TODAY'S TECHNICIAN™

SHOP MANUAL

FOR AUTOMOTIVE SUSPENSION & STEERING SYSTEMS

TODAY'S TECHNICIAN™

SHOP MANUAL FOR AUTOMOTIVE SUSPENSION & STEERING SYSTEMS

SIXTH EDITION

Mark Schnubel

CENGAGE Learning®

Australia • Brazil • Mexico • Singapore • United Kingdom • United States

CENGAGE Learning®

Today's Technician™: Automotive Suspension & Steering Systems, 6th Edition

Mark Schnubel

VP, General Manager, Skills and Planning: Dawn Gerrain

Director, Development, Global Product Management, Skills: Marah Bellegarde

Product Manager: Erin Brennan

Senior Product Development Manager: Larry Main

Senior Content Developer: Meaghan Tomaso

Product Assistant: Scott Royael

Marketing Manager: Linda Kuper

Market Development Manager: Jonathon Sheehan

Senior Production Director: Wendy Troeger

Production Manager: Mark Bernard

Content Project Manager: Christopher Chien

Art Director: GEX

Media Developer: Debbie Bordeaux

Cover Image(s): © cla78/Shutterstock

© 2015, 2011 Cengage Learning

WCN: 01-100-101

ALL RIGHTS RESERVED. No part of this work covered by the copyright herein may be reproduced, transmitted, stored, or used in any form or by any means graphic, electronic, or mechanical, including but not limited to photocopying, recording, scanning, digitizing, taping, Web distribution, information networks, or information storage and retrieval systems, except as permitted under Section 107 or 108 of the 1976 United States Copyright Act, without the prior written permission of the publisher.

For product information and technology assistance, contact us at
Cengage Learning Customer & Sales Support, 1-800-354-9706

For permission to use material from this text or product, submit all requests online at **www.cengage.com/permissions**
Further permissions questions can be e-mailed to
permissionrequest@cengage.com

Library of Congress Control Number: 2014930383

ISBN-13: 978-1-285-43814-6

ISBN-10: 1-285-43814-0

Cengage Learning
200 First Stamford Place, 4th Floor
Stamford, CT 06902
USA

Cengage Learning is a leading provider of customized learning solutions with office locations around the globe, including Singapore, the United Kingdom, Australia, Mexico, Brazil, and Japan. Locate your local office at: **www.cengage.com/global**

Cengage Learning products are represented in Canada by Nelson Education, Ltd.

To learn more about Cengage Learning Solutions, visit **www.cengage.com**

Purchase any of our products at your local college store or at our preferred online store **www.cengagebrain.com**

Notice to the Reader
Publisher does not warrant or guarantee any of the products described herein or perform any independent analysis in connection with any of the product information contained herein. Publisher does not assume, and expressly disclaims, any obligation to obtain and include information other than that provided to it by the manufacturer. The reader is expressly warned to consider and adopt all safety precautions that might be indicated by the activities described herein and to avoid all potential hazards. By following the instructions contained herein, the reader willingly assumes all risks in connection with such instructions. The publisher makes no representations or warranties of any kind, including but not limited to, the warranties of fitness for particular purpose or merchantability, nor are any such representations implied with respect to the material set forth herein, and the publisher takes no responsibility with respect to such material. The publisher shall not be liable for any special, consequential, or exemplary damages resulting, in whole or part, from the readers' use of, or reliance upon, this material.

Printed in the United States of America
2 3 4 5 6 7 18 17 16 15

CONTENTS

Photo Sequences ..... ***viii***

Job Sheets ..... ***ix***

Preface ..... ***xi***

CHAPTER 1 *Safety* ***1***

• Introduction 2 • Occupational Safety and Health Act 2 • Shop Hazards 2 • Shop Safety Rules 3 • Air Quality 5 • Shop Safety Equipment 6 • Shop Layout 9 • Safety in the Automotive Shop 10 • General Shop Safety 10 • Personal Safety 11 • Electrical Safety 13 • Gasoline Safety 13 • Fire Safety 14 • Vehicle Operation 16 • Housekeeping Safety 16 • Air Bag Safety 17 • Lifting and Carrying 20 • Hand Tool Safety 21 • Lift Safety 21 • Hydraulic Jack and Safety Stand Safety 22 • Power Tool Safety 23 • Compressed-Air Equipment Safety 23 • Cleaning Equipment Safety and Environmental Considerations 24 • Hazardous Waste Disposal 27 • General Hybrid Electric Vehicle Safety 30 • Case Study 34 • Summary 34 • Terms to Know 35 • ASE-Style Review Questions 36 • Job Sheets 37

CHAPTER 2 *Tools and Shop Procedures* ***45***

• Using Suspension and Steering Equipment 45 • Measuring Systems 45 • Basic Diagnostic Procedure 47 • Suspension and Steering Tools 48 • Hydraulic Pressing and Lifting Equipment 61 • Suspension and Steering Service, Diagnostic, and Measurement Tools 65 • Employer and Employee Obligations 76 • National Institute for Automotive Service Excellence (ASE) Certification 78 • Service Manuals 80 • Case Study 86 • Terms to Know 86 • Summary 87 • ASE-Style Review Questions 87 • Job Sheets 89

CHAPTER 3 *Tire and Wheel Servicing and Balancing* ***95***

• Tire Noises and Steering Problems 95 • Tire Rotation 97 • Tire and Wheel Service 98 • Tire and Wheel Service Precautions 100 • Wheel Rim Service 106 • Tire Remounting Procedure 107 • Diagnosing and Servicing Tire Pressure Monitoring Systems 108 • Tire and Wheel Runout Measurement 114 • Tread Wear Measurement 116 • Preliminary Wheel Balancing Checks 117 • Tire Inflation Pressure 118 • Static and Dynamic Wheel Balance Procedure 119 • Electronic Wheel Balancers with Lateral Force Measurement (LFM) and Radial Force Variation Capabilities 121 • On-Car Wheel Balancing 125 • Noise, Vibration and Harshness Diagnosis 126 • Case Study 132 • Terms to Know 132 • ASE-Style Review Questions 133 • ASE Challenge Questions 134 • Job Sheets 135

CHAPTER 4 *Wheel Bearing and Seal Service* ***147***

• Diagnosis of Bearing Defects 147 • Service and Adjustment of Tapered Roller Bearing-Type Wheel Bearings 148 • Wheel Hub Unit Diagnosis and Service 154 • Front Drive Axle Diagnosis 158 • Drive Axle Removal 158 • Special Procedures for Drive Axle Removal 160 • Front Wheel Bearing Hub Unit Removal and Replacement 161 • Rear-Axle Bearing and Seal Service, Rear-Wheel-Drive Cars 163 • Case Study 166 • Terms to Know 166 • ASE-Style Review Questions 167 • ASE Challenge Questions 169 • Job Sheets 171

CHAPTER 5 *Shock Absorber and Strut Diagnosis and Service* ***179***

• Shock Absorber Visual Inspection 179 • Shock Absorber Testing 180 • Shock Absorber Manual Test 181 • Air Shock Absorber Diagnosis and Replacement 183 • Shock Absorber Replacement 183 • Diagnosis of Front Spring and Strut Noise 183 • Strut Removal and Replacement 185 • Removal of Strut from Coil Spring 185 • Strut Disposal Procedure 187

• Installation of Coil Spring on Strut 187 • Installation of Strut-and-Spring Assembly in Vehicle 189 • Rear Strut Replacement 190 • Installing Strut Cartridge, Off-Car 190 • Installing Strut Cartridge, On-Car 193 • Diagnosis of Electronically Controlled Shock Absorbers 195 • Case Study 197 • Terms to Know 197 • ASE-Style Review Questions 197 • ASE Challenge Questions 199 • Job Sheets 201

CHAPTER 6 *Front Suspension System Service* 211

• Curb Riding Height Measurement 211 • Front Suspension Diagnosis and Service 212 • Control Arm Diagnosis and Service 225 • Removing and Replacing Longitudinally Mounted Torsion Bars 235 • Case Study 237 • Terms to Know 237 • ASE-Style Review Questions 237 • ASE Challenge Questions 238 • Job Sheets 239

CHAPTER 7 *Rear Suspension Service*..... 249

• Lower Control Arm and Ball Joint Diagnosis and Replacement 256 • Rear Leaf-Spring Diagnosis and Replacement 261 • Track Bar Diagnosis and Replacement 262 • Stabilizer Bar Diagnosis and Service 263 • Rear Suspension Tie Rod Inspection and Replacement 264 • Case Study 265 • Terms to Know 265 • ASE-Style Review Questions 265 • ASE Challenge Questions 266 • Job Sheets 267

CHAPTER 8 *Steering Column and Linkage Diagnosis and Service*... 273

• Air Bag Deployment Module, Steering Wheel, and Clock Spring Electrical Connector Removal and Replacement 273 • Steering Column Service 278 • Steering Column Removal and Replacement 278 • Collapsible Steering Column Inspection 279 • Tilt Steering Column Disassembly 280 • Tilt Steering Column Inspection and Parts Replacement 283 • Tilt Steering Column Assembly 284 • Steering Column Flexible Coupling and Universal Joint Diagnosis and Service 286 • Steering Column Diagnosis 286 • Rack and Pinion Steering Gear Tie Rod Service 290 • Steering Linkage Diagnosis and Service 291 • Case Study 300 • Terms to Know 300 • ASE-Style Review Questions 300 • ASE Challenge Questions 302 • Job Sheets 303

CHAPTER 9 *Four Wheel Alignment Procedure, Part 1*..... 317

• Wheel Alignment Preliminary Diagnosis and Inspection 317 • Four-Wheel Alignment with Computer Alignment Systems 323 • Wheel Alignment Screens 328 • Adjustment Screens 333 • Diagnostic Drawing and Text Screens 334 • Wheel Alignment Procedure 337 • Causes of Improper Rear Wheel Alignment 337 • Rear Suspension Adjustments 338 • Front Camber Adjustment 346 • Front Caster Adjustment Procedure 353 • Toe Adjustment 356 • Checking Toe Change and Steering Linkage Height 357 • Manual Steering Wheel Centering Procedure 359 • Terms to Know 361 • Case Study 362 • ASE-Style Review Questions 362 • ASE Challenge Questions 364 • Job Sheets 365

CHAPTER 10 *Diagnostic Alignment Angles and Frame Diagnosis*... 383

• Bent Front Strut Diagnosis 386 • Setback Measurement and Correction Procedure 386 • Steering Axis Inclination (SAI) Correction Procedure 389 • Rear Wheel Tracking Measurement with a Track Gauge 389 • Frame Diagnosis and Service 394 • Frame Diagnosis 395 • Checking Frame Alignment 396 • Measuring Unitized Body Alignment 402 • Case Study 408 • Terms to Know 410 • ASE-Style Review Questions 411 • ASE Challenge Questions 412 • Job Sheets 413

CHAPTER 11 *Computer-Controlled Suspension System Service*.... 421

• Preliminary Inspection of Computer-Controlled Suspension Systems 421 • Programmed Ride Control System Diagnosis 422 • Electronic Air Suspension Diagnosis and Service 424 • Servicing and Diagnosing Vehicle Dynamic Suspension Systems 432

• Diagnosis of Electronic Suspension Control Systems 437 • Scan Tool Diagnosis of Electronic Suspension Control 439 • Case Study 447 • Terms to Know 448 • ASE-Style Review Questions 449 • ASE Challenge Questions 450 • Job Sheets 451

CHAPTER 12 *Power Steering Pump Diagnosis and Service* 459

• Power Steering Pump Belt Service 459 • Power Steering Pump Fluid Service 467 • Power Steering Pump Diagnosis 469 • Power Steering Pump Service 471 • Inspecting and Servicing Power Steering Lines and Hoses 476 • Power Steering Hose Replacement 477 • Hybrid Electric Vehicle (HEV) and Electrohydraulic Power Steering (EHPS) Diagnosing and Servicing Procedures 478 • Case Study 487 • Terms to Know 487 • ASE-Style Review Questions 487 • ASE Challenge Questions 488 • Job Sheets 491

CHAPTER 13 *Rack and Pinion Steering Gear and Four Wheel Steering Diagnosis and Service* 497

• Manual or Power Rack and Pinion Steering Gear On-Car Inspection 497 • Manual or Power Rack and Pinion Steering Gear Removal and Replacement 500 • Manual Rack and Pinion Steering Gear Diagnosis and Service 505 • Power Rack and Pinion Steering Gear Diagnosis and Service 508 • Diagnosis of Power Steering, Steering Column, and Suspension Systems 518 • Diagnosis of Magnasteer Systems 519 • Diagnosis of Rack-Drive Electronic Power Steering 520 • Diagnosis of Column-Drive Electronic Power Steering 522 • Active Steering System Preliminary Diagnosis 525 • Electronic Four-Wheel Steering Diagnosis and Service 528 • Preliminary Inspection 529 • Quadrasteer Diagnosis 529 • Diagnosis of Four-Wheel Active Steering (4WAS) System 533 • Case Study 537 • Terms to Know 537 • ASE-Style Review Questions 538 • ASE Challenge Questions 539 • Job Sheets 541

CHAPTER 14 *Recirculating Ball Steering Gear Diagnosis and Service* 551

• Power Recirculating Ball Steering Gear Diagnosis 551 • Power Recirculating Ball Steering Gear Replacement 552 • Power Recirculating Ball Steering Gear Adjustments 553 • Power Recirculating Ball Steering Gear Oil Leak Diagnosis 557 • Power Recirculating Ball Steering Gear Seal Replacement 557 • Case Study 560 • Terms to Know 560 • ASE-Style Review Questions 560 • ASE Challenge Questions 561 • Job Sheets 563

***Appendix A ASE Practice Examination* 569**

***Appendix B Metric Conversions* 575**

***Appendix C Special Tool Suppliers* 576**

***Appendix D Manufacturer Websites* 577**

***Appendix E Suspension and Steering Professional Associations* 578**

***Glossary* 579**

***Index* 592**

PHOTO SEQUENCES

1. Typical Procedure for Removing Air Bag Module 18
2. Typical Procedure for Performing Four Wheel Alignment with a Computer Wheel Aligner 75
3. Typical Procedure for Dismounting and Mounting a Tire on a Wheel Assembly 109
4. Typical Off-Car Wheel Balancing Procedure 122
5. Typical Procedure for Adjusting Rear Wheel Bearings on a Front-Wheel-Drive Car 155
6. Typical Procedure for Measuring Front Wheel Hub Endplay—Integral, Sealed Wheel Bearing Hub Assemblies 157
7. Rear Shock Absorber Visual Inspection and Bounce Test 182
8. Typical Procedure for Removing and Replacing a MacPherson Strut 191
9. Typical Procedure for Measuring the Lower Ball Joint Horizontal Movement on a MacPherson Strut Front Suspension 216
10. Vertical Ball Joint Measurement 218
11. Typical Procedure for Measuring Front and Rear Curb Riding Height 252
12. Typical Procedure for Removing a Rear Strut-and-Spring Assembly on a Front-Wheel-Drive Car 257
13. Typical Procedure for Removing a Steering Wheel 276
14. Diagnosing, Removing, and Replacing an Outer Tie Rod End on a Vehicle with a Parallelogram Steering Linkage 295
15. Typical Procedure for Performing Four-Wheel Alignment with a Computer Wheel Aligner 330
16. Typical Procedure for Adjusting Rear Wheel Camber and Toe 342
17. Typical Procedure for Performing Front and Rear Suspension Alignment Adjustments 345
18. Typical Procedure for Front Wheel Turning Radius Measurement 384
19. Typical Procedure for Performing Frame Measurement, Plumb Bob Method 399
20. Performing Underhood Measurements 404
21. Inflate L/H Front Air Spring Using a Scan Tool 429
22. Reading Scan Tool Data on an Electronic Suspension Control System 442
23. Typical Procedure for Pressure Testing a Power Steering Pump 473
24. Observing Safety Precautions When Servicing Hybrid Electric Vehicles (HEVs) 486
25. Typical Procedure for Removing and Replacing a Power Rack and Pinion Steering Gear 503
26. Typical Procedure for Removing and Replacing Inner Tie Rod End, Power Rack, and Pinion Steering Gear 517
27. Preliminary Inspection, Four-Wheel Steering Diagnosis 530
28. Typical Procedure for Performing a Worm Shaft Bearing Preload Adjustment 554
29. Pitman Sector Shaft Lash Adjustment 556

JOB SHEETS

1. Demonstrate Proper Lifting Procedures 37
2. Locate and Inspect Shop Safety Equipment 39
3. Shop Housekeeping Inspection 41
4. Personnel and Shop Safety Assessment 43
5. Raise a Car with a Floor Jack and Support It on Safety Stands 89
6. Follow the Proper Procedure to Hoist a Car 91
7. Determine the Availability and Purpose of Suspension and Steering Tools 93
8. Tire Dismounting and Mounting 135
9. Tire and Wheel Runout Measurement 137
10. Off-Car Wheel Balancing 139
11. On-Car Wheel Balancing 141
12. Diagnose Tire and Wheel Vibration, Steering Pull, and Chassis Waddle 143
13. Inspect, Diagnose, and Calibrate Tire Pressure Monitoring Systems 145
14. Service Integral Wheel Bearing Hubs 171
15. Diagnose Wheel Bearings 173
16. Clean, Lubricate, Install, and Adjust Nonsealed Wheel Bearings 175
17. Remove Strut-and-Spring Assembly and Disassemble Strut and Spring 201
18. Assemble Strut and Spring and Install Strut-and-Spring Assembly 205
19. Install Strut Cartridge, Off-Car 209
20. Measure Lower Ball Joint Vertical and Horizontal Movement, Short-and-Long Arm Suspension Systems 239
21. Determine if the Upper or Lower Suspension Pivot Point Is a Primary Load Carrier or a Follower Joint and Measure Joint for Vertical and Horizontal Movement, in a Suspension System 241
22. Ball Joint Replacement 243
23. Steering Knuckle Removal, MacPherson Strut Front Suspension 247
24. Remove and Service Rear Suspension Strut and Coil Spring Assembly 267
25. Remove Rear Suspension Lower Control Arm and Ball Joint Assembly 269
26. Install Rear Suspension Lower Control Arm and Ball Joint Assembly 271
27. Remove and Replace Air Bag Inflator Module and Steering Wheel 303
28. Remove and Replace Steering Column 307
29. Inspect Manual or Power Rack and Pinion Steering Gear and Tie Rods 309
30. Diagnose, Remove, and Replace Idler Arm 313
31. Remove and Replace Outer Tie Rod End, Parallelogram Steering Linkage 315
32. Road Test Vehicle and Diagnose Steering Operation 365
33. Measure Front and Rear Wheel Alignment Angles with a Computer Wheel Aligner 367
34. Check Proper Steering Linkage Height by Measuring Toe Change 371
35. Center Steering Wheel 373
36. Adjust Front Wheel Alignment Angles 375

JOB SHEETS

37. Adjust Rear Wheel Alignment Angles 379

38. Frame Measurement, Plumb Bob Method 413

39. Inspect and Measure Front Cradle 417

40. Inspect and Weld Vehicle Frame 419

41. Diagnostic Procedures and Steps to Follow 451

42. Inspection and Preliminary Diagnosis of Computer-Controlled Suspension System 453

43. Adjust Vehicle Ride (Trim) Height with a Scan Tool 455

44. Perform an On-Demand Self-Test on a Vehicle Dynamic Suspension (VDS) System 457

45. Draining and Flushing Power Steering System 491

46. Testing Power Steering Pump Pressure 493

47. Measure and Adjust Power Steering Belt Tension and Alignment 495

48. Remove and Replace Manual or Power Rack and Pinion Steering Gear 541

49. Diagnose Power Rack and Pinion Steering Gear Oil Leakage Problems 545

50. Perform a Learn Rear Wheel Alignment Procedure on a Quadrasteer System 547

51. Quadrasteer Diagnostic System Check 549

52. Power Recirculating Ball Steering Gear Oil Leak Diagnosis 563

53. Adjust Power Recirculating Ball Steering Gear Worm Shaft Thrust Bearing Preload, Steering Gear Removed 565

54. Adjust Power Recirculating Ball Steering Gear Sector Lash, Steering Gear Removed 567

PREFACE

Thanks to the support the *Today's Technician* series has received from those who teach automotive technology, Cengage Learning, the leader in automotive-related textbooks, is able to live up to its promise to regularly provide new editions of texts of this series. We have listened and responded to our critics and our fans and present this new updated and revised sixth edition. By revising this series on a regular basis, we can respond to changes in the industry, changes in technology, changes in the certification process, and to the ever-changing needs of those who teach automotive technology.

We also listened to instructors who said something was missing or incomplete in the last edition. We responded to those and the results are included in this sixth edition.

The *Today's Technician* series features textbooks that cover all mechanical and electrical systems of automobiles and light trucks. Principally, the individual titles correspond to the certification areas for 2013 in areas of National Institute for Automotive Service Excellence (ASE) certification.

Additional titles include remedial skills and theories common to all of the certification areas and advanced or specific subject areas that reflect the latest technological trends.

This new edition, like the last, was designed to give students a chance to develop the same skills and gain the same knowledge that today's successful technician has. This edition also reflects the changes in the guidelines established by the National Automotive Technicians Education Foundation (NATEF) in 2013.

The purpose of NATEF is to evaluate technician training programs against standards developed by the automotive industry and recommend qualifying programs for certification (accreditation) by ASE. Programs can earn ASE certification upon the recommendation of NATEF. NATEF's national standards reflect the skills that students must master. ASE certification through NATEF evaluation ensures that certified training programs meet or exceed industry-recognized, uniform standards of excellence.

The technician of today and for the future must know the underlying theory of all automotive systems and be able to service and maintain those systems. Dividing the material into two volumes, a Classroom Manual and a Shop Manual, provides the reader with the information needed to begin a successful career as an automotive technician without interrupting the learning process by mixing cognitive and performance learning objectives into one volume.

The design of Cengage Learning's *Today's Technician* series was based on features that are known to promote improved student learning. The design was further enhanced by a careful study of survey results, in which the respondents were asked to value particular features. Some of these features can be found in other textbooks, whereas others are unique to this series.

Each Classroom Manual contains the principles of operation for each system and subsystem. The Classroom Manual also contains discussions on design variations of key components used by the different vehicle manufacturers. It also looks into emerging technologies that will be standard or optional features in the near future. This volume is organized to build upon basic facts and theories. The primary objective of this volume is to allow the reader to gain an understanding of how each system and subsystem operates. This understanding is necessary to diagnose the complex automobiles of today and tomorrow. Although the basics contained in the Classroom Manual provide the knowledge needed for diagnostics, diagnostic procedures appear only in the Shop Manual. An understanding of the underlying theories is also a requirement for competence in the skill areas covered in the Shop Manual.

A spiral-bound Shop Manual covers the "how-tos." This volume includes step-by-step instructions for diagnostic and repair procedures. Photo Sequences are used to illustrate some of the common service procedures. Other common procedures are listed and are accompanied with fine line drawings and photos that allow the reader to visualize and conceptualize the finest details of the procedure. This volume also contains the reasons for performing the procedures, as well as when that particular service is appropriate.

The two volumes are designed to be used together and are arranged in corresponding chapters. Not only are the chapters in the volumes linked together, the contents of the chapters are also linked. This linking of content is evidenced by marginal callouts that refer the reader to the chapter and page that the same topic is addressed in the other volume. This feature is valuable to instructors. Without this feature, users of other two-volume textbooks must search the index or table of contents to locate supporting information in the other volume. This is not only cumbersome but also creates additional work for an instructor when planning the presentation of material and when making reading assignments. It is also valuable to the students, with the page references they also know exactly where to look for supportive information.

Both volumes contain clear and thoughtfully selected illustrations. Many of which are original drawings or photos specially prepared for inclusion in this series. This means that the art is a vital part of each textbook and not merely inserted to increase the number of illustrations.

The page layout, used in the series, is designed to include information that would otherwise break up the flow of information presented to the reader. The main body of the text includes all of the "need-to-know" information and illustrations. In the wide side margins of each page are many of the special features of the series. Items that are truly "nice-to-know" information such as: simple examples of concepts just introduced in the text, explanations or definitions of terms that are not defined in the text, examples of common trade jargons used to describe a part or operation, and exceptions to the norm explained in the text. This type of information is placed in the margin, out of the normal flow of information. Many textbooks attempt to include this type of information and insert it in the main body of text; this tends to interrupt the thought process and cannot be pedagogically justified. By placing this information off to the side of the main text, the reader can select when to refer to it.

Jack Erjavec
Series Editor

Highlights of this Edition—Classroom Manual

The text was updated to include the latest technology in suspension and steering systems. Updated and expanded coverage of basic electrical theory and hybrid electric vehicle safety, hybrid vehicle steering systems, active steering systems, rear active steering (RAS), CAN bus networking, computer-controlled suspension systems, and adaptive cruise control systems. Expanded coverage of tires and wheels including tire plus sizing and rim offset considerations. As well as expanded alignment theory coverage and sequencing of information. The text also includes the latest technology in vehicle stability control systems, traction control systems, active roll control, lane departure warning (LDW) systems, collision mitigation systems, telematics, and tire pressure monitoring systems (TPMS).

The first chapter explains the design and purpose of basic suspension and steering systems. This chapter provides students with the necessary basic understanding of suspension and steering systems. The other chapters in the book allow the students to build upon their understanding of these basic systems.

The second chapter explains all the basic theories required to understand the latest suspension and steering systems described in the other chapters. Students must understand these basic theories to comprehend the complex systems explained later in the text.

The other chapters in the book are designed to be stand alone to allow reordering of topics covered to fit individual program needs and explain all the current model systems and components such as wheel bearings, tires and wheels, shock absorbers and struts, front and rear suspension systems, computer-controlled suspension systems, steering columns and linkages, power steering pumps, steering gears and systems, four-wheel steering systems, frames, and four-wheel alignment. All of the art pieces have been replaced with color photos and color diagrams throughout the text to improve visual concepts of suspension and steering systems and components.

Highlights of this Edition—Shop Manual

The chapters in the Shop Manual have been updated to explain the diagnostic and service procedures for the latest systems and components described in the Classroom Manual. Diagnostics is a very important part of an automotive technician's job. Therefore, proper diagnostic procedures are emphasized in the Shop Manual.

All Photo Sequences are now in color. These Photo Sequences illustrate the correct diagnostic or service procedure for a specific system or component. These Photo Sequences allow the students to visualize the diagnostic or service procedure. Visualization of these diagnostic and service procedures helps students to remember the procedures, and perform them more accurately and efficiently. The text covers the information required to pass an ASE test in Suspension and Steering Systems. New and updated Job Sheets have been created to meet current NATEF tasks.

Chapter 1 explains the necessary safety precautions and procedures in an automotive repair shop. General shop safety and the required shop safety equipment are explained in the text. The text describes safety procedures when operating vehicles and various types of automotive service equipment. Correct procedures for handling hazardous waste materials are detailed in the text.

Chapter 2 describes suspension and steering diagnostic and service equipment and the use of service manuals. This chapter also explains employer and employee obligations, and ASE certification requirements.

The other chapters in the text have been updated to explain the diagnostic and service procedures for the latest suspension and steering systems explained in the Classroom Manual. New job sheets related to the new systems and components have been added in the text. All of the art pieces have been replaced with color photos and color diagrams throughout the text to improve the student's visualization of diagnostic and service procedures.

Mark Schnubel

SHOP MANUAL

To stress the importance of safe work habits, the Shop Manual also dedicates one full chapter to safety. Other important features of this manual include:

PERFORMANCE-BASED OBJECTIVES

These objectives define the contents of the chapter and define what the student should have learned upon completion of the chapter. These objectives also correspond with the list of required tasks for NATEF certification. *Each NATEF task is addressed.*

Although this textbook is not designed to simply prepare someone for the certification exams, it is organized around the NATEF task list. These tasks are defined generically when the procedure is commonly followed and specifically when the procedure is unique for specific vehicle models. Imported and domestic model automobiles and light trucks are included in the procedures.

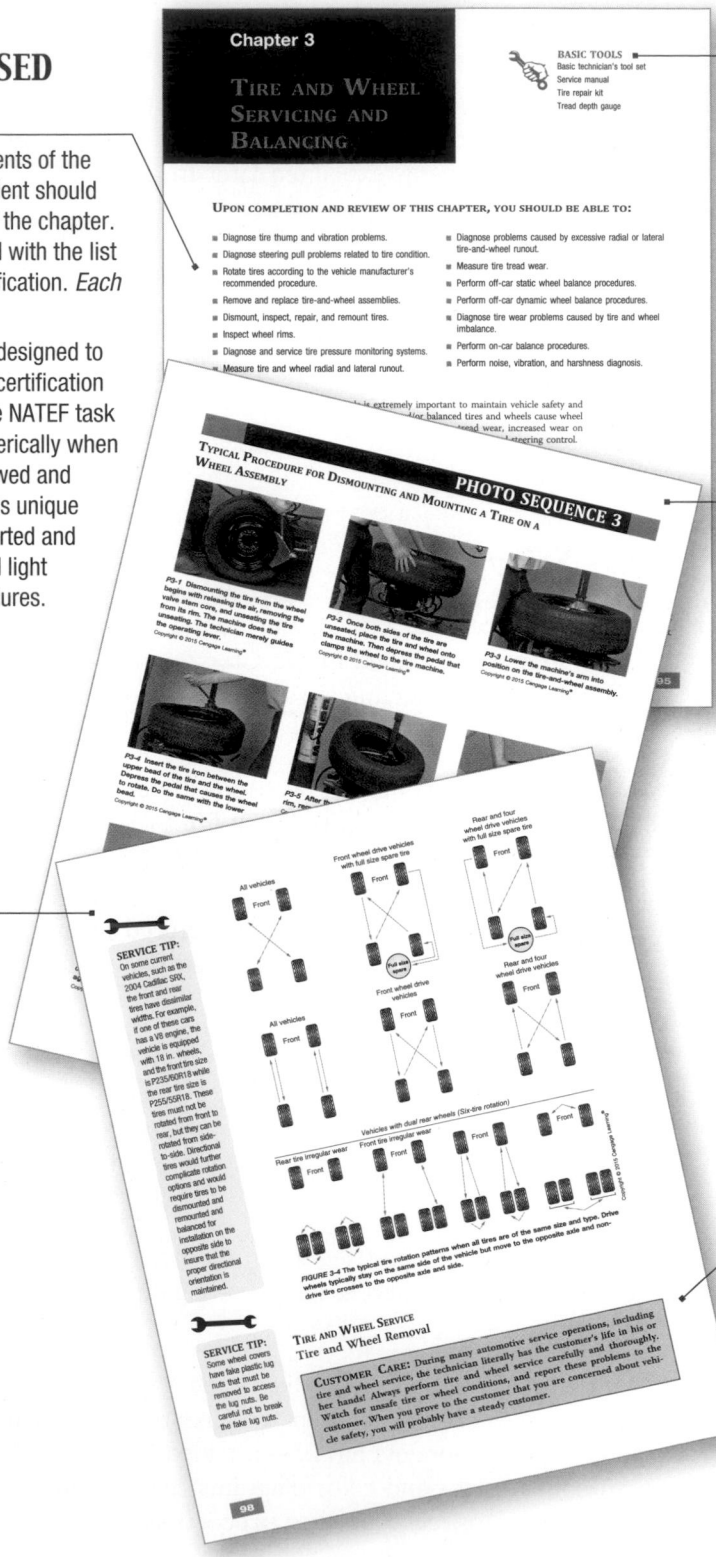

Chapter 3

TIRE AND WHEEL SERVICING AND BALANCING

BASIC TOOLS
Basic technician's tool set
Service manual
Tire repair kit
Tread depth gauge

UPON COMPLETION AND REVIEW OF THIS CHAPTER, YOU SHOULD BE ABLE TO:

- Diagnose tire thump and vibration problems.
- Diagnose steering pull problems related to tire condition.
- Rotate tires according to the vehicle manufacturer's recommended procedure.
- Remove and replace tire-and-wheel assemblies.
- Dismount, inspect, repair, and remount tires.
- Inspect wheel rims.
- Diagnose and service tire pressure monitoring systems.
- Measure tire and wheel radial and lateral runout.
- Diagnose problems caused by excessive radial or lateral tire-and-wheel runout.
- Measure tire tread wear.
- Perform off-car static wheel balance procedures.
- Perform off-car dynamic wheel balance procedures.
- Diagnose tire wear problems caused by tire and wheel imbalance.
- Perform on-car balance procedures.
- Perform noise, vibration, and harshness diagnosis.

TYPICAL PROCEDURE FOR DISMOUNTING AND MOUNTING A TIRE ON A WHEEL ASSEMBLY

PHOTO SEQUENCE 3

P3-1 Dismounting the tire from the wheel begins with releasing the air, removing the valve stem cores, and separating the tire from the rim. The machine does the unseating. The technician merely guides the operating lever.

P3-2 Once both sides of the tire are unseated, place the tire and wheel onto the machine. Then depress the pedal that clamps the wheel to the tire machine.

P3-3 Lower the machine's arm into position on the tire-and-wheel assembly.

P3-4 Insert the tire iron between the upper bead of the tire and the wheel. Depress the pedal that causes the wheel to rotate. Do the same with the lower bead.

SERVICE TIP: On some current vehicles, such as the 2004 Cadillac SRX, the front and rear tires have dissimilar widths. For example, if one of these cars has a V8 engine, the vehicle is equipped with 18 in. wheels, and the front tire size is P235/60R18 while the rear tire size is P255/55R18. These tires must not be rotated from front to rear, but they can be rotated from side-to-side. Directional tires would further complicate rotation options and would require tires to be dismounted and remounted and balanced for installation on the opposite side to insure that the proper directional orientation is maintained.

All vehicles
Front wheel drive vehicles with full size spare tire
Rear and four wheel drive vehicles with full size spare tire
Front wheel drive vehicles
Rear and four wheel drive vehicles
Vehicles with dual rear wheels (Six-tire rotation)
Rear tire irregular wear
Front tire irregular wear

FIGURE 3-4 The typical tire rotation patterns when all tires are of the same size and type. Drive wheels typically stay on the same side of the vehicle but move to the opposite axle and non-drive tire crosses to the opposite axle and side.

SERVICE TIP: Some wheel covers have fake plastic lug nuts that must be removed to access the lug nuts. Be careful not to break the fake lug nuts.

TIRE AND WHEEL SERVICE
Tire and Wheel Removal

CUSTOMER CARE: During many automotive service operations, including tire and wheel service, the technician literally has the customer's life in his or her hands! Always perform tire and wheel service carefully and thoroughly. Watch for unsafe tire or wheel conditions, and report these problems to the customer. When you prove to the customer that you are concerned about vehicle safety, you will probably have a steady customer.

98

BASIC TOOLS LISTS

Each chapter begins with a list of the basic tools needed to perform the tasks included in the chapter.

PHOTO SEQUENCES

Many procedures are illustrated in detailed Photo Sequences. These detailed photographs show the students what to expect when they perform particular procedures. They also can provide the student a familiarity with a system or type of equipment, which the school may not have.

SERVICE TIPS

Whenever a short-cut or special procedure is appropriate, it is described in the text. These tips are generally those things commonly done by experienced technicians.

CUSTOMER CARE

This feature highlights those little things a technician can do or say to enhance customer relations.

SPECIAL TOOLS LISTS

Whenever a special tool is required to complete a task, it is listed in the margin next to the procedure.

CROSS-REFERENCES TO THE CLASSROOM MANUAL

Reference to the appropriate page in the Classroom Manual is given whenever necessary. Although the chapters of the two manuals are synchronized, material covered in other chapters of the Classroom Manual may be fundamental to the topic discussed in the Shop Manual.

MARGINAL NOTES

These notes add "nice-to-know" information to the discussion. They may include examples or exceptions, or may give the common trade jargon for a component.

CAUTIONS AND WARNINGS

Throughout the text, warnings are given to alert the reader to potentially hazardous materials or unsafe conditions. Cautions are given to advise the student of things that can go wrong if instructions are not followed or if a nonacceptable part or tool is used.

TERMS TO KNOW LIST

Terms in this list can be found in the Glossary at the end of the manual.

CASE STUDIES

Case Studies concentrate on the ability to properly diagnose the systems. Beginning with Chapter 3, each chapter ends with a case study in which a vehicle has a problem, and the logic used by a technician to solve the problem is explained.

ASE-STYLE REVIEW QUESTIONS

Each chapter contains ASE-style review questions that reflect the performance-based objectives listed at the beginning of the chapter. These questions can be used to review the chapter as well as to prepare for the ASE certification exam.

ASE CHALLENGE QUESTIONS

Each technical chapter ends with five ASE challenge questions. These are not more review questions; rather they test the students' ability to apply general knowledge to the contents of the chapter.

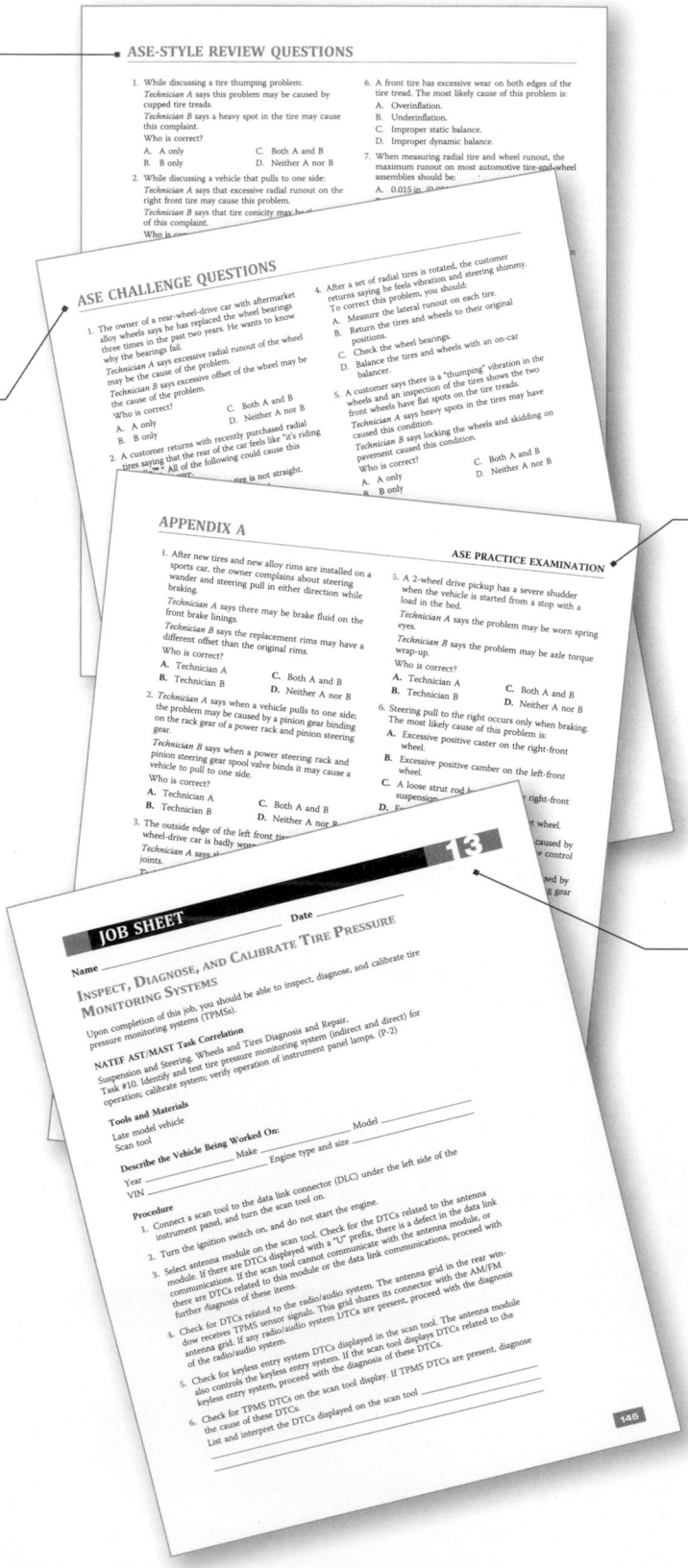

ASE-STYLE REVIEW QUESTIONS

1. While discussing a tire thumping problem:
Technician A says this problem may be caused by cupped tire treads.
Technician B says a heavy spot in the tire may cause this complaint.
Who is correct?
A. A only C. Both A and B
B. B only D. Neither A nor B

2. While discussing a vehicle that pulls to one side:
Technician A says that excessive radial runout on the right front tire may cause this problem.
Technician B says that tire conicity may be

6. A front tire has excessive wear on both edges of the tire tread. The most likely cause of this problem is:
A. Overinflation.
B. Underinflation.
C. Improper static balance.
D. Improper dynamic balance.

7. When measuring radial tire and wheel runout, the maximum runout on most automotive tire-and-wheel assemblies should be:
A. 0.015 in.

ASE CHALLENGE QUESTIONS

1. The owner of a rear-wheel-drive car with aftermarket alloy wheels says he has replaced the wheel bearings three times in the past two years. He wants to know why the bearings fail.
Technician A says excessive radial runout of the wheel may be the cause of the problem.
Technician B says excessive offset of the wheel may be the cause of the problem.
Who is correct?
A. A only C. Both A and B
B. B only D. Neither A nor B

2. A customer returns with recently purchased radial tires saying that the rear of the car feels like "it's riding
All of the following could cause this

4. After a set of radial tires is rotated, the customer returns saying he feels vibration and steering shimmy. To correct this problem, you should:
A. Measure the lateral runout on each tire.
B. Return the tires and wheels to their original positions.
C. Check the wheel bearings.
D. Balance the tires and wheels with an on-car balancer.

5. A customer says there is a "thumping" vibration in the wheels and an inspection of the tires shows the two front wheels have flat spots on the tire treads.
Technician A says heavy spots in the tires may have caused this condition.
Technician B says locking the wheels and skidding on pavement caused this condition.
Who is correct?
A. A only C. Both A and B
B. B only D. Neither A nor B

APPENDIX A

ASE PRACTICE EXAMINATION

1. After new tires and new alloy rims are installed on a sports car, the owner complains about steering wander and steering pull in either direction while braking.
Technician A says there may be brake fluid on the front brake linings.
Technician B says the replacement rims may have a different offset than the original rims.
Who is correct?
A. Technician A
B. Technician B C. Both A and B
D. Neither A nor B

2. *Technician A* says when a vehicle pulls to one side; the problem may be caused by a pinion gear binding on the rack gear of a power rack and pinion steering gear.
Technician B says when a power steering rack and pinion steering gear spool valve binds it may cause a vehicle to pull to one side.
Who is correct?
A. Technician A
B. Technician B C. Both A and B
D. Neither A nor B

5. A 2-wheel drive pickup has a severe shudder when the vehicle is started from a stop with a load in the bed.
Technician A says the problem may be worn spring eyes.
Technician B says the problem may be axle torque wrap-up.
Who is correct?
A. Technician A
B. Technician B C. Both A and B
D. Neither A nor B

6. Steering pull to the right occurs only when braking. The most likely cause of this problem is:
A. Excessive positive caster on the right-front wheel.
B. Excessive positive camber on the left-front wheel.
C. A loose strut rod
suspension

JOB SHEET 3

Name Date

INSPECT, DIAGNOSE, AND CALIBRATE TIRE PRESSURE MONITORING SYSTEMS

Upon completion of this job, you should be able to inspect, diagnose, and calibrate tire pressure monitoring systems (TPMSs).

NATEF AST/MAST Task Correlation
Suspension and Steering, Wheels and Tires Diagnosis and Repair, Task #10. Identify and test tire pressure monitoring system (indirect and direct) for operation; calibrate system; verify operation of instrument panel lamps. (P-2)

Tools and Materials
Late model vehicle
Scan tool

Describe the Vehicle Being Worked On:
Year Make Model
VIN Engine type and size

Procedure
1. Connect a scan tool to the data link connector (DLC) under the left side of the instrument panel, and turn the scan tool on.
2. Turn the ignition switch on, and do not start the engine.
3. Select antenna module on the scan tool. Check for the DTCs related to the antenna module. If there are DTCs displayed with a "U" prefix, there is a defect in the data link communications. If the scan tool cannot communicate with the antenna module, or there are DTCs related to this module or the data link communications, proceed with further diagnosis of these items.
4. Check for DTCs related to the radio/audio system. The antenna grid in the rear window receives TPMS sensor signals. This grid shares its connector with the AM/FM antenna grid. If any radio/audio system DTCs are present, proceed with the diagnosis of the radio/audio system.
5. Check for keyless entry system DTCs displayed in the scan tool. The antenna module also controls the keyless entry system. If the scan tool displays DTCs related to the keyless entry system, proceed with the diagnosis of these DTCs.
6. Check for TPMS DTCs on the scan tool display. If TPMS DTCs are present, diagnose the cause of these DTCs.
List and interpret the DTCs displayed on the scan tool

145

ASE PRACTICE EXAMINATION

A 50 question ASE practice exam, located in the appendix, is included to test students on the contents of the Shop Manual.

JOB SHEETS

Located at the end of each chapter, the Job Sheets provide a format for students to perform procedures covered in the chapter. A reference to the NATEF Task addressed by the procedure is referenced on the Job Sheet.

Features of the Classroom Manual include the following:

COGNITIVE OBJECTIVES

These objectives define the contents of the chapter and define what the student should have learned upon completion of the chapter. *Each topic is divided into small units to promote easier understanding and learning.*

MARGINAL NOTES

These notes add "nice-to-know" information to the discussion. They may include examples or exceptions, or may give the common trade jargon for a component.

A BIT OF HISTORY

This feature gives the student a sense of the evolution of the automobile. This feature not only contains nice-to-know information, but also should spark some interest in the subject matter.

AUTHOR'S NOTES

This feature includes simple explanations, stories, or examples of complex topics. These are included to help students understand difficult concepts.

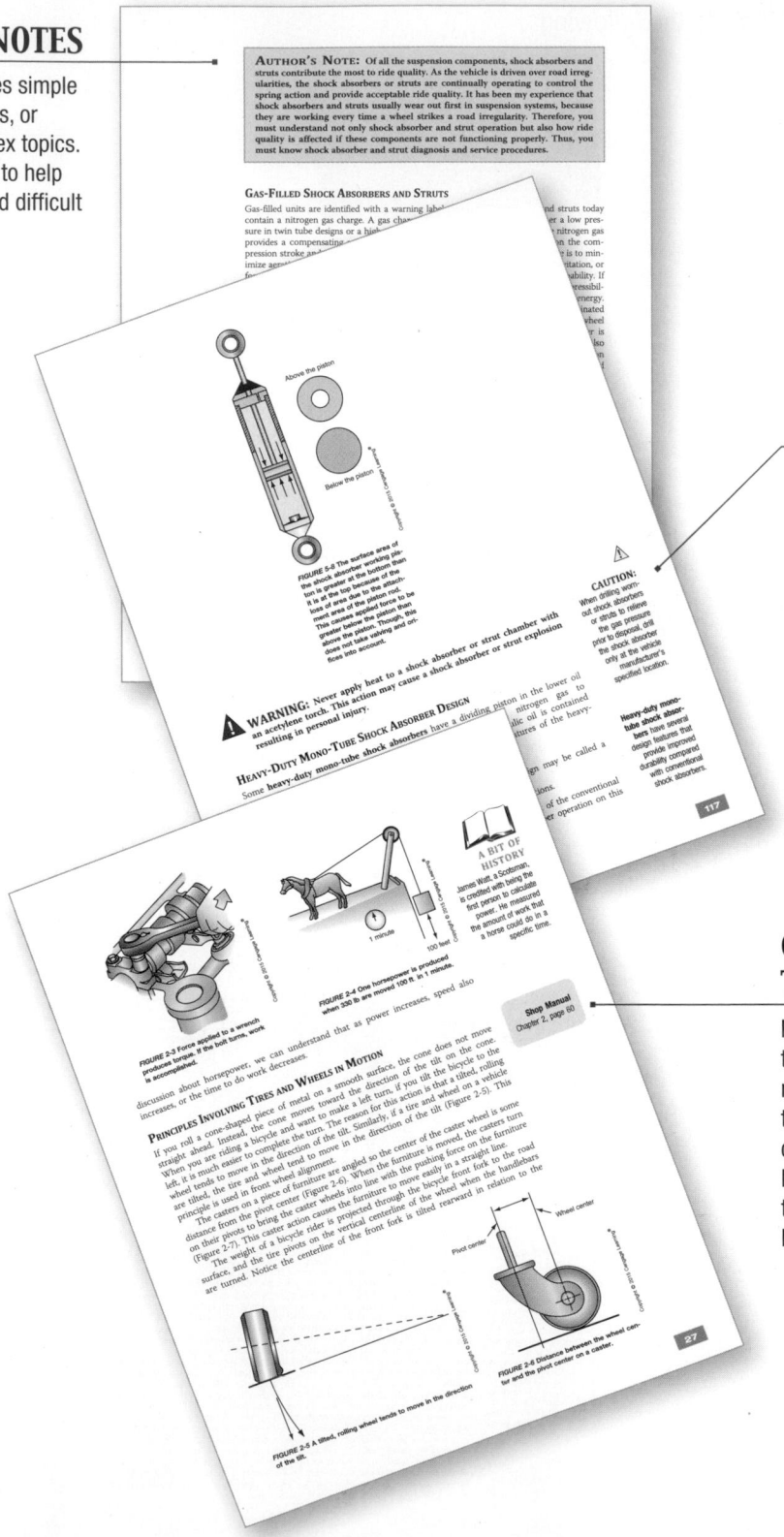

AUTHOR'S NOTE: Of all the suspension components, shock absorbers and struts contribute the most to ride quality. As the vehicle is driven over road irregularities, the shock absorbers or struts are continually operating to control the spring action and provide acceptable ride quality. It has been my experience that shock absorbers and struts usually wear out first in suspension systems, because they are working every time a wheel strikes a road irregularity. Therefore, you must understand not only shock absorber and strut operation but also how ride quality is affected if these components are not functioning properly. Thus, you must know shock absorber and strut diagnosis and service procedures.

GAS-FILLED SHOCK ABSORBERS AND STRUTS

WARNING: Never apply heat to a shock absorber or strut chamber with an acetylene torch. This action may cause a shock absorber or strut explosion resulting in personal injury.

HEAVY-DUTY MONO-TUBE SHOCK ABSORBER DESIGN

CAUTION: When drilling worn-out shock absorbers or struts, to relieve the gas pressure prior to disposal, drill the shock absorber only at the vehicle manufacturer's specified location.

Heavy-duty mono-tube shock absorbers have several design features that provide improved durability compared with conventional shock absorbers.

A BIT OF HISTORY

James Watt, a Scotsman, is credited with being the first person to calculate power. He measured the amount of work that a horse could do in a specific time.

Shop Manual Chapter 2, page 60

CAUTIONS AND WARNINGS

Throughout the text, warnings are given to alert the reader to potentially hazardous materials or unsafe conditions. Cautions are given to advise the student of things that can go wrong if instructions are not followed or if a nonacceptable part or tool is used.

CROSS-REFERENCES TO THE SHOP MANUAL

Reference to the appropriate page in the Shop Manual is given whenever necessary. Although the chapters of the two manuals are synchronized, material covered in other chapters of the Shop Manual may be fundamental to the topic discussed in the Classroom Manual.

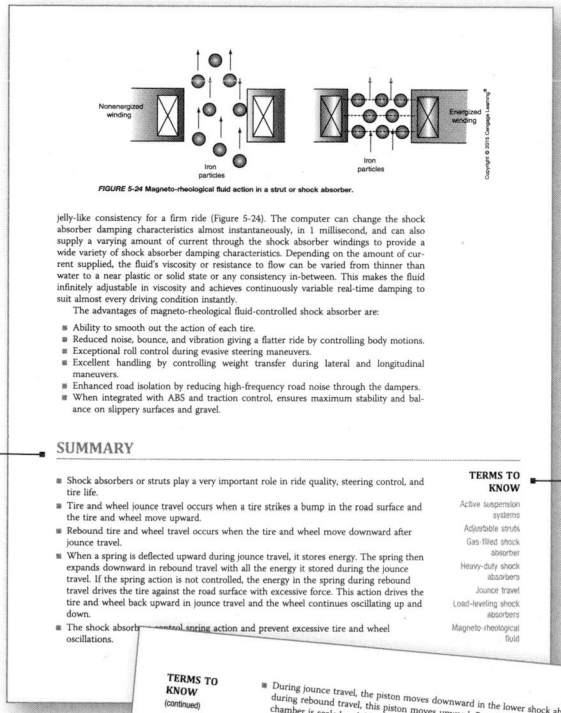

FIGURE 5-24 **Magneto-rheological fluid action in a strut or shock absorber.**

jelly-like consistency for a firm ride (Figure 5-24). The computer can change the shock absorber damping characteristics almost instantaneously, in 1 millisecond, and can also supply a varying amount of current through the shock absorber windings to provide a wide variety of shock absorber damping characteristics. Depending on the amount of current supplied, the fluid's viscosity or resistance to flow can be varied from thinner than water to a near plastic or solid state or any consistency in-between. This makes the fluid infinitely adjustable in viscosity and achieves continuously variable real-time damping to suit almost every driving condition instantly.

The advantages of magneto-rheological fluid-controlled shock absorber are:

- Ability to smooth out the action of each tire.
- Reduced noise, bounce, and vibration giving a flatter ride by controlling body motions.
- Exceptional roll control during evasive steering maneuvers.
- Excellent handling by controlling weight transfer during lateral and longitudinal maneuvers.
- Enhanced road isolation by reducing high-frequency road noise through the dampers.
- When integrated with ABS and traction control, ensures maximum stability and balance on slippery surfaces and gravel.

SUMMARY

- Shock absorbers or struts play a very important role in ride quality, steering control, and tire life.
- Tire and wheel jounce travel occurs when a tire strikes a bump in the road surface and the tire and wheel move upward.
- Rebound tire and wheel travel occurs when the tire and wheel move downward after jounce travel.
- When a spring is deflected upward during jounce travel, it stores energy. The spring then expands downward in rebound travel with all the energy it stored during the jounce travel. If the spring action is not controlled, the energy in the spring during rebound travel drives the tire against the road surface with excessive force. This action drives the tire and wheel back upward in jounce travel and the wheel continues oscillating up and down.
- The shock absorb[illegible] control spring action and prevent excessive tire and wheel oscillations.

TERMS TO KNOW

Active suspension systems
Adjustable struts
Gas-filled shock absorber
Heavy-duty shock absorbers
Jounce travel
Load-leveling shock absorbers
Magneto-rheological fluid

TERMS TO KNOW (continued)

Rebound travel
Shock absorbers
Shock absorber ratios
Spring insulator
Struts
Travel-sensitive struts
Upper strut mount

- During jounce travel, the piston moves downward in the lower shock absorber chamber; during rebound travel, this piston moves upward. Because the lower shock absorber chamber is sealed and filled with oil, this oil must flow past the piston during any piston movement.
- Valves and openings in the shock absorber piston provide precision control of the oil flow past the piston to control spring action. Shock absorber valves are matched to the amount of energy that may be stored in the spring.
- A nitrogen gas charge is located in the oil reservoir of many shock absorbers and struts to prevent oil cavitation or foaming, which provides more positive shock absorber action.
- Shock absorber ratio refers to the difference between the shock absorber control on the compression and extension cycle. Many shock absorbers provide more control on the extension cycle.
- Internal design is similar in shock absorbers and struts, but struts also support the coil spring.
- Most front struts are connected between the steering knuckle and the upper strut mount.
- Many rear struts are connected between the spindle and the upper support.
- A travel-sensitive shock absorber provides increased resistance to piston movement as the shock absorber is extended.
- Adjustable shock absorbers and struts have a manual adjustment that allows the technician or owner to adjust the strut orifice opening.
- Load-leveling shock absorbers have air pumped into the shock absorbers from an onboard compressor to maintain a specific rear suspension height regardless of the rear suspension load.
- Most front struts rotate on the upper strut mount bearing as the front wheels are turned.

REVIEW QUESTIONS

Short Answer Essays

1. Describe spring action during jounce and rebound travel.
2. Describe uncontrolled spring action without a shock absorber.
3. Explain shock absorber operation.
4. Describe the vehicle safety hazards created by worn-out shock absorbers.
5. Explain shock absorber ratio.
6. Explain the purpose of the nitrogen gas charge in shock absorbers and struts.
7. Explain the differences between heavy-duty mono-tube and conventional twin tube shock absorbers.
8. Identify the purposes of an upper strut mount.
9. Explain the purpose of strut-and-spring insulators.
10. Describe the purpose of the rubber spring bumper on a strut piston rod.

Fill-in-the-Blanks

1. Shock absorbers control spring action and prevent excessive spring _____.
2. Shock absorber design is matched to the _____ rate of the spring.
3. Modern shock absorbers are _____ damping devices that increase resistance the faster the suspension moves.
4. The nitrogen gas charge in a shock absorber prevents oil _____ and _____.
5. The single tube design in a heavy-duty shock absorber prevents excessive _____.
6. A shock absorber with a 70/30 ratio provides more control on the _____ cycle.
7. The lower end of many front struts is bolted to the steering _____.

128

SUMMARIES

Each chapter concludes with a summary of key points from the chapter. These are designed to help the reader review the chapter contents.

TERMS TO KNOW LIST

A list of new terms appears next to the Summary.

REVIEW QUESTIONS

Short answer essay, fill-in-the-blank, and multiple-choice questions are found at the end of each chapter. These questions are designed to accurately assess the student's competence in the stated objectives at the beginning of the chapter.

SUPPLEMENTS

Instructor Resources

The Instructor Resources, now available online, are a robust ancillary product that contains all preparation tools to meet any instructor's classroom needs. It includes chapter outlines in PowerPoint with images, video clips, and animations that coincide with each chapter's content coverage, chapter tests powered by Cognero with hundreds of test questions, a searchable Image Library with all photos and illustrations from the text, theory-based worksheets in Word that provide homework or in-class assignments, the Job Sheets from the Shop Manual in Word, a NATEF correlation chart, and an Instructor's Guide in electronic format.

To access these Instructor Resources, go to login.cengagebrain.com, and create an account or log into your existing account.

CourseMate

The all new CourseMate for *Today's Technician: Automotive Suspension & Steering Systems*, 6e, offers students and instructors access to important tools and resources, all in an online environment. The CourseMate includes an Interactive e-book of the core text, interactive quizzes, flashcards, as well as an Engagement Tracker tool for monitoring students' progress in the CourseMate product.

To learn more about this resource and access free demo CourseMate resources, go to **www.cengagebrain.com**, and search by this book's ISBN (9781285438108). To access CourseMate materials that you have purchased, go to login.cengagebrain.com, enter your access code, and create an account or log into your existing account.

REVIEWERS

The author and publisher would like to extend a special thanks to the instructors who reviewed this text and offered invaluable feedback:

Clay Brown
McHenry County College
Crystal Lake, IL

Steve Calvert
University of Northwestern Ohio
Lima, OH

Christopher J. Marker
University of Northwestern Ohio
Lima, OH

Chapter 1

SAFETY

UPON COMPLETION AND REVIEW OF THIS CHAPTER, YOU SHOULD BE ABLE TO:

- Recognize shop hazards and take the necessary steps to avoid personal injury or property damage.
- Describe the general shop rules that must be followed in an automotive shop.
- Explain the purposes of the Occupational Safety and Health Act.
- Explain two reasons why batteries are a shop hazard.
- Describe how long hair may be a shop hazard.
- Explain how incandescent trouble lights may be a shop hazard.
- Describe the harmful effects of carbon monoxide on the human body.
- Explain why asbestos dust in the shop air is a hazard.
- Describe the requirements for the location of safety equipment in the shop.
- Explain four different types of fires, and the fire extinguisher required for each type of fire.
- Describe three pieces of shop safety equipment other than fire extinguishers.
- Describe the main purposes of the Right-to-Know laws.
- Describe the purpose of MSDS and list the information that must be contained on these sheets.
- Explain the employer's responsibility regarding hazardous materials in the shop.
- Observe all general shop safety precautions.
- Demonstrate all precautions related to personal safety when working in the shop.
- Explain why smoking is dangerous in the shop.
- Explain why drug or alcohol use is dangerous in the shop.
- Describe basic electrical safety precautions.
- Explain gasoline safety precautions.
- Describe housekeeping safety precautions.
- Explain fire safety precautions.
- Describe the proper procedure for using a fire extinguisher.
- Describe the necessary precautions to maintain air bag safety.
- Describe the precautions to be observed when driving vehicles in the shop.
- Explain the proper procedure for lifting heavy objects.
- Describe the necessary steps for hand tool safety.
- Explain the necessary precautions when operating a vehicle lift.
- Describe hydraulic jack and safety stand safety precautions.
- Explain the necessary safety precautions when using power tools.
- Describe the precautions required to maintain compressed-air equipment safety.
- Explain the precautions and environmental concerns related to cleaning equipment safety.
- Describe the safety precautions related to working on a Hybrid electric vehicle.

Introduction

Safety is extremely important in the automotive shop! The knowledge and practice of safety precautions prevent serious personal injury and expensive property damage. Automotive students and technicians must be familiar with shop hazards and shop safety rules. The first step in providing a safe shop is learning about shop hazards and safety rules. The second, and most important, step in this process is applying your knowledge of shop hazards and safety rules while working in the shop. In other words, you must develop safe working habits in the shop from your understanding of shop hazards and safety rules. When shop employees have a careless attitude toward safety, accidents are more likely to occur; therefore, all shop personnel must develop a serious attitude toward safety. The result of this attitude is serious shop personnel who will learn and adopt all shop safety rules.

Shop personnel must be familiar with their rights regarding hazardous waste disposal. These rights are explained in the **Right-to-Know laws**. Shop personnel must also be familiar with the types of hazardous materials in the automotive shop and the proper disposal methods for these materials according to state and federal regulations.

Occupational Safety and Health Act

The **Occupational Safety and Health Act (OSHA)** was passed by the U.S. government in 1970. The purposes of this legislation are:

1. to assist and encourage the citizens of the United States in their efforts to ensure safe and healthful working conditions by providing research, information, education, and training in the field of occupational safety and health.
2. to ensure safe and healthful working conditions for working men and women by authorizing enforcement of the standards developed under the Act.

Because approximately 25 percent of workers are exposed to health and safety hazards on the job, the OSHA is necessary to monitor, control, and educate workers regarding health and safety in the workplace.

Shop Hazards

Service technicians and students encounter many hazards in an automotive shop. When these hazards are known, basic shop safety rules and procedures must be followed to avoid personal injury. Some of the hazards in an automotive shop include the following:

1. Flammable liquids, such as gasoline and paint, must be handled and stored properly in approved, closed containers to comply with safety regulations.
2. Flammable materials, such as oily rags, must be stored properly in closed containers to avoid a fire hazard.
3. Batteries contain a corrosive sulfuric acid solution and produce explosive hydrogen gas while charging.
4. Loose sewer and drain covers may cause foot or toe injuries.
5. Caustic liquids, such as those in hot cleaning tanks, are harmful to skin and eyes.
6. High-pressure air in the shop's compressed-air system can be very dangerous or fatal if it penetrates the skin and enters the bloodstream. High-pressure air released near the eyes may cause eye injury.
7. Frayed cords on electrical equipment and lights may result in severe electrical shock.
8. Hazardous waste material, such as batteries and caustic cleaning solutions, must be handled with adequate personal protection to avoid injury (Figure 1-1).
9. Carbon monoxide from vehicle exhaust is poisonous and potentially fatal.
10. Loose clothing or long hair may become entangled in rotating parts on equipment or vehicles, resulting in serious injury.

Copyright © 2015 Cengage Learning®

FIGURE 1-1 **Always wear recommended safety clothing and equipment when handling hazardous materials.**

11. Dust and vapors generated during some repair jobs are harmful. Asbestos dust, which may be released during brake lining service and clutch service, is a contributor to lung cancer.
12. High noise levels from shop equipment such as an air chisel may be harmful to the ears.
13. Oil, grease, water, or parts cleaning solutions on shop floors may cause someone to slip and fall, resulting in serious injury.
14. The incandescent bulbs used in some trouble lights may shatter if the light is dropped, igniting flammable materials in the area and causing a fire. Many insurance companies now require the use of trouble lights with fluorescent bulbs in the shop.

Shop Safety Rules

Applying basic shop rules helps prevent serious, expensive accidents. Failure to comply with shop rules may cause personal injury or expensive damage to vehicles and shop facilities. It is the responsibility of the employer and all shop employees to make sure that shop rules are understood and followed until these rules become automatic habits. The following basic shop rules should be observed:

1. Always wear safety glasses and other protective equipment that is required by a service procedure (Figure 1-2). For example, a brake parts washer must be used to avoid breathing asbestos dust into the lungs. Asbestos dust is a known cause of lung cancer. This dust is encountered in manual transmission clutch facings and brake linings.
2. Tie long hair securely behind your head, and do not wear loose or torn clothing.
3. Do not wear rings, watches, or loose hanging jewelry. If jewelry such as a ring, metal watchband, or chain makes contact between an electrical terminal and ground, the jewelry becomes extremely hot, resulting in severe burns.
4. Do not work in the shop while under the influence of alcohol or drugs.
5. Set the parking brake when working on a vehicle. If the vehicle has an automatic transmission, place the gear selector in park unless a service procedure requires another

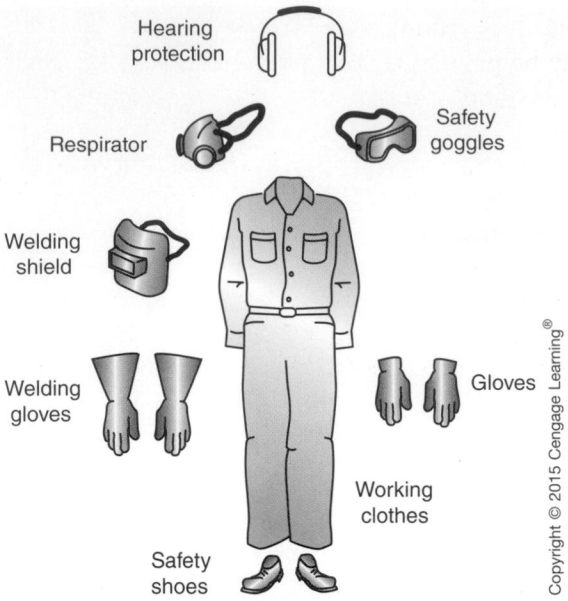

FIGURE 1-2 **Shop safety clothing and equipment, including safety goggles, respirator, welding shield, proper work clothes, ear protection, welding gloves, work gloves, and safety shoes.**

selector position. When the vehicle is equipped with a manual transmission, position the gear selector in neutral with the engine running or in reverse with the engine stopped.

6. Always connect a shop exhaust hose to the vehicle tailpipe, and be sure the shop exhaust fan is running. If it is absolutely necessary to operate a vehicle without a shop exhaust pipe connected to the tailpipe, open the large shop door to provide adequate ventilation. Carbon monoxide in the vehicle exhaust may cause severe headaches and other medical problems. High concentrations of carbon monoxide may result in death!
7. Keep hands, clothing, and wrenches away from rotating parts such as cooling fans. Remember that electric-drive fans may start rotating at any time, even with the ignition off.
8. Always leave the ignition switch off unless a service procedure requires another switch position.
9. Always follow the vehicle manufacturer's recommended procedure to disable the high-voltage electrical system and wait for 5 minutes before working on a **hybrid vehicle**. A switch under the steering column is pushed to the Off position to disable the high-voltage system on some hybrid vehicles. On other hybrid vehicles, the high-voltage disconnect switch is mounted at or near the high-voltage battery pack. On some hybrid vehicles, the high-voltage system retains voltage for 5 minutes after the disable switch is turned off.

A hybrid vehicle has a power train with two power sources. The most common type of hybrid vehicle has a gasoline engine and an electric drive motor(s).

10. Do not smoke in the shop. If the shop has designated smoking areas, smoke only in those areas.
11. Store oily rags and other discarded combustibles in covered metal containers designed for this purpose.
12. Always use the wrench or socket that fits properly on the bolt. Do not substitute metric for English wrenches, or vice versa.
13. Keep tools in good condition. For example, do not use a punch or chisel with a mushroomed end because when struck with a hammer, a piece of the mushroomed metal could break off, resulting in severe eye or other injury.

14. Do not leave power tools running and unattended.
15. Serious burns may be prevented by avoiding contact with hot metal components, such as exhaust manifolds and other exhaust system components, radiators, and some air-conditioning hoses.
16. When a lubricant such as engine oil is drained, always wear heavy plastic gloves because the oil could be hot enough to cause burns.
17. Prior to getting under a vehicle, be sure the vehicle is placed securely on safety stands.
18. Operate all shop equipment, including lifts, according to the equipment manufacturer's recommended procedure. Do not operate equipment unless you are familiar with the correct operating procedure.
19. Do not run or engage in horseplay in the shop.
20. Obey all state and federal fire, safety, and environmental regulations.
21. Do not stand in front of or behind vehicles.
22. Always place fender, seat, and floor mat covers on a customer's vehicle before working on the car.
23. Inform the shop foreman of any safety dangers, as well as suggestions for safety improvement.
24. Do not direct high-pressure air from an air gun toward human skin or near the eyes. High-pressure air may penetrate the skin and enter the bloodstream. Air in the bloodstream may be fatal! High-pressure air discharged near the eyes may cause serious eye damage.

Air Quality

Vehicle exhaust contains small amounts of carbon monoxide, which is a poisonous gas. Weak concentrations of carbon monoxide in the shop air may cause nausea and headaches; strong concentrations may be fatal. All shop personnel are responsible for air quality in the shop. Shop management is responsible for providing an adequate exhaust system that can remove exhaust fumes from the maximum-allowable number of vehicles that may be running in the shop at one time. Technicians should never run a vehicle in the shop unless a shop exhaust hose is installed on the tailpipe of the vehicle. The exhaust fan must be switched on to remove exhaust fumes.

If shop heaters or furnaces have restricted chimneys, they release carbon monoxide emissions into the shop air. Therefore, chimneys should be checked periodically for restriction and proper ventilation. Diesel exhaust contains some carbon monoxide, but particulate emissions are also present in the exhaust from these engines. Particulates are small carbon particles that can be harmful to the lungs.

Monitors are available to measure the level of carbon monoxide in the shop. Some of these monitors read the amount of carbon monoxide present in the shop air; others provide an audible alarm if the concentration of carbon monoxide exceeds the danger level.

The sulfuric acid solution in car batteries is a corrosive, poisonous liquid. If a battery is charged with a fast charger at a high rate for a period of time, the battery becomes hot, and the sulfuric acid solution begins to boil. Under this condition, the battery may emit strong sulfuric acid fumes that may be harmful to the lungs. If this happens, the battery charger should be turned off or the charging rate should be reduced considerably.

Some automotive clutch facings and brake linings contain asbestos. Never use an air hose to blow dirt from these components, because this action disperses asbestos dust into the shop where it may be inhaled by technicians and other people in the shop. A brake parts washer or a vacuum cleaner with special attachments must be used to clean the dust from these components. Even though technicians take every precaution to maintain air quality in the shop, some undesirable gases may still get into the air. For example, exhaust manifolds may get oil on them during an engine overhaul. When the engine is started and these manifolds get hot, the oil burns off the manifolds and pollutes the shop air with oil smoke. Adequate shop ventilation must be provided to take care of this type of air contamination.

Shop Safety Equipment

Fire Extinguishers

Fire extinguishers are one of the most important pieces of safety equipment. All shop personnel must know the location of each fire extinguisher in the shop. If you have to waste time looking for an extinguisher after a fire starts, the fire could get out of control before you get the extinguisher into operation. Fire extinguishers should be located where they are easily accessible at all times. A decal on each fire extinguisher identifies the type of chemical in the extinguisher and provides operating information (Figure 1-3). Shop personnel should be familiar with the following types of fires and fire extinguishers:

1. *Class A fires* are those involving ordinary combustible materials such as paper, wood, clothing, and textiles. **Multipurpose dry chemical fire extinguishers** are used on these fires.
2. *Class B fires* involve the burning of flammable liquids such as gasoline, oil, paint, solvents, and greases. These fires can also be extinguished with multipurpose dry chemical fire extinguishers. In addition, fire extinguishers containing halogen, or halon, may be used to extinguish class B fires. The chemicals in this type of extinguisher attach to the hydrogen, hydroxide, and oxygen molecules to stop the combustion process almost instantly. However, the resultant gases from the use of halogen-type extinguishers are very toxic and harmful to the operator of the extinguisher.
3. *Class C fires* involve the burning of electrical equipment such as wires, motors, and switches. These fires can be extinguished with multipurpose dry chemical fire extinguishers.
4. *Class D fires* involve the combustion of metal chips, turnings, and shavings. Dry chemical fire extinguishers are the only type of extinguisher recommended for these fires.

Some multipurpose dry chemical fire extinguishers may be used on Class A, B, C, or D fires. Additional information regarding which types of extinguishers are used for various types of fires is provided in Table 1-1.

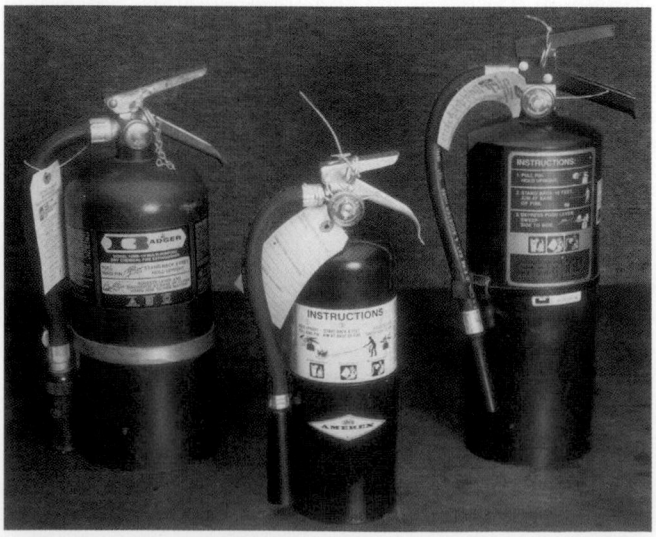

Copyright © 2015 Cengage Learning®

FIGURE 1-3 **Types of fire extinguishers.**

TABLE 1-1 **GUIDE TO FIRE EXTINGUISHER SELECTION**

	Class of fire	Typical fuel involved	Type of extinguisher
Class A Fires (green)	**For ordinary combustibles** Put out a class A fire by blowing its temperature or by coating the burning combustibles.	Wood Paper Cloth Rubber Plastics Rubbish Upholstery	Water* Foam* Multipurpose dry chemical
Class B Fires (red)	**For flammable liquids** Put out a class B fire by smothering it. Use an extinguisher that gives a blanketing, flame-interrupting effect; cover whole flaming liquid surface.	Gasoline Oil Grease Paint Lighter fluid	Foam* Carbon dioxide Halogenated agent Standard dry chemical Purple K dry chemical Multipurpose dry chemical
Class C Fires (blue)	**For electrical equipment** Put out a class C fire by shutting off power as quickly as possible and by always using a nonconducting extinguisher agent to prevent electric shock.	Motors Appliances Wiring Fuse boxes Switchboards	Carbon dioxide Halogenated agent Standard dry chemical Purple K dry chemical Multipurpose dry chemical
Class D Fires (yellow)	**For combustible metals** Put out a class D fire of metal chips, turnings, or shavings by smothering or coating with a specially designed extinguisher agent.	Aluminum Magnesium Potassium Sodium Titanium Zirconium	Dry power extinguisher and agents only

*Cartridge-operated water, foam, and soda-acid types of extinguishers are no longer manufactured. These extinguishers should be removed from service when they become due for their next hydrostatic pressure test.

Copyright © 2015 Cengage Learning®

Causes of Eye Injuries

Eye injuries can occur in various ways in the automotive shop. Some of the more common eye accidents are:

1. Thermal burns from excessive heat
2. Irradiation burns from excessive light such as from an arc welder
3. Chemical burns from strong liquids such as battery electrolyte
4. Foreign material in the eye
5. Penetration of the eye by a sharp object
6. A blow from a blunt object

Wearing safety glasses and observing shop safety rules will prevent most eye accidents.

Eyewash Fountains

If a chemical gets in your eyes, it must be washed out immediately to prevent a chemical burn. An eyewash fountain is the most effective way to wash the eyes. An eyewash fountain is similar to a drinking water fountain, but the eyewash fountain has water jets placed throughout the fountain top. Every shop should be equipped with some type of eyewash facility (Figure 1-4). Be sure you know the location of the eyewash fountain in the shop.

Safety Glasses and Face Shields

The mandatory use of eye protection, either safety glasses or a face shield, is one of the most important safety rules in an automotive shop. Face shields protect the face; safety

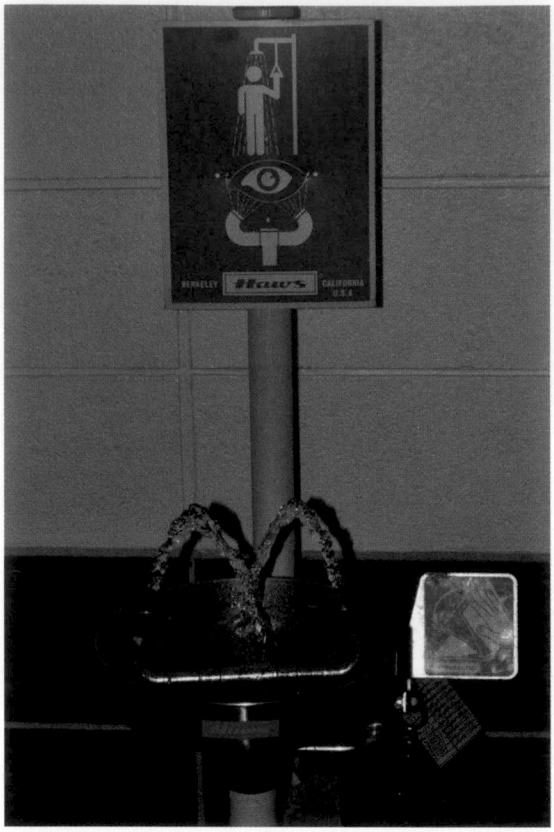

Copyright © 2015 Cengage Learning®

FIGURE 1-4 **Eyewash fountain.**

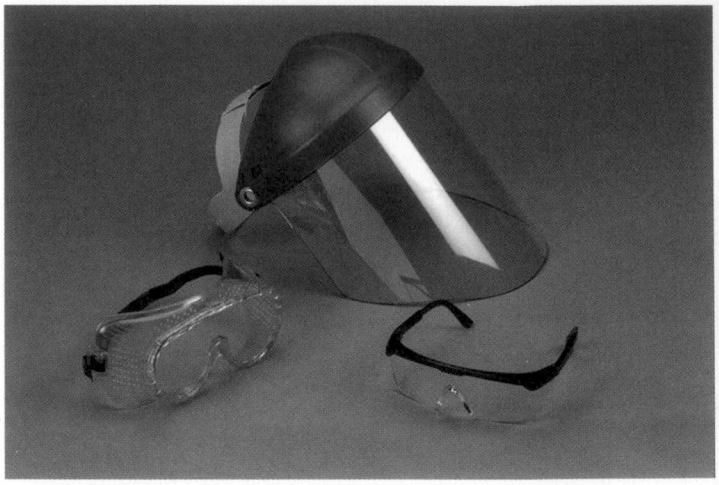

FIGURE 1-5 **Safety glasses with side protection must be worn in the automotive shop.**

glasses protect the eyes. When grinding, safety glasses must be worn, and a face shield can be worn. Many shop insurance policies require the use of eye protection in the shop. Some automotive technicians have been blinded in one or both eyes because they did not wear safety glasses. All safety glasses must be equipped with safety glass and should provide some type of side protection (Figure 1-5). When selecting a pair of safety glasses, they should feel comfortable on your face because if they are uncomfortable, you may remove them, leaving your eyes unprotected. A face shield should be worn when handling hazardous chemicals or when using an electric grinder or buffer (Figure 1-6).

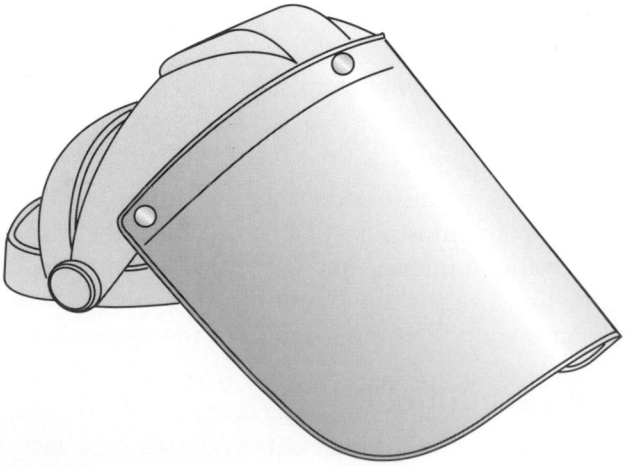

Copyright © 2015 Cengage Learning®

FIGURE 1-6 **Face shield.**

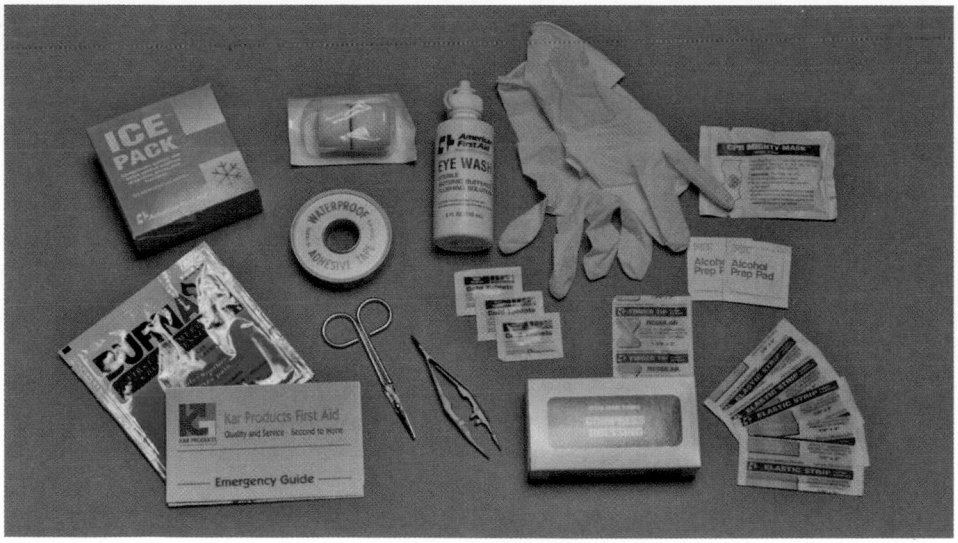

Copyright © 2015 Cengage Learning®

FIGURE 1-7 **First-aid kit.**

First-Aid Kits

First-aid kits should be clearly identified and conveniently located (Figure 1-7). These kits contain such items as bandages in a variety of sizes and ointment required for minor cuts. All shop personnel must be familiar with the location of first-aid kits. At least one of the shop personnel should have basic first-aid training. This person should be in charge of administering first aid and keeping first-aid kits filled.

Shop Layout

There are many different types of shops in the automotive service industry, including:

- New car dealers
- Independent repair shops
- Specialty shops
- Service stations
- Fleet shops

In any shop, the layout is important to maintain shop efficiency and contribute to safety. Every shop employee must be familiar with the location of all safety equipment in the shop. Shop layout includes bays for various types of repairs, equipment storage areas, and office locations. Some specific types of bays are electrical repair, wheel alignment and tires, and machining bays (Figure 1-8). Safety equipment such as fire extinguishers, first-aid kits, and eyewash fountains must be in easily accessible locations, and the location of each piece of safety equipment must be clearly marked. Areas such as the parts department and the parts cleaning area must be located so they are easily accessible from all areas of the shop. The service manager's office should also be centrally located. All shop personnel should familiarize themselves with the shop layout, especially the location of safety equipment. If you know the exact fire extinguisher locations, you may get an extinguisher into operation a few seconds faster. Those few seconds could make the difference between a fire that is quickly extinguished and one that gets out of control, causing extensive damage and personal injury!

The tools and equipment required for a certain type of work are stored in that specific bay. For example, the equipment for electrical and electronic service work is stored in the bay allotted to that type of repair. When certain bays are allotted to specific types of repair work, unnecessary equipment movement is eliminated. Each technician has his or her own

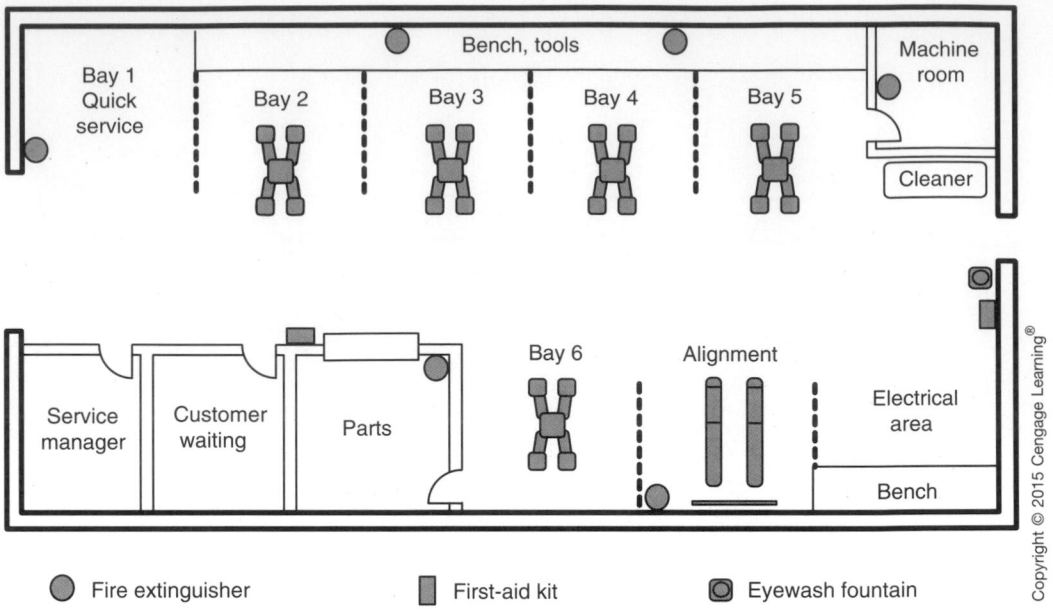

FIGURE 1-8 **Shop layout.**

tools on a portable roll cabinet that is moved to the vehicle being repaired. Special tools are provided by the shop, and these tools should be located on tool boards attached to the wall. Other shops may have a tool room where special tools are located. Adequate workbench space must be provided in those bays where bench work is required.

Safety in the Automotive Shop

Each person in an automotive shop must follow certain basic shop safety rules to remove the danger from shop hazards and prevent personal injury, vehicle damage, and property damage.

Technicians must understand shop safety as it applies to these categories:

1. General shop safety
2. Personal safety

General Shop Safety

When general shop safety rules are observed, personal injury and expensive property or vehicle damage may be avoided, and shop production is increased.

1. All sewer covers must fit properly and be kept securely in place.
2. When servicing brakes or clutches from manual transmissions, always clean asbestos dust from these components with an OSHA-approved **brake parts washer** (Figure 1-9). The brake parts washer contains a Greasoff® brake cleaning solution. The washer is placed under the wheel brake assembly to be washed, and a pump in the washer forces the cleaning solution through a hose and brush that is used to clean the brake components. Removed brake components may also be washed in the top tray on the washer.

A **brake parts washer** is used to wash brake or clutch components and to prevent asbestos particles from becoming airborne and entering the technician's lungs.

WARNING: Do not breathe asbestos dust, because this dust is a known contributor to lung cancer.

Copyright © 2015 Cengage Learning®

FIGURE 1-9 **Brake parts washer.**

3. Always use the correct tool for the job. For example, never strike a hardened steel component, such as a piston pin, with a steel hammer. This type of component may shatter, and fragments can penetrate eyes or skin.
4. Follow the car manufacturer's recommended service procedures.
5. Avoid working on a vehicle parked on an incline.
6. Never work under a vehicle unless the vehicle chassis is supported securely on safety stands.
7. When one end of a vehicle is raised, place wheel chocks on both sides of the wheels remaining on the floor.
8. Be sure that you know the location of shop first-aid kits, eyewash fountains, and fire extinguishers.
9. Collect oil, fuel, brake fluid, and other liquids in the proper safety containers.
10. Use only approved cleaning fluids and equipment. Do not use gasoline to clean parts.
11. Do not leave running equipment unattended.
12. Be sure the safety shields are in place on rotating equipment.
13. All shop equipment must have regularly scheduled maintenance and adjustment.
14. Some shops have safety lines around equipment. Always work within these lines when operating equipment.
15. Be sure that shop heating equipment is well-ventilated.
16. Do not run in the shop or engage in horseplay.
17. Post emergency phone numbers near the phone. These numbers should include doctor, ambulance, fire department, hospital, and police.
18. Do not place hydraulic jack handles where someone can trip over them.
19. Keep aisles clear of debris.

Personal Safety

Personal safety is the responsibility of each technician in the shop.

Always follow these safety practices:

1. Always use the correct tool for the job. If a wrong tool is used, it may slip and cause injury.
2. Follow the vehicle manufacturer's recommended service procedures.

3. Always wear eye protection, such as safety glasses with side protection or a face shield.
4. Wear protective gloves when cleaning parts in hot or cold tanks and when handling hot parts, such as exhaust manifolds.
5. Always wear 1,000V, Class 0 rated, high-voltage gloves to protect against electrical shock when servicing hybrid vehicles.
6. Regularly inspect high-voltage gloves for pin holes or other damage that could result in electric shock. Replace high-voltage gloves at the glove manufacturer's recommended intervals.
7. Before working on a hybrid vehicle, always follow the vehicle manufacturer's recommended service procedure to disable the high-voltage electrical system. On many hybrid vehicles, the high-voltage disconnect switch is located at or near the high-voltage battery pack. Turn the high-voltage disconnect switch Off to disable the high-voltage system.
8. After the high-voltage system is disabled on a hybrid vehicle, always wait for the time period specified by the vehicle manufacturer before servicing the vehicle. On some hybrid vehicles, the high-voltage system retains voltage for 5 minutes after the disable switch is turned off.
9. Some hybrid vehicles contain nickel–metal hydride batteries in the high-voltage battery pack. These batteries contain potassium hydroxide, a highly alkaline solution that damages human tissue. Always follow the vehicle manufacturer's recommended service procedures and wear the recommended protective equipment such as a face shield and gloves when servicing these batteries.
10. Do not smoke when working in the shop. A spark from a cigarette or lighter may ignite flammable materials in the work area.
11. When working on a running engine, keep hands and tools away from rotating parts. Remember that electric-drive fans may start rotating at any time.
12. Do not wear loose clothing, and keep long hair tied behind your head. Loose clothing or long hair is easily entangled in rotating parts.
13. Wear safety shoes or boots. Heavy-duty work boots or shoes with steel toe caps are best for working in the automotive shop. Footwear must protect against heavy falling objects, flying sparks, and corrosive liquids. Soles on footwear must protect against punctures by sharp objects. Athletic shoes and street shoes are not recommended in the shop.
14. Be sure that the shop has adequate ventilation. Carbon monoxide is odorless; do not expect to be able to smell it.
15. Make sure the work area has adequate lighting.
16. When servicing a vehicle, always apply the parking brake. Place the transmission in park with an automatic transmission, or neutral with a manual transmission.
17. Avoid working on a vehicle parked on an incline.
18. Never work under a vehicle unless the vehicle chassis is supported securely on safety stands.
19. Do not use electrical equipment, including trouble lights, with frayed cords.
20. Be sure the safety shields are in place on rotating equipment.
21. Before operating electric equipment, be sure the power cord has a ground connection.
22. When working in an area where noise levels are extreme, wear earplugs or ear covers.
23. Wear a respirator to protect your lungs when working in dusty conditions.
24. Do not use **incandescent bulb-type trouble lights**. Use **fluorescent trouble lights** (Figure 1-10).

Smoking, Alcohol, and Drugs in the Shop

Do not smoke when working in the shop. If the shop has designated smoking areas, smoke only in these areas. Do not smoke in customers' cars. A nonsmoker will not appreciate

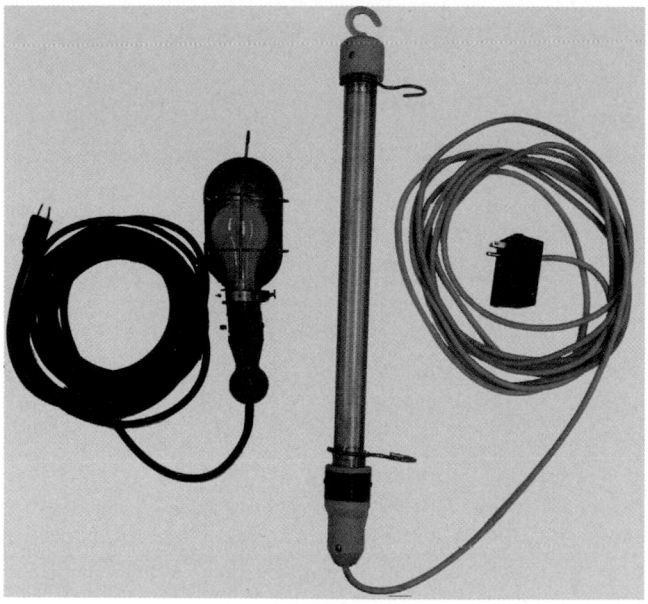

Copyright © 2015 Cengage Learning®

FIGURE 1-10 **Incandescent-type trouble light and fluorescent-type trouble light.**

cigarette odor in the car. A spark from a cigarette or lighter may ignite flammable materials in the workplace. The use of drugs or alcohol must be avoided while working in the shop. Even a small amount of drugs or alcohol affects reaction time. In an emergency situation, slow reaction time may cause personal injury. If a heavy object falls off the workbench, and your reaction time is slowed by drugs or alcohol, you may not get your foot out of the way in time to avoid a foot injury. When a fire starts in the workplace, and you are a few seconds slower getting a fire extinguisher into operation because of alcohol or drug use, it could make the difference between extinguishing a fire and having expensive fire damage.

The improper or excessive use of alcoholic beverages, inhalants, and/or drugs may be referred to as substance abuse.

Electrical Safety

In the automotive shop you will be using electric drills, shop lights, wheel balancers, and wheel aligners.

Observe the following electrical safety precautions on this equipment:

1. Frayed cords on electrical equipment must be replaced or repaired immediately.
2. All electrical cords from lights and electrical equipment must have a ground connection. The ground connector is the round terminal in a three-pronged electrical plug. Do not use a two-pronged adapter to plug in a three-pronged electrical cord. Three-pronged electrical outlets should be mandatory in all shops.
3. Do not leave electrical equipment running and unattended.

Gasoline Safety

Gasoline is a very explosive liquid! One exploding gallon of gasoline has a force equal to 14 sticks of dynamite. The expanding vapors from gasoline are extremely dangerous. These vapors are present even in cold temperatures. Vapors formed in gasoline tanks on cars are controlled, but vapors from a gasoline storage can may escape, resulting in a hazardous situation. Therefore, gasoline storage containers must be placed in a well-ventilated space.

Approved gasoline storage cans have a flash-arresting screen at the outlet (Figure 1-11). This screen prevents external ignition sources from igniting the gasoline within the can while the gasoline is being poured.

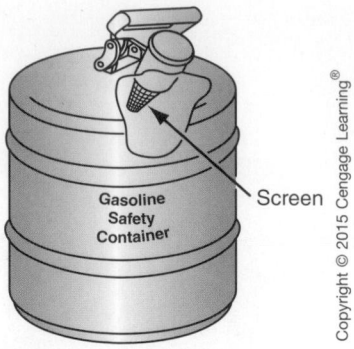

Copyright © 2015 Cengage Learning®

FIGURE 1-11 **Approved gasoline container.**

Follow these safety precautions regarding gasoline containers:

1. Always use approved gasoline containers that are painted red for proper identification.
2. Do not fill gasoline containers completely full. Always leave the level of gasoline at least one inch from the top of the container. This allows for expansion of the gasoline at higher temperatures. If gasoline containers are completely full, the gasoline will expand when the temperature increases. This expansion forces gasoline from the can and creates a dangerous spill.
3. If gasoline containers must be stored, place them in a well-ventilated area such as a storage shed. Do not store gasoline containers in your home or in the trunk of a vehicle.
4. When a gasoline container must be transported, be sure it is secured against upsets.
5. Do not store a partially filled gasoline container for long periods of time, because it may give off vapors and produce potential danger.
6. Never leave gasoline containers open except while filling or pouring gasoline from the container.
7. Do not prime an engine with gasoline while cranking the engine.
8. Never use gasoline as a cleaning agent.

A gasoline can may be called a jerry can.

Fire Safety

When fire safety rules are observed, personal injury and expensive fire damage to vehicles and property may be avoided. Follow these fire safety rules:

1. Familiarize yourself with the location and operation of all shop fire extinguishers.
2. If a fire extinguisher is used, report it to management so the extinguisher can be recharged.
3. Do not use any type of open-flame heater to heat the work area.
4. Do not turn on the ignition switch or crank the engine with a gasoline line disconnected.
5. Store all combustible materials such as gasoline, paint, and oily rags in approved safety containers.
6. Clean up gasoline, oil, or grease spills immediately.
7. Always wear clean shop clothes. Do not wear oil-soaked clothes.
8. Do not allow sparks and flames near batteries.
9. Be sure that welding tanks are securely fastened in an upright position.
10. Do not block doors, stairways, or exits.
11. Do not smoke when working on vehicles.
12. Do not smoke or create sparks near flammable materials or liquids.
13. Store combustible shop supplies such as paint in a closed steel cabinet.

14. Store gasoline in approved safety containers.
15. If a gasoline tank is removed from a vehicle, do not drag the tank on the shop floor.
16. Know the approved fire escape route from your classroom or shop to the outside of the building.
17. If a fire occurs, do not open doors or windows. This action creates extra draft, which makes the fire worse.
18. Do not pour water on a gasoline fire, because the water will make the fire worse.
19. Call the fire department as soon as a fire begins, and then attempt to extinguish the fire.
20. If possible, stand 6 to 10 feet from the fire and aim the fire extinguisher nozzle at the base of the fire with a sweeping action.
21. If a fire produces a lot of smoke in the room, remain close to the floor to obtain oxygen and avoid breathing smoke.
22. If the fire is too hot or the smoke makes breathing difficult, get out of the building.
23. Do not re-enter a burning building.
24. Keep solvent containers covered except when pouring from one container to another. When flammable liquids are transferred from bulk storage, the bulk container should be grounded to a permanent shop fixture, such as a metal pipe. During this transfer process, the bulk container should be grounded to the portable container (Figure 1-12). These ground wires prevent the buildup of a static electric charge, which could cause a spark and a disastrous explosion. Always discard or clean empty solvent containers, because fumes in these containers are a fire hazard.
25. Familiarize yourself with different types of fires and fire extinguishers, and know the type of extinguisher to use on each type of fire.

Using a Fire Extinguisher

Everyone working in the shop must know how to operate the fire extinguishers.

There are several different types of fire extinguishers, but their operation usually involves the following steps:

1. Get as close as possible to the fire without jeopardizing your safety.
2. Grasp the extinguisher firmly and aim the extinguisher at the fire.
3. Pull the pin from the extinguisher handle.

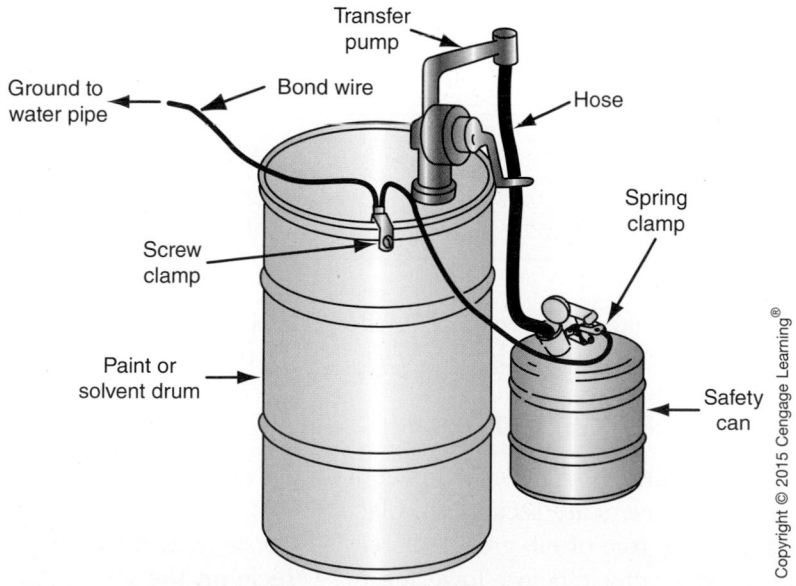

FIGURE 1-12 **Safe procedures for flammable liquid transfer.**

4. Squeeze the handle to dispense the contents of the extinguisher.
5. Direct the fire extinguisher nozzle at the base of the fire, and dispense the contents of the extinguisher with a sweeping action back and forth across the fire. Most extinguishers discharge their contents in 8 to 25 seconds.
6. Always be sure the fire is extinguished.
7. Always keep an escape route open behind you so a quick exit is possible if the fire gets out of control.

Vehicle Operation

When driving a customer's vehicle, observe the following precautions to prevent accidents and maintain good customer relations:

1. Prior to driving a vehicle, make sure the brakes are operating and fasten the safety belt.
2. Before you start the engine, check to be sure there is no person or object under the car.
3. If the vehicle is parked on a lift, be sure the lift is fully down and that the lift arms or components are not in contact with the vehicle chassis.
4. Before driving away, check to see if any objects are directly in front of or behind the vehicle.
5. Always drive slowly in the shop, and watch carefully for personnel and other moving vehicles.
6. Make sure the shop door is up high enough so there is plenty of clearance between the top of the vehicle and the door.
7. Watch the shop door to be certain that it is not coming down as you attempt to drive under the door.
8. If a road test is necessary, wear your seat belt, obey all traffic laws, and never drive in a reckless manner.
9. Do not squeal tires when accelerating or turning corners.

If customers observe that service personnel take good care of their car by driving carefully and by installing fender, seat, and floor mat covers, the service department's image is greatly enhanced in their eyes. These procedures impress upon the customers that shop personnel respect their car. Conversely, if grease spots are found on the upholstery or fenders after service work is completed, the customers will probably think the shop is careless, not only in car care but also in service work quality.

Housekeeping Safety

Customer Care: When customers see that you are concerned about their vehicle and that you operate a shop with excellent housekeeping habits, they will be impressed and will likely keep returning for service.

Careful housekeeping habits prevent accidents and increase worker efficiency. Good housekeeping also helps impress upon the customer that quality work is a priority in this shop.

Follow these housekeeping rules:

1. Keep aisles and walkways clear of tools, equipment, and other items.
2. Be sure all sewer covers are securely in place.
3. Keep floor surfaces free of oil, grease, water, and loose material.
4. Sweep up under a vehicle before lowering the vehicle on the lift.

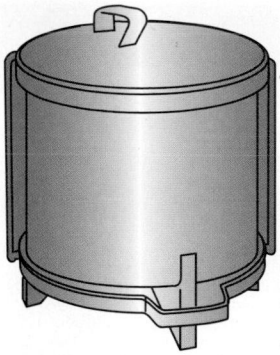

Copyright © 2015 Cengage Learning®

FIGURE 1-13 **Dirty shop towels or rags must be kept in an approved, closed container.**

Copyright © 2015 Cengage Learning®

FIGURE 1-14 **Store combustible materials in an approved safety cabinet.**

5. Proper trash containers must be conveniently located, and these containers should be emptied regularly.
6. Access to fire extinguishers must be unobstructed at all times, and fire extinguishers should be checked for proper charge at regular intervals.
7. Tools must be kept clean and in good condition.
8. When not in use, tools must be stored in their proper location.
9. Oily rags must be stored in approved, covered containers (Figure 1-13). A slow generation of heat occurs from the oxidation of oil on these rags. Heat may continue to be generated until the ignition temperature is reached. The oil and rags then begin to burn, causing a fire. This action is called spontaneous combustion. However, if the oily rags are in an airtight, approved container, there is not enough oxygen to cause burning.
10. Store paint, gasoline, and other flammable liquids in a closed steel cabinet (Figure 1-14).
11. Rotating components on equipment and machinery must have guards, and all shop equipment should have regular service and adjustment schedules.
12. Keep the workbenches clean. Do not leave heavy objects, such as used parts, on the bench after you are finished with them.
13. Keep parts and materials in their proper location.
14. When not in use, creepers must not be left on the shop floor. Creepers should be stored in a specific location.
15. The shop should be well lit, and all lights should be in working order.
16. Frayed electrical cords on lights or equipment must be replaced.
17. Walls and windows should be cleaned regularly.
18. Stairs must be clean, well lit, and free of loose material.

SERVICE TIP:
When you are finished with a tool, never set it on the customer's car. After using a tool, the best place for it is in your tool box or on the workbench. Many tools have been lost by leaving them on customers' vehicles.

Air Bag Safety

1. When service is performed on any **air bag system** component, always disconnect the negative battery cable, isolate the cable end, and wait for the amount of time specified by the vehicle manufacturer before proceeding with the necessary diagnosis or service. The average waiting period is two minutes, but some vehicle manufacturers specify up to 10 minutes. Photo Sequence 1 shows a typical procedure for removing an air bag module.

   On some recent-model vehicles, the air bag system is divided into different disabling and enabling zones for diagnostic and service purposes. When performing

An **air bag system** is designed to protect the driver and/or passengers in a vehicle collision.

PHOTO SEQUENCE 1

Typical Procedure for Removing Air Bag Module

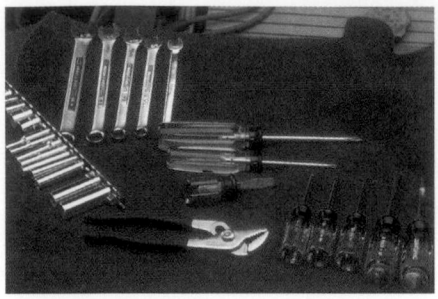

P1-1 **Tools required to remove the air bag module: safety glasses, seat covers, screwdriver set, torx driver set, battery terminal pullers, battery pliers, assorted wrenches, ratchet and socket set, and service manual.**

Copyright © 2015 Cengage Learning®

P1-2 **Place the seat and fender covers on the vehicle.**

Copyright © 2015 Cengage Learning®

P1-3 **Place the front wheels in the straight-ahead position, and turn the ignition switch to the lock position.**

Copyright © 2015 Cengage Learning®

P1-4 **Disconnect the negative battery cable.**

Copyright © 2015 Cengage Learning®

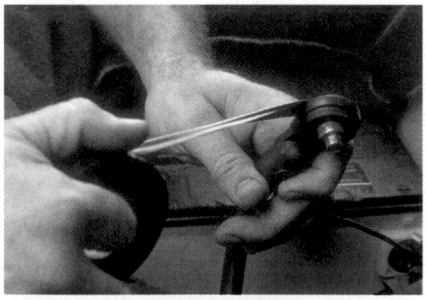

P1-5 **Tape the cable terminal to prevent accidental connection with the battery post. Note: A piece of rubber hose can be substituted for the tape.**

Copyright © 2015 Cengage Learning®

P1-6 **Remove the SIR fuse from the fuse box. Wait 10 minutes to allow the reserve energy to dissipate.**

Copyright © 2015 Cengage Learning®

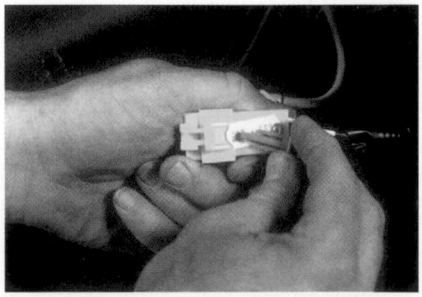

P1-7 **Remove the connector position assurance (CPA) from the yellow electrical connector at the base of the steering column.**

Copyright © 2015 Cengage Learning®

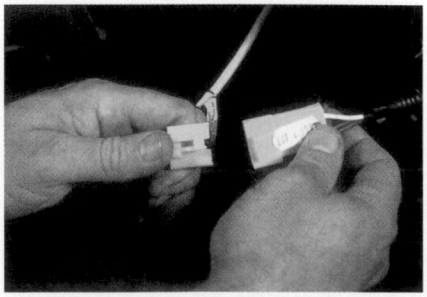

P1-8 **Disconnect the yellow two-way electrical connector.**

Copyright © 2015 Cengage Learning®

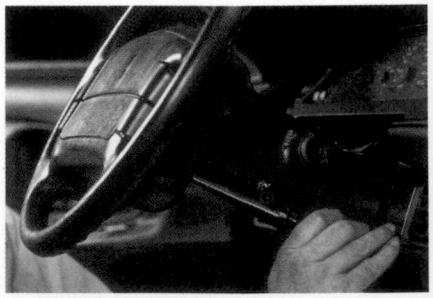

P1-9 **Remove the four bolts that secure the module from the rear of the steering wheel.**

Copyright © 2015 Cengage Learning®

P1-10 **Rotate the horn lead 1/4 turn and disconnect.**

Copyright © 2015 Cengage Learning®

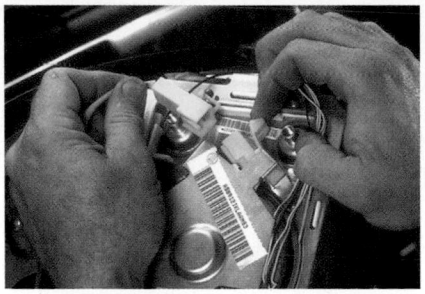

P1-11 **Disconnect the electrical connectors.**

Copyright © 2015 Cengage Learning®

P1-12 **Remove the module.**

Copyright © 2015 Cengage Learning®

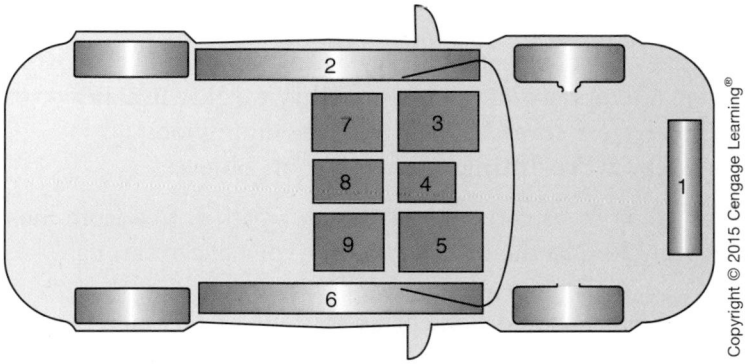

FIGURE 1-15 **Air bag system disabling and enabling zones.**

vehicle service on or near air bag system components, the vehicle manufacturer recommends disabling the air bag components, in the zone where the air bag components are located rather than disconnecting the vehicle battery (Figure 1-15). The air bag system components in each disabling zone are the following:

Zone 1–Front end air bag sensors
Zone 2–Driver/side impact sensor
Zone 3–Steering wheel air bag inflator module and coil
Zone 4–Air bag system sensing and diagnostic module (SDM)
Zone 5–Air bag system instrument panel module
Zone 6–Passenger/side impact sensor
Zone 7–Driver seat with L/F side impact module and seat belt pretensioner
Zone 8–Not used
Zone 9–Passenger seat with R/F side impact module and seat belt pretensioner

2. Replacement air bag system parts must have the same part number as the original part. Replacement parts of lesser or questionable quality must not be used. Improper or inferior components may result in inappropriate air bag deployment and injury to the vehicle occupants.
3. Do not strike or jar a sensor or an air bag system diagnostic monitor (ASDM). This may cause air bag deployment or make the sensor inoperative. Accidental air bag deployment may cause personal injury, and an inoperative sensor may result in air bag deployment failure, causing personal injury to vehicle occupants.

SERVICE TIP:
After disconnecting battery voltage from an air bag system, failure to wait for the specified time period before servicing the system may cause accidental air bag deployment.

4. All sensors and mounting brackets must be properly torqued to ensure correct sensor operation before an air bag system is powered up. If sensor fasteners do not have the proper torque, improper air bag deployment may result in injury to vehicle occupants.
5. When working on the electrical system on an air-bag-equipped vehicle, use only the vehicle manufacturer's recommended tools and service procedures. The use of improper tools or service procedures may cause accidental air bag deployment and personal injury. For example, do not use 12V or self-powered test lights when servicing the electrical system on an air-bag-equipped vehicle.
6. When handling a deployed air bag always wear safety glasses and shop gloves. Dispose of the deployed air bag through the normal refuse channels. Remove the shop gloves and wash your hands in mild soap and water.

WARNING: A deployed air bag may be partially covered with sodium hydroxide, which is a skin and eye irritant.

Lifting and Carrying

Many automotive service jobs require heavy lifting. Know your maximum weight-lifting ability, and do not attempt to lift more than this weight. If a heavy part exceeds your weight-lifting ability, have a coworker help with the lifting job.

Follow these steps when lifting or carrying an object:

1. If the object is going to be carried, be sure your path is free from loose parts or tools.
2. Position your feet close to the object; position your back reasonably straight for proper balance.
3. Your back and elbows should be kept as straight as possible. Continue to bend your knees until your hands reach the best lifting location on the object.
4. Be certain the container is in good condition. If a container falls apart during the lifting operation, parts may drop out of the container and result in foot injury or part damage.
5. Maintain a firm grip on the object; do not attempt to change your grip while lifting is in progress.
6. Straighten your legs to lift the object and keep the object close to your body. Use your leg muscles rather than your back muscles (Figure 1-16).
7. If you have to change the direction of travel, turn your whole body. Do not twist.

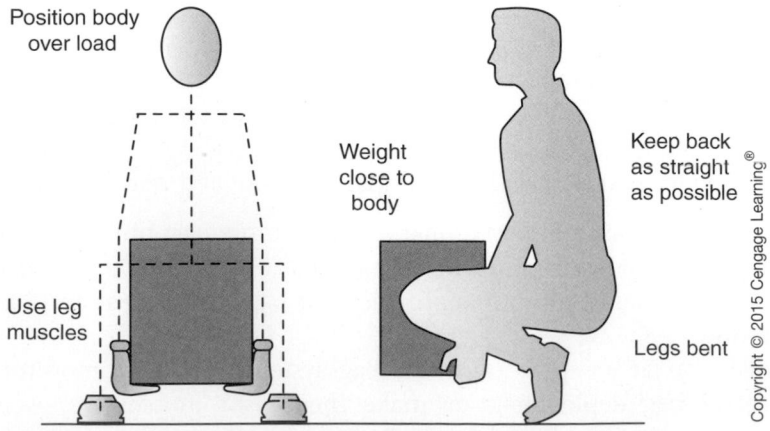

FIGURE 1-16 **Use your leg muscles, never your back, to lift heavy objects.**

8. Do not bend forward to place an object on a workbench or table. Position the object on the front surface of the workbench and slide it back. Do not pinch your fingers under the object while setting it on the front of the bench.
9. If the object must be placed on the floor or a low surface, bend your legs to lower the object. Do not bend your back forward, because this movement strains back muscles.
10. When a heavy object must be placed on the floor, place suitable blocks under the object to prevent jamming your fingers.

Hand Tool Safety

Many shop accidents are caused by improper use and care of hand tools.

Follow these safety steps when working with hand tools:

1. Maintain tools in good condition and keep them clean. Worn tools may slip and result in hand injury. If a hammer with a loose head is used, the head may fly off and cause personal injury or vehicle damage. If your hand slips off a greasy tool, it may cause some part of your body to hit the vehicle, causing injury.
2. Using the wrong tool for the job may damage the tool, fastener, or your hand if the tool slips. If you use a screwdriver as a chisel or pry bar, the blade may shatter, causing serious personal injury.
3. Use sharp-pointed tools with caution. Always check your pockets before sitting on the vehicle seat. A screwdriver, punch, or chisel in the back pocket may put an expensive tear in the upholstery. Do not lean over fenders with sharp tools in your pockets.
4. Tools that are intended to be sharp should be kept sharp. A sharp chisel, for example, will do the job faster with less effort.

Lift Safety

CAUTION:
When a vehicle is raised on a lift, the vehicle must be raised high enough to allow engagement of the lift locking mechanism.

Special precautions and procedures must be followed when a vehicle is raised on a lift.

Follow these steps for safe lift operation:

1. Always be sure the lift is completely lowered before driving a vehicle on or off the lift.
2. Do not hit or run over lift arms and adaptors when driving a vehicle on or off the lift. Have a coworker guide you when driving a vehicle onto the lift. Do not stand in front of a lift with the car coming toward you.
3. Be sure the lift pads contact the car manufacturer's recommended lifting points shown in the service manual. If the proper lifting points are not used, components under the vehicle such as brake lines or body parts may be damaged. Failure to use the recommended lifting points may cause the vehicle to slip off the lift, resulting in severe vehicle damage and personal injury.
4. Before a vehicle is raised or lowered, close the doors, hood, and trunk lid.
5. When a vehicle has been lifted a short distance off the floor, stop the lift and check the contact between the hoist lift pads and the vehicle to be sure the lift pads are still on the recommended lifting points.
6. When a vehicle has been raised, be sure the safety mechanism is in place to prevent the lift from dropping accidentally.
7. Prior to lowering a vehicle, always make sure there are no objects, tools, or people under the vehicle.
8. Do not rock a vehicle on a lift during a service job.
9. When a vehicle is raised, removal of some heavy components may cause vehicle imbalance. For example, because front-wheel-drive cars have the engine and transaxle at the front of the vehicle, these cars have most of their weight on the front end. Removing a

heavy rear-end component on these cars may cause the back end of the car to rise off the lift. If this happens, the vehicle could fall off the lift!

10. Do not raise a vehicle on a lift with people in the vehicle.
11. When raising pickup trucks and vans on a lift, remember these vehicles are higher than a passenger car. Be sure there is adequate clearance between the top of the vehicle and the shop ceiling or components under the ceiling.
12. Do not raise a four-wheel drive vehicle with a frame contact lift unless proper adaptors are used. Lifting a vehicle on a frame contact lift without the proper adaptors may damage axle joints.
13. Do not operate a front-wheel drive vehicle that is raised on a frame contact lift. This may damage the front drive axles.

Hydraulic Jack and Safety Stand Safety

WARNING: Always make sure the safety stand weight capacity rating exceeds the vehicle weight that you are planning to raise. A safety stand with insufficient weight capacity may collapse, resulting in vehicle damage and/or personal injury.

WARNING: Never lift a vehicle with a floor jack if the weight of the vehicle exceeds the rated capacity of the jack. If the vehicle weight exceeds the weight rating of the floor jack, the vehicle weight may collapse the floor jack, resulting in jack and vehicle damage and/or personal injury.

Accidents involving the use of floor jacks and safety stands may be avoided if these safety precautions are followed:

1. Never work under a vehicle unless safety stands are placed securely under the vehicle chassis and the vehicle is resting on these stands (Figure 1-17).
2. Prior to lifting a vehicle with a floor jack, be sure that the jack lift pad is positioned securely under a recommended lifting point on the vehicle. Lifting the front end of a vehicle with the jack placed under a radiator support may cause severe damage to the radiator and support.
3. Position the safety stands under a strong chassis member such as the frame or axle housing. The safety stands must contact the vehicle manufacturer's recommended lifting points.

Copyright © 2015 Cengage Learning®

FIGURE 1-17 **Safety stands.**

4. Because the floor jack is on wheels, the vehicle and safety stands tend to move as the vehicle is lowered from a floor jack onto the safety stands. Always be sure the safety stands remain under the chassis member during this operation, and be sure the safety stands do not tip. All the safety stand legs must remain in contact with the shop floor.
5. When the vehicle is lowered from the floor jack onto the safety stands, remove the floor jack from under the vehicle. Never leave a jack handle sticking out from under a vehicle, which may trip someone and cause an injury.

Power Tool Safety

Power tools use electricity, shop air, or hydraulic pressure as a power source. Careless operation of power tools may cause personal injury or vehicle damage.

Follow these steps for safe power tool operation:

1. Do not operate power tools with frayed electrical cords.
2. Be sure the power tool cord has a proper ground connection.
3. Do not stand on a wet floor while operating an electric power tool.
4. Always unplug an electric power tool before servicing the tool.
5. Do not leave a power tool running and unattended.
6. When using a power tool on small parts, do not hold the part in your hand. The part must be secured in a bench vise or with locking pliers.
7. Do not use a power tool on a job where the maximum capacity of the tool is exceeded.
8. Be sure that all power tools are in good condition; always operate these tools according to the tool manufacturer's recommended procedure.
9. Make sure all protective shields and guards are in position.
10. Maintain proper body balance while using a power tool.
11. Always wear safety glasses or a face shield.
12. Wear ear protection.
13. Follow the equipment manufacturer's recommended maintenance schedule for all shop equipment.
14. Never operate a power tool unless you are familiar with the tool manufacturer's recommended operating procedure, because serious accidents can occur.
15. Always make sure that the wheels are securely attached and in good condition on the electric grinder.
16. Keep fingers and clothing away from grinding and buffing wheels. When grinding or buffing a small part, hold the part with a pair of locking pliers.
17. Always make sure the sanding or buffing disc is securely attached to the sander pad.
18. Special heavy-duty sockets must be used on impact wrenches. If ordinary sockets are used on an impact wrench, they may break and cause serious personal injury.

Compressed-Air Equipment Safety

The shop air supply contains high-pressure air in the shop compressor and air lines. Serious injury or property damage may result from careless operation of compressed-air equipment.

Follow these steps to improve safety:

1. Never operate an air chisel unless the tool is securely connected to the chisel with the proper retaining device.
2. Never direct a blast of air from an air gun toward any part of your body. If air penetrates the skin and enters the bloodstream, it may cause very serious health problems and even death.

3. Safety glasses or a face shield should be worn for all shop tasks, including those tasks involving the use of compressed-air equipment.
4. Wear ear protection when using compressed-air equipment.
5. Always maintain air hoses and fittings in good condition. If an end suddenly blows off an air hose, the hose will whip around, causing possible personal injury.
6. Do not direct compressed air against the skin, because it may penetrate the skin, especially through small cuts or scratches, and enter the bloodstream, causing death or serious health complications. Use only OSHA-approved air gun nozzles.
7. Do not use an air gun to blow debris off clothing or hair.
8. Do not clean the workbench or floor with compressed air. This action may blow very small parts toward your skin or into your eye. Small parts blown by compressed air may also cause vehicle damage. For example, if the car in the next stall has the air cleaner removed, a small part may find its way into the carburetor or throttle body. When the engine is started, this part will likely be pulled into a cylinder by engine vacuum, and the part will penetrate through the top of a piston.
9. Never spin bearings with compressed air because the bearing will rotate at extremely high speed. This may damage the bearing or cause it to disintegrate, causing personal injury.
10. All pneumatic tools must be operated according to the tool manufacturer's recommended operating procedure.
11. Follow the equipment manufacturer's recommended maintenance schedule for all compressed-air equipment.

Cleaning Equipment Safety and Environmental Considerations

Cleaning Equipment Safety

All technicians are required to clean parts during their normal work routines. Face shields and protective gloves must be worn while operating cleaning equipment. In most states, environmental regulations require that the runoff from steam cleaning must be contained in the steam cleaning system. This runoff cannot be dumped into the sewer system. Because it is expensive to contain this runoff in the steam cleaning system, the popularity of steam cleaning has decreased. The solution in hot and cold cleaning tanks may be caustic, and contact between this solution and skin or eyes must be avoided. Parts cleaning often creates a slippery floor, and care must be taken when walking in the parts cleaning area. The floor in this area should be cleaned frequently. When the cleaning solution in hot or cold cleaning tanks is replaced, environmental regulations require that the old solution be handled as hazardous waste. Use caution when placing aluminum or aluminum alloy parts in a cleaning solution. Some cleaning solutions will damage these components. Always follow the cleaning equipment manufacturer's recommendations.

Parts Washers with Electromechanical Agitation

Some parts washers provide electromechanical agitation of the parts to provide improved cleaning action (Figure 1-18). These parts washers may be heated with gas or electricity. Various water-based **hot cleaning tank** solutions are available, depending on the type of metals being cleaned. For example, Kleer-Flo Greasoff® No. 1 powdered detergent is available for cleaning iron and steel. Nonheated electromechanical parts washers are also available, and these washers use cold cleaning solutions such as Kleer-Flo Degreasol® formulas.

A **hot cleaning tank** uses a heated solution to clean metal parts.

Many cleaning solutions, such as Kleer-Flo Degreasol® 99R, contain no ingredients listed as hazardous by the Environmental Protection Agency's Resource Conservation and Recovery Act (RCRA). This cleaning solution is a blend of sulfur-free hydrocarbons,

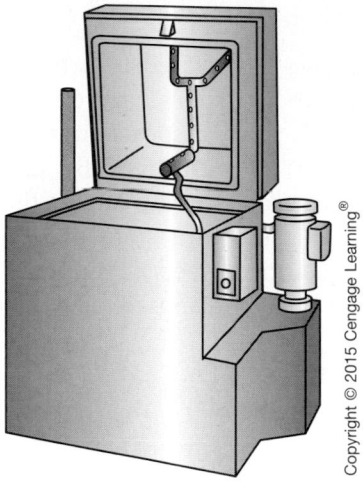

Copyright © 2015 Cengage Learning®

FIGURE 1-18 **Parts washer with electromechanical agitation.**

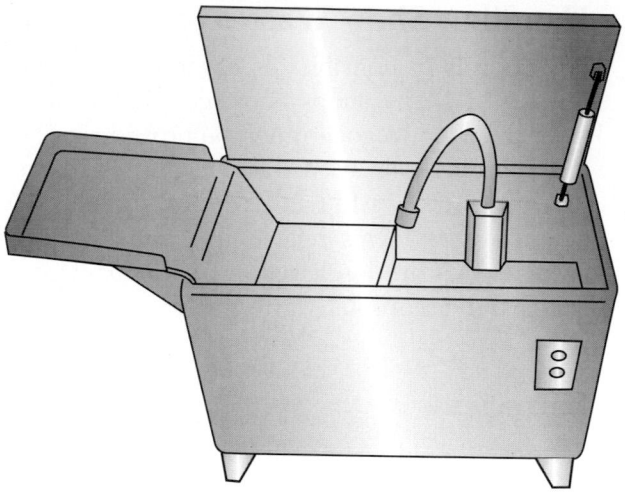

Copyright © 2015 Cengage Learning®

FIGURE 1-19 **Cold parts washer with agitated immersion tank.**

wetting agents, and detergents. Degreasol® 99R does not contain aromatic or chlorinated solvents, and it conforms to California's Rule 66 for clean air. Always use the cleaning solution recommended by the equipment manufacturer.

Cold Parts Washer with Agitation Immersion Tank

Some parts washers have an agitator immersion chamber under the shelves that provides thorough parts cleaning. Folding work shelves provide a large upper cleaning area with a constant flow of solution from the dispensing hose (Figure 1-19). This **cold parts washer** operates on Degreasol® 99R cleaning solution.

A **cold parts washer** uses a nonheated solution to clean metal parts.

Aqueous Parts Cleaning Tank

The **aqueous parts cleaning tank** uses a water-based, environmentally friendly cleaning solution, such as Greasoff® 2, rather than traditional solvents. The immersion tank is heated and agitated for effective parts cleaning (Figure 1-20). A sparger bar pumps a constant flow of cleaning solution across the surface to push floating oils away, and an integral skimmer removes these oils. This action prevents floating surface oils from redepositing on cleaned parts.

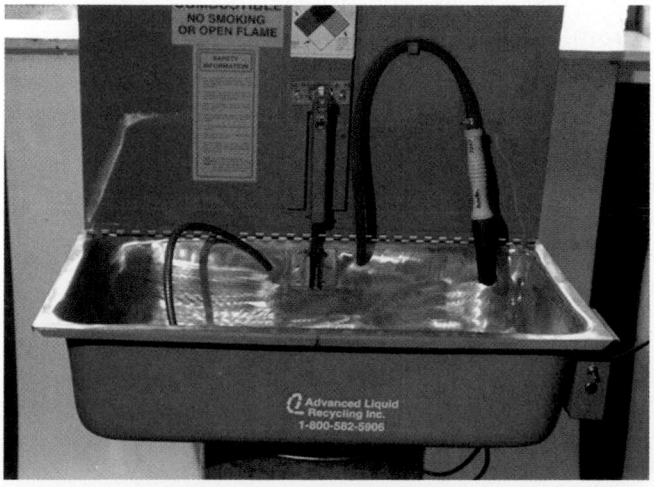

Copyright © 2015 Cengage Learning®

FIGURE 1-20 **Aqueous parts cleaning tank.**

Handling Shop Wastes

The shop is responsible for hazardous waste until such waste is delivered to a hazardous waste site. Many shops contract a hazardous waste hauler to transport hazardous waste from the shop to government-approved recyclers or hazardous waste disposal sites. Always hire a properly licensed waste hauler, and have a written contract with them. Be sure you know how the waste hauler disposes of shop wastes. The hazardous waste hauler fills out the necessary forms related to waste disposal and communicates with state and federal agencies that are in charge of hazardous waste disposal regulations. Keep all shipping bills from your hazardous waste hauler to prove you have recycled or disposed of hazardous waste material.

Batteries

Batteries should be recycled by shipping them to a reclaimer or back to the battery distributor. When defective batteries are stored on-site, they should be kept in watertight, acid-resistant containers. Acid residue from batteries is hazardous because it is corrosive and may contain lead and other toxins. Inspect defective batteries for cracks and leaks. Cover spilled battery acid with baking soda or lime to neutralize it, then clean up and dispose of any other hazardous material.

Oil

Used oil is commonly hauled to an oil recycling facility. Place oil drip pans under a vehicle with oil leaks so oil does not drip onto the storage area. In some states it is legal to burn used oil in a commercial space heater. State and local authorities must be contacted regarding regulations and permits.

Oil Filters

Used oil filters should be allowed to drain into an appropriate drip pan for 24 hours. After the draining process, oil filters should be squashed and recycled.

Solvents

Parts cleaning equipment that uses hazardous cleaning chemicals should be replaced with cleaning equipment that uses water-based degreasers. If hazardous chemicals are used in cleaning equipment, these chemicals must be recycled or disposed of as hazardous waste. Evaporation from spent cleaning chemicals is a contributor to ozone depletion and smog formation. Spent cleaning chemicals should be placed in closed, labeled containers and stored in drip pans or in diked areas. The storage area for waste materials should be covered to prevent rain from washing contaminants from stored materials into the groundwater. This storage area may have to be fenced and locked if vandalism is a possibility.

Liquids

Engine coolant should be collected and recycled in an approved coolant recycling machine. Other liquids such as brake fluid and transmission fluid should be labeled and stored in the same area as solvents. Used brake fluid or transmission fluid should be recycled or disposed of as hazardous waste material.

Shop Towels

When dirty shop towels are stored on-site, they should be placed in closed containers that are clearly marked "Contaminated shop towels only." Shop towels should be cleaned by a laundry service that has the capability to treat the wastewater generated by cleaning these towels.

Refrigerants

When servicing automotive air-conditioning systems, it is illegal to vent refrigerants to the atmosphere. Certified equipment must be used to recover and recycle the refrigerant and to recharge air-conditioning systems. This service work must be performed by an EPA-certified technician.

The EPA approves a Refrigerant Recovery and Recycling Review and Quiz supplied by the National Institute for Automotive Service Excellence (ASE). Upon successful completion of the ASE review and quiz, a technician is certified to service mobile air-conditioning refrigerant systems. The ASE review and quiz meets Section 609 regulations in the Clean Air Act Amendments.

Hazardous Waste Disposal

Hazardous waste materials in automotive shops are chemicals or components, which the shop no longer needs. These materials pose a danger to the environment and to people if they are disposed off in ordinary trash cans or sewers. However, no material is considered hazardous waste until the shop has finished using it and is ready to dispose it off. The **Environmental Protection Agency (EPA)** publishes a list of hazardous materials, which is included in the Code of Federal Regulations. Waste is considered hazardous if it is included in the EPA list of hazardous materials, or if it has one or more of these characteristics:

The **Environmental Protection Agency (EPA)** is the federal agency responsible for air and water quality in the United States.

1. **Reactive**. Any material that reacts violently with water or other chemicals is considered hazardous. If a material releases cyanide gas, hydrogen sulphide gas, or similar gases when exposed to low-pH acid solutions, it is hazardous.
2. **Corrosive**. If a material burns the skin or dissolves metals and other materials, it is considered hazardous.
3. **Toxic**. Materials are hazardous if they leach one or more of eight heavy metals in concentrations greater than 100 times the primary drinking water standard.
4. **Ignitable**. A liquid is hazardous if it has a flashpoint below 140°F (60°C). A solid is hazardous if it ignites spontaneously.

A material that is **reactive** reacts with some other chemicals and gives off gas(es) during the reaction.

Federal and state laws control the disposal of hazardous waste materials. Every shop employee must be familiar with these laws. Hazardous waste disposal laws include the **Resource Conservation and Recovery Act (RCRA)**. This law states that hazardous material users are responsible for hazardous materials from the time they become waste until the proper waste disposal is completed. Many automotive shops hire an independent hazardous waste hauler to dispose of hazardous waste material (Figure 1-21). The shop owner or manager should have a written contract with the hazardous waste hauler. Rather than hauling hazardous waste material to an approved hazardous waste disposal site, a shop may choose to recycle the material in the shop; therefore, hazardous waste material must be properly and safely stored. The user is responsible for the transportation of this material until it arrives at an approved hazardous waste disposal site and is processed according to the law.

A material that is **corrosive** causes another material to be gradually worn away by chemical action.

A **toxic** substance is poisonous to animal or human life.

The RCRA controls these types of automotive waste:

1. Paint and body repair products waste
2. Solvents for parts and equipment cleaning
3. Batteries and battery acid
4. Mild acids used for metal cleaning and preparation
5. Waste oil, engine coolants, and antifreeze
6. Air-conditioning refrigerants
7. Engine oil filters

A substance that is **ignitable** can be ignited spontaneously or by another source of heat or flame.

Copyright © 2015 Cengage Learning®

FIGURE 1-21 **Hazardous waste hauler.**

Never, under any circumstances, use these methods to dispose of hazardous waste material:

1. Pour hazardous wastes on weeds to kill them.
2. Pour hazardous wastes on gravel streets to prevent dust.
3. Throw hazardous wastes in a dumpster.
4. Dispose of hazardous wastes anywhere but an approved disposal site.
5. Pour hazardous wastes down sewers, toilets, sinks, or floor drains.

The Right-to-Know laws state that employees have a right to know when the materials they use at work are hazardous. The Right-to-Know laws started with the **Hazard Communication Standard** published by the Occupational Safety and Health Administration (OSHA) in 1983. This document was originally intended for chemical companies and manufacturers that required employees to handle hazardous materials in their work situation. Currently, most states have established their own Right-to-Know laws. Meanwhile, federal courts have decided to apply these laws to all companies, including automotive service shops. Under the Right-to-Know laws, the employer has three responsibilities regarding the handling of hazardous materials by its employees.

First, all employees must be trained about the types of hazardous materials they will encounter in the workplace. Employees must be informed about their rights under legislation regarding the handling of hazardous materials. All hazardous materials must be properly labeled, and information about each hazardous material must be posted on **safety data sheets (SDS)**, which are available from the manufacturer (Figure 1-22). A major change that occurred under OSHA's HazCom 2012 regulations has to do with Material Safety Data Sheets. OSHA's current requirement aligns itself with the UN's Globally Harmonized System of Classification and Labeling of Chemicals (GHS) Safety Data Sheet (SDS) preparation requirements. What has historically been called a Material Safety Data Sheet, or MSDS, will now be referred to as Safety Data Sheets, or SDS. In Canada, MSDS may be called **workplace hazardous materials information systems (WHMIS)**.

Safety data sheets (SDS) provide all the necessary data about hazardous materials.

The employer has a responsibility to place MSDS where they are easily accessible by all employees. The MSDS provide extensive information about hazardous materials such as:

1. Chemical name.
2. Physical characteristics.

MATERIAL SAFETY DATA SHEET

PRODUCT NAME: KLEAN-A-KARB (aerosol) #- HPMS 102068
PRODUCT: 5078, 5931, 6047T
(page 1 of 2)

1. Ingredients	CAS #	ACGIH TLV	OSHA PEL	OTHER LIMITS	%
Acetone	67-64-1	750 ppm	750 ppm		2–5
Xylene	1330-20-7	100 ppm	100 ppm		68–75
2-Butoxy ethanol	111-76-2	25 ppm	25 ppm	(skin)	3–5
Methanol	67-56-1	200 ppm	200 ppm		3–5
Detergent	-	NA	NA		0–1
Propane	74-98-6	NA	1000 ppm		10–20
Isobutane	75-28-5	NA	NA	1000 ppm	10–20

2. PHYSICAL DATA : (without propellent)

Specific Gravity : 0.865
Vapor Pressure : ND
% Volatile : >99
Boiling Point : 176°F Initial
Evaporation Rate : Moderately Fast
Freezing Point : ND
Vapor Density : ND
Solubility: Partially soluble in water
pH :NA
Appearance and Odor: A clear colorless liquid, aromatic odor

3. FIRE AND EXPLOSION DATA

Flashpoint : –40°F
Method : TCC
Flammable Limits propellent
LEL: 1.8 UEL: 9.5
Extinguishing Media: CO2, dry chemical, foam
Unusual Hazards : Aerosol cans may explode when heated above 120°F.

4. REACTIVITY AND STABILITY

Stability : Stable
Hazardous decomposition products: CO2, carbon monoxide (thermal)

5. PROTECTION INFORMATION

Ventilation : Use mechanical means to ensure vapor concentration is below TLV.
Respiratory: Use self-contained breathing apparatus above TLV.

Gloves : Solvent resistant
Eye and Face: Safety Glasses
Other Protective Equipment: Not normally required for aerosol product usage

Copyright © 2015 Cengage Learning®

FIGURE 1-22 **Safety data sheets (SDS) inform employees about hazardous materials.**

3. Protective equipment required for handling.
4. Explosion and fire hazards.
5. Other incompatible materials.
6. Health hazards such as signs and symptoms of exposure, medical conditions aggravated by exposure, and emergency and first-aid procedures.

7. Safe handling precautions.
8. Spill and leak procedures.

Second, the employer has a responsibility to make sure that all hazardous materials are properly labeled. The label information must include health, fire, and reactivity hazards posed by the material and the protective equipment necessary to handle the material. The manufacturer must supply all warning and precautionary information about hazardous materials, and this information must be read and understood by the employee before handling the material.

Third, employers are responsible for maintaining permanent files regarding hazardous materials. These files must include information on hazardous materials in the shop, proof of employee training programs, and information about accidents such as spills or leaks of hazardous materials. The employer's files must also include proof that employees' requests for hazardous material information such as SDS have been met. A general Right-to-Know compliance procedure manual must be maintained by the employer.

General Hybrid Electric Vehicle Safety

Because the hybrid electric vehicle (HEV) system can use voltages in excess of 300V (both DC and AC), it is vital the service technician be familiar with, and follow, all safety precautions. Many high-voltage electric vehicles today use a high-voltage electric air-conditioning compressor as part of the refrigerant system. Failure to perform the correct procedures can result in electrical shock, battery leakage, an explosion, or even death. The following are some general service precautions to be aware of:

- Test the integrity of the rubber insulating gloves (electrical lineman gloves) prior to use.
- Wear high-voltage (HV) insulating gloves when disconnecting the service plug and use the leather top glove during heavy or abrasive service procedures.

WARNING: Do not attempt to test or service the system for 5 minutes after the high-voltage service plug is removed. At least 5 minutes is required to discharge the high-voltage capacitors inside the inverter module.

- Never cut the orange HV power cables. The wire harness, terminals, and connectors of the high-voltage system are identified by orange. In addition, HV components may have a "High Voltage" caution label attached to them.
- Use insulating tools.
- Do not wear metallic objects that may cause electrical shorts.
- Wear protective safety goggles when inspecting the high-voltage battery.

WARNING: Be sure to use the proper safety equipment when working on any high-voltage system. Failure to do so may result in a serious or fatal injury.

- Follow the service information diagnostic procedures.
- Never open HV components.
- Before touching any of the HV system wires or components, wear insulating gloves, make sure the HV service plug is removed, and disconnect the auxiliary battery.
- Remove the service disconnect prior to performing a resistance check.

- Remove the service plug prior to disconnecting or reconnecting any HV connections or components.
- Isolate any HV wires that have been removed with insulation tape.
- Properly torque the high-voltage terminals.

WARNING: When the vehicle has been left unattended, recheck that the service disconnect has not been reinstalled, by a well-meaning associate.

SPECIAL TOOLS

DMM capable of reading 400V AC/DC

Insulating gloves

Insulating tape

When working on an HEV, always assume the HV system is live until you have proven otherwise, you can never be too safe. Your first mistake may be your last! If the vehicle has been driven into the service department, you know that the HV system was energized because most HEVs do not move without the HV system operating.

It is critical that the proper tools be used when working on the HV system. These include protective hand tools and a digital multimeter (DMM) with an insulation test function. The meter must be capable of checking for insulation up to 1,000V and measuring resistance at over 1.1 mega ohms. In addition, the DMM insulation test function is used to confirm proper insulation of the HV system components after a repair is performed.

Whenever possible use the one-hand rule when servicing the HV system (Figure 1-23). The one-hand rule means working with only one hand while servicing the HV systems so that in the event of an electric shock the high-voltage will not pass through your body. It is important to follow this rule when performing the HV check out procedure because confirmation of HV system power down has not been proven yet.

Insulated Glove Integrity Test

The rubber isolating gloves (lineman gloves), which the technician must wear for protection while serving the HV system, are your first line of defense when it comes to preventing contact with energized electrical components and must be tested for integrity before they are used. Also, pay attention to the date code on the gloves, gloves have an expiration date, they do not last forever. In addition for heavy services use the leather protective glove on top of the isolating gloves.

Not just any gloves will do, lineman gloves must meet current ASTM D120 specifications and NFPA 70E standards. These requirements are enforced by OSHA as part of their

Copyright © 2015 Cengage Learning®

FIGURE 1-23 **It is recommended to follow the one-hand rule when testing any high-voltage system.**

Copyright © 2015 Cengage Learning®

FIGURE 1-24 **Insulated rubber isolation gloves rated Class "0" must be worn when working on high-voltage system.**

CFR 1910.137 regulation. These standards dictate testing, retesting, and manufacturing criteria for lineman gloves. For HV vehicles the lineman gloves must meet Class "0" requirements of a rating of 1,000V alternate current (AC) (Figure 1-24). Electrical protective gloves are categorized by the amount of voltage both AC and direct current (DC) have been proof-tested to. In addition, the technician should wear rubber-soled shoes, cotton clothing, and safety glasses with side shields as part of their personnel protective equipment. Remove all jewelry and make sure metal zippers are not exposed. Always have a second set of isolating gloves available and let someone in the shop know their location. When preparing to work on any HV vehicle, let associates know in the event they must come to your aid.

OSHA regulations require that all insulating gloves must be electrically tested before first issue and retested every 6 months thereafter by a test laboratory. For this reason most shops will discard and replace gloves after 6 months. Any unused gloves after 12 months must be retested or discarded. The manufacturers and/or suppliers of insulating gloves can assist in providing test laboratory locations for retest certification of the glove if you prefer over replacement. It is recommended that gloves be stored out of direct sunlight and away from sources of ozone (i.e., electric motors). They should be stored in a glove bag flat, never folded, and hung up rather than laid down on a flat surface.

WARNING: Do not use insulating gloves that have been used for over 6 month or are new in the package but older than 12 months unless they have been recertified by a licensed test laboratory. Perform a daily or prior to use safety inspection of insulating gloves before working on any HV vehicle.

Daily or prior to use safety inspection of rubber insulating gloves procedure:

1. Visually inspect rubber gloves prior to use for cracks, tears, holes, signs of ozone damage, possible chemical contact, and signs of abrasion or after any situation that may have caused damage to the gloves.
2. OSHA requires a glove air inflator test.
3. Blow air into glove to inflate them and seal the opening by folding the base of the glove.
4. Slowly role the base of the glove toward the fingers to increase the pressure (Figure 1-25).

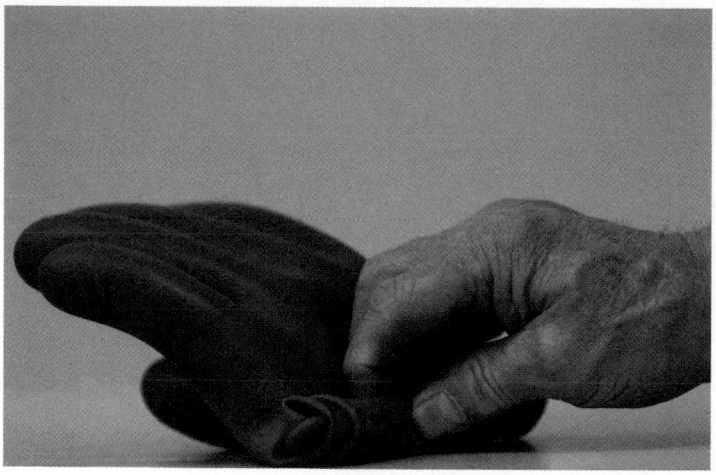

Copyright © 2015 Cengage Learning®

FIGURE 1-25 **Insulated rubber isolation must be tested for leaks before working on high-voltage system.**

5. Look and feel for pin holes on all surface sides.
6. If a leak or damage is detected, discard the glove. For extra precaution the glove should be rendered unusable (i.e., cut in half).

High-Voltage Service Plug

The HEV is equipped with a high-voltage service plug that disconnects the HV battery from the system. Usually this plug is located near the battery (Figure 1-26). Prior to disconnecting the high-voltage service plug, the vehicle must be turned off. Some manufactures also require the negative terminal of the auxiliary battery be disconnected. Once the high-voltage service plug is removed, the high-voltage circuit is shut off at the intermediate position of the HV battery.

The high-voltage service plug assembly contains a safety interlock reed switch. The reed switch is opened when the clip on the high-voltage service plug is lifted. The open reed switch turns off power to the service main relay (SMR). The main fuse for the high-voltage circuit is inside the high-voltage service plug assembly.

However, never assume that the high-voltage circuit is off. The removal of the high-voltage service plug does not disable the individual high-voltage batteries. Use a DMM to verify that 0V are in the system before beginning service. When testing the circuit for voltage, set the voltmeter to the 400 VDC scale.

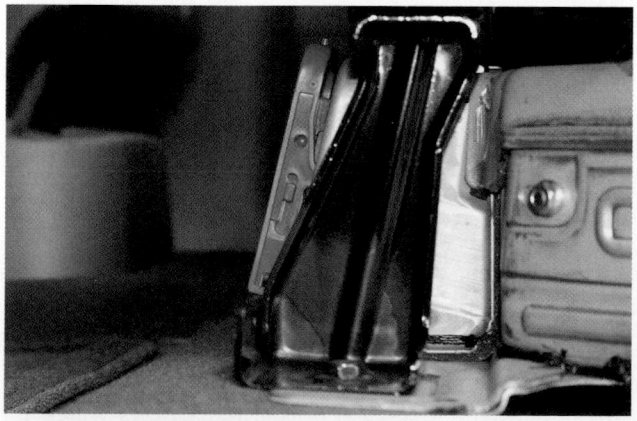

Copyright © 2015 Cengage Learning®

FIGURE 1-26 **The high-voltage service plug is generally located near the HV battery.**

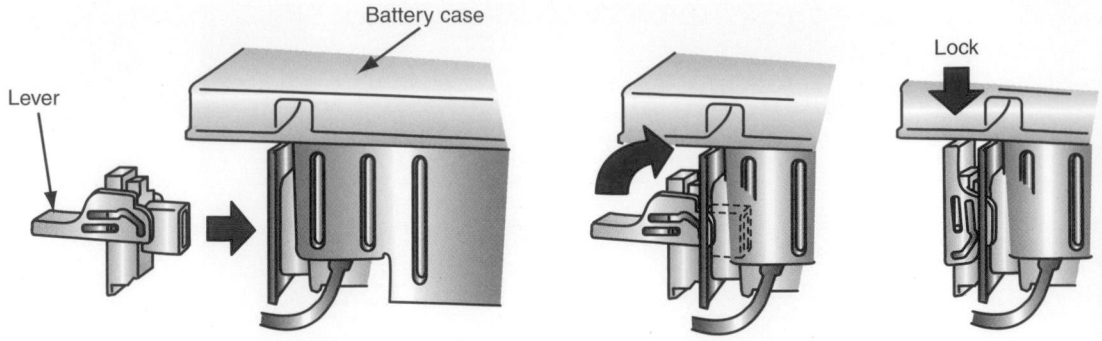

Copyright © 2015 Cengage Learning®

FIGURE 1-27 **The high-voltage service plug lever must be placed in the down and locked position after installation.**

After the high-voltage service plug is removed, a minimum of 5 minutes must pass before beginning service on the system. This is required to discharge the high voltage from the condenser in the inverter circuit.

To install the high-voltage service plug, make sure the lever is locked in the DOWN position (Figure 1-27). Slide the plug into the receptacle, and lock it in place by lifting the lever upward. Once it is locked in place, it closes the reed switch returning power to the system.

CASE STUDY

A technician raised a Grand Marquis on a lift to perform an oil and filter change and a chassis lubrication. This lift was a twin-post-type with separate front and rear lift posts. On this type of lift, the rear wheels must be positioned in depressions in the floor to position the rear axle above the rear lift arm. Then the front lift post and arms must be moved forward or rearward to position the front lift arms under the front suspension. The front lift arms must also be moved inward or outward so they are lifting on the vehicle manufacturer's specified lifting points.

The technician carefully positioned the front lift post and arms properly, but forgot to check the position of the rear tires in the floor depressions. The car was raised on the lift, and the technician proceeded with the service work. Suddenly there was a loud thump, and the rear of the car bounced up and down! The rear lift arms were positioned against the floor of the trunk rather than on the rear axle, and the lift arms punched through the floor of the trunk, narrowly missing the fuel tank. The technician was extremely fortunate the car did not fall off the lift, which would have resulted in more severe damage. If the rear lift arms had punctured the fuel tank, a disastrous fire could have occurred! Luckily, these things did not happen.

The technician learned a very important lesson about lift operation. Always follow all the recommended procedures in the lift operator's manual! The trunk floor was repaired at no cost to the customer, and fortunately, the shop and the vehicle escaped without major damage.

SUMMARY

- The U.S. Occupational Safety and Health Act (OSHA) of 1970 ensures safe and healthful working conditions and authorizes enforcement of safety standards.
- Many hazardous materials and conditions can exist in an automotive shop, including flammable liquids and materials, corrosive acid solutions, loose sewer covers, caustic liquids, high-pressure air, frayed electrical cords, hazardous waste materials, carbon monoxide, improper clothing, harmful vapors, high noise levels, and spills on shop floors.

- Material safety data sheets (MSDS) provide information regarding hazardous materials, labeling, and handling.
- The danger regarding hazardous conditions and materials may be avoided by eliminating shop hazards and applying the necessary shop rules and safety precautions.
- The automotive shop owner/management must supply the necessary shop safety equipment, and all shop personnel must be familiar with the location and operation of this equipment. Shop safety equipment includes gasoline safety cans, steel storage cabinets, combustible material containers, fire extinguishers, eyewash fountains, safety glasses and face shields, first-aid kits, and hazardous waste disposal containers.
- General shop safety includes following all safety rules and precautions to ensure a safe working environment and reduce on-the-job injuries.
- Maintaining adequate personal safety requires technicians to wear protective clothing such as proper footwear with steel toe caps, and safety glasses or a face shield. Technicians must avoid wearing loose fitting clothing that may become entangled in rotating equipment or components.
- To provide adequate personal safety, technicians must avoid wearing watches, jewelry, or rings.
- Shop safety requires a properly ventilated shop with an adequate exhaust removal system to avoid carbon monoxide gas in the shop.
- Shop safety equipment such as first-aid kits, eyewash fountains, and fire extinguishers must be clearly marked and easily accessible.
- Electrical cords must have a ground connection, and frayed electrical cords must be repaired or replaced immediately.
- Gasoline must be stored in clearly marked, approved gasoline containers.
- Fire safety in the shop includes storing combustible materials in approved, covered safety containers, avoiding sparks and flames near combustible materials, and quickly cleaning up gasoline, oil, or grease spills.
- To maintain shop safety, technicians must be familiar with the location and operation of fire extinguishers.
- Technicians must follow proper driving rules when driving vehicles in the shop.
- Proper lifting procedures must be followed to avoid personal injury when lifting heavy objects.
- Safe operating procedures must be followed when using hand tools and operating shop equipment such as lifts, hydraulic jacks, power tools, compressed air equipment, and cleaning equipment.
- Hybrid system can use voltage in excess of 300V, both AC and DC; it is critical that you as a service technician be familiar with and follow all safety and service precautions.
- Always remove the high-voltage service plug prior to servicing any HV system and place the plug, such as on your tool box, where it cannot be accidentally reinstalled by someone else.
- Do not attempt to test or service the high-voltage system of an HEV for 5 minutes after the high-voltage service plug is removed.
- Confirm that the high-voltage circuits have 0V using a DMM before performing any service procedure on the high-voltage system of an HEV. Never assume that the HV circuit is off.
- Test the integrity of your lineman gloves (isolation gloves) prior to servicing any high-voltage system. Do not use linemen gloves that fail a leak test, visual inspection, or are over 6 months old.

TERMS TO KNOW

Air bag system
Approved gasoline storage cans
Aqueous parts cleaning tank
Brake parts washer
Cold parts washer
Corrosive
Environmental Protection Agency (EPA)
Fluorescent trouble lights
Hazard Communication Standard
Hot cleaning tank
Hybrid vehicle
Ignitable
Incandescent bulb-type trouble lights
Multipurpose dry chemical fire extinguishers
Occupational Safety and Health Act (OSHA)
Reactive
Resource Conservation and Recovery Act (RCRA)
Right-to-Know laws
Safety data sheets (SDS)
Toxic
Workplace hazardous materials information systems (WHMIS)

ASE-STYLE REVIEW QUESTIONS

1. Breathing carbon monoxide may cause:
   A. Arthritis
   B. Cancer
   C. Impaired vision
   D. Headaches

2. While discussing shop rules:
   Technician A says the wire harness, terminals, and connectors of the high-voltage system are identified by red.
   Technician B says to remove the service plug prior to disconnecting or reconnecting any HV connections or components.
   Who is correct?
   A. A only
   B. B only
   C. Both A and B
   D. Neither A nor B

3. *Technician A* says that OSHA is a state agency responsible for Occupational Standards for Heating and Air Conditioning.
   Technician B says that an OSHA poster must be displayed in the employees' common area, such as the break room.
   Who is correct?
   A. A only
   B. B only
   C. Both A and B
   D. Neither A nor B

4. While discussing personal safety:
   Technician A says rings and jewelry may be worn in the automotive shop.
   Technician B says some electric-drive cooling fans may start rotating at any time.
   Who is correct?
   A. A only
   B. B only
   C. Both A and B
   D. Neither A nor B

5. While lifting heavy objects in the automotive shop:
   A. Bend your back to pick up the heavy object.
   B. Place your feet as far as possible from the object.
   C. Bend forward to place the object on the workbench.
   D. Straighten your legs to lift an object off the floor.

6. All of the following statements are true concerning high-voltage system safety, EXCEPT:
   A. Turn the power switch to the off position prior to performing a resistance check.
   B. Do not attempt to test or service the system for 5 minutes after the high-voltage service plug is removed.
   C. Test lineman gloves for damage and leaks prior to use.
   D. Disconnect the motor generators prior to turning the ignition off.

7. While operating hydraulic equipment safely in the automotive shop remember that:
   A. Safety stands have a maximum weight capacity.
   B. The driver's door should be open when raising a vehicle on a lift.
   C. A lift does not require a safety mechanism to prevent lift failure.
   D. Four-wheel drive vehicles should be lifted on a frame contact lift.

8. While discussing shop hazards:
   Technician A says high-pressure air from an air gun may penetrate the skin.
   Technician B says air in the bloodstream may be fatal.
   Who is correct?
   A. A only
   B. B only
   C. Both A and B
   D. Neither A nor B

9. While discussing hazardous material disposal:
   Technician A says certain types of hazardous waste material can be poured down a floor drain.
   Technician B says a shop is responsible for hazardous waste materials from the time they become waste until the proper waste disposal is completed.
   Who is correct?
   A. A only
   B. B only
   C. Both A and B
   D. Neither A nor B

10. When working in an automotive shop, hazardous conditions must be avoided by:
    A. Storing oily rags in open containers.
    B. Storing flammable liquids such as paint in metal safety cabinet.
    C. Leaving sewer covers loose for quick access.
    D. Cleaning brake parts with an air gun.

JOB SHEET 1

Name _____ **Date** _____

Demonstrate Proper Lifting Procedures

Upon completion of this job sheet, you should be able to follow the proper procedure when lifting heavy objects.

Tools and Materials

A heavy object with a weight that is within the weight-lifting capability of the technician.

Procedure

Task Completed

Have various members of the class demonstrate proper weight-lifting procedures by lifting an object off the shop floor and placing it on the workbench. Other class members are to observe and record any improper weight-lifting procedures.

1. If the object is to be carried, be sure your path is free from loose parts or tools. ☐
2. Position your feet close to the object, and position your back reasonably straight for proper balance. ☐
3. Your back and elbows should be kept as straight as possible. Continue to bend your knees until your hands reach the best lifting location on the object to be lifted. ☐
4. Be certain the container is in good condition. If a container falls apart during the lifting operation, parts may drop out of the container, resulting in foot injury or part damage. ☐
5. Maintain a firm grip on the object; do not attempt to change your grip while lifting is in progress. ☐
6. Straighten your legs to lift the object, keeping the object close to your body. Use leg muscles rather than back muscles. ☐
7. If you have to change direction of travel, turn your whole body. Do not twist. ☐
8. Do not bend forward to place an object on a workbench or table. Position the object on the front surface of the workbench and slide it back, watching your fingers under the object to avoid pinching them. ☐
9. If the object must be placed on the floor or a low surface, bend your legs to lower the object. Do not bend your back forward because this movement strains back muscles. ☐
10. When a heavy object must be placed on the floor, put suitable blocks under the object to prevent jamming your fingers under the object. ☐

11. List any improper weight-lifting procedures observed during the weight-lifting demonstrations.

1. _____
2. _____
3. _____
4. _____
5. _____
6. _____

Instructor's Response _____

_____

_____

JOB SHEET 2

Name _____ **Date** _____

Locate and Inspect Shop Safety Equipment

Upon completion of this job sheet, you should be familiar with the location of shop safety equipment and know if this equipment is serviced properly.

Procedure

Task Completed

1. Fire extinguishers: Are the fire extinguishers tagged to indicate they have been checked or serviced recently? ☐ Yes ☐ No

   Draw a basic diagram of the shop layout in the space below, and mark the locations of the fire extinguishers and fire exits.

2. Eyewash fountain or shower: Is the eyewash fountain or shower operating properly? ☐ Yes ☐ No

   Mark the location of the eyewash fountain or shower on the shop layout diagram in step 1.

3. First-aid kits: Are the first-aid kits properly stocked with supplies? ☐ Yes ☐ No

   Mark the location of the first-aid kits on the shop layout diagram in step 1.

4. Electrical shut-off box: Mark the location of the shop electrical shut-off box on the shop layout diagram in step 1. ☐

5. Are the trash containers equipped with proper covers? ☐ Yes ☐ No

   Mark the location of the trash containers on the shop layout diagram in step 1.

6. Metal storage cabinet: Mark the location of metal storage cabinet(s) for combustible materials on the shop layout diagram in step 1. ☐

Instructor's Response _____

_____

_____

JOB SHEET 3

Name _____ **Date** _____

Shop Housekeeping Inspection

Upon completion of this job sheet, you should be able to apply shop housekeeping rules in your shop.

Procedure

Task Completed

When students from another class of Automotive Technology are working in the shop, evaluate their shop housekeeping procedures using the 16 shop housekeeping procedures. List all the improper shop housekeeping procedures that you observed in the space provided at the end of the job sheet.

1. Keep aisles and walkways clear of tools, equipment, and other items. ☐
2. Be sure all sewer covers are securely in place. ☐
3. Keep floor surfaces free of oil, grease, water, and loose material. ☐
4. Proper trash containers must be conveniently located, and these containers should be emptied regularly. ☐
5. Access to fire extinguishers must be unobstructed at all times; fire extinguishers should be checked for proper charge at regular intervals. ☐
6. Tools must be kept clean and in good condition. ☐
7. When not in use, tools must be stored in their proper location. ☐
8. Oily rags and other combustibles must be placed in proper, covered metal containers. ☐
9. Rotating components on equipment and machinery must have guards. All shop equipment should have regular service and adjustment schedules. ☐
10. Benches and seats must be kept clean. ☐
11. Keep parts and materials in their proper location. ☐
12. When not in use, creepers must not be left on the shop floor. Creepers should be stored in a specific location. ☐
13. The shop should be well-lit, and all lights should be in working order. ☐
14. Frayed electrical cords on lights or equipment must be replaced. ☐
15. Walls and windows should be cleaned regularly. ☐

Task Completed

☐

16. Stairs must be clean, well lit, and free of loose material.
Observed improper shop housekeeping procedures:

1. _____
2. _____
3. _____
4. _____
5. _____
6. _____

Instructor's Response _____

_____

_____

JOB SHEET 4

Name _____ **Date** _____

Personnel and Shop Safety Assessment

As a service professional, one of your first concerns should be safety. Upon completion of this job sheet you should have an increased awareness of personnel and shop safety item. As you take these personnel assessment and survey your shop answering the following questions, you will learn to evaluate your workplace and personnel safety.

Procedure

Give a brief description following each step. Always wear eye protection when working or walking around a service facility.

1. Before you evaluate your work place you must first evaluate yourself. Are you dressed for work? ☐ Yes ☐ No

   a. If yes, why do you believe your attire is appropriate?

   _____

   _____

   b. If no, what must you correct to be properly attired?

   _____

   _____

   Are your safety glasses OSHA approved? ☐ Yes ☐ No

   c. Do you have side shields? ☐ Yes ☐ No

   Are you wearing leather boots or shoes with oil resistant soles? ☐ Yes ☐ No

   d. Does your footwear have steel toes? ☐ Yes ☐ No

   Is your shirt tucked into your pants? ☐ Yes ☐ No

   If you have long hair, is it tied back or under a hat? ☐ Yes ☐ No

Next carefully inspect your shop, noting any potential hazards.

Note: A hazard is not necessarily a safety violation but in an area of which you must be aware (i.e., pothole in parking lot).

Are there safety areas marked around grinders and other machinery?
☐ Yes ☐ No

What is the air pressure in the shop set at? _____

Where are the tools stored in the shop?

_____

_____

What recommendations would you make to improve the tool storage?

_____

_____

Have you been instructed on proper use of shop vehicle hoist (lift)?
☐ Yes ☐ No

e. If not, ask your instructor to demonstrate vehicle lift use.

Where is the first-aid kit located? _____

Where is the eyewash station located? _____

List the location of the exits. _____

What is the emergency evacuation plan and where are you to assemble once you evacuate the building?

_____

_____

_____

_____

_____

Where are the MSDS located? _____

2. Does the shop have a hazardous spill response kit? ☐ Yes ☐ No

List any additional questions your instructor would like you to answer:

_____

_____

_____

_____

_____

_____

_____

_____

_____

_____

_____

Instructor's Response _____

_____

_____

Chapter 2

Tools and Shop Procedures

Upon completion and review of this chapter, you should be able to:

- Diagnose steering and suspension problems using a basic diagnostic procedure.
- Raise a vehicle with a floor jack and lower the vehicle so it is supported on safety stands.
- Raise a vehicle with a lift.
- Test power steering pump pressure with an appropriate pressure gauge and valve assembly.
- Measure ball joint wear with a dial indicator designed for this purpose.
- Remove and replace a coil spring on a strut using a coil spring compressor tool.
- Use a tire changer to dismount and mount a tire.
- Use a scan tool to diagnose a computer-controlled suspension system.
- Balance tire and wheel assemblies with an electronic wheel balancer.
- Perform a four wheel alignment with a computer wheel aligner.
- Fulfill employee obligations when working in the shop.
- Accept job responsibility for each job completed in the shop.
- Describe the ASE automotive technician testing and certification process including the eight areas of certification.

Using Suspension and Steering Equipment

Learn to use equipment properly the first time. When improper procedures are learned and become a habit, it is more difficult to break these wrong habits. Using service and test equipment will help you to work more safely and efficiently, and in the long term, this should improve your income. We will discuss the proper use of some of the most common service and test equipment used by suspension and steering technicians.

Measuring Systems

Two systems of weights and measures are commonly used in the United States. One system is the **U.S. customary (USC) system**, which is commonly referred to as the English system. Well-known measurements for length in the USC system are the inch, foot, yard, and mile. In this system, the quart and gallon are measurements for volume; ounce, pound, and ton are measurements for weight. A second system of weights and measures is called the **international system (IS)**, also known as the metric system.

In the USC system, the basic linear measurement is the yard; in the metric system, it is the meter. Each unit of measurement in the metric system is related to the other metric units by a factor of 10. Thus, every metric unit can be multiplied or divided by 10 (or 100

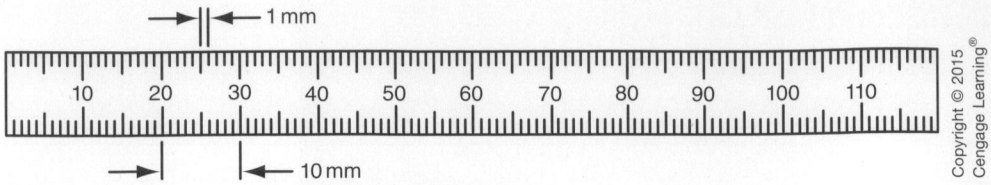

FIGURE 2-1 **Metric tape graduated in millimeters.**

or 1,000) to obtain larger units (multiples) or smaller units (submultiples). For example, the meter maybe divided by 100 to obtain centimeters (1/100 meter) or by 1,000 to obtain millimeters (1/1,000 meter).

The U.S. government passed the Metric Conversion Act in 1975 in an attempt to move American industry and the general public to accept and adopt the metric system. The automotive industry has adopted the metric system, and in recent years most bolts, nuts, and fittings on vehicles have been changed to metric. During the early 1980s, some vehicles had a mix of English and metric bolts. Imported vehicles have used the metric system for many years. Although the automotive industry has changed to the metric system, the general public in the United States has been slow to convert from the USC system to the metric system. One of the factors involved in this change is cost. What would it cost to change every highway distance and speed sign in the United States to read kilometers? Probably hundreds of millions, or even billions, of dollars.

Service technicians must be able to work with both the USC and the metric system. One meter (m) in the metric system is equal to 39.37 inches (in.) in the USC system. A metric tape measure may be graduated in millimeters, and 10 millimeters = 1 centimeter (Figure 2-1).

Some common equivalents between the metric and USC systems are:

1 meter (m) = 39.378 inches
1 centimeter (cm) = 0.3937 inch
1 millimeter (mm) = 0.03937 inch
1 inch (in.) = 2.54 centimeter
1 inch = 25.4 millimeter

In the USC system, phrases such as 1/8 of an inch are used for measurements. The metric system uses a set of prefixes. For example, in the word kilometer the prefix kilo means 1,000, indicating that there are 1,000 in a kilometer. Common prefixes in the metric system include:

NAME	SYMBOL	MEANING
mega	M	one million
kilo	k	one thousand
hecto	h	one hundred
deca	da	ten
deci	d	one tenth of
centi	c	one hundredth of
milli	m	one thousandth of
micro	μ	one millionth of

Measurement of Mass

In the metric system mass is measured in grams, kilograms, or tonnes: 1,000 grams (g) = kilogram (kg). In the USC system mass is measured in ounces, pounds, or tons. When converting pounds to kilograms, 1 pound = 0.453 kilograms.

Measurement of Length

In the metric system length is measured in millimeters, centimeters, meters, or kilometers: 10 millimeters (mm) = 1 centimeter (cm). In the USC system length is measured in inches, feet, yards, or miles.

When distance conversions are made between the two systems some of the following conversion factors are used:

1 inch = 25.4 millimeters
1 foot = 30.48 centimeters
1 yard = 0.91 meters
1 mile = 1.60 kilometers

Measurement of Volume

In the metric system volume is measured in milliliters, cubic centimeters, and liters: 1 cubic centimeter = 1 milliliter. If a cube has a length, depth, and height of 10 centimeters (cm), the volume of the cube is 10 cm × 10 cm × 10 cm = 1,000 cm^3 = 1 liter. When volume conversions are made between the two systems, 1 cubic inch = 16.38 cubic centimeters. If an engine has a displacement of 350 cubic inches, 350 × 16.38 = 5,733 cubic centimeters, and 5,733 ÷ 1,000 = 5.7 liters.

Basic Diagnostic Procedure

One of the most important parts of a technician's job is diagnosing automotive problems. Each year more vehicle systems are controlled by increasingly sophisticated electronic systems, and this makes accurate diagnosis even more important.

The following basic diagnostic procedure may be used to diagnose various automotive problems:

1. Be absolutely sure the problem is identified. Obtain all the information you can from the customer; for example, politely ask the customer to describe the exact symptoms of the problem. Ask the customer at what vehicle speed, engine temperature, and atmospheric temperature the problem occurs. From your discussion, find out when, where, and how the problem occurs.
2. Verify the customer complaint. If necessary, road test the vehicle under the same conditions as the customer described to verify the problem.
3. Think of the possible causes of the problem based on your own experience in diagnosing similar problems on other vehicles.
4. Consult original equipment manufacturer (OEM) or generic technical service bulletin (TSB) information regarding service procedures, parts replacement, or computer reprogramming that are designed to correct the problem.
5. Perform the necessary diagnostic tests to locate the exact cause of the complaint. Always begin the diagnostic tests with the quickest, easiest test, and if the problem is not located, proceed with the more complicated, time-consuming tests. During your diagnosis watch for other problems with the vehicle that may give trouble in the near future. If there are any other potential problems such as worn belts and radiator hoses, advise the customer about them and attempt to obtain the customer's approval to repair these problems. If the customer does not approve the repair of these potential problems, describe them on the work order. If we repair a car based on a complaint on a customer's vehicle, and two weeks later the customer brings the vehicle back with another serious complaint, the customer may assume that the repairs were not completed thoroughly.

6. After the cause of the complaint is definitely located, advise the customer about the extent and cost of the repairs. After the customer approves the necessary repairs, perform the appropriate service work on the vehicle.
7. Road test the vehicle if necessary to be sure the customer's complaint is eliminated.

CAUTION:
Do not hit the seal case with a hammer. This action will damage the seal.

Suspension and Steering Tools

Customer Care: Some automotive service centers have a policy of performing some minor service as an indication of their appreciation to the customer. This service may include cleaning all the windows and/or vacuuming the floors before the car is returned to the customer.

Although this service involves more labor costs for the shop, it may actually improve profits over a period of time. When customers find their windows cleaned and/or the floors vacuumed, it impresses them with the quality of work you do and the fact you care about their vehicle. They will likely return for service, besides telling their friends about the quality of service your shop performs.

CAUTION:
Always be sure the seal goes into the housing squarely and evenly. If the seal does not go squarely into the housing, the outer seal case may be distorted, and this condition may cause an oil leak around the seal housing.

Seal Drivers

WARNING: When using any suspension and steering tools, the vehicle manufacturer's recommended procedure in the service manual must be followed. Improper use of tools may lead to personal injury.

Seal drivers are designed to fit squarely against the seal case and inside the seal lip. A soft hammer is used to tap the seal driver and drive the seal straight into the housing. Some tool manufacturers market a seal driver kit with drivers to fit many common seals (Figure 2-2).

Bearing Pullers

A variety of **bearing pullers** are available to pull different sizes of bearings in various locations (Figure 2-3). Some bearing pullers are slide-hammer-type, whereas others are screw-type.

SERVICE TIP:
When installing a seal, the garter spring must face toward the flow of lubricant.

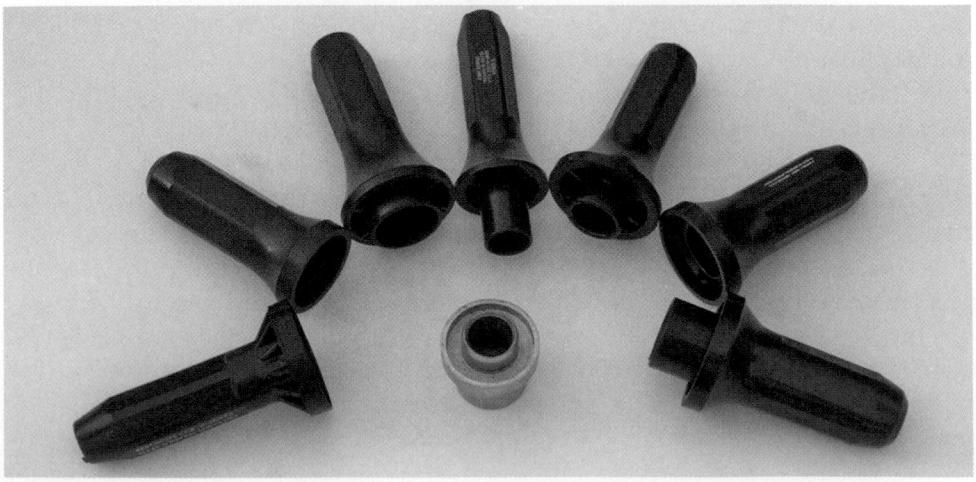

Copyright © 2015 Cengage Learning®

FIGURE 2-2 **Seal drivers.**

Seal drivers are available in various diameters to fit squarely against the outside edge of different-size seals. The seal driver handle is tapped with a soft hammer to install the seal.

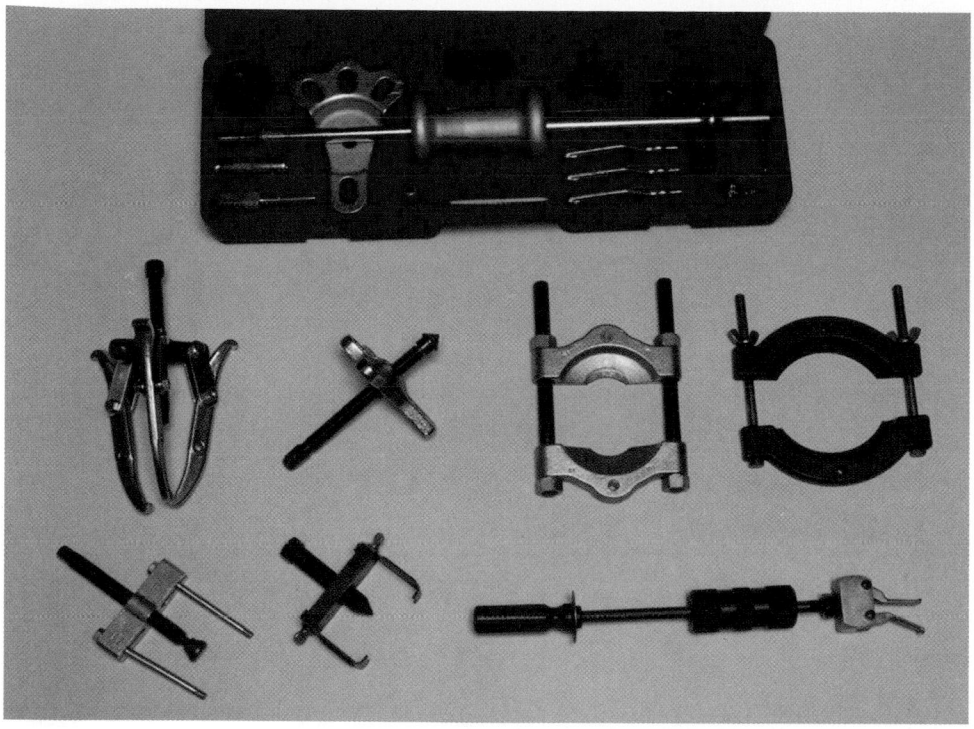

Copyright © 2015 Cengage Learning®

FIGURE 2-3 **Bearing pullers.**

Copyright © 2015 Cengage Learning®

FIGURE 2-4 **Axle puller.**

Bearing pullers are designed to fit over the outer diameter of the bearing or through the center opening in the bearing, pulling the bearing from its mounting shaft.

Axle pullers are usually slide-hammer type. These pullers attach to the axle studs to remove the axle.

Axle Pullers

Axle pullers are used to pull rear axle shafts in rear-wheel-drive vehicles. Most rear axle pullers are slide-hammer-type (Figure 2-4).

Stethoscope

A **stethoscope** is a mechanical device that amplifies sound and diagnoses the source of noises such as bearing noise. The stethoscope pickup is placed on the suspected noise source, and the ends of the two arms are placed in the technician's ears (Figure 2-5). Electronic stethoscopes provide improved sound amplification compared with mechanical stethoscopes and can locate the causes of mechanical noises, air leaks, and wind whistles.

A **stethoscope** is placed in the technician's ears, and the pickup is placed on or near the suspected noise source to diagnose noise problems.

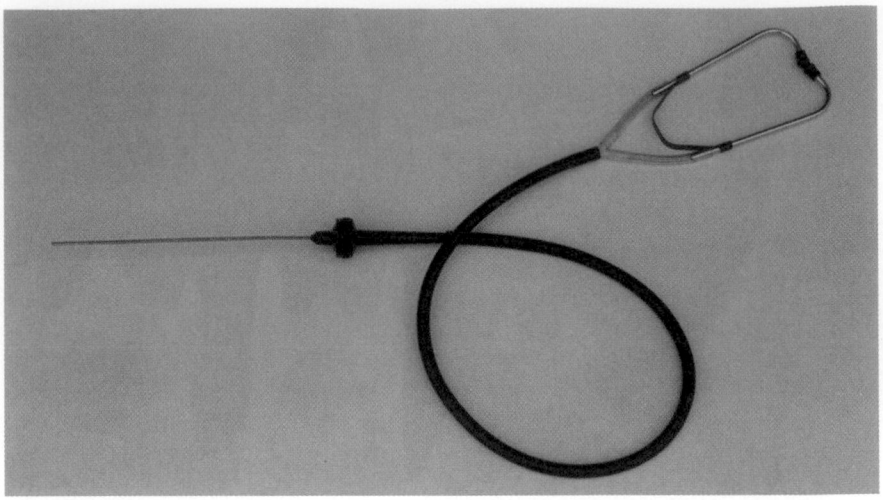

Copyright © 2015 Cengage Learning®

FIGURE 2-5 **Stethoscope.**

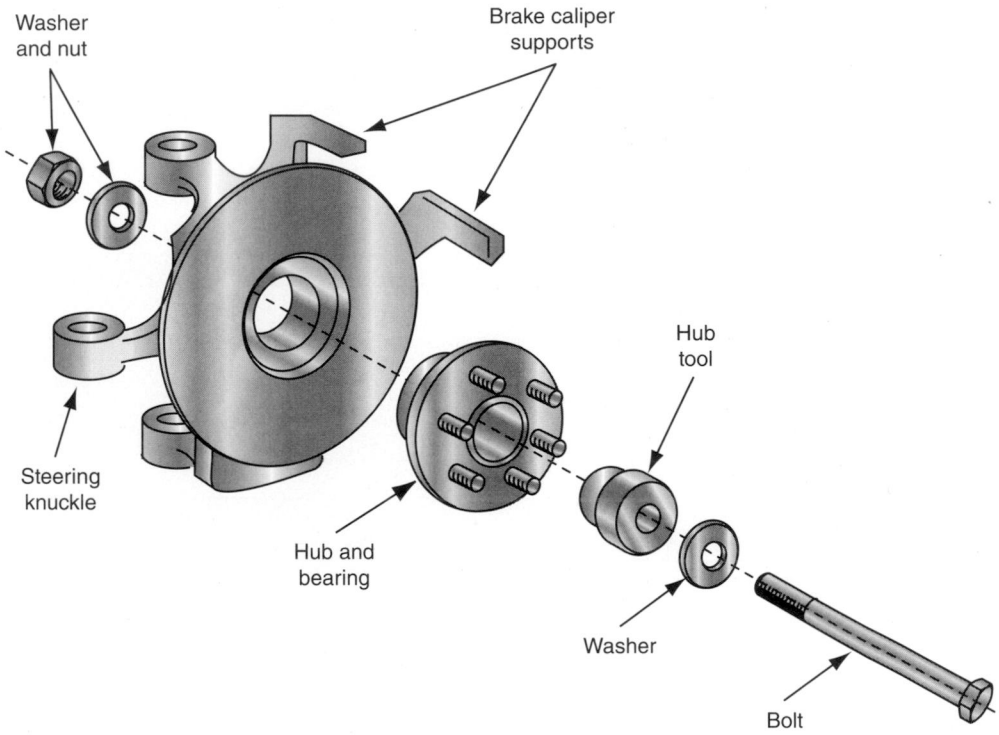

Copyright © 2015 Cengage Learning®

FIGURE 2-6 **Front bearing hub tool.**

SERVICE TIP:
Dial indicators must be kept clean and dry for accuracy and long life.

Front Bearing Hub Tool

Front bearing hub tools are designed to remove and install front wheel bearings on front-wheel-drive cars (Figure 2-6). These bearing hub tools are usually designed for a specific make of vehicle.

Dial Indicator

Dial indicators are used for measuring in many different locations. In the suspension and steering area, dial indicators are used for measurements, such as tire runout and ball joint movement (Figure 2-7). Dial indicators have many different attaching devices to connect the indicator to the component to be measured.

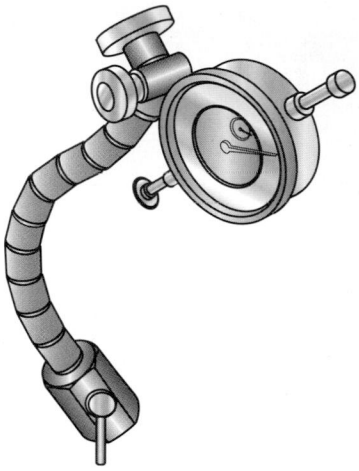

Copyright © 2015 Cengage Learning®

FIGURE 2-7 **Dial indicator designed for ball joint measurement.**

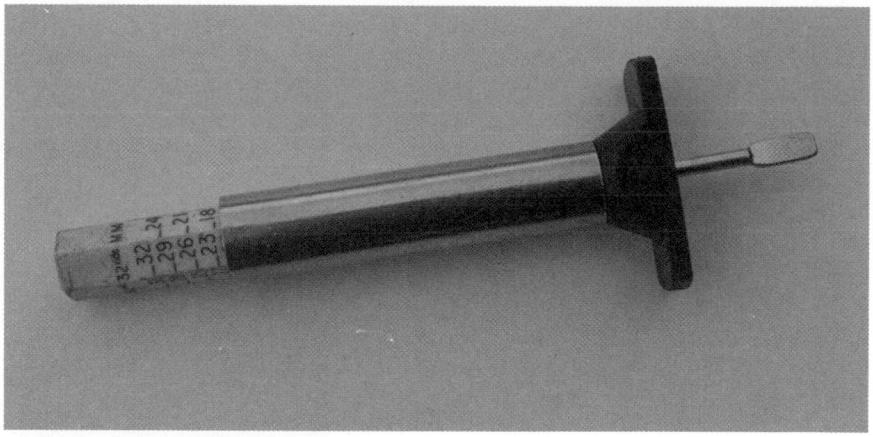

Copyright © 2015 Cengage Learning®

FIGURE 2-8 **Tire tread depth gauge.**

Tire Tread Depth Gauge

A **tire tread depth gauge** measures tire tread depth. This measurement should be taken at three or four locations around the tire's circumference. The lowest reading is considered to be the tread depth (Figure 2-8). Do not place the gauge tip on a wear indicator bar. This gauge is essential when making tire warranty adjustments.

Tire Changer

A **tire changer** is operated pneumatically to dismount and mount tires from the rims.

WARNING: Never operate a tire changer until you are familiar with its operation because this may lead to serious personal injury. Your instructor will explain and demonstrate the operation of this equipment. You should read the equipment operator's manual and use this equipment under the instructor's supervision until you are familiar with it.

Tire changers are used to dismount and mount tires (Figure 2-9). These changers may be used on most common tire sizes. A wide variety of tire changers are available, and each one

Copyright © 2015 Cengage Learning®

FIGURE 2-9 **Typical tire machine.**

Copyright © 2015 Cengage Learning®

FIGURE 2-10 **Typical road force balancer.**

operates differently. Always follow the procedure in the equipment operator's manual and the directions provided by your instructor.

Wheel Balancer

WARNING: Using a wheel balancer before you are familiar with its operation may result in serious personal injury and property damage.

Electronic wheel balancers indicate the required weight location on the rim to obtain proper static and dynamic wheel balance.

Electronic wheel balancers are used in most automotive repair shops (Figure 2-10). Do not attempt to use this equipment until you have studied wheel balance theory. Your instructor will explain and demonstrate the use of this equipment before you attempt to use it. This equipment should be used under the supervision of your instructor until you are familiar with it.

Machinist's Rule

A **machinist's rule** performs many accurate measurements in the shop. Most machinist's rules are graduated in inches and millimeters (Figure 2-11).

Air Chisel or Hammer

An air chisel is an electrically operated tool to which various chisels, cutters, and punches may be attached to complete cutting and riveting jobs. The air hammer operates the

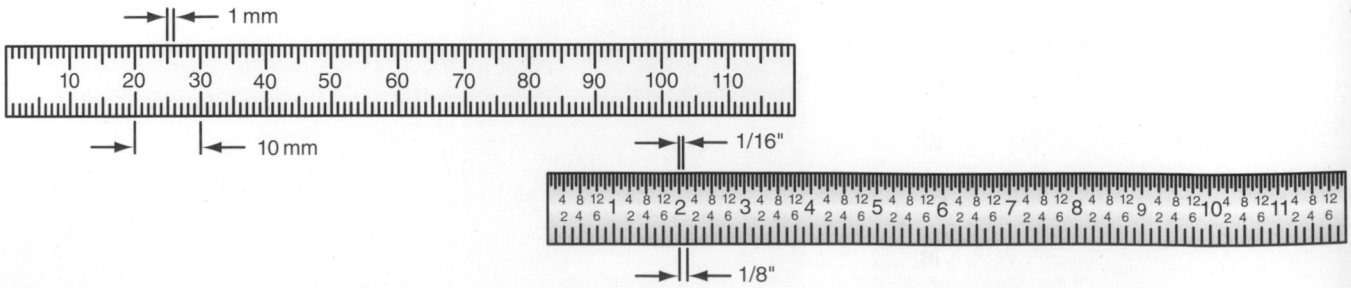

FIGURE 2-11 **Graduations on a typical machinist's rule.**

attached tool with a very fast back-and-forth action (Figure 2-12). Different tools that may be attached to the air chisel are illustrated in Figure 2-13.

WARNING: Always be sure the tool being used is properly attached to the air chisel. Improper tool attachment may result in serious personal injury.

Copyright © 2015 Cengage Learning®

FIGURE 2-12 **Air chisel.**

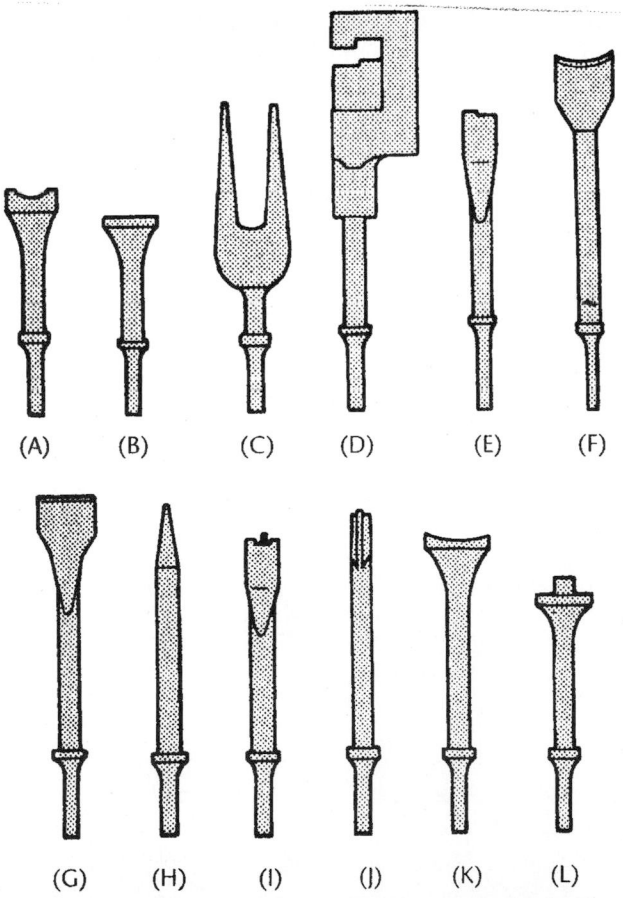

Copyright © 2015 Cengage Learning®

FIGURE 2-13 **Air chisel accessories: (A) universal joint and tie rod end tool, (B) smoothing hammer, (C) ball joint separator, (D) panel crimper, (E) shock absorber chisel, (F) tailpipe cutter, (G) scraper, (H) tapered punch, (I) edging tool, (J) rubber bushing splitter, (K) bushing remover, and (L) bushing installer.**

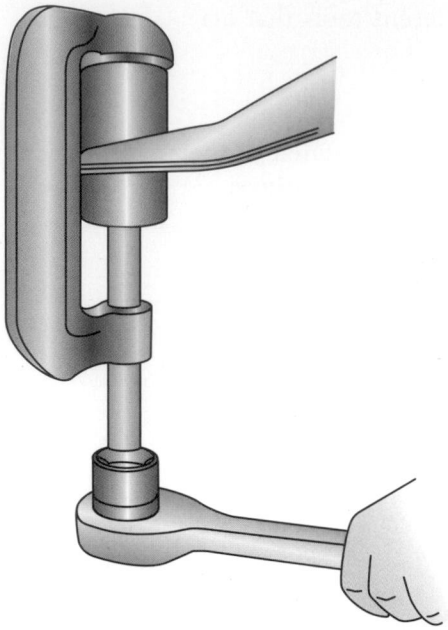

Copyright © 2015 Cengage Learning®

FIGURE 2-14 **Ball joint removal and installation tool.**

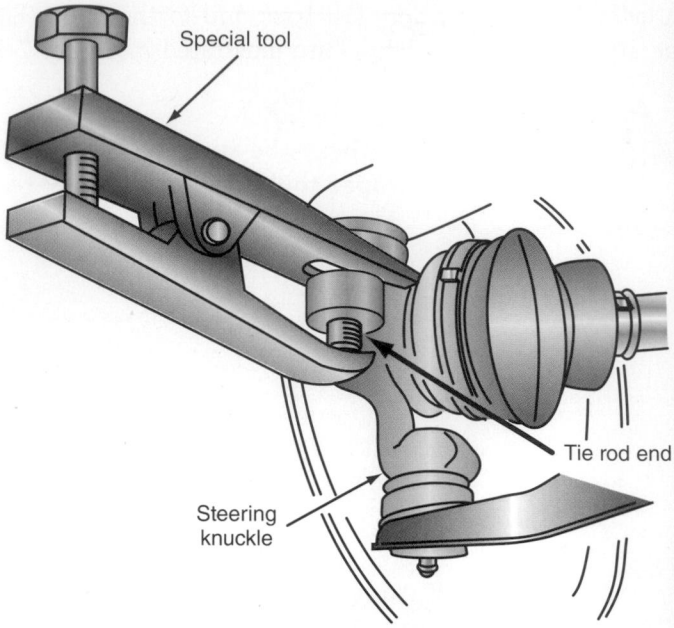

FIGURE 2-15 **Tie rod end and ball joint puller.**

Ball Joint Removal and Installation Tools

Ball joint removal and installation tools remove and install pressed-in ball joints on front suspension systems (Figure 2-14). The size of the removal and pressing tool adapter must match the size of the ball joint.

Tie Rod End and Ball Joint Puller

Some car manufacturers recommend a **tie rod end and ball joint puller** to remove tie rod ends and pull ball joint studs from the steering knuckle (Figure 2-15).

Control Arm Bushing Tools

A variety of **control arm bushing tools** are available to remove and replace control arm bushings. The bushing removal and installation tool must match the size of the control arm bushing (Figure 2-16).

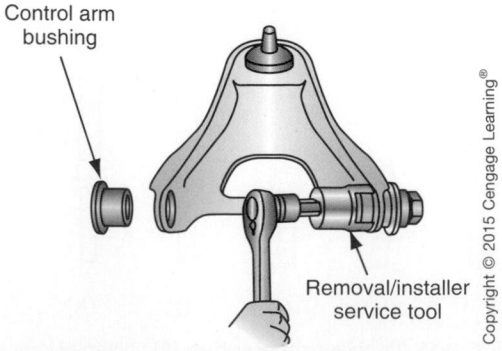

Copyright © 2015 Cengage Learning®

FIGURE 2-16 **Control arm bushing removal and replacement tools.**

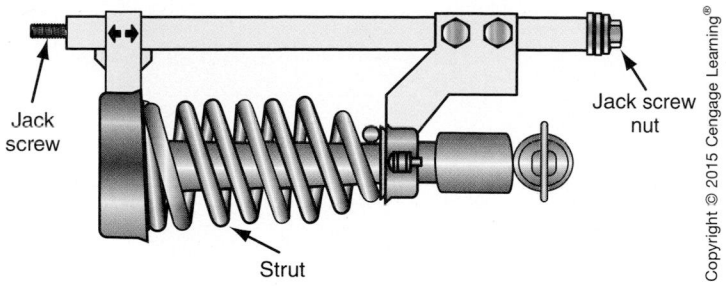

FIGURE 2-17 **MacPherson strut coil spring compressor tool.**

Coil Spring Compressor Tool

CAUTION: The vehicle manufacturer's and equipment manufacturer's recommended procedures must be followed for each type of spring compressor tool.

WARNING: There is a tremendous amount of energy in a compressed coil spring. Never disconnect any suspension component that will suddenly release this tension because this may result in serious personal injury and vehicle or property damage.

Many types of **coil spring compressor tools** are available to the automotive service industry (Figure 2-17). These tools compress the coil spring and hold it in the compressed position while removing the strut from the coil spring or performing other suspension work. Each type of front suspension system requires a different type of spring compressor tool. The vehicle manufacturer's and equipment manufacturer's recommended procedure must be followed.

Power Steering Pressure Gauge

WARNING: The power steering pump delivers extremely high pressure during the pump pressure test. Always follow the recommended test procedure in the vehicle manufacturer's service manual to avoid personal injury during this test.

A **power steering pressure gauge** is used to test the power steering pump pressure (Figure 2-18). Because the power steering pump delivers extremely high pressure during this test, the recommended procedure in the vehicle manufacturer's service manual must be followed.

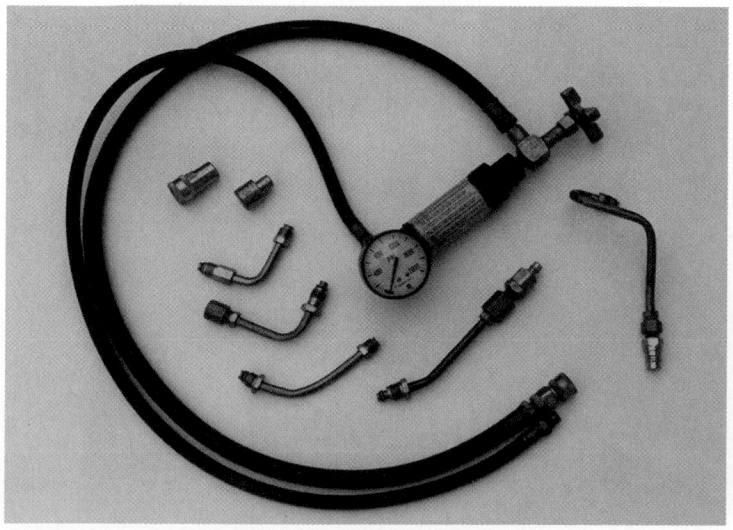

FIGURE 2-18 **Power steering pressure gauge.**

Copyright © 2015 Cengage Learning®

FIGURE 2-19 **Pitman arm puller.**

Pitman Arm Puller

WARNING: Never strike a puller with a hammer when it is installed and tightened because this may result in personal injury.

A **pitman arm puller** is a heavy-duty puller that removes the pitman arm from the pitman shaft (Figure 2-19).

Vacuum Hand Pump

A **vacuum hand pump** creates a vacuum to test vacuum-operated components and hoses (Figure 2-20).

Torque Wrenches and Torque Sticks

A torque wrench is required to tighten fasteners to a specified torque. Tightening fasteners to the specified torque is extremely important. If fastener torque is less than specified, the component retained by the fastener(s) may loosen, resulting in severe component or vehicle damage. When fasteners are tightened to a torque above the specified value, the component may become warped, resulting in improper component operation. Torque wrenches can be beam, click, or dial-type (Figure 2-21). A beam-type torque wrench bends as torque is applied, and a beam on the wrench points to the torque applied on a

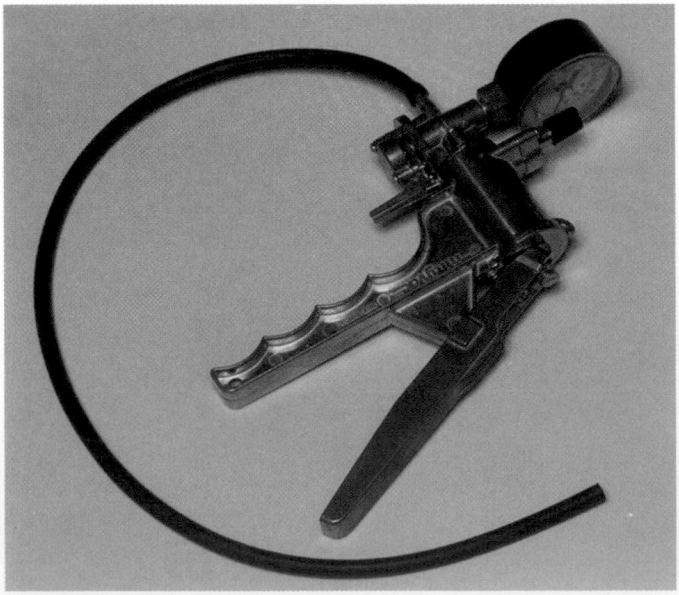

Copyright © 2015 Cengage Learning®

FIGURE 2-20 **Vacuum hand pump.**

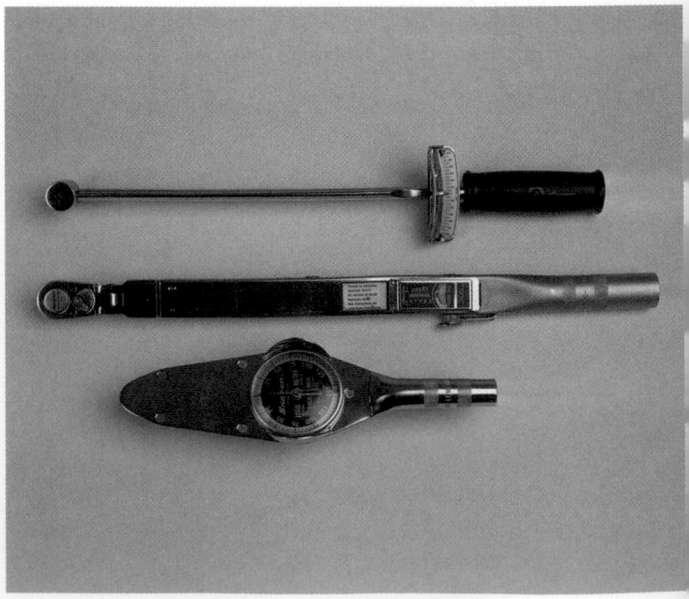

FIGURE 2-21 **Various types of torque wrenches.**

scale attached to the wrench handle. On a click-type torque wrench, the handle is rotated to set the specified torque on a scale adjacent to the handle. When the fastener is tightened to the torque setting, the torque wrench provides an audible click. On a dial-type torque wrench, the torque applied to the fastener is indicated on the dial. Digital-type torque wrenches that indicate the torque reading on a digital display are also available.

Some tool manufactures provide torque sticks for tightening fasteners to the specified torque with an impact wrench. One end of the torque stick has a ½ in. drive opening, and a socket is attached to the opposite end. A spring steel, heat-treated shaft is connected between the ½ in. drive opening and the socket end of the torque stick. Torque sticks are calibrated to twist at a specific torque with each blow of an impact wrench, and thus prevent further tightening of the fastener. Torque sticks are available in various USC and metric sizes and are usually sold in color-coded sets. The torque stick color indicates the specified torque at which they twist. For example, a yellow torque stick twists at 65 ft. lbs. Some torque sticks are a ½ in. extension that fits between the impact wrench drive and a thin-wall impact socket (Figure 2-22).

SERVICE TIP:
Using a torque stick allows the use of an impact wrench when tightening fasteners, and this speeds up the service operation. However, the use of torque sticks may not be as accurate as using a torque wrench. It is a good practice to double check the accuracy of a torque stick by checking the fastener torque with a torque wrench after it is tightened with a torque stick.

Turning Radius Gauge

Turning radius gauge turntables are placed under the front wheels during a wheel alignment. The top plate in the turning radius gauge rotates on the bottom plate to allow the front wheels to be turned during a wheel alignment. A degree scale and a pointer on the gauge indicate the number of degrees the front wheels are turned (Figure 2-23). If the car has four-wheel steering (4WS), the turning radius gauges are also placed under the rear wheels during a wheel alignment.

Plumb Bob

A **plumb bob** is a metal weight with a tapered end (Figure 2-24). This weight is suspended on a string. Plumbers use a plumb bob to locate pipe openings directly below each other at the top and bottom of partitions. Some vehicle manufacturers recommend checking vehicle frame measurements with a plumb bob.

Tram Gauge

A **tram gauge** is a long, straight graduated bar with an adjustable pointer at each end (Figure 2-25). The tram gauge performs frame and body measurements.

Copyright © 2015 Cengage Learning®

GURE 2-22 **Torque stick.**

Copyright © 2015 Cengage Learning®

FIGURE 2-23 **Turing radius gauge.**

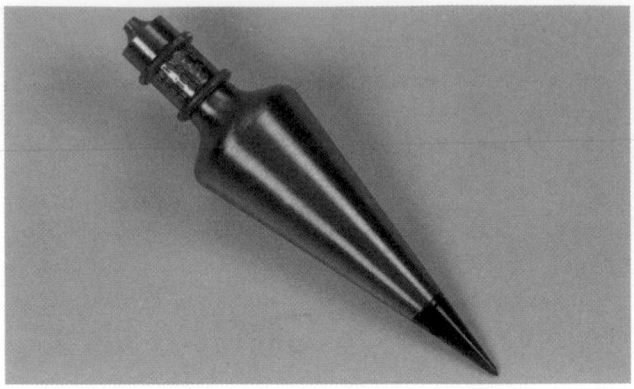

Copyright © 2015 Cengage Learning®

FIGURE 2-24 **Plumb bob.**

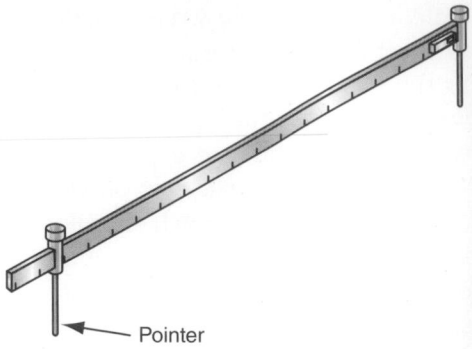

FIGURE 2-25 **Tram gauge.**

SERVICE TIP:
Magnetic wheel alignment gauge mounting surfaces must be clean with no metal burrs.

Magnetic Wheel Alignment Gauge

Each **magnetic wheel alignment gauge** contains a strong magnet that holds the gauge securely on the front wheel hubs. The magnetic wheel alignment gauge measures some of the front suspension alignment angles (Figure 2-26).

Rim Clamps

When the wheel hub is inaccessible to the magnetic alignment gauge, an adjustable **rim clamp** may be attached to each front wheel. The magnetic gauges may be attached to the rim clamp (Figure 2-27). Rim clamps are also used on computer wheel aligners.

Brake Pedal Jack

A **brake pedal jack** must be installed between the front seat and the brake pedal to apply the brakes while checking some front wheel alignment angles (Figure 2-28).

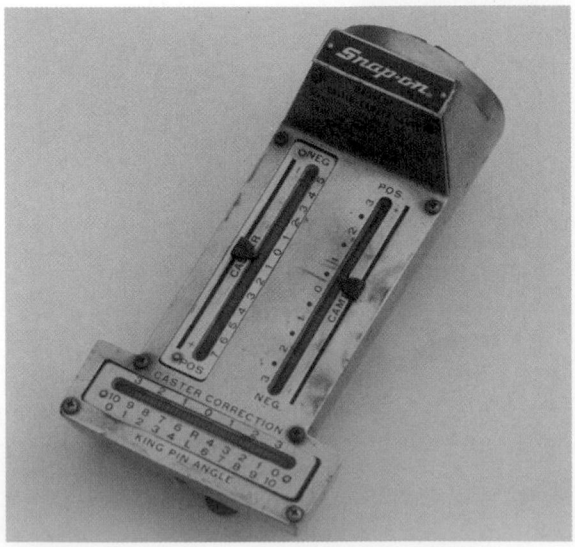

Copyright © 2015 Cengage Learning®

FIGURE 2-26 **Magnetic wheel alignment gauge.**

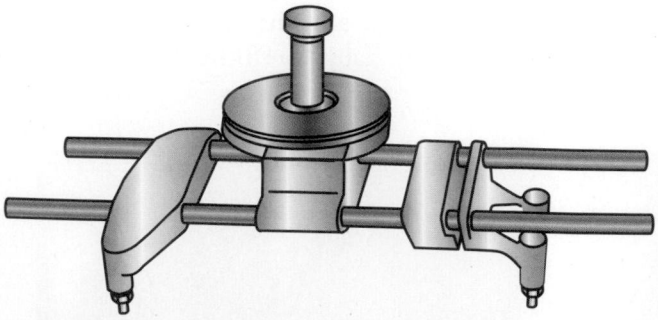

Copyright © 2015 Cengage Learning®

FIGURE 2-27 **Rim clamps.**

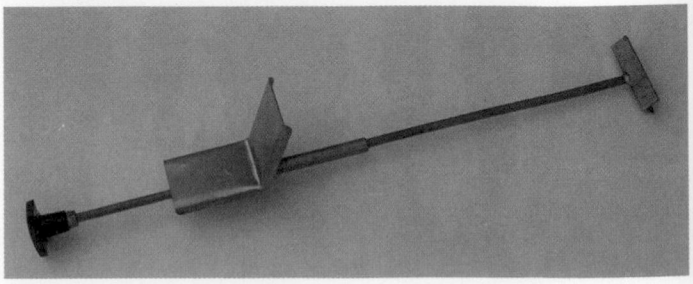

FIGURE 2-28 **Brake pedal jack.**

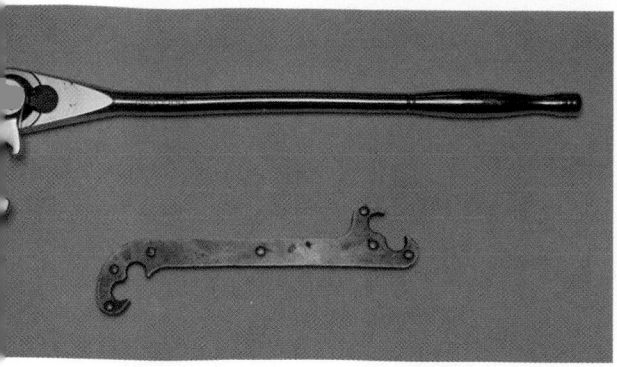

Copyright © 2015 Cengage Learning®

E 2-29 Tie rod sleeve adjusting tool.

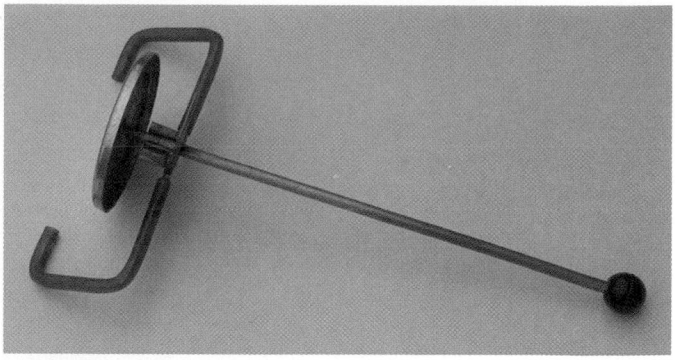

Copyright © 2015 Cengage Learning®

FIGURE 2-30 **Steering wheel locking tool.**

Tie Rod Sleeve Adjusting Tool

A **tie rod sleeve adjusting tool** rotates the tie rod sleeves and performs some front wheel adjustments (Figure 2-29).

CAUTION:
Do not use anything except a tie rod adjusting tool to adjust the tie rod sleeves. Tools, such as a pipe wrench, will damage the sleeves.

Steering Wheel Locking Tool

A **steering wheel locking tool** locks the steering wheel while performing some front suspension service (Figure 2-30).

Toe Gauge

A **toe gauge** is a long, straight, graduated bar that measures front wheel toe (Figure 2-31).

Track Gauge

Some **track gauges** use a fiber-optic alignment system to measure front wheel toe and to determine if the rear wheels are tracking directly behind the front wheels. The front and rear fiber-optic gauges may be connected to the wheel hubs or to rim clamps attached to the wheel rims (Figure 2-32). A remote light source in the main control is sent through fiber-optic cables to the wheel gauges. A strong light beam between the front and rear wheel units informs the technician if the rear wheel tracking is correct.

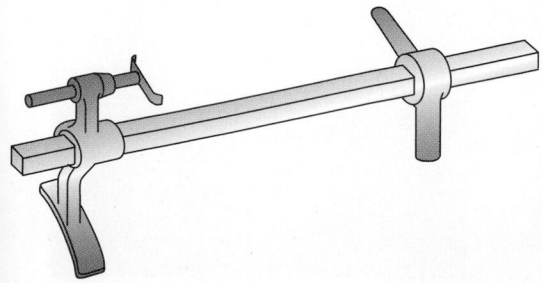

Copyright © 2015 Cengage Learning®

FIGURE 2-31 **Toe gauge.**

Copyright © 2015 Cengage Learning®

FIGURE 2-32 **Fabric optic track and toe gauge.**

Copyright © 2015 Cengage Learning®

FIGURE 2-33 **Computer wheel aligner.**

Computer Four Wheel Aligner

Many automotive shops are equipped with a **computer four wheel aligner** (Figure 2-33). These wheel aligners perform all front and rear wheel alignment angles quickly and accurately.

Scan Tool

A variety of **scan tools** are available for diagnosing automotive computer systems (Figure 2-34). These testers obtain fault codes and perform other diagnostic functions on computer-controlled suspension systems.

Bench Grinder

WARNING: Always wear a face shield when using a bench grinder. Failure to observe this precaution may cause personal injury.

Copyright © 2015 Cengage Learning®

FIGURE 2-34 **Scan tool for diagnosing computer-controlled suspension systems.**

Copyright © 2015 Cengage Learning®

URE 2-35 Bench grinder.

Copyright © 2015 Cengage Learning®

FIGURE 2-36 Bench grinder accessories.

WARNING: When grinding small components on a grinding wheel, wire brush wheel, or buffing wheel, always hold these components with a pair of vise grips to avoid injury to fingers and hands.

WARNING: Grinding and buffing wheels on bench grinders must be mounted on the grinder according to the manufacturer's instructions. Grinding and buffing wheels must be retained with the manufacturer's specified washers and nuts, and the retaining nuts must be tightened to the specified torque. Personal injury may occur if grinding and buffing wheels are not properly attached to the bench grinder.

Bench grinders usually have a grinding wheel and a wire wheel brush driven by an electric motor (Figure 2-35). The grinding wheel may be replaced with a grinding disc containing several layers of synthetic material (Figure 2-36). A buffing wheel may be used in place of the wire wheel brush. The grinding wheel may be used for various grinding jobs and deburring. A buffing wheel is most commonly used for polishing. Most bench grinders have grinding wheels and wire brush wheels that are 6 to 10 in. (15.24 to 25.4 cm) in diameter. Bench grinders must be securely bolted to the workbench.

Hydraulic Pressing and Lifting Equipment

Hydraulic Press

A **hydraulic press** uses a hydraulic cylinder and ram to remove and install precision-fit components from their mounting location.

WARNING: When operating a hydraulic press, always be sure that the components being pressed are supported properly on the press bed with steel supports. Improperly supported components may slip, resulting in personal injury.

WARNING: When using a hydraulic press, never operate the pump handle if the pressure gauge exceeds the maximum pressure rating of the press. If this maximum pressure is exceeded, some part of the press may suddenly break, causing severe personal injury.

WARNING: Be sure the safety cage is in place around the press to prevent personal injury.

When two components have a tight precision fit between them, a hydraulic press is used to either separate these components or press them together. The hydraulic press rests on the shop floor, and an adjustable steel-beam bed is retained to the lower press frame with heavy steel pins. A hydraulic cylinder and ram is mounted on the top part of the press with the ram facing downward toward the press bed (Figure 2-37). The component being pressed is placed on the press bed with appropriate steel supports. A hand-operated hydraulic pump is mounted on the side of the press. When the handle is pumped, hydraulic fluid is forced into the cylinder, and the ram is extended against the component on the press bed to complete the pressing operation. A pressure gauge on the press indicates the pressure applied from the hand pump to the cylinder. The press frame is designed for a certain maximum pressure, and this pressure must not be exceeded during hand pump operation.

Floor Jack

WARNING: The maximum lifting capacity of the floor jack is usually written on the jack decal. Never lift a vehicle that exceeds the jack lifting capacity. This action may cause the jack to break or collapse, resulting in vehicle damage or personal injury.

A **floor jack** uses hydraulic pressure supplied to a hydraulic cylinder, ram, and lift pad to lift one end or one corner of a vehicle.

A floor jack is a portable unit mounted on wheels. The lifting pad on the jack is placed under the chassis of the vehicle, and the jack handle is operated with a pumping action (Figure 2-38), forcing fluid into a hydraulic cylinder in the jack. Then this cylinder extends to force the jack lift pad upward and lift the vehicle. Always be sure that the lift pad is positioned securely under one of the car manufacturer's recommended lifting points. To release the hydraulic pressure and lower the vehicle, the handle or release lever must be turned slowly. Do not leave the jack handle where someone can trip over it.

Vehicle Lift (Hoist)

A **vehicle lift** raises a vehicle so the technician can work under it. The lift arms must be placed under the car manufacturer's recommended lifting points prior to raising a vehicle. Twin posts are used on some lifts, whereas other lifts have a single post (Figure 2-39). Some lifts have an electric motor, which drives a hydraulic pump to create fluid pressure

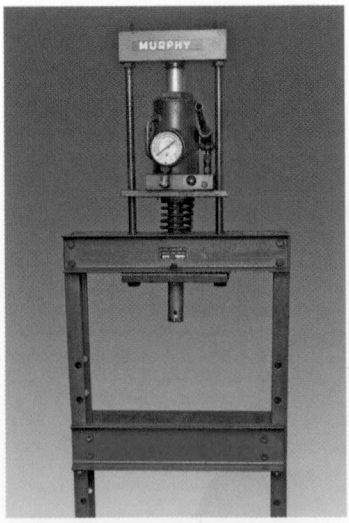

Copyright © 2015 Cengage Learning®

FIGURE 2-37 **Hydraulic press.**

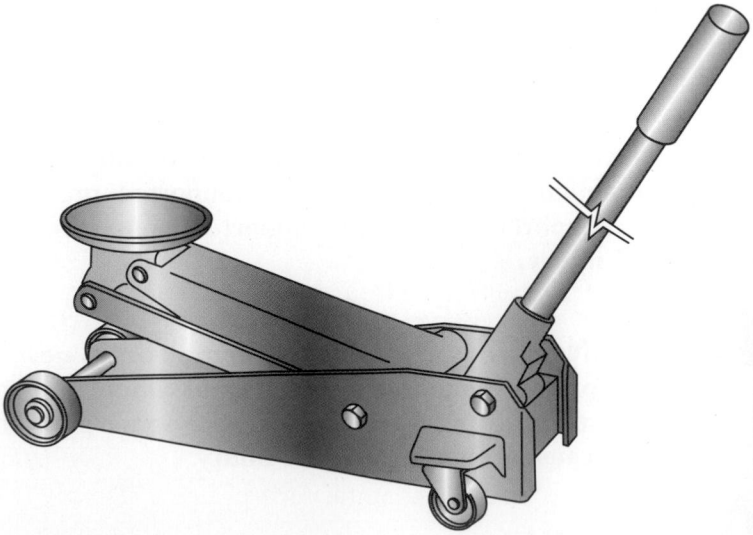

Copyright © 2015 Cengage Learning®

FIGURE 2-38 **Hydraulic floor jack.**

Copyright © 2015 Cengage Learning®

RE 2-39 **Vehicle lifts are used to raise a vehicle.**

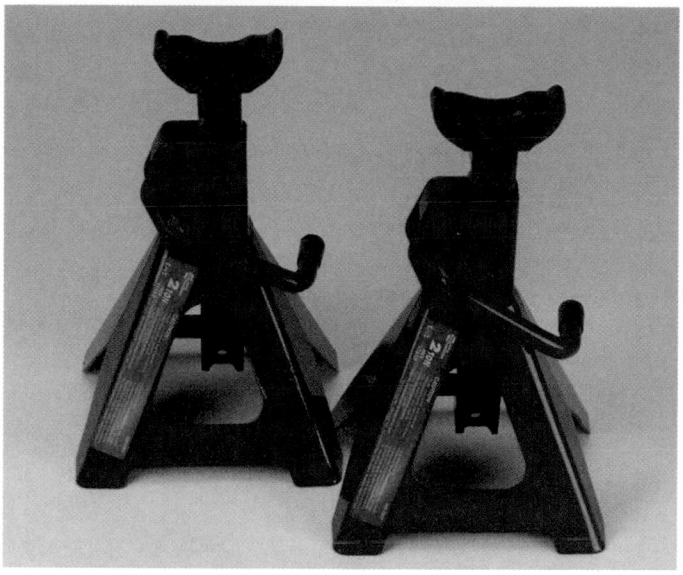

Copyright © 2015 Cengage Learning®

FIGURE 2-40 **Typical safety stands.**

and force the lift upward. Other lifts use air pressure from the shop air supply to force the lift upward. If shop air pressure is used, it is applied to fluid in the lift cylinder. A control lever or switch is placed near the lift, which supplies shop air pressure to the lift cylinder and turns on the lift pump motor. Always be sure that the safety lock is engaged after the lift is raised. When the safety lock is released, a release lever is operated slowly to lower the vehicle.

CAUTION: Always be sure that the lift arms are securely positioned under the car manufacturer's recommended lifting points before raising the vehicle. These lifting points are shown in the service manual. If the lift arms are not positioned on the proper lift points, chassis components may be damaged, or the vehicle may slide off the lift, resulting in vehicle damage and/or personal injury.

Using a Floor Jack and Safety Stands

A floor jack is often used to raise a vehicle off the floor. The vehicle is then lowered onto **safety stands** to allow service work to be performed under the vehicle (Figure 2-40).

Follow this procedure to lift a vehicle with a floor jack and then support it on safety stands:

1. Check the weight ratings of the floor jack and the safety stands; be sure these ratings exceed the weight to be lifted and supported.

WARNING: If the vehicle weight to be lifted and supported exceeds the weight rating of the floor jack or safety stands, the floor jack or safety stands may collapse, allowing the vehicle to drop suddenly, which may cause personal injury and/or vehicle damage.

2. Be sure the vehicle is parked on a level surface. Place wheel blocks behind the rear wheels of the vehicle if you are lifting the front end. Place these wheel blocks in front of the front wheels if you are lifting the rear end.
3. Determine the vehicle manufacturer's specified lift point from the vehicle service manual, and position the floor jack lift pad under this lift point. Pump the floor jack handle until the lift pad contacts the lift point on the vehicle.
4. Look under the vehicle and be sure the floor jack lift pad is contacting the specified lift point. If necessary, lower the lift pad and move it so it contacts the lift point properly.
5. Pump the floor jack handle until the vehicle is at the desired height. Place safety stands adjusted to the appropriate height under the vehicle manufacturer's lift points. These lift points will be a strong chassis member such as the frame or axle housing.

A vehicle lift may be called a hoist.

Safety stands may be called jack stands.

CAUTION:
The maximum capacity of the vehicle lift is placed on an identification plate. Never lift a vehicle that is heavier than the maximum capacity of the lift. Lifting a vehicle with a weight greater than the lift weight rating may bend the lift or cause the vehicle to slip off the lift, resulting in vehicle damage and/or personal injury.

6. Because the floor jack is on casters, the vehicle and floor jack tend to move slightly as the vehicle is lowered onto safety stands. Slowly operate the hydraulic pressure release on the floor jack to lower the vehicle onto the safety stands. As the vehicle is lowered, be sure the safety stands do not tip. All four legs on both safety stands must remain in contact with the floor surface. Be sure the safety stands are contacting the vehicle lift points properly, and double-check to be sure the safety stand legs are all contacting the floor.

WARNING: Never work under a vehicle unless it is securely supported on safety stands. If a vehicle is supported only by a floor jack, the jack may allow the vehicle to drop, resulting in personal injury.

7. After the vehicle is supported properly on safety stands, lower and remove the floor jack to allow more room for under-car service.

WARNING: Never under any condition leave a floor jack handle down so it is near the floor surface; someone may trip over the handle, resulting in personal injury.

CAUTION:
Lifting or supporting a vehicle at a location other than the vehicle manufacturer's specified lift point may cause damage to the vehicle chassis and other components.

Using a Vehicle Lift

A vehicle lift raises a vehicle to allow a technician to perform under-car repairs.

Follow this procedure to raise a vehicle on a lift:

1. Determine both the weight rating of the lift and the vehicle weight to be lifted. Be sure the weight rating of the lift exceeds the vehicle's weight.
2. Before driving a vehicle on a lift, be sure the lift arms are positioned properly out of the path of the vehicle. The lift arms must also be positioned to provide unobstructed clearance between the lift and the vehicle chassis.
3. After driving the vehicle on the lift, be sure the vehicle is properly positioned in relation to the lift. Make sure the vehicle doors, hood, and trunk lid are closed. Do not raise a vehicle with any person inside the vehicle.
4. Determine the specified vehicle lift points from the vehicle service manual (Figure 2-41).
5. Position the lift supports so they are directly under the specified vehicle lift points. Operate the lift so the lift supports contact the specified lift points.

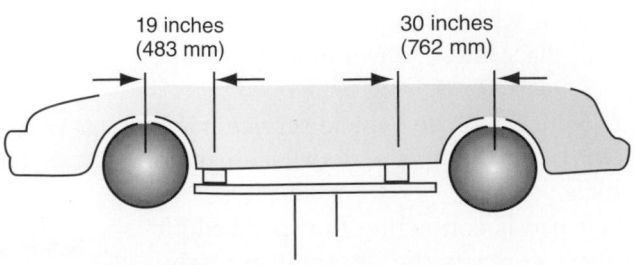

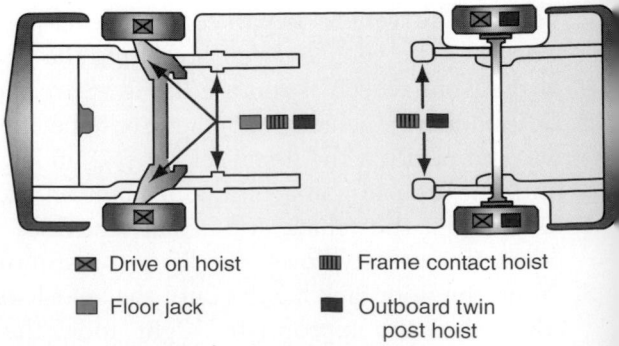

FIGURE 2-41 **Lift points for a typical unibody vehicle.**

WARNING: If the lift supports are not in proper contact with the specified vehicle lift points, the vehicle may slip off the lift, resulting in personal injury and vehicle damage.

CAUTION: When driving a vehicle on a lift, do not run over lift arms, adapters, or axle supports. This action may damage the lift, tires, or vehicle chassis components.

6. Operate the lift a short distance off the shop floor and stop the lift. Look under the vehicle to be sure the lift supports are in full contact with the specified lift points. If the lift supports are not in full contact with the specified lift points, lower the vehicle onto the floor and reposition the lift supports.
7. Raise the vehicle to the desired height. Be sure the lift safety mechanism is engaged.

WARNING: If heavy components are removed from one end of a vehicle, the vehicle may become unstable on the lift. Therefore, to prevent this action, chain a strong chassis component to the lift on the opposite end of the vehicle before removing heavy components from the other end.

CAUTION: If the lift supports are not in proper contact with the specified vehicle lift points, chassis components such as brake lines may be damaged.

8. Before lowering the lift, remove all tools, toolboxes, and equipment from under the vehicle, and sweep up discarded parts from under the vehicle.
9. Be sure nobody is standing under the vehicle before lowering it.
10. Lower the vehicle slowly until it is resting on the shop floor and be sure the lift is completely lowered.
11. Before driving the vehicle off the lift, be sure there is proper clearance between all parts of the lift and the vehicle chassis.

Suspension and Steering Service, Diagnostic, and Measurement Tools

Using a Power Steering Pressure Gauge

The power steering pump pressure test is one of the most important indications of power steering pump condition.

Follow these steps to use the power steering pressure gauge to test power steering pump pressure:

A **power steering pressure gauge** is connected in the power steering system to test power steering pump pressure.

1. Check the power steering belt condition and if it is cracked, damaged, or oil soaked, replace the belt. If a V-belt is bottomed in the pulley, excessive wear on the sides of a V-belt is indicated. This condition also requires belt replacement.
2. Measure the power steering belt tension with a belt tension gauge. If this tension is less than specified, adjust the power steering belt tension.
3. Be sure the engine is warmed up and the power steering fluid is at normal operating temperature.
4. With the engine shut off, check the fluid level in the power steering pump reservoir. If necessary, add the vehicle manufacturer's specified fluid until the proper level is obtained.
5. Listen to the power steering pump with the engine idling. A growling noise from the pump may indicate aeration of the power steering fluid in the reservoir or air in the power steering system. If air is indicated in the system, turn the steering wheel several times fully in each direction to bleed air from the system before proceeding with the power steering pump pressure test.
6. Shut the engine off and remove the high-pressure hose from the power steering pump. Connect the power steering pressure gauge and valve assembly to the fitting on the end

Classroom Manual
Chapter 2, page 32

Copyright © 2015 Cengage Learning®

FIGURE 2-42 **Pressure gauge connections to power steering.**

of the high-pressure hose and to the fitting on the pump from which the high-pressure hose was removed (Figure 2-42). Be sure the valve on the power steering pressure gauge is fully open.

7. With the engine idling, close the power steering gauge valve for no more than 10 seconds and observe the power steering pump pressure on the gauge. Turn the gauge valve fully open. If the power steering pump pressure is less than the vehicle manufacturer's specified pressure, replace the power steering pump.

WARNING: Wear a face shield and protective gloves during the power steering pressure test. Power steering fluid and components become hot during this test.

CAUTION:
Do not allow the power steering fluid to become too hot during the pump pressure test. Excessive fluid temperature reduces pump pressure, resulting in false test results.

WARNING: If the gauge valve is closed for more than 10 seconds during the power steering pressure test, excessive power steering pump pressure may rupture power steering hoses. Hot fluid spraying from a ruptured hose may burn anyone near the vehicle.

8. Shut the engine off and remove the power steering pressure gauge. Tighten the high-pressure hose fitting to the specified torque. Start the engine and check for leaks at this fitting. Be sure the power steering reservoir is filled to the proper level.

Using a Dial Indicator

A dial indicator performs precision measurements on some suspension components such as ball joints. A special dial indicator with a stiff, flexible attaching bracket is available for measuring ball joint wear. When measuring ball joint wear, the end of this flexible bracket is clamped to a stable suspension component with a pair of vise grips. The dial indicator must be mounted so there is no movement in the indicator mounting. A dial indicator contains a movable plunger that is positioned against the component to be measured. The pointer on the dial indicator registers the amount of plunger movement. Each rotation of

Copyright © 2015 Cengage Learning®

FIGURE 2-43 **The dial indicator pointer rotation indicates the amount of plunger movement.**

the pointer represents 0.100 in. (2.54 mm) of plunger movement (Figure 2-43). The dial indicator plunger must be positioned against the component to be measured so the dial indicator is preloaded approximately 0.250 in. (6.35 mm). When the dial indicator is preloaded properly, about one-half of the plunger should be outside the dial indicator. After the preload procedure is completed, the dial indicator face must be rotated until the 0 position on the scale is aligned with the pointer. Always look squarely at the indicator face. Looking at the dial indicator face from an angle may provide an inaccurate reading.

The following is a typical procedure for measuring ball joint wear on a short-and-long arm suspension system with the coil spring positioned between the lower control arm and the chassis:

1. Lift the front of the vehicle with a floor jack positioned under the manufacturer's specified lift point, and position a safety stand under the lower control arm. Lower the vehicle so the lower control arm and vehicle weight rests on the safety stand. The tire must remain several inches off the floor.
2. Clamp the end of the dial indicator's flexible bracket onto the lower control arm.
3. Position the dial indicator plunger against the lower side of the steering knuckle next to the nut on the ball joint stud. The dial indicator plunger should be positioned at a 90 degree angle with the ball joint stud (Figure 2-44).
4. Preload and zero the dial indicator. Be sure the dial indicator is clamped securely to the lower control arm without any movement in the indicator mounting.

Copyright © 2015 Cengage Learning®

FIGURE 2-44 **Dial indicator installed to measure vertical ball joint movement.**

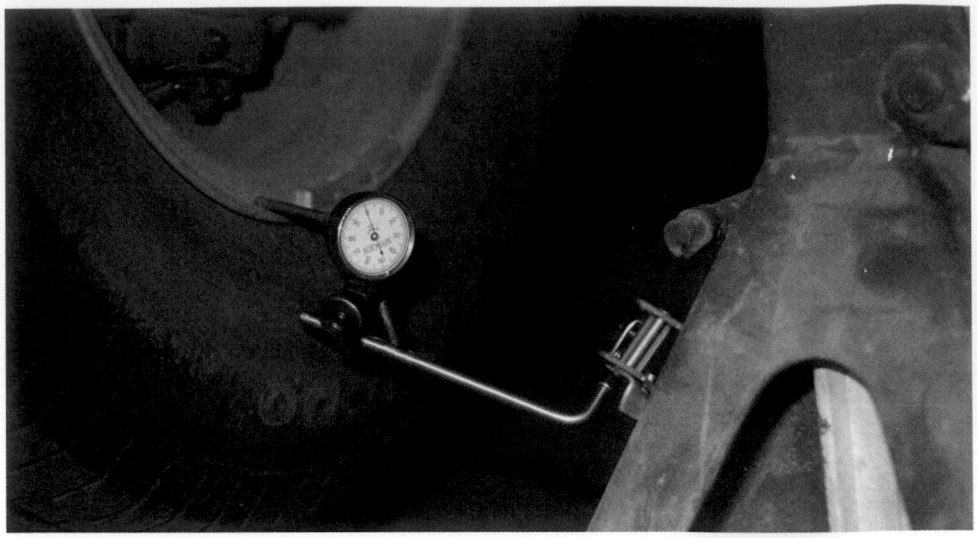

Copyright © 2015 Cengage Learning®

FIGURE 2-45 **Dial indicator installed to measure horizontal joint movement.**

SERVICE TIP: The procedure for measuring ball joint wear varies depending on the type of suspension system and the type of ball joint mounting. Always consult the vehicle manufacturer's service manual.

5. Position a long steel bar under the tire, and lift upward on the bar while an assistant observes the movement on the dial indicator pointer. If the dial indicator pointer movement exceeds the maximum-specified ball joint vertical movement, the ball joint must be replaced.
6. Be sure the front wheel bearings are adjusted properly. Position the dial indicator plunger against the inner edge of the wheel rim (Figure 2-45). Preload and zero the dial indicator. Grasp the top and bottom of the tire, and attempt to tip the tire out at the bottom and in at the top. Release the tire, and repeat this procedure several times while an assistant observes the dial indicator. If the dial indicator pointer movement exceeds the maximum-specified ball joint horizontal movement, the ball joint must be replaced.

Using a Coil Spring Compressing Tool

A coil spring compressing tool is required to compress the coil spring before removing the spring from the strut on a MacPherson strut suspension system.

Follow this procedure for using a coil spring compressing tool to remove a spring from a strut:

1. With the strut-and-coil-spring assembly removed from the vehicle, install the coil spring in the compressing tool according to the tool manufacturer's or vehicle manufacturer's recommended procedure.

CAUTION: If the coil spring has an enamel-type coating, tape the spring in the areas where the compressor tool contacts the spring. If the compressor tool chips this coating, the spring may break prematurely.

WARNING: Unless the coil spring is compressed, never loosen the nut on the strut rod that retains the upper strut mount on the strut and coil spring. Loosening this nut will suddenly release the coil spring tension, resulting in personal injury.

2. Adjust the compressing arms on the spring compressing tool so the arms contact the coils farthest away from the center of the spring (Figure 2-46).

WARNING: Always use the coil spring compressor tool recommended by the vehicle manufacturer. Some compressor tools are designed to work only on specific strut-and-coil-spring assemblies. If the compressor tool does not fit properly on the coil spring, the tool may slip off the spring, suddenly releasing the coil spring tension, which may result in personal injury.

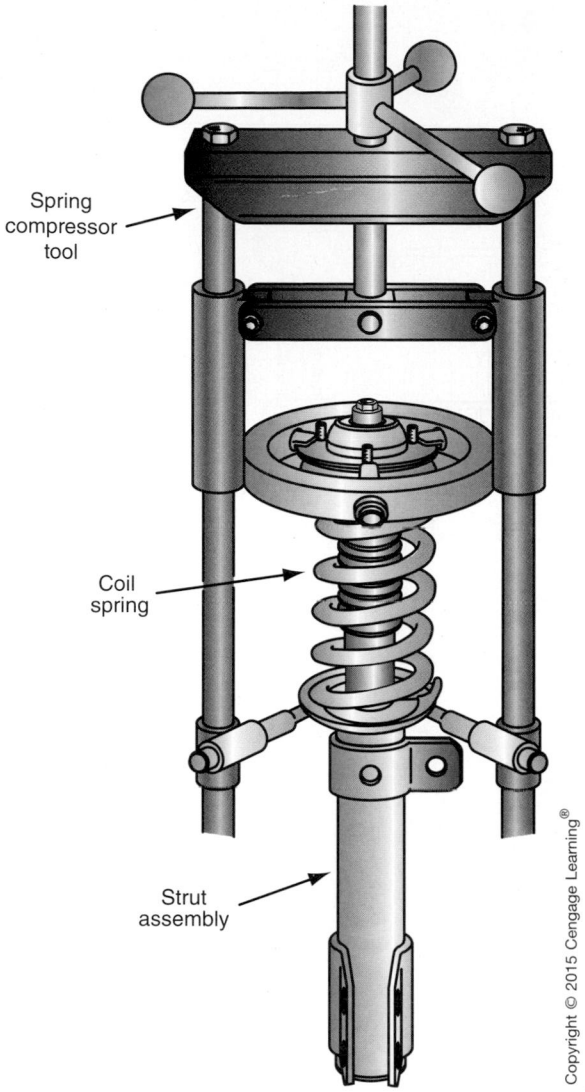

FIGURE 2-46 **Coil spring and strut assembly mounted in a spring compressing tool.**

3. Turn the handle on top of the compressor tool until all the spring tension is removed from the upper strut mount.
4. Loosen and remove the strut rod nut in the center of the upper strut mount (Figure 2-47). Be sure all the spring tension is removed from the upper strut mount before loosening this nut.
5. Remove the upper strut mount assembly, mount bearing, and then remove the upper spring seat and insulator.
6. Rotate the handle on the spring compressing tool to release all the tension on the coil spring, and then remove the spring.
7. Remove the dust shield and jounce bumper from the strut rod, and then remove the lower spring insulator (Figure 2-48).
8. Replace all worn or defective parts, and be sure the strut rod is fully extended. Install the lower insulator on the lower strut spring seat, and be sure the insulator is properly positioned on the seat.
9. Install the spring bumper on the strut rod.
10. Place the coil spring properly in the spring compressing tool, and rotate the handle on the spring compressing tool to compress the coil spring.

CAUTION:
Never clamp the lower end of the strut in a vise, which may cause internal strut damage.

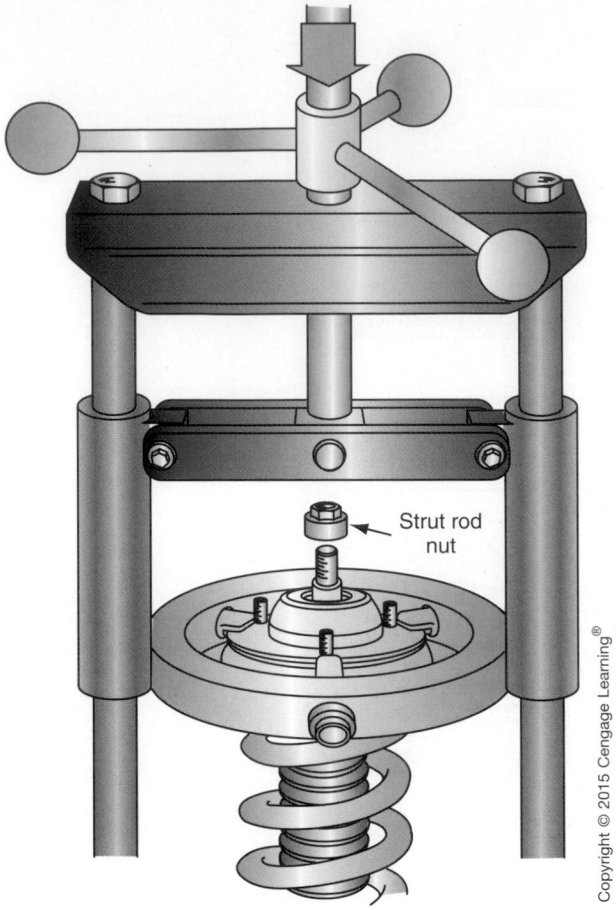

FIGURE 2-47 **After the compressing tool is operated to remove all the spring tension, remove the strut rod nut.**

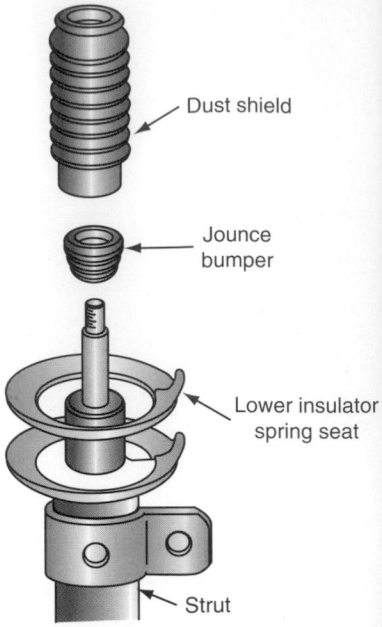

FIGURE 2-48 **Removal of the dust shield, jounce bumper, and lower spring insulator.**

SERVICE TIP:
The tire dismounting and mounting procedure varies depending on the type of tire changer. Always follow the dismounting and mounting procedure recommended by the tire changer manufacturer.

11. Install the strut into the coil spring, and be sure the lower insulator is properly positioned on the coil spring.
12. Install the upper spring insulator and upper strut mount, and be sure these components are properly positioned.
13. Install the strut rod nut and tighten this nut to the specified torque.
14. Rotate the handle on the spring compressing tool to slowly release all the spring tension, and remove the strut and spring assembly from the tool.

CAUTION:
Always use the tire changer attachments recommended by the tire changer manufacturer. For example, specific bead removal and installation attachments may be recommended for certain types of wheel rims.

Using a Tire Changer

A tire changer dismounts and mounts tires on wheel rims without damaging the rims. Most tire changers are operated by shop air pressure.

WARNING: If the tire changer manufacturer's recommended procedure for dismounting and mounting tires is not followed, personal injury and/or rim damage may occur.

The following is a typical tire dismounting and mounting procedure:

1. Position the tire and wheel assembly in the specified location on the tire changer (Figure 2-49). Be sure the tire and wheel assembly is properly positioned and secured.
2. Remove the valve cap and valve core, and be sure the tire is completely deflated.

Copyright © 2015 Cengage Learning®

FIGURE 2-49 **Typical tire changer.**

3. Position the bead unseating tool to unseat the outer bead, and operate the changer to unseat this bead.
4. Operate the changer to unseat the bead on the inner side of the rim.
5. Position the bead removal tool to remove the outer bead over the rim, and operate the changer to remove this bead over the rim.
6. Position the bead removal tool to remove the inner bead over the rim, and operate the changer to remove this bead over the rim.
7. After the beads are removed over the rim, remove the tire from the changer.
8. Clean dirt and rust from the rim sealing flanges with a wire brush, and apply a coating of rubber compound to the bead area of the tire.
9. Install the tire on top of the rim in the desired position. Many tires have a dot painted on the tire that should be aligned with the valve stem when the tire is mounted on the rim. Therefore, this dot must be on the outer side of the tire prior to mounting.
10. Position the bead installation tool to install the inner bead over the rim, and operate the changer to install this bead over the rim.
11. Position the bead installation tool to install the outer bead over the rim, and operate the changer to install this bead over the rim.
12. Rotate the tire on the rim to align the painted dot with the valve stem, and install the valve core. Be sure the tire is centered properly on the rim.
13. Inflate the tire to the specified tire pressure, install the valve cap, and remove the tire and wheel from the tire changer.

Using a Scan Tool

A scan tool performs an electronic diagnosis of computer-controlled suspension systems and other electronic systems. Various test modules may be plugged into the scan tool, depending on the vehicle and system being diagnosed (Figure 2-50). The scan tool cable is connected to the data link connector (DLC) under the dash. On 1996 and newer vehicles

SERVICE TIP:
Many shops have a policy of always installing a new valve stem when repairing or replacing a tire, which prevents problems with cracked or defective valve stems.

SERVICE TIP:
Aligning the painted dot on the tire sidewall ensures proper tire balancing with a minimum amount of balance weight.

SERVICE TIP:
The valve core may be left out of the valve stem and the tire partially inflated. This action allows the air pressure to enter the tire quicker and seat the beads against the rim flanges. After the beads are seated, allow most of the air to escape from the tire, and then install the valve core.

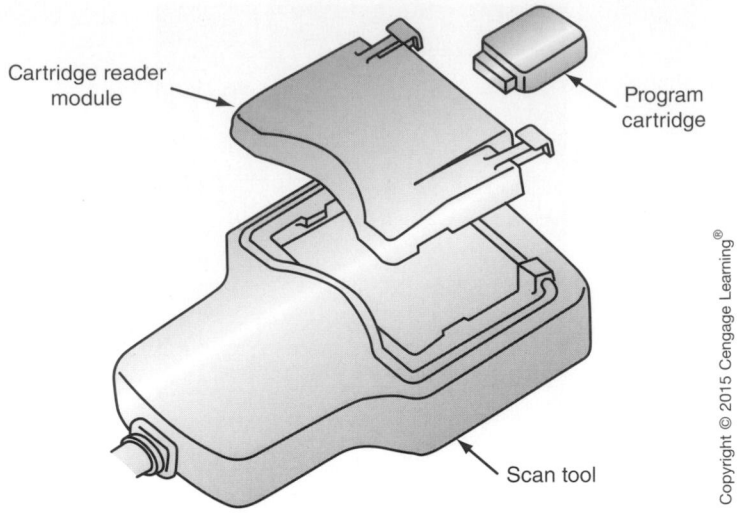

FIGURE 2-50 **Scan tool and related modules.**

CAUTION:
Never connect or disconnect the scan tool data cable with the ignition switch on. This action may damage expensive electronic components on the vehicle or the scan tool.

with on-board diagnostic II (OBD II) systems, the scan tool is powered from a terminal in the DLC. On most pre-1996 vehicles, a power cable on the scan tool has to be connected to a 12V power source such as the cigarette lighter.

After the scan tool is connected, the vehicle make, model year, and engine size usually has to be selected on the scan tool display. On vehicles with several different computer systems, the technician must select the computer system to be diagnosed from the list of computer systems on the vehicle. For example, if the technician wants to diagnose the continuously variable road sensing suspension system (CVRSS), then RSS must be selected from the list of computer systems displayed on the scan tool display. After the RSS selection, the scan tool display will ask the technician to select diagnostic trouble codes (DTCs), data, inputs, outputs, or clear codes.

On OBD II vehicles, the DTCs contain five digits. A typical DTC from a CVRSS is C1712. The "C" indicates this DTC is in the chassis group, and the "1" informs the technician this is a manufacturer specific code. A common DTC dictated by Society of Automotive Engineers (SAE) standards is indicated if the second digit is a "0." The digits 712 in the DTC indicate an open circuit in the left front (LF) damper actuator circuit. On a CVRSS, the damper actuators are solenoids in each shock absorber or strut. Diagnostic trouble codes (DTCs) indicate a defect in a certain electrical area, but they do not indicate a defect in a specific component. For example, the C1712 DTC indicates there is an open circuit in the LF damper actuator solenoid or in the connecting wires between this solenoid and the suspension computer. The technician usually has to use a volt-ohm meter to locate the exact cause of the DTC. Diagnostic trouble codes (DTCs) may be identified on the scan tool display as CURRENT or HISTORY. A CURRENT DTC is one that is present at the time of testing. A HISTORY DTC represents an intermittent defect that occurred in the past, which has since disappeared.

If "data" is selected, the scan tool display indicates data related to the CVRSS. This data includes readings from the input sensors and readings such as battery voltage or vehicle speed. When "inputs" is selected, the scan tool displays readings from the input sensors in the CVRSS. In the "output" parameter the scan tool displays readings from the outputs in the CVRSS. The output displays may be a voltage reading or solenoids displayed as being "on" or "off." If "clear codes" is selected on the scan tool display, the DTCs are erased from the computer memory.

Depending on the computer-controlled suspension system being diagnosed, the scan tool may be used to perform different functions. For example, on some computer-controlled air suspension systems, the scan tool may be used to command the suspension computer to fill and vent each air spring. This test mode may be used to check the operation of system components. The scan tool may be used to flash program the suspension computer memory. This mode installs new software in the computer as directed by service bulletins from the original equipment manufacturer (OEM). On some computer-controlled suspension systems, such as the system on the 2002 Lincoln Blackwood, the scan tool is used to adjust the suspension ride height.

Using an Electronic Wheel Balancer

The most popular type of wheel balancer in automotive service shops is the electronic balancer. Most electronic wheel balancers have an electric motor that spins the tire and wheel assembly at moderate speed during the balance procedure. On some electronic wheel balancers, the tire and wheel assembly is spun by hand for balance purposes. Certain preliminary checks must be performed on the tire and wheel assembly before installing this assembly on an electronic wheel balancer.

These preliminary checks include the following:

1. Clean all mud and debris from the tire and wheel assembly.
2. Remove all old wheel weights from the rim.

WARNING: Improperly installed wheel weights may fly off the rim when the wheel is spun on the balancer, resulting in personal injury.

3. Remove objects such as stones from the tire tread.

WARNING: Stones or other objects in the tire tread may fly out when the wheel is spun on the balancer, resulting in personal injury.

4. Inspect the tire tread and sidewall for defects, such as splits, cuts, chunks out of the tread or sidewall, and bulges indicating ply separation. Replace tires with these defects.
5. Inflate the tire to the specified pressure.

The tire and wheel assembly must be installed on the wheel balancer according to the instructions of the wheel balancer manufacturer. After the tire and wheel assembly is installed securely on the balancer, slowly rotate the tire by hand and listen for objects such as balls of rubber rolling inside the tire. Objects inside the tire must be removed, because they make wheel balancing difficult or impossible.

Classroom Manual
Chapter 2, page 28

WARNING: If the tire and wheel assembly is not securely installed on the wheel balancer, the wheel may come loose and fly off the balancer during the balance procedure, resulting in personal injury.

Use a dial indicator to measure vertical and lateral tire and wheel runout. If tire and wheel runout is excessive, wheel balancing may be difficult. Enter the wheel diameter, width, and offset in the wheel balancer. The balancer must have this information to calculate the required amount of wheel weight and the weight position. Lower the safety hood on the wheel balancer, and activate the balancer to spin the wheel. Apply the brake on the balancer to stop the wheel. Install the indicated amount of wheel weight in the proper position as indicated on the wheel balancer display. Be sure the weight(s) are securely attached to the wheel rim.

Using a Computer Four Wheel Aligner

If a vehicle has a frame and a live axle rear suspension system, it is not as likely to require rear wheel alignment as a unibody vehicle with a semi-independent or independent rear suspension system. Because of increased four wheel alignment requirements on unibody

A **computer four wheel aligner** uses a wheel sensor mounted on each wheel to measure all the front and rear wheel alignment angles.

cars, most shops that offer wheel alignment service are equipped with a computer four wheel aligner. Before performing a wheel alignment, the technician must complete a preliminary vehicle inspection. The purpose of the preliminary inspection is to locate any defective or worn components that would make wheel alignment inaccurate.

A preliminary wheel alignment inspection includes these checks:

1. Be sure the vehicle has the normal curb weight that it has when the driver is operating the vehicle. Make sure the gas tank is full, and check for excessive mud adhered to the underside of the chassis.
2. Be sure the tires are inflated to the specified tire pressure, and inspect the tires for excessive wear, bulges, cuts, and splits. Be sure the tires are all the same size.
3. Check for the specified front and rear suspension height. Because the suspension height affects many of the wheel alignment angles, this height must be corrected before performing a wheel alignment if it is not within specifications.
4. Check the steering wheel free play. If this free play is more than specified, inspect the steering linkages and tie rod ends for excessive wear. The excessive steering wheel free play must be corrected before a wheel alignment, because this problem affects front wheel toe and steering quality.
5. Be sure the shock absorbers and struts are in satisfactory condition.
6. Check the front and rear wheel bearing adjustments, and correct these adjustments as necessary. Loose wheel bearing adjustments affect some of the wheel alignment angles.
7. Check the ball joints for excessive wear. Worn ball joints affect some of the wheel alignment angles.
8. Inspect the front and rear, upper and lower control arms for damage, and inspect all the control arm bushings for wear. Bent control arms or worn control arm bushings affect wheel alignment angles.
9. Inspect the front and rear stabilizer bars and bushings. Worn stabilizer bushings affect ride quality and vehicle handling.

After the preliminary inspection is completed, and all the necessary components have been replaced or adjusted, drive the vehicle onto the wheel alignment rack.

Then follow this procedure to complete the four wheel alignment with a computer wheel aligner:

1. Follow all instructions provided in the wheel aligner operator's manual.
2. Be sure the front wheels are centered on the wheel aligner turntables.
3. Be sure the rear wheels are positioned properly on the slip plates.
4. Mount the wheel sensor units on each wheel.
5. Select the vehicle make and model year on the wheel aligner screen.
6. On the wheel aligner preliminary inspection screen, check the condition of each item. Most of these items are mentioned in the previous preliminary inspection procedure.
7. Display the ride height screen on the wheel aligner, and be sure the front and rear ride height is within specifications.
8. Display the wheel runout screen of the wheel aligner, and compensate for wheel runout as directed.
9. Display the turning angle screen on the wheel aligner. Apply the brakes with a brake pedal depressor as directed on the screen, and perform the turning angle check.
10. Display the front and rear wheel alignment angles on the screen. Most computer four wheel aligners mark the alignment angles that are not within specifications.
11. Display the adjustment screen on the wheel aligner, and perform the necessary front and rear suspension adjustments to bring all the alignment angles within specifications. Photo Sequence 2 illustrates a typical procedure for performing a four wheel alignment with a computer wheel aligner.

SERVICE TIP: The wheel runout compensation procedure varies, depending on the make and year of the wheel aligner.

PHOTO SEQUENCE 2

Typical Procedure for Performing Four Wheel Alignment with a Computer Wheel Aligner

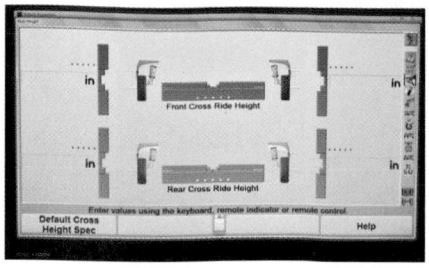

P2-1 **Display the ride height screen. Check the tire condition for each tire on the tire condition screen.**

Copyright © 2015 Cengage Learning®

P2-2 **Position the vehicle on the alignment rack.**

Copyright © 2015 Cengage Learning®

P2-3 **Make sure the front tires are positioned properly on the turntables.**

Copyright © 2015 Cengage Learning®

P2-4 **Position the rear wheels on the slip plates.**

Copyright © 2015 Cengage Learning®

P2-5 **Attach the wheel units.**

Copyright © 2015 Cengage Learning®

P2-6 **Select the vehicle make and model year.**

Copyright © 2015 Cengage Learning®

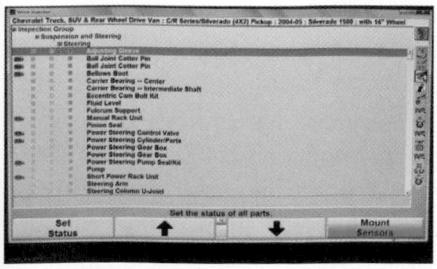

P2-7 **Check the items on the screen during the preliminary inspection.**

Copyright © 2015 Cengage Learning®

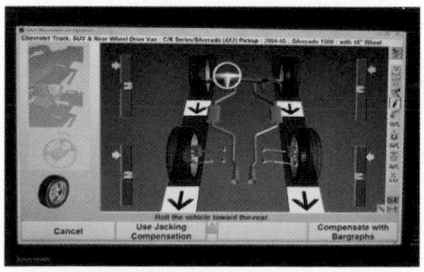

P2-8 **Display the wheel runout compensation screen.**

Copyright © 2015 Cengage Learning®

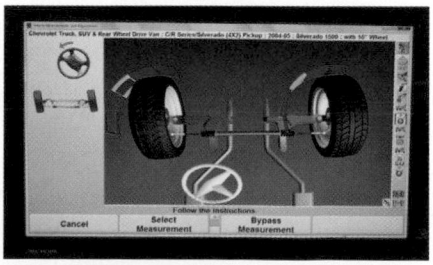

P2-9 **Display the turning angle screen and perform the turning angle check.**

Copyright © 2015 Cengage Learning®

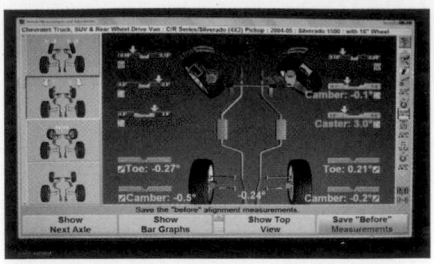

P2-10 **Display the front and rear wheel alignment angle screen.**
Copyright © 2015 Cengage Learning®

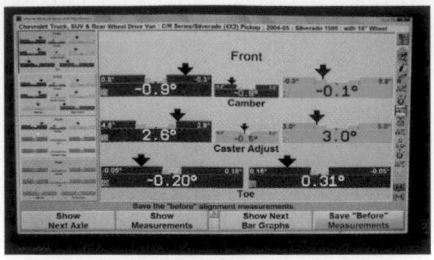

P2-11 **Display the adjustment screen.**
Copyright © 2015 Cengage Learning®

Employer and Employee Obligations

The ever increasing electronics content on today's vehicles also requires that technicians are familiar with the latest electronics technology. To be a successful automotive technician, you must be committed to lifelong training. There are many different ways to obtain this training, but it is absolutely essential.

Automotive training may be obtained by these methods:

1. Obtain bulletins, service manuals, and training information from original equipment manufacturers (OEMs); independent parts and component manufacturers; and independent suppliers of service manuals and training books. After the information is obtained, it is essential that you read and study it.
2. Join an organization dedicated to supplying information to automotive service technicians such as the International Automotive Technician's Association (IATA).
3. Join an Internet organization such as the International Automotive Technicians Network where you can communicate with other technicians and obtain answers to service problems.
4. Download information available on the Internet from automotive equipment manufacturers. Many of these manufacturers provide operator's manuals and other information on their equipment for downloading purposes.
5. Attend satellite training seminars available from some independent automotive training organizations and OEMs.
6. Attend training seminars in your location sponsored by equipment and parts manufacturers or OEMs.
7. Attend training seminars sponsored by the automotive department at your local college.
8. Attend National Institute for Automotive Service Excellence (ASE) certification classes.

Successful automotive technicians must be committed to lifelong training, but their employers must also be committed to assisting them in obtaining the necessary training. This assistance may be financial, providing time off work to attend training seminars or arranging the necessary training programs.

When you begin employment, you enter into a business agreement with your employer. A business agreement involves an exchange of goods or services that have value. Although the automotive technician may not have a written agreement with his or her employer, the technician exchanges time, skills, and effort for money paid by the employer. Both the employee and the employer have obligations.

The automotive technician's obligations include the following:

1. Productivity: As an automotive technician you have a responsibility to your employer to make the best possible use of time on the job. Each job should be done in a reasonable length of time. Employees are paid for their skills, effort, and time.

2. Quality: Each repair job should be a quality job! Work should never be done in a careless manner. Nothing improves customer relations like quality workmanship.
3. Teamwork: The shop staff are a team, and everyone including technicians and management personnel is a team member. You should cooperate with and care about other team members. Each member of the team should strive for harmonious relations with fellow workers. Cooperative teamwork helps improve shop efficiency, productivity, and customer relations. Customers may be "turned off" by bickering between shop personnel.
4. Honesty: Employers and customers expect and deserve honesty from automotive technicians. Honesty creates a feeling of trust among technicians, employers, and customers.
5. Loyalty: As an employee, you are obliged to act in the best interests of your employer, both on and off the job.
6. Attitude: Employees should maintain a positive attitude at all times. As in other professions, automotive technicians have days when it may be difficult to maintain a positive attitude. For example, there will be days when the technical problems on a certain vehicle are difficult to solve. However, a negative attitude certainly will not help the situation! A positive attitude has a positive effect on the job situation as well as on the customer and employer.
7. Responsibility: You are responsible for your conduct on the job and your work-related obligations. These obligations include always maintaining good workmanship and customer relations. Attention to details such as always placing fender, seat, and floor mat covers on customer vehicles prior to driving or working on the vehicle greatly improves customer relations.
8. Following directions: All of us like to do things "our way." Such action, however, may not be in the best interests of the shop; as an employee, you have an obligation to follow the supervisor's directions.
9. Punctuality and regular attendance: Employees have an obligation to be on time for work and to be regular in attendance on the job. It is very difficult for a business to operate successfully if it cannot count on its employees to be on the job at the appointed time.
10. Regulations: Automotive technicians should be familiar with all state and federal regulations pertaining to their job situation, such as the Occupational Safety and Health Act (OSHA) and hazardous waste disposal laws. In Canada, employees should be familiar with workplace hazardous materials information systems (WHMIS).

Employer to employee obligations include:

1. Wages: The employer has a responsibility to inform the employee regarding the exact amount of financial remuneration they will receive and when they will be paid.
2. Fringe benefits: A detailed description of all fringe benefits should be provided by the employer. These benefits may include holiday pay, pension plans, and sickness and accident insurance.
3. Working conditions: A clean, safe workplace must be provided by the employer. The shop must have adequate safety equipment and first-aid supplies. Employers must be certain that all shop personnel maintain the shop area and equipment to provide adequate safety and a healthy workplace atmosphere.
4. Employee instruction: Employers must provide employees with clear job descriptions and be sure that each worker is aware of his or her obligations.
5. Employee supervision: Employers should inform their workers regarding the responsibilities of their immediate supervisors and other management personnel.
6. Employee training: Employers must make sure that each employee is familiar with the safe operation of all the equipment that they are required to use in their job situation. Since automotive technology is changing rapidly, employers should provide regular update training for their technicians. Under the Right-to-Know Laws, employers are required to inform all employees about hazardous materials in the shop. Employees should be familiar with material safety data sheets (MSDSs), which detail the labeling and handling of hazardous waste, and the health problems if exposed to hazardous waste.

Job Responsibilities

An automotive technician has specific responsibilities regarding each job performed on a customer's vehicle.

These job responsibilities include:

1. Do every job to the best of your ability. There is no place in the automotive service industry for careless workmanship! Automotive technicians and students must realize they have a very responsible job.
2. Treat customers fairly and honestly on every repair job. Do not install parts that are unnecessary to complete the repair job.
3. Use published specifications; do not guess at adjustments.
4. Follow the service procedures in the service manual provided by the vehicle manufacturer or an independent manual publisher.
5. When the repair job is completed, always be sure the customer's complaint has been corrected.
6. Do not be too concerned with work speed when you begin working as an automotive technician. Speed comes with experience.

Liable Responsibilities

During many repair jobs you, as a student or technician working on a customer's vehicle, actually have the customer's life and the vehicle safety in your hands. For example, if you are doing a brake job and leave the nuts loose on one wheel, that wheel may fall off the vehicle at high speed. This could result in serious personal injury for the customer and others, plus extensive vehicle damage. If this type of disaster occurs, the individual who worked on the vehicle and the shop may be involved in a very expensive legal action. As a student or technician working on customer vehicles, you are responsible for the safety of every vehicle that you work on! Even when careless work does not create a safety hazard, it leads to dissatisfied customers, who often take their business to another shop. Nobody benefits when that happens.

National Institute for Automotive Service Excellence (ASE) Certification

The National Institute for Automotive Service Excellence (ASE) has provided voluntary testing and certification of automotive technicians on a national basis for many years. The image of the automotive service industry has been enhanced by the ASE certification program. More than 415,000 technicians now have current certifications and work in a wide variety of automotive service shops.

ASE provides certification in eight areas of automotive repair:

1. Engine repair
2. Automatic transmissions/transaxles
3. Manual drivetrain and axles
4. Suspension and steering
5. Brakes
6. Electrical systems
7. Heating and air conditioning
8. Engine performance

A technician may take the ASE test and become certified in any or all of the eight areas. When a technician passes an ASE test in one of the eight areas, an Automotive Technician's shoulder patch is issued by ASE. Technicians who pass all eight tests receive

Copyright © 2015 Cengage Learning®

FIGURE 2-51 **ASE certification shoulder patches worn by Automotive Technicians and Master Technicians.**

a Master Technician's shoulder patch (Figure 2-51). Retesting at five-year intervals is required to remain certified. The certification test in each of the eight areas contains 40 to 80 multiple-choice questions. The test questions are written by a panel of automotive service experts from various areas of automotive service, including automotive instructors, service managers, automotive manufacturers' representatives, test equipment representatives, and certified technicians. The test questions are pretested and checked for quality by a national sample of technicians. On an ASE certification test, approximately 45 to 50 percent of the questions are Technician A and Technician B format, and the multiple-choice format is used in 40 to 45 percent of the questions. Less than 10 percent of ASE certification questions are an EXCEPT format in which the technician selects one incorrect answer out of four possible answers. ASE regulations demand that each technician must have two years of working experience in the automotive service industry prior to taking a certification test or tests. However, relevant formal training may be substituted for one year of working experience. Contact ASE for details regarding this substitution. The contents of the suspension and steering test are listed in Table 2-1.

ASE also provides certification tests in automotive specialty areas, such as Parts Specialist; Advanced Engine Performance Specialist; Alternate Fuels, Light Vehicle—Compressed Natural Gas; Machinist, Cylinder Head Specialist; Machinist, Cylinder Block Specialist; and Machinist, Assembly Specialist.

Shops that employ ASE-certified technicians display an official **ASE blue seal of excellence**. This blue seal increases the customer's awareness of both the shop's commitment to quality service and the competency of its certified technicians.

TABLE 2-1 **SUSPENSION AND STEERING TEST SUMMARY**

Content area	Questions in test	Percentage of test
A. Steering systems diagnosis and repair 1. Steering columns and manual steering gears (3) 2. Power-assisted steering units (4) 3. Steering linkage (3)	10	25%
B. Suspension systems diagnosis and repair 1. Front suspensions (6) 2. Rear suspensions (5) 3. Miscellaneous service (2)	13	33%
C. Wheel alignment diagnosis, adjustment, and repair	12	30%
D. Wheel and tire diagnosis and repair	5	12%
TOTAL	**40**	**100%**

Copyright © 2015 Cengage Learning®

Repair Orders (R.O.)

Repair orders may vary depending on the shop, but repair orders usually have this basic information:

1. Customer's name, address, and phone number(s)
2. Customer's signature
3. Vehicle make, model, year, and color
4. Vehicle identification number (VIN)
5. Vehicle mileage
6. Engine displacement
7. Date and time
8. Service writers code number
9. Work order number
10. Labor rate
11. Estimate of repair costs
12. Accurate and concise description of the vehicle problem.

In many shops the repair orders are completed on a computer terminal, and the computer may automatically write the vehicle repair history on the repair order if the vehicle has a previous repair history in the shop computer system.

The repair order informs the technician regarding the problem(s) with the vehicle. In many shops the technician has to enter a starting time on the repair order. This may be done on a computer terminal or by inserting the work order into the time clock. The technician may also have to enter his code number on the work order to indicate who worked on the vehicle. The technician's code number on the work order is also used to pay the technician for the repair job. The technician must diagnose and repair the problem(s) indicated on the repair order. When the technician obtains parts from the parts department to complete the repair, the technician must present the order number. The parts personnel enter the parts and the cost on the repair order. In some shops the technician is required to enter an accurate description of the completed repairs on the work order. For example, the description of the problem on the work order may be "A/C system inoperative." If the technician replaced the A/C compressor fuse to correct the problem, he or she may enter, "Diagnosed A/C electrical defects and found a blown A/C fuse. Replaced the fuse and operated the A/C system. A/C system operation is normal." Some shops require the service writer or shop foreman to sign the work order when the repair job is successfully completed. The work order is routed back to the accountant who calculates all the charges on the work order including the appropriate taxes. Some shops add a miscellaneous charge on the work order. A typical miscellaneous charge is 10 percent of the total charges on the work order. This miscellaneous charge is to cover the cost of small items such as bolts, cotter pins, grease, lubricants, and sealers that are not entered separately on the work order.

In many shops the technicians work on a flat rate basis. In these shops the technician is paid a flat rate for each repair. In dealership the flat rate time is set by the vehicle manufacturer. Independent shops use generic flat rate manuals published by firms such as Mitchell Publications. If the flat rate time is 2 hours for completing a specific vehicle repair, the customer is charged for 2 hours, and the technician is paid for 2 hours even though he or she completed the repair in 1.5 hours. Conversely, if the technician takes 2.5 hours to complete the job, the technician is only paid for 2 hours and the customer is charged for 2 hours. The flat rate time is usually entered on the work order.

Service Manuals

The service manual is one of the most important tools for today's technician. It provides information concerning component identification, service procedures, specifications, and diagnostic information. In addition, the service manual provides information concerning

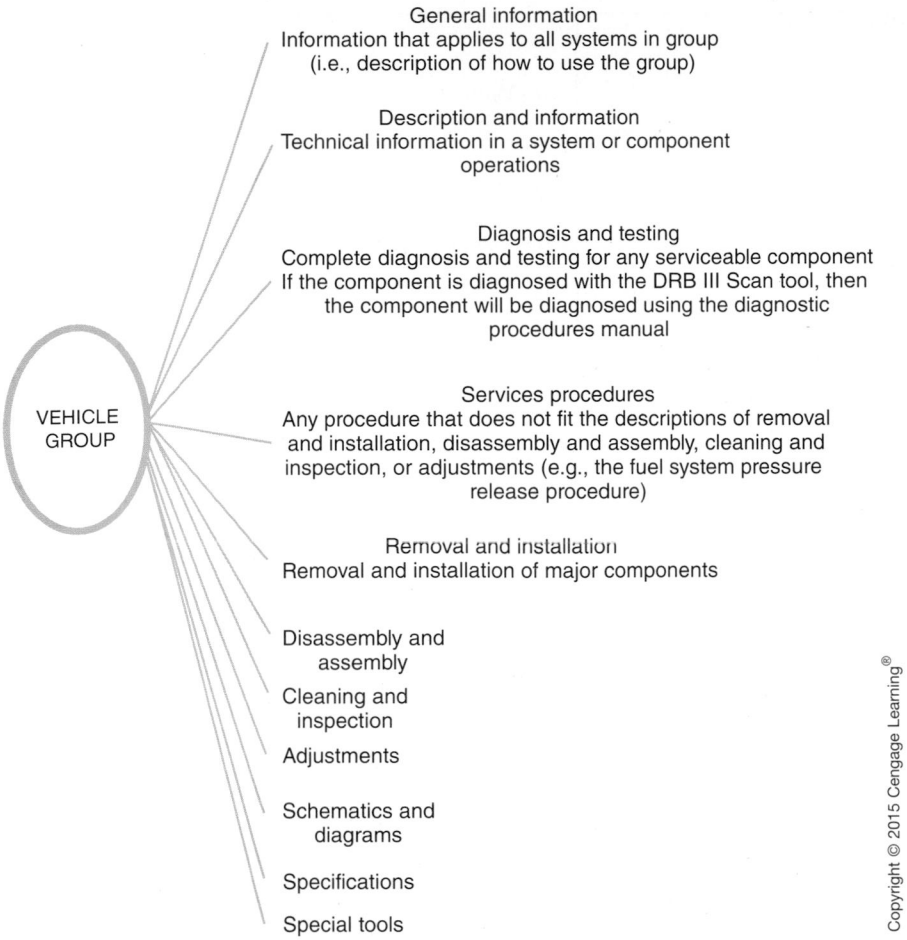

FIGURE 2-52 **Uniform service manual layout.**

wiring harness connections and routing, component location, and fluid capacities. Service manuals can be supplied by the vehicle manufacturer or through aftermarket suppliers.

The service manual provides an explanation of the vehicle identification number (VIN). The VIN information is essential when ordering parts. Most service manuals published by vehicle manufacturers now have a standard format (Figure 2-52). The service manual usually provides illustrations to guide the technician through the service operation (Figure 2-53). Always use the correct manual for the vehicle and system being serviced. Follow each step in the service procedure. Do not skip steps! Measurements such as torque, end play, and clearance specifications are located in or near the service manual text or procedural information. Specification tables are usually provided at the end of the procedural information or component area (Figure 2-54).

Because the service manual is divided into a number of main component and system areas, a table of contents is provided at the front of the manual to provide quick access to the desired information. Each component area or system is covered in a section of the service manual (Figure 2-55). At the beginning of each section in the service manual, a smaller table of contents guides the technician to the information regarding the specific system or component being serviced. The service manual may be divided into several volumes, because of the extensive amount of information required to service today's vehicles. Diagnostic information in each section of the service manual is usually provided in **diagnostic procedure charts** (Figure 2-56). The test results obtained in a specific diagnostic step guide the technician to the next appropriate step.

Service and parts information can also be provided through computer services (Figure 2-57). Computerized service information may be provided on computer discs or compact discs (CDs), which are easier to store and access. Computers may also be connected to a

Ball Joint Boot Replacement

NOTE: The upper control arm ball joint, lower control arm ball joint, and knuckle upper ball joint are attached with the boot retainer to improve the sealing efficiency of the boot.

1. Remove the set ring and boot.

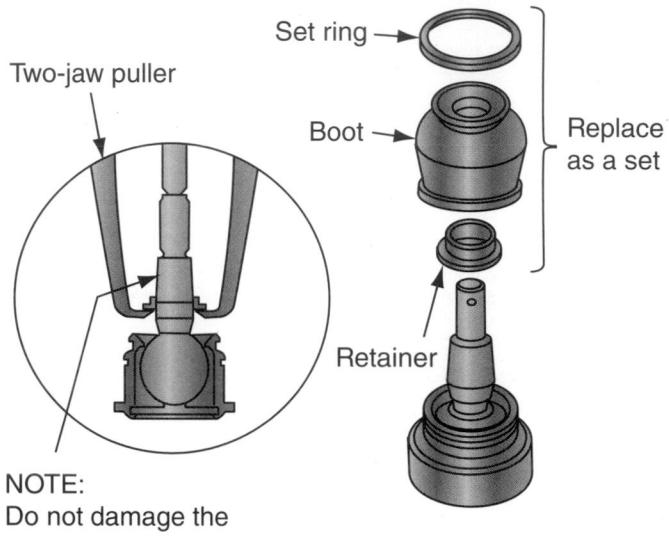

NOTE:
Do not damage the tapered section of the ball pin with the bearing puller.

2. Remove the retainer.

NOTE: The knuckle lower ball joint does not have a retainer.

3. Pack the interior of the boot and lip with grease.

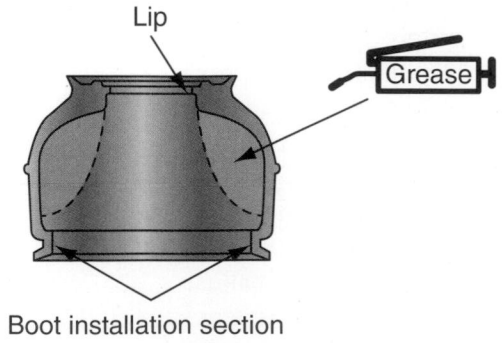

CAUTION: Do not contaminate the boot installation section with grease.

4. Wipe the grease off the sliding surface of the ball pin, and pack with fresh grease.

5. Insert the new retainer lightly into the ball joint pin.

NOTE: When installing the ball joint, press the retainer into the ball joint pin.

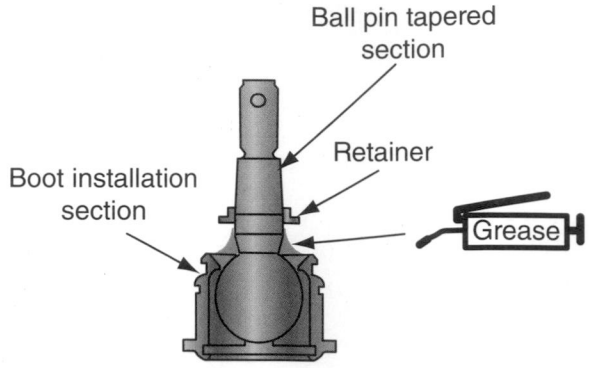

CAUTION:

Keep grease off the boot installation section and the tapered section of the ball pin.

Do not allow dust, dirt, or other foreign materials to enter the boot.

6. Install the boot in the groove of the boot installation section securely, then bleed air.

7. Adjust the special tool with the adjusting bolt until the end of the tool aligns with the groove on the boot.

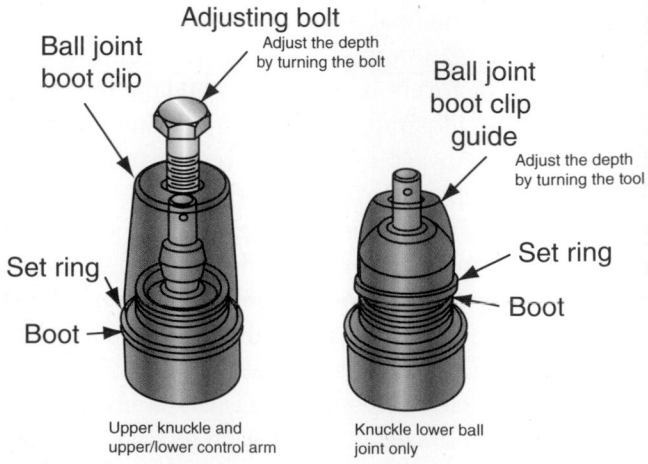

8. Slide the set ring over the tool and into position.

CAUTION: After installing the boot, check the ball pin tapered section and threads for grease contamination and wipe them if necessary.

FIGURE 2-53 **Illustrations in the service manual that guide the technician through service procedures.**

		Tire size	Pressure	
			Front	Rear
Cold tire inflation pressure	For all roads including full rated loads	P195/70R14	220 kPa (2.2 kgf/cm², 32 psi)	240 kPa (2.4 kgf/cm², 34 psi)
		P205/65R15	220 kPa (2.2 kgf/cm², 32 psi)	240 kPa (2.4 kgf/cm², 34 psi)
	Optional inflation for reduced loads (1 to 4 passengers)	P195/70R14	180 kPa (1.8 kgf/cm², 26 psi)	180 kPa (1.8 kgf/cm², 26 psi)
		P205/65R15	180 kPa (1.8 kgf/cm², 26 psi)	180 kPa (1.8 kgf/cm², 26 psi)
Vehicle height	Tire size		Height	
			Front	Rear
	P195/70R14		210 mm (8.27 in.)	270 mm (10.63 in.)
	P205/65R15		213 mm (8.39 in.)	276 mm (10.87 in.)
Front wheel alignment	Toe-in (total)		0° +/- 0.2° (0 +/- 2 mm, 0 +/- 0.08 in.)	
	Wheel angle	Tire size	Inside wheel	Outside wheel
		P195/70R14	37°20' +/- 2°	32°15'
		P205/65R15	36°00' +/- 2°	31°20'
	Camber		-0°35' +/- 45'	
	Cross camber		45' or less	
	Caster		1°05' +/- 45'	
	Cross caster		45' or less	
	Steering axis inclination		13°00' +/- 45'	
Rear wheel alignment	Toe-in (total)		0.4° +/- 0.2° (4 +/- 2 mm, 0.16 +/- 0.08 in.)	
	Camber		-0°15' +/- 45'	
	Cross camber		45' or less	

Copyright © 2015 Cengage Learning®

FIGURE 2-54 **Specification table.**

central database to obtain service information. Using the mouse, light pen, computer keyboard, or touch-sensitive screen, the technician selects choices from a series of menus on the computer monitor. When the desired information is accessed, it may be printed out for detailed study. Modern automotive service information is presently computer or Web-based.

Service procedures may be modified by the vehicle manufacturer at any time. Service bulletins provide up-to-date corrections for the service manuals. If a significant number of corrections are required, a second edition of the manual may be published. When service information is provided on CDs, the CDs are updated frequently to provide the latest information.

CUSTOMER CARE: When advising customers regarding when the work will be completed on their car, it is a good idea to estimate a longer time than you anticipate. For example, if you expect it will take 3 hours to repair the vehicle, advise the customer it will be ready in 4 hours. This allows extra time for diagnosing difficult problems or road testing the vehicle. It can be very frustrating for customers when they come to pick up their vehicle at the appointed time, and the vehicle is not ready.

TABLE OF CONTENTS	SECTION NUMBER
GENERAL INFO. AND LUBE	
General Information	0A
Maintenance and Lubrication	0B
HEATING AND AIR COND.	
Heating and Vent. (non-A/C)	1A
Air-Conditioning System	1B
V-5 A/C Compressor Overhaul	1D3
BUMPERS AND FRONT BODY PANELS	
Bumpers (See 10-4)	
Fr. End Body Panels (See 10-5)	
STEERING, SUSPENSION, TIRES, AND WHEELS	
Diagnosis	3
Wheel Alignment	3A
Power Steering Gear & Pump	3B1
Front Suspension	3C
Rear Suspension	3D
Tires and Wheels	3E
Steering Col. On-Vehicle Service	3F
Steering Col. – Std. Unit Repair	3F1
Steering Col. – Tilt, Unit Repair	3F2
DRIVE AXLES	
Drive Axles	4D
BRAKES	
General Info. – Diagnosis and On-Car Service	5
Compact Master Cylinder	5A1
Disc Brake Caliper	5B2
Drum Brake - Anchor Plate	5C2
Power Brake Booster Assembly	5D2
ENGINES	
General Information	6
2.0 Liter I-4 Engine	6A1
3.1 Liter V6 Engine	6A3
Cooling System	6B
Fuel System	6C
Engine Electrical – General	6D
Battery	6D1
Cranking System	6D2
Charging System	6D3
Ignition System	6D4
Engine Wiring	6D5
Driveability & Emissions – Gen.	6E
Driveability & Emissions – TBI	6E2
Driveability & Emissions – PFI	6E3
Exhaust System	6F

TABLE OF CONTENTS	SECTION NUMBER
TRANSAXLE	7A
Auto.Transaxle On-Car Serv.	3T40-HD
Auto. Trans. – Hydraulic Diagnosis	
Auto. Trans. – Unit Repair	3T40
Man. Trans. On-Car Service	7B
5-Sp. 5TM40 Man. Trans. Unit Repair	7B1
5-Sp. Isuzu Man. Trans. Unit Repair	7B2
Clutch	7C
CHASSIS ELECTRICAL, INSTRUMENT PANEL & WIPER/WASHER	
Electrical Diagnosis	
Lighting and Horns	8A
Instrument Panel and Console	8B
Windshield Wiper/Washer	8C
	8E5
ACCESSORIES	
Audio System	9A
Cruise Control	9B
Engine Block Heater	9C
BODY SERVICE	
General Body Service	10-1
Stationary Glass	10-2
Underbody	10-3
Bumpers	10-4
Body Front End	10-5
Doors	10-6
Rear Quarters	10-7
Body Rear End	10-8
Roof & Convertible Top	10-9
Seats	10-10
Safety Belts	10-11
Body Wiring	10-12
Unibody Collision Repair	11-1
Welded Panel Replacement	11-2
INDEX	
Alphabetical Index	

Copyright © 2015 Cengage Learning®

FIGURE 2-55 **The table of contents directs you to the major systems and component areas in the service manual.**

EPS INDICATOR LIGHT DOES NOT COME ON

The EPS indicator light does not come on when the ignition switch is turned on (II).

Note: All indicator lights except the charging system indicator light and malfunction indicator lamp (MIL) will not come on when the BACK-UP LIGHTS/ALTERNATOR (15a) fuse is blown.

Check the BACK-UP LIGHTS/ALTERNATOR (15a) fuse in the under-dash fuse box.

Is the fuse OK?

NO → Replace the fuse and recheck.

Note: reinstall the fuse if it is OK.

YES

Check for an open in the IG1 circuit:
1. Disconnect the gauge assembly 30-pin connector.
2. Turn the ignition switch ON (II).
3. Measure the voltage between the terminal No. 13 and body ground.

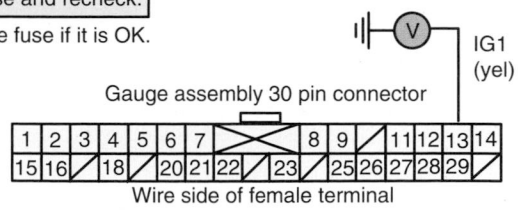

Is there battery voltage?

NO → Repair open in the wire between the BACK-UP LIGHTS/ALTERNATOR (15a) fuse and the gauge assembly. Replace the under-dash fuse box (open circuit inside the fuse box).

YES

Check the EPS indicator light bulb in the gauge assembly.

Is the bulb OK?

NO → Replace the EPS indicator light bulb.

YES

Check the gauge assembly:
1. Connect the gauge assembly 30-pin connector.
2. Connect the gauge assembly 30-pin connector terminal No. 5 to body ground.

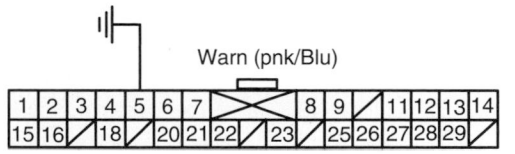

Does the EPS indicator light come on?

NO → Replace the printed circuit board in the gauge assembly.

YES

(To next page)

Copyright © 2015 Cengage Learning®

FIGURE 2-56 **Typical diagnostic chart.**

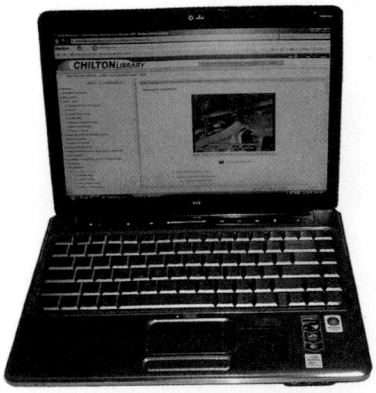

Copyright © 2015 Cengage Learning®

FIGURE 2-57 **Computers are replacing printed service manuals in many shops.**

TERMS TO KNOW

ASE blue seal of excellence
Axle pullers
Ball joint removal and installation tools
Bearing pullers
Brake pedal jack
Coil spring compressor tool
Computer four wheel aligner
Control arm bushing tools
Diagnostic procedure charts
Dial indicators
Electronic wheel balancers
Floor jack
Hydraulic press
International system (IS)
Machinist's rule
Magnetic wheel alignment gauge
Pitman arm puller
Plumb bob
Power steering pressure gauge
Rim clamp
Safety stands
Scan tools
Seal drivers
Steering wheel locking tool
Stethoscope
Tie rod sleeve adjusting tool
Tie rod end and ball joint puller
Tire tread depth gauge
Toe gauge
Track gauges
Tram gauge
Turning radius gauge turntables
U.S. customary (USC) system
Vacuum hand pump
Vehicle lift

CASE STUDY

A technician was removing and replacing the alternator of a General Motors car. After installing the replacement alternator and connecting the alternator battery wire, she proceeded to install the alternator belt. The rubber boot was still removed from the alternator battery terminal. While installing this belt, her wristwatch expansion bracelet made electrical contact from the alternator battery terminal to ground on the alternator housing. Even though the alternator battery wire is protected with a fuse link, which melted, a high current flowed through the wristwatch bracelet. This heated the bracelet to a very high temperature and severely burned the technician's arm.

The technician forgot two safety rules:

1. Never wear jewelry, such as watches and rings, while working in an automotive shop.
2. Before performing electrical work on a vehicle, disconnect the negative battery cable. If the vehicle is equipped with an air bag, wait the specified time after this cable is disconnected.

CASE STUDY

A customer brought a Cavalier into the shop with a rough idle complaint. The customer informed the service salesman that he had this problem since the car was purchased new, and the problem was always caused by an open exhaust gas recirculation (EGR) valve. The customer said the (EGR) valve had been replaced three times under warranty, and he had also paid for an EGR valve replacement, and one EGR valve cleaning, after the warranty had expired.

The technician connected a scan tool to the DLC, and verified that a diagnostic trouble code (DTC) was present indicating the EGR valve was stuck open. Next the technician checked for technical service bulletins (TSBs) related to this problem in the shop database. One TSB indicated that the vehicle manufacturer had revised powertrain control module (PCM) software that caused the PCM to pulse the EGR valve briefly each time the PCM closed the EGR valve to remove carbon buildup on the EGR valve pintle. Using the appropriate equipment the technician installed the updated software in the PCM, and this permanently corrected the problem.

Previous technicians who serviced this vehicle had forgotten, or neglected, to check the TSB database for the cause of this problem. As a result, both the vehicle manufacturer and the customer spent a considerable amount of money for EGR valve replacements, when computer reprogramming was required to correct the problem.

CASE STUDY

A technician had just replaced the engine in a Ford vehicle and was performing final adjustments. In this shop, the cars were parked in the work bays at an angle on both sides of the shop. With the engine running at fast idle, the automatic transmission suddenly slipped into reverse. The car went backward across the shop and collided with a car in one of the electrical repair bays. Both vehicles were damaged to a considerable extent. Fortunately, no personnel were injured.

This technician forgot to apply the parking brake while working on the vehicle!

SUMMARY

- Each unit in the metric system of measurements can be multiplied or divided by 10 to obtain larger or smaller units.
- A stethoscope amplifies sound to help the technician identify the source of abnormal noises.
- A dial indicator is a precision measuring device that measures movement or wear on various components in thousands of an inch.
- A tire changer is a pneumatically operated machine that is used to dismount and mount tires on rims.
- An electronic wheel balancer indicates static and dynamic unbalance in tire and wheel assemblies. The electronic wheel balancer also indicates the mounting location and size of wheel weight(s) required to provide a balanced tire and wheel assembly.
- A spring compressor tool compresses a coil spring so the spring and other suspension components can be safely removed.
- A power steering pressure gauge is used to test power steering pump pressure and determine the condition of the power steering system.
- Turning radius gauges are placed under the front wheels during a wheel alignment.
- A magnetic wheel alignment gauge may be attached to the outer surface of the front wheel hubs to measure certain wheel alignment angles.
- A computer four wheel aligner is used to electronically measure the alignment angles on the front and rear suspensions.
- A scan tool is connected to the data link connector (DLC) under the dash to diagnose various electronic systems.
- A vehicle lift is used to raise vehicles off the floor so under-car service may be performed.

ASE-STYLE REVIEW QUESTIONS

1. While discussing systems of weights and measures:
   Technician A says the international system is called the metric system.
   Technician B says every unit in the metric system can be divided or multiplied by 10.
   Who is correct?
   A. A only
   B. B only
   C. Both A and B
   D. Neither A nor B

2. While discussing the metric system:
   Technician A says that 1 inch is equal to 2.54 millimeters.
   Technician B says that 1 meter is equal to 36.37 inches.
   Who is correct?
   A. A only
   B. B only
   C. Both A and B
   D. Neither A nor B

3. When diagnosing a vehicle operational problem, the first step is to:
   A. Think of the possible causes of the problem.
   B. Perform diagnostic tests to locate the exact cause of the problem.
   C. Consult TSB information to find the cause of the problem.
   D. Verify the customer complaint.

4. When testing power steering pump pressure, the power steering gauge valve should be closed for a maximum of:
   A. 10 seconds.
   B. 20 seconds.
   C. 25 seconds.
   D. 35 seconds.

5. While discussing employer and employee responsibilities:

   Technician A says employers are required to inform their employees about hazardous materials in the shop.

   Technician B says that employers have no obligation to inform their employees about the safe operation of shop equipment.

   Who is correct?

   A. A only
   B. B only
   C. Both A and B
   D. Neither A nor B

6. When connecting a scan tool to diagnose a computer-controlled suspension system on a 2014 vehicle:

   A. The scan tool power cord is connected to the cigarette lighter socket.
   B. The scan tool is connected to the DLC generally located under the dash.
   C. The ignition switch should be on when disconnecting the scan tool.
   D. On OBD II vehicles the 7-digit DTCs are displayed on the scan tool.

7. All these statements about suspension and steering tools are true EXCEPT:

   A. A plumb bob may be used for frame measurement.
   B. A tram gauge may be used to measure front wheel alignment.
   C. A scan tool may be used to diagnose electronically controlled suspension systems.
   D. A brake pedal jack may be used to apply the brakes during a wheel alignment.

8. When using a vehicle lift:

   A. The lift arms may be positioned on any strong chassis member.
   B. The safety catch must be engaged after the vehicle is raised on a lift.
   C. The vehicle hood should be open when raising a vehicle on a lift.
   D. The vehicle weight may exceed the maximum capacity of the lift.

9. When taking ASE certification tests:

   A. A technician must pass four of the eight ASE certification tests to receive a Master Technician's shoulder patch.
   B. Each ASE certification test contains 35 to 45 questions.
   C. An ASE certification test is available in suspension and steering.
   D. Retesting at three-year intervals is required to maintain ASE certification.

10. When taking ASE certification tests:

    A. A technician must have four years of automotive repair experience before taking an ASE certification test.
    B. ASE may allow relevant training to be substituted for one year of work experience.
    C. On any ASE certification test, 90 percent of the questions are Technician A and Technician B type.
    D. Employers who employ ASE-certified technicians may display a green ASE seal of excellence.

JOB SHEET 5

Name _____ **Date** _____

Raise a Car with a Floor Jack and Support It on Safety Stands

Upon completion of this job sheet, you should be able to raise the front and rear of a vehicle with a floor jack and support the vehicle on safety stands.

NATEF MAST/AST Correlation

Required Supplemental Tasks, Shop and Personal Safety: 3. Identify and use proper placement of floor jacks and jack stands.

Tools and Materials

A car
Four safety stands
Hydraulic floor jack

Describe the Vehicle Being Worked on:

Year _____ Make _____ Model _____
VIN _____ Engine type and size _____

This job sheet must be completed under the supervision of your instructor!

Procedure

Task Completed

1. Determine the proper vehicle floor jack lifting points in the car manufacturer's service manual. Proper front lifting point _____ Proper rear lifting point _____
2. Be sure the car is parked on a level shop floor surface with the parking brake applied. ☐
3. Place the floor jack under the proper lifting point on the front of the vehicle, and raise the floor jack pad until it contacts the lifting point. ☐
4. Release the parking brake and be sure the transmission is in neutral. Is the parking brake released? ☐ Yes ☐ No

   Is the transmission in neutral? ☐ Yes ☐ No
5. Operate the floor jack and raise the vehicle to the desired height. Place safety stands under the proper support points on the vehicle chassis or suspension. ☐
6. Very slowly operate the release lever on the hydraulic floor jack to slowly lower the vehicle until the support points lightly contact the safety stands. Stop lowering the floor jack. ☐
7. Be sure the safety stands contact the proper support points on the vehicle, and check to be sure all the safety stand legs contact the shop floor evenly.

   Are the safety stands contacting the proper support points on the vehicle?
   ☐ Yes ☐ No

Task Completed

Are all safety stand legs contacting the floor evenly? ☐ Yes ☐ No

Instructor check _____

☐ 8. If the answer to both questions in step 7 is yes, very slowly operate the release lever on the hydraulic floor jack to slowly lower the vehicle until the vehicle weight is completely supported on the safety stands. Be sure all the safety stand legs are contacting the shop floor evenly, and then remove the floor jack.

☐ 9. Place the floor jack under the proper lifting point on the rear of the vehicle, and raise the floor jack pad until it contacts the lifting point.

10. Operate the floor jack, and raise the vehicle to the desired height. Place safety stands under the proper support points on the rear of the vehicle chassis or suspension.

Are the safety stands properly positioned? ☐ Yes ☐ No

Instructor check _____

☐ 11. Very slowly operate the release lever on the hydraulic floor jack to slowly lower the vehicle until the rear support points lightly contact the safety stands. Stop lowering the floor jack.

12. Be sure the safety stands contact the proper rear support points on the vehicle, and check to be sure all the safety stand legs contact the shop floor evenly.

Are the safety stands contacting the proper support points on the vehicle?
☐ Yes ☐ No

Are all the safety stand legs contacting the floor evenly? ☐ Yes ☐ No

Instructor check _____

☐ 13. If the answer to both questions in step 12 is yes, very slowly operate the release lever on the hydraulic floor jack to slowly lower the rear of the vehicle until the vehicle weight is completely supported on the safety stands. Be sure all the safety stand legs are contacting the shop floor evenly; then remove the floor jack.

Instructor's Response _____

_____

_____

JOB SHEET 6

Name _____ **Date** _____

Follow the Proper Procedure to Hoist a Car

Upon completion of this job sheet, you should be able to raise and lower a car on a hoist.

CAUTION: Do not raise a four-wheel-drive vehicle with a frame contact lift because this may damage axle joints.

NATEF MAST/AST Correlation

Required Supplemental Tasks, Shop and Personal Safety: 4. Identify and use proper procedures for safe lift operation.

Tools and Materials

Car
Lift with enough capacity to hoist the car

Describe the Vehicle Being Worked on:

Year _____ Make _____ Model _____
VIN _____ Engine type and size _____

Procedure

Task Completed

1. Always be sure the lift is completely lowered before driving the car onto the lift. ☐
2. Do not hit or run over lift arms and adaptors when driving a car onto the lift. ☐
3. Have a coworker guide you when driving a car onto the lift.

WARNING: Do not stand in front of a lift with the car coming toward you. This action may result in personal injury.

4. Be sure the lift pads on the lift are contacting the car manufacturer's recommended lifting points shown in the service manual. ☐
   Is the vehicle properly positioned on lift? ☐ Yes ☐ No
   Recommended front lifting points: ☐ Right side ☐ Left side
   Recommended rear lifting points: ☐ Right side ☐ Left side
   Are all four lift pads contacting the recommended lifting points? ☐ Yes ☐ No
   Instructor check _____
5. Be sure the doors and hood are closed, and be sure there are no people in the car. ☐
6. When a car is lifted a short distance off the floor, stop the lift and check the contact between the lift pads and the car chassis to be sure the lift pads are still on the recommended lifting points. ☐
7. Be sure there is adequate clearance between the top of the vehicle and the shop ceiling or components under the ceiling. ☐

Task Completed

☐ 8. When a car is raised on a lift, be sure the safety mechanism is in place to prevent the lift from dropping accidentally.

Is the lift safety mechanism in place? ☐ Yes ☐ No

Instructor check_____

List one precaution that must be observed when a front-wheel-drive vehicle is raised on a lift, and explain the reason for this precaution.

_____

_____

☐ 9. Prior to lowering a car on a lift, always make sure there are no tools, objects, or people under the vehicle.

WARNING: Do not rock a car on a lift during a service job. This action may cause the car to fall off the lift, resulting in personal injury and vehicle damage.

WARNING: When a car is raised on a lift, removal of some heavy components may cause car imbalance on the lift, which may cause the car to fall off the lift, resulting in personal injury and vehicle damage.

☐ 10. Be sure the lift is lowered completely and no lift components are contacting the vehicle before backing the vehicle off the lift.

Instructor's Response _____

_____

_____

CAUTION:
If the proper lifting points are not used, components under the vehicle such as brake lines or body parts may be damaged. Failure to use the recommended lifting points may cause the car to slip off the lift, resulting in severe vehicle damage and personal injury.

JOB SHEET 7

Name _____ **Date** _____

Determine the Availability and Purpose of Suspension and Steering Tools

NATEF MAST/AST Correlation

Required Supplemental Tasks, Tools and Equipment: 1. Identify tools and their usage in automotive applications.

Procedure

Locate the following tools in your shop or tool room, and explain the purpose of each tool.

1. Stethoscope: available? ☐ Yes ☐ No
   Location _____
   Purpose _____
2. Dial indicator for measuring ball joint wear: available? ☐ Yes ☐ No
   Location _____
   Purpose _____
3. Ball joint removal and installing tools: available? ☐ Yes ☐ No
   Location _____
   Purpose _____
4. Coil spring compressing tool: available? ☐ Yes ☐ No
   Location _____
   Purpose _____
5. Power steering pressure gauge: available? ☐ Yes ☐ No
   Location _____
   Purpose _____
6. Vacuum hand pump: available? ☐ Yes ☐ No
   Location _____
   Purpose _____
7. Turning radius gauge: available? ☐ Yes ☐ No
   Location _____
   Purpose _____
8. Brake pedal jack: available? ☐ Yes ☐ No
   Location _____
   Purpose _____
9. Steering wheel locking tool: available? ☐ Yes ☐ No
   Location _____
   Purpose _____

10. Scan tool: available? ☐ Yes ☐ No

Location _____

Purpose _____

Instructor's Response _____

_____

_____

Chapter 3

Tire and Wheel Servicing and Balancing

BASIC TOOLS
Basic technician's tool set
Service manual
Tire repair kit
Tread depth gauge

Upon completion and review of this chapter, you should be able to:

- Diagnose tire thump and vibration problems.
- Diagnose steering pull problems related to tire condition.
- Rotate tires according to the vehicle manufacturer's recommended procedure.
- Remove and replace tire and wheel assemblies.
- Dismount, inspect, repair, and remount tires.
- Inspect wheel rims.
- Diagnose and service tire pressure monitoring systems.
- Measure tire and wheel radial and lateral runout.
- Diagnose problems caused by excessive radial or lateral tire and wheel runout.
- Measure tire tread wear.
- Perform off-car static wheel balance procedures.
- Perform off-car dynamic wheel balance procedures.
- Diagnose tire wear problems caused by tire and wheel imbalance.
- Perform on-car balance procedures.
- Perform noise, vibration, and harshness diagnosis.

Proper servicing of tires and wheels is extremely important to maintain vehicle safety and provide normal tire life. Improperly serviced and/or balanced tires and wheels cause wheel vibration and shimmy problems, resulting in excessive tire tread wear, increased wear on suspension and steering components, and decreased vehicle stability and steering control.

Tire Noises and Steering Problems

Diagnosis of Tire Noises

Uneven tread surfaces may cause tire noises that seem to originate elsewhere in the vehicle. These noises may be confused with differential noise. Differential noise usually varies with acceleration and deceleration, whereas tire noise remains more constant in relation to these forces. Tire noise is most pronounced on smooth asphalt road surfaces at speeds of 15 to 45 mph (24 to 72 km/h).

Tire thump may be defined as a pounding noise caused by tire and wheel rotation.

Tire Thump and Vibration

When tire thump and tire vibration are present, check these items:

1. Cupped tire treads
2. Excessive tire **radial runout**
3. Manufacturing defects such as heavy spots, weak spots, or tread chunking
4. Incorrect wheel balance

Tire vibration is a fast shaking of the tire that is transferred to the chassis and passenger compartment.

SERVICE TIP:
Tire noise varies with road surface conditions, whereas differential noise is not affected when various road surfaces are encountered.

FIGURE 3-1 **Tire conicity.**

Radial runout refers to a tire out-of-round condition.

Steering pull is the tendency of the steering to gradually pull to the right or left when the vehicle is driven straight ahead on a reasonably smooth, straight road.

Steering Pull

A vehicle should maintain the straight-ahead forward direction on smooth, straight road surfaces without excessive steering wheel correction by the driver. If the steering gradually pulls to one side on a smooth, straight road surface, a tire, steering, or suspension defect is present. Tires of different types, sizes, designs, or inflation pressures on opposite sides of a vehicle cause **steering pull**. Sometimes a tire manufacturing defect occurs in which the belts are wound off center on the tire. This condition is referred to as **tire conicity**. A cone-shaped object rolls in the direction of its smaller diameter. Similarly, a tire with conicity tends to lead, or pull, to one side, which causes the vehicle to follow the action of the tire (Figure 3-1). In the trade this is often referred to as a radial pull.

Because tire conicity cannot be diagnosed by a visual inspection, it must be diagnosed by switching the two front tires and reversing the front and rear tires (Figure 3-2).

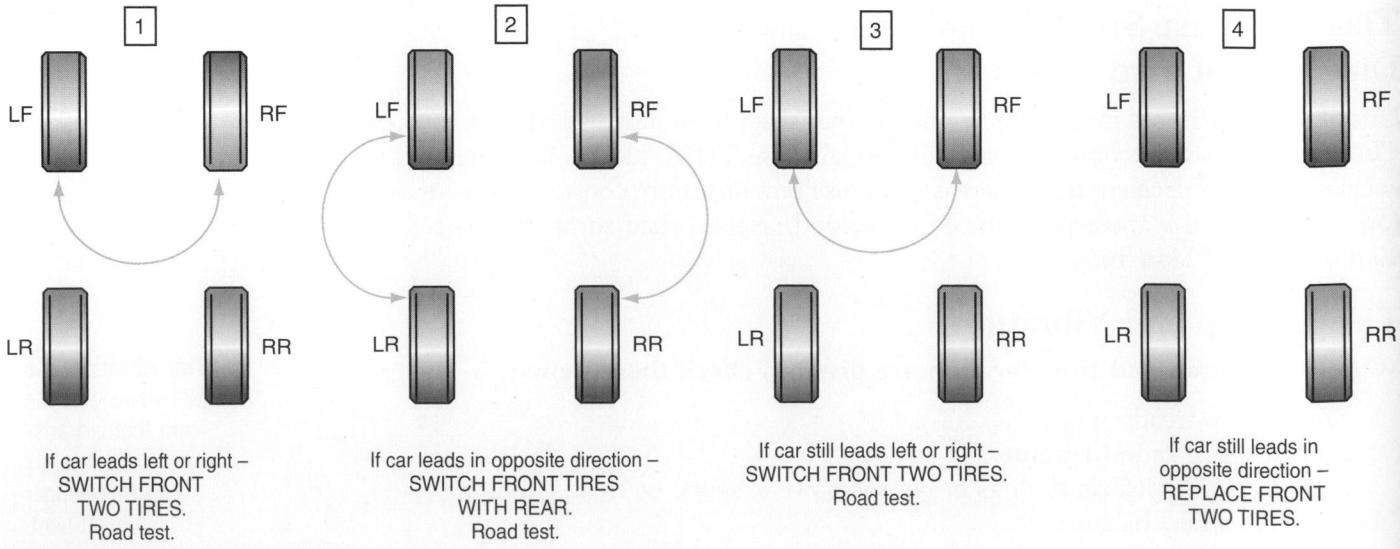

FIGURE 3-2 **Tire conicity diagnosis.**

Incorrect front suspension alignment angles as well as worn or damaged suspension components may also cause steering pull.

Some wheel balancers with force variation capabilities sense and indicate tire conicity defects. These wheel balancers have a roller that presses against the tire tread during the wheel balance procedure, and this roller senses tire conicity.

SERVICE TIP:
Tire conicity is not visible. It can be diagnosed only by changing the tire and wheel position.

Tire Rotation

Driving habits determine tire life to a large extent. Severe brake applications, high-speed driving, turning at high speeds, rapid acceleration and deceleration, and striking curbs are just a few driving habits that shorten tire life. Most car manufacturers recommend tire rotation at specified intervals to obtain maximum tire life, generally every 6,000 miles (9,656 km). The exact tire rotation procedure depends on the model year, the type of tires, and whether the vehicle has a conventional spare or a compact spare (Figure 3-3). Tire rotation procedures do not include the compact spare. The vehicle manufacturer provides tire rotation information in the owner's manual and service manual. As an example of the typical tire rotation pattern, when all tires and wheels are of the same size and type (Figure 3-4 tire rotation), the drive wheel assemblies typically stay on the same side of the vehicle but move to the opposite axle and the non-drive wheel assemblies cross to the opposite axle and side.

Classroom Manual
Chapter 3, page 66

When tires and wheels are installed on a vehicle, it is very important that the wheel nuts are torqued to manufacturer's specifications in the proper sequence (Figure 3-5). Do not use an impact wrench when tightening wheel nuts to the specified torque.

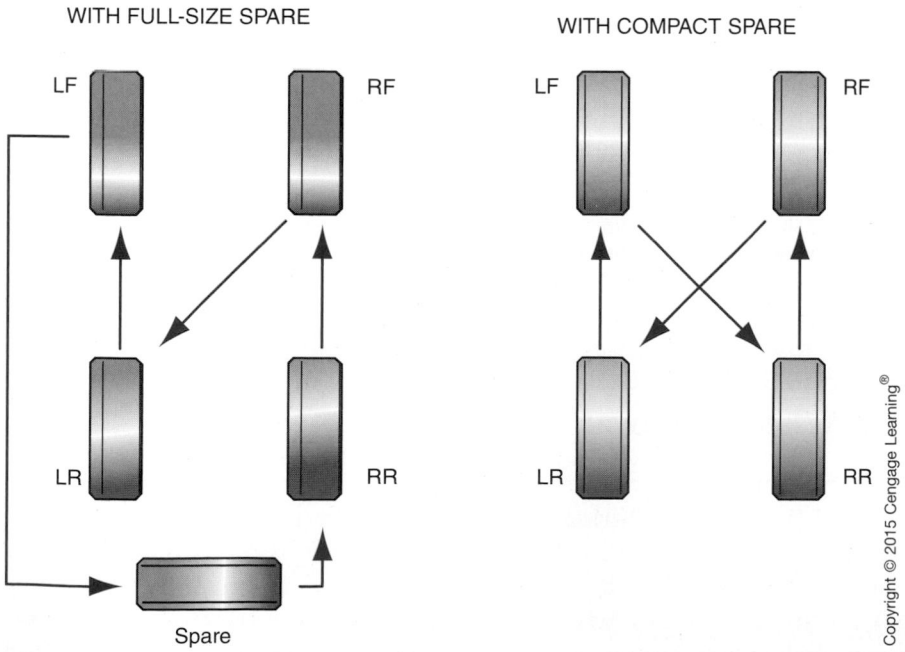

FIGURE 3-3 **Radial tire rotation procedure.**

SERVICE TIP:
On some current vehicles, such as the 2004 Cadillac SRX, the front and rear tires have dissimilar widths. For example, if one of these cars has a V8 engine, the vehicle is equipped with 18 in. wheels, and the front tire size is P235/60R18 while the rear tire size is P255/55R18. These tires must not be rotated from front to rear, but they can be rotated from side-to-side. Directional tires would further complicate rotation options and would require tires to be dismounted and remounted and balanced for installation on the opposite side to insure that the proper directional orientation is maintained.

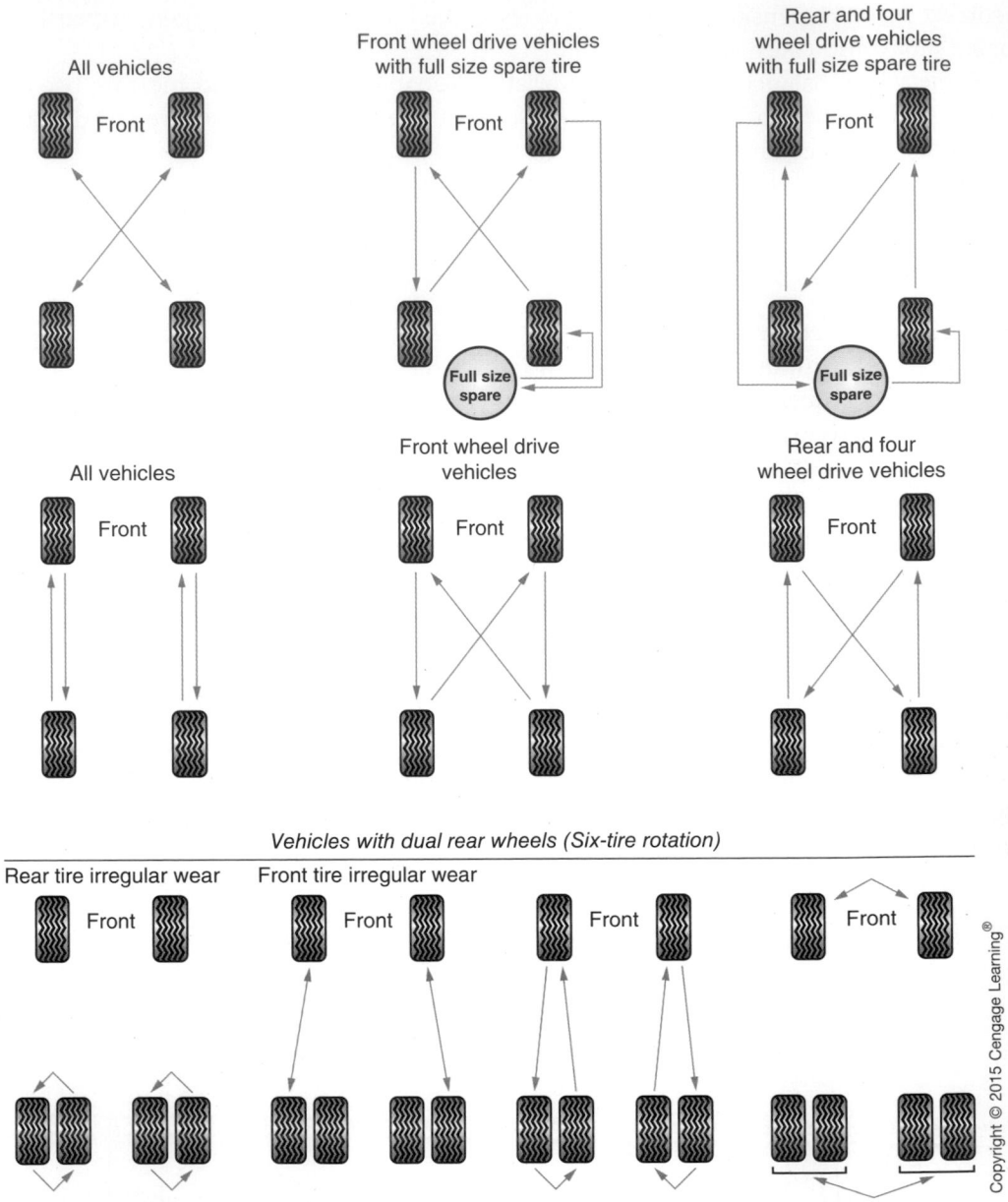

FIGURE 3-4 **The typical tire rotation patterns when all tires are of the same size and type. Drive wheels typically stay on the same side of the vehicle but move to the opposite axle and non-drive tire crosses to the opposite axle and side.**

SERVICE TIP:
Some wheel covers have fake plastic lug nuts that must be removed to access the lug nuts. Be careful not to break the fake lug nuts.

Tire and Wheel Service

Tire and Wheel Removal

CUSTOMER CARE: During many automotive service operations, including tire and wheel service, the technician literally has the customer's life in his or her hands! Always perform tire and wheel service carefully and thoroughly. Watch for unsafe tire or wheel conditions, and report these problems to the customer. When you prove to the customer that you are concerned about vehicle safety, you will probably have a steady customer.

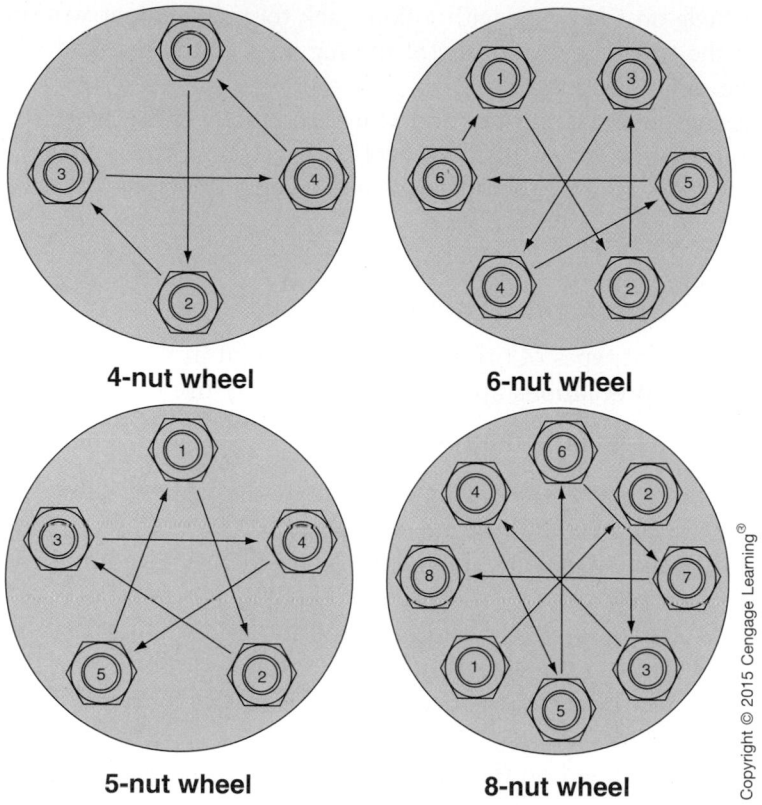

FIGURE 3-5 **Wheel nut tightening sequence.**

When it is necessary to remove a tire and wheel assembly, follow these steps:

1. Remove the wheel cover. If the vehicle is equipped with **antitheft locking wheel covers**, the lock bolt for each wheel cover is located behind the ornament in the center of the wheel cover. A special key wrench is supplied to the owner for ornament and lock bolt removal. If the customer's key wrench has been lost, a master key is available from the vehicle dealer.
2. Loosen the wheel lug nuts about one-half turn, but do not remove the wheel nuts. Some vehicles are equipped with **antitheft wheel nuts**. A special lug nut key is supplied to the vehicle owner. This lug nut key has a hex nut on the outer end and a special internal projection that fits in the wheel nut opening. Install the lug nut key on the lug nuts, and connect the lug nut wrench on the key hex nut to loosen the lug nuts.

Antitheft locking wheel covers prevent unauthorized personnel from removing the wheel covers.

Antitheft wheel nuts prevent unauthorized personnel from removing the wheel nuts.

WARNING: Before the vehicle is raised on a hoist, be sure that the hoist is lifting on the car manufacturer's recommended lifting points. If the hoist is not lifting on the car manufacturer's recommended lift points, chassis components may be damaged, and the vehicle may slip off the hoist, resulting in personal injury.

WARNING: If the vehicle is lifted with a floor jack, place safety stands under the suspension or frame, and lower the vehicle onto the safety stands. Then remove the floor jack from under the vehicle. If the vehicle is not supported properly on safety stands, the vehicle may suddenly drop, resulting in personal injury.

CAUTION: If heat is used to loosen a rusted wheel, the wheel and/or wheel bearings may be damaged.

3. Raise the vehicle on a hoist or with a floor jack to a convenient working level.
4. Chalk mark the tire, wheel, and one of the lug nuts so the tire and wheel can be reinstalled in the same position.
5. Remove the lug nuts and the tire and wheel assembly. If the wheel is rusted and will not come off, hit the inside of the wheel with a large rubber mallet. Do not hit the wheel with a steel hammer, because this action could damage the wheel. Do not heat the wheel.

Tire and Wheel Service Precautions

There are many different types of tire changing equipment in the automotive service industry. However, specific precautions apply to the use of any tire changing equipment.

These precautions include the following:

1. Before you operate any tire changing equipment, always be absolutely certain that you are familiar with the operation of the equipment.
2. When operating tire changing equipment, always follow the equipment manufacturer's recommended procedure.
3. Always deflate a tire completely before attempting to dismount the tire.
4. Clean the bead seats on the wheel rim before mounting the tire on the wheel rim.
5. Lubricate the outer surface of the tire beads with rubber lubricant before mounting the tire on the wheel rim.
6. When the tire is mounted on the wheel rim, be sure the tire is positioned evenly on the wheel rim.
7. While inflating a tire, do not stand directly over the tire. An air hose extension allows the technician to stand back from the tire during the inflation process.
8. Do not overinflate tires.
9. When mounting tires on cast aluminum alloy wheel rims or cast magnesium alloy wheel rims, always use the tire changing equipment manufacturer's recommended tools and procedures.
10. When mounting or dismounting run-flat tires, be sure the tire changing equipment is compatible with these tires and wheels.

CAUTION:
Never use petroleum-based lubricants on tire beads. This action damages the beads.

CAUTION:
If hand tools or tire irons are used to dismount tires, tire bead and wheel rim damage may occur.

Tire Dismounting

Always use a tire changer to dismount tires. Do not use hand tools or tire irons for this purpose. Various types of tire changers are available. When servicing tires it is very important that the tire changing equipment will mount and dismount run-flat tires, low-profile tires, and tires mounted on alloy wheels without damaging the wheel rims or tires. Some tire changers have the features illustrated in Figure 3-6 and described in the following explanation:

1. A floating 3-point articulated arm has a polymer mount/dismount head on the end of the arm. This type of mount/dismount head goes between the tire bead and the rim when dismounting or mounting the tire (Figure 3-7). The polymer head will not damage alloy wheel rims. A mechanical bead pusher with a nylon roller facilitates handling stiff, low-profile tires.
2. A spring-loaded wheel centering support centers the wheel rim in the clamping system.
3. A rubber-pad clamping system eliminates steel-jaw clamping and wheel rim damage. The rim clamping system easily adjusts for internal or external rim clamping (Figure 3-8).
4. A multi-piston air motor in the tire changer drives the clamp assembly and wheel rim in either direction up to 15 revolutions per minute (rpm). The technician controls the

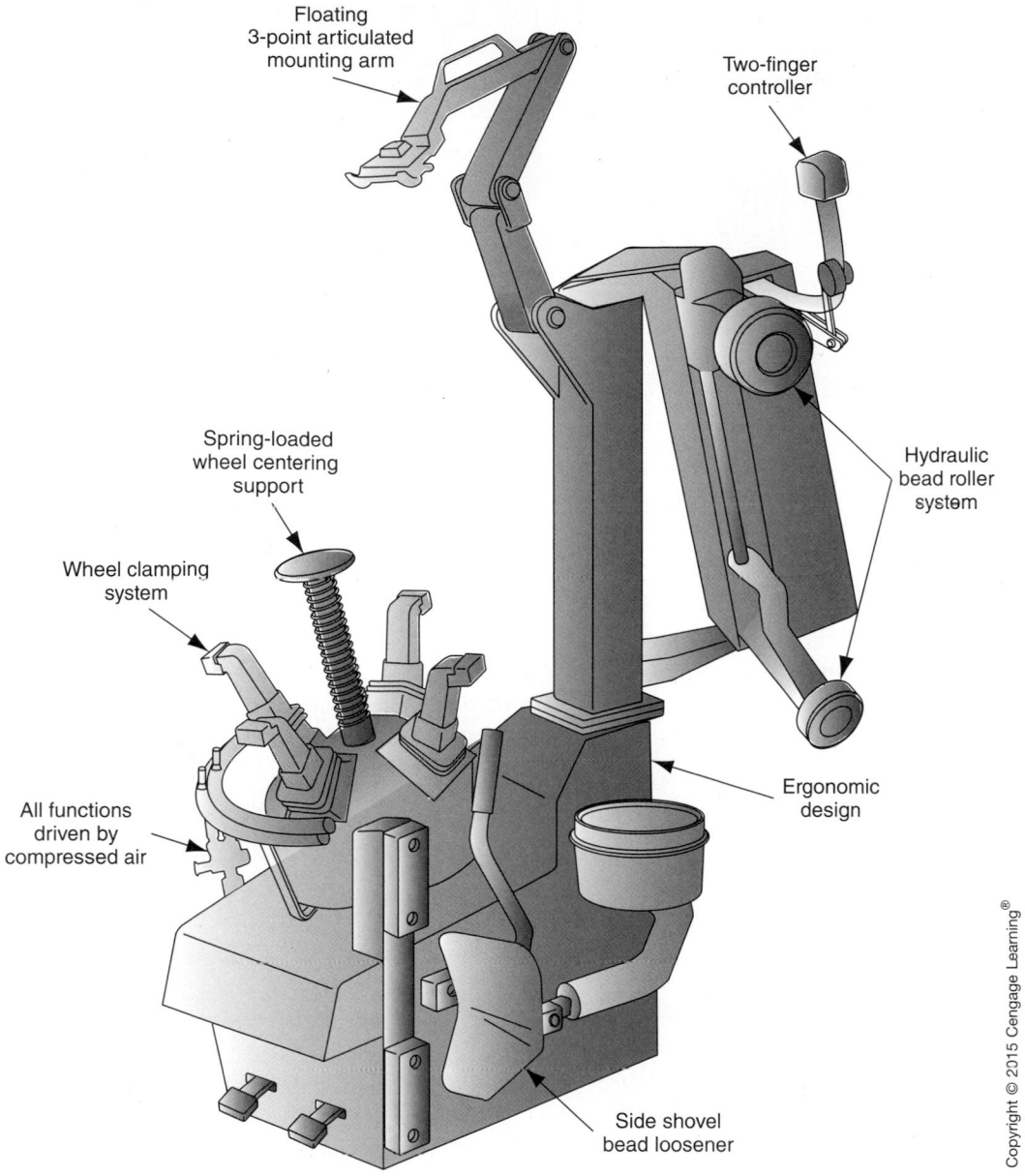

Copyright © 2015 Cengage Learning®

FIGURE 3-6 **Tire changer.**

clamp and wheel rim rotation by foot-operated pedals on the front of the tire changer. This type of clamping system allows the technician to rotate the wheel rim inside the tire to correct excessive wheel and tire runout conditions.

5. The technician uses two fingers to operate control switches that control the upper and lower bead rollers by air-over-hydraulic pressure.
6. The nonmetal bead rollers are placed against the tire next to the rim, and the technician uses the control switches to supply pressure to the rollers (Figure 3-9). The clamp and wheel assembly is rotated to loosen the upper and lower beads.
7. The tire changer is designed so the complete bead loosening, tire dismounting, and tire mounting operation can be performed from one position, which saves time and physical energy.
8. A side shovel bead-loosening tool is an optional feature that facilitates dismounting and mounting tires on special applications such as motorcycles and all-terrain vehicles (ATVs).

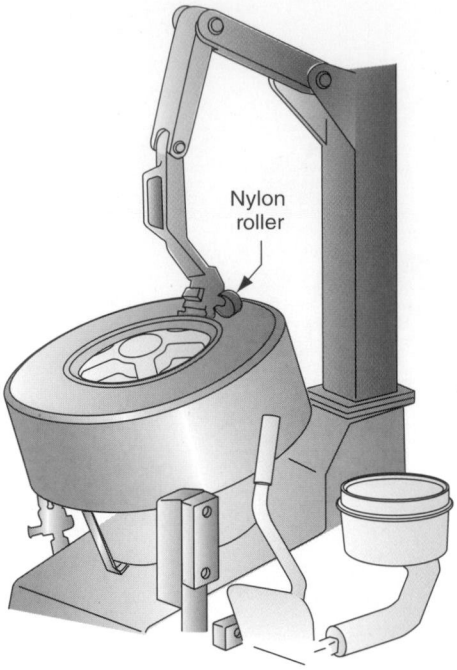

Copyright © 2015 Cengage Learning®

FIGURE 3-7 **Articulating arm with polymer head and mechanical bead pusher.**

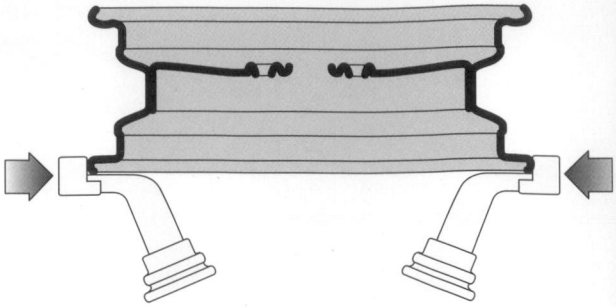

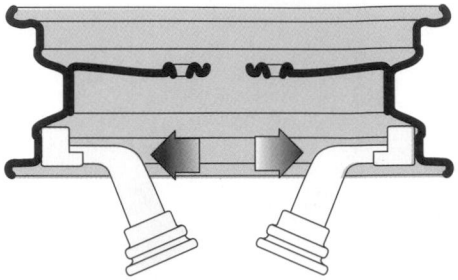

FIGURE 3-8 **Rim clamping system provides internal or external rim clamping action.**

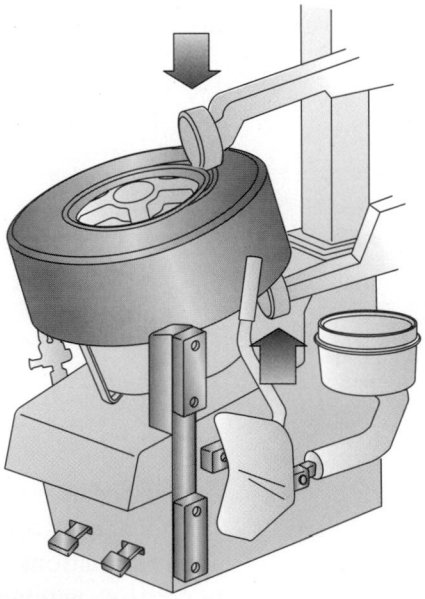

Copyright © 2015 Cengage Learning®

FIGURE 3-9 **Bead loosening rollers.**

Other tire changers have similar features but a different tire and wheel assembly mounting (Figure 3-10). The tire and wheel assembly is mounted on a clamping bell that centers the wheel on the bell. A threaded center post is mounted in the center of the clamping bell. After the tire and wheel assembly is placed on the clamping bell, a tapered cone is threaded onto the center post to retain the tire and wheel assembly. The nonmetal bead rollers are used to loosen the upper and lower beads. On alloy wheels the polymer head on the articulating arm is used to dismount and mount the tire. On steel wheels a steel mount tool may be placed

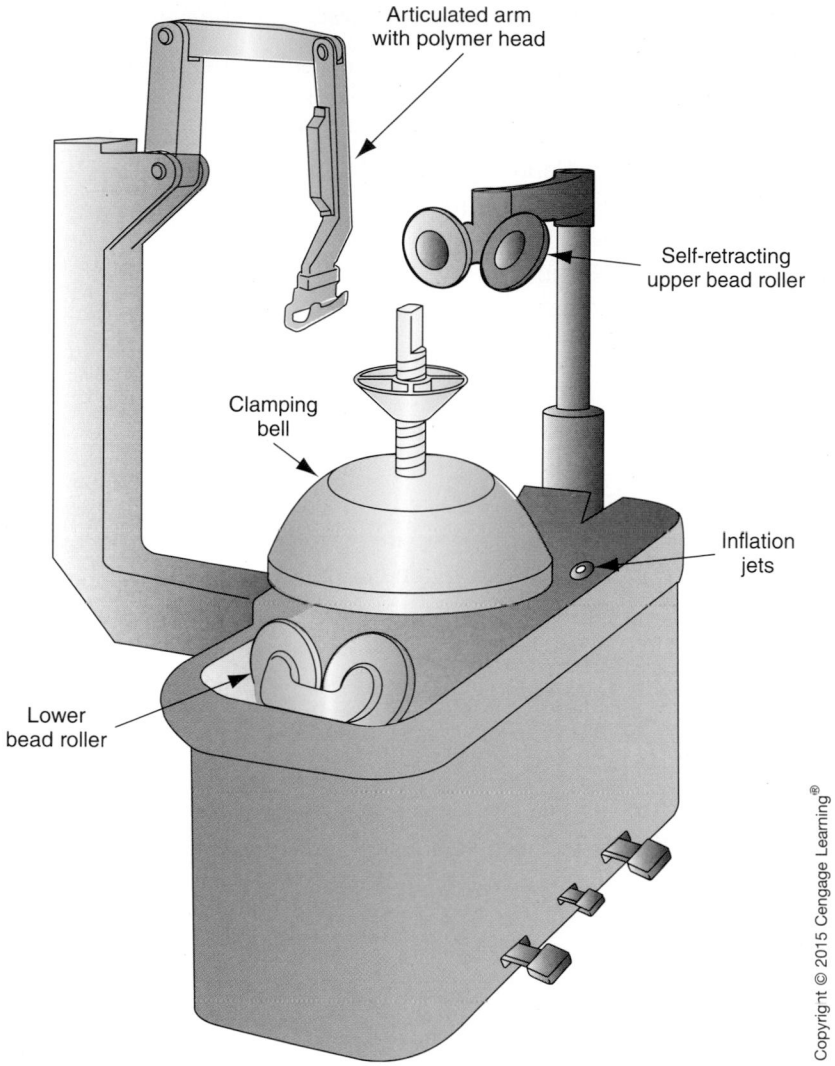

FIGURE 3-10 **Tire changer with horizontal clamping mechanism.**

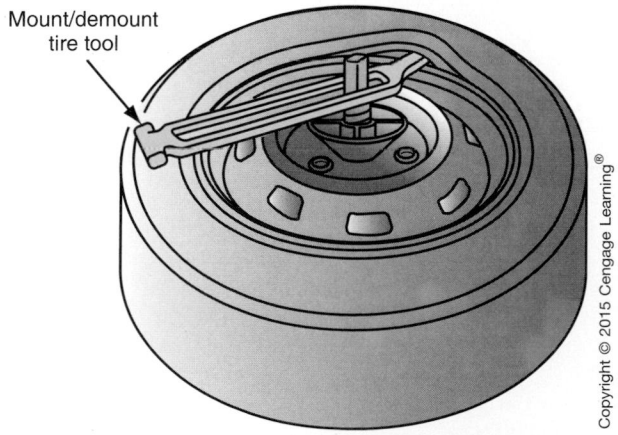

FIGURE 3-11 **Steel mount/demount tool for steel rims.**

through the center post and under the tire bead, and the tire and wheel assembly is rotated to dismount or mount the tire (Figure 3-11). Many tire changers have inflation jets in the clamping mechanism to supply air pressure to the inside of the tire and inflate the tire after the tire is mounted properly.

SPECIAL TOOL
Tire changer

Classroom Manual
Chapter 3, page 68

CAUTION:
The use of sealants or cord plugs when repairing PAX tires will immediately void the tire warranty.

A typical tire dismounting procedure follows:

1. Remove the valve core and be sure the tire is completely deflated.
2. Place the wheel and tire on the tire changer with the narrow bead ledge facing upward.
3. Follow the operating procedure recommended by the manufacturer of the tire changer to force the tire bead inward and separate it from the rim on both sides.
4. Push one edge of the top bead into the drop center of the rim.
5. Place the tire changer's head or lever between the bead and the rim on the opposite side of the rim from where the bead is in the drop center.
6. Operate the tire changer to rotate the head or lever and move the bead over the top of the rim.
7. Repeat steps 4, 5, and 6 to move the lower bead over the top of the rim.

PAX Tire Service

PAX tires are usually serviced at Michelin-authorized PAX tire dealers. Some car dealerships that sold cars with original equipment PAX tires may be authorized Michelin PAX dealers. Other car dealerships will only provide complete PAX tire and wheel replacements. If PAX tires are punctured in the tread area, they are repairable.

Special tire changing equipment is required to dismount and mount PAX tires. An accessory kit and a different press arm may be installed on some tire changers to upgrade the changer to PAX capabilities. A PAX accessory kit and PAX press arm are available for the Coats 9024E tire changer (Figure 3-12). This accessory kit (Figure 3-13) contains the following items:

1. Quick release hub nut
2. Centering cones
3. Removable center shaft
4. Cushioned PAX riser
5. Dual roller tool
6. Tool holders

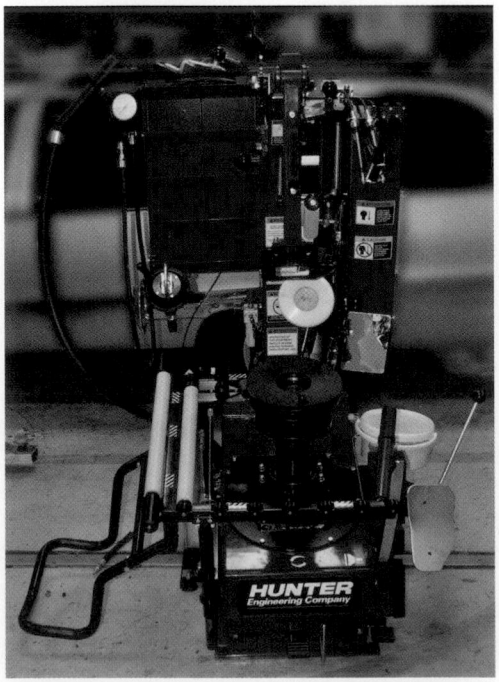

Copyright © 2015 Cengage Learning®

FIGURE 3-12 **Tire changer adaptable to PAX.**

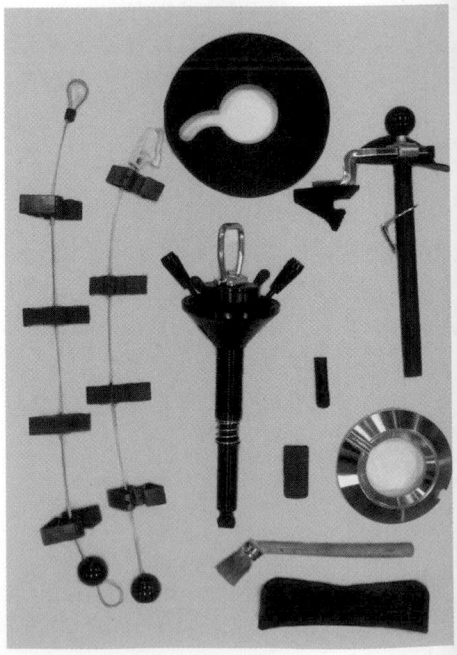

FIGURE 3-13 **Tire changer accessory kit for PAX tire services.**

7. PAX bead lift tool
8. Bead flip tool
9. Plain bead lock tool
10. Spring-loaded bead lock tool
11. Plastic protectors
12. Tapered roller

When dismounting and mounting PAX tires, follow the procedure specified by the tire changer manufacturer. All PAX tires have a tire pressure monitoring system (TPMS). Service precautions related to TPMS systems must be observed when servicing PAX tires and wheels. When a PAX tire is dismounted, always inspect the deflation ring mounted on the wheel. If the ring is damaged, it must be replaced.

CAUTION:
If a tire changer without PAX capabilities is used to attempt PAX tire dismounting and mounting, tire and wheel damage may occur.

Tire Inspection and Repair

To find a leak in a tire and wheel, inflate the tire to the pressure marked on the sidewall, and then submerge the tire and wheel in a tank of water. An alternate method of leak detection is to sponge soapy water on the tire and wheel. Bubbles will appear wherever the leak is located in the tire or wheel. Mark the leak location in the tire or wheel rim with a crayon, and mark the tire at the valve stem location so the tire can be reinstalled in the same position on the wheel to maintain proper balance.

A puncture is the most common cause of a tire leak, and many punctures can be repaired satisfactorily. Do not attempt to repair punctures over 1/4 in. in diameter. Punctures in the sidewalls or on the tire shoulders should not be repaired. The repairable area in belted bias-ply tires is approximately the width of the belts (Figure 3-14). The belts in radial tires are wider than those in bias-ply tires. The repairable area in radial tires is also the width of the belts. Because compact spare tires have thin treads, do not attempt to repair these tires.

Inspect the tire; do not repair a tire with any of the following defects, signs of damage, or excessive wear:

1. Tires with the wear indicators showing
2. Tires worn until the fabric or belts are exposed
3. Bulges or blisters
4. **Ply separation**
5. Broken or cracked beads
6. Cuts or cracks anywhere in the tire

When **ply separation** occurs, the tire plys are pulled apart. This condition often appears as a bulge on the tire surface.

Because most vehicles are equipped with tubeless tires, we will discuss this type of tire repair. If the cause of the puncture, such as a nail, is still in the tire, remove it from the tire. Most punctures can be repaired from inside the tire with a service plug or vulcanized patch service kit. The instructions of the tire service kit manufacturer should be followed, but we will discuss three common tire repair procedures.

Plug Installation Procedure

1. Buff the area around the puncture with a wire brush or wire buffing wheel.
2. Select a plug slightly larger than the puncture opening, and insert the plug in the eye of the insertion tool.
3. Wet the plug and the insertion tool with vulcanizing fluid.
4. While holding and stretching the plug, pull the plug into the puncture from the inside of the tire (Figure 3-15). The head of the plug should contact the inside of the tire. If the plug pulls through the tire, repeat the procedure.
5. Cut the plug off 1/32 in. from the tread surface. Do not stretch the plug while cutting.

CAUTION:
Radial tire patches should have arrows that must be positioned parallel to the radial plies to provide proper adhesion.

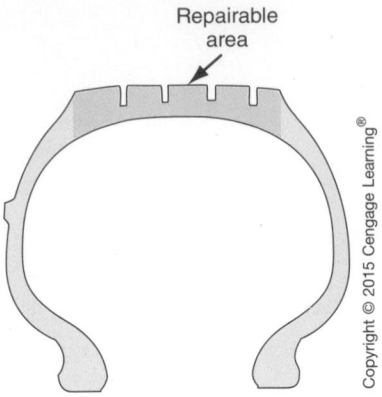

FIGURE 3-14 **Repairable area on bias-ply and belted bias-ply tires.**

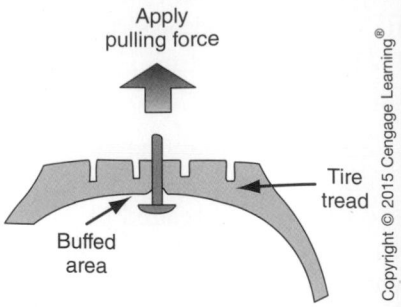

FIGURE 3-15 **Plug installation procedure.**

CAUTION:
The use of abrasive cleaners, alkaline-base detergents, or caustic agents on aluminum or magnesium wheel rims may cause discoloration or damage to the protective coating.

Cold Patch Installation Procedure

1. Buff the area around the puncture with a wire brush or buffing wheel.
2. Apply vulcanizing fluid to the buffed area and allow it to dry until it is tacky.
3. Peel the backing from the patch, and apply the patch over the puncture. Center the patch over the puncture.
4. Run a stitching tool back and forth over the patch to improve bonding.

Hot Patch Installation Procedure

1. Buff the area around the puncture with a wire brush or buffing wheel.
2. Apply vulcanizing fluid to the buffed area, if required.
3. Peel the backing from the patch and install the patch so it is centered over the puncture on the inside of the tire. Many hot patches are heated with an electric heating element clamped over the patch. This element should be clamped in place for the amount of time recommended by the equipment or patch manufacturer.
4. After the heating element is removed, allow the patch to cool for a few minutes and be sure the patch is properly bonded to the tire.

Classroom Manual
Chapter 3, page 69

Wheel Rim Service

Steel rims should be spray cleaned with a water hose. Aluminum or magnesium wheel rims should be cleaned with a mild soap and water solution, and rinsed with clean water. The use of abrasive cleaners, alkaline-base detergents, or caustic agents may damage aluminum or magnesium wheel rims. Clean the rim bead seats on these wheel rims thoroughly with the mild soap and water solution. The rim bead seats on steel wheel rims should be cleaned with a wire brush or coarse steel wool.

CAUTION:
Steel wheel rims must not be welded, heated, or peened with a ball-peen hammer. These procedures may weaken the rim and create a safety hazard.

Steel wheel rims should be inspected for excessive rust and corrosion, cracks, loose rivets or welds, bent or damaged bead seats, and elongated lug nut holes. Aluminum or magnesium wheel rims should be inspected for damaged bead seats, elongated lug nut holes, cracks, and porosity. If any of these conditions are present on either type of wheel rim, replace the wheel rim.

Many shops always replace the tire valve assembly when a tire is repaired or replaced. This policy helps prevent future problems with tire valve leaks. The inner end of the valve may be cut off with a pair of diagonal pliers, and then the outer end may be pulled from the rim. Coat the new valve with rubber tire lubricant and pull it into the rim opening with a special puller screwed onto the valve threads.

Wheel Rim Leak Repair

A wheel rim leak may be repaired if the leak is not caused by excessive rust on a steel rim, and the rim is in satisfactory condition.

Follow these steps for wheel rim leak repair:

1. Use #80-grit sandpaper to thoroughly clean the area around the leak on the tire side of the rim.
2. Use a shop towel to remove any grit from the leak area.
3. Be sure the wheel rim is at room temperature, and apply a heavy coating of silicone rubber sealer over the leak area.
4. Spread the sealer over the entire sanded area with a putty knife.
5. Allow the sealer to cure for six hours before remounting the tire.

CAUTION: Installing an inner tube to correct leaks in a tubeless tire or wheel rim is not an approved procedure.

Tire Remounting Procedure

1. Be sure the wheel rim bead seats are thoroughly cleaned.
2. Coat the tire beads and the wheel rim bead seats with rubber tire lubricant.
3. Secure the wheel rim on the tire changer with the narrow bead ledge facing upward, and place the tire on top of the wheel rim with the bead on the lower side of the tire in the drop center of the wheel rim.
4. Use the tire changer head or lever under the tire bead to install the tire bead over the wheel rim. Always operate the tire changer with the manufacturer's recommended procedure.
5. Repeat steps 3 and 4 to install the upper bead over the wheel rim.
6. Rotate the tire on the wheel rim until the crayon mark is aligned with the valve stem. This mark was placed on the tire prior to dismounting.

WARNING: When a bead expander is installed around the tire, never exceed 10 psi (69 kPa) pressure in the tire. A higher pressure may cause the expander to break and fly off the tire, causing serious personal injury or property damage.

WARNING: When a bead expander is not used, never exceed 40 psi (276 kPa) tire pressure to move the tire beads out tightly against the wheel rim. A higher pressure may blow the tire bead against the rim with excessive force, and this action could burst the rim or tire, resulting in serious personal injury or property damage.

WARNING: While inflating a tire, do not stand directly over the tire. In this position, serious injury could occur if the tire or wheel rim flies apart.

Follow the recommended procedure supplied by the manufacturer of the tire changer to inflate the tire. This procedure may involve the use of a bead expander installed around the center of the tire tread to expand the tire beads against the wheel rim. If a bead expander is used, inflate the tire to 10 psi (69 kPa) to move the beads out tightly against the wheel rim. Never exceed this pressure with a bead expander installed on the tire. Always observe the circular marking around the tire bead as the tire is inflated. This mark should be centered around the wheel rim (Figure 3-16). Always observe both beads while a tire is inflated. If the circular mark around the tire bead is not centered on the rim, deflate

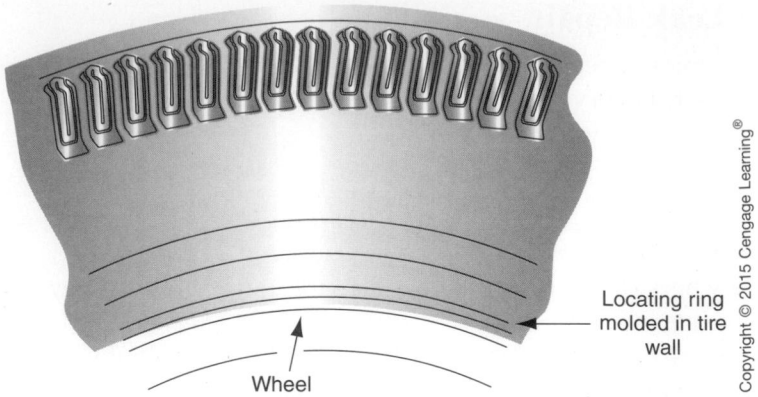

FIGURE 3-16 **The circular ring around the tire bead must be centered on the wheel rim.**

the tire and center it on the wheel rim. If a bead expander is not used, never exceed 40 psi (276 kPa) when moving tire beads out tightly against the rim. When either tire bead will not move out tightly against the wheel rim with 40 psi (276 kPa) tire pressure, deflate the tire and center it on the wheel rim again.

Photo Sequence 3 shows a typical procedure for dismounting and mounting a tire on a wheel assembly.

Diagnosing and Servicing Tire Pressure Monitoring Systems

Visual Inspection

If the TPMS warning light is illuminated and/or a warning message is displayed in the message center, the first step in diagnosing the system is a visual inspection. During this inspection be sure each tire is inflated to the specified pressure. Make sure all four or five tires are the size specified by the vehicle manufacturer, and that each wheel contains a tire pressure sensor. Inspect the wiring harness connection to the TPMS module for loose or corroded connections and damaged wires.

Classroom Manual
Chapter 3, page 78

Scan Tool Diagnosis

The vehicle manufacturer's recommended diagnostic procedure varies considerably depending on the vehicle model year. Most systems are diagnosed using a scan tool with TPMS capabilities. The following is a typical scan tool diagnosis. If the visual inspection does not reveal any defects, use a scan tool to proceed with the diagnostic system check.

Diagnostic System Check

The diagnostic system check provides the following information:

1. Identification of the control modules in the TPMS system.
2. Indication of the ability of the system control modules to communicate through the serial data circuit.
3. Indication of any diagnostic trouble codes (DTCs) stored in the system control modules.

Follow these steps to complete the diagnostic system check.

1. Connect a scan tool to the data link connector (DLC) under the left side of the instrument panel, and turn the scan tool on.

PHOTO SEQUENCE 3

Typical Procedure for Dismounting and Mounting a Tire on a Wheel Assembly

P3-1 **Dismounting the tire from the wheel begins with releasing the air, removing the valve stem core, and unseating the tire from its rim. The machine does the unseating. The technician merely guides the operating lever.**

Copyright © 2015 Cengage Learning®

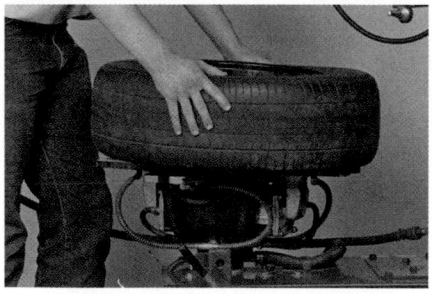

P3-2 **Once both sides of the tire are unseated, place the tire and wheel onto the machine. Then depress the pedal that clamps the wheel to the tire machine.**

Copyright © 2015 Cengage Learning®

P3-3 **Lower the machine's arm into position on the tire and wheel assembly.**

Copyright © 2015 Cengage Learning®

P3-4 **Insert the tire iron between the upper bead of the tire and the wheel. Depress the pedal that causes the wheel to rotate. Do the same with the lower bead.**

Copyright © 2015 Cengage Learning®

P3-5 **After the tire is totally free from the rim, remove the tire.**

Copyright © 2015 Cengage Learning®

P3-6 **Prepare the wheel for the mounting of the tire by using a wire brush to remove all dirt and rust from the sealing surface. Apply rubber compound to the bead area of the tire.**

Copyright © 2015 Cengage Learning®

P3-7 **Place the tire onto the wheel and lower the arm into place. As the machine rotates the wheel, the arm will force the tire over the rim. After the tire is completely over the rim, install the air ring over the tire. Activate it to seat the tire against the wheel.**

Copyright © 2015 Cengage Learning®

P3-8 **Reinstall the valve stem core and inflate the tire to the recommended inflation.**

Copyright © 2015 Cengage Learning®

2. Turn the ignition switch on, and do not start the engine.
3. Select antenna module on the scan tool. Check for the DTCs related to the antenna module. If there are DTCs displayed with a "U" prefix, there is a defect in the data link communications. If the scan tool cannot communicate with the antenna module, or there are DTCs related to this module or the data link communications, proceed with further diagnosis of these items.
4. Check for DTCs related to the radio/audio system. The antenna grid in the rear window receives TPMS system sensor signals. This grid shares its connector with the AM/FM antenna grid. If any radio/audio system DTCs are present, proceed with the diagnosis of the radio/audio system.
5. Check for keyless entry system DTCs displayed in the scan tool. The antenna module also controls the keyless entry system. If the scan tool displays DTCs related to the keyless entry system, proceed with the diagnosis of these DTCs.
6. Check for TPMS system DTCs on the scan tool display. If TPMS system DTCs are present, diagnose the cause of these DTCs.

TPMS System Data Display

TPMS data display may be selected on the scan tool display. The data display is very useful when diagnosing the TPMS system and sensors. The following data displays may be selected on the scan tool:

1. Battery voltage—The scan tool displays the amount of battery voltage supplied to the antenna module.
2. LF pressure sensor ID—The scan tool displays an eight-digit number or an asterisk. The eight-digit number is a unique LF sensor identification number, and this display indicates the number has been learned by the antenna module. If an asterisk is displayed, the sensor ID number has not been learned by the antenna module.
3. LF pressure sensor mode—This mode display may indicate **stationary, wake, drive, remeasure, learn, low bat**, or **N/A** (not available). If **stationary** is displayed, the sensor roll switch is open and the sensor has sent a stationary message that only occurs every 60 minutes from the previous stationary transmission. A **wake** display indicates the sensor has detected an initial roll switch closure, and the sensor is changing from the stationary mode to the drive mode. A **drive** display indicates the vehicle speed is above 20 mph (32 km/h), and the sensor roll switch has been closed for a minimum of 10 seconds. In this mode the sensor transmits every 60 seconds. If **remeasure** is displayed, the sensor has detected a 1.6 psi (11 kPa) pressure change. A **learn** display indicates the sensor has been activated by a low-frequency voltage signal that occurs during a sensor relearn procedure. A **low battery** display indicates the internal sensor battery has low voltage and sensor replacement is necessary.
4. LF tire pressure—The scan tool displays the actual tire pressure between 0 and 51 psi (0 to 344 kPa).
5. LF tire pressure sensor status—In this mode the scan tool displays **valid** or **invalid**. A **valid** display indicates the specified LF tire pressure is present. If **invalid** is displayed, the LF tire pressure is not within specifications and DTC C0750 is currently set in the antenna module memory. This DTC indicates a defective LF sensor or faulty LF sensor circuit.

CAUTION:
Do not remove the valve stem core to relieve the tire pressure. If the valve stem core is inadvertently removed from the valve stem, a new nickel-plated valve core must be installed. Failure to use a nickel-plated valve core will result in corrosion and possible loss of tire pressure.

The same data from each wheel sensor may be displayed on the scan tool. Other DTCs include C0755 indicating a defective LR sensor, C0760 representing a faulty RF sensor, and C0765 indicating a defect in the RR sensor. A current DTC indicates the fault causing the DTC is present at the time of diagnosis. The DTCs vary depending on the vehicle make and model year. A history DTC represents a fault that occurred in the past, but it is not present at the time of diagnosis. The vehicle speed must be above 20 mph (32 km/h) for the antenna module to run the DTC check. A DTC is set in the antenna module memory if any wheel sensor does not enter the drive mode or does not transmit any data for 10 minutes or more. When a DTC is set in the antenna module memory, the following actions are taken:

1. A DTC is stored in memory.
2. The driver information center (DIC) displays a SERVICE TIRE MONITOR message.
3. The DIC indicates row of dashes in place of the suspect tire pressure display.

A current DTC is cleared from memory when the fault causing the DTC is corrected. A history DTC is cleared after 100 consecutive fault-free ignition cycles. A scan tool may be used to clear DTCs.

CAUTION:
The tire pressure sensors are not serviceable. If a sensor is defective, it must be replaced.

Sensor Replacement

Vehicles with a TPMS may require a tire pressure adjustment in relation to atmospheric temperature. For example, if the atmospheric temperature decreases 30°F when a vehicle is parked outside overnight, the tire pressure decreases 3 psi. This pressure decrease may activate the TPMS warning system. Refer to the vehicle manufacturer's specifications for tire inflation pressure at various atmospheric temperatures.

Proper valve stem and sensor removal procedures must be followed to avoid damage to the sensor: To remove a valve stem and sensor, follow these steps:

1. Before removing the tire and wheel assembly from the vehicle, mark the wheel rim in relation to one of the wheel studs.
2. Position the tire and wheel assembly with the valve stem at the 6 o'clock position.
3. Remove the sensor retention nut and push the valve stem and sensor into the tire.
4. Place the wheel and tire on the turntable of the tire changer so the valve stem is positioned 270° from the mounting/demounting arm on the tire changer (Figure 3-17).
5. Loosen both tire beads away from the wheel rim.

CAUTION:
Tire pressure sealing products may render the TPMS sensor inoperative. If this occurs, remove all of the tire sealing material and replace the sensor.

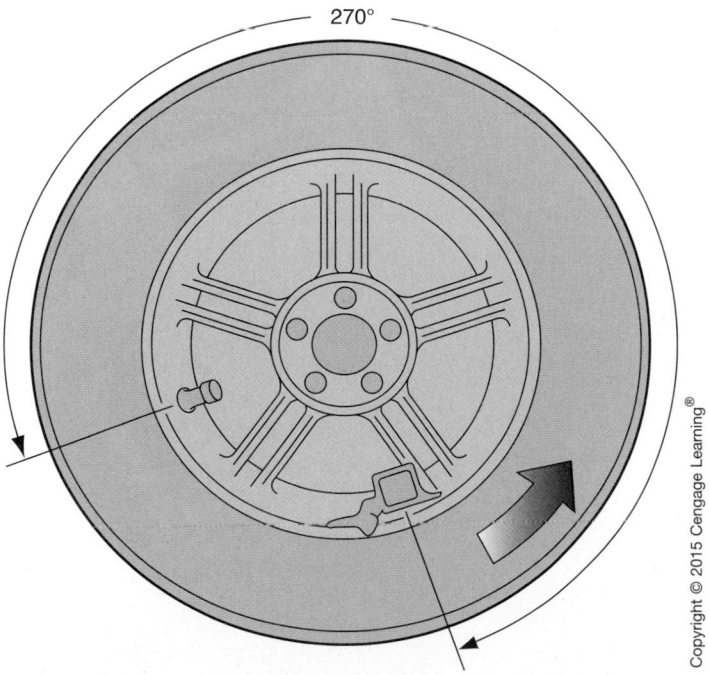

FIGURE 3-17 **Proper tire and wheel position on a tire changer.**

CAUTION:
Each time a TPMS sensor is removed, a new grommet, retention nut, and valve cap must be installed. Replace the valve core if it has been removed or damaged.

6. Place index marks on the tire beside the valve stem opening and the wheel weight positions.
7. Lubricate the outer tire bead, and dismount the outer bead over the wheel rim.
8. Lift up on the outer tire bead, and remove the TPMS sensor and valve stem assembly.
9. Install the new sensor in the valve stem opening, and install a new grommet and retention nut.
10. Tighten the retention nut to 71 in.-lbs (8 Nm).
11. Lubricate and mount the outer tire bead.
12. Inflate the tire to the specified pressure.
13. Remove the tire and wheel from the tire changer and install this assembly in the proper position on the wheel studs. Tighten the wheel retaining nuts in the proper sequence to the specified torque.

Sensor Learning Procedure with Magnetic Tool

If a TPMS sensor or component is serviced, a sensor learning procedure must be performed. There are a number of different sensor learning procedures depending on the vehicle make and model year. The sensor learning procedure usually involves the use of a magnetic tool or a scan tool.

Follow these steps to complete the sensor learning procedure with a magnetic tool:

SPECIAL TOOL
Magnetic learn tool

1. Starting with the ignition switch off, cycle the ignition switch on and off three times, and on the third cycle leave the ignition switch in the on position. Do not wait more than two seconds between switch cycles.
2. Press and release the brake pedal.
3. Repeat the ignition switch cycling procedure as explained in step 1. Upon completion of this procedure, the horn should sound once to indicate successful entry to the learn mode.
4. After the horn sounds, a TRAIN LEFT FRONT TIRE message should appear in the instrument panel message center.
5. Place the special magnetic tool on the valve stem of the left front tire (Figure 3-18). When the TPMS module recognizes the left front sensor, the horn sounds momentarily.

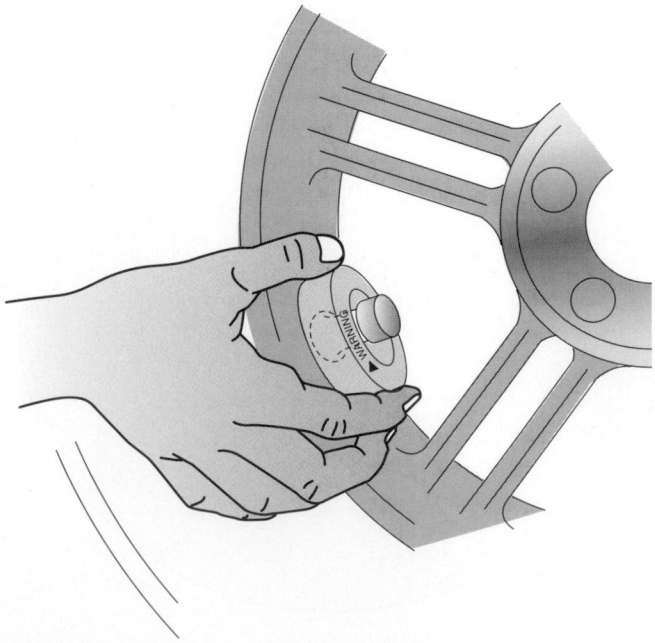

Copyright © 2015 Cengage Learning®

FIGURE 3-18 **Magnetic learning tool for TPMS sensors.**

WARNING: The special magnetic tool may adversely affect magnetically sensitive devices such as heart pacers, and this action may result in personal injury!

6. Repeat step 5 at the right front, right rear, left rear, and spare wheels.
7. If the learn procedure fails on any wheel, the horn sounds twice, and a TIRES NOT LEARNED-REPEAT message appears in the message center. If this action occurs, the learn procedure must be repeated from step 1.

Sensor Learning Procedure with Scan Tool

Follow these steps to complete the sensor learning procedure with a scan tool:

1. Connect a scan tool to the DLC.
2. Turn on the ignition switch, and do not start the engine.
3. Apply the parking brake.
4. Select Special Functions on the scan tool.
5. Select Sensor Learn Mode on the scan tool, and press the Enter key.
6. Press the On soft key. A horn chirp should sound to indicate the sensor learn mode is enabled.
7. Starting with the LF tire, increase or decrease the tire pressure for 5–8 seconds or until a horn chirp sounds. The horn chirp may occur before the 5- to 8-second time period, or up to 30 seconds after this time period.

WARNING: If you are increasing tire pressure during the learning procedure, never inflate a tire above the vehicle manufacturer's maximum specified tire pressure. This action may cause personal injury and tire damage.

8. After the horn chirp sounds, repeat step 7 on the other 3 or 4 sensors in the following order:
   (a) RF
   (b) RR
   (c) LR
   (d) Spare tire (if applicable)

If a horn chirp is not heard after 35 seconds for any of the 4 sensors, turn off the ignition switch and exit the learn mode on the scan tool. Repeat the sensor learning procedure from step 4.

9. After the learning procedure has been completed on all the sensors, a double horn chirp sounds to indicate the learning procedure is completed on all the sensors.
10. Turn off the ignition switch and disconnect the scan tool.
11. Inflate all the tires to the specified pressure.

Sensor Learning Procedure with Keyless Entry Transmitter

On some vehicles, the keyless entry remote transmitter may be used to complete the wheel sensor learning procedure without the use of any other equipment. Follow these steps to perform a sensor learning procedure using the keyless entry remote transmitter:

1. Turn on the ignition switch.
2. Apply the parking brake.
3. Use the keyless entry remote transmitter to lock and unlock the doors three times.

4. Simultaneously press the lock and unlock buttons on the keyless entry remote transmitter until a horn chirp sounds.
5. Starting with the LF tire, increase or decrease the tire pressure for 5–8 seconds, or until a horn chirp sounds. The horn chirp may sound before the 5–8 seconds is completed, or up to 35 seconds after this time period.
6. After the horn chirp sounds on the LF tire, follow the procedure in step 5 on the other 3 or 4 sensors in the following order:
   (a) RF
   (b) RR
   (c) LR
   (d) Spare (if applicable)

If a horn chirp does not sound after 35 seconds on any of the tires, turn off the ignition switch and exit the learn mode on the scan tool. Repeat the procedure starting with step 1.

7. After all the sensors have been learned, a double horn chirp sounds to indicate all the sensors have been learned.
8. Turn off the ignition switch and disconnect the scan tool.
9. Inflate all the tires to the specified tire pressure.

Tire and Wheel Runout Measurement

Radial tire runout refers to excessive variations in the tread circumference.

Ideally, a tire and wheel assembly should be perfectly round. However, this condition is rarely achieved. A tire and wheel assembly that is out-of-round is said to have radial runout. If the radial runout exceeds manufacturer's specifications, a vibration may occur because the radial runout causes the spindle to move up and down (Figure 3-19). A defective tire with a variation in stiffness may also cause this up-and-down spindle action.

Wheel balancers with force variation capabilities have a roller that is pressed against the tire tread during the wheel balance procedure. This roller senses and indicates stiffness variation in a tire.

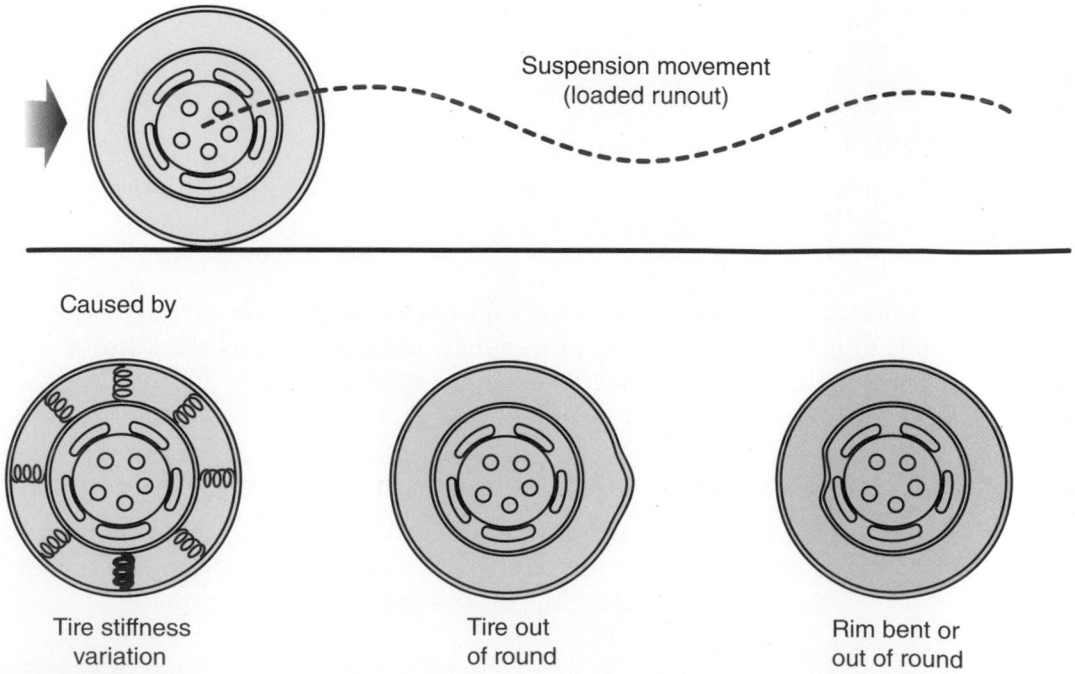

Copyright © 2015 Cengage Learning®

FIGURE 3-19 **Vertical tire and wheel vibrations caused by radial tire or wheel runout, or variation in tire stiffness.**

Copyright © 2015 Cengage Learning®

FIGURE 3-20 **Measuring tire radial runout.**

A dial indicator gauge may be positioned against the center of the tire tread as the tire is rotated slowly to measure radial runout (Figure 3-20). Radial runout of more than 0.060 in. (1.5 mm) will cause vehicle shake. If the radial runout is between 0.045 in. to 0.060 in. (1.1 mm to 1.5 mm), vehicle shake may occur. These are typical radial runout specifications. Always consult the vehicle manufacturer's specifications. Mark the highest point of radial runout on the tire with chalk, and mark the valve stem position on the tire.

If the radial tire runout is excessive, dismount the tire and check the runout of the wheel rim with a dial indicator positioned against the lip of the rim while the rim is rotated (Figure 3-21). Use chalk to mark the highest point of radial runout on the wheel rim. Radial wheel runout should not exceed 0.035 in. (0.9 mm), whereas the maximum lateral wheel runout is 0.045 in. (1.1 mm). If the highest point of wheel radial runout coincides

SPECIAL TOOL

Tire runout gauge

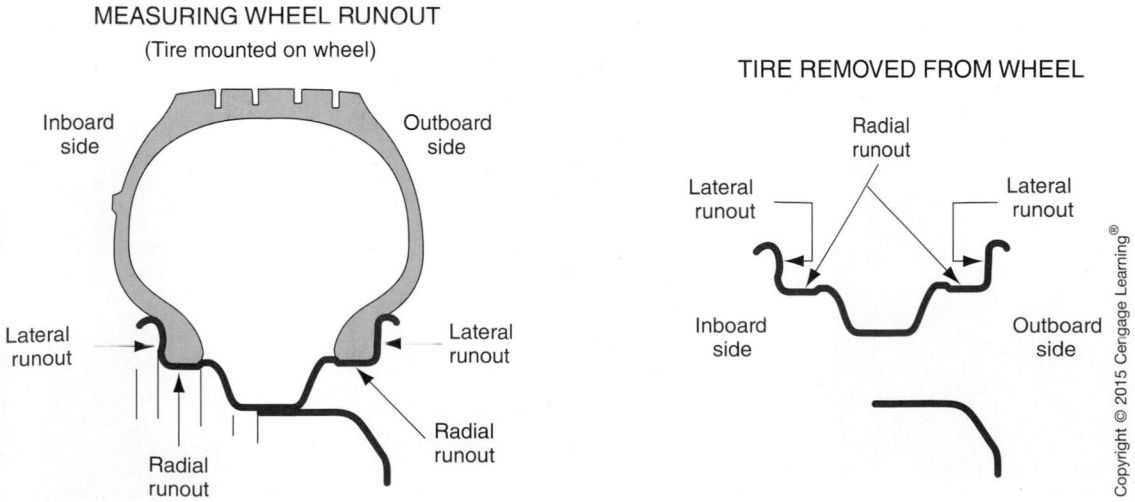

FIGURE 3-21 **Measuring wheel radial and lateral runout.**

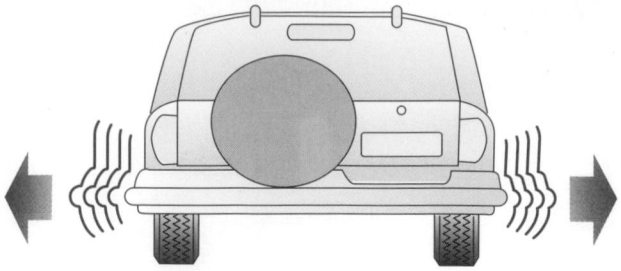

Tire waddle often caused by

- Steel belt not straight within tire
- Excessive lateral runout

Copyright © 2015 Cengage Learning®

FIGURE 3-22 **Chassis waddling action caused by lateral tire or wheel runout, or a defective tire with a belt that is not straight.**

Side-to-side tire and chassis movement caused by excessive lateral tire runout may be called tire or chassis waddle.

with the chalk mark from the highest point of maximum tire radial runout, the tire may be rotated 180° on the wheel to reduce radial runout. Tires or wheels with excessive runout are usually replaced.

Lateral tire runout may be measured with a dial indicator located against the sidewall of the tire. Excessive lateral runout causes the tire to waddle as it turns, and this waddling sensation may be transmitted to the passenger compartment (Figure 3-22). A chassis waddling action may also be caused by a defective tire in which the belt is not straight. If the lateral runout exceeds 0.050 in. (1.27 mm) off vehicle or 0.060 in. (1.52 mm) on vehicle, wheel shake problems will occur on the vehicle. Chalk mark the tire and wheel at the highest point of lateral runout. When the tire runout is excessive, the tire should be removed from the wheel, and the wheel lateral runout should be measured with a dial indicator positioned against the edge of the wheel as the wheel is rotated. Tires or wheels with excessive lateral runout should be replaced.

Classroom Manual
Chapter 3, page 65

Tread Wear Measurement

On most tires, the **tread wear indicators** appear as wide bands across the tread when tread depth is worn to 2/32 in. (1.6 mm). Most tire manufacturers recommend tire replacement when the tread wear indicators appear across two or more tread grooves at three locations around the tire (Figure 3-23). If tires do not have wear indicators, a tread depth gauge may be used to measure the tread depth (Figure 3-24). The tread depth

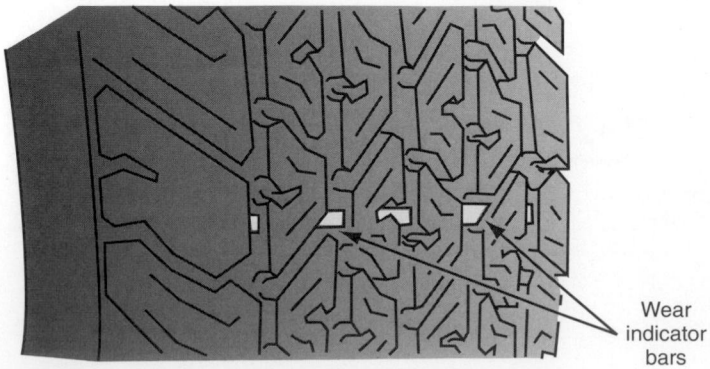

Copyright © 2015 Cengage Learning®

FIGURE 3-23 **Tire tread wear indicators.**

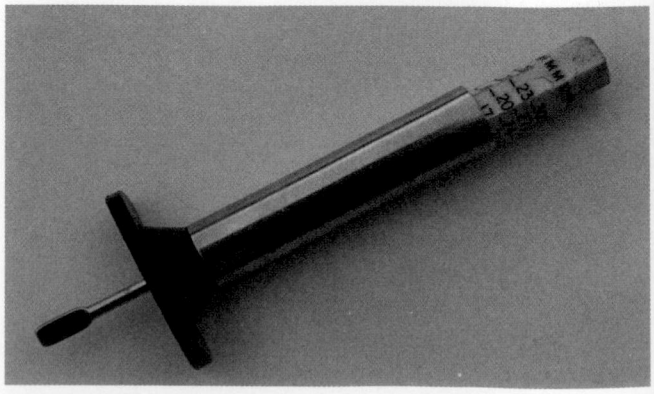

FIGURE 3-24 **Tread depth gauge.**

gauge reads in 32nds of an inch. Tires with 2/32 in. of tread depth or less should be replaced.

Preliminary Wheel Balancing Checks

Classroom Manual
Chapter 3, page 86

These preliminary checks should be completed before a tire and wheel are balanced:

1. Check for objects in the tire tread.
2. Check for objects inside the tire.
3. Inspect the tread and sidewall.
4. Check the inflation pressure.
5. Measure the tire and wheel runout.
6. Check the wheel bearing adjustment.
7. Check for mud collected on the inside of the wheel and wash the tire and wheel assembly.
8. Inspect the wheel rim for damage and excessive rust.

WARNING: On many wheel balancers, the tire and wheel are spun at high speed during the dynamic balance procedure. Be sure that all wheel weights are attached securely, and check for other loose objects on the tire and wheel, such as stones in the tread. If loose objects are detached from the tire or wheel at high speed, they may cause serious personal injury or property damage.

WARNING: On the type of wheel balancer that spins the tire and wheel at high speed during the dynamic balance procedure, always attach the tire and wheel assembly securely to the balancer. Follow the equipment manufacturer's recommended wheel mounting procedure. If the tire and wheel assembly becomes loose on the balancer at high speed, serious personal injury or property damage may result.

WARNING: Prior to spinning a tire and wheel at high speed on a wheel balancer, always lower the protection shield over the tire. This shield provides protection in case anything flies off the tire or wheel.

All of the items on the preliminary check list influence wheel balance or safety. Therefore, it is extremely important that the preliminary checks be completed. Since a tire and wheel assembly is rotated at high speed during the dynamic balance procedure, it is very important that objects such as stones be removed from the treads. Centrifugal force may dislodge objects from the treads and cause serious personal injury. For this reason, it is also extremely important that the old wheel weights be removed from the wheel prior to balancing and the new weights attached securely to the wheel during the balance procedure.

Follow these preliminary steps before attempting to balance a wheel and tire:

1. When off-car wheel balancers are used, the tire and wheel must be removed from the vehicle and installed on the balancer. All mud, dust, and debris must be washed from the wheel after it is removed.

2. Objects inside a tire, such as balls of rubber, make balance impossible. When the tire and wheel assembly is mounted on the balancer, be absolutely sure that the wheel is securely tightened on the balancer. As the tire is rotated slowly, listen for objects rolling inside the tire. If such objects are present, they must be removed prior to wheel balancing.
3. The tire should be inspected for tread and sidewall defects before the balance procedure. These defects create safety hazards, and they may influence wheel balance. For example, tread chunking makes the wheel balancing difficult.

Tire Inflation Pressure

A tire depends on correct inflation pressure to maintain its correct shape and support the vehicle weight.

Excessive inflation pressure causes the following problems:

1. Excessive center tread wear (Figure 3-25)
2. Hard ride
3. Damage to tire carcass

When tires are underinflated, the following problems will be evident:

1. Excessive wear on each side of the tread (Figure 3-23)
2. Hard steering
3. Wheel damage
4. Excessive heat buildup in the tire and possible severe tire damage with resultant hazardous driving

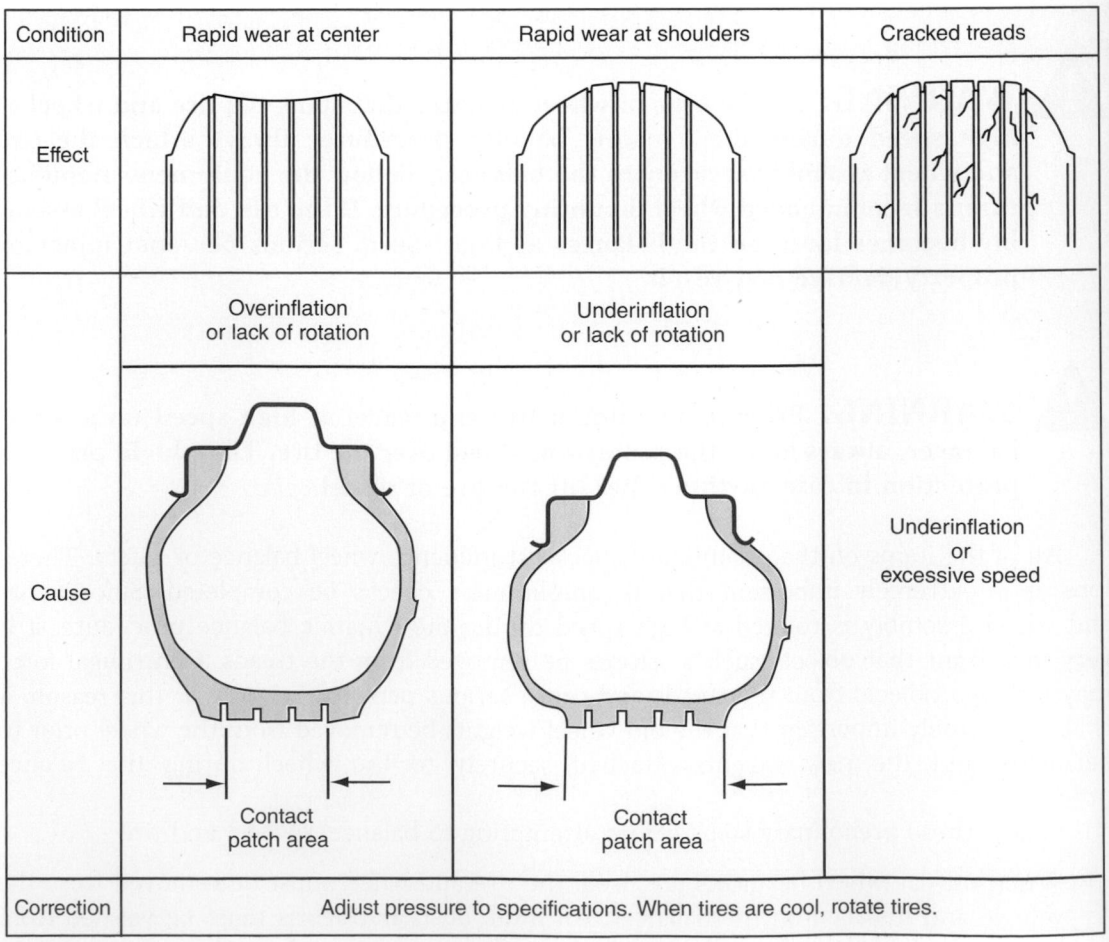

FIGURE 3-25 **Tire tread wear caused by underinflation or overinflation.**

The tires should be inflated to the car manufacturer's recommended pressure prior to the balance procedure. Loose wheel bearings allow lateral wheel shaking and simulate an imbalanced wheel condition when the vehicle is in motion. Therefore, wheel bearing adjustments should be checked when wheel balance conditions are diagnosed.

Static and Dynamic Wheel Balance Procedure

The off-vehicle electronic wheel balancer is presently the most common type in use (Figure 3-26). When using this type of wheel balancer, each tire and wheel assembly is removed from the vehicle and installed on the wheel balancer. A special cone centers the wheel rim on the balancer shaft, and a large nut retains the wheel and tire assembly on the balancer. Tire and wheel assemblies must always be mounted on the balancer using the service procedures recommended by the balancer manufacturer. Most electronic wheel balancers have an electric motor that spins the tire and wheel assembly at moderate speed during the balance procedure. A safety cover on the balancer must be lowered over the tire and wheel assembly when the tire and wheel assembly is spun on the balancer. This hood protects the technician from any debris that might be dislodged from the tire tread. On some electronic wheel balancers, the tire and wheel assembly is spun by hand during the balance procedure.

SPECIAL TOOL
Off-car electronic wheel balancer

Static balance refers to the balance of a stationary wheel.

Dynamic wheel balance refers to proper balance of the tire and wheel assembly during tire and wheel rotation.

After the tire and wheel assembly is mounted properly on the balancer, the technician must enter the wheel diameter, wheel width, and rim offset in the balancer computer. A special caliper may be used to measure the wheel width. A special tool on the balancer may be extended so it contacts the flange on the side of the wheel rim next to the balancer. This tool provides the required wheel offset measurement that indicates the position of the wheel on the balancer shaft. An electronic wheel balancer performs **static balance** and **dynamic wheel balance** calculations simultaneously. The following is a typical wheel balance procedure:

1. Complete all the preliminary balance checks mentioned previously in this chapter.
2. Be sure the tire and wheel assembly is mounted on the balancer using the balancer manufacturer's recommended mounting procedure.

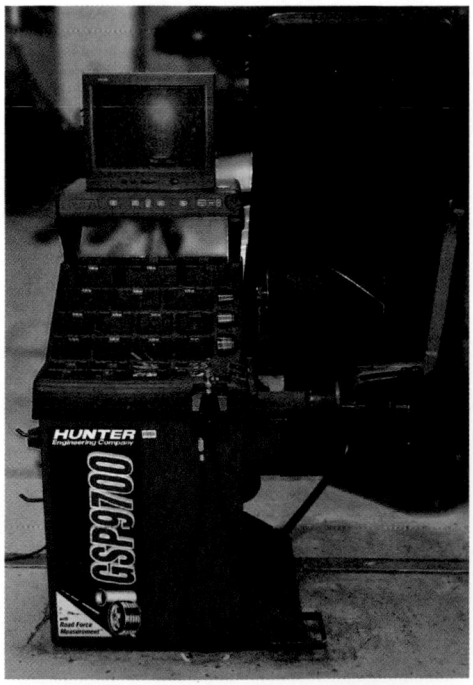

Copyright © 2015 Cengage Learning®

FIGURE 3-26 **Electronic wheel balancer.**

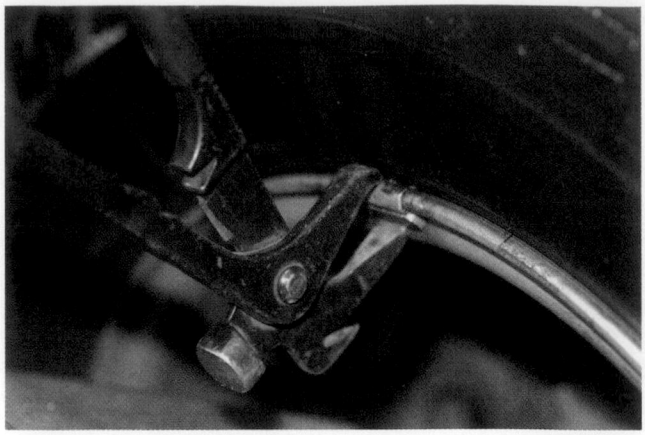

Copyright © 2015 Cengage Learning®

FIGURE 3-27 **Wheel weights are removed and installed with special wheel weight pliers.**

3. Use a pair of wheel weight pliers to remove all the old wheel weights from the wheel rim (Figure 3-27).
4. Enter the wheel diameter, width, and offset in the balancer computer.
5. Be sure the safety hood is lowered over the tire and wheel assembly, and activate the balancer control to spin the wheel and tire assembly.
6. Operate the balancer brake to slow and stop the wheel, and observe the balancer display screen to determine the size and location of wheel weight(s) required on the wheel rim.
7. Install the correct wheel weights in the locations indicated on the balancer screen. Use that hammer head on the wheel weight pliers to install the weights.
8. Spin the wheel again on the balancer; the balancer screen should indicate a balanced tire and wheel assembly.
9. Remove the wheel and tire assembly from the balancer, and install this assembly on the vehicle using the alignment marks placed on the tire wheel and hub stud prior to wheel removal.
10. Tighten the wheel lug nuts to the specified torque in the proper sequence.

SERVICE TIP:
Some wheel rims such as magnesium rims require the use of stick-on wheel weights.

SERVICE TIP:
Wheel weights are rated in ounces or grams.

On some pre electronic wheel balancers, a static balance procedure is performed by allowing the wheel and tire assembly to rotate by gravity on the wheel balancer. A **heavy spot** in the tire rotates the tire and wheel assembly until the heavy spot is at the bottom. The necessary wheel balance weights were then added at the top of the wheel 180° from the heavy spot. Equal wheel weights were installed on each side of the wheel (Figure 3-28). If a wheel has dynamic unbalance, and the heavy spot is on the outside

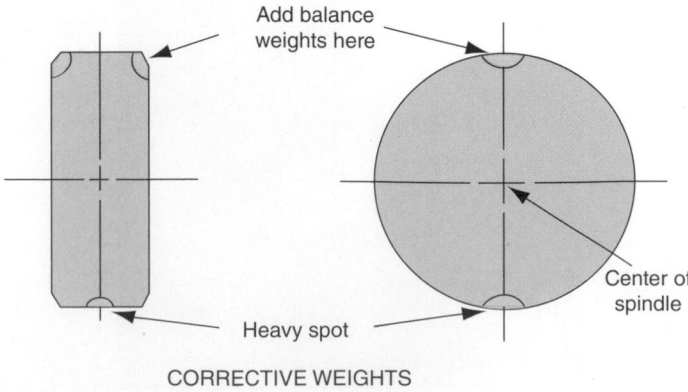

Copyright © 2015 Cengage Learning®

FIGURE 3-28 **Static wheel balance procedure.**

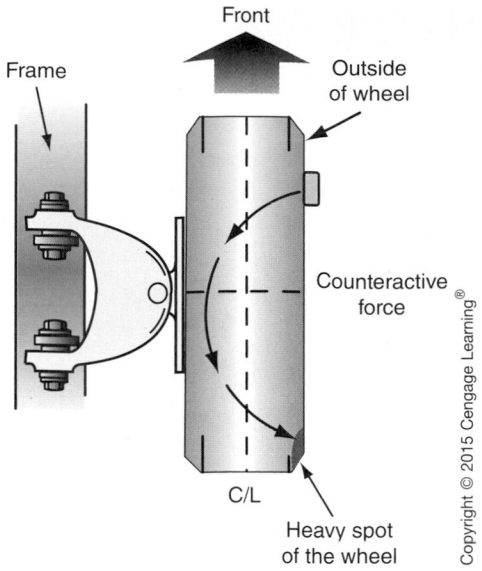

FIGURE 3-29 **Dynamic wheel balance procedure with heavy spot on the outside edge of the tread.**

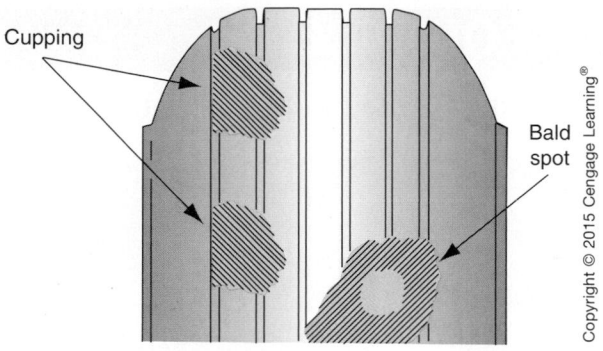

FIGURE 3-30 **Cupped tire tread wear with bald spots on the tread caused by improper static or dynamic wheel balance.**

edge of the tire tread, the correct size of wheel weight is installed on the rim 180° from the heavy spot (Figure 3-29). Improper static or dynamic wheel balance causes cupped tire tread wear and bald spots on the tread (Figure 3-30). Incorrect static wheel balance results in wheel tramp, and improper dynamic wheel balance causes wheel shimmy. When using an electronic wheel balancer, the technician does not have to determine the wheel weight size or location because the balancer provides this information to the technician.

The Center for Environmental Health (CEH) in California together with Chrysler Group and three major wheel weight manufacturers reached an agreement to end the use of lead wheel weights in California by December 31, 2009. It is estimated that lead wheel weights cause 500,000 lbs of lead to be placed in the environment each year from wheel weights breaking off vehicle wheels. Lead wheel weights will be replaced with a metal or material that is more environmentally friendly. Be sure you are installing the type of wheel weights required by the legislation in your state.

Photo Sequence 4 illustrates a typical off-car wheel balance procedure.

Electronic Wheel Balancers with Lateral Force Measurement (LFM) and Radial Force Variation Capabilities

Improper wheel balance is only one cause of wheel vibration. Wheel vibration may also be caused by tire and/or wheel runout or stiffness variation in the tire sidewalls. Some wheel balancers now have the capability to measure wheel and tire runout and stiffness or force variation in the tire sidewalls. Tire and wheel rim runout may be measured with a dial indicator, but this operation is very time consuming. Many tires marketed today have a paint dot or mark on the sidewall that indicates either the high side or low side of the force variation in the tire. This paint dot is aligned with the valve stem when the tire is mounted on the wheel rim. Using this system, the technician is assuming the paint dot is placed at the high point of radial force, and the valve stem hole is at the low point of runout on the rim. However, this is not always true, and a tire mounted with this tire position in relation to the wheel rim may cause vibration even though it is properly balanced. Other tires may have a slight conicity problem that causes steering pull on the vehicle.

PHOTO SEQUENCE 4

Typical Off-Car Wheel Balancing Procedure

P4-1 **Complete all the preliminary wheel balance checks.**

Copyright © 2015 Cengage Learning®

P4-2 **Follow the wheel balancer manufacturer's recommended procedure to mount the wheel and tire assembly securely on the electronic wheel balancer. Some wheel balancers have a centering check in the programming procedure. If the tire and wheel are not mounted properly, tire and wheel balance will be inaccurate.**

Copyright © 2015 Cengage Learning®

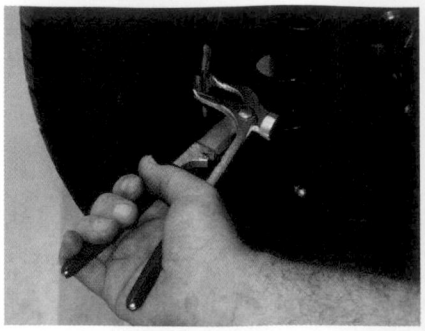

P4-3 **Use a pair of wheel weight pliers to remove all the wheel weights from the wheel rim.**

Copyright © 2015 Cengage Learning®

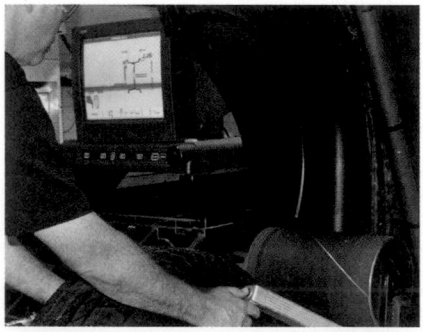

P4-4 **Enter the wheel diameter, width, and offset on the wheel balancer screen.**

Copyright © 2015 Cengage Learning®

P4-5 **Lower the safety hood over the wheel and tire on the wheel balancer.**

Copyright © 2015 Cengage Learning®

P4-6 **Activate the wheel balancer control to spin the wheel and tire assembly.**

Copyright © 2015 Cengage Learning®

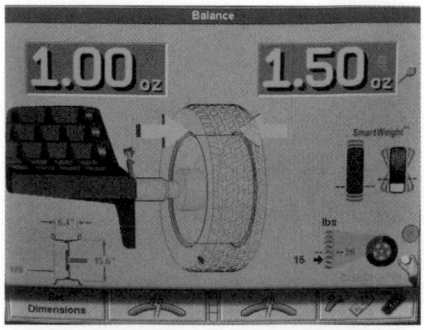

P4-7 **Stop the wheel and tire assembly. Observe the wheel balancer display screen to determine the correct size and location of the required wheel weights.**

Copyright © 2015 Cengage Learning®

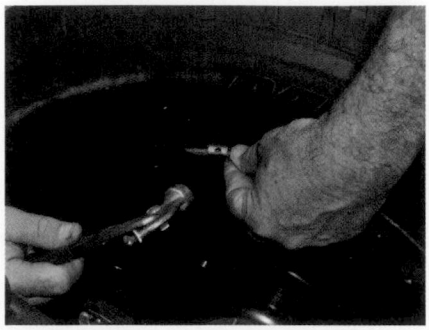

P4-8 **Install the correct size wheel weights in the location(s) indicated on the wheel balancer screen.**

Copyright © 2015 Cengage Learning®

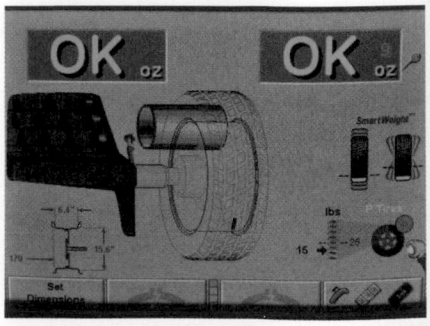

P4-9 **Activate the wheel balancer control to spin the wheel again, and observe the wheel balancer display to confirm that it indicates a balanced wheel and tire assembly.**

Copyright © 2015 Cengage Learning®

P4-10 **Stop the wheel, lift the safety hood, and remove the wheel and tire assembly from the balancer.**

Copyright © 2015 Cengage Learning®

An electronic wheel balancer with radial force variation capabilities has a roller that is forced against the tire tread with considerable force during the balance procedure. This type of balancer provides wheel balance following the usual procedure. The roller forced against the tire allows the balancer to measure tire conicity that results in lateral force and steering pull (Figure 3-31). As each tire and wheel assembly is balanced, the balancer display indicates a number for each assembly (Figure 3-32). The balancer display also indicates the position on the vehicle where each numbered tire and wheel assembly should be installed to provide the least amount of steering pull and the best straight-ahead steering stability (Figure 3-33). The wheel balancer with LFM capabilities provides a much faster and more accurate method for diagnosing and correcting steering pull caused by tire conicity compared to the previous method of rotating tires to try and correct this problem.

The balancer also measures lateral and radial tire and wheel runout, and indicates if the runout is rim or tire related. This type of balancer also measures force variation in the tire (Figure 3-34). The balancer display indicates the amount of tire and tire and wheel assembly road force. The display also indicates the necessary tire and wheel service to correct the excessive road force. Using the balancer display, a chalk mark may be placed on the tire at the location of the high force variation, and a second chalk mark on the rim 180° away from the first chalk mark (Figure 3-35). The tire and wheel assembly may be installed on a tire changer, and the rim may be rotated in relation to the tire to correct the force variation.

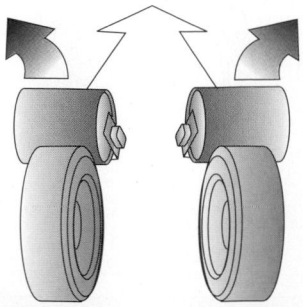

Copyright © 2015 Cengage Learning®

FIGURE 3-31 **Tire conicity sensed by the roller on the balancer causes lateral force and steering pull.**

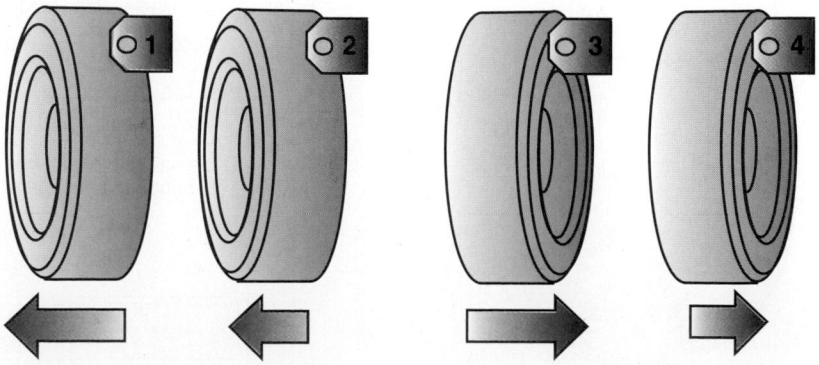

Copyright © 2015 Cengage Learning®

FIGURE 3-32 **The balancer assigns a number to each tire and wheel assembly.**

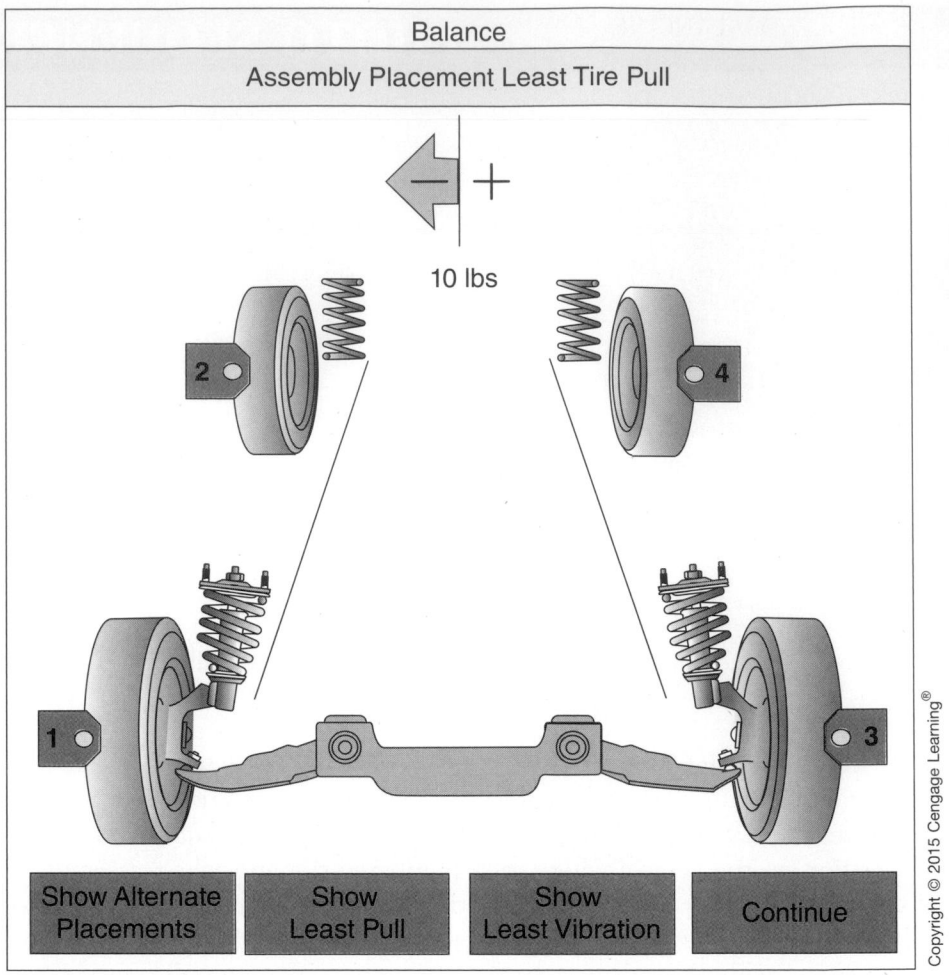

FIGURE 3-33 **The balancer display indicates the location on the vehicle where each numbered tire should be installed.**

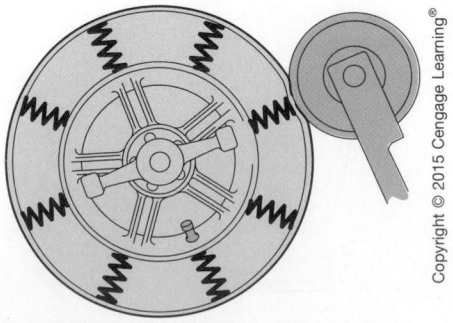

FIGURE 3-34 **The roller on the balancer senses force variation in the tire.**

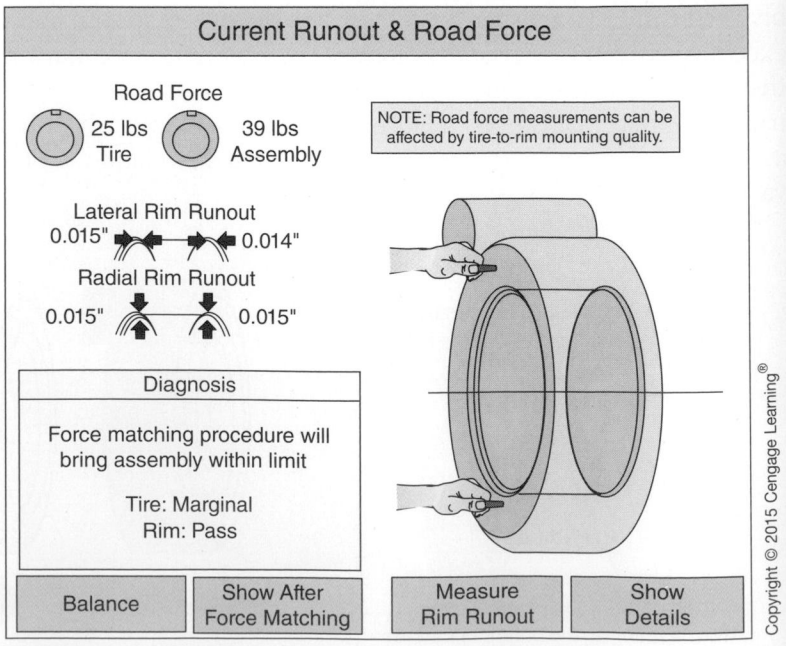

FIGURE 3-35 **The balancer display indicates the necessary correction for excessive force variation.**

On-Car Wheel Balancing

Though only performed by a few select shops anymore, on-car wheel balancing was once the preferred method of wheel balancing. On-car wheel balancers contain a drum driven by an electric motor. This drum is positioned against the tire on the vehicle, which allows the electric motor to rotate the wheel. Many on-car balancers have a strobe light with a meter and an electronic vibration sensor. Off-car wheel balancing and on-car wheel balancing are a complementary combination for fine-tuning wheel balance. For example, a wheel vibration problem may still exist after an off-car balance procedure. If this problem occurs, the on-car balancer may be used to correct the problem. The on-car balance procedure corrects imbalance problems in all rotating components, including brake drums or rotors.

SPECIAL TOOL
On-car wheel balancer

When a front wheel on a rear-wheel-drive car is balanced, use this procedure:

1. Perform the preliminary checks listed previously in this chapter.
2. Raise the wheel being balanced 5 in. (12 cm) off the floor and be sure the chassis is supported on safety stands so the wheel drops downward.
3. Install the electronic vibration sensor between the lower control arm and the floor.
4. Chalk mark a reference mark on the outer sidewall of the tire.
5. Spin the wheel just fast enough to produce vibration on the front bumper.
6. When vibration causes strobe light flashing, move the balancer drum away from the tire and allow the tire to spin freely.
7. Shine the strobe light around the tire sidewall and note the chalk-mark position.
8. Note the pointer position on the meter.
9. Use the balancer's brake plate to slow and stop the wheel.
10. Rotate the wheel until the chalk mark is in the exact position where it appeared under the strobe light. The heavy spot is now at the bottom of the wheel and the balancing weight should be attached 180° from the heavy spot. Install the amount of weight indicated on the meter.
11. Spin the wheel again (Figure 3-36). If the wheel balance is satisfactory, the meter pointer will read in the balanced position. When the pointer does not indicate a balanced wheel, shine the strobe light on the tire sidewall. If the installed wheel weight is at the 12 o'clock position, additional weight is required, whereas the 6 o'clock weight position indicates excessive weight. A 3 o'clock or 9 o'clock weight position may be corrected by moving the weight 1 in. (2.5 cm) toward the 12 o'clock position.

Additional on-car wheel balancer precautions include:

1. Do not spin the wheel at excessive speeds.
2. Do not spin the front wheels on a front-wheel-drive vehicle with the floor jack under the chassis and the suspension dropped downward. Under this condition, severe angles

Copyright © 2015 Cengage Learning®

FIGURE 3-36 **On-car wheel balancer.**

exist in the front drive axle joints, and these joints may be damaged if the wheels are rotated with the balancer. Place the floor jack under the lower control arm to raise the wheel.

CAUTION:
On a rear-wheel-drive vehicle when a rear wheel is rotated with an on-car balancer, do not allow the speed indicated on the speedometer to exceed 35 mph (56 km/h), because the wheel speed is much faster than the speed indicated on the speedometer.

Rear Wheel Balancing

If an on-car balancer is used on the rear wheels of a rear-wheel-drive car, the technician must determine if the vehicle has a conventional or a limited slip differential. With the transmission in park, or in gear with a manual transmission, rotate one rear wheel by hand. If the vehicle has a limited slip differential, the rear wheels will not rotate, whereas a free-turning wheel indicates a conventional differential.

When the vehicle has a conventional differential, use the same front wheel balance procedure on the rear wheels. Raise only the wheel to be balanced from the floor, and do not allow the speed indicated on the speedometer to exceed 35 mph (56 km/h). When the other rear tire is resting on the floor, the speed of the wheel being balanced is 70 mph (112 km/h).

If the vehicle has a limited slip differential, proceed as follows for on-car rear wheel balancing:

Classroom Manual
Chapter 3, page 85

1. Raise the vehicle and place safety stands under the frame to support the vehicle weight.
2. Lift the rear axle housing with a floor jack to reduce the universal joint angles, but do not take the vehicle weight off the safety stands.
3. Remove the wheel that is not being balanced. Install and torque the wheel nuts on the brake drum to hold the drum in place.
4. Balance the opposite wheel using the same procedure as for the front wheels.
5. Leave the balanced wheel on the vehicle and install and balance the other wheel. Be sure to torque the wheel nuts properly. Do not allow rear wheel speed to exceed 55 mph (90 km/h) on the speedometer.

Noise, Vibration and Harshness Diagnosis

Classroom Manual
Chapter 3, page 91

The technician must have a logical sequence for vibration diagnosis. This problem may be diagnosed using the same basic diagnostic procedure that is used for diagnosing defects in other parts of the vehicle.

A basic diagnostic procedure follows:

1. Question the customer regarding the vibration problem, and obtain as much information as possible from this source.
2. Be sure the complaint is identified. Road test the vehicle if necessary.
3. Think of the possible causes of the vibration.
4. Perform the necessary inspection and test procedures to locate the cause of the vibration.
5. Complete the necessary repairs to correct the problem(s) identified in the inspection and test procedures.
6. Be sure the original customer complaint is eliminated. If necessary, road test the vehicle.

When questioning the customer regarding the vibration problem, the technician should obtain the answers to these questions:

1. At what vehicle speed is the vibration most noticeable?
2. Is the vibration felt in the vehicle, and in what area of the vehicle is the vibration present? (For example, left front, right rear, under the vehicle floor.)
3. Can the customer hear the vibration?

4. Does engine load, such as hard acceleration, affect the vibration?
5. Does vehicle load, such as the number of passengers or weight in the trunk, affect the vibration?
6. Does the vibration occur in all transmission gear ranges?
7. When did the customer first notice the vibration? Did the vibration occur after some other service work was performed on the vehicle?

Basic Road Test

The purpose of a road test is to verify the vibration complaint and determine any operating conditions under which the vibration changes or is eliminated refer to Table 3-1. A road test should also determine if the vibration is related to engine speed or vehicle road speed. During any road test for vibration diagnosis, if the vehicle does not have an in-dash tachometer, connect a tachometer to the ignition system, and install the leads through one of the windows so the tachometer may be placed in view of the driver. Connect the power cord from an **electronic vibration analyzer (EVA)** to the cigarette lighter socket, and attach the EVA vibration sensor to the area where the suspected vibration is located (Figure 3-37).

The **electronic vibration analyzer (EVA)** is a tester that helps to locate the source of a vibration.

When it is determined that the vibration is related to engine speed or vehicle road speed, the technician has to perform special test procedures to locate the cause of the vibration in one of these areas:

1. Engine, engine or transmission mount, clutch disc and pressure plate (manual transmission), flexplate and torque converter (automatic transmission)
2. Propeller shaft, transmission output shaft, differential pinion flange (rear-wheel-drive vehicles), front drive axles (front-wheel-drive vehicles)
3. Tires, wheels, hubs, and brake rotors

Special Road Tests

Slow Acceleration Test. All of the special road tests help the technician to locate the cause of the vibration. A tire and wheel inspection should be completed before any road test. Defects discovered during the tire inspection may explain the cause of the vibration that occurs during a road test.

TABLE 3-1 **TIRE AND WHEEL DIAGNOSIS**

Problem	Symptoms	Possible Causes
Tire thump	Thumping noise while driving	Cupped tire treads Excessive radial tire or wheel runout Tire defects Improper wheel balance Variation in tire stiffness
Chassis waddling	Lateral rear chassis oscillations while driving	Excessive rear tire or wheel lateral runout
Steering pull	Steering pulls to one side while driving straight ahead	Tire conicity Unmatched tires side-to-side Improper wheel alignment angles
Wheel vibration	A vibration that occurs at certain speeds and may be felt on the steering wheel or throughout the chassis	Improper wheel balance Excessive lateral tire runout Variation in tire stiffness Tire defects
Excessive wear on tire treads	Cupped tire treads Wear on tread center Equal wear on tread edges	Improper wheel balance Overinflation Underinflation

Copyright © 2015 Cengage Learning®

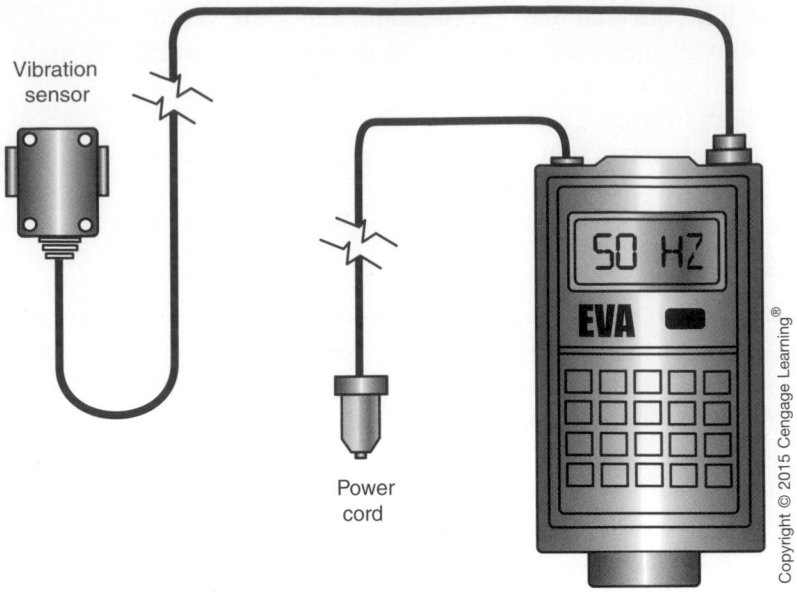

FIGURE 3-37 **Electronic vibration analyzer (EVA).**

Inspect the tires and wheel for these defects:

1. Check the inflation pressure in each tire.
2. Check for unusual tire wear such as cupping, excessive tread wear, and flat or bald spots. Inspect the tire sidewalls and tread area for bulges and ply separation. These tire defects may cause vibration, slapping noise, and tire growl or howl.
3. Inspect the rims for bends, damage, and excessive corrosion.

Perform the **slow acceleration test** on a smooth level road. During this test, accelerate the vehicle from a stop to legal highway speed. Check for any vibration that matches the customer's description. If any vibration is present, record the frequency, vehicle speed and engine speed when the vibration occurred.

Neutral Coast-Down Test. Perform the **neutral coast-down test** on a smooth, level road. Accelerate the vehicle to a speed slightly higher than the speed at which the vibration occurs. Shift the transmission into neutral and allow the vehicle to coast down through the speed range at which the vibration occurs. If the vibration still occurs in neutral, the vibration is vehicle-speed sensitive, and the engine, propeller shaft, flexplate, and torque converter have been eliminated as causes of the vibration. When the vibration occurs with the transmission in neutral, the vibration diagnosis should concentrate on the drive axle assemblies and the tire and wheel assemblies. Off-car or on-car balance procedures are performed to correct balance problems in brake drums or rotors and tires and wheels.

Downshift Test. To complete the **downshift test**, drive the vehicle on a smooth, level road at the speed when the vibration is present. Note the engine rpm. Decelerate and reduce the vehicle speed until the vibration is no longer present. Manually downshift the transmission into the next lowest gear, and accelerate the vehicle until the engine is running at the previously recorded rpm. If the vibration returns at the same engine rpm, the engine, propeller shaft, flexplate and torque converter (automatic transmission) or clutch disc and pressure plate (manual transmission) are the most likely causes of the vibration.

Neutral Run-Up Test. Perform the **neutral run-up test** when the customer complains about vibration at idle speed or as a follow-up to the downshift test. To perform the neutral run-up test, slowly increase the engine rpm from idle. Note the engine rpm and frequency at which vibration occurs. If vibration occurs, the cause of the vibration is related to the engine, flexplate and torque converter (automatic transmission) or clutch disc and pressure plate (manual transmission).

Brake Torque Test. The **brake torque test** may be used to identify engine vibrations that were not present during the downshift and neutral run-up tests. This test may also be used when diagnosing vibrations that are sensitive to engine load.

To perform the brake torque test, follow this procedure:

1. Firmly apply the parking brake.
2. Block the front wheels.
3. Firmly apply the brake pedal.
4. Shift transmission into drive.
5. Slowly increase the engine speed to 1,200 rpm and check for vibrations.
6. Note the engine rpm when vibration occurred.

CAUTION:
Do not increase the engine rpm above 1,200, and limit this test to 10 seconds to protect the transmission from overheating.

If engine vibrations occur during the brake torque test, but they did not occur during the downshift and neutral run-up tests, the engine and/or transmission mounts are a prime suspect for the cause of the vibration.

Steering Input Test. The **steering input test** may be performed to diagnose vibrations in suspension components. To perform this test, drive the vehicle through sweeping turns at the speed when the vibration occurs. Drive the vehicle when turning in one direction and then perform some turns in the opposite direction. If the vibration changes during these turns, the vibration is likely caused by defective wheel bearings, tire tread, or front drive axle shafts and joints.

Standing Acceleration Test. The **standing acceleration test** is used to diagnose a vibration that may be called a launch shudder. To perform the standing acceleration test bring the vehicle to a complete stop. Place the transmission in drive and remove your foot from the brake pedal. Accelerate the vehicle to 30 to 40 mph (48 to 64 km/h), and check for vibrations that match the customer's complaint. If a shudder-type vibration occurs during the test, the cause of this vibration could be incorrect curb riding height, defective front drive axle joints, defective engine or transmission mounts, or faulty exhaust hangers or mounts.

Vibration Diagnosis with the Electronic Vibration Analyzer (EVA)

The EVA is helpful in locating the source of vibration problems.

Follow these steps to connect the EVA:

1. Be sure the correct software cartridge is securely installed in the bottom of the EVA.
2. Connect the vibration sensor cord to input A or B on the EVA. Position the release button facing downward on the lower side of this connector. Push the connector into the input port until the connector clicks and locks in place.
3. Follow these steps to perform the sensor calibration procedure:
   - **a.** Lay the sensor on a flat stationary surface, and be sure the side of the sensor marked UP is facing upward.
   - **b.** Plug the EVA power cord into a 12 V power supply, such as the cigarette lighter.
   - **c.** Press the "up" arrow key.
   - **d.** Press number 2 on the key pad three times until the word BURNING appears, followed by a request to turn the sensor over.
   - **e.** Turn the sensor over and press any key on the keypad to begin calibration. The calibration procedure takes approximately 20 seconds, and the display returns to the active mode when calibration is completed.

CAUTION:
When installing the sensor connector in the EVA input terminal, do not twist the sensor connector. Do not remove this connector when the EVA is in use. This action may damage the EVA.

4. Two vibration sensors can be connected to ports A and B on the EVA at the same time, and the A/B button on the tester can be used to switch from one sensor to the other. Place the sensor(s) on any component that is suspected of being a vibration source. Each sensor has a 20-ft. cord, so it may be placed anywhere on the vehicle. Each sensor contains a magnet for attachment to metal surfaces. For attachment to nonmagnetic surfaces, use putty or a hook and loop fastener. Vibrations are usually felt in an up-and-down direction. Because the vibration sensor is directionally sensitive, this sensor must be mounted on the lower part of the suspected component with the marking on the sensor facing upward (Figure 3-38). When repeating a vibration test on the same component, always be sure the vibration sensor is mounted in the exact same location.

Hertz or rpm are displayed on the left side of the EVA display. Either of these values may be selected by pressing the "RPM/Hertz" button on the tester. The selected value is displayed at the top of the left column on the display. A bar graph of the suspected source of the vibration is shown in the center of the display. When the AUTO mode is selected, the suspected vibration source is shown in the display. If the AUTO mode is not in use, the bar graph is indicated in the center of the display (Figure 3-39). The amplitude of each frequency indicated on the left of the screen is displayed on the right side of the screen. These amplitudes are displayed in Gs of acceleration force. Higher amplitudes indicate stronger vibration forces. The EVA displays up to three of the most dominant input frequencies in descending order of amplitude strength.

The technician may press the averaging (AVG) button to select this mode. In the AVG mode, the EVA displays the average of multiple vibration readings taken over a period of time. The EVA is usually operated in the AVG mode to minimize distractions caused by a sudden vibration that is not related to the main vibration source. These distractions could be caused by the vehicle wheels striking road irregularities. When the EVA is operating in the AVG mode and the AUTO Mode, "A" is displayed along the top of the screen. The AUTO mode is designed to be used with the vibration diagnosis tables in the appropriate service manual. If the EVA is operating in the AVG mode and the MANUAL mode, AVG is shown at the top of the screen. When the EVA is operating in the non-AVG mode and the AUTO mode, "I" is indicated at the top of the display. The top row on the display also indicates "A" or "B," representing the vibration sensor input port.

When the FREEZE button is pressed, the EVA displays FRZ on the top line of the display. In this mode the EVA will store data for a short time period when the RECORD button is pressed during various vibration tests. The technician may press the PLAY BACK button to display the recorded data. The technician presses the EXIT button to return the EVA to normal operation.

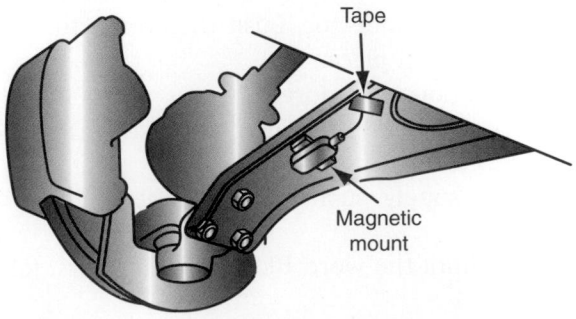

NOTE:
Sensor is directionally sensitive, marking on sensor must be facing upward.

Copyright © 2015 Cengage Learning®

FIGURE 3-38 **Vibration sensor mounting.**

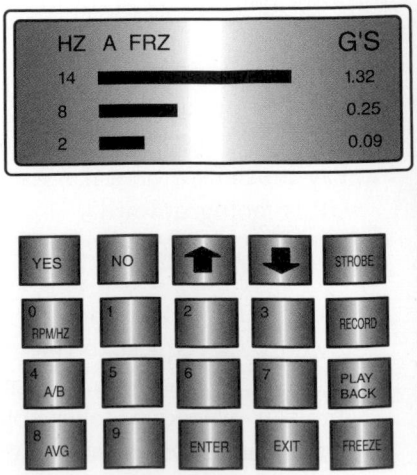

Copyright © 2015 Cengage Learning®

FIGURE 3-39 **Electronic vibration analyzer (EVA) display and keypad.**

The EVA has a strobe light trigger wire that may be used with an inductive pickup timing light to balance rotating components such as propeller shafts.

Calculating Vibration Frequencies of Various Components

Calculating Tire/Wheel Frequency. The first step in calculating tire/wheel frequency is to divide the vehicle speed by five-mile-per-hour increments. There are 12 increments of 5 miles per hour, if the vehicle speed is 60 mph. The next step is to use the chart in Figure 3-40 to determine how many times a wheel and tire rotate per second at 5 mph. As indicated in this chart, if the tire size is 195/70/14, the tires revolve at 1.17 times per second for every 5 mph of vehicle speed. To calculate the tire/wheel frequency at 60 mph, multiply the number of five-mile-per-hour increments, 12, by the number of times the tire and wheel revolve in one second per 5 mph increment, 1.17. Therefore we have the calculation $12 \times 1.17 = 14.04$, and this size of tire and wheel on a vehicle traveling at 60 mph produces a vibration of approximately 14 hertz (Hz). With a vibration sensor attached to the lower control arm, if the EVA displays a vibration of 14 Hz, the tire and wheel mounted on the knuckle attached to the control arm have a vibration problem. This vibration problem could also be caused by a rotor or hub.

With a vibration of 14 Hz, a second-order vibration is 28 Hz, a third-order vibration is 42 Hz, and a fourth-order vibration is 56 Hz. On a front-wheel-drive vehicle, a worn constant velocity drive axle joint may cause a vibration in the fifth order of the tire/wheel vibration, and a defective tripod drive axle joint may result in a vibration of the third order of the tire/wheel vibration. On a rear-wheel-drive-vehicle if the tire/wheel frequency is 14 and the rear axle ratio is 3.42:1, the frequency of the drive shaft is $14 \times 3.42 = 47.88$.

Tire Size	Tire Frequency at 5 mph	Tire Size	Tire Frequency at 5 mph
P145/80/13	1.31	215/70/15	1.08
155/80/13	1.27	215/75/15	1.05
175/65/14	1.26	215/75/16	1.01
185/65/14	1.23	215/85/16	0.95
185/70/14	1.20	220/75/15	1.03
185/75/14	1.16	225/45/17	1.16
195/60/15	1.20	225/55/17	1.08
195/65/15	1.16	225/60/16	1.09
195/70/14	1.17	225/70/15	1.06
195/75/14	1.13	225/70/16	1.02
195/75/15	1.09	225/75/15	1.02
205/55/15	1.21	225/75/16	0.99
205/55/16	1.16	235/70/15	1.04
205/60/14	1.22	235/75/15	1.00
205/60/15	1.17	235/75/16	0.97
205/65/15	1.14	245/70/15	1.02
205/70/14	1.14	245/70/16	0.98
205/70/15	1.10	245/75/16	0.95
205/75/14	1.11	275/40/17	1.13
205/75/15	1.07	295/40/20	0.99
215/50/17	1.14	335/35/17	1.10
215/55/16	1.14	30X9.5/15	0.98
215/65/15	1.11	31X10.5/15	0.95
215/65/16	1.07		

Copyright © 2015 Cengage Learning®

FIGURE 3-40 **Tire/wheel frequencies.**

Calculating Engine and Engine Firing Frequency. To calculate the engine frequency, divide the engine rpm by 60 seconds per minute. For example, if the engine rpm is 3,000, the engine frequency is 3,000 ÷ 60 = 50 revolutions per second, or 50 Hz. To determine the engine firing frequency, divide the number of engine cylinders by two, and multiply the answer by the engine frequency. Let us assume we are working on a V6 engine running at 3,000 rpm. Divide the number of cylinders by two, and so 6 ÷ 2 = 3. Then multiply this answer by the engine frequency of 50 to calculate the firing frequency. Therefore, 3 × 50 = 150 is the engine firing frequency. Because a cylinder in a four-cycle engine expels exhaust only once every two revolutions, a single cylinder frequency is one-half the engine frequency. Therefore, in our example if the engine frequency is 50, a single cylinder frequency is 25. This is the frequency of exhaust pulses in a single cylinder and also the frequency of the camshaft that turns at one-half the engine speed.

Calculating Accessory Drive Frequencies. To calculate the frequency of the belt-driven accessories, divide the diameter of the crankshaft pulley by the diameter of the pulley on the belt-driven accessory and multiply the answer by the engine frequency. Let us assume we are still working on the engine in our previous example, which is a V6 engine with a frequency of 50 running at 3,000 rpm. With a vibration sensor mounted on the engine the EVA displays a vibration of 160 Hz. The diameter of the crankshaft pulley is 8 inches, the diameter of the power steering pulley is 5 inches, the diameter of the A/C compressor pulley is 4.5 inches, and the diameter of the alternator pulley is 2.5 inches.

Therefore, we have the following calculations:

Power steering pump frequency: 8 ÷ 5 = 1.6 × 50 = 80
A/C compressor frequency: 8 ÷ 4.5 = 1.7 × 50 = 85
Alternator frequency: 8 ÷ 2.5 = 3.2 × 50 = 160

When we compare the vibration frequency displayed on the EVA to the belt-driven accessory frequencies, the alternator frequency matches the frequency on the EVA, indicating the alternator is the source of the vibration.

TERMS TO KNOW

Antitheft locking wheel covers
Antitheft wheel nuts
Brake torque test
Downshift test
Dynamic wheel balance
Electronic vibration analyzer (EVA)
Heavy spot
Lateral tire runout
Neutral coast-down test
Neutral run-up test
Ply separation
Radial runout
Slow acceleration test
Standing acceleration test
Static balance
Steering input test
Steering pull
Tire conicity
Tire thump
Tire vibration
Tread wear indicators

CASE STUDY

A customer complains about severe front end waddling on a 2013 Cadillac CTS equipped with steel-belted radial tires, which have 11/32" of tread depth. The technician questions the customer further regarding this problem and learns that the problem occurs only after the car has sat overnight, and it lasts only for approximately six blocks when the customer starts driving the car.

The technician test drove the vehicle but was unable to duplicate the customer complaint. The technician carefully checks all steering and suspension components, wheel bearing adjustment, and tire runout. No problems are found during these checks, and the technician asks the customer to leave the car in the shop overnight. The technician installs the front tires on the rear of the car and rotates the rear tires to the front, and the car is left in the shop overnight. A road test the next morning indicates rear end waddling, which proves the problem must be caused by a tire problem such as a radial cord ply imperfection, since moving the tires original position was the only thing changed on the car. The technician removed the two rear tires and checked them on a Road Force tire balancer once they cooled down again. The tire that had been moved to the left rear position was found to have excessive road force variation, which was beyond the limits of match mounting. The technician replaced the defective tire and checks the balance of the other three tires. The customer reports later that the problem has been completely corrected.

ASE-STYLE REVIEW QUESTIONS

1. While discussing a tire thumping problem:
   Technician A says this problem may be caused by cupped tire treads.
   Technician B says a heavy spot in the tire may cause this complaint.
   Who is correct?
   A. A only
   B. B only
   C. Both A and B
   D. Neither A nor B

2. While discussing a vehicle that pulls to one side:
   Technician A says that excessive radial runout on the right front tire may cause this problem.
   Technician B says that tire conicity may be the cause of this complaint.
   Who is correct?
   A. A only
   B. B only
   C. Both A and B
   D. Neither A nor B

3. While discussing tire noise:
   Technician A says if it is tire noise the noise will vary with road surface conditions.
   Technician B says that if it is tire noise the noise will remain constant when the vehicle is accelerated and decelerated.
   Who is correct?
   A. A only
   B. B only
   C. Both A and B
   D. Neither A nor B

4. While discussing tire wear:
   Technician A says that static imbalance causes feathered tread wear.
   Technician B says that dynamic imbalance causes cupped wear and bald spots on the tire tread.
   Who is correct?
   A. A only
   B. B only
   C. Both A and B
   D. Neither A nor B

5. While discussing on-car wheel balancing:
   Technician A says that during an on-car wheel balancing procedure on a rear wheel of a rear-wheel-drive car, the speed indicated on the speedometer should not exceed 65 mph (105 km/h).
   Technician B says that the speed indicated on the speedometer on this car must not exceed 35 mph (56 km/h).
   Who is correct?
   A. A only
   B. B only
   C. Both A and B
   D. Neither A nor B

6. A front tire has excessive wear on both edges of the tire tread. The most likely cause of this problem is:
   A. Overinflation.
   B. Underinflation.
   C. Improper static balance.
   D. Improper dynamic balance.

7. When measuring radial tire and wheel runout, the maximum runout on most automotive tire and wheel assemblies should be:
   A. 0.015 in. (0.038 mm).
   B. 0.025 in. (0.635 mm).
   C. 0.045 in. (1.143 mm).
   D. 0.070 in. (1.77 mm).

8. When measuring lateral wheel runout with the tire demounted from the wheel, the maximum runout on most automotive wheels is:
   A. 0.020 in. (0.508 mm).
   B. 0.030 in. (0.762 mm).
   C. 0.045 in. (1.143 mm).
   D. 0.055 in. (1.397 mm).

9. All of these statements about improper wheel balance are true EXCEPT:
   A. Dynamic imbalance may cause wheel shimmy.
   B. Dynamic imbalance may cause steering pull in either direction.
   C. Static imbalance causes wheel tramp.
   D. Static imbalance causes rapid wear on suspension components.

10. When diagnosing wheel balance problems:
    A. Balls of rubber inside the tire have no effect on wheel balance.
    B. Loose wheel bearing adjustment may simulate improper static wheel balance.
    C. Improper dynamic wheel balance may be caused by a heavy spot in the center of the tire tread.
    D. After a tire patch is installed, the tire and wheel may be improperly balanced.

ASE CHALLENGE QUESTIONS

1. The owner of a rear-wheel-drive car with aftermarket alloy wheels says he has replaced the wheel bearings three times in the past two years. He wants to know why the bearings fail.

   Technician A says excessive radial runout of the wheel may be the cause of the problem.

   Technician B says excessive offset of the wheel may be the cause of the problem.

   Who is correct?

   A. A only
   B. B only
   C. Both A and B
   D. Neither A nor B

2. A customer returns with recently purchased radial tires saying that the rear of the car feels like "it's riding on Jello™." All of the following could cause this problem EXCEPT:

   A. The radial belt of a rear tire is not straight.
   B. The wheel is improperly mounted.
   C. Excessive lateral wheel runout.
   D. Excessive radial tire runout.

3. A customer says the new tires he just purchased vibrate. The installer says he balanced the wheels and tires with a conventional electronic balancer before placing them on the car.

   Technician A says one of the tires may have a conicity problem.

   Technician B says one of the tires may have a force variation problem.

   Who is correct?

   A. A only
   B. B only
   C. Both A and B
   D. Neither A nor B

4. After a set of radial tires is rotated, the customer returns saying he feels vibration and steering shimmy. To correct this problem, you should:

   A. Measure the lateral runout on each tire.
   B. Return the tires and wheels to their original positions.
   C. Check the wheel bearings.
   D. Balance the tires and wheels with an on-car balancer.

5. A customer says there is a "thumping" vibration in the wheels and an inspection of the tires shows the two front wheels have flat spots on the tire treads.

   Technician A says heavy spots in the tires may have caused this condition.

   Technician B says locking the wheels and skidding on pavement caused this condition.

   Who is correct?

   A. A only
   B. B only
   C. Both A and B
   D. Neither A nor B

JOB SHEET 8

Name _____ **Date** _____

Tire Dismounting and Mounting

Upon completion of this job sheet, you should be able to demount and mount tires with or without a tire pressure monitor sensor.

NATEF AST/MAST Task Correlation

Suspension and Steering. Wheels and Tires Diagnosis and Repair,

Task #F-6. Dismount, inspect, and remount tire on wheel; balance wheel and tire assembly (static and dynamic). (P-1)

Task #F-7. Dismount, inspect, and remount tire on wheel equipped with tire pressure monitoring system sensor. (P-2)

Task #F-11. Demonstrate knowledge of steps required to remove and replace sensors in a tire pressure monitoring system. (P-1)

Tools and Materials

Tire changer
Tire and wheel assembly

Procedure

1. Remove the valve core to release all the air pressure from the tire. Chalk mark the tire at the valve stem opening in the wheel so the tire may be reinstalled in the same position to maintain proper wheel balance.
   Is all the air pressure released from the tire? ☐ Yes ☐ No
   Is the tire chalk marked at the valve stem location in the wheel?
   Instructor check _____

SERVICE TIP: The following is a generic tire demounting and mounting procedure.

2. Guide the operating lever on the tire changer, at 90° to valve stem to avoid damage to tire pressure sensor, if equipped, to unseat both tire beads. Are both tire beads unseated? ☐ Yes ☐ No

3. Place the tire and wheel assembly properly on the tire changer. Is the tire and wheel assembly positioned properly on the tire changer? ☐ Yes ☐ No

WARNING: Do not proceed to dismount the tire unless the tire and wheel assembly is securely attached to the tire changer. This action may cause personal injury.

4. Press the pedal on the tire changer that clamps the wheel to the changer. Is the wheel clamped properly to the tire changer? ☐ Yes ☐ No

5. Lower the arm on the tire changer into position on the tire and wheel assembly. Is the tire changer arm positioned properly on the tire and wheel assembly?
   ☐ Yes ☐ No

Task Completed

SERVICE TIP: Different makes of tire changers have various ways of tire bead seating on the wheel. Some tire changers have an air inflation ring that is positioned on top of the wheel and tire. Other tire changers have air jets in each clamping jaw that hold the wheel on the tire changer. Air pressure from these jets helps force the tire beads against the wheel sealing surfaces.

6. Insert the tire iron properly between the upper tire bead and the wheel. Be sure the tire iron is properly positioned. Depress the tire changer pedal that causes the wheel to rotate. This rotation moves the top bead out over the wheel. Is the top tire bead above the wheel rim? ☐ Yes ☐ No

7. Lift the tire up and install the tire iron between the lower bead and the wheel. Be sure the tire iron is properly positioned. Depress the tire changer pedal that causes the wheel to rotate. This rotation moves the lower bead out over the wheel. Is the lower bead above the wheel rim? ☐ Yes ☐ No

8. Remove the tire from the wheel.

9. Remove, inspect, and reinstall tire pressure monitor sensor from wheel.

10. Use a wire brush to remove rust and debris from the tire bead sealing areas on the wheel rim. Are the tire bead sealing areas on the wheel properly cleaned?
☐ Yes ☐ No

11. Apply rubber lubricant to both tire beads. Are both tire beads lubricated?
☐ Yes ☐ No

12. Place the tire onto the wheel, and lower the tire changer arm into position. Depress the tire changer lever that causes the wheel to rotate. This action causes the lower tire bead over the wheel. Repeat this process to install the upper bead over the wheel. Is the tire properly installed on the wheel? ☐ Yes ☐ No

WARNING: Do not proceed to mount the tire unless the wheel is securely attached to the tire changer. This action may cause personal injury.

13. Rotate the tire on the wheel so the chalk mark placed on the tire in step 8 is aligned with the valve stem opening in the wheel. Is the chalk mark on the tire aligned with the valve stem position? ☐ Yes ☐ No

☐

14. Install and tighten the valve core in the valve stem.

SERVICE TIP: Some tire changers have an air hose and air pressure gauge designed into the tire changer. On other tire changers a shop air hose must be used to inflate the tire.

WARNING: When inflating a tire, do not stand or lean over the tire and wheel. If the tire bead comes off the rim or the rim cracks, personal injury may occur. When using a shop air hose to inflate the tire, install an extension on the air hose.

15. Supply air to the air ring or air jets on the tire changer to move the tire beads outward against the wheel. Are the tire beads moved outward against the wheel?
☐ Yes ☐ No

16. Connect an air supply hose to the tire valve stem. When using a shop air hose, connect an extension to the air hose. Inflate the tire to the specified pressure.
Specified tire pressure _____
Actual tire pressure _____

Instructor's Response _____

JOB SHEET 9

Name _____ **Date** _____

Tire and Wheel Runout Measurement

Upon completion of this job sheet, you should be able to measure radial and lateral tire and wheel runout.

NATEF AST/MAST Task Correlation

Suspension and Steering. Wheels and Tires Diagnosis and Repair,
Task #F-4. Measure wheel, tire, axle flange, and hub runout; determine necessary action. (P-2)

SERVICE TIP: The tire and wheel may be installed on an off-car wheel balancer to measure tire and wheel runout.

Tools and Materials

Dial indicator
Wheel balancer
Tire and wheel assembly
Wheel diameter _____
Manufacturer and type of tire _____

_____
Task Completed

Procedure

1. Mount the tire and wheel assembly properly on the wheel balancer and position a dial indicator against the center of the tire tread as the tire is rotated slowly to measure radial runout.
   Radial runout _____
   Specified radial runout _____
2. Mark the highest point of radial runout on the tire with chalk, and mark the valve stem position on the tire. ☐
3. If the radial tire runout is excessive, demount the tire, and check the runout of the wheel rim with a dial indicator positioned against the lip of the rim while the rim is rotated.
   Wheel radial runout _____
   Specified wheel radial runout _____
4. Use chalk to mark the highest point of radial runout on the wheel rim. ☐
5. If the highest point of wheel radial runout coincides with the chalk mark from the highest point of maximum tire radial runout, the tire may be rotated 180° on the wheel to reduce radial runout. Tires or wheels with excessive runout are usually replaced.
   Does the highest point of tire radial runout coincide with the highest point of wheel radial runout? ☐ Yes ☐ No

   Instructor's Response _____
   State the required action to correct excessive radial runout and the reason for this action.
   _____
   _____

Task Completed

6. Position a dial indicator located against the sidewall of the tire to measure lateral runout.

   Tire lateral runout _____

   Specified tire lateral runout _____

☐ 7. Chalk mark the tire and wheel at the highest point of lateral runout.

☐ 8. If the tire runout is excessive, the tire should be demounted from the wheel and the wheel lateral runout measured.

9. If the tire runout was excessive, measure the wheel lateral runout with a dial indicator positioned against the edge of the wheel as the wheel is rotated.

   Wheel lateral runout _____

   Specified wheel lateral runout _____

   State the required action to correct excessive wheel lateral runout and the reason for this action. _____

   _____

Instructor's Response _____

_____

_____

JOB SHEET 10

Name _____ **Date** _____

Off-Car Wheel Balancing

Upon completion of this job sheet, you should be able to perform off-car wheel balancing procedures.

NATEF AST/MAST Task Correlation

Suspension and Steering. Wheels and Tires Diagnosis and Repair,

Task #F-6. Dismount, inspect, and remount tire on wheel; balance wheel and tire assembly (static and dynamic). (P-1)

Tools and Materials

Electronic off-car wheel balancer
Wheel weights
Tire and wheel
Wheel diameter _____
Manufacturer and type of tire _____

Describe the Vehicle Being Worked On:

Year _____ Make _____ Model _____
VIN _____ Engine type and size _____

_____
Task Completed

Procedure

1. Wash mud and debris from the tire and wheel.
2. Mount the tire and wheel assembly on the wheel balancer with the wheel balancer manufacturer's recommended mounting components. ☐
3. Tighten the tire and wheel retaining nut on the wheel balancer to the specified torque. Tire and wheel retaining nut on the wheel balancer tightened to the specified torque? ☐ Yes ☐ No
   Instructor's Check _____
4. Remove stones and other objects from the tire tread.
   Are stones and other objects removed from the tire tread? ☐ Yes ☐ No
   Instructor's Check _____
5. Check for objects inside tire.
   Are there objects inside the tire? ☐ Yes ☐ No
   Instructor's Check _____
6. Remove all the old wheel weights from the wheel. Are the old wheel weights removed? ☐ Yes ☐ No
   Instructor's Check _____

Task Completed

7. Inspect the tread and sidewall.
List defective tread and sidewall conditions. _____

_____

List the types of tire tread wear and state the causes of this wear.

_____

_____

Based on your inspection of the tire, list the required tire service.

_____

_____

8. Check tire inflation pressure.
Tire inflation pressure _____
Specified tire inflation pressure _____

9. Measure tire and wheel runout.
Radial tire and wheel runout _____
Specified tire and wheel radial runout _____
Lateral tire and wheel runout _____
Specified lateral tire and lateral wheel runout _____
Recommended tire and wheel service. _____

10. Lower the safety hood on the wheel balancer.
Is the safety hood lowered? ☐ Yes ☐ No

11. Enter the required information on wheel balancer keypad, such as wheel diameter, wheel width, and wheel offset or distance. Is the required wheel information entered in the wheel balancer? ☐ Yes ☐ No
Instructor's Check _____

12. If the wheel balancer has a mode selector, select the desired balance mode.
Balance mode selected _____

13. Spin the wheel on the balancer, then apply the balancer brake to stop the wheel.

14. Observe the balancer display indicating the necessary amount of wheel weights and the required location of these weights. Install the required wheel weights in the locations indicated on the balancer. Be sure the wheel weights are securely attached to the rim. Are the wheel weight(s) securely attached to the rim at the locations indicated by the wheel balancer? ☐ Yes ☐ No
☐ Instructor's Check _____

15. Spin the tire and wheel assembly on the balancer again to be sure this assembly is properly balanced. Does the wheel balancer indicate proper wheel balance?
☐ Yes ☐ No
If the answer to this question is no, repeat the balance procedure.

Instructor's Response _____

_____

_____

JOB SHEET 11

Name _____ **Date** _____

On-Car Wheel Balancing

Upon completion of this job sheet, you should be able to perform on-car wheel balancing procedures.

NATEF AST/MAST Task Correlation

Suspension and Steering. Wheels and Tires Diagnosis and Repair,

Task #F-6. Dismount, inspect, and remount tire on wheel; balance wheel and tire assembly (static and dynamic). (P-1)

Tools and Materials

On-car wheel balancer
Car
Wheel weights

Describe the Vehicle Being Worked On:

Year _____ Make _____ Model _____

VIN _____ Engine type and size _____

Procedure

Task Completed

1. Wash mud and debris from the wheel and tire. ☐
2. Raise the wheel being balanced 5 in. (12 cm) off the floor; be sure the chassis is securely supported on safety stands. ☐
3. Remove stones and other objects from the tire tread.
   Are all objects removed from the tire tread? ☐ Yes ☐ No
   Instructor's Check _____
4. Inspect the tire tread and sidewall condition.
   Tread and sidewall condition _____
   _____
   Types of tire tread wear _____
   _____
   Causes of tire tread wear _____
   _____
   _____
   Based on your inspection of the tire tread and sidewall, state the necessary tire service.
   _____
   _____
5. Check the wheel bearing adjustment and adjust to specifications if necessary.
   Wheel bearing adjustment: ☐ Satisfactory ☐ Unsatisfactory

CAUTION:
Do not spin the front wheels on a front-wheel-drive vehicle with the floor jack under the chassis and the suspension dropped downward. Under this condition, severe angles exist in the front drive axle joints, and these joints may be damaged if the wheels are rotated with the balancer. Place the floor jack under the lower control arm to raise the wheel.

CAUTION:
To obtain proper wheel balance, spin the wheel only to the speed where the vibration occurs. Do not spin the wheel at excessive speeds.

6. Install the electronic vibration sensor between the lower control arm and the floor.
   Is the vibration sensor properly installed? ☐ Yes ☐ No
   Instructor's Check _____
7. Chalk mark a reference mark on the outer sidewall of the tire.
8. Spin the wheel just fast enough to produce vibration on the front bumper.
9. When vibration causes strobe light flashing, move the balancer drum away from the tire and allow the tire to spin freely.
10. Shine the strobe light around the tire sidewall and note the chalk-mark position.
    Chalk-mark clock position _____
11. Note the pointer position on the meter.
    Meter pointer position _____
12. Use the balancer's brake plate to slow and stop the wheel.
13. Rotate the wheel until the chalk mark is in the exact position where it appeared under the strobe light. The heavy spot is now at the bottom of the wheel and the balancing weight should be attached 180° from the heavy spot. Install the amount of weight indicated on the meter.
    Amount of wheel weight installed _____
14. Spin the wheel again. If the wheel balance is satisfactory, the meter pointer will read in the balanced position. If the pointer does not indicate a balanced wheel, shine the strobe light on the tire sidewall. If the installed wheel weight is at the 12 o'clock position, additional weight is required. A 6 o'clock weight position indicates excessive weight. A 3 o'clock or 9 o'clock weight position may be corrected by moving the weight 1 in. (2.5 cm) toward the 12 o'clock position.
    Tire and wheel balance: ☐ Satisfactory ☐ Unsatisfactory
    If the wheel balance is unsatisfactory, state the action required to balance tire and wheel.
    _____
    _____

Instructor's Response _____
_____
_____

JOB SHEET 12

Name _____ **Date** _____

Diagnose Tire and Wheel Vibration, Steering Pull, and Chassis Waddle

Upon completion of this job sheet, you should be able to diagnose tire and wheel vibration problems, steering pull, and chassis waddle.

NATEF AST/MAST Task Correlation

Suspension and Steering. Wheels and Tires Diagnosis and Repair,
Task #F-2. Diagnose wheel/tire vibration, shimmy, and noise; determine necessary action. (P-2)

Tools and Materials

Late model vehicle

Describe the Vehicle Being Worked On:

Year _____ Make _____ Model _____
VIN _____ Engine type and size _____

Procedure

1. Drive the vehicle at 10 mph (16 km/h), 20 mph (32 km/h), 30 mph (48 km/h), 40 mph (64 km/h), 50 mph (80 km/h), 60 mph (96.5 km/h). Maintain each vehicle speed long enough to diagnose tire, steering, and chassis problems. Do not exceed the legal speed limit. Did the vehicle indicate any of the following problems at these various speeds?

2. Chassis waddle. Speed at which waddle occurred _____
   Did the waddle occur at the front or the rear of the vehicle? _____
   List the possible causes of waddle and the necessary diagnostic steps and service procedures to correct the problem.
   _____
   _____
   _____

3. Tire thump and vibration. Speed at which tire thump or vibration occurred _____
   At which corner of the vehicle did the thump and vibration occur. _____
   List the possible causes of tire thump and vibration and the necessary diagnostic steps and service procedures to correct the problem.
   _____
   _____
   _____

4. Wheel shimmy. Speed at which wheel shimmy occurred. _____

List the possible cause of wheel shimmy, and the necessary diagnostic steps and service procedures to correct the problem.

_____

_____

_____

5. Steering pull. Speed at which steering pull occurred. _____

Direction of steering pull. _____

List the possible causes of steering pull, and the necessary diagnostic steps and service procedures to correct the problem.

_____

_____

_____

Instructor's Response _____

_____

_____

JOB SHEET 13

Name _____ **Date** _____

Inspect, Diagnose, and Calibrate Tire Pressure Monitoring Systems

Upon completion of this job, you should be able to inspect, diagnose, and calibrate tire pressure monitoring systems (TPMSs).

NATEF AST/MAST Task Correlation

Suspension and Steering. Wheels and Tires Diagnosis and Repair,
Task #F-10. Identify and test tire pressure monitoring system (indirect and direct) for operation; calibrate system; verify operation of instrument panel lamps. (P-2)

Tools and Materials

Late model vehicle
Scan tool

Describe the Vehicle Being Worked On:

Year _____ Make _____ Model _____
VIN _____ Engine type and size _____

Procedure

1. Connect a scan tool to the data link connector (DLC) under the left side of the instrument panel, and turn the scan tool on.
2. Turn the ignition switch on, and do not start the engine.
3. Select antenna module on the scan tool. Check for the DTCs related to the antenna module. If there are DTCs displayed with a "U" prefix, there is a defect in the data link communications. If the scan tool cannot communicate with the antenna module, or there are DTCs related to this module or the data link communications, proceed with further diagnosis of these items.
4. Check for DTCs related to the radio/audio system. The antenna grid in the rear window receives TPMS sensor signals. This grid shares its connector with the AM/FM antenna grid. If any radio/audio system DTCs are present, proceed with the diagnosis of the radio/audio system.
5. Check for keyless entry system DTCs displayed in the scan tool. The antenna module also controls the keyless entry system. If the scan tool displays DTCs related to the keyless entry system, proceed with the diagnosis of these DTCs.
6. Check for TPMS DTCs on the scan tool display. If TPMS DTCs are present, diagnose the cause of these DTCs.
   List and interpret the DTCs displayed on the scan tool _____
   _____
   _____

TPMS Learning Procedure

1. Connect a scan tool to the DLC.
2. Turn on the ignition switch, and do not start the engine.
3. Apply the parking brake.
4. Select Special Functions on the scan tool.
5. Select Sensor Learn Mode on the scan tool, and press the Enter key.
6. Press the On soft key. A horn chirp should sound to indicate the sensor learn mode is enabled.
7. Starting with the LF tire, increase or decrease the tire pressure for 5–8 seconds or until a horn chirp sounds. The horn chirp may occur before the 5- to 8-second time period, or up to 30 seconds after this time period.

WARNING: If you are increasing tire pressure during the learning procedure, never inflate a tire above the vehicle manufacturer's maximum specified tire pressure. This action may cause personal injury and tire damage.

8. After the horn chirp sounds, repeat step 7 on the other 3 or 4 sensors in the following order:
   (a) RF
   (b) RR
   (c) LR
   (d) Spare tire (if applicable)

If a horn chirp is not heard after 35 seconds for any of the 4 sensors, turn off the ignition switch and exit the learn mode on the scan tool. Repeat the sensor learning procedure from step 4.

9. After the learning procedure has been completed on all the sensors, a double horn chirp sounds to indicate the learning procedure is completed on all the sensors.
10. Turn off the ignition switch and disconnect the scan tool.
11. Inflate all the tires to the specified pressure.
    Learning procedure completed ? ☐ Yes ☐ No
    If the answer is no, state the reason for the test not being completed _____
    _____

Instructor's Response _____
_____
_____

Chapter 4

Wheel Bearing and Seal Service

BASIC TOOLS

Basic technician's tool set
Service manual
Inch-pound torque wrench
Foot-pound torque wrench
Fine-toothed round and flat files
Wheel bearing grease
Differential lubricant

Upon completion and review of this chapter, you should be able to:

- Diagnose bearing defects.
- Clean and repack wheel bearings.
- Reassemble and adjust wheel bearings.
- Remove and replace wheel bearing seals.
- Diagnose wheel bearing problems.
- Diagnose problems in wheel bearing hub units.
- Remove and replace front drive axles.
- Remove and replace wheel bearing hub units.
- Remove and replace rear-axle bearings on rear-wheel-drive cars.

Technicians must accurately diagnose wheel bearing problems to avoid repeat bearing failures, and thus provide customer satisfaction. Accurate wheel bearing service procedures are essential to maintain vehicle safety! Improper wheel bearing service may cause brake problems, steering complaints, and premature bearing failure. Improper wheel bearing service may even cause a wheel to fly off a vehicle, resulting in personal injury and vehicle damage. A knowledge of drive axle removal is necessary to service the front wheel bearings on front-wheel-drive vehicles, because the drive axle must be removed from the front hub to service the wheel bearings. Similarly, technicians must also understand rear-axle bearing service procedures on rear-wheel-drive vehicles.

Diagnosis of Bearing Defects

Bearings are designed to provide long life, but there are many causes of premature bearing failure. If a bearing fails, the technician must decide if the bearing failure was caused by normal wear or if the bearing failed prematurely. For example, if a front wheel bearing fails on a car that is two years old with an original odometer reading of 30,000 *miles* (48,280 *kilometers*), experience tells us the bearing failure is premature, because front wheel bearings normally last for a much longer mileage period, often in excess of 100,000 *miles* (160,934 *kilometers*). Always listen to the customer's complaints, and obtain as much information as possible from the customer. Ask the customer specific questions about abnormal or unusual vehicle noises and operation. If a bearing fails prematurely, there must be some cause for the failure.

Bearing galling refers to metal smears on the ends of the rollers.

The causes of premature bearing failure are:

1. Lack of lubrication
2. Improper type of lubrication
3. Incorrect endplay adjustment (where applicable)
4. Misalignment of related components, such as shafts or housings
5. Excessive bearing load
6. Improper installation or service procedures

7. Excessive heat
8. Dirt or contamination

When a bearing fails prematurely, the technician must correct the cause of this failure to prevent the new bearing from failing. Refer to Table 4-1 for additional information on noise and vibration diagnosis. The types of bearing failures and the necessary corrective service procedures are provided in Figure 4-1 and Figure 4-2. **Bearing fatigue spalling** appears as flaking of surface metal on bearing rollers and races. **Bearing brinelling** shows up as straight-line indentations on the races and rollers. **Bearing smears** appear as metal loss in a circular, blotched pattern around the bearing races and rollers. **Bearing frettage** shows up as a fine, corrosive wear pattern around the bearing races and rollers. This wear pattern is circular on the races.

Bearing abrasive step wear is a fine circular wear pattern on the ends of the rollers.

Bearing etching appears as a loss of material on the bearing rollers and races. Bearing surfaces are gray or grayish black.

Bearing indentations are surface depressions on the rollers and races.

The first indication of bearing failure is usually a howling noise while the bearing is rotating. The howling noise will likely vary depending on the bearing load. A front wheel bearing usually provides a more noticeable howl when the vehicle is turning a corner, because this places additional thrust load on the bearing. A defective rear-axle bearing usually provides a howling noise that is more noticeable at lower speeds. The howling noise is more noticeable when driving on a narrow street with buildings on each side, because the noise vibrates off the nearby buildings. A rear-axle bearing noise is present during acceleration and deceleration, because the vehicle weight places a load on the bearing regardless of the operating condition. The rear-axle bearing noise may be somewhat more noticeable during deceleration because there is less engine noise at that time.

Service and Adjustment of Tapered Roller Bearing-Type Wheel Bearings

Cleaning Bearings

Classroom Manual
Chapter 4, page 98

WARNING: Always wear safety goggles when working in the shop.

WARNING: Do not aim compressed air directly toward any part of your body. Compressed air can penetrate human flesh and enter the blood stream with very serious consequences.

WARNING: Do not spin the bearing at high speed with compressed air. Bearing damage or disintegration may result. Bearing disintegration may cause serious personal injury.

Bearing brinelling may be caused by the continuous vibration when transporting new vehicles by rail or truck from the factory to the dealership.

CAUTION: Never wash a sealed bearing or a bearing that is shielded on both sides because solvent may enter the bearing and destroy the lubricant in the bearing, resulting in very short bearing life.

TABLE 4-1 **DIAGNOSTIC TABLE**

Problem	Symptoms	Possible causes
Noise in drive train, front-wheel-drive car.	Clicking noise while turning a corner at low speeds.	Worn outer front drive axle joint.
Vibration in front drive train, front-wheel-drive car.	Vibration while decelerating at 35 to 45 mph, 56 to 72 km/h.	Worn inner drive axle joint.
Noise in front suspension.	Growling noise in front suspension, most noticeable when turning a corner or driving at low speeds.	Worn wheel bearings.
Lack of directional stability.	Steering wanders to either side when driving straight ahead.	Excessive front wheel bearing endplay, improper wheel bearing adjustment.
Noise in drive train, rear-wheel-drive car.	Constant growling or clicking noise in rear axle, most noticeable when driving at low speeds or down a narrow street.	Worn rear-axle bearing.

Copyright © 2015 Cengage Learning®

TAPERED ROLLER BEARING DIAGNOSIS

Consider the following factors when diagnosing bearing condition:
1. General condition of all parts during disassembly and inspection.
2. Classify the failure with the aid of the illustrations.
3. Determine the cause.
4. Make all repairs following recommended procedures.

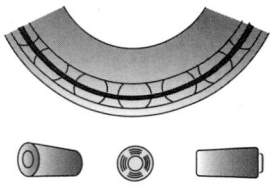

ABRASIVE STEP WEAR

Pattern on roller ends caused by fine abrasives. Clean all parts and housings, check seals and bearings, and replace if leaking, rough, or noisy.

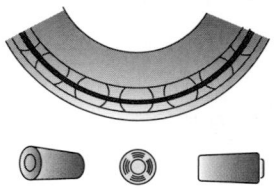

GALLING

Metal smears on roller ends due to overheating, lubricant failure, or overload. Replace bearing, check seals, and check for proper lubrication.

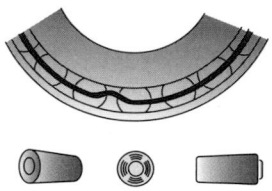

BENT CAGE

Cage damaged due to improper handling or tool usage. Replace bearing.

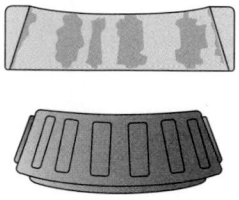

ABRASIVE ROLLER WEAR

Pattern on races and rollers caused by fine abrasives. Clean all parts and housings, check seals and bearings, and replace if leaking, rough, or noisy.

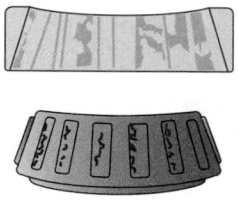

ETCHING

Bearing surfaces appear gray or grayish black in color with related etching away of material usually at roller spacing. Replace bearings, check seals, and check for proper lubrication.

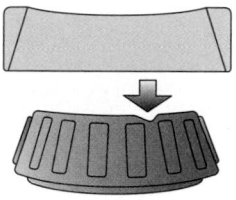

BENT CAGE

Cage damaged due to improper handling or tool usage. Replace bearing.

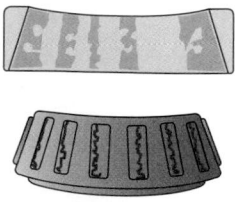

INDENTATIONS

Surface depressions on race and rollers caused by hard particles of foreign material. Clean all parts and housings. Check seals, and replace bearings if rough or noisy.

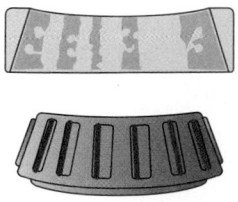

BRINELLING

Surface indentations in raceway caused by rollers either under impact loading or vibration while the bearing is not rotating. Replace bearing if rough or noisy.

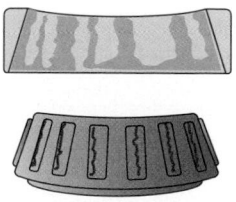

MISALIGNMENT

Outer race misalignment due to foreign object. Clean related parts and replace bearing. Make sure races are properly sealed.

Copyright © 2015 Cengage Learning®

FIGURE 4-1 **Bearing failures and corrective procedures.**

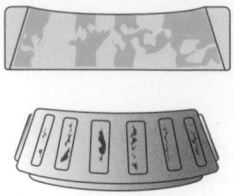

FATIGUE SPALLING

Flaking of surface metal resulting from fatigue. Replace bearing, clean all related parts.

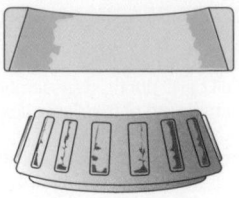

STAIN DISCOLORATION

Discoloration can range from light brown to black caused by incorrect lubricant or moisture. Reuse bearings if stains can be removed by light polishing or if no evidence of overheating is observed. Check seals and related parts for damage.

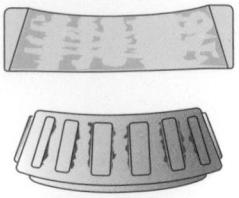

CAGE WEAR

Wear around outside diameter of cage and roller pockets caused by abrasive material and inefficient lubrication. Clean related parts and housings. Check seals and replace bearings.

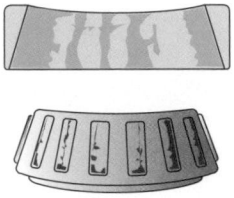

HEAT DISCOLORATION

Heat discoloration can range from faint yellow to dark blue, resulting from overload or incorrect lubricant. Excessive heat can cause softening of races or rollers. To check for loss of temper on races or rollers, a simple file test may be made. A file drawn over a tempered part will grab and cut metal, whereas, a file drawn over a hard part will glide readily with no metal cutting. Replace bearings if overheating damage is indicated. Check seals and other parts.

FRETTAGE

Corrosion set up by small relative movement of parts with no lubrication. Replace bearings. Clean related parts. Check seals and check for proper lubrication.

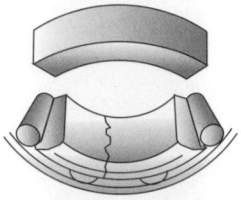

CRACKED INNER RACE

Race cracked due to improper fit, cocking, or poor bearing seats. Replace bearing and correct bearing seats.

SMEARS

Smearing of metal due to slippage. Slippage can be caused by poor fits, lubrication, overheating, overloads, or handling damage. Replace bearings, clean related parts, and check for proper fit and lubrication.

FIGURE 4-2 **Bearing failures and corrective procedures, continued.**

Copyright © 2015 Cengage Learning®

Copyright © 2015 Cengage Learning®

FIGURE 4-3 **Cleaning a bearing with solvent.**

Copyright © 2015 Cengage Learning®

FIGURE 4-4 **Wrapping a bearing in waterproof paper.**

WARNING: Never strike a bearing with a ball peen hammer. This action will damage the bearing and the bearing may shatter, causing severe personal injury.

Two separate tapered roller bearings are used in the front wheel hubs of many rear-wheel-drive cars. The rear wheel hubs on some front-wheel-drive cars have the same type of bearings. Similar service and adjustment procedures apply to these tapered roller bearings.

The outer bearing race on a tapered roller bearing may be called a bearing cup.

These bearings should be cleaned, inspected, and packed with wheel bearing grease at the vehicle manufacturer's recommended service intervals. Pry the grease seal out of the inner hub opening with a seal removal tool, and discard the seal. This seal should always be replaced when the bearings are serviced. Do not attempt to wash sealed bearings or bearings that are shielded on both sides. If a bearing is sealed on one side, it may be washed in solvent and repacked with grease.

Bearings may be placed in a tray and lowered into a container of clean solvent. A brush may be used to remove old grease from the bearing (Figure 4-3). The bearings may be dried with compressed air after the cleaning operation. Be sure the shop air supply is free from moisture, which causes rust formation in the bearing. After all the old grease has been cleaned from the bearing, rinse the bearing in clean solvent and dry it thoroughly with compressed air.

Classroom Manual
Chapter 4, page 98

When bearing cleaning is completed, bearings should be inspected for the defects illustrated in Figure 4-1 and Figure 4-2 If any of these conditions are present on the bearing, replacement is necessary. Tapered roller bearings and their matching outer races must be replaced as a set. If the bearing installation is not done immediately, cover the bearings with a protective lubricant and wrap them in waterproof paper (Figure 4-4). Be sure to identify the bearings, or lay them in order, so you reinstall them in their original location. Do not clean bearings or races with paper towels. If you are using a shop towel for this purpose, be sure it is lint free. Lint from shop towels or paper towels may contaminate the bearing. Bearing races and the inner part of the wheel hub should be thoroughly cleaned with solvent and dried with compressed air. Inspect the seal mounting area in the hub for metal burrs. Remove any burrs with a fine round file.

SPECIAL TOOL
Bearing driver

Bearing races must be replaced if any of the defects described in Figure 4-1 and Figure 4-2 are found. The proper bearing race driving tool must be used to remove the bearing races (Figure 4-5). If a driver is not available for the bearing races, a long brass punch and hammer may be used to drive the races from the hub. When a hammer and punch are used for this purpose, be careful not to damage the hub inner surface with the punch.

Small dirt particles left behind the outer bearing race cause race misalignment and premature bearing failure.

The new bearing races should be installed in the hub with the correct bearing race driving tool (Figure 4-6). When bearings and races are replaced, be sure they are the same as the original bearings. The part numbers should be the same on the old bearings and the replacement bearings.

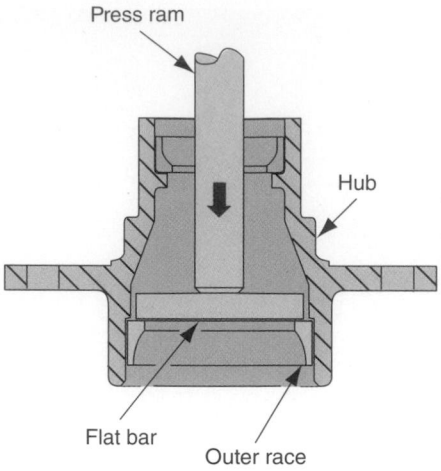

Copyright © 2015 Cengage Learning®

FIGURE 4-5 **Bearing race removal.**

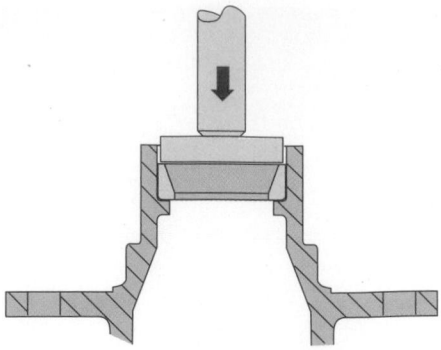

Copyright © 2015 Cengage Learning®

FIGURE 4-6 **Bearing race installation.**

CAUTION:
Cleanliness is very important during wheel bearing service. Always maintain cleanliness of hands, tools, work area, and all related bearing components. One small piece of dirt in a bearing will cause bearing failure.

Inspect the bearing and seal mounting surfaces on the spindle. Small metal burrs may be removed from the spindle with a fine-toothed file. If the spindle is severely scored in the bearing or seal mounting areas, spindle replacement is necessary.

Bearing Lubrication and Assembly

After the bearings and races have been cleaned and inspected, the bearings should be packed with grease. Always use the vehicle manufacturer's specified wheel bearing grease. Vehicle manufacturers usually recommend a lithium-based wheel bearing grease. Place a lump of grease in the palm of one hand and grasp the bearing in the other hand. Force the widest edge of the bearing into the lump of grease, and squeeze the grease into the bearing. Continue this process until grease is forced into the bearing around the entire bearing circumference. Place a coating of grease around the outside of the rollers, and apply a light coating of grease to the races. A bearing packing tool may be used to force grease into the bearings rather than using the hand method. Bearing packers may be hand operated or pressure operated (Figure 4-7).

CAUTION:
Always keep grease containers covered when not in use. Uncovered grease containers are easily contaminated with dirt and moisture.

Place some grease in the wheel hub cavity and position the inner bearing in the hub (Figure 4-8). Check the fit of the new bearing seal on the spindle and in the hub. The seal lip must fit snugly on the spindle, and the seal case must fit properly in the hub opening. The part number on the old seal and the replacement seal should be the same. Be sure the seal is installed in the proper direction with the garter spring and higher part of the lip toward the lubricant in the hub. The new inner bearing seal must be installed in the hub with a suitable seal driver (Figure 4-9). Place a light coating of wheel bearing grease on the spindle and slide the hub assembly onto the spindle. Install the outer wheel bearing and be sure there is

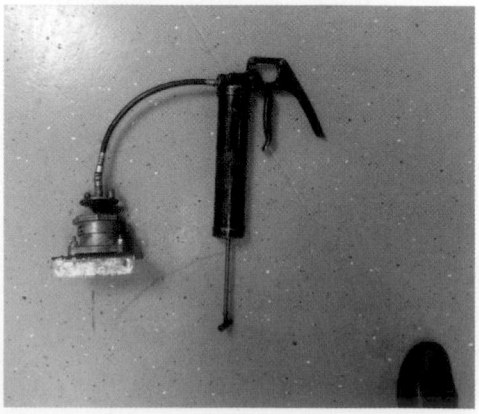

Copyright © 2015 Cengage Learning®

FIGURE 4-7 **Mechanical wheel bearing packer.**

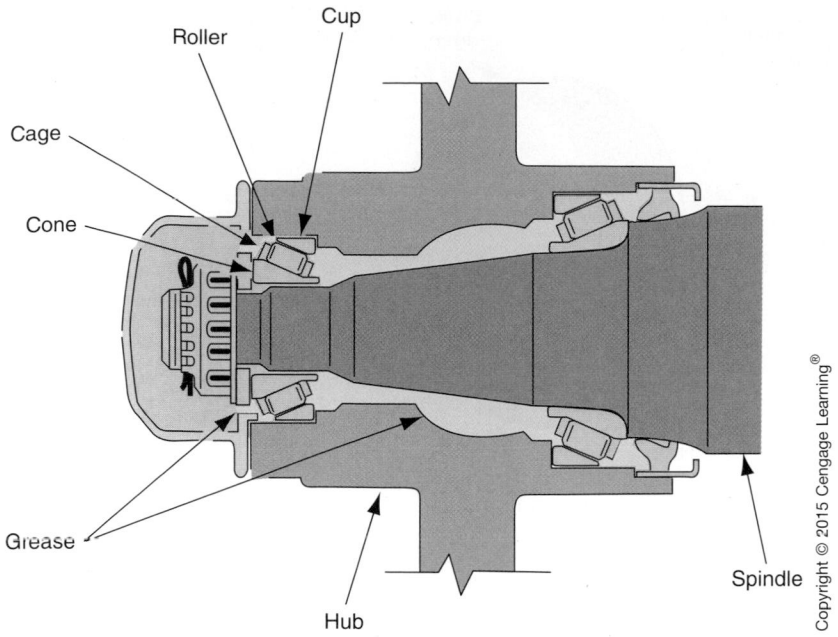

FIGURE 4-8 **Wheel bearing lubrication.**

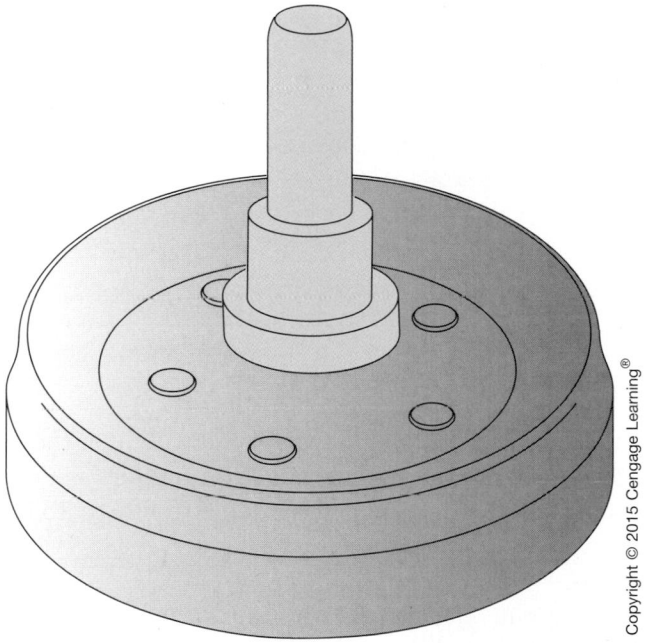

FIGURE 4-9 **Seal installation.**

SERVICE TIP:
When a lip seal is installed, the garter spring should always face toward the flow of lubricant.

SPECIAL TOOL
Seal driver

adequate lubrication on the bearing and race. Be sure the washer and nut are clean and install these components on the spindle (Figure 4-10). Tighten the nut until it is finger tight.

Photo Sequence 5 shows a typical procedure for adjusting rear wheel bearings on a front-wheel-drive car.

Wheel Bearing Adjustment with Two Separate Tapered Roller Bearings in the Wheel Hub

Wheel bearing endplay is the amount of horizontal wheel bearing hub movement. If a bearing has a preload condition, a slight tension is placed on the bearing. Loose front wheel bearing adjustment results in lateral front wheel movement and reduced directional stability. If the wheel bearing adjusting nut is tightened excessively, the bearings may

Classroom Manual
Chapter 4, page 102

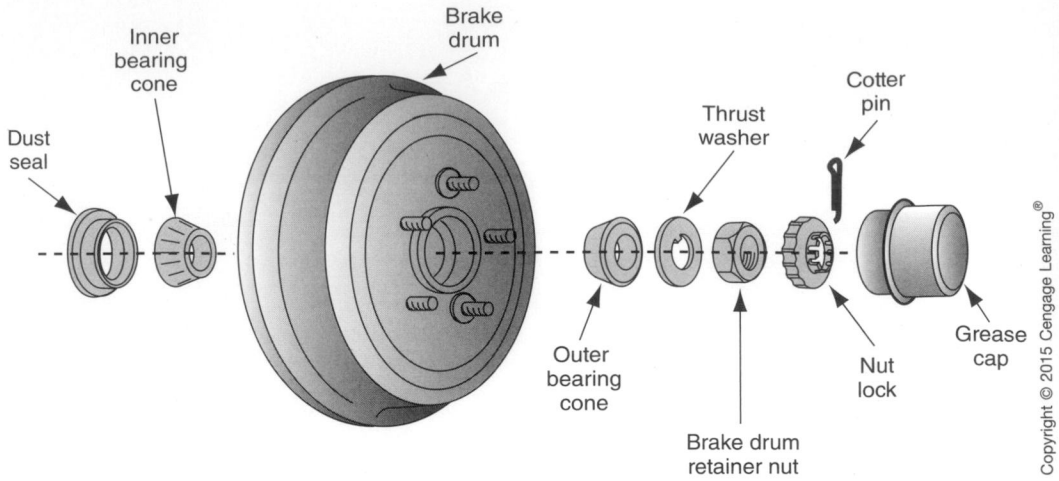

FIGURE 4-10 **Installation of wheel bearings and related components.**

overheat, resulting in premature bearing failure. The bearing adjustment procedure may vary depending on the make of vehicle. Always follow the procedure in the vehicle manufacturer's service manual.

> **CUSTOMER CARE: Never sell a customer automotive service that is not required on his or her car. Selling preventive maintenance, however, is a sound business practice and may save a customer some future problems. An example of preventive maintenance is selling a cooling system flush when the cooling system is not leaking but the manufacturer's recommended service interval has elapsed. If customers find out they were sold some unnecessary service, they will probably never return to the shop. They will likely tell their friends about their experience, and that kind of advertising the shop can do without.**

A typical bearing adjustment procedure follows:

1. With the hub and bearings assembled on the spindle, tighten the adjusting nut to 17 to 25 ft.-lb. (23 to 34 Nm) while the hub is rotated in the forward direction.
2. Loosen the adjusting nut 1/2 turn and retighten it to 10 to 15 in.-lb (1.0 to 1.7 Nm). This specification varies depending on the make of vehicle. Always use the manufacturer's specifications.
3. Position the adjusting nut retainer over the nut so the retainer slots are aligned with the cotter pin hole in the spindle.
4. Install a new cotter pin, and bend the ends around the retainer flange.
5. Install the grease cap, and make sure the hub and drum rotate freely.

After a wheel bearing adjustment is performed, the hub must have an endplay of 0.001 in. to 0.005 in. (0.0254 mm to 0.127 mm). A dial indicator may be mounted with the stem positioned against the outer edge of the hub to perform this measurement. When the hub is pulled outward and pushed inward, the specified endplay must be displayed on the dial indicator. If the endplay is not correct, the wheel bearing adjustment must be repeated.

WHEEL HUB UNIT DIAGNOSIS AND SERVICE

When wheel bearings and hubs are an integral assembly or unitized hub assembly (Figure 4-11), the bearing endplay should be measured with a dial indicator stem mounted

PHOTO SEQUENCE 5

Typical Procedure for Adjusting Rear Wheel Bearings on a Front-Wheel-Drive Car

P5-1 **Always make sure the car is positioned safely on a lift before working on the vehicle.**

Copyright © 2015 Cengage Learning®

P5-2 **Remove the dust cap from the wheel hub.**

Copyright © 2015 Cengage Learning®

P5-3 **Remove the cotter pin and nut retainer from the bearing adjusting nut.**

Copyright © 2015 Cengage Learning®

P5-4 **Tighten the bearing adjusting nut to 17 to 25 ft.-lb.**

Copyright © 2015 Cengage Learning®

P5-5 **Loosen the bearing adjusting nut one-half turn.**

Copyright © 2015 Cengage Learning®

P5-6 **Tighten the bearing adjusting nut to 10 to 15 in.-lb.**

Copyright © 2015 Cengage Learning®

P5-7 **Position the adjusting nut retainer over the adjusting nut so the slots are aligned with the holes in the nut and spindle.**

Copyright © 2015 Cengage Learning®

P5-8 **Install a new cotter pin and bend the ends around the retainer flange.**

Copyright © 2015 Cengage Learning®

P5-9 **Install the dust cap and be sure the hub rotates freely.**

Copyright © 2015 Cengage Learning®

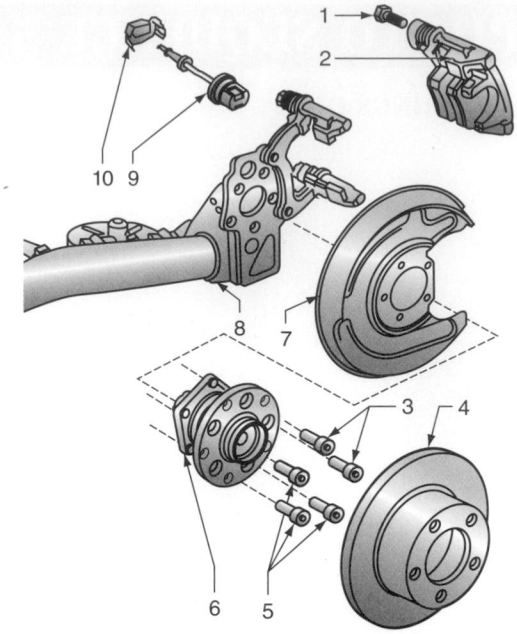

1. Bolt
2. Caliper
3. Socker head bolt
4. Rotor
5. Socket head bolt
6. Wheel hub w/bearing
7. Backing plate
8. Axle beam
9. ABS speed sensor
10. Bracket

Copyright © 2015 Cengage Learning®

FIGURE 4-11 **A typical non-serviceable integral hub bearing assembly.**

Copyright © 2015 Cengage Learning®

FIGURE 4-12 **Wheel bearing diagnosis on vehicle.**

SPECIAL TOOL

Dial indicator

Classroom Manual

Chapter 4, page 102

against the hub. If the endplay exceeds 0.005 in. (0.127 mm) as the hub is moved in and out, the hub and bearing assembly should be replaced. This specification is typical, but the vehicle manufacturer's specifications must be used. Hub and bearing replacement is also necessary if the bearing is rough or noisy. Integral-type bearing and hub assemblies are used on the front and rear wheels on some front-wheel-drive cars. Photo Sequence 6 shows a typical procedure for measuring front wheel hub endplay.

When removable front wheel bearings are mounted in the steering knuckle, the wheel bearings may be checked with the vehicle raised on the hoist and a dial indicator positioned against the outer wheel rim lip as shown in Figure 4-12.

When the wheel is moved in and out, the maximum bearing movement on the dial indicator should be as follows:

0.020 in. (0.508 mm) for 13-in. (33-cm) wheels
0.023 in. (0.584 mm) for 14-in. (35.5-cm) wheels
0.025 in. (0.635 mm) for 15-in. (38-cm) wheels

If the bearing movement is excessive, check the hub nut torque before replacing the bearing. When this torque is correct and bearing movement is excessive, the bearing should be replaced.

When a wheel is removed to service the wheel bearings, proper balance must be maintained between the wheel and tire and the hub. Therefore, the tire, wheel, hub stud, and brake rotor should be chalk marked prior to removal.

Service of the integral hub assembly is relatively straight forward but does require the removal of the brake caliper and rotor assembly if the vehicle is equipped with disk brakes.

First remove the wheel assembly.

Remove the disc brake caliper mounting bolts and secure with a piece of mechanics' wire to avoid damage to the brake hose.

PHOTO SEQUENCE 6

Typical Procedure for Measuring Front Wheel Hub Endplay—Integral, Sealed Wheel Bearing Hub Assemblies

P6-1 **Be sure the vehicle is properly positioned on a lift before the wheel bearing hub endplay measurement is performed. The vehicle should be properly positioned on a lift with the lift raised to a comfortable working height for performing this measurement.**

Copyright © 2015 Cengage Learning®

P6-2 **Remove the wheel cover and dust cap.**

Copyright © 2015 Cengage Learning®

P6-3 **Attach a magnetic dial indicator base securely to the inside of the fender at the lower edge of the wheel opening. Position the dial indicator stem against the vertical wheel surface as close as possible to the top wheel stud, and preload the dial indicator stem.**

Copyright © 2015 Cengage Learning®

P6-4 **Zero the dial indicator pointer.**

Copyright © 2015 Cengage Learning®

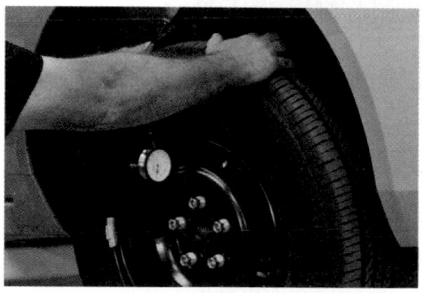

P6-5 **Grasp the top of the tire with both hands. Push and pull on the top of the tire without rotating the tire, and note the dial indicator readings with the tire pushed inward and the tire pulled outward. The difference between the two readings is the wheel hub endplay. Repeat this procedure twice to verify the endplay reading.**

Copyright © 2015 Cengage Learning®

P6-6 **Maximum wheel bearing endplay should be 0.005 in. (0.127 mm). If the endplay measurement is not correct, wheel bearing hub replacement is necessary.**

Copyright © 2015 Cengage Learning®

P6-7 **Remove the dial indicator and install the dust cap and wheel cover.**

Copyright © 2015 Cengage Learning®

Remove the brake rotor.

If a wheel speed sensor for the antilock brake system (ABS) is integral in the front wheel hubs, disconnect the wheel speed sensor connectors. If the wheel speed sensor is not integrated into the hub it will be necessary to remove it from the steering knuckle to avoid damage.

If this is the drive axle hub it will be necessary to remove the drive axle retaining nut.

Remove the hub-to-knuckle bolts, and place the transaxle in park.

Use the proper puller to remove the wheel-bearing hub from the drive axle.

Install the new hub and bearing assembly on the drive axle splines, and install a new drive axle nut. Tighten the drive axle nut to pull the hub onto the splines. Do not tighten this nut to the specified torque at this time.

Place the transaxle in neutral and install the wheel bearing hub bolts. Tighten these bolts to the specified torque.

Install the wheel sensor and connectors, rotors, and calipers. Tighten the caliper mounting bolts to the specified torque. Place a large punch between the rotor fins and the caliper, and tighten the drive axle nut to the specified torque.

Install the wheels in their original positions, and tighten the wheel nuts to the specified torque.

Lower the vehicle onto the shop floor, and tighten the new drive axle nut to the specified torque. Install a new cotter pin if used instead of a new locking nut.

Road test the car and listen for abnormal wheel bearing noises.

Front Drive Axle Diagnosis

On many front-wheel-drive vehicles, the front drive axles must be removed before the wheel hub unit or steering knuckle and bearing can be detached. Therefore, we will discuss front drive axle diagnosis and removal. Because drive axle noises may be confused with front wheel bearing noise, a brief discussion of drive axle noises and problems may be helpful. A defective inner drive axle joint usually causes a vibration when the vehicle is decelerating at 35 to 45 miles per hour (mph), or 56 to 72 kilometers per hour (km/h). A worn inner drive axle joint may also cause vibration during acceleration. When an outer drive axle joint is worn, a clicking noise is heard during a hard turn below 20 mph (32 km/h). To determine which drive axle has the defective joint, lift the vehicle on a hoist, and allow the front wheels to drop down. This action will position the axle joints at a different angle than when the car is driven on the road. Lift the lower control arms one at a time with a floor jack, and place the transmission in drive to simulate the driving conditions that provided the vibration or noise. If the vibration or noise occurs with one lower control arm lifted, that side has the defective drive axle joint.

Drive Axle Removal

A **circlip** is a split, circular clip mounted in a groove on the inner drive axle joint extension to retain the drive axle in the transaxle on some front-wheel-drive-vehicles.

Many drive axles have a **circlip** on the inner joint extension that holds the inner joint into the differential side gear. Drive axle systems vary depending on the vehicle. Follow the drive axle removal procedure in the vehicle manufacturer's service manual.

A general front drive axle removal procedure follows:

1. Loosen the front wheel nuts and hub nuts.
2. Lift the vehicle on a hoist and be sure the hoist safety mechanism is in place; then remove the front wheels and tires.
3. Remove the brake calipers and rotors. Connect a piece of wire from the calipers to a suspension or chassis component. Do not allow the calipers to hang on the end of the brake line.
4. Install protective drive axle boots if these are supplied by the car manufacturer.

5. Remove the ball joint to steering knuckle clamp bolt (Figure 4-13).
6. Pry the ball joint stud from the steering knuckle.
7. Pull the inner axle from the transaxle (Figure 4-14); do not allow the axle to drop down at a severe angle.
8. Remove the hub nut and washer, and separate the outer axle joint from the wheel hub. Some outer axle joint splines are slightly spiraled. On this type of outer axle joint, a special puller is required to separate the axle joint from the wheel (Figure 4-15).
9. Remove the drive axle from the chassis.

Reverse the drive axle removal procedure for front drive axle installation.

SPECIAL TOOL

Front wheel bearing hub pulling and installing tools

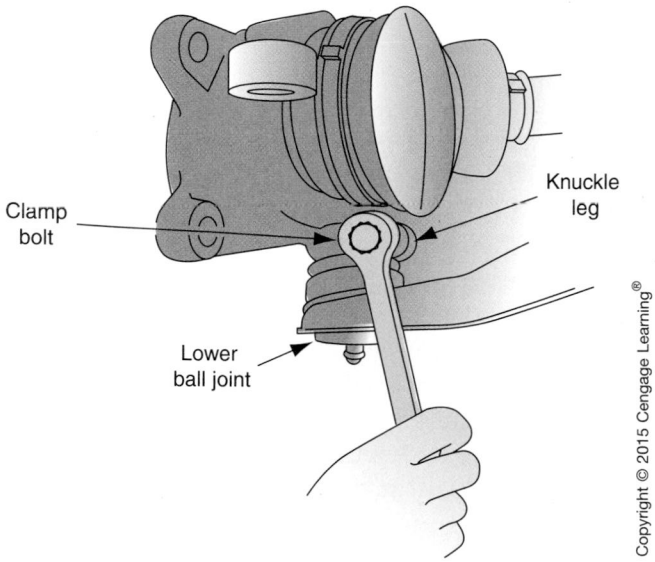

FIGURE 4-13 **Removal of ball joint to steering knuckle clamp bolt.**

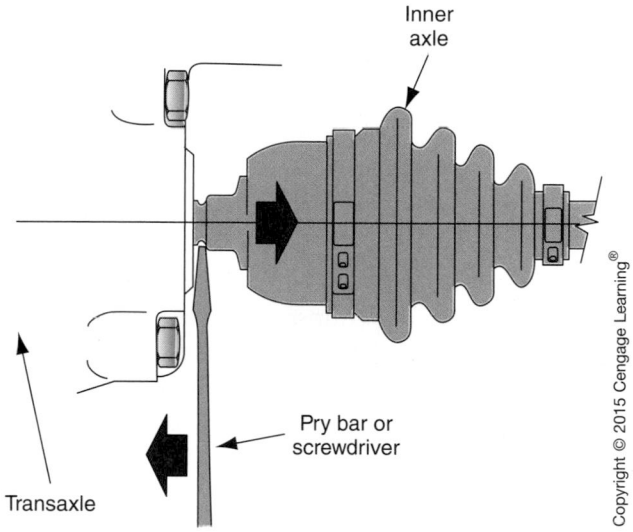

FIGURE 4-14 **Removal of inner axle joint from transaxle.**

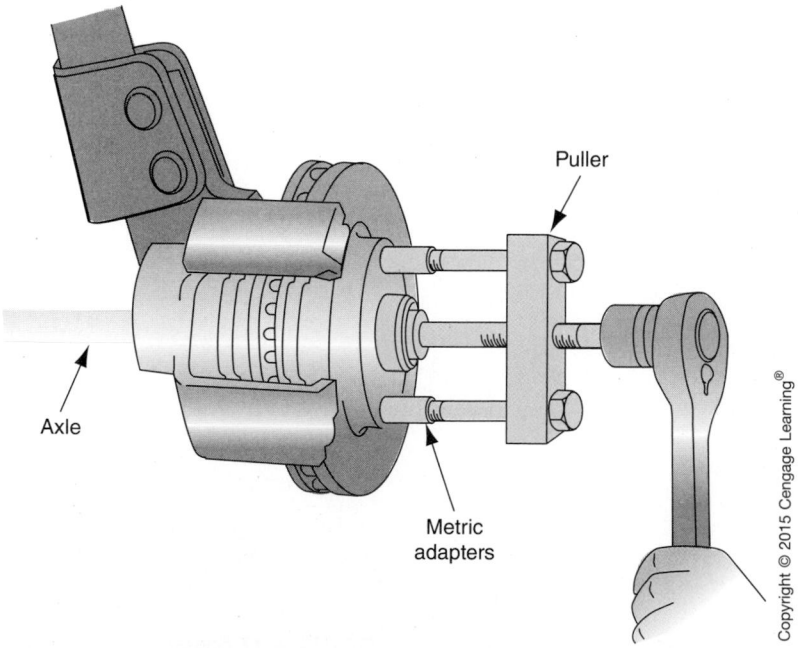

FIGURE 4-15 **Removal of outer axle joint from front wheel hub.**

Special Procedures for Drive Axle Removal

Some car manufacturers recommend removal of the strut from the steering knuckle rather than removal of the ball joint when the drive axles are removed. This type of suspension system has a threaded nut on the ball joint stem to hold the ball joint into the steering knuckle. If an eccentric camber adjustment bolt is positioned in the strut, the bolt head position should be marked in relation to the strut before the bolt is removed (Figure 4-16).

On these suspension systems, the brake calipers and brake line clamps should be removed before the drive axles. Some car manufacturers supply a slide-hammer-type puller to remove the inner axle joints from the transaxle.

On Chrysler manual and automatic transaxles, the speedometer gear must be removed from the right differential extension housing before the right drive axle is removed (Figure 4-17).

Early model Chrysler transaxles had circlip retainers, which held the inner drive axle joints into the differential side gears. The differential cover had to be removed and these circlips had to be compressed with needle nose pliers before the drive axles could be removed from the differential side gears. The drive axles had to be rotated until a flat area on the axle and the circlip ends were visible (Figure 4-18).

The inner joint housings did not have tripod springs on early models. Later model Chrysler drive axles may be pulled from the differential side gears without collapsing the circlips. To determine which type of circlip is used, grasp the inner joint housing and try to pull it out of the transaxle. If the inner joint housing moves outward and springs back into the transaxle, the joint has a later model circlip. When the inner joint is solid in the transaxle, the older model circlip, which requires compressing before removal, is used.

On Ford automatic transaxles (ATX), the left inner axle joint is inset into the transaxle housing. In this location, it is impossible to pry the axle joint from the transaxle. Therefore, the right axle joint must be removed first. A special tool is then used to drive the left inner axle joint from the differential side gear (Figure 4-19).

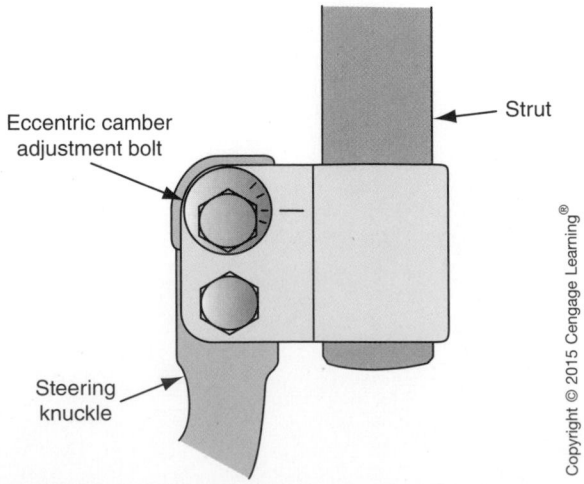

FIGURE 4-16 **Marking eccentric strut bolt before removal.**

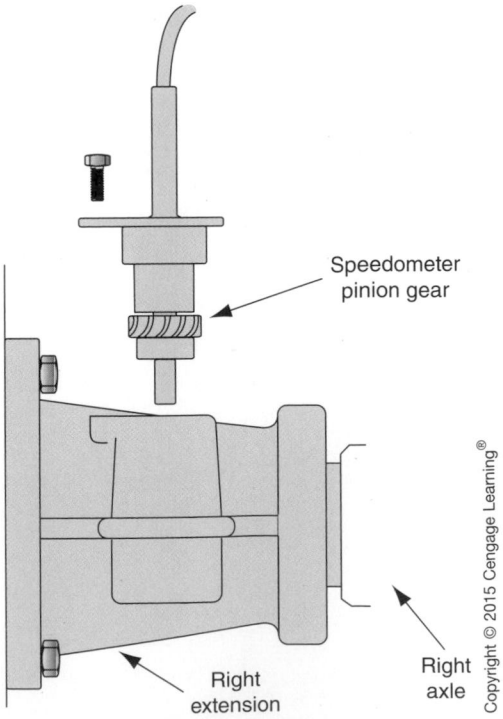

FIGURE 4-17 **Speedometer drive gear removal before right axle removal.**

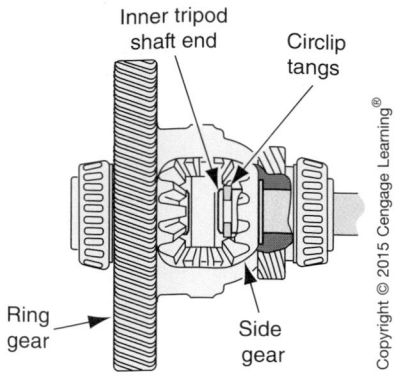

FIGURE 4-18 **Compressible circlips in early model Chrysler transaxles.**

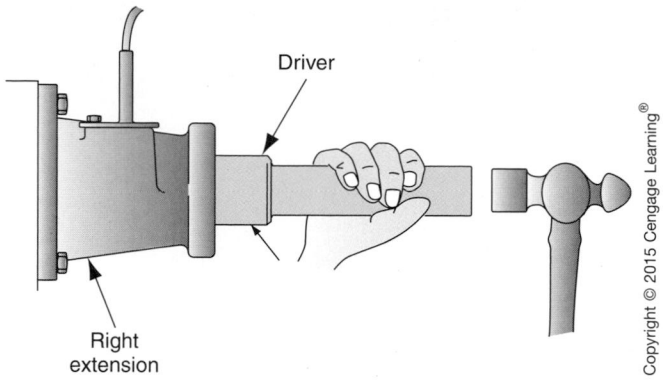

FIGURE 4-19 **Removal of left inner drive axle joint on Ford automatic transaxles.**

If both left and right drive axles are removed at the same time on Ford transaxles, the differential side gears may become dislocated. When this occurs, the differential must be removed to realign the side gears. The special tool for driving out the left inner axle joint may be left in place to support the side gears on ATX models. On manual transaxles (MTX), shipping plugs T81P-1177-B should be installed in each side gear when the drive axles are removed.

SERVICE TIP:
Do not remove the tool until the left inner axle joint is reinstalled in the Ford ATX. If the tool is removed without installing the left inner axle joint, the side gears may fall out of place. If this action occurs, the Ford ATX may have to be disassembled to install the side gears properly.

Front Wheel Bearing Hub Unit Removal and Replacement

The front wheel bearing removal and replacement procedure varies depending on the vehicle and the type of front wheel bearing. Always follow the front wheel bearing removal and replacement procedure in the manufacturer's service manual. The following procedure applies to front wheel bearing units that are pressed into the steering knuckle.

When front wheel bearing replacement is necessary, the steering knuckle must be removed and the wheel hub must be pressed from the bearing with a special tool (Figure 4-20).

A special puller is used to remove and replace the wheel bearing in the knuckle (Figure 4-21 and Figure 4-22).

The wheel hub must be pulled into the wheel bearing with a special tool (Figure 4-23). The proper driving tool is used to install the seal behind the bearing in the knuckle (Figure 4-24).

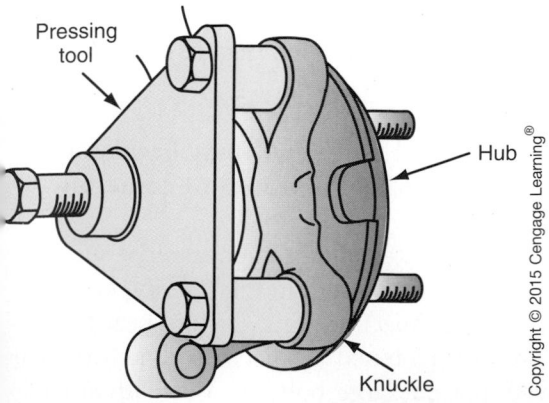

FIGURE 4-20 **Wheel hub removal from bearing.**

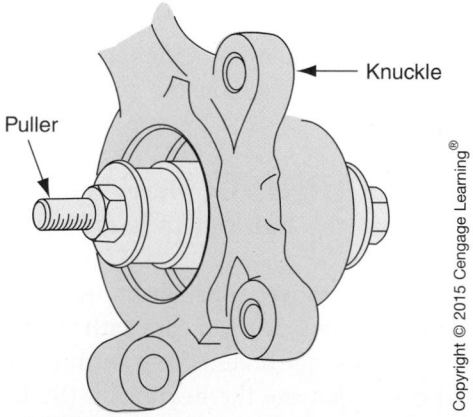

FIGURE 4-21 **Wheel bearing removal from knuckle.**

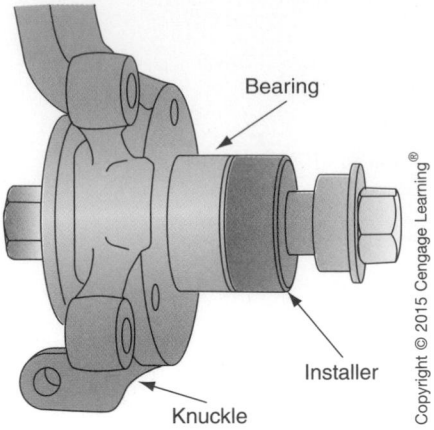

FIGURE 4-22 **Wheel bearing installation in knuckle.**

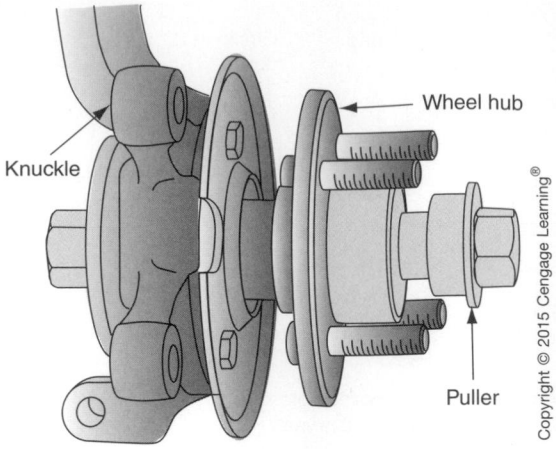

FIGURE 4-23 **Wheel hub installation in wheel bearing.**

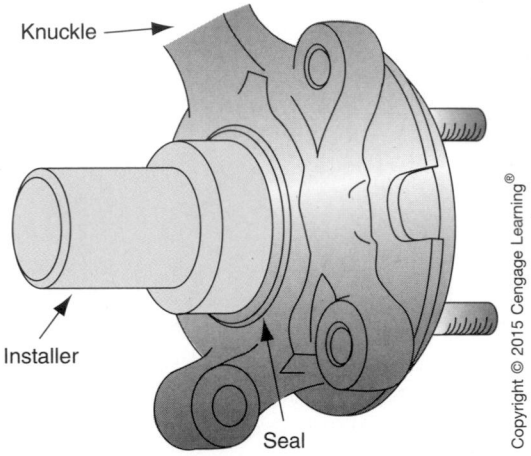

FIGURE 4-24 **Seal installation in steering knuckle.**

CAUTION:
Never use an impact wrench to tighten a hub nut. This action may cause wheel bearing damage.

When two separate roller bearings are mounted in the steering knuckle, the bearing races must be driven from the knuckle with a hammer and punch. These bearings must be lubricated with wheel bearing grease prior to installation, as described earlier in this chapter. When the wheel bearings are removed, all wheel bearing seals must be replaced. A staked-type hub nut must be replaced if it is removed.

On these front-wheel-drive cars, the hub nut torque applies the correct adjustment on the front wheel bearings. Therefore, this torque is extremely important. With the brakes applied, the hub nut should be tightened to the specified torque (Figure 4-25). When the hub nut is torqued to specifications, the nut lock and cotter pin should be installed (Figure 4-26).

Classroom Manual
Chapter 4, page 101

WARNING: Never reuse a cotter pin. A used cotter pin may break, allowing the hub nut to loosen. This may allow the wheel and hub to come off.

SPECIAL TOOL
Technician's stethoscope

After the wheel is installed, the wheel nuts should be tightened in sequence to the specified torque (Figure 4-27). On cars with the front wheel bearings mounted in the steering knuckles, never move a car unless the front hub nuts are torqued to specifications. Lack of bearing preload could damage the bearings if the hub nuts are not tightened to specifications. If the car must be moved when the drive axles are removed, place a large bolt and nut with suitable washers through the front wheel bearing and tighten the nut to specifications.

Copyright © 2015 Cengage Learning®

FIGURE 4-25 **Hub nut torquing.**

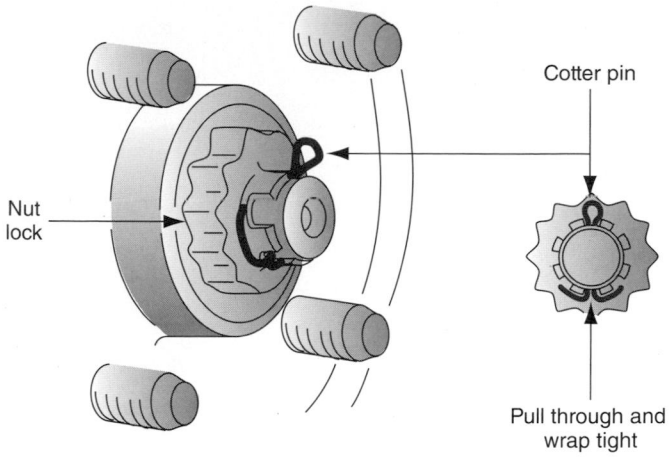

Copyright © 2015 Cengage Learning®

FIGURE 4-26 **Nut lock and cotter pin installation.**

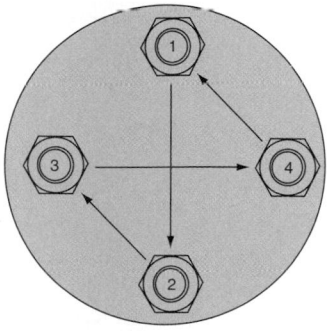

4-nut wheel

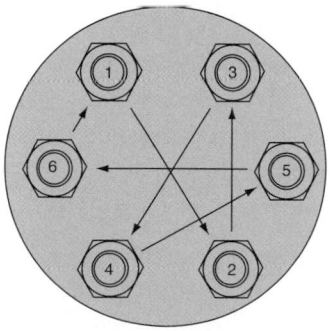

6-nut wheel

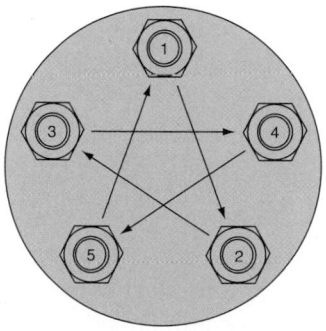

5-nut wheel

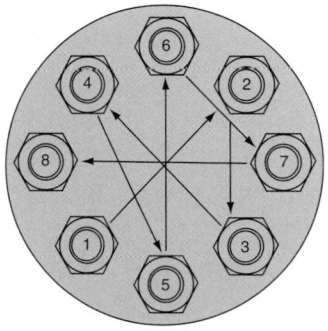

8-nut wheel

Copyright © 2015 Cengage Learning®

FIGURE 4-27 **Wheel nut tightening sequence.**

Rear-Axle Bearing and Seal Service, Rear-Wheel-Drive Cars

WARNING: Use extreme caution when diagnosing problems with a vehicle raised on a hoist and the engine running with the transmission in drive. Keep away from rotating wheels, drive shafts, or drive axles.

Rear-axle bearing noise may be diagnosed with the vehicle raised on a hoist. Be sure the hoist safety mechanism is engaged after the vehicle is raised on the hoist. With the engine running and the transmission in drive, operate the vehicle at a moderate speed of 35 to 45 mph (56 to 72 km/h) and listen with a **stethoscope** placed on the rear axle housing directly over the axle bearings. If grinding or clicking noises are heard, the bearing must be replaced.

Many axle shafts in rear-wheel-drive cars have a roller bearing and seal at the outer end (Figure 4-28). These axle shafts are often retained in the differential with **"C" locks** that must be removed before the axles.

"C" locks are split, circular, metal rings that fit in rear axle grooves to retain the rear axles in the differential on some rear-wheel-drive vehicles.

The rear-axle bearing removal and replacement procedure varies depending on the vehicle make and model year. Always follow the rear-axle bearing removal and replacement procedure in the manufacturer's service manual.

A typical rear axle shaft removal and replacement procedure on a rear-wheel-drive car with "C" lock axle retainers is as follows:

1. Loosen the rear wheel nuts and chalk mark the rear wheel position in relation to the rear axle studs.
2. Raise the vehicle on a hoist and make sure the hoist safety mechanism is in place.

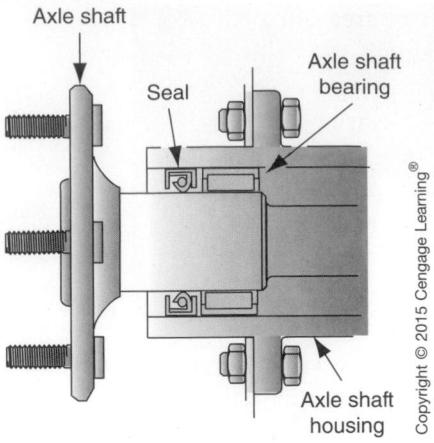

FIGURE 4-28 **Rear axle roller bearing and seal, rear-wheel-drive car.**

SPECIAL TOOLS

Seal puller

Bearing puller

3. Remove the rear wheels and brake drums, or calipers and rotors.
4. Place a drain pan under the differential and remove the differential cover. Discard the old lubricant.
5. Remove the differential lock bolt, pin, pinion gears, and shaft (Figure 4-29).
6. Push the axle shaft inward and remove the axle "C" lock.
7. Pull the axle from the differential housing.

Reverse the axle removal procedure to reinstall the axle. Always use a new differential cover gasket, and fill the differential to the bottom of the filler plug opening with the manufacturer's recommended lubricant. Be sure all fasteners, including the wheel nuts, are tightened to the specified torque.

A typical axle bearing and seal removal procedure follows:

1. Remove the axle seal with a seal puller.
2. Use the proper bearing puller to remove the axle bearing (Figure 4-30).
3. Clean the axle housing seal and bearing mounting area with solvent and a brush. Next clean this area with compressed air.
4. Check the seal and bearing mounting area in the housing for metal burrs and scratches. Remove any burrs or irregularities with a fine-toothed round file.
5. Wash the axle shaft with solvent and blow it dry with compressed air.

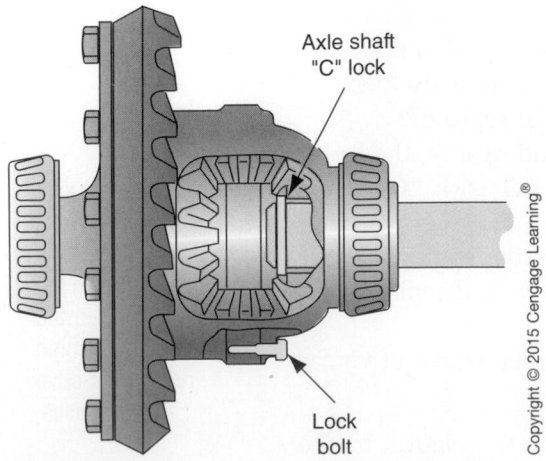

FIGURE 4-29 **Rear axle "C" lock, lock bolt, and pinion gears.**

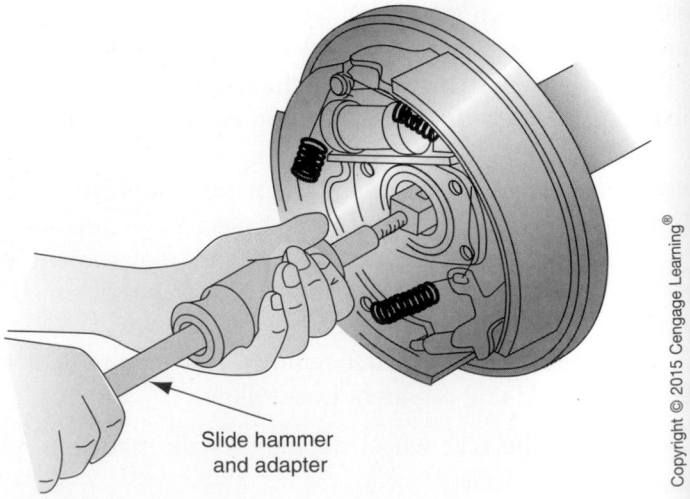

FIGURE 4-30 **Rear-axle bearing puller.**

6. Check the bearing contact area on the axle for roughness, pits, and scratches. If any of these conditions are present, the axle must be replaced.
7. Be sure the new bearing fits properly on the axle and in the housing. Install the new bearing with the proper bearing driver (Figure 4-31). The bearing driver must apply pressure to the outer race that is pressed into the housing.
8. Be sure the new seal fits properly on the axle shaft and in the housing. Make sure the garter spring on the seal faces toward the differential. Use the proper seal driver to install the new seal in the housing (Figure 4-32).
9. Lubricate the bearing, seal, and bearing surface on the axle with the manufacturer's specified differential lubricant.
10. Reverse the rear axle removal procedure to reinstall the rear axle.
11. Be sure all fasteners are tightened to the specified torque.

Classroom Manual
Chapter 4, page 104

WARNING: Never use an acetylene torch to heat axle bearings or adaptor rings during the removal and replacement procedure. The heat may cause fatigue in the steel axle and the axle may break suddenly, causing the rear wheel to fall off. This action will likely result in severe vehicle damage and personal injury.

Some rear axles have a sealed bearing that is pressed onto the axle shaft and held in place with a retainer ring. These rear axles usually do not have "C" locks in the differential. A retainer plate is mounted on the axle between the bearing and the outer end of the axle.

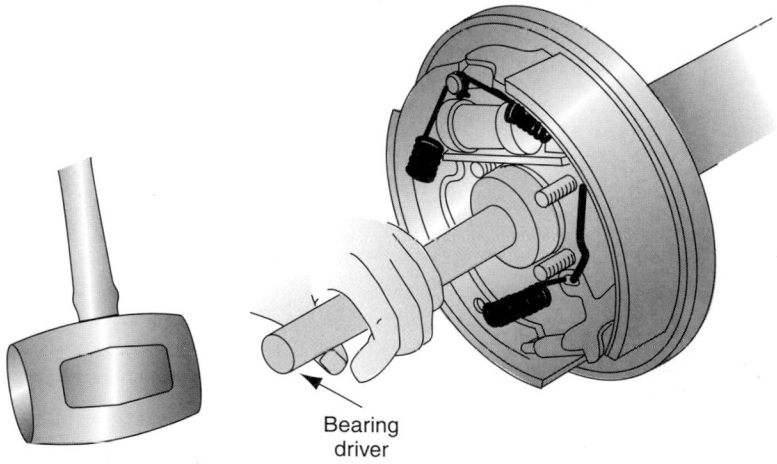

Copyright © 2015 Cengage Learning®

FIGURE 4-31 **Rear-axle bearing driver.**

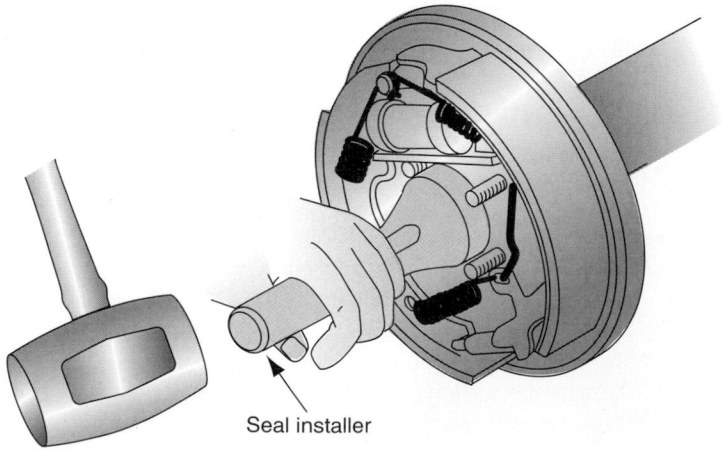

Copyright © 2015 Cengage Learning®

FIGURE 4-32 **Installing rear axle seal.**

SPECIAL TOOL
Axle puller

Classroom Manual
Chapter 4, page 104

This plate is bolted to the outer end of the differential housing. After the axle retainer plate bolts are removed, a slide-hammer-type puller is attached to the axle studs to remove this type of axle. When this type of axle bearing is removed, the adaptor ring must be split with a hammer and chisel while the axle is held in a vise. Do not heat the adaptor ring or the bearing with an acetylene torch during the removal or installation process. After the adaptor ring is removed, the bearing must be pressed from the axle shaft, and the bearing must not be reused. A new bearing and adaptor ring must be pressed onto the axle shaft. The bearing removal and replacement procedure is shown in Figure 4-33.

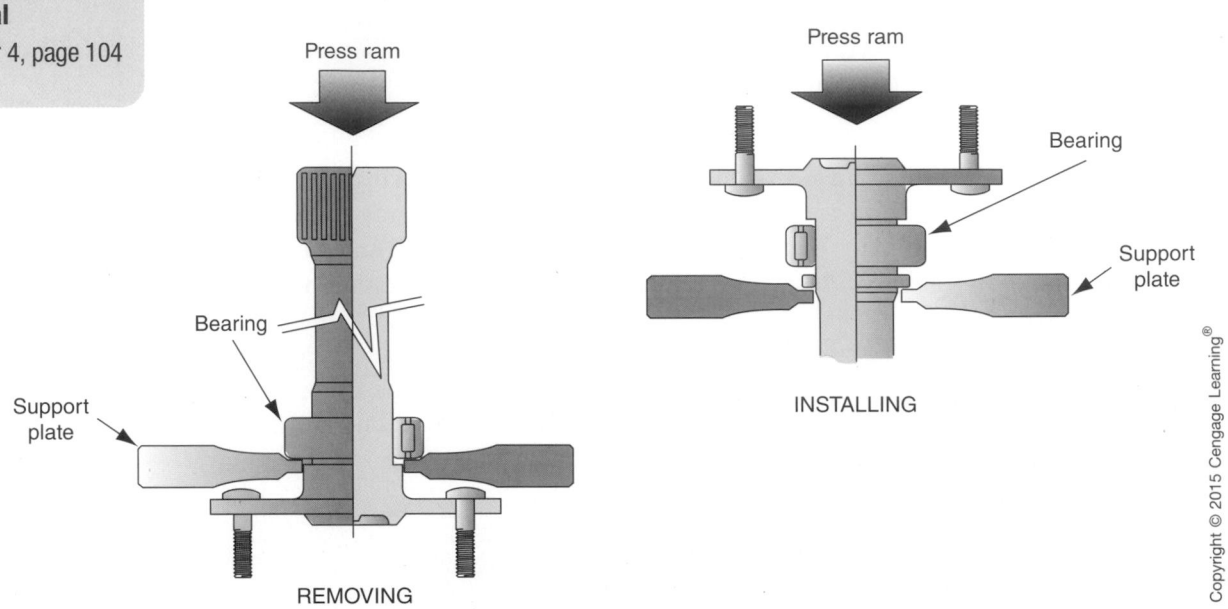

FIGURE 4-33 **Axle bearing and adaptor ring removal and replacement.**

CASE STUDY

A customer complains of a bearing noise in the right front wheel of his Lincoln Town Car. He says the right front outer wheel bearing has been replaced twice in the last year, and this is the third failure. The technician asks the customer about the mileage intervals between bearing replacements, and the customer indicates that the wheel bearing has lasted about 8,000 miles (12,800 kilometers) each time it has been replaced. The technician finds out from the customer that no other work was done on the car each time the bearing was replaced.

When the technician removes the right front wheel and hub, the outer bearing rollers and races are badly scored. After cleaning both bearings, races, and hub, the technician closely examines the outer bearing race. It shows an uneven wear pattern, which indicates misalignment. The technician removes the outer bearing race and finds a small metal burr behind the bearing race. This burr caused race misalignment and excessive wear on the race and rollers. The technician removes the burr with a fine-toothed file. The inner bearing and race have indentation wear because metal particles from the outer bearing contaminated the lubricant in the hub. The technician removes the inner bearing race and cleans the hub and spindle thoroughly. He replaces both bearings and the inner seal and repacks the bearings and hub with grease. After reinstalling the hub, the technician carefully adjusts the bearings to the manufacturer's specifications and tightens the wheel nuts to the specified torque. A road test indicates that the bearing noise has been eliminated.

One small metal burr caused this customer a considerable amount of unnecessary expense. This experience proves that a technician's diagnostic capability is extremely important!

TERMS TO KNOW

Bearing brinelling
Bearing fatigue spalling
Bearing frettage
Bearing smears
Circlip
"C" locks
Stethoscope

ASE-STYLE REVIEW QUESTIONS

1. While discussing defective bearings:

   Technician A says brinelling appears as indentations across the bearing races.

   Technician B says brinelling occurs while the bearing is rotating.

   Who is correct?

   A. A only
   B. B only
   C. Both A and B
   D. Neither A nor B

2. While discussing bearing defects:

   Technician A says misalignment wear on a front wheel bearing could be caused by a metal burr behind one of the bearing races.

   Technician B says misalignment wear on a front wheel bearing could be caused by a bent front spindle.

   Who is correct?

   A. A only
   B. B only
   C. Both A and B
   D. Neither A nor B

3. While discussing front wheel bearing service on a vehicle with the bearing hub unit pressed into the steering knuckle:

   Technician A says the front wheel bearings may be damaged if this front-wheel-drive vehicle is moved without the hub nuts torqued to specifications.

   Technician B says the hub nut torque supplies the correct wheel bearing adjustment.

   Who is correct?

   A. A only
   B. B only
   C. Both A and B
   D. Neither A nor B

4. While discussing front wheel bearing service on a front-wheel-drive car with two separate tapered roller bearings mounted in the front steering knuckles:

   Technician A says the hub nut torque supplies the correct bearing adjustment.

   Technician B says that the brake should be applied while the front hub nuts are torqued.

   Who is correct?

   A. A only
   B. B only
   C. Both A and B
   D. Neither A nor B

5. A front-wheel-drive vehicle has 14-in. (35.5-cm) rims and front wheel bearings mounted in the steering knuckles. The vehicle is lifted on a hoist, and a dial indicator is positioned against the outer rim lip. The total inward and outward rim movement is 0.035 in (0.889 mm).

   Technician A says that the wheel bearing may require replacement.

   Technician B says that the hub nut torque should be checked prior to bearing replacement.

   Who is correct?

   A. A only
   B. B only
   C. Both A and B
   D. Neither A nor B

6. Which of the following statements about wheel bearings is correct?

   A. Sealed bearings should be washed in solvent.
   B. High-pressure air may be used to spin the bearings.
   C. A bent bearing cage may be caused by improper tool use.
   D. Bearing overload may cause bearing cage wear.

7. All these statements about hub seal service are true EXCEPT:

   A. The garter spring must face toward the flow of lubricant.
   B. A ball peen hammer should be used to install the seal.
   C. Seal contact area on the spindle must be clean and free from metal burrs.
   D. The outer edge of the seal case should be coated with sealant.

8. Which of the following statements about servicing press-in rear-axle bearings on a rear-wheel-drive car is correct?

   A. These bearings may be reused after they are removed from the axle shaft.
   B. The bearing adaptor ring should be removed by splitting it with a hammer and chisel.
   C. A cutting torch may be used to cut the bearing off the axle shaft.
   D. An acetylene torch may be used to heat the new adaptor ring prior to installation.

9. A front-wheel-drive vehicle has two serviceable tapered roller bearings in each rear wheel hub. Which of the following statements about adjusting these wheel bearings is correct?
   A. The adjusting nut should be tightened to 17 to 25 ft.-lb (23 to 24 Nm), backed off 1/2 turn, and then tightened to 10 to 15 in.-lb (1.0 to 1.7 Nm).
   B. The adjusting nut should be tightened to 40 ft.-lb (54 Nm), backed off 1 turn, and then tightened to 10 ft.-lb (13.5 Nm).
   C. The adjusting nut should be tightened to 50 ft.-lb (67.5 Nm), backed off 3/4 turn, and then tightened to 10 to 15 in.-lb (1.0 to 1.7 Nm).
   D. The wheel and hub should not be rotated while adjusting the wheel bearings.

10. A unitized front wheel bearing hub that is bolted to the steering knuckle has 0.010 in (0.254 mm) of hub endplay. The proper repair procedure is to:
   A. Repack and readjust the wheel bearings in the hub.
   B. Tighten the hub nut to the specified torque.
   C. Inspect the drive axle and hub splines, and replace the worn components.
   D. Replace the wheel bearing hub assembly.

ASE CHALLENGE QUESTIONS

1. The customer says her front-wheel-drive car makes "a moaning noise" in a turn. Which of the following could cause this problem?
   A. Outer front wheel bearing.
   B. Rear-axle bearing.
   C. Differential gear noise.
   D. Transaxle output shaft bearing.

2. The customer complains of a "whining noise" in the back of her rear-wheel-drive car when driving between 30 and 40 miles per hour.
   Technician A says a good way to diagnose this problem is on a hoist with a stethoscope.
   Technician B says a good way to diagnose this problem is on the road with a microphone.
   Who is correct?
   A. A only
   B. B only
   C. Both A and B
   D. Neither A nor B

3. Upon inspecting a noisy front wheel bearing from a 4WD sport utility vehicle, brinelling damage to the outer bearing race was noticed.
   Technician A says the damage was probably caused by tightening the bearing nut with an impact wrench.
   Technician B says the damage was probably caused by driving the vehicle through deep, muddy water.
   Who is correct?
   A. A only
   B. B only
   C. Both A and B
   D. Neither A nor B

4. While discussing the cause of overheating damage of a front wheel bearing:
   Technician A says overheating may be caused by insufficient or incorrect bearing lubricant.
   Technician B says overheating may be caused by overtightening the wheel bearing nut on installation.
   Who is correct?
   A. A only
   B. B only
   C. Both A and B
   D. Neither A nor B

5. The customer says her front-wheel-drive car has a noise and vibration when decelerating and is most noticeable under 50 miles per hour. Which would be the most probable cause of this problem?
   A. Defective outer front wheel bearing.
   B. Defective inner front wheel bearing.
   C. Defective outer drive axle joint.
   D. Defective inner drive axle joint.

JOB SHEET 14

Name _____ **Date** _____

SERVICE INTEGRAL WHEEL BEARING HUBS

Upon completion of this job sheet, you should be able to measure integral wheel bearing hub endplay, and remove and replace integral wheel bearing hubs.

NATEF AST/MAST Task Correlation:

Suspension and Steering. related Suspension and Steering Service,
Task #D-2. Remove, inspect, and service or replace front and rear wheel bearings. (P-1)

Tools and Materials

A front-wheel-drive vehicle with integral-type front wheel bearing hubs
A floor jack and safety stands, or a lift for raising the vehicle
Dial indicator
Integral wheel bearing hub puller
Torque wrench

Describe the Vehicle Being Worked On:

Year _____ Make _____ Model _____
VIN _____ Engine type and size _____

Procedure

_____
Task Completed

1. Road test the vehicle and listen for any abnormal wheel bearing noises; check for steering looseness or wander that may be caused by loose wheel bearings.

Type of abnormal noise	Possible causes
a. _____	a. _____
b. _____	b. _____
c. _____	c. _____

Steering wander and looseness
☐ Yes ☐ No

2. Use a dial indicator to measure the endplay on the front wheel bearing hubs with the wheel installed on the hub.
   Endplay left side _____ Endplay right side _____
   Specified endplay _____
   Necessary repairs _____

3. Loosen and remove the cotter pin, lock nut, and drive axle nut before raising the vehicle. ☐

4. Raise the vehicle on a lift and remove the front wheels. ☐

5. Remove the brake calipers and brake rotors. Use a length of wire to tie the calipers to the chassis. ☐

Task Completed

6. Use a dial indicator to measure the endplay directly on the front wheel bearing hubs.
   Endplay left side _____ Endplay right side _____
   Specified endplay _____
   Necessary repairs _____

☐ 7. If a wheel speed sensor for the antilock brake system (ABS) is integral in the front wheel hubs, disconnect the wheel speed sensor connectors.

☐ 8. Remove the hub-to-knuckle bolts, and place the transaxle in park.

☐ 9. Use the proper puller to remove the wheel bearing hub from the drive axle.

☐ 10. Install the new hub and bearing assembly on the drive axle splines, and install a new drive axle nut. Tighten the drive axle nut to pull the hub onto the splines. Do not tighten this nut to the specified torque at this time.

11. Place the transaxle in neutral and install the wheel bearing hub bolts. Tighten these bolts to the specified torque.
    Specified wheel bearing hub bolt torque _____
    Actual wheel bearing hub bolt torque _____

12. Install the wheel sensor connectors, rotors, and calipers. Tighten the caliper mounting bolts to the specified torque. Place a large punch between the rotor fins and the caliper, and tighten the drive axle nut to the specified torque.
    Specified caliper bolt torque _____
    Actual caliper bolt torque _____
    Specified drive axle nut torque _____
    Actual drive axle nut torque _____

13. Install the wheels in their original positions, and tighten the wheel nuts to the specified torque.
    Specified wheel nut torque _____
    Actual wheel nut torque _____

14. Lower the vehicle onto the shop floor, and tighten the drive axle nut to the specified torque. Install the lock nut and a new cotter pin.
    Specified drive axle nut torque _____
    Actual drive axle nut torque _____
    Are the lock nut and cotter pin properly installed and tightened?
    Instructor's Check _____

15. Road test the car and listen for abnormal wheel bearing noises.
    Wheel bearing noise: ☐ Satisfactory ☐ Unsatisfactory

Instructor's Response _____

_____

_____

JOB SHEET 15

Name _____ **Date** _____

Diagnose Wheel Bearings

Upon completion of this job sheet, you should be able to diagnose wheel bearings.

NATEF AST/MAST Task Correlation:

Suspension and Steering. related Suspension and Steering Service,
Task #D-2. Remove, inspect, and service or replace front and rear wheel bearings. (P-1)

Tools and Materials

A vehicle with wheel bearing noise or bearing noise from another source

Describe the Vehicle Being Worked On:

Year _____ Make _____ Model _____
VIN _____ Engine type and size _____

Procedure

Task Completed

1. Road test the vehicle, listen for any abnormal wheel noises, and check for steering looseness and wander that may be caused by loose wheel bearings. ☐
2. Wheel bearing noise: ☐ Satisfactory ☐ Unsatisfactory
3. Is the bearing noise coming from the front or rear of the vehicle? ☐ Front ☐ Rear
4. Is this noise more noticeable when turning a corner at low speeds? ☐ Yes ☐ No
5. Is this noise more noticeable during acceleration? ☐ Yes ☐ No
6. Is this noise more noticeable during deceleration? ☐ Yes ☐ No
7. Is this noise more noticeable when driving at a steady speed? ☐ Yes ☐ No
   If the answer to this question is yes, state the speed when the bearing noise is most noticeable. _____
8. Is this noise most noticeable when driving down a narrow street at a steady low speed? ☐ Yes ☐ No
9. State the exact cause of the bearing noise, and explain the reason for your diagnosis.
   _____
   _____
   _____

Instructor's Response _____
_____
_____

JOB SHEET 16

Name _____ **Date** _____

Clean, Lubricate, Install, and Adjust Nonsealed Wheel Bearings

Upon completion of this job sheet, you should be able to clean, lubricate, install, and adjust nonsealed wheel bearings.

NATEF AST/MAST Task Correlation:

Suspension and Steering. related Suspension and Steering Service,
Task #D-2. Remove, inspect, and service or replace front and rear wheel bearings. (P-1)

Tools and Materials

A vehicle with tapered roller-type front wheel bearings or a front-wheel-drive car with tapered roller-type rear wheel bearings
A floor jack and safety stands, or a lift for raising the vehicle
Wheel bearing grease
Cleaning solution
Seal drivers
Torque wrench

Describe the Vehicle Being Worked On:

Year _____ Make _____ Model _____
VIN _____ Engine type and size _____

Procedure — Task Completed

1. Raise the vehicle on a lift with the chassis supported on the lift. ☐
2. Chalk mark the wheels in relation to the studs; remove the wheels, wheel hubs, and brake drums or rotors. ☐
3. Remove the inner hub seal and wheel bearings from the hub. ☐
4. Clean the hub and bearings with an approved cleaning solvent.
   Are the hub and bearings properly cleaned? ☐ Yes ☐ No
   Instructor's Check _____
5. Inspect the wheel bearings, bearing cones, and hubs.

Type of defects	Necessary replacement parts
a. _____	a. _____
b. _____	b. _____
c. _____	c. _____

6. Inspect all bearing-related conditions such as lubrication, alignment, spindle condition, and hub condition. State the cause(s) of the bearing defects listed in step 5.

   a. _____

   b. _____

   c. _____

7. Repack the wheel bearings and hub with the car manufacturer's specified wheel bearing grease.

   Are the wheel bearings and hub repacked with the manufacturer's recommended grease? ☐ Yes ☐ No

   Instructor's Check _____

8. Install a new inner hub seal with the proper seal driver. Lubricate the seal lips with a light coating of the car manufacturer's recommended wheel bearing grease.

   Is the new inner hub seal properly installed? ☐ Yes ☐ No

   Instructor's Check _____

9. Inspect, clean, and lubricate the spindles and seal contact area.

   Spindle and seal contact condition: ☐ Satisfactory ☐ Unsatisfactory

   If unsatisfactory, state necessary repairs. _____

   Is the spindle and seal contact area properly cleaned and lubricated? ☐ Yes ☐ No

   Instructor's Check _____

10. Install the wheel hubs, wheel bearings, washer, and retaining nut. Adjust the wheel bearing retaining nut to the specified initial torque.

    Specified initial wheel bearing torque _____

    Actual initial wheel bearing torque _____

11. Back off the wheel bearing retaining nut the specified amount.

    Specified portion of a turn to back off the wheel bearing retaining nut _____

    Actual portion of a turn the wheel bearing retaining nut is backed off _____

12. Tighten the wheel bearing retaining nut to the final specified torque.

    Specified final wheel bearing retaining nut torque _____

    Actual final wheel bearing retaining nut torque _____

13. Install a new cotter pin through the wheel bearing retaining nut and spindle opening.

    Is the new cotter pin properly installed? ☐ Yes ☐ No

    Instructor's Check _____

14. Install the wheel bearing dust covers. Install the wheels in their original position, and tighten the wheel nuts to the specified torque.

    Specified wheel nut torque _____

    Actual wheel nut torque _____

15. Road test the car and listen for abnormal wheel bearing noises.
Wheel bearing noise ☐ Satisfactory ☐ Unsatisfactory

Instructor's Response _____

_____

_____

Chapter 5

Shock Absorber and Strut Diagnosis and Service

BASIC TOOLS

Basic technician's tool set
Service manual
Hydraulic floor jack
Jack stands
Center punch

Upon completion and review of this chapter, you should be able to:

- Perform a visual shock absorber inspection.
- Perform a shock absorber bounce test, and determine shock absorber condition.
- Determine shock absorber condition from a manual shock absorber test.
- Remove and replace shock absorbers.
- Diagnose shock absorber and strut noise complaints.
- Remove and replace front and rear struts.
- Remove struts from coil springs.
- Install coil springs on struts.
- Follow the vehicle manufacturer's recommended strut disposal procedure.
- Perform off-car strut cartridge replacement procedures.
- Perform on-car strut cartridge replacement procedures.
- Diagnose electrically controlled shock absorbers.

Shock absorbers and struts must be in good condition to provide satisfactory ride quality and maintain vehicle safety. Worn-out shock absorbers and struts may cause excessive chassis oscillations and harsh ride quality with resulting passenger discomfort. On a severely rough road surface, the chassis oscillations caused by worn-out shock absorbers and struts may contribute to a loss of steering control, resulting in a vehicle collision. Therefore, technicians must be familiar with shock absorber and strut diagnosis and service procedures.

SERVICE TIP: During a visual shock absorber inspection, check the rebound, or strikeout, bumpers on the control arms or chassis. If the rebound bumpers are severely worn, the shock absorbers may be worn out.

Shock Absorber Visual Inspection

Bolt Mounting

Shock absorbers should be inspected for loose mounting bolts and worn mounting bushings. If these components are loose, rattling noise is evident, and replacement of the bushings and bolts is necessary.

Bushing Condition

In some shock absorbers, the bushing is permanently mounted in the shock, and the complete unit must be replaced if the bushing is worn. When the mounting bushings are worn, the shock absorber will not provide proper spring control.

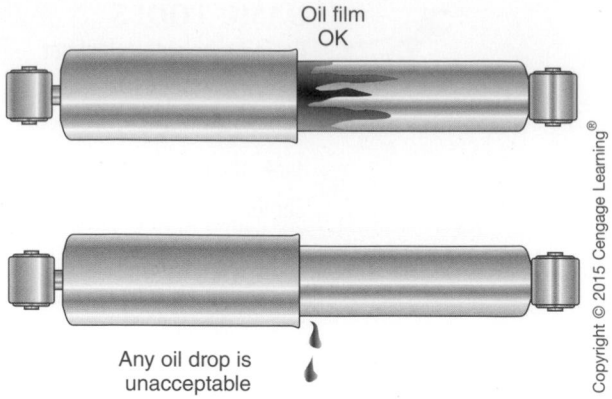

FIGURE 5-1 **Shock absorber and strut oil leak diagnosis.**

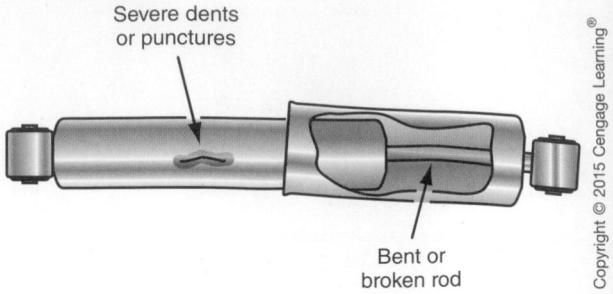

FIGURE 5-2 **Damaged shock absorber inspection.**

Struts are similar internally to shock absorbers, but struts also support the steering knuckle. In many applications, the coil spring is mounted on the strut.

Oil Leakage

Shock absorbers and **struts** should be inspected for oil leakage. A slight oil film on the lower oil chamber is acceptable. Any indication of oil dripping is not acceptable, and unit replacement is necessary (Figure 5-1).

Shock absorbers and struts should be inspected visually for physical damage, such as a bent condition and severe dents or punctures. When any of these conditions are present, unit replacement is required (Figure 5-2).

Shock Absorber Testing

Classroom Manual
Chapter 5, page 111

Customer Care: Always be willing to spend a few minutes explaining problems, including safety concerns, regarding the customer's vehicle. When customers understand why certain repairs are necessary, they feel better about spending the money. For example, if you explain that worn-out shock absorbers cause excessive chassis oscillations and may result in loss of steering control on irregular road surfaces, the customer is more receptive to spending the money for shock absorber replacement.

A shock absorber may be called a shock.

Although many technicians still rely on the bounce test described below to test the condition of the hydraulic shock absorber or strut, it is not the most reliable method of diagnosis because it does not simulate all the road force conditions that today's velocity sensitive shock absorbers are exposed to under all driving conditions. The test may uncover a noise from a shock absorber or a strut that have completely failed but does not insure that the suspension system is still capable of providing vehicle stability and designed damping control.

All modern hydraulic shock absorbers and struts are velocity sensitive and react to the piston travel speed. There are various areas that can wear internally on a shock absorber. The top seal can wear creating an external leak and the piston and cylinder wall can wear causing an internal leak (Figure 5-3). But a more common point of wear or failure is the damping valves in the piston assembly. The best way to test a shock absorber's performance is to go on a test drive and watch for:

- Brake dive on sudden stops
- Acceleration squat

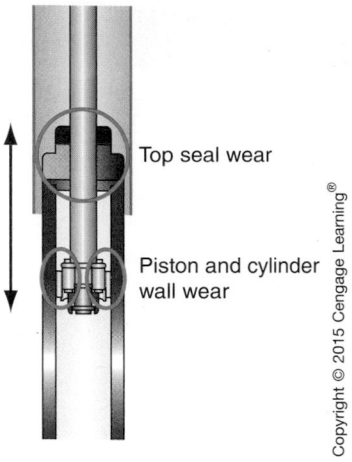

FIGURE 5-3 **Shock absorbers can develop wear at the top seal creating an external leak, or at the piston seal or bore creating an internal leak.**

- Body roll and swerve
- Ride comfort under all driving conditions

When the **bounce test** is performed, the bumper is pushed downward with considerable weight applied on each corner of the vehicle. The bumper is released after this action, and one free upward bounce should stop the vertical chassis movement if the shock absorber or strut provides proper spring control. Shock absorber replacement is required if more than one free upward bounce occurs. The shock absorber bounce test is illustrated in Photo Sequence 7. Refer to Table 5-1 for addition diagnostic information.

SERVICE TIP: A road test should be completed to test for shock absorber noise. If the gas leaks out of a shock absorber, oil will leak past the piston and the shock valve may knock (slap) against the oil. This knocking noise is heard when driving the vehicle over small bumps. This noise cannot be duplicated during a bounce test.

SERVICE TIP: On twin I-beam front suspension systems, front wheel camber may be noticeably different after a bounce test.

Shock Absorber Manual Test

A **manual test** may be performed on shock absorbers. When this test is performed, disconnect the lower end of the shock, and move the shock up and down as rapidly as possible. A satisfactory shock absorber should offer a strong, steady resistance to movement on the entire compression and rebound strokes. The amount of resistance may be different on the compression stroke compared with the rebound stroke. If a loss of resistance is experienced during either stroke, shock replacement is essential.

Some defective shock absorbers or struts may have internal clunking, clicking, and squawking noises, or binding conditions. When these shock absorber noises or conditions are experienced, shock absorber or strut replacement is necessary.

TABLE 5-1 **SHOCK ABSORBER AND STRUT DIAGNOSIS**

Problem	Symptoms	Possible Causes
Harsh ride quality	Excessive chassis bouncing Excessive vertical wheel oscillations	Worn-out shock absorbers or struts
Shock absorber or strut oil leaks	Oil dripping from shock absorber or strut	Worn-out seal in shock absorber or strut, damaged lower shock absorber or strut housing
Rattling noise	Rattling noise in chassis when driving	Worn-out shock absorber or strut mountings
Chattering noise	Chattering noise when turning the front wheels to the right or left	Worn-out upper strut mount and bearing assembly
Worn, damaged rebound bumpers	Chassis bottoming out	Worn-out shock absorbers or struts, low ride height

Copyright © 2015 Cengage Learning®

PHOTO SEQUENCE 7

Rear Shock Absorber Visual Inspection and Bounce Test

P7-1 **Raise the vehicle on a lift.**

Copyright © 2015 Cengage Learning®

P7-2 **Inspect the rear shock absorbers for oil leaks and damage such as dents and punctures.**

Copyright © 2015 Cengage Learning®

P7-3 **Grasp the lower shock absorber tube and attempt to move the shock absorber vertically and horizontally to check for looseness and wear on the shock absorber bushing and mounting bolt.**

Copyright © 2015 Cengage Learning®

P7-4 **Grasp the upper shock absorber cover and attempt to move it vertically and horizontally to check for looseness and wear on the shock absorber mounting bushing and mounting bolt.**

Copyright © 2015 Cengage Learning®

P7-5 **Disconnect the lower shock absorber mounting and grasp the lower end of the shock absorber. Pull the shock absorber downward on the extension stroke and push the lower end of the shock absorber upward on the compression stroke. The shock absorber should offer resistance to movement in relation to the shock absorber ratio.**

Copyright © 2015 Cengage Learning®

P7-6 **Install the lower shock absorber mounting and tighten the mounting bolt to the specified torque.**

Copyright © 2015 Cengage Learning®

P7-7 **Lower the vehicle onto the shop floor, and move the vehicle off the lift.**

Copyright © 2015 Cengage Learning®

P7-8 **Using your knee, press down on the rear bumper with considerable weight, and suddenly remove your knee from the bumper. Count the number of free chassis oscillations before the chassis stops bouncing upward and downward.**

Copyright © 2015 Cengage Learning®

Air Shock Absorber Diagnosis and Replacement

Air shock absorbers are similar to conventional shocks except they have air pressure applied to them to control chassis curb height. Some air shock absorbers must be pressurized with the shop air hose. This type of unit contains a valve for inflation purposes. Other air shock absorbers are inflated by an onboard compressor with interconnecting plastic lines between the compressor and the shocks. Shock absorber lines must be inspected for breaks, cracks, and sharp bends. If any of these defects are present, the line must be replaced. The shock absorber lines must be secured to the chassis, and they must not rub against other components.

When air shock absorbers slowly lose their air pressure and reduce the curb riding height, shock replacement is required. Before removing an air shock, relieve the air pressure in the shock.

CAUTION: Gas-filled shock absorbers will extend when disconnected.

Upward wheel movement is referred to as wheel jounce, and downward wheel movement is called rebound.

Shock Absorber Replacement

WARNING: Never apply heat to the lower shock absorber or strut chamber with an acetylene torch. Excessive heat may cause a shock absorber or strut explosion, which could result in personal injury.

When shock absorber replacement is necessary, follow this procedure:

1. Before replacing rear shock absorbers, lift the vehicle on a hoist and support the rear axle on safety stands so the shock absorbers are not fully extended.
2. When a front shock absorber must be changed, lift the front end on the vehicle with a floor jack, then place safety stands under the lower control arms. Lower the vehicle onto the safety stands and remove the floor jack.
3. Disconnect the upper shock mounting nut and grommet.
4. Remove the lower shock mounting nut or bolts, and remove the shock absorber.
5. Reverse steps 1 through 4 to install the new shock absorber and grommets.
6. With the full vehicle weight supported on the suspension, tighten the shock mounting nuts to the specified torque.

Diagnosis of Front Spring and Strut Noise

Strut chatter may be heard when the steering wheel is turned with the vehicle not moving or moving at low speed. To verify the location of this chattering noise, place one hand on a front coil spring while someone turns the steering wheel. If strut chatter is present, the spring binds and releases as it turns. This condition is caused by the upper spring seat binding against the strut bearing mount. A revised spring seat is available to correct this problem on some models.

Classroom Manual
Chapter 5, page 119

A noise that occurs on sharp turns or during front suspension jounce may be caused by one of the following problems:

1. Interference between the upper strut rebound stop and the upper mount or **strut tower**.
2. Interference between the coil spring and the tower (Figure 5-4).
3. Interference between the coil spring and the upper mount (Figure 5-5).

A **strut tower** is a raised, circular, reinforced area inboard of the front fenders, which supports the upper end of the strut and coil spring assembly.

On some models, these coil spring interference problems may be corrected by installing upper coil spring spacers on top of the coil spring. Spring removal from the strut is required to install these spacers.

On many front suspension systems the strut is bolted to the upper end of the knuckle. When the front wheels are turned, the strut and coil spring rotate together and pivot on the upper strut mount bearing. In this type of suspension, a defective upper strut mount may cause strut chatter and noise.

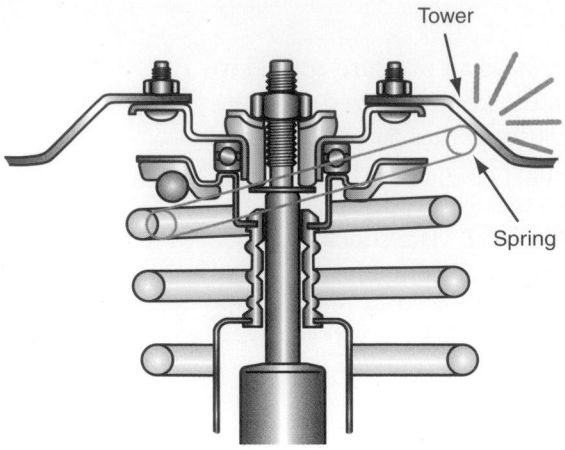

Copyright © 2015 Cengage Learning®

FIGURE 5-4 **Coil spring to upper strut tower interference.**

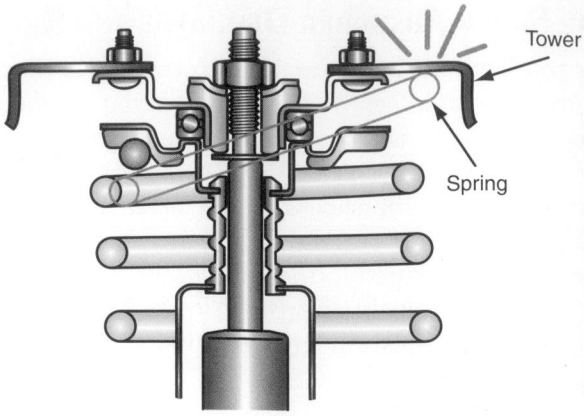

Copyright © 2015 Cengage Learning®

FIGURE 5-5 **Coil spring to upper mount interference.**

On other front suspension systems, the strut extends downward and the lower ball joint retains the lower end of the strut to the lower control arm (Figure 5-6). The front hub is bolted to a flange on the knuckle. A bearing assembly is mounted in the lower spring seat. When the front wheels are turned, the strut rotates on the bearing in the lower spring seat, but the spring and upper strut mount do not rotate. Therefore, in this type of suspension system, strut chatter may be caused by a worn or defective lower spring seat bearing.

If the lateral movement of the strut rod and nut above the strut tower exceeds 3/16 in. (4.76 mm), the upper strut bearing and mount assembly should be replaced.

Worn **spring insulators** or broken coil springs cause a rattling noise on road irregularities. Broken coil springs result in reduced curb riding height and harsh riding.

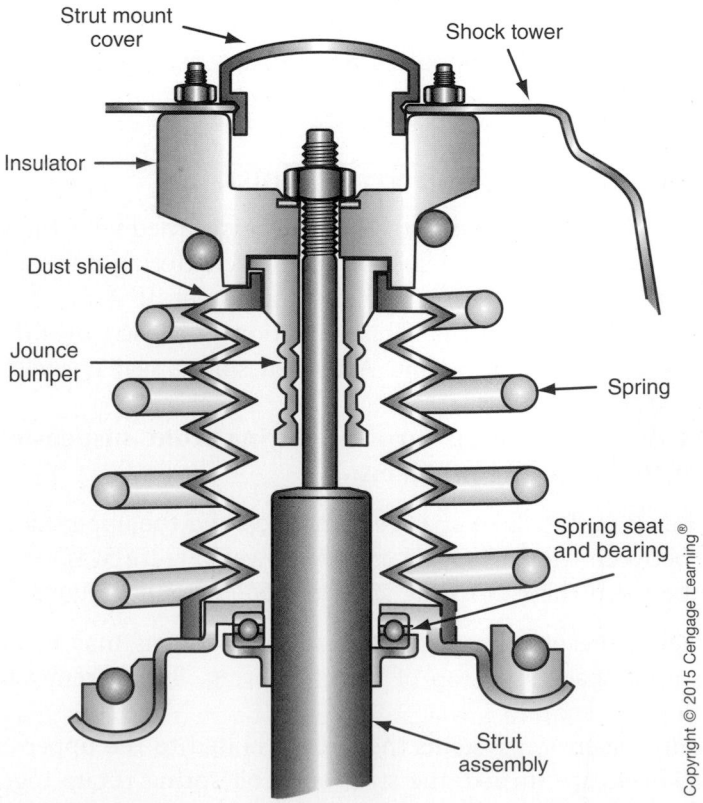

Copyright © 2015 Cengage Learning®

FIGURE 5-6 **Strut assembly with bearing mounted in the lower spring seat.**

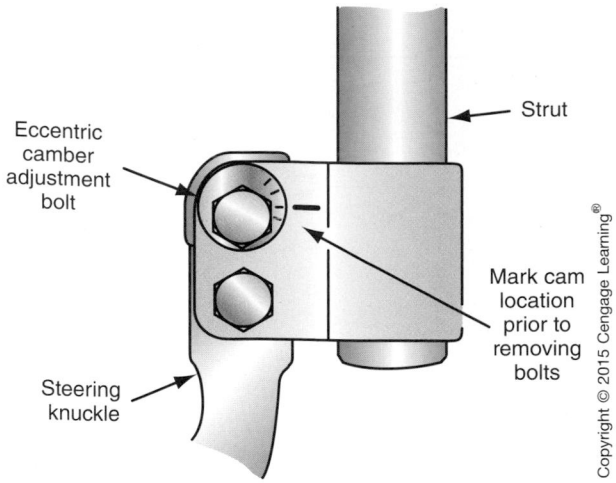

FIGURE 5-7 **Camber bolt marking for strut removal.**

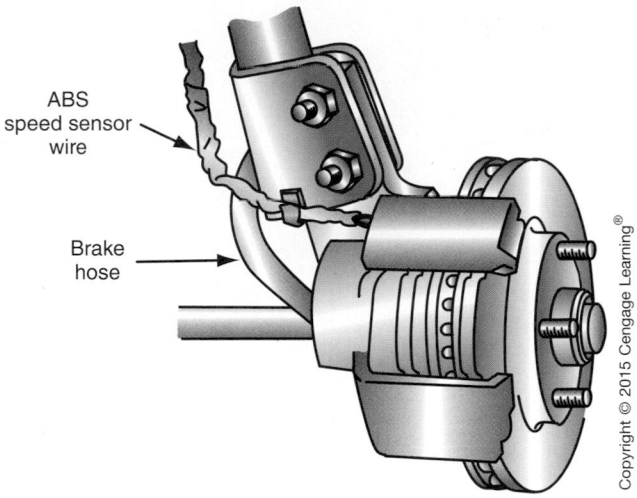

FIGURE 5-8 **Brake line and ABS wheel speed sensor wire removed from the strut.**

Strut Removal and Replacement

Before a front strut-and-spring assembly is removed, the strut must be removed from the steering knuckle, and the top strut mount bolts must be removed from the strut tower. If an **eccentric camber bolt** is used to attach the strut to the knuckle, always mark the bolt head in relation to the strut and reinstall the bolt in the same position (Figure 5-7).

An **eccentric camber bolt** has an oblong head, which provides inward or outward steering knuckle movement in relation to the strut as this bolt is rotated.

Always follow the vehicle manufacturer's recommended procedure in the service manual for removal of the strut-and-spring assembly.

A typical procedure for strut-and-spring assembly removal follows:

1. Raise the vehicle on a hoist or floor jack. If a floor jack is used to raise the vehicle, lower the vehicle onto safety stands placed under the chassis so the lower control arms and front wheels drop downward. Remove the floor jack from under the vehicle.
2. Remove the brake line and antilock brake system (ABS) wheel speed sensor wire from clamps on the strut (Figure 5-8). In some cases, the clamps may also have to be removed from the strut.
3. Remove the strut-to-steering knuckle retaining bolts, and remove the strut from the knuckle (Figure 5-9).
4. Remove the upper strut mounting bolts on top of the strut tower, and remove the strut-and-spring assembly (Figure 5-10).

Classroom Manual
Chapter 5, page 120

Removal of Strut from Coil Spring

Classroom Manual
Chapter 5, page 119

WARNING: Always use a coil spring compressing tool according to the tool or vehicle manufacturer's recommended service procedure. Be sure the tool is properly installed on the spring. If a coil spring slips off the tool when the spring is compressed, severe personal injury or property damage may occur.

SPECIAL TOOL
Coil spring compressing tool

WARNING: Never loosen the upper strut mount retaining nut on the end of the strut rod unless the spring is compressed enough to remove all spring tension from the upper strut mount. If this nut is loosened with spring tension on the upper mount, this mount becomes a very dangerous projectile that may cause serious personal injury or property damage.

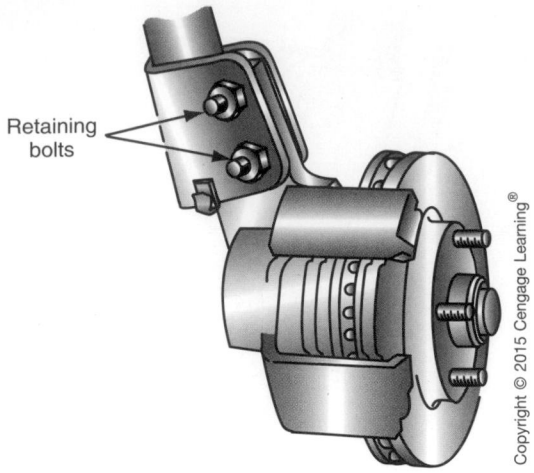

FIGURE 5-9 **Removing strut-to-steering knuckle retaining bolts.**

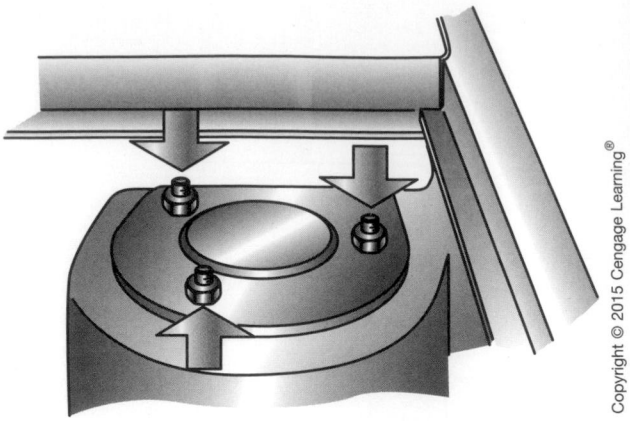

FIGURE 5-10 **Removing upper strut mounting bolts on top of strut tower.**

Spring compressing tools are used to compress a coil spring and relieve the spring tension on the upper strut mount to allow spring removal from the strut.

The coil spring must be compressed with a special tool before the strut can be removed. All the tension must be removed from the upper spring seat before the upper strut piston rod nut is loosened. Many different **spring compressing tools** are available, and they must always be used according to the manufacturer's recommended procedure. *If the coil spring has an enamel-type coating, tape the spring where the compressing tool contacts the spring.* The spring may break prematurely if this coating is chipped.

A typical procedure for removing a strut from a coil spring follows:

1. Install the coil spring and strut assembly in the spring compressing tool according to the tool or vehicle manufacturer's recommended procedure.
2. Adjust the compressing arms on the spring compressing tool so the arms contact the coils farthest away from the center of the spring (Figure 5-11).
3. Turn the handle on top of the compressing tool until all the spring tension is removed from the upper strut mount (Figure 5-11).
4. Loosen and remove the nut on the upper strut rod (Figure 5-12). Be sure all the spring tension is removed from the upper strut mount before loosening this nut.

CAUTION:
Never clamp the lower shock absorber or strut chamber in a vise with excessive force. This action may distort the lower chamber and affect piston movement in the shock absorber or strut.

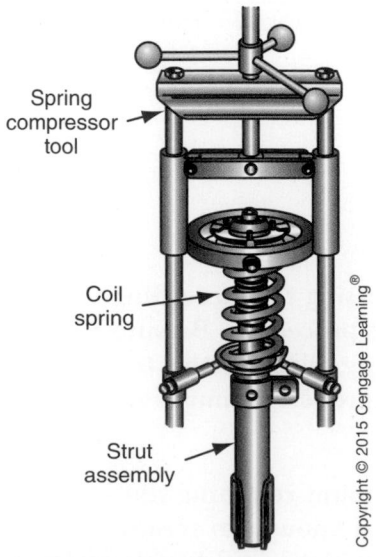

FIGURE 5-11 **Coil spring and strut assembly mounted in a spring compressing tool.**

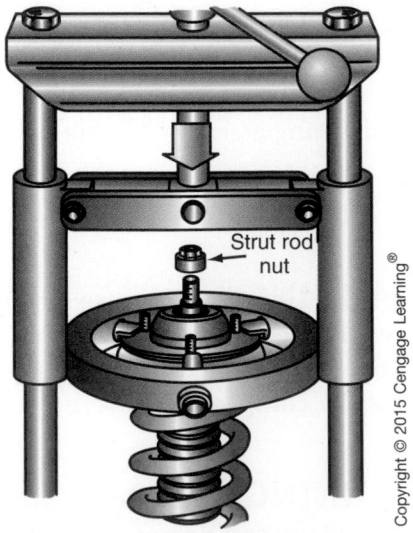

FIGURE 5-12 **After the compressing tool is operated to remove all the spring tension, remove the strut rod nut.**

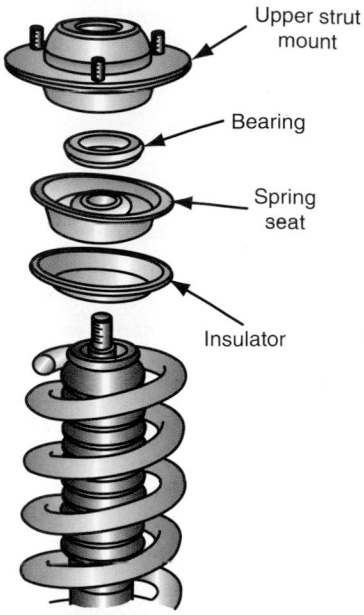

Copyright © 2015 Cengage Learning®

FIGURE 5-13 **Removal of the upper strut mount, bearing, upper spring seat, and insulator.**

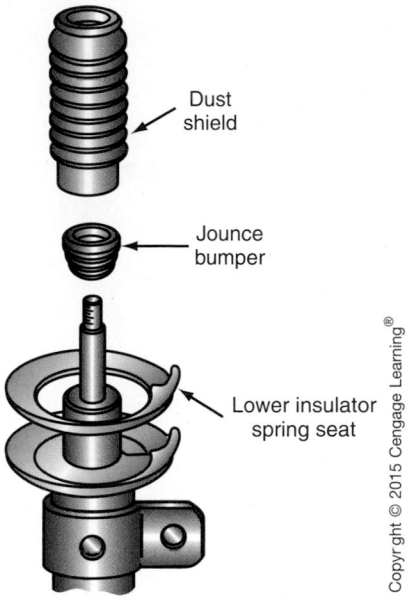

Copyright © 2015 Cengage Learning®

FIGURE 5-14 **Removal of the dust shield, jounce bumper, and lower spring insulator.**

5. Remove the upper strut mount assembly and mount bearing, and then remove the upper spring seat and insulator (Figure 5-13).
6. Rotate the handle on the spring compressing tool to release all the tension on the coil spring, and then remove the spring.
7. Remove the dust shield and jounce bumper from the strut rod, and then remove the lower spring insulator (Figure 5-14).

SERVICE TIP: Because of high labor costs, it is often more economical for the customer if the technician replaces a strut and coil spring as a complete assembly rather than replacing these components individually. Many automotive parts suppliers market struts and coil spring as complete assemblies.

Strut Disposal Procedure

WARNING: Always follow the vehicle manufacturer's recommended procedure for strut disposal. Do not throw gas-filled shock absorbers or struts in a fire of any kind or in a dumpster. If the vehicle manufacturer recommends drilling the strut to release the gas charge, drill the strut at the manufacturer's recommended location.

The following is a typical strut drilling procedure that is performed prior to strut disposal:

1. Fully extend the strut rod.
2. Center punch the strut at the manufacturer's recommended drilling location (Figure 5-15).
3. Drill a small hole at the center-punched position.

Installation of Coil Spring on Strut

A typical procedure for installing a coil spring on a strut follows:

1. Install the lower insulator on the lower strut spring seat, and be sure the insulator is properly seated (Figure 5-16).
2. Install the spring bumper on the strut rod (Figure 5-17).

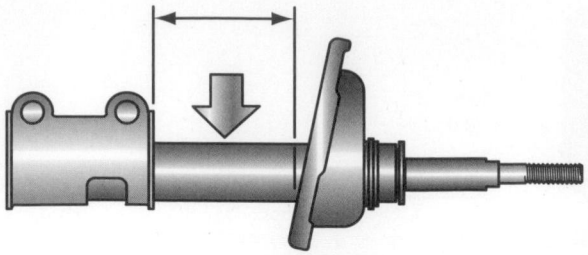

Copyright © 2015 Cengage Learning®

FIGURE 5-15 **Strut drilling location.**

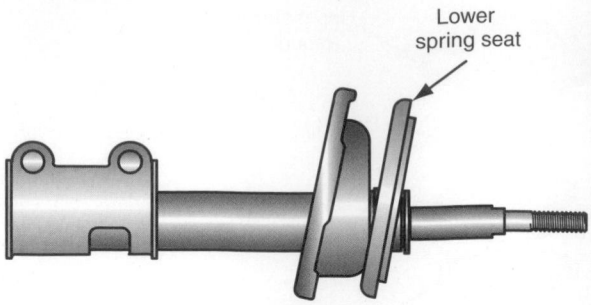

Copyright © 2015 Cengage Learning®

FIGURE 5-16 **Insulator installation on lower spring seat.**

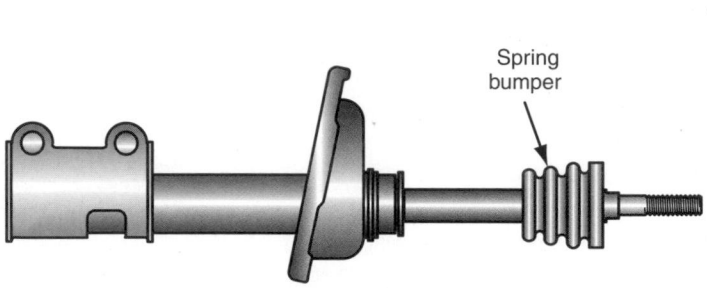

Copyright © 2015 Cengage Learning®

FIGURE 5-17 **Spring bumper installation on strut piston rod.**

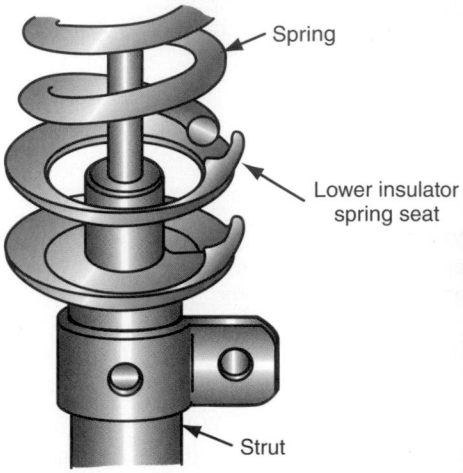

Copyright © 2015 Cengage Learning®

FIGURE 5-18 **Installing strut in coil spring with proper lower spring seat alignment.**

3. With the coil spring compressed in the spring compressing tool, install the strut in the spring (Figure 5-18). Be sure the spring is properly seated on the lower insulator spring seat.
4. Be sure the strut piston rod is fully extended and install the upper insulator on top of the coil spring.
5. Install the upper strut mount on the upper insulator (Figure 5-19).
6. Be sure the spring, upper insulator, and upper strut mount are properly positioned and seated on the coil spring and strut piston rod (Figure 5-20).
7. Install the strut rod nut, and tighten this nut to the specified torque (Figure 5-21).
8. Rotate the upper strut mount until the lowest bolt in this mount is aligned with the tab on the lower spring seat (Figure 5-22).

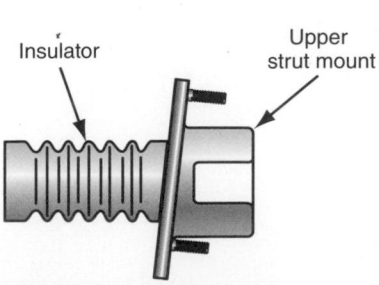

Copyright © 2015 Cengage Learning®

FIGURE 5-19 **Upper insulator and upper strut mount installed on coil spring.**

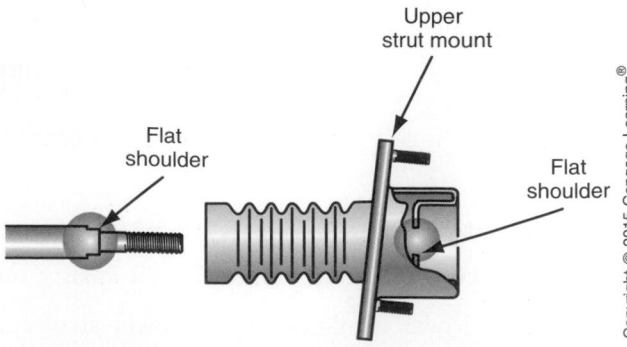

Copyright © 2015 Cengage Learning®

FIGURE 5-20 **Upper strut mount properly positioned on strut piston rod.**

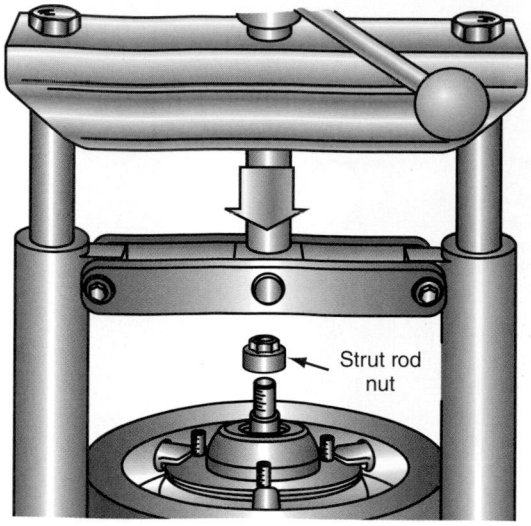

Copyright © 2015 Cengage Learning®

FIGURE 5-21 **Installing strut rod nut.**

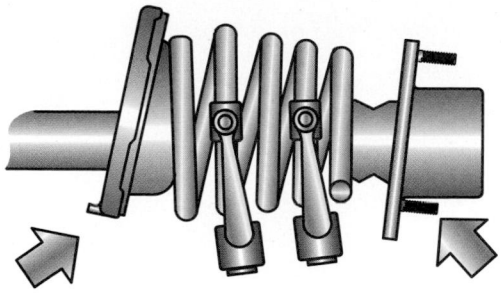

Copyright © 2015 Cengage Learning®

FIGURE 5-22 **Aligning lowest bolt on upper strut mount with tab on lower spring seat.**

9. Gradually loosen the handle on the compressing tool until all the spring tension is released from the tool, and remove the strut-and-spring assembly from the tool.

Installation of Strut-and-Spring Assembly in Vehicle

A typical installation procedure for a strut-and-spring assembly follows:

1. Install the strut-and-spring assembly with the upper strut mounting bolts extending through the bolt holes in the strut tower. Tighten the nuts on the upper strut mounting bolts to the specified torque (Figure 5-23).
2. Install the lower end of the strut in the steering knuckle to the proper depth, and tighten the strut-to-knuckle retaining bolts to the specified torque (Figure 5-24). If one of the strut-to-knuckle bolts is an eccentric camber bolt, be sure the eccentric is aligned with the mark placed on the strut during the removal procedure.
3. Install the brake hose in the clamp on the strut. Place the ABS wheel speed sensor wire in the strut clamp if the vehicle is equipped with ABS (Figure 5-25).

Photo Sequence 8 shows a typical procedure for removing and replacing a MacPherson strut.

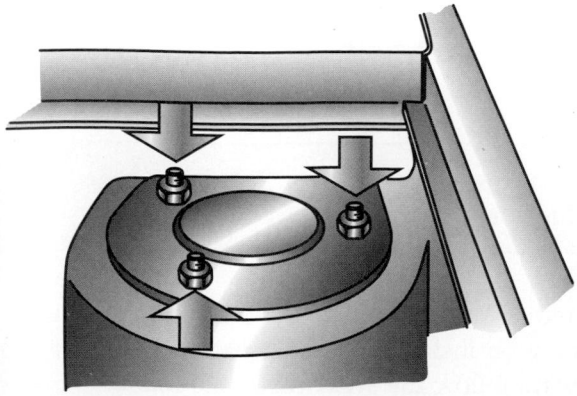

Copyright © 2015 Cengage Learning®

FIGURE 5-23 **Nuts installed on upper strut mount bolts.**

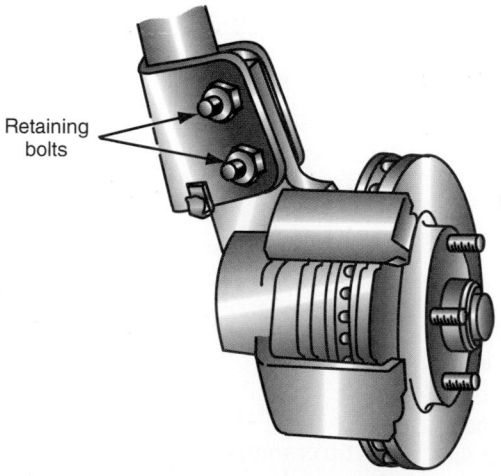

Copyright © 2015 Cengage Learning®

FIGURE 5-24 **Lower end of strut installed in steering knuckle.**

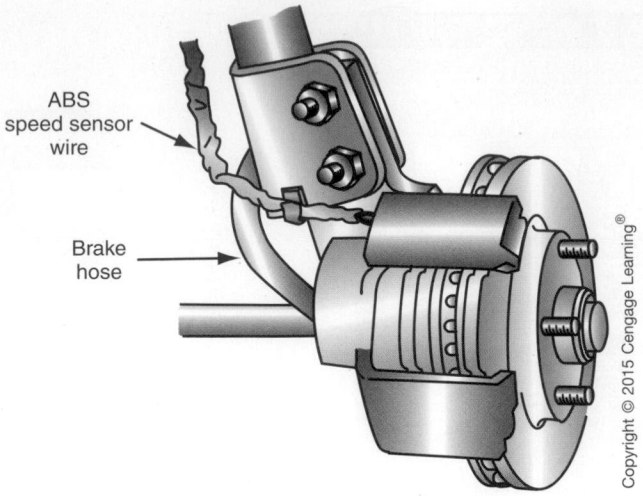

FIGURE 5-25 **Brake hose and ABS wire installed in strut clamps.**

Rear Strut Replacement

The rear strut replacement procedure varies depending on the type of rear suspension. Always follow the vehicle manufacturer's recommended procedure outlined in the appropriate service manual.

A typical rear strut replacement procedure follows:

1. Lift the vehicle with a floor jack and lower the vehicle onto safety stands placed under the chassis to support the vehicle weight.
2. Place the floor jack under the lower control arm and operate the jack to support some of the spring tension.
3. Remove the tire and wheel assembly.
4. Remove the strut-to-spindle bolts.
5. Pull upward on the strut to remove the strut from the spindle. If necessary, lower the floor jack slightly to remove the strut from the spindle.
6. Disconnect the upper strut mount from the chassis, and remove the strut.
7. When the new strut or mount is installed, reverse steps 1 through 6. Tighten all the bolts to the specified torque.
8. Check vehicle alignment.

The rear coil springs may be removed from the rear struts using the same basic procedure for spring removal from the front struts.

Installing Strut Cartridge, Off-Car

Customer Care: Check the cost of the strut cartridges versus the price of new struts. Give customers the best value for their repair dollar!

A **strut cartridge** contains the inner working part of the strut, which may be installed in the outer housing of the old strut.

Many struts are a sealed unit, and thus rebuilding is impossible. However, some manufacturers supply a replacement cartridge that may be installed in the strut housing after the strut has been removed from the vehicle. Always follow the **strut cartridge** manufacturer's recommended installation procedure.

PHOTO SEQUENCE 8

Typical Procedure for Removing and Replacing a MacPherson Strut

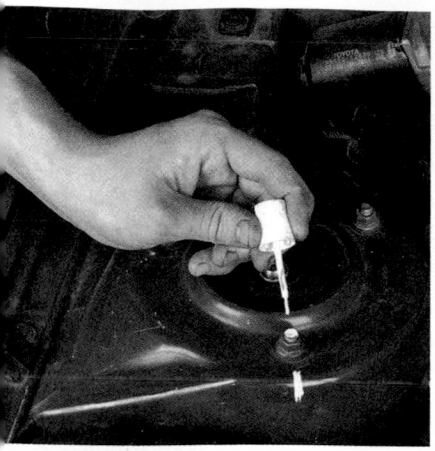

P8-1 **The top of the strut assembly is mounted directly to the chassis of the car. Prior to loosening the strut-to-chassis bolts, scribe alignment marks on the strut bolts and the chassis.**

Copyright © 2015 Cengage Learning®

P8-2 **With the top strut bolts or nuts removed, raise the car to a working height. It is important that the car be supported on its frame and not on its suspension components.**

Copyright © 2015 Cengage Learning®

P8-3 **Remove the wheel assembly. The strut is accessible from the wheel well after the wheel is removed.**

Copyright © 2015 Cengage Learning®

P8-4 **Remove the bolt that fastens the brake line or hose to the strut assembly.**

Copyright © 2015 Cengage Learning®

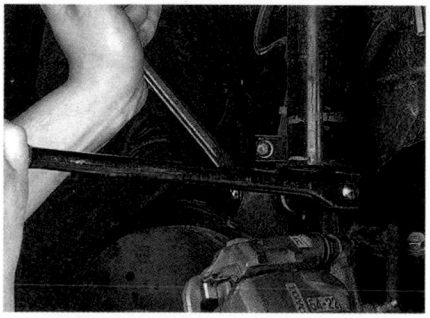

P8-5 **Remove the strut-to-steering knuckle bolts.**

Copyright © 2015 Cengage Learning®

P8-6 **Support the steering knuckle with wire and remove the strut assembly from the car.**

Copyright © 2015 Cengage Learning®

P8-7 **Install the strut assembly into the proper type spring compressor. Then compress the spring until all spring tension is removed from the upper strut mount. Loosen and remove the strut rod nut.**

Copyright © 2015 Cengage Learning®

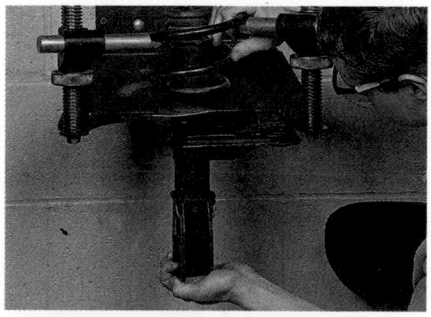

P8-8 **Remove the old strut assembly from the spring and install the new strut. Compress the spring to allow for reassembly and tighten the strut rod nut.**

Copyright © 2015 Cengage Learning®

P8-9 **Reinstall the strut assembly into the car. Make sure all bolts are properly tightened and in the correct locations.**

Copyright © 2015 Cengage Learning®

SPECIAL TOOL
Pipe cutter

The following is a typical off-car strut cartridge installation procedure:

1. Install a bolt and two nuts in the upper strut-to-knuckle mounting bolt hole. Place a nut on the inside and outside of the strut flange.
2. Clamp this bolt in a vise to hold the strut.
3. Locate the line groove near the top of the strut body, and use a pipe cutter installed in this groove to cut the top of the strut body.
4. After the cutting procedure, remove the strut piston assembly from the strut (Figure 5-26).
5. Remove the strut from the vise and dump the oil from the strut.
6. Place the special tool supplied by the vehicle manufacturer or cartridge manufacturer on top of the strut body. Strike the tool with a plastic hammer until the tool shoulder contacts the top of the strut body. This action removes burrs from the strut body and places a slight flare on the body.

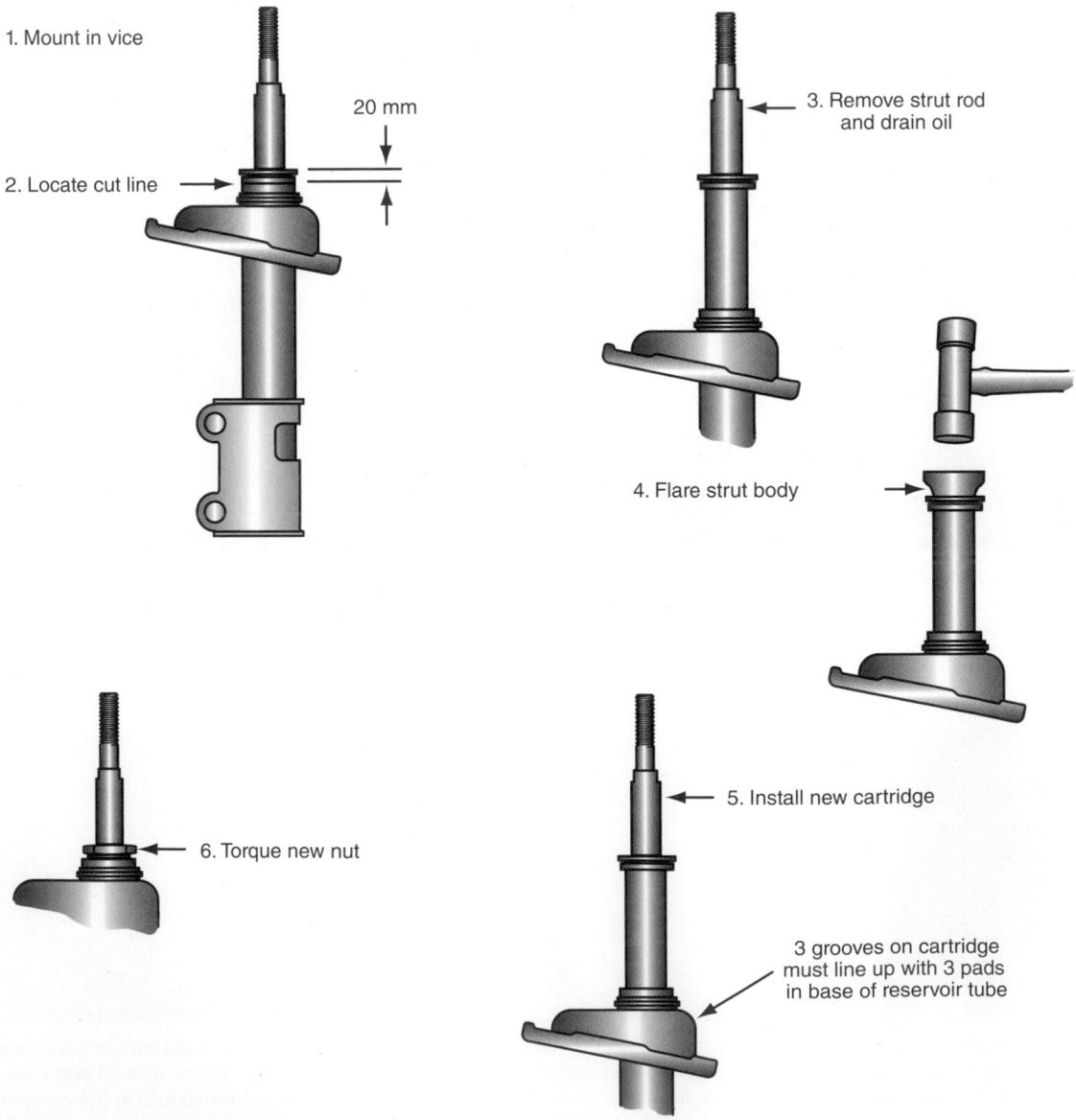

Copyright © 2015 Cengage Learning®

FIGURE 5-26 **Installation of strut cartridge.**

7. Remove the tool from the strut body.
8. Install the required amount of oil in the strut, place the new cartridge in the strut body, and turn the cartridge until it settles into indentations in the bottom of the strut body.
9. Place the new nut over the cartridge.
10. Using a special tool supplied by the vehicle or cartridge manufacturer, tighten the nut to the specified torque.
11. Move the strut piston rod in and out several times to check for proper strut operation.

Installing Strut Cartridge, On-Car

WARNING: If a vehicle is hoisted or lifted in any way during an on-car strut cartridge replacement, the coil spring may fly off the strut, causing vehicle damage and personal injury.

On some vehicles, the front strut cartridge may be removed and replaced with the strut installed in the vehicle. Always consult the vehicle manufacturer's service manual for the proper strut service procedure.

A typical on-car strut cartridge installation procedure follows:

1. Remove the nuts from the upper strut mount and remove the strut mount cover (Figure 5-27).
2. Use the special tool supplied by the vehicle manufacturer and a number 50 torx bit to remove the nut from the strut piston rod (Figure 5-28). Never lift or hoist the vehicle once this nut is removed.
3. Remove the bushing from the upper strut mount (Figure 5-29).
4. Thread the special tool supplied by the vehicle manufacturer onto the top of the strut piston rod, and push this rod downward into the strut (Figure 5-30). Remove this tool and then remove the jounce bumper retainer (Figure 5-31).
5. Remove the jounce bumper (Figure 5-32).

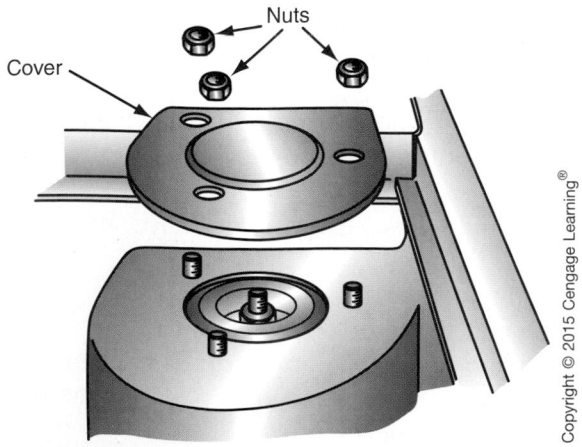

FIGURE 5-27 **Removing upper strut mount nuts and cover.**

FIGURE 5-28 **Removing nut from strut piston rod.**

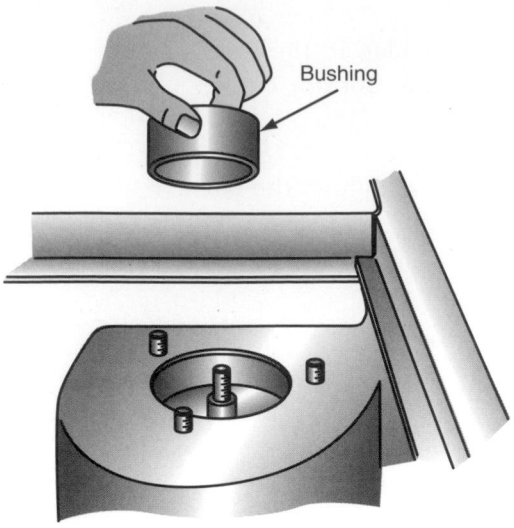

Copyright © 2015 Cengage Learning®

FIGURE 5-29 **Removing upper strut mount bushing.**

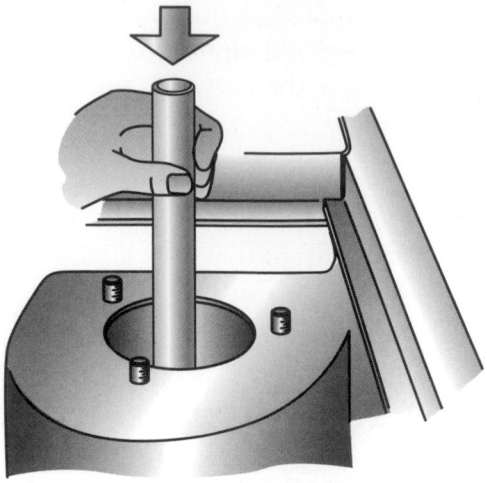

Copyright © 2015 Cengage Learning®

FIGURE 5-30 **Pushing strut piston rod downward.**

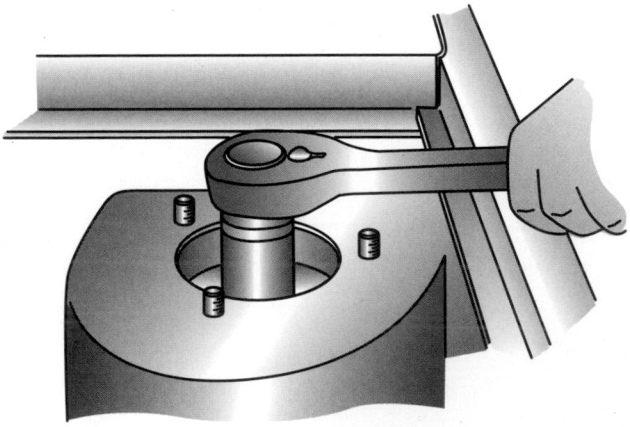

Copyright © 2015 Cengage Learning®

FIGURE 5-31 **Removing jounce bumper retainer.**

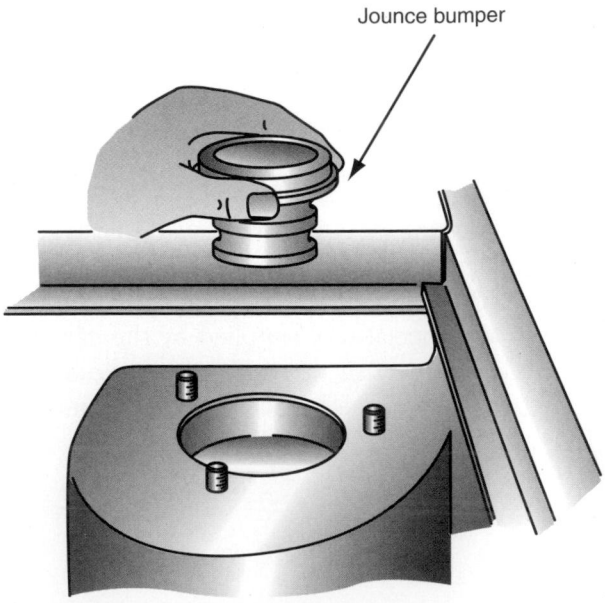

Copyright © 2015 Cengage Learning®

FIGURE 5-32 **Removing jounce bumper.**

6. Thread the special tool onto the strut piston rod, and reextend this rod (Figure 5-33). Use the special tool supplied by the vehicle manufacturer to remove the closure nut on top of the strut (Figure 5-34).
7. Grasp the top of the strut piston rod and remove the strut valve mechanism (Figure 5-35).
8. Remove the oil from the strut tube with a hand-operated suction pump.
9. Install the new strut cartridge in the strut, and tighten the closure nut to the specified torque (Figure 5-36). Reverse steps 1 through 6 to complete the strut cartridge replacement. Place a light coating of the vehicle manufacturer's recommended engine oil on the upper strut mount bushing prior to bushing installation.

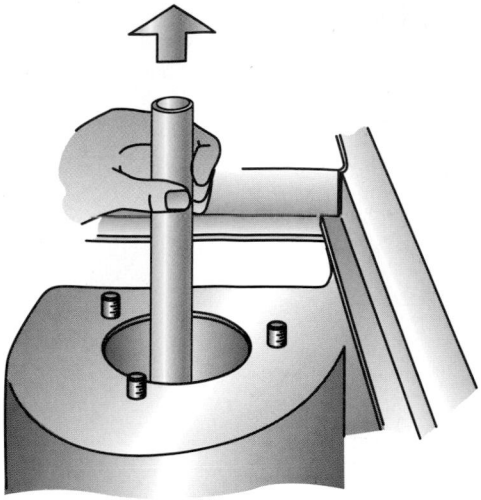

Copyright © 2015 Cengage Learning®

FIGURE 5-33 **Reextending strut piston rod.**

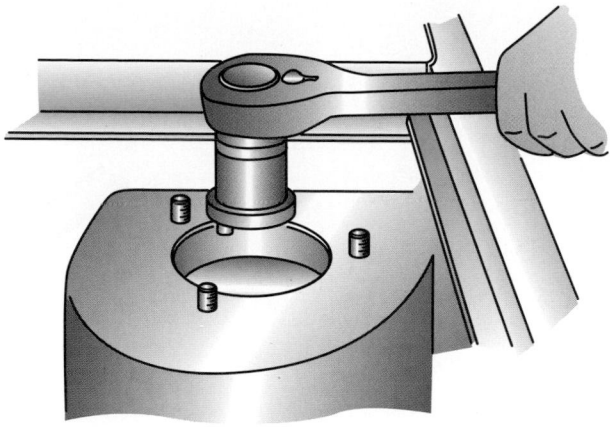

Copyright © 2015 Cengage Learning®

FIGURE 5-34 **Removing strut closure nut.**

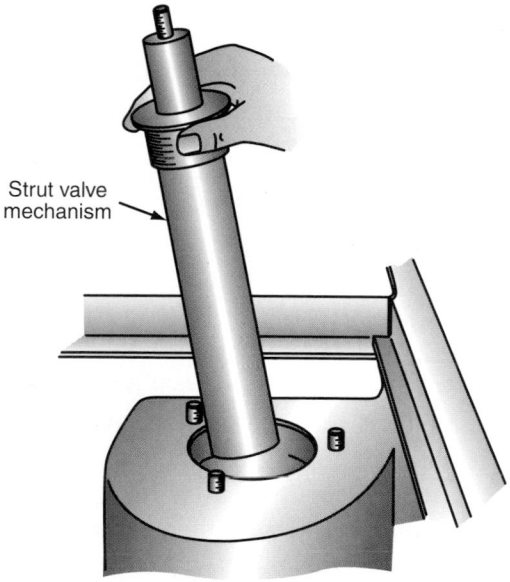

Copyright © 2015 Cengage Learning®

FIGURE 5-35 **Removing strut valve mechanism.**

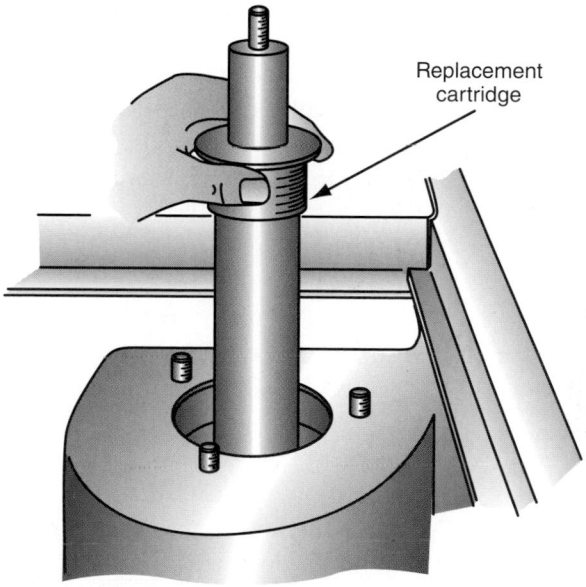

Copyright © 2015 Cengage Learning®

FIGURE 5-36 **Installing new strut cartridge.**

Diagnosis of Electronically Controlled Shock Absorbers

The actuators on electronically controlled shock absorbers can be removed by pushing inward simultaneously on the two actuator retaining tabs and lifting the actuator off the top of the strut (Figure 5-37).

An actuator may be called a solenoid.

WARNING: If the strut piston rod nut and actuator mounting bracket are removed, do not move or raise the vehicle. This action releases the coil spring tension and may result in personal injury and vehicle damage.

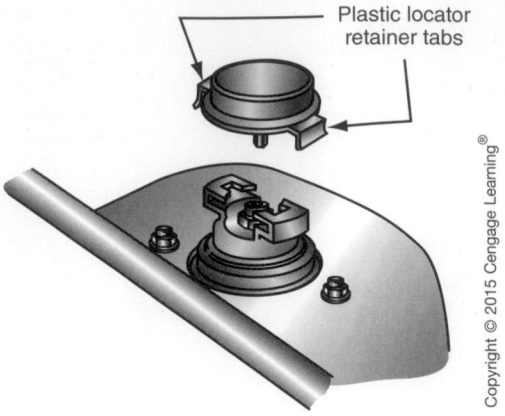

FIGURE 5-37 **Electronically controlled strut actuator.**

Follow these steps to diagnose the electronic actuator:

1. With the actuator removed from the strut and the actuator wiring harness connected, turn the ignition switch on. Move the ride control switch to the auto position and wait 5 seconds.
2. Move the ride control switch to the firm position, and wait 5 seconds. If the actuator control tube on the bottom of the actuator rotated, the actuator is operating. If the actuator control tube did not rotate, proceed with the actuator tests.
3. With the ride control switch in the firm position, place matching H's beside the control tube and the actuator.
4. With the ride control switch in the auto position, place matching S's beside the control tube and the actuator.
5. Turn the ignition switch off and disconnect the actuator electrical connector. The actuator control tube may be rotated with a small screwdriver.
6. Connect a pair of ohmmeter leads to the position sense wire and the signal return wire. The position sense wire is white or white with a colored tracer, and the signal return wire is yellow with a black tracer. With the actuator in the S position, the position sense should be closed and the ohmmeter should indicate less than 10 ohms.
7. Rotate the actuator to the H position. Under this condition, the position sense switch should be open and the ohmmeter should indicate over 1,000 ohms. If the ohmmeter readings are not within specifications, replace the actuator.
8. Connect the ohmmeter leads to the position sense wire and soft power terminal in the actuator electrical connector. The position sense wire is white or white with a colored tracer, and the soft power wire is tan with a red tracer. If the ohmmeter indicates over 1,000 ohms, the actuator is satisfactory. If the ohmmeter reading is below 10 ohms, replace the actuator.
9. Connect the ohmmeter leads from the signal return terminal in the wiring harness side of the actuator connector to a chassis ground. The ohmmeter should indicate less than 10 ohms. If the ohmmeter reading is higher than specified, check the signal return wire and the programmed ride control (PRC) module.

SERVICE TIP: Wiring harness colors vary depending on the vehicle make and model year. The wiring colors in the following steps are based on Ford vehicles. If you are working on a different make of vehicle, refer to the wire colors in the vehicle manufacturer's service information.

CASE STUDY

A customer complained about a squeaking noise in the rear suspension of a 2009 Dodge Caravan. The customer said the noise occurred during normal driving at lower speeds. The technician lifted the car on a hoist and made a check of all rear suspension bushings in shock absorbers, track bar, and trailing arms. The spring insulators and all suspension bolts were checked visually. All of these items were in good condition, and there was no evidence of a squeaking noise as the chassis was bounced gently. Because the exact source of chassis noise can sometimes be difficult to locate, the technician performed a visual check of bushings, insulators, and fasteners in the front suspension. No problems were found in the front suspension.

One of the first requirements for successful automotive diagnosis is to obtain as much information as possible from the customer. The customer with the squeaking rear suspension in a Dodge Caravan lived in a part of the country where cold temperatures occur in the winter. This complaint occurred in January. The technician questioned the customer further about the conditions when the squeaking suspension noise was heard, and the customer revealed that the noise occurred when the temperature was severely cold. The customer also indicated that the noise disappeared at warmer temperatures.

The technician informed the customer that the only way to find the exact cause of this annoying squeak was to leave the car on the lot at the shop all night and check it first thing in the morning when it was colder. The customer complied with this suggestion, and the technician drove the car into the shop immediately the next morning. The squeaking noise occurred as the car was driven across the parking lot. The technician lifted the car on a hoist and listened with a stethoscope at each rear suspension bushing as a coworker gently bounced the rear suspension. No squeaking noise was heard at any of the rear suspension bushings. However, when the stethoscope pickup was placed on the left rear shock absorber, the squeaking noise was loud and clear. The shock absorber was quickly removed, and the squeaking noise was gone when the rear chassis was bounced gently. Replacement of the left rear shock absorber corrected this complaint and made the customer happy.

TERMS TO KNOW

Bounce test
Eccentric camber bolt
Manual test
Spring compressing tools
Spring insulators
Struts
Strut cartridge
Strut chatter
Strut tower

ASE-STYLE REVIEW QUESTIONS

1. A slight oil film appears on the lower shock absorber oil chamber, and the shock absorber performs satisfactorily on a road test:

   Technician A says the shock absorber is satisfactory.

   Technician B says the shock absorber may contain excessive pressure.

   Who is correct?

   A. A only
   B. B only
   C. Both A and B
   D. Neither A nor B

2. While discussing a shock absorber and strut bounce test:

   Technician A says the shock absorber is satisfactory if the bumper makes two free upward bounces.

   Technician B says the bumper must be pushed downward with considerable force.

   Who is correct?

   A. A only
   B. B only
   C. Both A and B
   D. Neither A nor B

3. While discussing a shock absorber manual test:

   Technician A says the shock absorber's resistance to movement should be jerky and erratic.

   Technician B says the shock absorber's resistance to movement may be greater on the rebound stroke compared with the compression stroke.

   Who is correct?

   A. A only
   B. B only
   C. Both A and B
   D. Neither A nor B

4. When the front wheels are turned on a vehicle equipped with front struts, the left front coil spring provides a chattering action and noise:

   Technician A says the strut has internal defects, and strut replacement is necessary.

   Technician B says the upper strut bearing and mount is defective.

   Who is correct?

   A. A only
   B. B only
   C. Both A and B
   D. Neither A nor B

5. While discussing enamel-coated coil springs:
   Technician A says if the enamel-type coating on a coil spring is chipped, the spring may break prematurely.
   Technician B says coil springs should be taped in the compressing tool contact areas to prevent chipping of the enamel-type coating.
   Who is correct?
   A. A only
   B. B only
   C. Both A and B
   D. Neither A nor B

6. During service procedures, gas-filled shock absorbers may:
   A. Be heated with an acetylene torch.
   B. Be thrown in an incinerator.
   C. Extend when disconnected.
   D. Retract when disconnected and extended.

7. During wheel jounce, a thumping noise is heard in the right front suspension, but the strut bounce test is satisfactory. All of these defects could cause the problem EXCEPT:
   A. A worn upper strut mount.
   B. Worn spring insulators.
   C. A broken coil spring.
   D. A fluid leak in the strut.

8. When removing and replacing a front strut on a MacPherson strut suspension:
   A. The cam-type strut-to-knuckle bolt should be marked in relation to the strut.
   B. The strut rod nut may be loosened with spring tension applied to the upper strut mount.
   C. The lower strut chamber may be clamped in a vise to support the strut.
   D. The spring compressor must be installed before the strut is removed from the vehicle.

9. When assembling and installing a strut and coil spring in a MacPherson strut front suspension:
   A. The upper spring insulator should be installed between the upper strut mount and the strut tower.
   B. The lower spring insulator should be installed between the coil spring and the spring bumper.
   C. On some vehicles, the upper strut mount must be properly aligned in relation to the lower spring seat on the strut.
   D. Final compressing of the coil spring may be done by tightening the nut on the strut rod.

10. During an on-car strut cartridge replacement:
   A. Oil must be removed from the strut with a hand-operated suction pump.
   B. The front suspension may be lifted with a floor jack.
   C. The strut rod must be pulled upward before removing the jounce bumper.
   D. The new strut cartridge must be filled with the manufacturer's specified fluid.

ASE CHALLENGE QUESTIONS

1. A customer says his MacPherson strut front suspension is making a chattering noise when the steering wheel is turned hard to the left. He says he also feels a "kind of vibration" in the steering wheel when the chatter is heard. To correct this problem, you should:
   A. Replace the front struts.
   B. Check the lower strut spring seating.
   C. Check the stabilizer links.
   D. Check the upper bearing and strut mounting.
2. A faulty shock absorber or strut may cause:
   A. Steering pull.
   B. Suspension bottoming.
   C. Lateral chassis oscillations.
   D. Excessive steering free play.
3. The front rebound bumpers on a coil spring suspension system are badly worn and damaged.
   Technician A says the shock absorbers may be worn out.
   Technician B says the coil springs may be weak.
   Who is correct?
   A. A only
   B. B only
   C. Both A and B
   D. Neither A nor B
4. A manual test of a shock absorber shows a stronger resistance to rebound than compression.
   Technician A says the shock is defective.
   Technician B says the shock has a 50/50 ratio.
   Who is correct?
   A. A only
   B. B only
   C. Both A and B
   D. Neither A nor B
5. *Technician A* says oil leaking from a strut or shock requires shock or strut replacement.
   Technician B says a slight film of oil on a strut or shock is OK if it performs satisfactorily.
   Who is correct?
   A. A only
   B. B only
   C. Both A and B
   D. Neither A nor B

JOB SHEET 17

Name _____ **Date** _____

Remove Strut-and-Spring Assembly and Disassemble Strut and Spring

Upon completion of this job sheet, you should be able to remove a strut-and-spring assembly from a car and disassemble the strut and spring.

NATEF AST/MAST Task Correlation

Suspension and Steering. Suspension Systems Diagnosis and Repair,

Task #C-10. Inspect, remove and install strut cartridge or assembly, strut coil spring, insulators (silencers), and upper strut bearing mount. (P-3)

Tools and Materials

Front-wheel-drive car
Floor jack
Safety stands
Coil spring compressor

Describe the Vehicle Being Worked On:

Year _____ Make _____ Model _____

VIN _____ Engine type and size _____

Procedure

1. With the vehicle parked on the shop floor, perform a strut bounce test.

   Based on the bounce test results, state the strut condition, and give the reason for your diagnosis.

   _____

   _____

2. Raise the vehicle on a hoist or with a floor jack. If a floor jack is used to raise the vehicle, lower the vehicle onto safety stands placed under the chassis so the lower control arms and front wheels drop downward. Remove the floor jack from under the vehicle.

   Is the vehicle securely supported on safety stands? ☐ Yes ☐ No

   Instructor check _____

3. Remove the brake line and antilock brake system (ABS) wheel speed sensor wire from clamps on the strut. In some cases, these clamps may have to be removed from the strut.

   Is the brake line and ABS wheel speed sensor wire disconnected? ☐ Yes ☐ No

   Instructor check _____

Task Completed

4. Punch mark the cam bolt in relation to the strut, remove the strut-to-steering knuckle retaining bolts, and remove the strut from the knuckle.
   Is the cam bolt marked in relation to the strut? ☐ Yes ☐ No
   Is the strut removed from the knuckle? ☐ Yes ☐ No
   Instructor check _____
   State the reason for marking the cam bolt in relation to the strut.
   _____
   _____

☐ 5. Remove the upper strut mounting bolts on top of the strut tower; remove the strut-and-spring assembly.

CAUTION: Never clamp the lower shock absorber or strut chamber in a vise with excessive force. This action may distort the lower chamber and affect piston movement in the shock absorber or strut.

WARNING: Always use a coil spring compressing tool according to the tool or vehicle manufacturer's recommended service procedure. Be sure the tool is properly installed on the spring. If a coil spring slips off the tool when the spring is compressed, severe personal injury or property damage may occur.

WARNING: Never loosen the upper strut mount retaining nut on the end of the strut rod unless the spring is compressed enough to remove all spring tension from the upper strut mount. If this nut is loosened with spring tension on the upper mount, this mount becomes a very dangerous projectile that may cause serious personal injury or property damage.

6. Install the spring compressing tool on the coil spring according to the tool or vehicle manufacturer's recommended procedure.
   Is the compressing tool properly installed on the strut-and-spring assembly?
   ☐ Yes ☐ No
   Instructor check _____

☐ 7. Turn the nut on top of the compressing tool until all the spring tension is removed from the upper strut mount.
   Is all the spring tension removed from the upper strut mount? ☐ Yes ☐ No
   Instructor check _____

☐ 8. Install a bolt and two nuts in the upper strut-to-knuckle mounting bolt holes. Install a nut on each side of the strut flange. Clamp this bolt securely in a vise to hold the strut-and-spring assembly and the compressing tool.

CAUTION: If the coil spring has an enamel-type coating and the compressing tool contacts the coil spring, tape the spring where the compressing tool contacts the spring.

9. Use the bar on the spring compressing tool to hold the strut-and-spring assembly from turning, and loosen the nut on the upper strut mount. Be sure all the spring tension is removed from the upper strut mount before loosening this nut.
   Is all the spring tension removed from the upper strut mount? ☐ Yes ☐ No
   Is the upper strut mount retaining nut loosened? ☐ Yes ☐ No
   Instructor check _____

10. Remove the nut, upper strut mount, upper insulator, coil spring, spring bumper, and lower insulator.

Task Completed ☐

11. Inspect the strut, upper strut mount, coil spring, spring insulators, and spring bumper. On the basis of this inspection, list the necessary strut-and-spring service, and give the reasons for your diagnosis.

_____

_____

_____

Instructor's Response _____

_____

_____

JOB SHEET 18

Name _____ **Date** _____

Assemble Strut and Spring and Install Strut-and-Spring Assembly

Upon completion of this job sheet, you should be able to assemble a strut and spring and install a strut-and-spring assembly in a car.

NATEF AST/MAST Task Correlation

Suspension and Steering. Suspension Systems Diagnosis and Repair,

Task #C-10. Inspect, remove and install strut cartridge or assembly, strut coil spring, insulators (silencers), and upper strut bearing mount. (P-3)

Tools and Materials

Front-wheel-drive car
Floor jack
Safety stands
Coil spring compressor

Describe the Vehicle Being Worked On:

Year _____ Make _____ Model _____

VIN _____ Engine type and size _____

Procedure

Task Completed

1. Install a bolt in the upper strut-to-knuckle retaining bolt, and clamp this bolt in a vise to hold the strut, spring, and compressing tool as in the disassembly procedure. ☐
2. Install the lower insulator on the lower strut spring seat and be sure the insulator is properly seated.
   Is the lower insulator properly seated? ☐ Yes ☐ No
   Instructor check _____
3. Install the spring bumper on the strut rod.
   Is the spring bumper properly installed? ☐ Yes ☐ No
   Instructor check _____
4. With the coil spring compressed in the spring compressing tool, install the spring on the strut. Be sure the spring is properly seated on the lower insulator spring seat.
   Is the coil spring properly seated on the lower insulator? ☐ Yes ☐ No
   Instructor check _____

Task Completed

5. Be sure the strut piston rod is fully extended and install the upper insulator on top of the coil spring.
   Is the strut rod fully extended? ☐ Yes ☐ No
   Is the upper insulator properly installed? ☐ Yes ☐ No
   Instructor check _____

☐ 6. Install the upper strut mount on the upper insulator.

7. Be sure the spring, upper insulator, and upper strut mount are properly positioned and seated on the coil spring and strut piston rod.
   Are the spring, upper insulator, and upper strut mount properly positioned?
   ☐ Yes ☐ No
   Instructor check _____

8. Use the compressing tool bar to hold the strut and spring from turning, then tighten the strut piston rod nut to the specified torque.
   Specified strut piston rod nut torque _____
   Actual strut piston rod nut torque _____

9. Rotate the upper strut mount until the lowest bolt in this mount is aligned with the tab on the lower spring seat.
   Is the upper strut mount properly aligned? ☐ Yes ☐ No
   Instructor check _____

☐ 10. Gradually loosen the nut on the compressing tool until all the spring tension is released from the tool, and remove the tool from the spring.

11. Install the strut-and-spring assembly with the upper strut mounting bolts extending through the bolt holes in the strut tower. Tighten the nuts on the upper strut mounting bolts to the specified torque.
   Specified upper strut mount nut torque _____
   Actual upper strut mount nut torque _____

12. Install the lower end of the strut in the steering knuckle to the proper depth. Align the punch marks on the cam bolt and strut that were placed on these components during disassembly, and tighten the strut-to-knuckle retaining bolts to the specified torque.
   Is the cam bolt properly positioned? ☐ Yes ☐ No
   Specified strut-to-knuckle bolt torque _____
   Actual strut-to-knuckle bolt torque _____
   Instructor check _____

13. Install the brake hose in the clamp on the strut. Place the ABS wheel speed sensor wire in the strut clamp if the vehicle is equipped with ABS.
   Is the brake hose properly installed and tightened? ☐ Yes ☐ No
   Is the ABS wheel speed sensor properly installed and tightened? ☐ Yes ☐ No
   Instructor check _____

14. Raise the vehicle with a floor jack, remove the safety stands, and lower the vehicle onto the shop floor.

Task Completed ☐

Instructor's Response _____

_____

_____

JOB SHEET 19

Name _____ **Date** _____

INSTALL STRUT CARTRIDGE, OFF-CAR

Upon completion of this job sheet, you should be able to remove and replace a strut cartridge with the strut removed from the car.

NATEF AST/MAST Task Correlation

Suspension and Steering. Suspension Systems Diagnosis and Repair,

Task #C-10. Inspect, remove and install strut cartridge or assembly, strut coil spring, insulators (silencers), and upper strut bearing mount. (P-3)

Tools and Materials

Front strut removed from car
Pipe cutter
Torque wrench

Describe the Vehicle Being Worked On:

Year _____ Make _____ Model _____

VIN _____ Engine type and size _____

Procedure

_____
Task Completed

1. List the strut conditions and suspension operating conditions that indicate a strut should be serviced or replaced.

   _____

   _____

   _____

   _____

2. Install a bolt and two nuts in the upper strut-to-knuckle mounting bolt hole. Place a nut on the inside and outside of the strut flange. ☐

3. Clamp this bolt in a vise to hold the strut.

4. Locate the line groove near the top of the strut body, and use a pipe cutter installed in this groove to cut the top of the strut body. ☐

5. After the cutting procedure, remove the strut piston assembly from the strut.
   Is the strut piston assembly removed from the strut? ☐ Yes ☐ No
   Instructor check _____

6. Remove the strut from the vise and dump the oil from the strut. ☐

Task Completed

7. Place the special tool supplied by the vehicle manufacturer or cartridge manufacturer on top of the strut body and strike the tool with a plastic hammer until the tool shoulder contacts the top of the strut body. This action removes burrs from the strut body and places a slight flare on the body.
   Is the strut body properly flared? ☐ Yes ☐ No
   Instructor check _____

☐ 8. Remove the tool from the strut body.

9. Place the new cartridge in the strut body and turn the cartridge until it settles into indentations in the bottom of the strut body.
   Is the new strut cartridge properly seated in strut body indentations? ☐ Yes ☐ No
   Instructor check _____

☐ 10. Place the new nut over the cartridge.

11. Using a special tool supplied by the vehicle or cartridge manufacturer, tighten the nut to the specified torque.
   Specified strut nut torque _____
   Actual strut nut torque _____

12. Move the strut piston rod in and out several times to check for proper strut operation.
   Strut operation: ☐ Satisfactory ☐ Unsatisfactory

Instructor's Response _____

_____

_____

Chapter 6

Front Suspension System Service

BASIC TOOLS

Basic technician's tool set
Service manual
Floor jack
Safety stands
Pry bar
3/8″ electric drill and drill bits

Upon completion and review of this chapter, you should be able to:

- Measure curb riding height.
- Diagnose and correct curb riding height problems.
- Adjust torsion bars.
- Diagnose front suspension noise and body sway.
- Remove and replace ball joints, and check ball joint condition.
- Remove and replace steering knuckles, and check knuckle condition.
- Remove and replace lower control arms, and check control arm and bushing condition.
- Remove and replace coil springs, and check spring and insulator condition.
- Remove and replace upper control arms, and check control arm and bushing condition.
- Remove and replace control arm bushings.
- Inspect and replace rebound bumpers.
- Diagnose, remove, and replace stabilizer bars.
- Diagnose, remove, and replace strut rods.
- Diagnose, remove, and replace leaf springs.
- Replace torsion bars, and check torsion bar condition.

Proper front suspension system service is extremely important to provide adequate vehicle safety and to maintain ride comfort and normal tire life. If worn or loose front suspension system components are ignored, steering control may be adversely affected, which may result in loss of steering control and an expensive collision. Defective front suspension components, such as worn out shock absorbers or broken springs, may cause rough riding that results in driver and passenger discomfort. Other worn front suspension components, such as worn ball joints and control arm bushings, cause improper alignment angles that causes excessive front tire wear. Therefore, technicians must be familiar with front suspension service. Diagnostic Table 6-1 at the end of the chapter contains symptoms and causes of common front suspension issues.

Curb riding height is the distance from specific chassis locations to the road surface.

Directional stability refers to the tendency of the vehicle steering to remain in the straight-ahead position when driven straight ahead on a reasonably smooth, level road surface.

Steering effort is the amount of effort required to turn the steering wheel.

Curb Riding Height Measurement

Regular inspection and proper maintenance of suspension systems is extremely important for maintaining vehicle safety. The **curb riding height** is determined mainly by spring condition. Other suspension components, such as control arm bushings, affect curb riding height if they are worn. Since incorrect curb riding height affects most of the other suspension angles, this measurement is critical.

Reduced curb riding height on the front suspension may cause decreased **directional stability**. If the curb riding height is reduced on one side of the front suspension, the steering may pull to one side. Reduced rear suspension height increases **steering effort** and causes rapid steering wheel return after turning a corner. Harsh riding occurs when the curb riding height is less than specified. The curb riding height must be measured at the vehicle manufacturer's specified location, which varies depending on the type of suspension system.

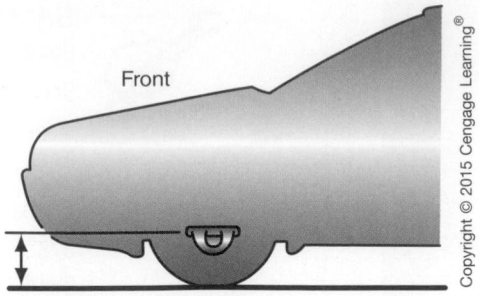

FIGURE 6-1 **Curb riding height measurement, front suspension.**

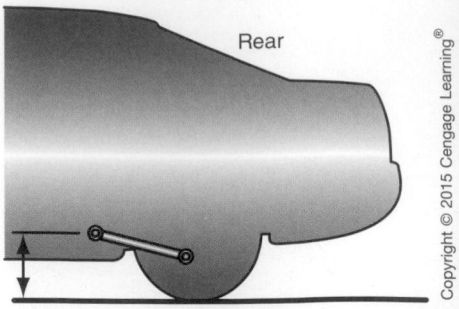

FIGURE 6-2 **Curb riding height measurement, rear suspension.**

When the vehicle is on a level floor or on an alignment rack, measure the curb riding height from the floor to the center of the lower control arm mounting bolt on both sides of the front suspension (Figure 6-1). On the rear suspension system, measure the curb riding height from the floor to the center of the strut rod mounting bolt (Figure 6-2).

If the curb riding height is less than specified, the control arms and bushings should be inspected and replaced as necessary. When the control arms and bushings are in normal condition, the reduced curb riding height may be caused by sagged springs that require replacement.

Front Suspension Diagnosis and Service

Torsion Bar Adjustment

Customer Care: Each time under-car service is performed, make a quick inspection of suspension and steering components. Advise the customer regarding any necessary repairs. This procedure often obtains additional work for the shop, and the customer will be impressed that you are interested in his or her car and personal safety.

SPECIAL TOOL
Machinist's rule

On torsion bar front suspension systems, the torsion bars may be adjusted to correct the curb riding height. The curb riding height must be measured at the location specified by the vehicle manufacturer. On some torsion bar front suspension systems, the curb riding height is measured with the vehicle on a lift and the tires supported on the lift. Measure the distance from the center of the front lower control arm bushing to the lift. Then measure the distance from the lower end of the front spindle to the lift (Figure 6-3). The difference between these two readings is the curb riding height.

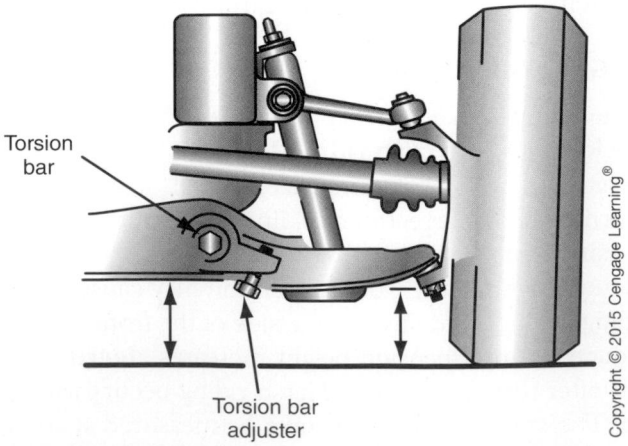

FIGURE 6-3 **Curb riding height measurement, torsion bar front suspension.**

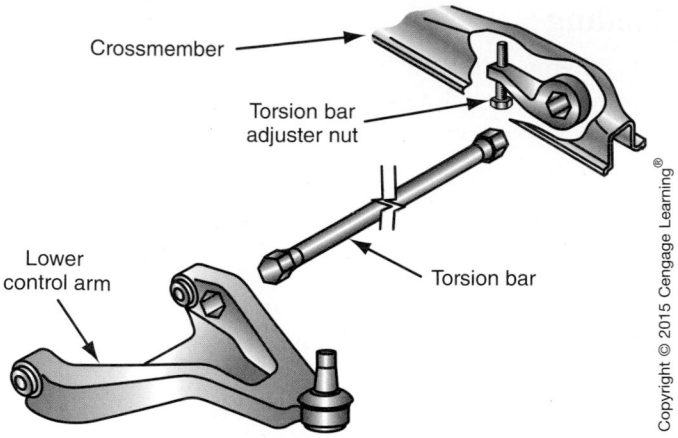

FIGURE 6-4 **Curb riding height adjustment, torsion bar front suspension.**

If the curb riding height is not correct on a torsion bar front suspension, the torsion bar anchor adjusting bolts must be rotated until the curb riding height equals the vehicle manufacturer's specifications (Figure 6-4).

Inspecting Ball Joints

CUSTOMER CARE: Regular chassis lubrication at the vehicle manufacturer's recommended service interval is one of the keys to long ball joint life. Always advise the customer of this fact.

Classroom Manual
Chapter 6, page 136

Wear Indicators. Some ball joints have a grease fitting installed in a floating retainer. The grease fitting and retainer may be used as a **ball joint wear indicator**. With the vehicle weight resting on the wheels, grasp the grease fitting and check for movement (Figure 6-5).

A **ball joint wear indicator** allows the technician to check ball joint wear by visibly inspecting the ball joint.

Some car manufacturers recommend ball joint replacement if any grease fitting movement is present. In some other ball joints, the grease fitting retainer extends a short distance through the ball joint surface (Figure 6-6). On this type of joint, replacement is necessary if the grease fitting shoulder is flush with or inside the ball joint cover.

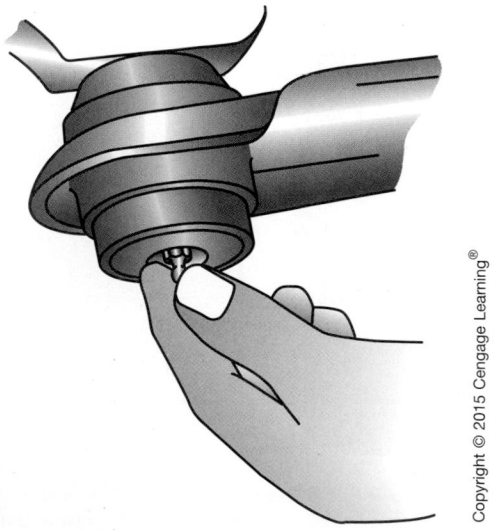

FIGURE 6-5 **Ball joint grease fitting wear indicator.**

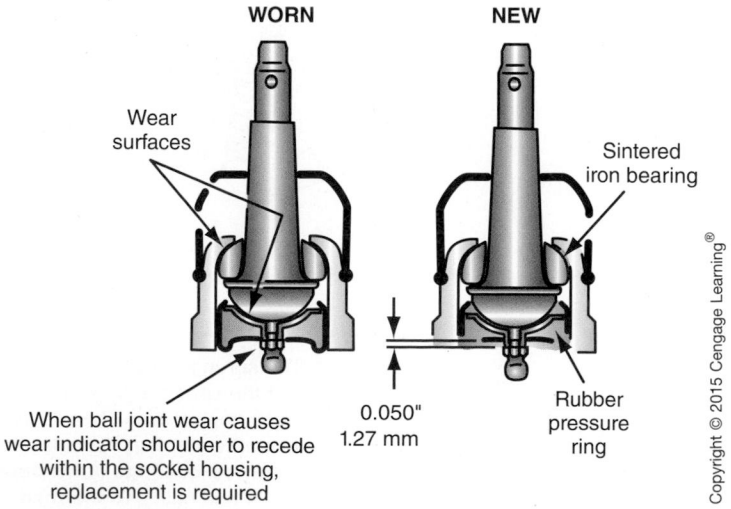

FIGURE 6-6 **Ball joint wear indicator with grease fitting extending from ball joint surface.**

Ball Joint Unloading

On many suspension systems, ball joint looseness is not apparent until the weight has been removed from the joint. To check for movement in a ball joint you must first determine if the joint is a load carrying or a non-load carrying joint. Then you can determine where to jack up the vehicle to properly inspect the joint (Figure 6-7). It is important to follow the lines of force to determine which joints are primary load carrying joints and which are followers. Start at the top of the spring and follow the lines of force as shown in Figure 6-8. The sprung weight of the vehicle is sitting on top of the spring. It is transmitted down to the lower control arm. Then through the lower ball joint to the steering knuckle. Next, the weight is transferred to the spindle and wheel bearings down through the wheel and tire and finally to the road surface. The upper ball joint is a follower joint.

When the coil spring is positioned between the lower control arm and the chassis, such as in a short-and-long-arm, modified MacPherson, or double wishbone suspension system, place a floor jack near the outer end of the lower control arm and raise the tire off the floor (Figure 6-9). This will transfer the weight off the lower control arm and the jack to the floor relieving pressure on the lower ball joint. Be sure the rebound bumper is not in contact with the control arm or frame.

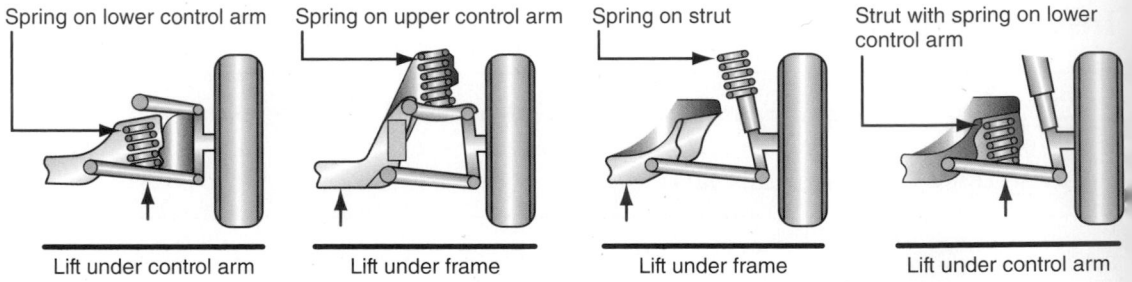

Lifting points for unloading different types of suspensions

FIGURE 6-7 **To check for movement in a ball joint you must first determine if the joint is a load carrying or a non-load carrying joint. Then you can determine where to jack up the vehicle to properly inspect the joint.**

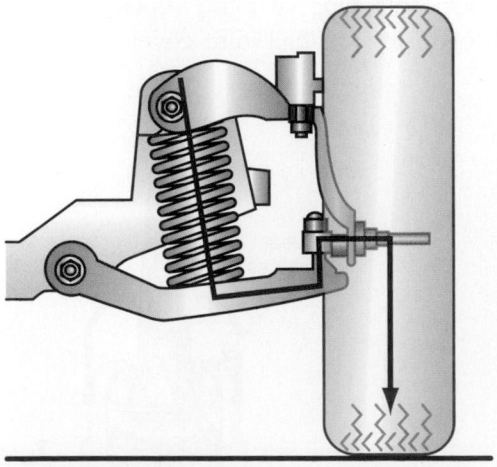

Copyright © 2015 Cengage Learning®

FIGURE 6-8 **The sprung weight of the vehicle is sitting on top of the spring. It is transmitted down to the lower control arm. Then through the lower ball joint to the steering knuckle. Next to the spindle and wheel bearings down through the wheel and tire and finally to the road surface.**

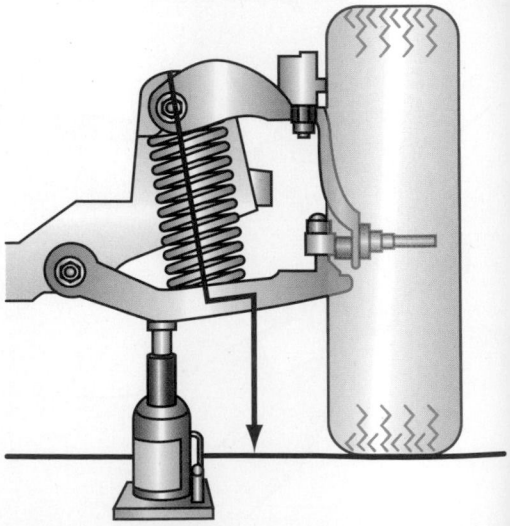

FIGURE 6-9 **Raise the vehicle under the lower control arm to relieve pressure on the lower ball joint for proper inspection.**

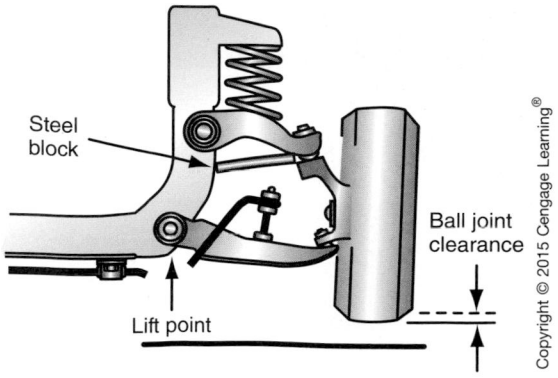

FIGURE 6-10 **Floor jack position to check ball joint wear with spring between the upper control arm and the chassis.**

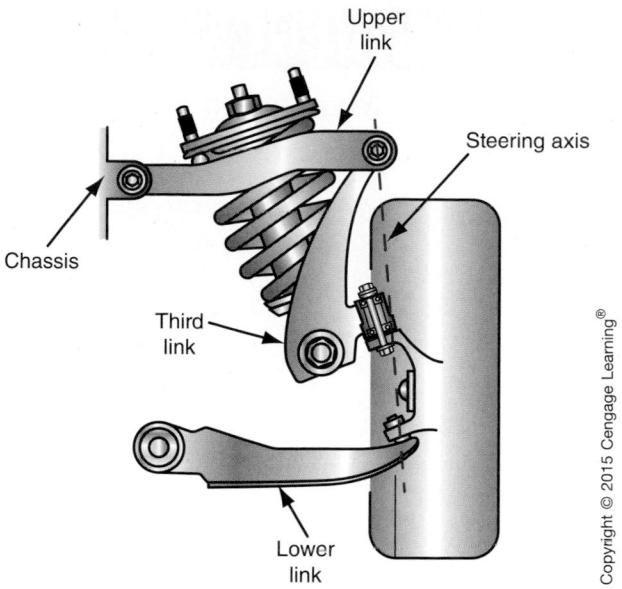

FIGURE 6-11 **The lower ball joint in this multilink suspension system is a non-load primary load carrying joint.**

When the coil springs are positioned between the upper control arm and the chassis, place a steel block between the upper control arm and the frame. With this block in place, raise the tire off the floor with a floor jack under the front crossmember (Figure 6-10). In this example, the upper ball joint is the primary load carrying joint.

On a multilink suspension system the spring load is carried by the lower spring seat on the strut, the third link and the upper pivot point on the steering knuckle. In the multilink suspension system in Figure 6-11, the lower ball joint is a follower. When the tire is lifted off the floor, it is not necessary to unload this type of ball joint.

The lower ball joint on a MacPherson strut front suspension is a non-load carrying joint and should be checked for radial movement with a dial indicator when the tire is lifted off the floor (Figure 6-12). Since the spring load is carried by the upper and lower spring seats when the tire is lifted off the floor, it is not necessary to unload this type of ball joint. Photo Sequence 9 shows a typical procedure for measuring the lower ball joint radial movement on a MacPherson strut front suspension. In this suspension system the

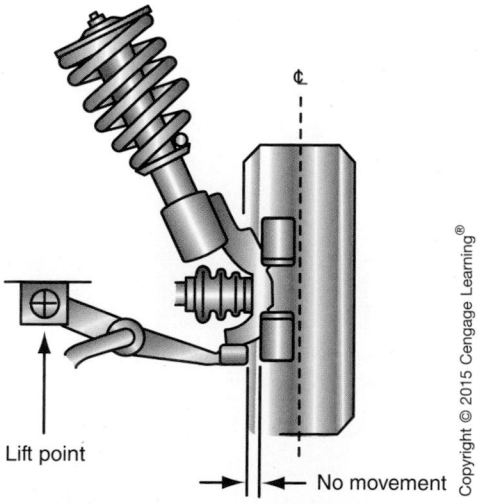

FIGURE 6-12 **Ball joint wear measurement on MacPherson strut front suspension.**

PHOTO SEQUENCE 9

Typical Procedure for Measuring the Lower Ball Joint Horizontal Movement on a MacPherson Strut Front Suspension

P9-1 **Raise the front suspension with a floor jack and place safety stands under the chassis at the vehicle manufacturer's recommended lifting points.**

Copyright © 2015 Cengage Learning®

P9-2 **Grasp the front tire at the top and bottom and rock the tire inward and outward while a coworker visually checks for movement in the front wheel bearing. If there is movement in the front wheel bearing, adjust or replace the bearing.**

Copyright © 2015 Cengage Learning®

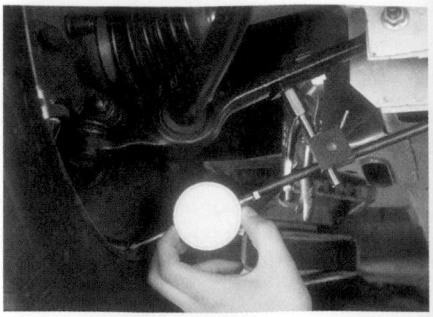

P9-3 **Position a dial indicator against the inner edge of the rim at the bottom. Preload and zero the dial indicator.**

Copyright © 2015 Cengage Learning®

P9-4 **Grasp the bottom of the tire and pull outward.**

Copyright © 2015 Cengage Learning®

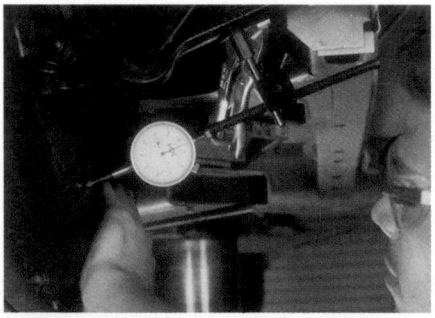

P9-5 **With the tire held outward, read the dial indicator.**

Copyright © 2015 Cengage Learning®

P9-6 **Push the bottom of the tire inward and be sure the dial indicator reading is zero. Adjust the dial indicator as required.**

Copyright © 2015 Cengage Learning®

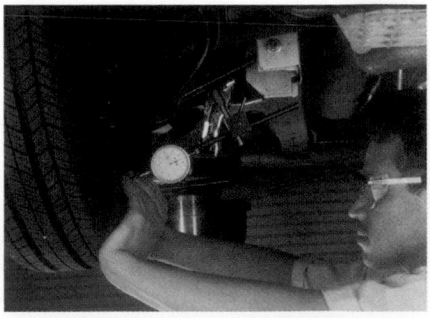

P9-7 **Grasp the bottom of the tire and pull outward.**

Copyright © 2015 Cengage Learning®

P9-8 **With the tire held in this position, read the dial indicator.**

Copyright © 2015 Cengage Learning®

P9-9 **If the dial indicator reading is more than specified, replace the lower ball joint.**

Copyright © 2015 Cengage Learning®

lower ball joint is the non-load carrier, but the upper strut mount is the primary load carrying pivot point. It is important to check the upper strut mount for proper pivoting and excessive play in the assembly. This is an often overlooked part of the suspension inspection. A binding upper strut mount can lead to a memory steer condition, while a mount with excessive movement can cause a rattle over bumps.

SPECIAL TOOL

Dial indicator for ball joint measurement

Ball Joint Vertical Measurement

The vehicle manufacturer may provide ball joint vertical and horizontal tolerances. A dial indicator is one of the most accurate ball joint measuring devices (Figure 6-13). Always install the dial indicator at the vehicle manufacturer's recommended location for ball joint measurement. When measuring the **ball joint vertical movement** in a compression-loaded ball joint, attach the dial indicator to the lower control arm and position the dial indicator stem on the lower side of the steering knuckle beside the ball joint stud (Figure 6-14). Depress the dial indicator stem approximately 0.250 in. (6.35 mm) and zero the dial indicator. Place a pry bar under the tire and pry straight upward while observing the vertical ball joint movement on the dial indicator. If this movement is more than specified, ball joint replacement is necessary. Photo Sequence 10 illustrates the vertical ball joint measurement procedure.

Ball joint vertical movement refers to up and down movement in a ball joint.

On a tension-loaded ball joint, clean the top end of the lower ball joint stud, and position the dial indicator stem against the top end of this stud. Depress the dial indicator plunger approximately 0.250 in. (6.35 mm) and zero the dial indicator. Lift upward with a pry bar under the tire and observe the dial indicator reading. If the vertical ball joint movement exceeds the manufacturer's specifications, ball joint replacement is required.

Camber angles are an imaginary line through the centerline of the tire and wheel in relation to the true vertical centerline of the tire.

Ball Joint Horizontal Measurement

Worn ball joints cause improper **camber angles** and **caster angles**, which result in reduced directional stability and tire tread wear. Connect the dial indicator to the lower control arm of the ball joint being checked and position the dial indicator stem against the edge of the wheel rim (Figure 6-15).

Caster angles are an imaginary line through the upper and lower ball joint centers in relation to the true vertical tire centerline viewed from the side.

Be sure the front wheel bearings are adjusted properly prior to the ball joint horizontal measurement. While an assistant grasps the top and bottom of the raised tire and attempts to move the tire and wheel horizontally in and out, observe the reading on the dial indicator (Figure 6-16).

Copyright © 2015 Cengage Learning®

FIGURE 6-13 **Dial indicator designed for ball joint measurement.**

Copyright © 2015 Cengage Learning®

FIGURE 6-14 **Dial indicator installed to measure vertical ball joint movement on a compression-loaded ball joint.**

PHOTO SEQUENCE 10

VERTICAL BALL JOINT MEASUREMENT

P10-1 **Place a floor jack under a front, lower control arm, and raise the lower control arm until the front tire is approximately 4 in. (10 cm) off the floor.**

Copyright © 2015 Cengage Learning®

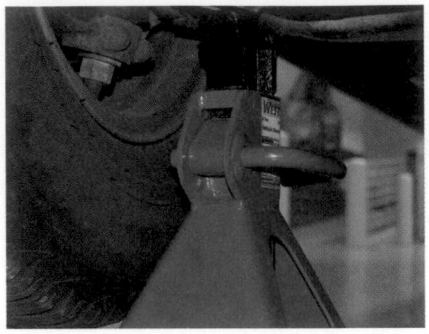

P10-2 **Lower the lower control arm until it is securely supported on a safety stand, and remove the floor jack.**

Copyright © 2015 Cengage Learning®

P10-3 **Attach the magnetic base of a dial indicator for measuring ball joints to the safety stand.**

Copyright © 2015 Cengage Learning®

P10-4 **Position the dial indicator stem against the steering knuckle beside the ball joint stud.**

Copyright © 2015 Cengage Learning®

P10-5 **Preload the dial indicator stem approximately 0.250 in. (6.35 mm), and zero the dial indicator scale.**

Copyright © 2015 Cengage Learning®

P10-6 **Position a pry bar under the tire and lift on the tire and wheel assembly while a coworker observes the dial indicator. Record the ball joint vertical movement indicated on the dial indicator.**

Copyright © 2015 Cengage Learning®

P10-7 **Repeat step 6 several times to confirm an accurate dial indicator reading.**

Copyright © 2015 Cengage Learning®

P10-8 **Compare the ball joint vertical movement indicated on the dial indicator to the vehicle manufacturer's specifications for ball joint wear. If the ball joint vertical movement exceeds specifications, ball joint replacement is necessary.**

Copyright © 2015 Cengage Learning®

Copyright © 2015 Cengage Learning®

FIGURE 6-15 **Dial indicator installed to measure radial ball joint movement.**

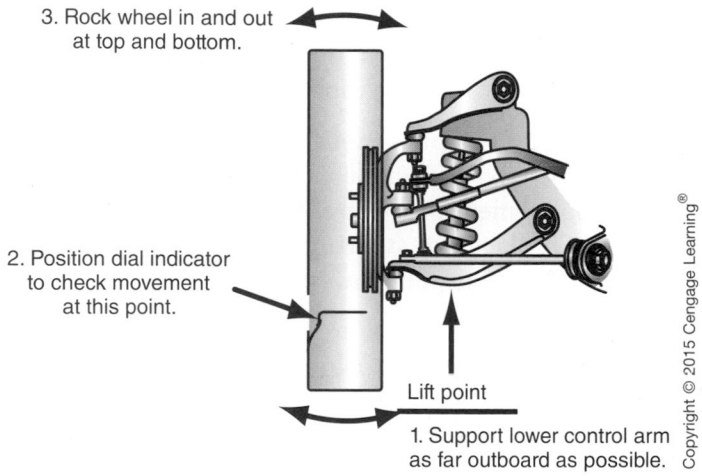

Copyright © 2015 Cengage Learning®

FIGURE 6-16 **Measuring radial ball joint movement.**

Diagnosis of Ride Harshness

When diagnosing a ride harshness condition, check these measurements and components:

1. Curb riding height—insufficient curb riding height causes ride harshness
2. Excessive vehicle load
3. Worn shock absorbers or struts
4. Broken or weak springs
5. Worn suspension bushings, such as strut mounts, control arm bushings, and shock absorber mounting bushings
6. Wheel alignment—excessive positive caster on front wheels causes ride harshness

Noise Diagnosis

A squeaking noise in the rear suspension may be caused by a suspension bushing or a defective strut or shock absorber.

Checking Ball Joints, Twin I-Beam Axles

Follow these steps to check the ball joints on a twin I-beam axle:

Classroom Manual
Chapter 6, page 160

1. Lift the front end of the vehicle with a floor jack and place safety stands near the outer end of the I-beams.
2. Lower the vehicle onto the safety stands.
3. While an assistant grasps the wheel at the bottom and moves the wheel in and out, watch for movement between the lower part of the axle jaw and the spindle lower arm. If this movement exceeds 1/32 in., replace the lower ball joint.
4. As an assistant grasps the top of the wheel and moves the wheel in and out, watch for movement between the upper axle jaw and the spindle upper arm. If this movement exceeds 1/32 in., replace the upper ball joint.

Front Steering Knuckle Diagnosis and Service, Front-Wheel-Drive Vehicle

Front Steering Knuckle Diagnosis. Steering looseness may be caused by a worn tie rod end opening or ball joint opening in the steering knuckle. If these knuckle openings are worn, reduced directional stability may be experienced. Since a bent steering knuckle affects front suspension alignment angles, this problem may cause reduced directional stability and tire tread wear. Many steering arm and knuckle assemblies must be replaced as a unit. If the steering arms are bent, knuckle replacement is required.

Steering Knuckle Removal and Replacement. The steering knuckle replacement procedure varies depending on the type of front suspension. Always follow the vehicle manufacturer's steering knuckle replacement procedure in the service manual.

Follow these steps to replace the steering knuckle on a front-wheel-drive vehicle with a MacPherson strut front suspension system:

1. Remove the wheel cover and loosen the front wheel nuts and the drive axle nut.
2. Lift the vehicle chassis on a hoist and allow the front suspension to drop downward. Remove the front wheel, brake caliper, brake rotor, and drive axle nut. Tie the brake caliper to a suspension component. Do not allow the caliper to hang on the end of the brake hose.
3. Remove the inner end of the drive axle from the transaxle with a pulling or prying action. On some Chrysler products, the differential cover must be removed and axle circlips compressed before the drive axle is removed from the transaxle.
4. Remove the outer end of the drive axle from the steering knuckle. On some early model Ford Escorts and Lynx, a puller is required for this operation.
5. Be sure the vehicle weight is supported on the hoist with the front suspension dropped downward. Disconnect the lower ball joint nuts from the mounting bolts in the lower control arm (Figure 6-17).
6. Remove the cotter pin from the outer tie rod nut (Figure 6-18). Remove the outer tie rod nut, and then use a puller to disconnect the tie rod end from the steering knuckle (Figure 6-19).
7. If an eccentric cam is used on one of the strut-to-knuckle bolts, mark the cam and bolt position in relation to the strut, and remove the strut-to-knuckle bolts.
8. Remove the knuckle from the strut, and lift the knuckle out of the chassis (Figure 6-20).
9. Pry the dust deflector from the steering knuckle with a large flat-blade screwdriver (Figure 6-21).
10. Use a puller to remove the ball joint from the steering knuckle (Figure 6-22). Check the ball joint and tie rod end openings in the knuckle for wear and out-of-round. Replace the knuckle if these openings are worn or out-of-round.

SPECIAL TOOL
Ball joint puller

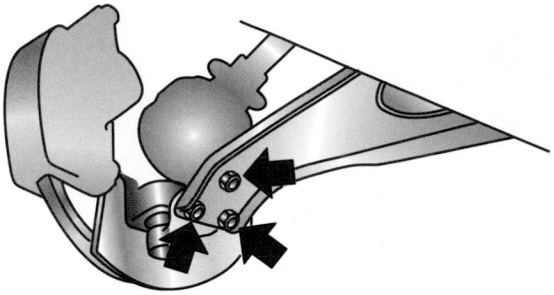

Copyright © 2015 Cengage Learning®

FIGURE 6-17 **Removing nuts from lower ball joint mounting bolts under the lower control arm.**

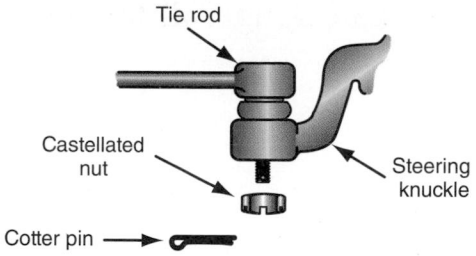

Copyright © 2015 Cengage Learning®

FIGURE 6-18 **Removing the cotter pin and nut from the outer tie rod end.**

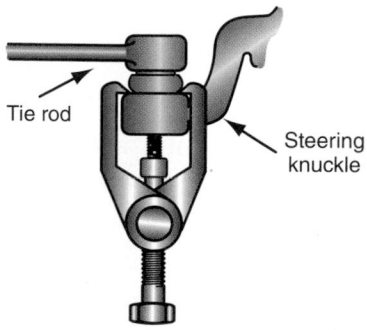

Copyright © 2015 Cengage Learning®

FIGURE 6-19 **Removing the outer tie rod end from the steering knuckle.**

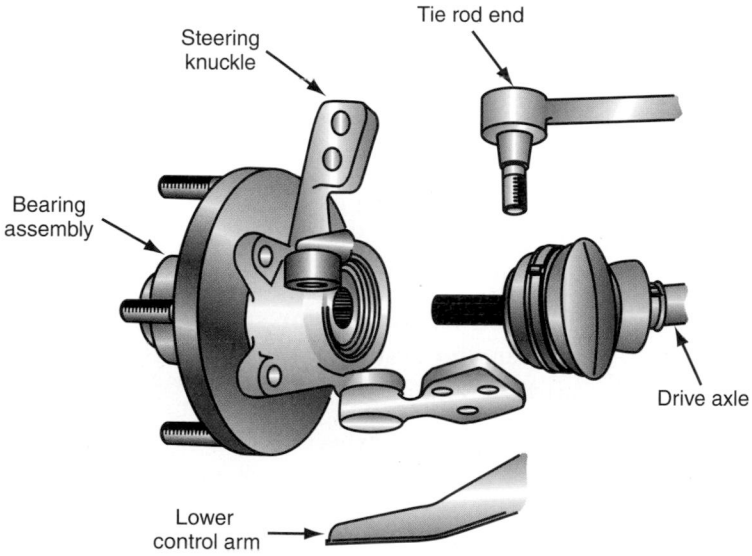

Copyright © 2015 Cengage Learning®

FIGURE 6-20 **Removing the steering knuckle from the lower control arm.**

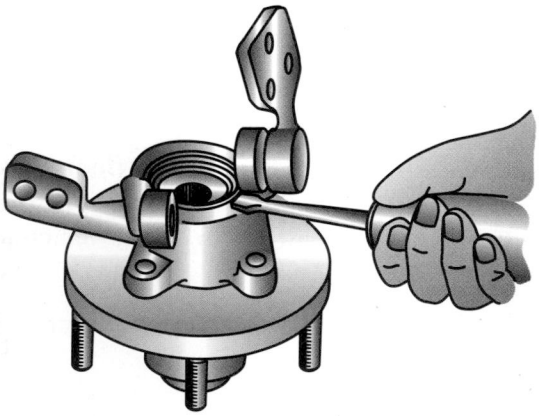

Copyright © 2015 Cengage Learning®

FIGURE 6-21 **Removing dust deflector from the steering knuckle.**

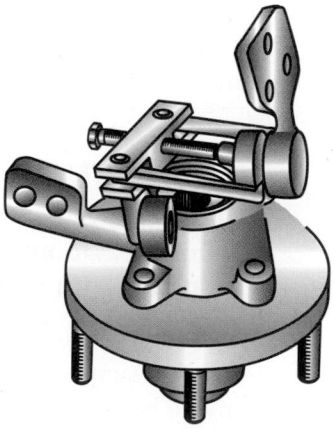

Copyright © 2015 Cengage Learning®

FIGURE 6-22 **Removing ball joint from the steering knuckle.**

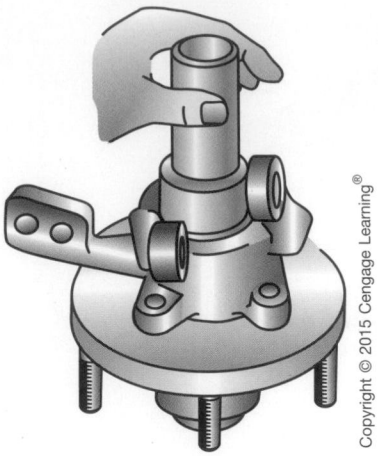

FIGURE 6-23 **Installing dust deflector in the steering knuckle.**

11. Use the proper driving tool to reinstall the dust deflector in the steering knuckle (Figure 6-23).
12. Reverse steps 1 through 10 to reinstall the knuckle. Service or replace the wheel bearing as required. (Refer to Chapter 4 for wheel bearing service.) Torque all nuts to specifications and install cotter pins as required.

Front Steering Knuckle Replacement, Rear-Wheel-Drive Vehicle

Proceed as follows for steering knuckle replacement on short-and-long arm front suspension systems with the coil spring positioned between the lower control arm and the frame:

1. Remove the wheel cover and loosen the wheel nuts.
2. Lift the vehicle with a floor jack and place safety stands under the chassis so the front suspension drops downward. Lower the vehicle onto the safety stands and remove the floor jack.
3. Remove the front wheel, brake caliper, and the brake rotor and hub with the wheel bearings. Attach the brake caliper to a chassis component with a piece of wire. Do not allow the caliper to hang on the end of the brake hose.
4. Remove the outer tie rod cotter pin and nut, and remove the outer tie rod end from the steering arm.
5. Remove the cotter pins from the upper and lower ball joint nuts and loosen, but do not remove the nuts.

WARNING: During the knuckle replacement procedure, the floor jack supports the spring tension when the ball joints are disconnected from the knuckle. Do not lower the floor jack. Chain the coil spring to the lower control arm as a safety precaution.

SPECIAL TOOL
Ball joint spreader tool

6. Support the lower control arm with a floor jack, and raise the floor jack until the spring tension is supported by the jack.
7. Remove both ball joint nuts, and remove the ball joint studs from the steering knuckle. A threaded spreader tool may be used to push the upper and lower ball joints from the knuckle (Figure 6-24 and Figure 6-25).
8. Pull upward on the upper control arm to remove the upper ball joint stud from the knuckle. Remove the knuckle from the lower ball joint stud. Check the ball joint and tie rod end openings in the knuckle for wear and out-of-round. Replace the knuckle if these conditions are present.

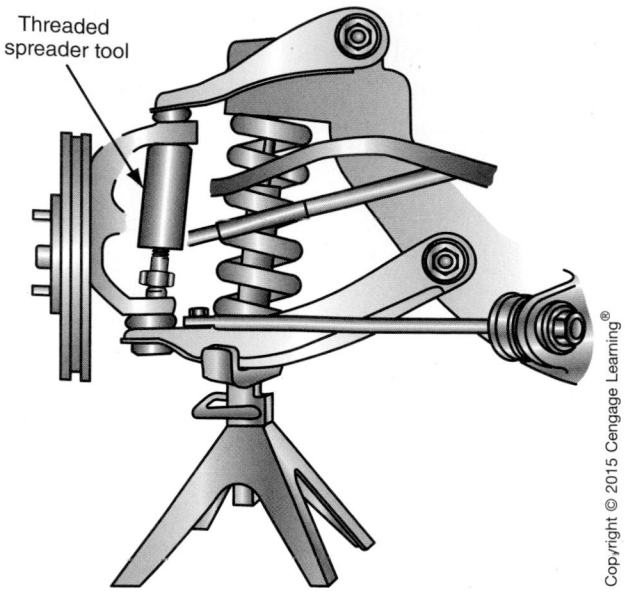

FIGURE 6-24 **Removing the lower ball joint from the steering knuckle.**

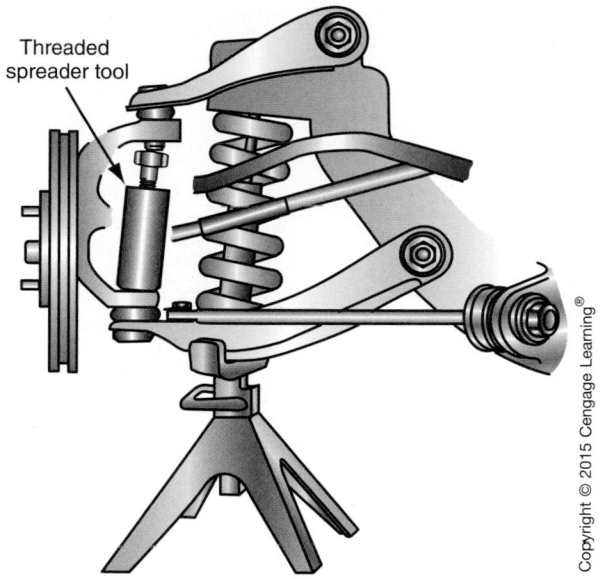

FIGURE 6-25 **Removing the upper ball joint from the steering knuckle.**

9. Reverse steps 1 through 8 to install the steering knuckle. Tighten the ball joint nuts and the outer tie rod nut to the specified torque, and install the cotter pins in these nuts. Adjust the front wheel bearings, and tighten the wheel nuts to the specified torque.

SERVICE TIP: Some ball joints have a tapered stud, and a nut is threaded onto the top of this stud to retain the ball joint in the steering knuckle. Other ball joints have a straight stud, and a pinch bolt extends through the steering knuckle and a notch in the side of the ball joint stud to hold the ball joint in the steering knuckle.

Ball Joint Replacement

Results of Ball Joint Wear. Worn ball joints affect steering angles and cause reduced directional stability and excessive tire tread wear.

Ball Joint Removal and Replacement. Ball joints may be pressed, bolted (or riveted), or threaded into the control arms (Figure 6-26). The ball joint replacement procedure varies depending on the type of suspension and the method of ball joint attachment. Always follow the vehicle manufacturer's recommended ball joint replacement procedure in the service manual.

The following are typical steps that apply to all three methods of ball joint attachment:

1. Remove the wheel cover and loosen the wheel nuts.
2. Lift the vehicle with a floor jack and place safety stands under the chassis so the front suspension is allowed to drop downward. Lower the vehicle onto the safety stands and remove the floor jack.

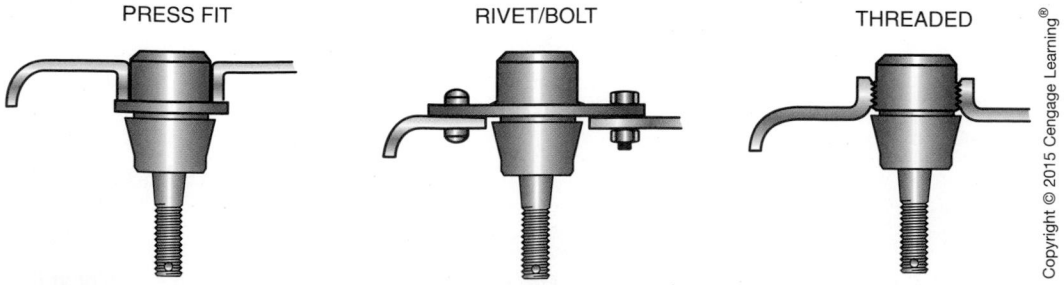

FIGURE 6-26 **Methods of ball joint attachment.**

SPECIAL TOOL

Ball joint pressing tools

3. Remove the wheel and place a floor jack under the outer end of the lower control arm. Operate the floor jack and raise the lower control arm until the ball joints are unloaded. Remove other components, such as the brake caliper, rotor, and drive axle, as required to gain access to the ball joints.
4. Remove the cotter pin in the ball joint or joints that requires replacement, and loosen, but do not remove the ball joint stud nuts.
5. Loosen the ball joint stud tapers in the steering knuckle. A threaded expansion tool is available for this purpose.
6. Remove the ball joint nut and lift the knuckle off the ball joint stud. Block or tie up the knuckle and hub assembly to access the ball joint.
7. If the ball joint is riveted to the control arm, drill and punch out the rivets, and bolt the new ball joint to the control arm (Figure 6-27).
8. If the ball joint is pressed into the lower control arm, remove the ball joint dust boot, and use a pressing tool to remove and replace the ball joint (Figure 6-28 and Figure 6-29).

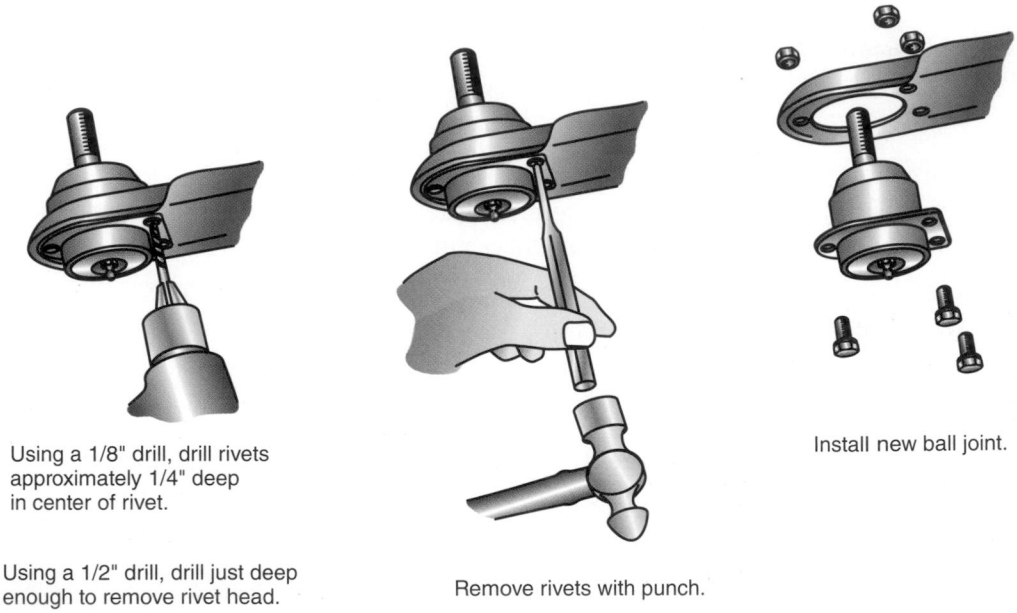

Copyright © 2015 Cengage Learning®

FIGURE 6-27 **Replacing the riveted ball joint.**

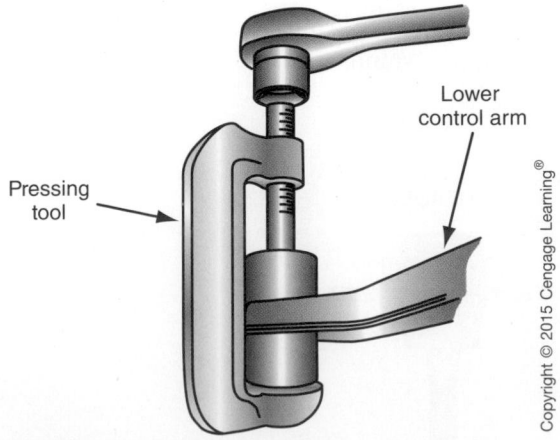

Copyright © 2015 Cengage Learning®

FIGURE 6-28 **Removing the pressed-in ball joint from the lower control arm.**

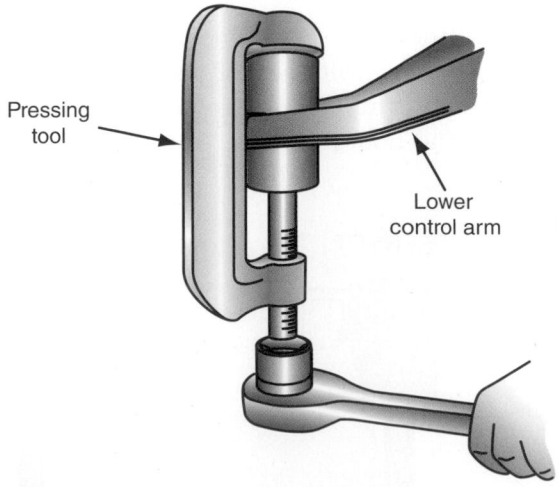

Copyright © 2015 Cengage Learning®

FIGURE 6-29 **Installing the pressed-in ball joint in the lower control arm.**

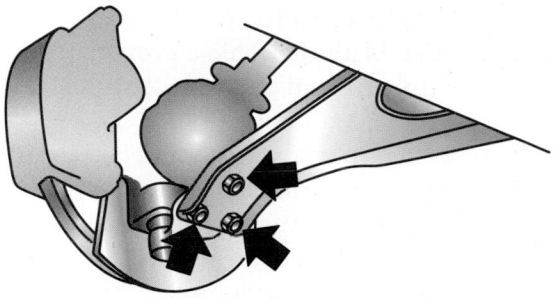

Copyright © 2015 Cengage Learning®

FIGURE 6-30 **Installing the ball joint retaining bolts and nuts in the lower control arm.**

9. If the ball joint housing is threaded into the control arm, use the proper size socket to remove and install the ball joint. The replacement ball joint must be torqued to the manufacturer's specifications. If a minimum of 125 ft.-lbs of torque cannot be obtained, the control arm threads are damaged and control arm replacement is necessary.
10. If the ball joint is bolted to the lower control arm, install the new ball joint and tighten the bolt and nuts to the specified torque (Figure 6-30).
11. Clean and inspect the ball joint stud tapered opening in the steering knuckle. If this opening is out-of-round or damaged, the knuckle must be replaced.
12. Check the fit of the ball joint stud in the steering knuckle opening. This stud should fit snugly in the opening and only the threads on the stud should extend through the knuckle. If the ball joint stud fits loosely in the knuckle tapered opening, either this opening is worn or the wrong ball joint has been supplied.
13. Install the ball joint stud in the steering knuckle opening, making sure the stud is straight and centered. Install the stud nut and tighten this nut to the specified torque. Install a new cotter pin through the stud and nut. Do not loosen the nut to align the nut and stud openings.
14. Reassemble the components that were removed in step 3. Make sure the wheel nuts are tightened to the specified torque.
15. After ball joint replacement, the front suspension alignment should be checked.

CAUTION:
Never back off a ball joint stud nut to align the cotter pin openings in the nut and stud. This action may cause the ball joint stud to become loose in the opening where it is mounted. Always tighten the nut to the next hole to install the cotter pin.

Control Arm Diagnosis and Service

Control Arm Diagnosis and Replacement

Upper and lower control arms should be inspected for deterioration, cracks, worn bushings and bent conditions (Figure 6-31). The rubber should be intact and in good condition all the way around the bushing. A polished area on the shaft indicates bushing wear and movement exists. If the control arm bushings are worn, steering is erratic, especially on irregular road surfaces. Worn control arm bushings may cause a rattling noise while

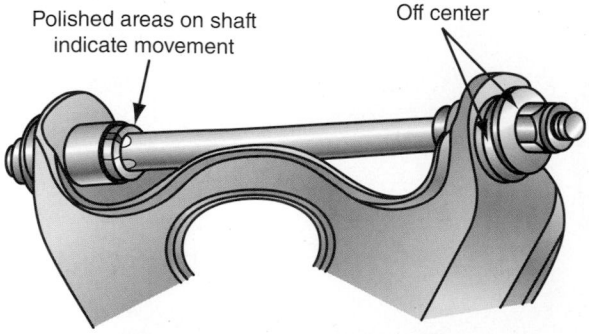

Copyright © 2015 Cengage Learning®

Bad upper control arm bushings

FIGURE 6-31 **Upper and lower control arms should be inspected for cracks and worn and off center bushings.**

driving on irregular road surfaces. Dry or worn control arm bushings may cause a squeaking noise on irregular road surfaces. With the vehicle properly lifted, use a pry bar between the mounting frame and the control arm and rock the control arm in and out and look and feel for bushing movement.

Caster and camber angles on the front suspension are altered by worn upper and lower control arm bushings. Incorrect caster or camber angles may cause the vehicle to pull to one side. Tire wear may be excessive when the camber angle is not within specifications.

The upper or lower control arm removal and replacement procedure varies depending on the type of suspension. Always follow the recommended procedure in the vehicle manufacturer's service manual.

Lower Control Arm Replacement, MacPherson Strut Suspension

SPECIAL TOOL

Tie rod end puller

The following is a lower control arm replacement procedure for a MacPherson strut front suspension:

1. Remove the wheel cover and loosen the wheel nuts and drive axle nut.
2. Lift the front of the vehicle with a floor jack and place jack stands under the chassis. Lower the vehicle onto the safety stands and allow the front suspension to drop downward.
3. Remove the front wheel and the front fender apron seal (Figure 6-32). Remove the drive axle nut (Figure 6-33).
4. Remove the cotter pin and nut from the outer tie rod end, and use a puller to remove this tie rod end from the steering arm (Figure 6-34).
5. Remove the stabilizer brackets from the lower control arm (Figure 6-35).
6. Remove the lower ball joint nuts and bolts from the lower control arm (Figure 6-36).
7. Remove the drive axle from the axle hub (Figure 6-37), and secure the drive axle to a suspension component with a piece of wire (Figure 6-38). Do not allow the drive axle to hang downward.
8. Remove the two attaching bolts at the front side of the lower control arm (Figure 6-39).
9. Remove the bolt and nut on the rear side of the lower control arm (Figure 6-40), and remove the lower arm bushing stopper from the lower arm shaft.

Classroom Manual
Chapter 6, page 152

Reverse steps 1 through 9 to install the replacement control arm. Tighten all bolts and nuts to the specified torque, and install cotter pins as required.

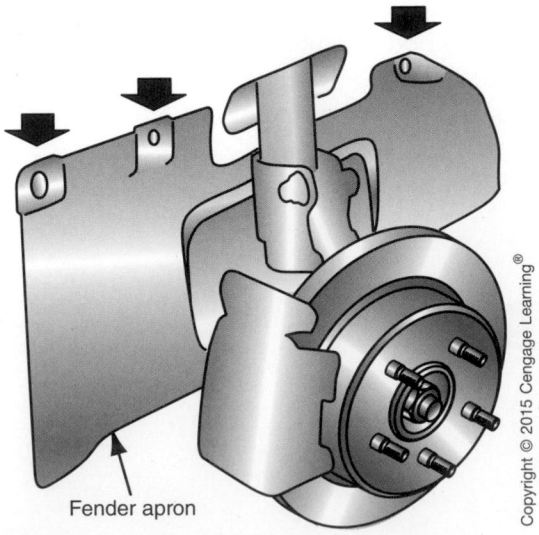

FIGURE 6-32 **Removing the front fender apron seal.**

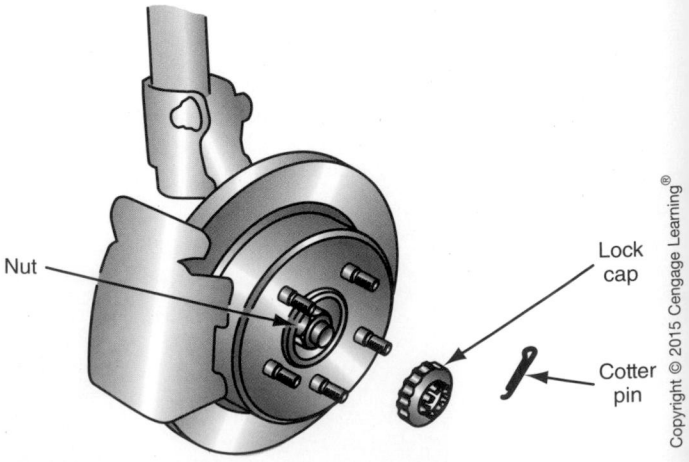

FIGURE 6-33 **Removing the drive axle cotter pin, lock, cap, and nut.**

Copyright © 2015 Cengage Learning®

FIGURE 6-34 **Removing the outer tie rod end.**

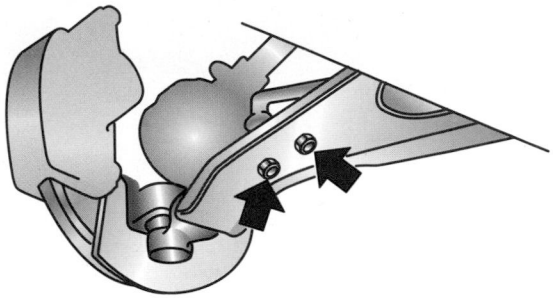

Copyright © 2015 Cengage Learning®

FIGURE 6-35 **Removing stabilizer brackets from the lower control arm.**

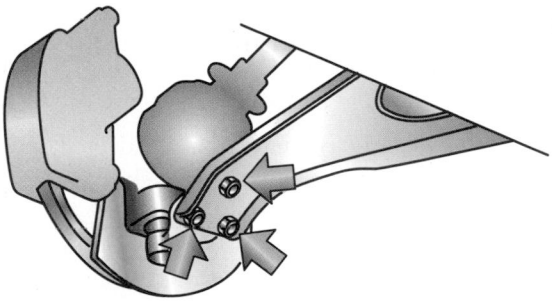

Copyright © 2015 Cengage Learning®

FIGURE 6-36 **Removing the lower ball joint bolts and nuts from the lower control arm.**

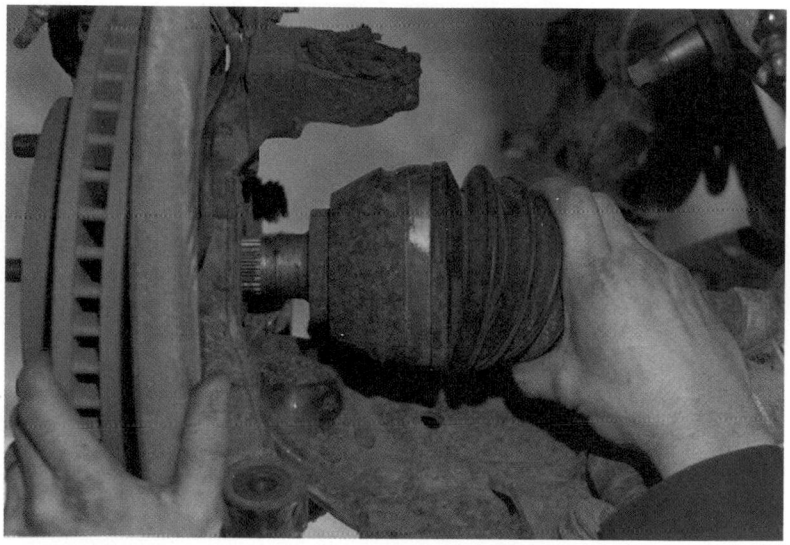

Copyright © 2015 Cengage Learning®

FIGURE 6-37 **Removing the drive axle from the axle hub.**

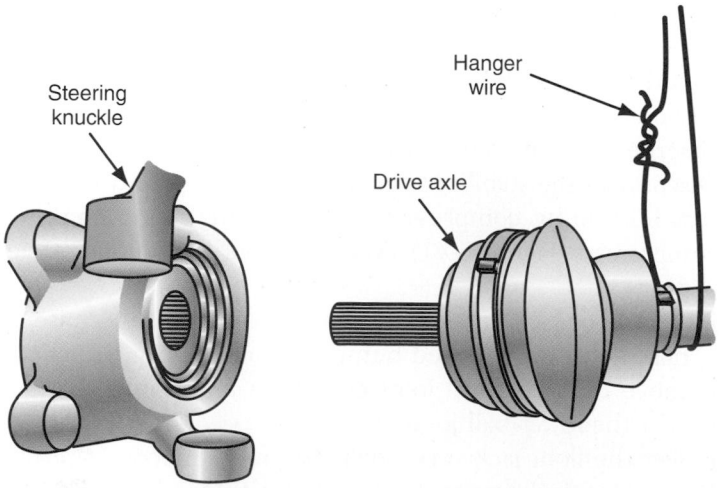

Copyright © 2015 Cengage Learning®

FIGURE 6-38 **Securing the drive axle with a piece of wire.**

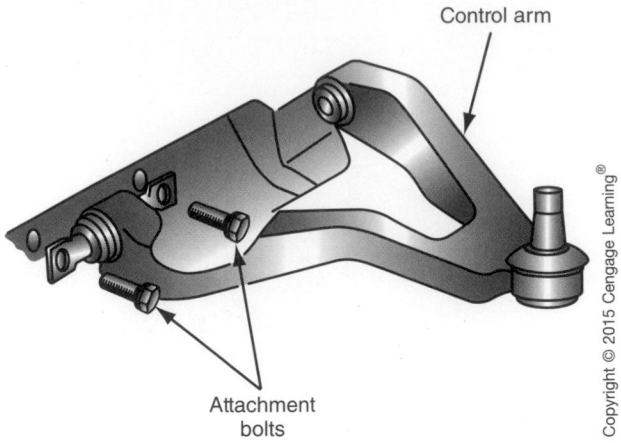

FIGURE 6-39 **Removing the bolts from the front side of the lower control arm.**

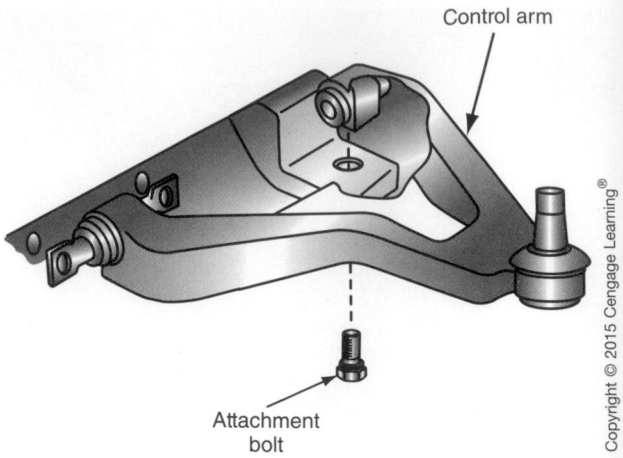

FIGURE 6-40 **Removing the bolt and nut from the rear side of the lower control arm.**

Lower Control Arm and Spring Replacement, Short-and-Long Arm Suspension

WARNING: During control arm and spring replacement, the coil spring tension is supported by a compressing tool or floor jack. Always follow the vehicle manufacturer's recommended control arm and spring replacement procedures very carefully. Serious personal injury or property damage may occur if the spring tension is released suddenly.

Broken coil springs may cause a rattling noise while driving on irregular road surfaces. Weak or broken coil springs also reduce curb riding height. A rattling noise while driving on road irregularities may also be caused by worn or broken spring insulators. Since weak or broken coil springs affect front suspension alignment angles, this problem may cause reduced directional stability, excessive tire wear, and harsh riding.

Follow this procedure when a lower control arm and/or spring is replaced on a vehicle with a short-and-long arm suspension system with the coil springs positioned between the lower control arm and the frame:

1. Lift the vehicle on a hoist until the tires are a short distance off the floor, and allow the front suspension to drop downward. An alternate method is to lift the vehicle with a floor jack, and then support the chassis securely on safety stands so the front suspension drops downward.
2. Disconnect the lower end of the shock absorber. On some applications, the shock absorber must be removed.
3. Disconnect the stabilizer bar from the lower control arm.
4. Install a spring compressor and turn the spring compressor bolt until the spring is compressed (Figure 6-41). Make sure all the spring tension is supported by the compressing tool.
5. Place a floor jack under the lower control arm, and raise the jack until the control arm is raised and the rebound bumper is not making contact.
6. Remove the lower ball joint cotter pin and nut, and use a threaded expansion tool to loosen the lower ball joint stud.
7. Lower the floor jack very slowly to lower the control arm and coil spring.
8. Disconnect the lower control arm inner mounting bolts, and remove the lower control arm.

SPECIAL TOOL

Spring compressing tool, short-and-long arm suspension

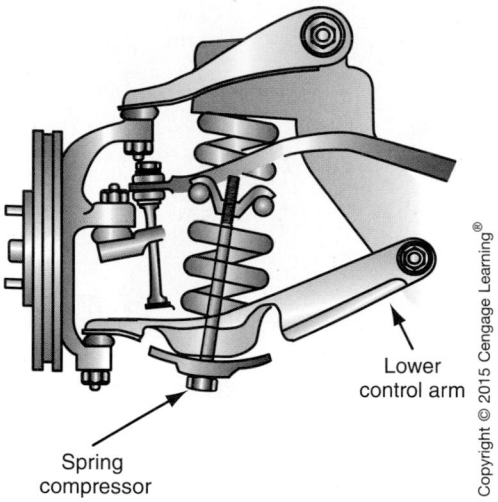

FIGURE 6-41 **Spring compressing tool installed on a spring in a short-and-long arm suspension system.**

9. Rotate the compressing tool bolt to release the spring tension, and remove the spring from the control arm.
10. Inspect the lower control arm for a bent condition or cracks. If either of these conditions is present, replace the control arm. Visually inspect all control arm bushings. Loose or worn bushings must be replaced. Visually inspect upper and lower spring insulators for cracks and wear, and inspect the spring seat areas in the chassis and lower control arm. Worn or cracked spring insulators must be replaced.
11. Reverse steps 1 through 9 to install the lower control arm. Be sure the coil spring and insulators are properly seated in the lower control arm and in the upper spring seat.

NOTE: See Chapter 5 for strut and spring service.

Upper Control Arm Removal and Replacement, Short-and-Long Arm Suspension System

Proceed as follows when replacing an upper control arm on a short-and-long arm suspension system with the front coil springs located between the lower control arm and the frame:

1. Remove the wheel cover and loosen the wheel nuts.
2. Lift the vehicle with a floor jack and install safety stands under the chassis. Lower the vehicle onto the safety stands so the front suspension drops downward.
3. Remove the front wheel and tire.
4. Place a floor jack under the lower control arm and raise this arm. Make sure the rebound bumper is not making contact.
5. Remove the cotter pin and loosen, but do not remove the upper ball joint nut.
6. Use a threaded expansion tool to loosen the ball joint stud in the control arm. Make sure the floor jack is supporting the spring tension.
7. Remove the ball joint nut and the inner control arm mounting bolts. If shims are located on the inner control arm mounting bolts, note the shim position.
8. Remove the control arm from the chassis.
9. Visually inspect the upper control arm for a bent or cracked condition. Inspect the control arm bushing openings for wear, and replace the control arm if it is bent, cracked, or worn in the bushing areas. Inspect the control arm shaft and bushings for wear, and replace all worn parts (Figure 6-42).

CAUTION:
Since the floor jack is supporting the spring tension, do not lower the floor jack until the ball joint is reconnected and the ball joint nut is tightened to the specified torque.

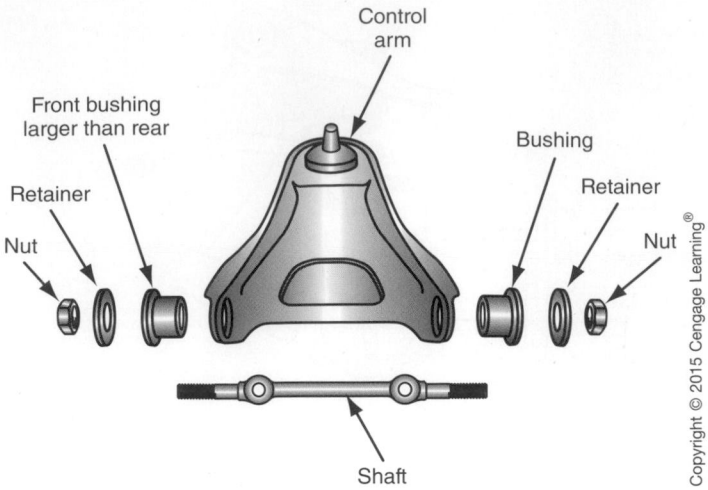

FIGURE 6-42 **Upper control arm components.**

SERVICE TIP: Excessively worn rebound bumpers indicate worn-out shock absorbers, reduced curb riding height, or driving continually on irregular road surfaces.

10. Install the control arm and place the original number of shims on the inner mounting bolts. Tighten these bolts to the manufacturer's specifications.
11. Install the ball joint stud in the steering knuckle, and tighten the stud nut to the specified torque.
12. Install the upper ball joint cotter pin and the front wheel. Tighten the wheel nuts to the specified torque.

The front suspension alignment angles should be checked after upper control arm replacement.

Rebound bumpers limit the maximum lower control arm movement to prevent coil spring bottoming. Rebound bumpers may be called strikeout bumpers.

Rebound Bumpers

Rebound bumpers are usually bolted to the lower control arm or to the chassis. Inspect the rebound bumpers for cracks, wear, and flattened conditions (Figure 6-43). Damaged rebound bumpers may be caused by sagged springs and insufficient curb riding height or worn-out shock absorbers and struts. If the rebound bumpers must be replaced, remove the mounting bolts and the bumper. Install the new bumper and tighten the mounting bolts to the specified torque.

SPECIAL TOOL

Control arm bushing removal, replacement, and flaring tools

Front Lower Control Arm Bushing Removal and Replacement

All suspension bushings should be checked periodically for wear, looseness, and deterioration. These bushings are important for providing quiet suspension operation and preventing the transmission of suspension vibration to the chassis and passenger compartment. When a bushing is contained in a steel sleeve, press only on the outer sleeve during the

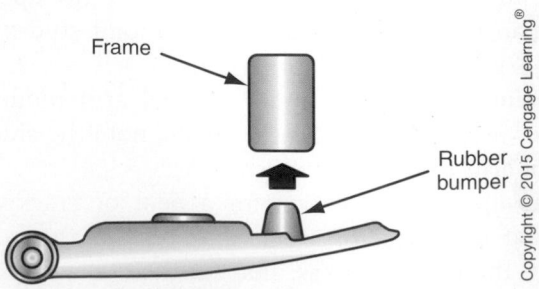

FIGURE 6-43 **Rebound bumper.**

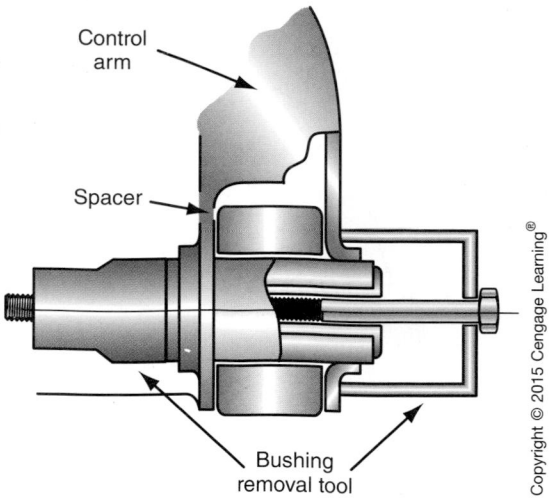

FIGURE 6-44 **Removing the front control arm bushing.**

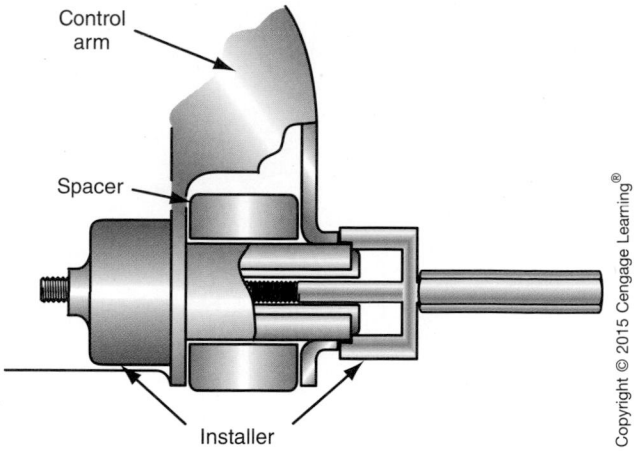

FIGURE 6-45 **Installing the front control arm bushing.**

removal and replacement procedure. The control arm bushing removal and replacement procedure varies depending on the type of suspension system. Always follow the vehicle manufacturer's recommended procedure in the service manual. Special bushing removal and replacement tools are required for control arm bushing replacement. A front control arm bushing removal tool and spacer is used to remove the control arm bushing (Figure 6-44). The spacer is installed between the bushing support lugs to prevent distorting these lugs during the bushing removal process.

A different tool is required to install the new front control arm bushing (Figure 6-45). The same spacer that is used during the bushing removal is installed between the bushing lugs during bushing installation.

After the front control arm bushing is installed, a bushing flaring tool is used to flare the bushing (Figure 6-46). Notice that a spacer is installed between the bushing lugs during the flaring process. Flaring retains the bushing securely (Figure 6-47).

The rear control arm bushing is replaced with the same procedure as the front control arm bushing. Some of the same special tools are used for rear control arm bushing replacement (Figure 6-48 and Figure 6-49).

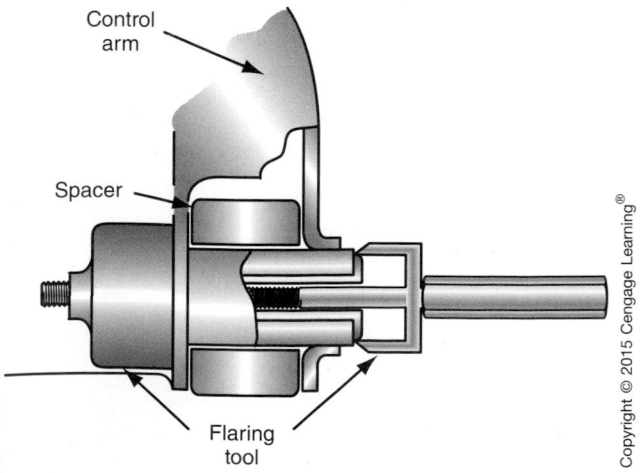

FIGURE 6-46 **Flaring the front control arm bushing.**

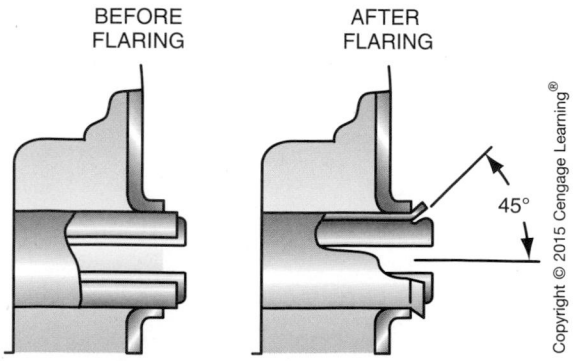

FIGURE 6-47 **Front control arm bushing after flaring.**

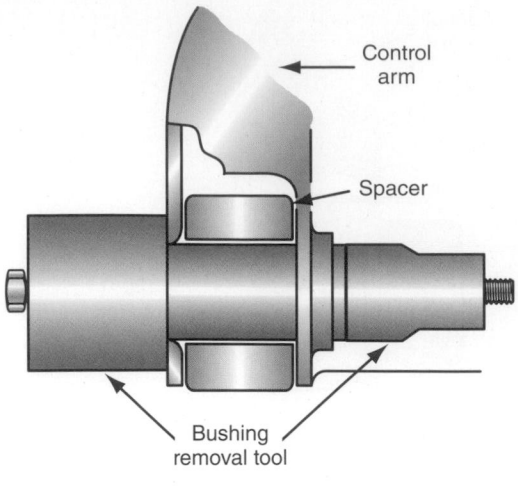

FIGURE 6-48 **Removing the rear control arm bushing.**

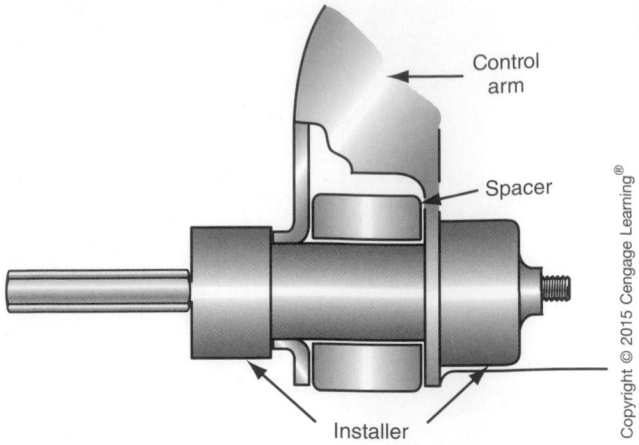

FIGURE 6-49 **Installing the rear control arm bushing.**

Stabilizer Bar Diagnosis and Replacement

A **stabilizer bar** may be referred to as a sway bar. It reduces body sway when one wheel strikes a road irregularity.

Worn **stabilizer bar** mounting bushings, grommets, or mounting bolts cause a rattling noise as the vehicle is driven on irregular road surfaces. A weak stabilizer bar or worn bushings and grommets cause harsh riding and excessive body sway while driving on irregular road surfaces. Worn or very dry stabilizer bar bushings may cause a squeaking noise on irregular road surfaces. All stabilizer bar components should be visually inspected for wear. Stabilizer bar removal and replacement procedures vary depending on the vehicle. Always follow the vehicle manufacturer's recommended procedure in the service manual.

The following is a typical stabilizer bar removal and replacement procedure:

Classroom Manual
Chapter 6, page 150

1. Lift the vehicle on a hoist and allow both sides of the front suspension to drop downward as the vehicle chassis is supported on the hoist.
2. Remove the mounting bolts at the outer ends of the stabilizer bar and remove the bushings, grommets, brackets, or spacers (Figure 6-50).
3. Remove the mounting bolts in the center area of the stabilizer bar.
4. Remove the stabilizer bar from the chassis.
5. Visually inspect all stabilizer bar components, such as bushings, bolts, and spacer sleeves. Replace the stabilizer bar, grommets, bushings, brackets, or spacers as required.

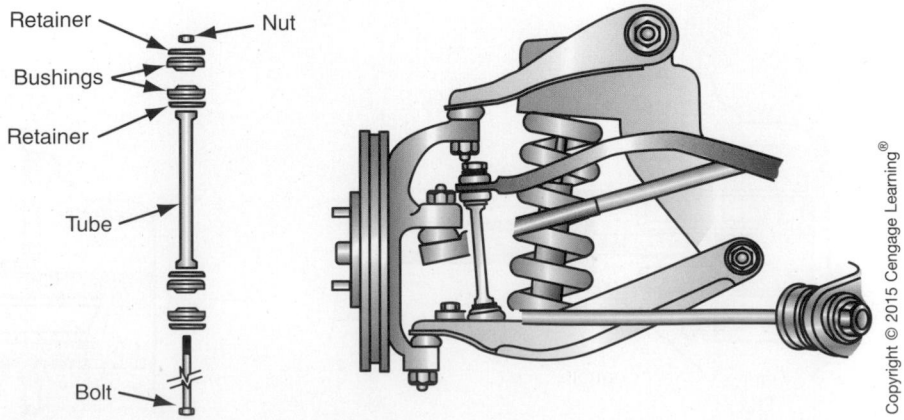

FIGURE 6-50 **Stabilizer bar components.**

Split bushings may be removed over the stabilizer bar. Bushings that are not split must be pulled from the bar.

6. Reverse steps 2 through 4 to install the stabilizer bar. Make sure all stabilizer bar components are installed in the original position, and tighten all fasteners to the specified torque.

Some vehicle manufacturers specify that stabilizer bars must have equal distances between the outer bar ends and the lower control arms. Always refer to the manufacturer's recommended measurement procedure. If this measurement is required, adjust the nut on the outer stabilizer bar mounting bolt until equal distances are obtained between the outer bar ends and the lower control arms. Worn grommets cause these distances to be unequal.

Strut Rod Diagnosis and Replacement

A **strut rod** prevents fore-and-aft lower control arm movement.

Some front suspension systems have a **strut rod** connected from the lower control arm to the frame. Rubber grommets isolate the rod from the chassis components. Worn strut rod grommets may allow the lower control arm to move rearward or forward. This movement changes the caster angle, which may affect steering quality and cause the vehicle to pull to one side. A bent strut rod also causes steering pull. Worn strut rod grommets or loose mounting bolts cause a rattling noise while driving on irregular road surfaces. Inspect the strut rod grommets visually for wear and deterioration. With the vehicle lifted on a hoist, grasp the strut rod firmly and apply vertical and horizontal force to check the rod and grommets for movement. Worn grommets must be replaced. If a strut rod is bent, replace the rod.

Follow these steps for strut rod replacement:

1. Lift the vehicle on a hoist.
2. Remove the strut rod nut from the front end of the rod.
3. Remove the strut rod bolts from the lower control arm.
4. Pull the strut rod rearward to remove the rod (Figure 6-51).
5. Remove the bushings from the opening in the chassis.
6. Visually inspect the strut rod, bushings, washers, and retaining bolts. Replace all worn parts. Reverse steps 1 through 5 to reinstall the strut rod. Tighten the strut rod nut and bolts to the specified torque.

Since strut rod and bushing conditions affect front suspension alignment, check front suspension alignment after strut rod service.

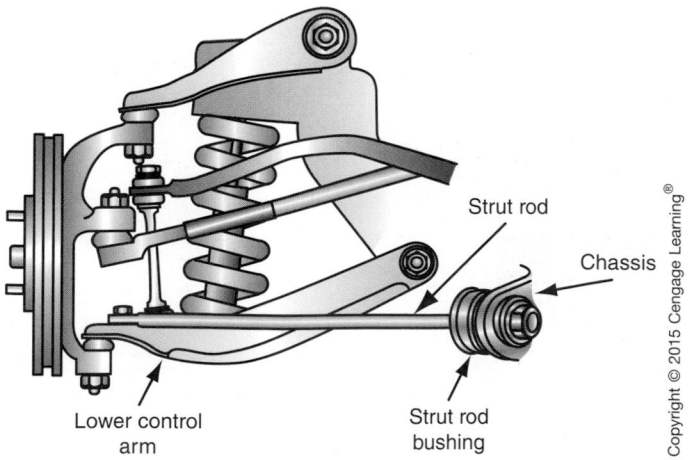

FIGURE 6-51 **Strut rod.**

Front Leaf Spring Inspection and Replacement

Front leaf springs are used in some truck suspension systems. These springs are mounted longitudinally with a spring on each side of the suspension. Many leaf springs have plastic silencers between the spring leafs. If these silencers are worn out, creaking and squawking noises will be heard when the vehicle is driven over road irregularities at low speeds.

When the silencers require checking or replacement, lift the vehicle with a floor jack and support the frame on safety stands so the suspension hangs downward. With the vehicle weight no longer applied to the springs, the spring leafs may be pried apart with a pry bar to remove and replace the silencers. On some leaf springs, the clamps must be removed to replace the silencers.

Leaf springs should be inspected for a sagged condition, which causes the curb riding height to be less than specified. Leaf springs should also be visually inspected for broken leafs, broken center bolts, and worn shackles or bushings. Weak or broken leaf springs affect front suspension alignment angles and cause excessive tire tread wear, reduced directional stability, and harsh riding. A rattling noise while driving on irregular road surfaces may be caused by worn shackles or bushings. A broken center bolt usually allows one side of the front suspension to move rearward in relation to the other side. This action may cause steering pull to one side. Worn shackles and bushings lower curb riding height and cause a rattling noise on road irregularities.

The following is a typical front leaf spring replacement procedure:

1. Lift the front end of the vehicle with a floor jack under the front axle and place safety stands under the frame. Lower the vehicle weight onto the safety stands, but leave the floor jack under the front axle to support some of the front suspension weight.
2. Remove the nuts from the spring U-bolts, and remove the U-bolts and lower spring plate. If the spring plate is attached to the shock absorber, this plate may be left on the shock absorber and moved out of the way. If shims are positioned between the spring plate and the front axle, be sure to note the position and the number of shims. These shims must be reinstalled in their original position because they set the front wheel caster angle.
3. Be sure the floor jack is lowered enough to relieve the vehicle weight from the springs.
4. Remove the front shackle assembly.
5. Remove the rear spring mounting bolt, and remove the spring from the chassis (Figure 6-52).
6. Check all the spring hangers, bolts, bushings, and shackle plates for wear, and replace as required. Springs with broken or sagged leafs are usually sent to a spring rebuilding shop for repair.
7. Check the spring center bolt to be sure it is not broken.
8. Reverse steps 1 through 5 to install the front springs. Tighten all bolts to the manufacturer's specifications.

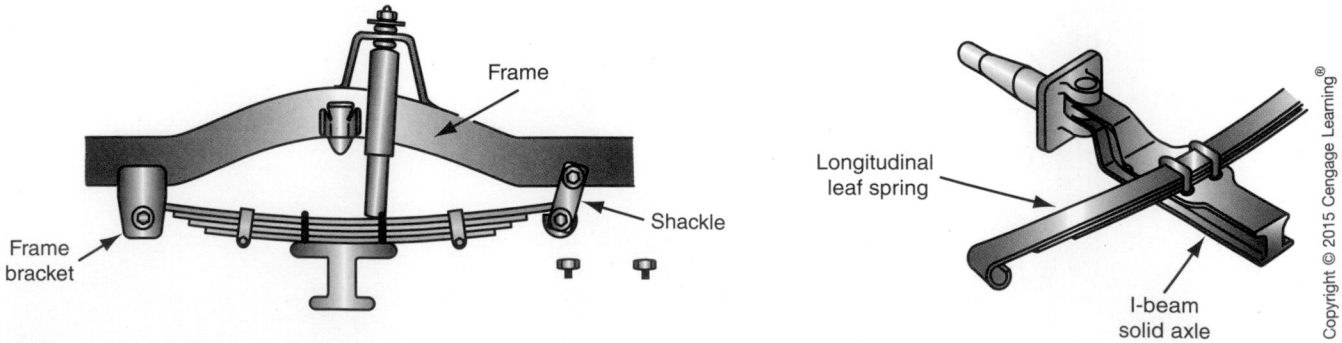

FIGURE 6-52 **Front leaf spring.**

Removing and Replacing Longitudinally Mounted Torsion Bars

Worn torsion bar components, such as pivot cushion bushings and control arm bushings, result in harsh riding and suspension noise when driving on road irregularities. A worn torsion bar hex and anchor may cause a rattling noise while driving on irregular road surfaces. Weak torsion bars or those with worn bushings or anchors cause reduced curb riding height, which may result in reduced directional stability and excessive tire tread wear. Always follow the vehicle manufacturer's recommended procedure in the service manual for torsion bar removal and replacement.

The following is a typical torsion bar removal and replacement procedure:

WARNING: Some torsion bar front suspension systems on sport utility vehicles (SUVs) are combined with an air suspension system. When servicing these systems, the air suspension switch in the rear jack storage area must be shut off before performing any suspension service or hoisting, jacking, or towing the vehicle. If this procedure is not followed, personal injury and vehicle damage may occur.

1. Raise the vehicle on a lift with the tires supported on the lift.
2. Remove the torsion bar cover plate bolts and the cover plate (Figure 6-53).
3. Measure and record the distance from the lip on the head of the torsion bar adjusting bolt to the casting surface this bolt is threaded into (Figure 6-54). When reinstalled, this bolt must be adjusted to this same measurement.
4. Install the torsion bar tool and adapters (Figure 6-55). Tighten this tool until the torsion bar adjuster lifts off the adjustment bolt.

CAUTION: The torsion bar adjusting bolt is coated with a dry adhesive. This bolt must be replaced if it is backed off or removed. Failure to follow this procedure may cause the bolt to loosen during operation, resulting in improper wheel alignment, reduced steering control, increased tire wear, and reduced ride quality.

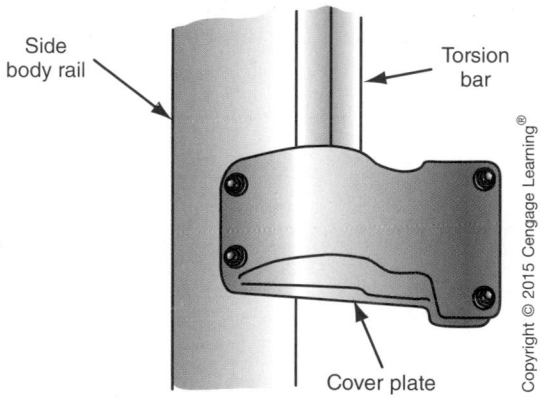

FIGURE 6-53 **Torsion bar cover plate.**

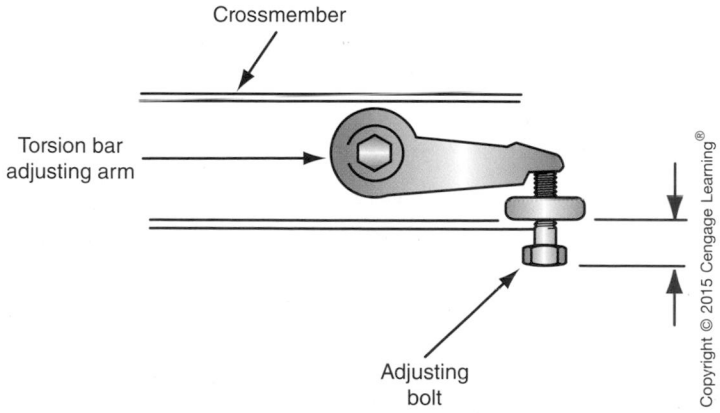

FIGURE 6-54 **Measuring the torsion bar adjusting bolt position.**

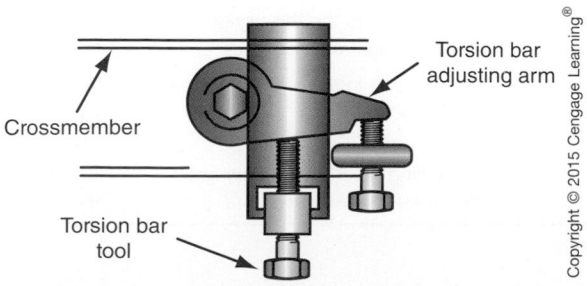

FIGURE 6-55 **Torsion bar tool and adapters.**

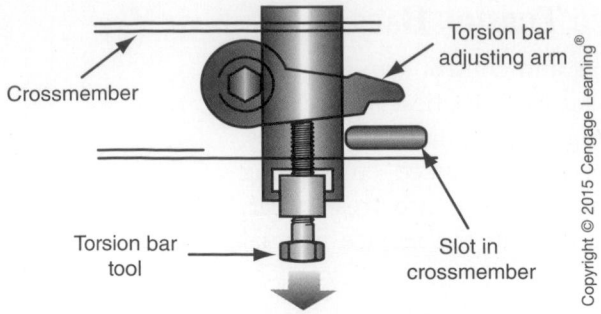

FIGURE 6-56 **Loosening the torsion bar tool.**

FIGURE 6-57 **Alignment marks placed on the torsion bar and adjuster.**

5. Remove the torsion bar adjustment bolt and nut.
6. Loosen the torsion bar tool to remove all the tension from the torsion bar (Figure 6-56).
7. Use the end of a screwdriver to place matching alignment marks on the torsion bar and the adjuster so these components may be reassembled in the same position. Remove the torsion bar insulator. Pull the torsion bar to the rear to remove this bar from the lower control arm.
8. Position the torsion bar in the lower control arm.
9. Install the torsion bar adjuster on the torsion bar with the alignment marks properly aligned (Figure 6-57).
10. Install the torsion bar tool and adapters. Tighten the torsion bar tool until a new adjustment bolt can be installed. Tighten this bolt until the measurement obtained in step 3 is obtained between the head of the bolt and the casting the bolt is threaded into.
11. Install the torsion bar cover plate, and tighten the plate retaining bolts to the specified torque.
12. Measure the suspension riding height as explained previously this chapter. If the suspension riding height is not within specifications, adjust the torsion bar adjusting bolt to obtain the specified riding height.

TABLE 6-1 **FRONT SUSPENSION DIAGNOSIS**

Problem	Symptoms	Possible Causes
Low riding height	Harsh ride quality, worn strikeout bumpers	Weak springs, worn control arm bushings, bent control arms, improper torsion bar adjustment
Steering wander, excessive edge wear on front tires	Erratic steering control when driving straight ahead	Worn ball joints, worn suspension or steering components, improper wheel alignment
Suspension noise	Rattling or squeaking noise when driving on road irregularities	Worn or dry stabilizer bar links and bushings, worn shock absorber or strut rod bushings
Improper steering control	Steering pull to the left or right when driving straight ahead	Improper wheel alignment, worn control arm bushings, bent control arms
Improper steering control while braking	Steering pull to the left or right only when braking	Worn strut rod bushings, loose strut rod mounting, worn lower control arm bushings
Leaf spring noise	Squeaking noise when driving on road irregularities	Worn silencers between the spring leafs

Copyright © 2015 Cengage Learning®

CASE STUDY

A customer complained about steering pull to the left while braking on a 2009 Chrysler 300. The technician questioned the customer about other symptoms, but the customer stated that the steering did not pull while driving, and the car had no other problems. Further questioning of the customer revealed that extensive brake work had been done in an attempt to correct the problem. The front brake pads had been replaced and the front brake rotors had been turned.

While performing a road test, the technician discovered the car had a definite pull to the left while braking, but the steering was normal while driving. During the road test, no other problems were evident. The technician removed the front wheels and brake rotors. A careful examination of the brake linings and rotor surfaces indicated these components were in excellent condition. The technician checked the rotor surfaces to be sure they were both in the same condition. The pistons in both front calipers moved freely. An inspection of the brake lines and hoses did not reveal any visible problems, and a pressure check at each front brake caliper indicated equal pressures at both front wheels during a brake application.

Next, the technician made a visual inspection of the steering and suspension components and discovered a worn, loose left front tension rod bushing. This tension rod was replaced, and the strut rod nut tightened to the specified torque. A road test of the car indicated no steering pull during brake application.

TERMS TO KNOW

Ball joint vertical movement
Ball joint wear indicator
Camber angles
Caster angles
Curb riding height
Directional stability
Rebound bumpers
Stabilizer bar
Steering effort
Strut rod

ASE-STYLE REVIEW QUESTIONS

1. While discussing curb riding height:
   Technician A says worn control arm bushings reduce curb riding height.
   Technician B says incorrect curb riding height affects most other front suspension angles.
   Who is correct?
   A. Technician A
   B. Technician B
   C. Both A and B
   D. Neither A nor B

2. The shoulder on the ball joint grease fitting is inside the ball joint cover:
   Technician A says the ball joint should be replaced.
   Technician B says a longer grease fitting should be installed.
   Who is correct?
   A. Technician A
   B. Technician B
   C. Both A and B
   D. Neither A nor B

3. While discussing ball joint unloading on a short-and-long arm suspension system with the coil springs between the lower control arm and the chassis:
   Technician A says a steel spacer should be installed under the upper control arm.
   Technician B says a floor jack should be placed under the lower control arm to unload the ball joints.
   Who is correct?
   A. Technician A
   B. Technician B
   C. Both A and B
   D. Neither A nor B

4. While discussing ball joint radial measurement:
   Technician A says the dial indicator should be positioned against the top of the ball joint stud.
   Technician B says the front wheel bearing play does not affect the ball joint radial measurement if measured at the wheel rim.
   Who is correct?
   A. Technician A
   B. Technician B
   C. Both A and B
   D. Neither A nor B

5. While discussing ball joint installation:
   Technician A says the ball joint nut may be backed off to install the cotter pin.
   Technician B says only the ball joint threads should extend above the opening in the steering knuckle.
   Who is correct?
   A. Technician A
   B. Technician B
   C. Both A and B
   D. Neither A nor B

6. When installing a new threaded ball joint in a lower control arm, the technician can only torque the ball joint to 90 ft.-lbs. The necessary repair for this problem is to:
   A. Weld the ball joint into the control arm.
   B. Place Loctite on the ball joint threads.
   C. Replace the lower control arm.
   D. Install a larger diameter ball joint.

7. All of these defects could result in worn-out rebound bumpers EXCEPT:
   A. Sagged springs.
   B. Worn-out shock absorbers.
   C. Continual driving on rough roads.
   D. Curb riding height more than specified.
8. When removing and replacing control arm bushings:
   A. If the bushing is contained in a steel sleeve, press on the rubber bushing.
   B. A spacer should be installed between the control arm support lugs during bushing removal and installation.
   C. The same tool is used for bushing removal and replacement.
   D. Bushing flaring is necessary to expand the bushing.
9. A car experiences excessive body sway when cornering, but there is no abnormal noise in the suspension. The most likely cause of this problem is:
   A. Stabilizer bar bushing is missing.
   B. A weak stabilizer bar.
   C. Stabilizer bar grommets are worn out.
   D. A broken stabilizer bar.
10. While driving a vehicle over a speed bump a squawking noise is heard from the front suspension system. All of the following could cause this EXCEPT:
   A. A broken sway bar link.
   B. A lower control arm bushing.
   C. A front sway bar bushing.
   D. An upper control arm bushing.

ASE CHALLENGE QUESTIONS

1. The customer says her truck with front torsion bar suspension pulls to the left. A cursory check of the vehicle shows that the right front tire height is too low. The next step to diagnose the condition would be:
   A. Check the ball joints.
   B. Check the torsion bar.
   C. Check the strut rod bushings.
   D. Check the stabilizer bar bushings.
2. A customer says her front-wheel-drive car has a steering wander problem on irregular road surfaces. All of the following could cause this problem EXCEPT:
   A. Worn stabilizer bar bushings.
   B. Worn ball joints.
   C. Worn strut rod bushings.
   D. Worn control arm bushings.
3. While diagnosing a customer's complaint of steering instability on a MacPherson strut front suspension with wear indicator-type ball joints, an inspection of the ball joints shows the grease fittings to be solid.
   Technician A says no movement of the fitting means the ball joint is good.
   Technician B says ball joint wear may not be apparent with the weight on the joint.
   Who is correct?
   A. A only
   B. B only
   C. Both A and B
   D. Neither A nor B
4. A customer says her rear-wheel-drive car has excessive body sway while cornering. In diagnosing this problem, which of the following components should you check first?
   A. Control arm bushings.
   B. Strut rod bushings.
   C. Stabilizer bar bushings.
   D. Shock absorber bushings.
5. *Technician A* says a bent strut rod can cause a car to pull to one side.
   Technician B says a deteriorated strut rod bushing can cause steering and braking problems.
   Who is correct?
   A. A only
   B. B only
   C. Both A and B
   D. Neither A nor B

JOB SHEET 20

Name _____ **Date** _____

Measure Lower Ball Joint Vertical and Horizontal Movement, Short-and-Long Arm Suspension Systems

Upon completion of this job sheet, you should be able to measure ball joint wear and determine the necessary ball joint service.

NATEF AST/MAST Task Correlation:

Suspension and Steering. Suspension Systems Diagnosis and Repair,
Task #C-5. Inspect, remove and install upper and/or lower ball joints (with or without wear indicators). (P-2)

Tools and Materials

Ball joint dial indicator
Floor jack
Safety stands

Describe the Vehicle Being Worked on:

Year _____ Make _____ Model _____
VIN _____ Engine type and size _____

Procedure — Task Completed

1. Raise the front of the vehicle with a floor jack lift pad positioned on the specified lifting point. ☐
2. Install safety stands near the outer ends of the lower control arms. Lower the floor jack so the control arms are supported on the safety stands. Remove the floor jack. ☐
3. Attach a dial indicator for ball joint measurement to the lower control arm, and position the dial indicator stem against the lower end of the steering knuckle next to the ball joint retaining nut if the suspension has a lower compression-loaded ball joint. When the suspension has a tension-loaded ball joint, place the indicator stem against the top of the ball joint stud.
   Type of ball joint
   Dial indicator stem position
4. Preload the dial indicator stem 0.250 in. (6.35 mm), and zero the dial indicator.
   Is the dial indicator preloaded and placed in the zero position? ☐ Yes ☐ No
   Instructor check _____
5. Place a pry bar under the front tire and lift straight upward on the pry bar while a coworker observes the dial indicator. ☐

Task Completed

☐ 6. Compare the reading on the dial indicator with the vehicle manufacturer's specifications. If the vertical ball joint movement exceeds specifications, replace the ball joint.

Specified ball joint vertical movement

Actual ball joint vertical movement

Necessary ball joint service

☐ 7. Be sure the front wheel bearings are properly adjusted.

8. Attach the dial indicator to the lower control arm, and position the dial indicator against the inner edge of the wheel rim. Preload the dial indicator 0.250 in. (6.35 mm) and place the dial in the zero position.

Is the dial indicator preloaded and placed in the zero position? ☐ Yes ☐ No

Instructor check _____

☐ 9. Grasp the tire at the top and bottom, and try to rock the tire inward and outward while a coworker observes the dial indicator.

☐ 10. Compare the reading on the dial indicator to the vehicle manufacturer's specifications. If the horizontal ball joint movement exceeds specifications, replace the ball joint.

Specified ball joint horizontal movement

Actual ball joint horizontal movement

☐ 11. Repeat the measurements in steps 3 through 10 on the opposite side of the front suspension.

12. Based on your ball joint measurements, state all the necessary ball joint service and explain the reasons for your diagnosis.

_____

_____

_____

Instructor's Response _____

_____

_____

JOB SHEET 21

Name _____ **Date** _____

Determine if the Upper or Lower Suspension Pivot Point Is a Primary Load Carrier or a Follower Joint and Measure Joint for Vertical and Horizontal Movement, in a Suspension System

Upon completion of this job sheet, you should be able to measure ball joint or pivot point wear and determine the necessary joint service.

NATEF AST/MAST Task Correlation:

Suspension and Steering. Suspension Systems Diagnosis and Repair,
Task #C-5. Inspect, remove and install upper and/or lower ball joints (with or without wear indicators). (P-2)

Tools and Materials

Ball joint dial indicator
Floor jack
Safety stands

Describe the Vehicle Being Worked on:

Year _____ Make _____ Model _____
VIN _____ Engine type and size _____

Procedure

1. Determine the load carrying ball-joint or pivot point.
   Upper
   Lower
   Both
   Position the floor jack correctly for testing. Describe: _____
2. Install safety stands and remove the floor jack.
3. Locate and record the specification for axial play:
   Load-carrying: _____ Follower: _____
4. Locate and record the specification for lateral play:
   Load-carrying: _____ Follower: _____
5. Attach a dial indicator for lower ball joint measurement to the lower control arm, and position the dial indicator stem against the lower end of the steering knuckle next to the ball joint retaining nut if the suspension has a lower compression-loaded ball joint. When the suspension has a tension-loaded ball joint, place the indicator stem against the top of the ball joint stud.

Task Completed

6. Preload the dial indicator stem 0.250 in. (6.35 mm), and zero the dial indicator. Is the dial indicator preloaded and placed in the zero position? ☐ Yes ☐ No
Instructor check _____

☐ 7. Place a pry bar under the front tire and lift straight upward on the pry bar while a coworker observes the dial indicator.

8. Grasp the tire at the top and bottom, and try to rock the tire inward and outward while a coworker observes the dial indicator.

9. Repeat the measurements in steps 4 through 9 for all four pivot points of the front suspension. Record your results of the inspect both load carrying and follower joints in the following table:

Ball Joint	Left	Pass/Fail	Right	Pass/Fail
Load-Carrying Axial Play				
Load-Carrying Lateral Play				
Follower Axial Play				
Follower Lateral Play				

☐ 10. Compare the reading on the dial indicator with the vehicle manufacturer's specifications. If the vertical ball joint movement exceeds specifications, replace the ball joint.

11. Based on your ball joint measurements, state all the necessary ball joint service and explain the reasons for your diagnosis.

_____

_____

_____

Instructor's Response _____

_____

_____

JOB SHEET 22

Name _____ **Date** _____

Ball Joint Replacement

Upon completion of this job sheet, you should be able to remove and replace ball joints.

NATEF AST/MAST Task Correlation:

Suspension and Steering. Suspension Systems Diagnosis and Repair,
Task #C-5. Inspect, remove and install upper and/or lower ball joints (with or without wear indicators). (P-2)

Tools and Materials

Floor jack
Ball joint loosening tool
Safety stands
Torque wrench

Describe the Vehicle Being Worked on:

Year _____ Make _____ Model _____
VIN _____ Engine type and size _____

Procedure — Task Completed

1. Remove the wheel cover and loosen the wheel nuts. ☐
2. Lift the vehicle with a floor jack and place safety stands under the chassis so the front suspension is allowed to drop downward. Lower the vehicle onto the safety stands and remove the floor jack. ☐
3. Remove the wheel and place a floor jack under the outer end of the lower control arm. Operate the floor jack and raise the lower control arm until the ball joints are unloaded.
   Are the ball joints properly unloaded? ☐ Yes ☐ No
   Instructor check _____
4. Remove other components, such as the brake caliper, rotor, and drive axle, as required to gain access to the ball joints. ☐
5. Remove the cotter pin in the ball joint or joints that require replacement, and loosen, but do not remove, the ball joint stud nuts.
   Is the ball joint stud nut loosened? ☐ Yes ☐ No
   Instructor check _____
6. Loosen the ball joint stud tapers in the steering knuckle. A threaded expansion tool is available for this purpose.
   Are the ball joint stud tapers loosened? ☐ Yes ☐ No
   Instructor check _____
7. Remove the ball joint nut and lift the knuckle off the ball joint stud. Block or tie up the knuckle and hub assembly to access the ball joint. ☐

Task Completed

☐ 8. If the ball joint is riveted to the control arm, drill and punch out the rivets and bolt the new ball joint to the control arm.

☐ 9. If the ball joint is pressed into the lower control arm, remove the ball joint dust boot and use a pressing tool to remove and replace the ball joint.

10. If the ball joint housing is threaded into the control arm, use the proper size socket to remove and install the ball joint. The replacement ball joint must be torqued to the manufacturer's specifications. If a minimum of 125 ft.-lbs of torque cannot be obtained, the control arm threads are damaged and control arm replacement is necessary.

Ball joint torque, threaded ball joint _____

Control arm condition, threaded ball joint: ☐ Satisfactory ☐ Unsatisfactory

11. If the ball joint is bolted to the lower control arm, install the new ball joint and tighten the bolt and nuts to the specified torque.

Ball joint bolt torque _____

☐ 12. Clean and inspect the ball joint stud tapered opening in the steering knuckle. If this opening is out-of-round or damaged, the knuckle must be replaced.

Condition of ball joint stud opening in the steering knuckle:

☐ Satisfactory ☐ Unsatisfactory

13. Check the fit of the ball joint stud in the steering knuckle opening. This stud should fit snugly in the opening and only the threads on the stud should extend through the knuckle. If the ball joint stud fits loosely in the knuckle tapered opening, either this opening is worn or the wrong ball joint has been supplied.

Ball joint stud smooth tapered area appearing above steering knuckle surface:

☐ Yes ☐ No

If the answer to this question is yes, state the necessary repairs and explain the reason for your diagnosis.

_____

_____

Ball joint stud fit in the steering knuckle opening:

☐ Satisfactory ☐ Unsatisfactory

If the ball joint stud fit in the steering knuckle opening is unsatisfactory, state the necessary repairs and explain the reason for your diagnosis.

_____

_____

CAUTION:
Never back off a ball joint stud to align the cotter pin openings in the nut and stud. This may cause the ball joint stud to loosen, resulting in a suspension failure. Always tighten the nut to the next hole to install the cotter pin.

14. Install the ball joint stud in the steering knuckle opening, making sure the stud is straight and centered. Install the stud nut and tighten this nut to the specified torque. Install a new cotter pin through the stud and nut. Do not loosen the nut to align the nut and stud openings.

Specified ball joint nut torque _____

Actual ball joint nut torque _____

Is a new cotter pin properly installed? ☐ Yes ☐ No

Instructor check _____

15. Reassemble the components that were removed in step 4. Make sure the wheel nuts are tightened to the specified torque.

Specified wheel nut torque _____

Actual wheel nut torque _____

Instructor's Response _____

_____

_____

JOB SHEET 23

Name _____ **Date** _____

Steering Knuckle Removal, MacPherson Strut Front Suspension

Upon completion of this job sheet, you should be able to remove the steering knuckle and determine knuckle condition on McPherson strut front suspension systems.

NATEF AST/MAST Task Correlation:

Suspension and Steering. Suspension Systems Diagnosis and Repair,
Task #C-6. Inspect, remove and install steering knuckle assemblies. (P-3)

Tools and Materials

Floor jack
Torque wrench
Safety stands
Driving tools
Tire-rod end puller

Procedure

Task Completed

1. Remove the wheel cover and loosen the front wheel nuts and the drive axle nut. ☐

2. Lift the vehicle chassis on a hoist and allow the front suspension to drop downward. Remove the front wheel, brake caliper, brake rotor, and drive axle nut. Tie the brake caliper to a suspension component; do not allow the caliper to hang on the end of the brake hose.
   Is the vehicle weight supported on hoist with front suspension dropped downward?
   ☐ Yes ☐ No
   Instructor check _____

3. Remove the inner end of the drive axle from the transaxle with a pulling or prying action. On some Chrysler products, the differential cover must be removed and axle circlips compressed prior to drive axle removal from the transaxle.
   Is the inner end of drive axle removed from transaxle? ☐ Yes ☐ No
   Instructor check _____

4. Remove the outer end of the drive axle from the steering knuckle and hub. On some early model Ford Escorts and Lynx, a puller is required for this operation.
   Is the outer end of drive axle removed from steering knuckle and hub?
   ☐ Yes ☐ No
   Instructor check _____

5. Be sure the vehicle weight is supported on the hoist with the front suspension dropped downward. Remove the ball joint retaining nuts in the lower control arm.
   Are the lower ball joint retaining nuts removed? ☐ Yes ☐ No
   Instructor check _____

Task Completed

6. Remove the cotter pin from the tie rod nut. Remove the outer tie rod nut and use a puller to disconnect the tie rod end from the steering knuckle.
   Is the outer tie rod end removed? ☐ Yes ☐ No
   Instructor check _____

7. If an eccentric cam is used on one of the strut-to-knuckle bolts, mark the cam and bolt position in relation to the strut and remove the strut-to-knuckle bolts.
   Is the eccentric camber marked in relation to the strut? ☐ Yes ☐ No
   Instructor check _____

☐ 8. Remove the knuckle from the strut and lift the knuckle out of the chassis.

☐ 9. Pry the dust deflector from the steering knuckle with a large flat-blade screwdriver.

10. Use a puller to remove the ball joint from the steering knuckle. Check the ball joint and tie rod end openings in the knuckle for wear and out-of-round. Replace the knuckle if these openings are worn or out-of-round.
    Condition of ball joint opening in the knuckle:
    ☐ Satisfactory ☐ Unsatisfactory
    Condition of tie rod end opening in the knuckle:
    ☐ Satisfactory ☐ Unsatisfactory
    State the necessary steering knuckle and related repairs, and explain the reason for your diagnosis.
    _____
    _____

☐ 11. Use the proper driving tool to reinstall the dust deflector in the steering knuckle.

Instructor's Response _____
_____
_____

Chapter 7

Rear Suspension Service

BASIC TOOLS

Basic technician's tool set
Service manual
Machinist's rule
Floor jack
Safety stands
Transmission jack
Foot-pound torque wrench
Pry bar

Upon completion and review of this chapter, you should be able to:

- Diagnose rear suspension noises.
- Diagnose rear suspension sway and lateral movement.
- Measure and correct rear suspension curb riding height.
- Remove and replace rear coil springs.
- Inspect rear springs, insulators, and seats.
- Inspect strut or shock absorber bushings and upper mount, and replace strut cartridge.
- Remove, inspect, and replace lower control arms.
- Inspect, remove, and replace rear ball joints.
- Inspect, remove, and replace suspension adjustment links.
- Diagnose, remove, and replace rear leaf springs.
- Diagnose, remove, and replace stabilizer bars.
- Diagnose, remove, and replace track bars.
- Inspect, remove, and replace tie rods.

Rear suspension system diagnosis and service is extremely important to maintain vehicle safety, ride quality, and tire life. Many rear suspension components may create a safety hazard if they are not serviced properly. For example, if wheel nuts or spindle nuts are not tightened properly, a wheel assembly may come off the vehicle with disastrous results. Improper rear suspension curb riding height, worn struts or shock absorbers, and sagged or broken springs will result in harsh ride quality. Rear wheel toe and camber must be adjusted to specifications to provide normal tire tread life.

Diagnosis of Ride Harshness

When diagnosing a ride harshness condition, check these measurements and components:

1. Curb riding height—insufficient curb riding height causes ride harshness
2. Excessive vehicle load
3. Worn shock absorbers or struts
4. Broken or weak springs
5. Worn suspension bushings, such as strut mounts, control arm bushings, and shock absorber mounting bushings
6. Wheel alignment—excessive positive caster on front wheels causes ride harshness

Noise Diagnosis

A squeaking noise in the rear suspension may be caused by a suspension bushing or a defective strut or shock absorber.

If a rattling noise occurs in the rear suspension, check these components:

1. Worn or missing suspension bushings, such as control arm bushings, track bar bushings, stabilizer bar bushings, trailing arm bushings, and strut rod bushings
2. Worn strut or shock absorber bushings or mounts
3. Defective struts or shock absorbers
4. Broken springs or worn spring insulators
5. Broken exhaust system hangers or improperly positioned exhaust system components

Special tools can be used to locate noises at any location on the vehicle. A Chassis Ear tool has six clamp-on sensors that are miniature microphones (Figure 7-1). These six sensors are connected to a switchbox that is also connected to a headset worn by the technician. The technician can rotate the switch on the switchbox to connect any sensor to the headset. The sensors can be mounted on various suspected noise sources. When the control on the switchbox is rotated, the technician can determine which sensor is connected to the noise source.

Body sway is leaning of the chassis to one side.

An electronic stethoscope can be used to locate the source of a noise. This tool has one pickup that is placed on or near a suspected noise source. The pickup is connected to an amplifier box, and the headset is also connected to this box (Figure 7-2). When the pickup is placed on the source of a noise, the noise is amplified in the headset.

SERVICE TIP:
Proper curb riding height must be maintained to provide normal steering quality and tire wear.

Sway and Lateral Movement Diagnosis

Excessive **body sway**, or roll, on road irregularities may be caused by a weak stabilizer bar or loose stabilizer bar bushings. If lateral movement is experienced on the rear of the chassis, the track bar or track bar bushings may be defective. Lateral movement is sideways movement.

Curb Riding Height Measurement

Classroom Manual
Chapter 7, page 188

Regular inspection and proper maintenance of suspension systems are extremely important to maintain vehicle safety. The curb riding height is determined mainly by spring condition. Other suspension components, such as control arm bushings, affect curb riding height if they are worn. *Since incorrect curb riding height affects most of the other suspension angles, this measurement is critical.* Reduced rear suspension height increases steering effort and causes rapid steering wheel return after turning a corner. Harsh riding occurs when the curb riding height is less than specified. The curb riding height must be measured at the

FIGURE 7-1 **Chassis Ear tool for locating noise sources.**

Copyright © 2015 Cengage Learning®

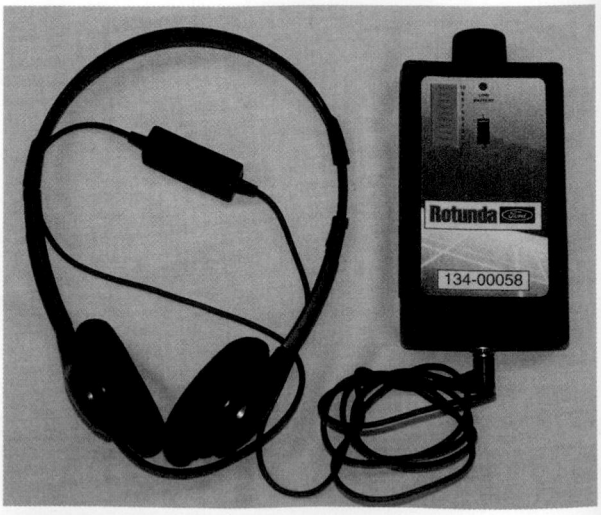

FIGURE 7-2 **Electronic stethoscope.**

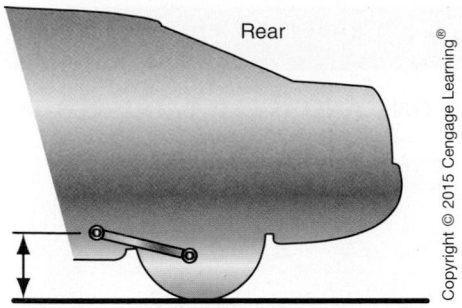

FIGURE 7-3 **Curb riding height measurement, rear suspension.**

vehicle manufacturer's specified location, which varies depending on the type of vehicle and suspension system. On some vehicles, the curb riding height on the rear suspension is measured from the floor to the center of the strut rod mounting bolt when the vehicle is on a level floor or an alignment rack (Figure 7-3). Photo Sequence 11 shows a typical procedure for measuring front and rear curb riding height.

Rear Strut, Coil Spring, and Upper Mount Diagnosis and Service

WARNING: When the rear coil springs are mounted on the struts, never loosen the upper strut nut until the entire spring tension is removed from the upper support with a spring compressor. If this nut is removed with spring tension on the upper support, the spring tension turns the spring and the upper support into a dangerous projectile, which may cause serious personal injury and/or property damage.

Weak springs cause harsh riding and reduced curb riding height. Broken springs or spring insulators cause a rattling noise while driving on road irregularities. Worn-out struts or shock absorbers cause excessive chassis oscillations and harsh riding. (Refer to Chapter 5 for strut and shock absorber service.) Loose or worn strut or shock absorber bushings cause a rattling noise on road irregularities. The rear coil spring removal and replacement procedure varies depending on the type of rear suspension system. Always follow the vehicle manufacturer's recommended procedure in the service manual.

The following is a typical rear strut and spring removal and replacement procedure on a MacPherson strut independent rear suspension system with the coil springs mounted on the struts:

1. Remove the rear seat and the package trim tray (Figure 7-4).
2. Remove the wheel cover, and loosen the wheel nuts.
3. Lift the vehicle with a floor jack, and lower the chassis onto safety stands so the rear suspension is allowed to drop downward.
4. Place a floor jack lift pad under the rear spindle on the side where the strut and spring removal is taking place. Raise the floor jack to support some of the suspension system weight (Figure 7-5).
5. Remove the rear wheel, disconnect the nut from the small spring in the lower arm (Figure 7-6), and remove the brake hose and antilock brake system (ABS) wire from the strut (Figure 7-7).
6. Remove the stabilizer bar link from the strut (Figure 7-8), and loosen the strut-to-spindle mounting bolts (Figure 7-9).
7. Remove the three upper support nuts under the package tray trim (Figure 7-10), and lower the floor jack to remove the strut from the knuckle (Figure 7-11). Remove the strut from the chassis.

SERVICE TIP:
If the plastic coating on a coil spring is chipped, the spring may break prematurely. The spring may be taped in the compressing tool contact areas to prevent chipping.

PHOTO SEQUENCE 11

Typical Procedure for Measuring Front and Rear Curb Riding Height

P11-1 **Check the trunk for extra weight.**

Copyright © 2015 Cengage Learning®

P11-2 **Check the tires for normal inflation pressure.**

Copyright © 2015 Cengage Learning®

P11-3 **Park the car on a level shop floor or alignment rack.**

Copyright © 2015 Cengage Learning®

P11-4 **Find the vehicle manufacturer's specified curb riding height measurement locations in the service manual.**

Copyright © 2015 Cengage Learning®

P11-5 **Measure and record the right front curb riding height.**

Copyright © 2015 Cengage Learning®

P11-6 **Measure and record the left front curb riding height.**

Copyright © 2015 Cengage Learning®

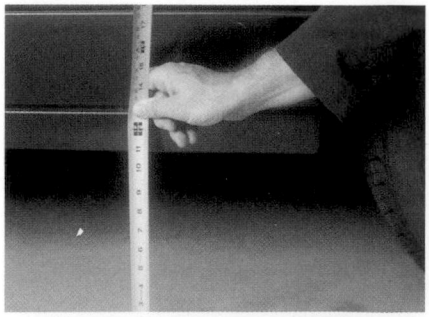

P11-7 **Measure and record the right rear curb riding height.**

Copyright © 2015 Cengage Learning®

P11-8 **Measure and record the left rear curb riding height.**

Copyright © 2015 Cengage Learning®

P11-9 **Compare the measurement results to the specified curb ride height in the service information.**

Copyright © 2015 Cengage Learning®

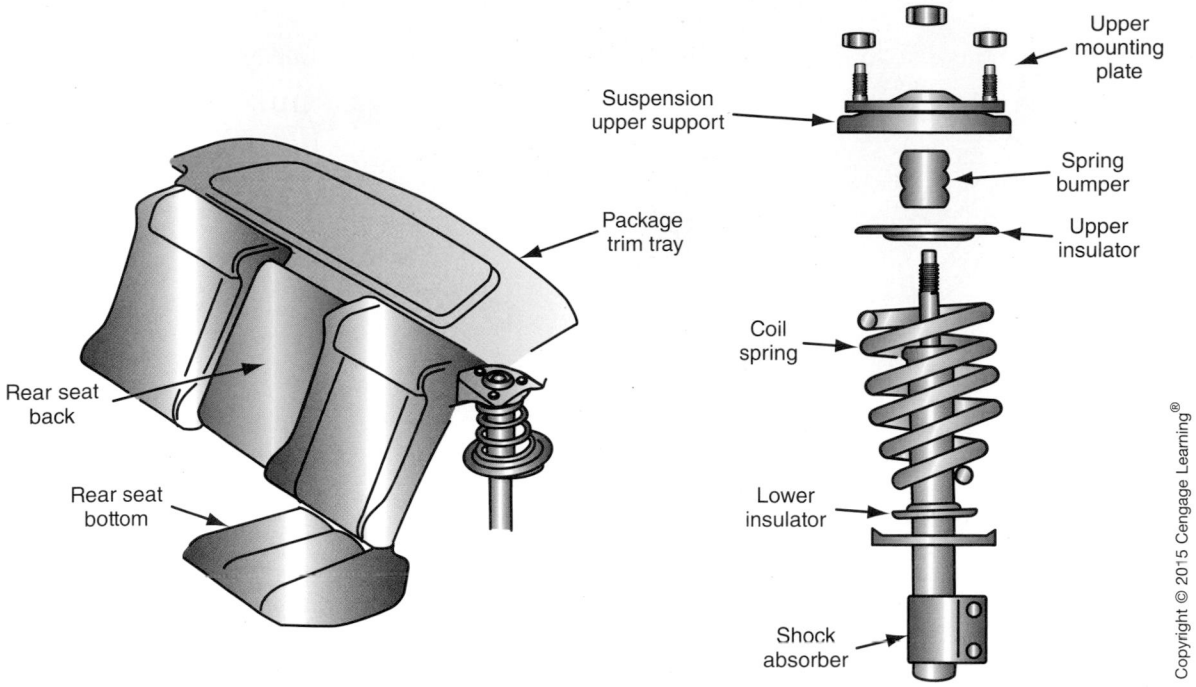

FIGURE 7-4 **Rear suspension with rear seat and package trim tray.**

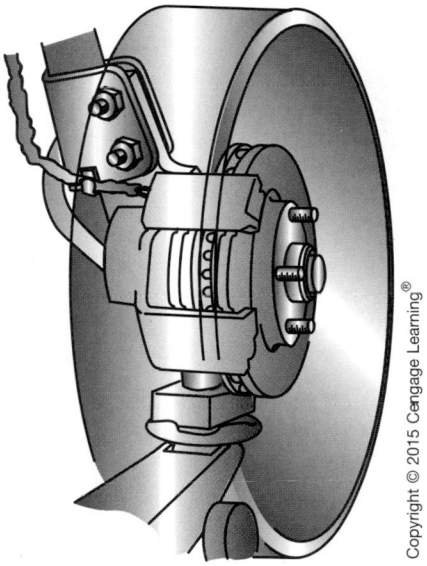

FIGURE 7-5 **Floor jack supporting some of the rear suspension weight.**

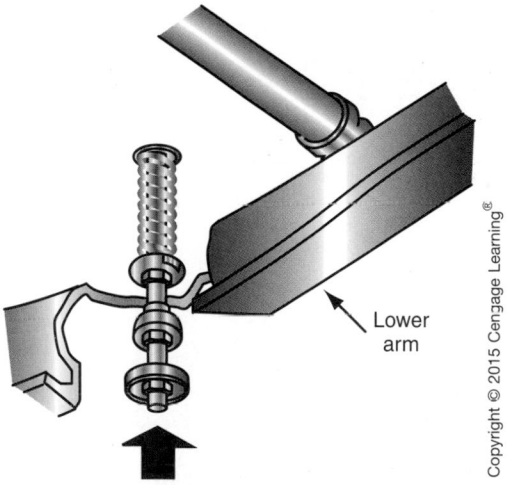

FIGURE 7-6 **Disconnecting nut from small spring in the lower arm.**

8. Following the vehicle or equipment manufacturer's recommended procedure, install a spring compressing tool on the coil spring, and tighten the tool until the entire spring tension is removed from the upper support (Figure 7-12).
9. Operate the compressing tool to remove the entire spring tension from the upper strut mount and then remove the strut rod nut (Figure 7-13).
10. Remove the strut rod nut, upper support, and upper insulator.
11. Remove the strut from the lower end of the spring.
12. If the spring is to be replaced, rotate the compressing tool handle until the entire spring tension is removed from the compressing tool, and then remove the spring from the tool.

SPECIAL TOOL

Coil spring compressor

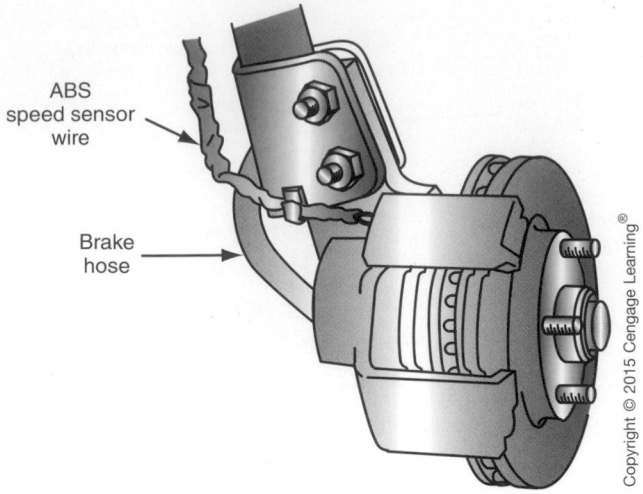

FIGURE 7-7 **Disconnecting brake hose and ABS wire from the rear strut.**

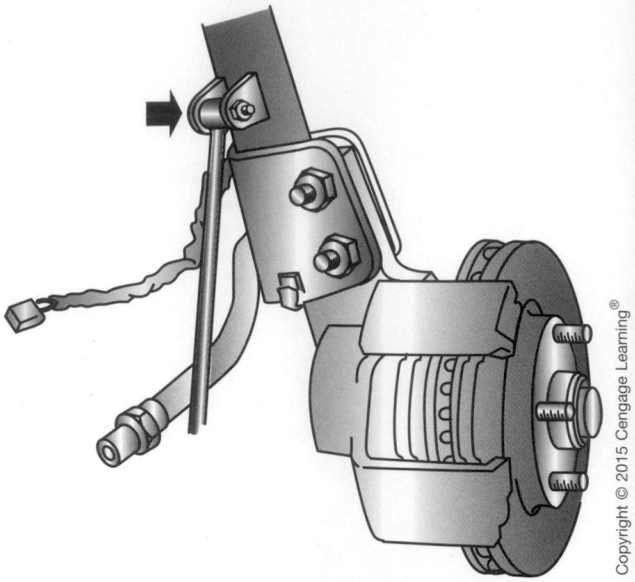

FIGURE 7-8 **Removing stabilizer bar from the rear strut.**

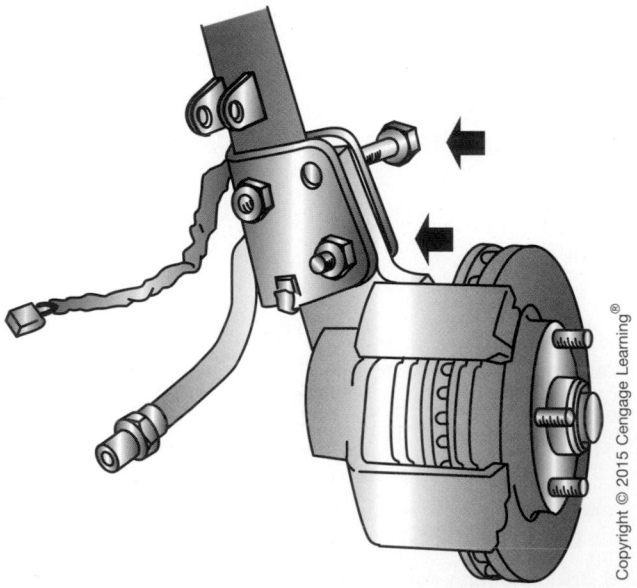

FIGURE 7-9 **Loosening strut-to-rear spindle bolts.**

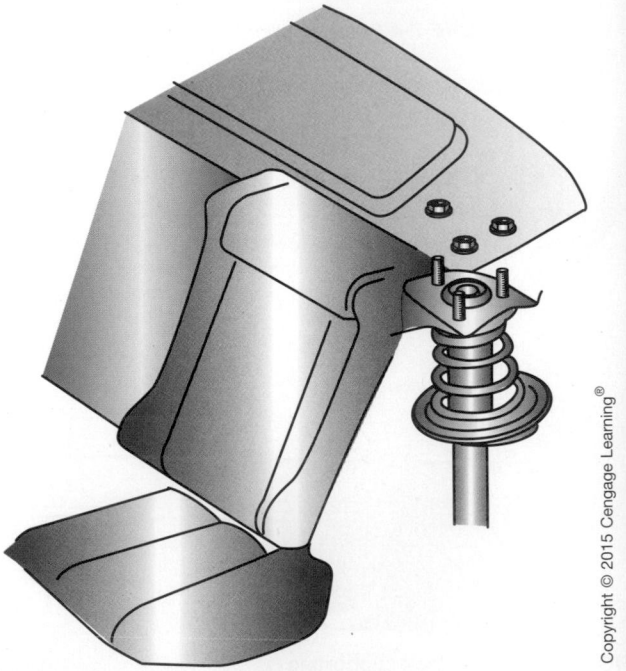

FIGURE 7-10 **Removing upper mount nuts.**

Classroom Manual
Chapter 7, page 176

13. Inspect the lower insulator and spring seat on the strut (Figure 7-14). If the spring seat is warped or damaged, replace the strut. A new cartridge may be installed in some rear struts. (The strut cartridge replacement procedure is explained in Chapter 5.)
14. Visually inspect the upper mount, insulator, and spring bumper. If any of these components are damaged, worn, or distorted, replacement is necessary. Worn upper mounts and insulators or damaged spring seats cause suspension noise while driving on road irregularities. Assemble the spring bumper, upper mount, and insulator (Figure 7-15), and then compress the coil spring in the compressing tool. Assemble the strut and related components into the spring (Figure 7-16). Tighten all fasteners to the specified torque. Photo Sequence 12 illustrates the proper procedure for removing a rear strut and spring.

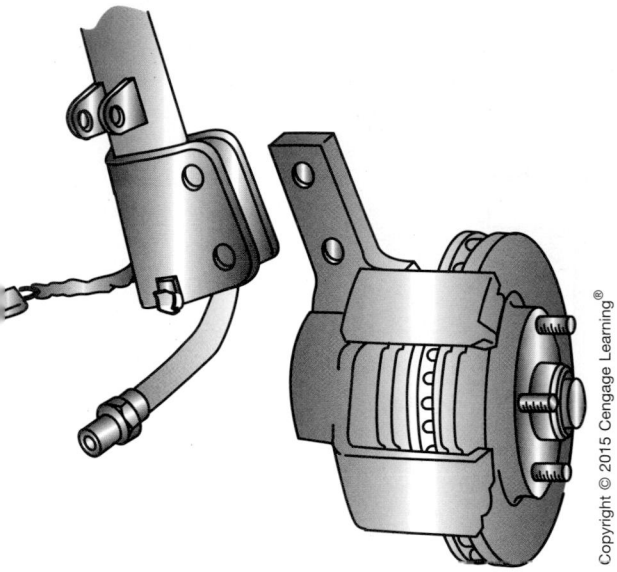

FIGURE 7-11 **Removing the strut from the rear spindle.**

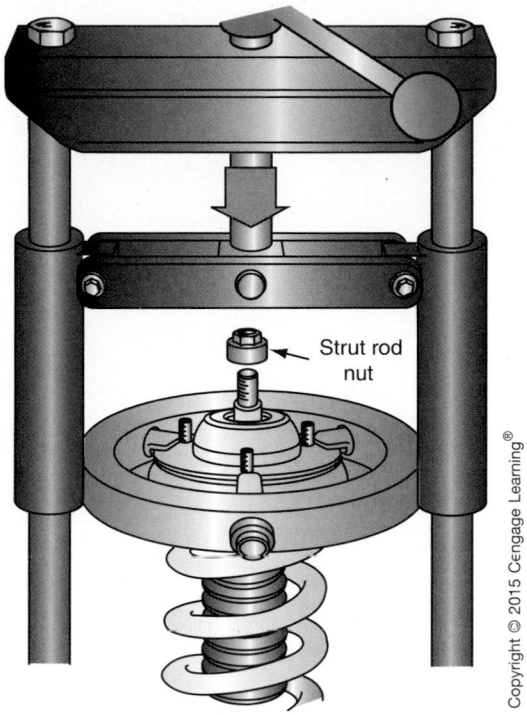

FIGURE 7-12 **A spring compressing tool is used to compress the coil spring.**

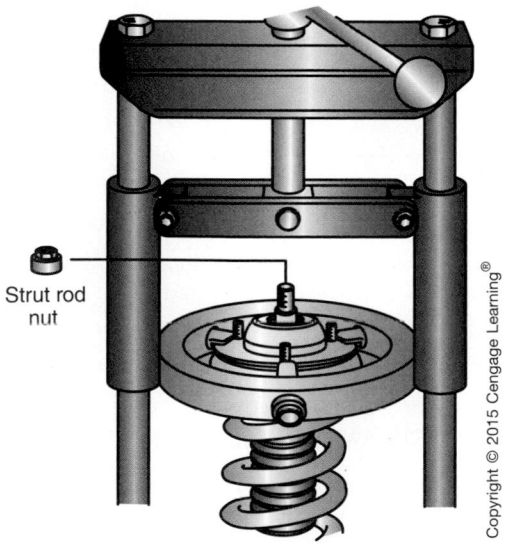

FIGURE 7-13 **After the compressing tool is operated to remove the entire spring tension, remove the strut rod nut.**

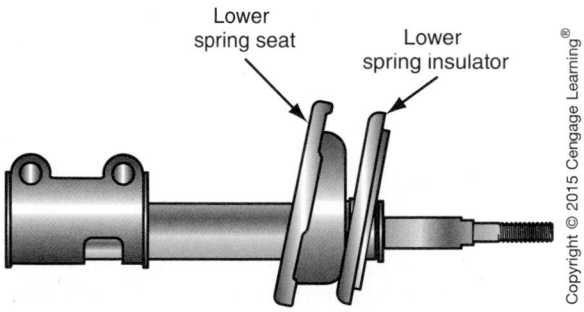

FIGURE 7-14 **Inspecting lower spring seat and insulator.**

FIGURE 7-15 **Assembly of rear strut, spring bumper, spring, upper mount, and insulator.**

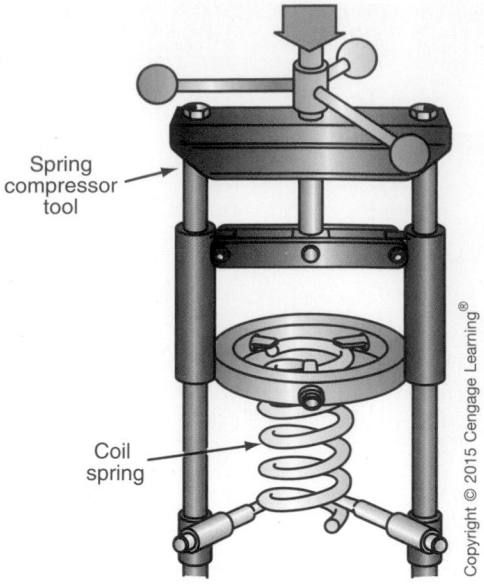

FIGURE 7-16 **Compressing the coil spring in the compressing tool prior to strut-and-spring assembly.**

CUSTOMER CARE: When talking to customers, always remember the two *P*s—*pleasant* and *polite*. There may be many days when we do not feel like being pleasant and polite. Perhaps we have several problem vehicles in the shop with symptoms that are difficult to diagnose and correct. Some service work may be behind schedule, and customers may be irate because their vehicles are not ready on time. However, we should always remain pleasant and polite with customers. Our attitude does much to make the customer feel better and realize their business is appreciated. A customer may not feel very happy about an expensive repair bill, but a pleasant attitude on our part may help improve the customer's feelings. When the two *P*s are remembered by service personnel, customer relations are enhanced, and the customer will return to the shop. Conversely, if service personnel have a grouchy, indifferent attitude, customers may be turned off and take their business to another shop.

LOWER CONTROL ARM AND BALL JOINT DIAGNOSIS AND REPLACEMENT

Worn bushings on the lower control arms may cause incorrect rear wheel camber or toe, which results in rear tire wear and steering pull. Bent lower control arms must be replaced. When ball joints with **wear indicators** are in normal condition, there is 0.050 in. (1.27 mm) between the grease nipple shoulder and the cover. If the ball joint is worn, the grease nipple shoulder is flush with or inside the cover (Figure 7-17). A worn ball joint causes improper rear wheel toe and/or camber, which may result in tire tread wear or steering pull.

Wear indicators show ball joint wear by the position of the grease fitting in the ball joint.

The lower control arm removal and replacement procedure varies depending on the vehicle and the type of suspension. Always follow the vehicle manufacturer's recommended procedure in the service manual.

The following is a typical lower control arm removal procedure:

1. Lift the vehicle on a hoist with the chassis supported on the hoist and the control arms dropped downward. The vehicle may be lifted with a floor jack, and the chassis supported on safety stands.
2. Remove the tire and wheel assembly.

PHOTO SEQUENCE 12

Typical Procedure for Removing a Rear Strut-and-Spring Assembly on a Front-Wheel-Drive Car

P12-1 **Remove the left rear wheel cover, and loosen the wheel nuts on the left rear wheel.**

Copyright © 2015 Cengage Learning®

P12-2 **Raise the vehicle on a frame-contact lift and chalk mark one of the left rear wheel studs in relation to the wheel, and remove the wheel nuts followed by the wheel and tire assembly.**

Copyright © 2015 Cengage Learning®

P12-3 **Remove the left rear brake caliper, and use a piece of wire to suspend the caliper from a chassis member. Do not allow the caliper to hang from the flexible brake hose.**

Copyright © 2015 Cengage Learning®

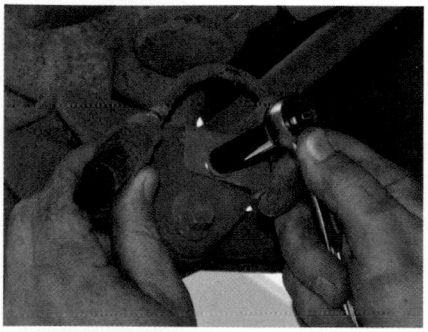

P12-4 **If the vehicle is equipped with antilock brakes, remove the left rear speed sensor wire, routing tube, and bracket from the trailing arm.**

Copyright © 2015 Cengage Learning®

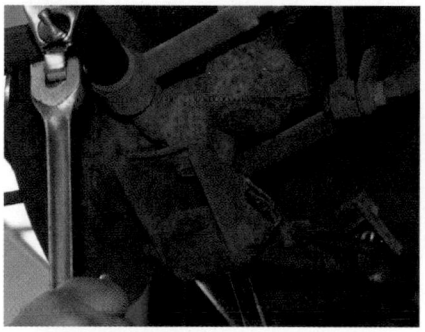

P12-5 **Remove both forward and rearward lateral lower link-to-spindle attaching bolts by installing a thin open-end wrench on the hex of the attaching link stud to prevent the stud from turning and then remove the lateral link-to-spindle attaching bolts.**

Copyright © 2015 Cengage Learning®

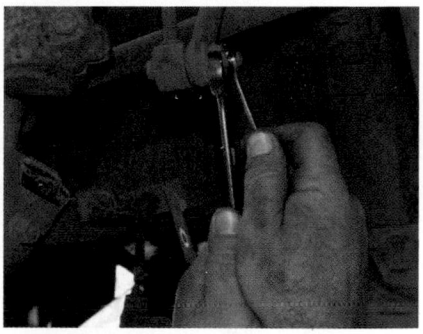

P12-6 **Use an open-end wrench to hold the left rear stabilizer bar attaching link stud from turning, and remove the stabilizer bar attaching link nut. Disconnect the stabilizer bar from the attaching link stud.**

Copyright © 2015 Cengage Learning®

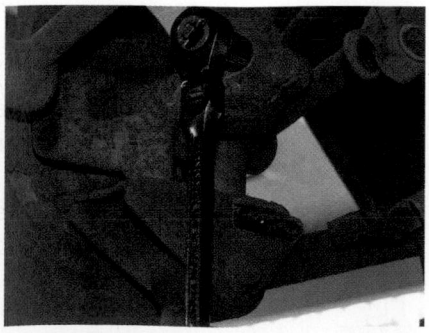

P12-7 **Remove the nut from the strut-to-spindle pinch bolt, and remove the pinch bolt.**

Copyright © 2015 Cengage Learning®

P12-8 **Tap a center punch into the slot in the lower end of the strut to spread the strut opening slightly. Be sure the inner end of the punch does not contact and puncture the strut.**

Copyright © 2015 Cengage Learning®

P12-9 **Use a hammer to tap the top of the spindle and drive it downward off the end of the strut. Allow the left rear spindle and assembled components to hang from the trailing arm.**

Copyright © 2015 Cengage Learning®

P12-10 **Lower the vehicle on the lift so the disconnected components on the left rear suspension and the other tires are a short distance off the shop floor. Open the trunk and remove the dust cap on top of the left rear strut opening. Loosen the left rear strut mounting nuts.**

Copyright © 2015 Cengage Learning®

P12-11 **Have a coworker hold the left rear strut-and-spring assembly and remove the strut mounting nuts. Remove the left rear strut from the vehicle.**

Copyright © 2015 Cengage Learning®

Classroom Manual
Chapter 7, page 176

SPECIAL TOOL
Lower control arm support tool

3. Remove the stabilizer bar from the knuckle bracket.
4. Remove the parking brake cable retaining clip from the lower control arm.
5. If the car has electronic level control (ELC), disconnect the height sensor link from the control arm.
6. Install a special tool to support the lower control arm in the bushing areas (Figure 7-18).
7. Place a transmission jack under the special tool and raise the jack enough to remove the tension from the control arm bushing retaining bolts. If the car was lifted with a floor jack and supported on safety stands, place a floor jack under the special tool.
8. Place a safety chain through the coil spring and around the lower control arm.
9. Remove the bolt from the rear control arm bushing.
10. Be sure the jack is raised enough to relieve the tension on the front bolt in the lower control arm and remove this bolt.
11. Lower the jack slowly and allow the control arm to pivot downward. When all the tension is released from the coil spring, remove the safety chain, coil spring, and insulators.

Check the coil spring for distortion and proper free length. If the spring has a vinyl coating, check this coating for scratches and nicks. Check the spring insulators for cracks and wear.

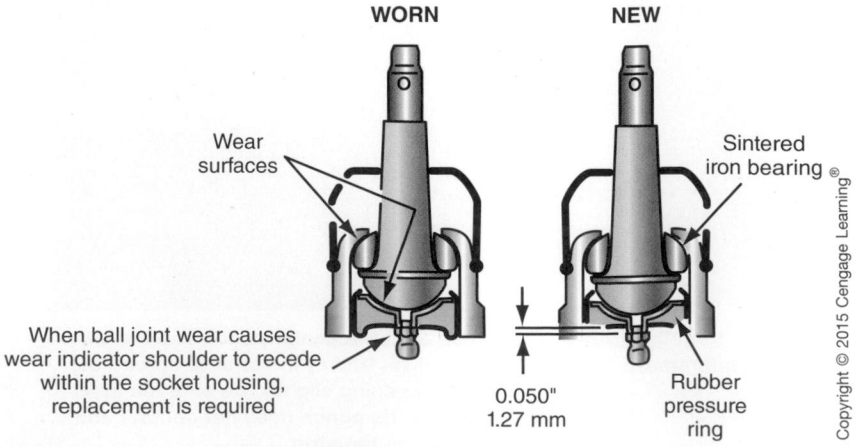

FIGURE 7-17 **Ball joint wear indicator.**

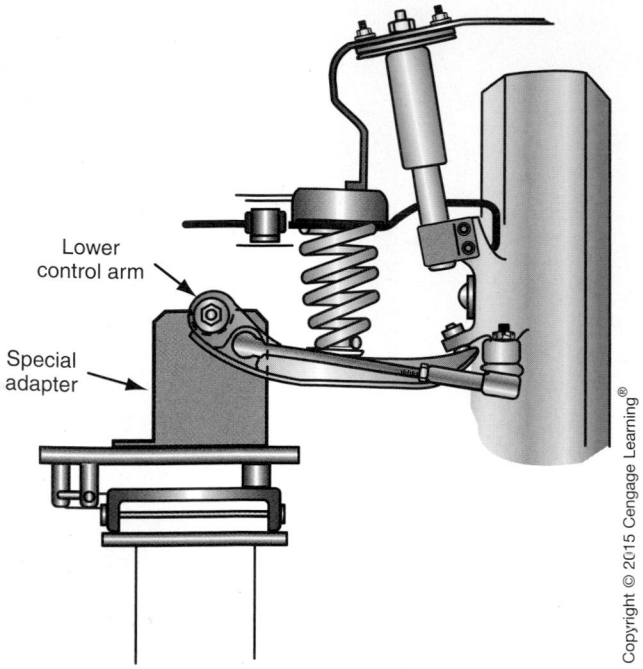

FIGURE 7-18 **Special tool installed to support the inner end of the lower control arm.**

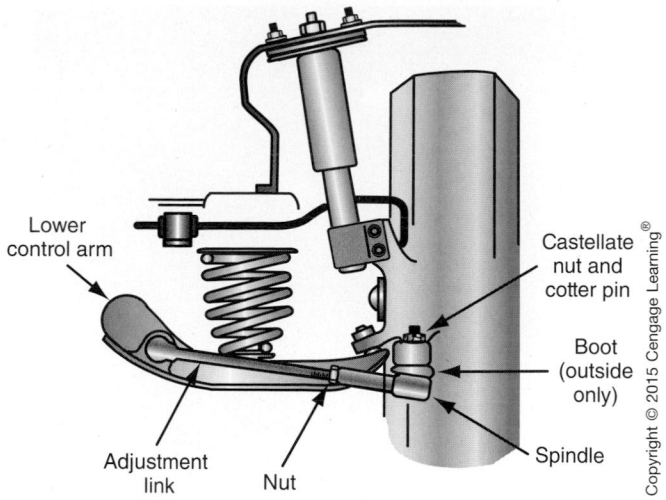

FIGURE 7-19 **Removing the suspension adjustment link from the lower control arm.**

After the coil spring is removed, follow these steps to remove the lower control arm:

1. Remove the nut on the inner end of the **suspension adjustment link**, and disconnect this link from the lower control arm (Figure 7-19).
2. Remove the cotter pin from the ball joint nut, and loosen, but do not remove, the nut from the ball joint stud.
3. Use a special ball joint removal tool to loosen the ball joint in the knuckle.
4. Remove the ball joint nut and the lower control arm (Figure 7-20).

A **suspension adjustment link** may be connected from the rear knuckle to the lower control arm to adjust rear wheel toe.

Check the lower control arm for bends, distortion, and worn bushings. A special ball joint pressing tool is used to press the ball joint from the lower control arm (Figure 7-21), and the same pressing tool with different adapters is used to press the new ball joint into the control arm (Figure 7-22).

Remove the nut and cotter pin on the outer end of the suspension adjustment link, and use a special puller to remove this link from the knuckle (Figure 7-23). Remove the

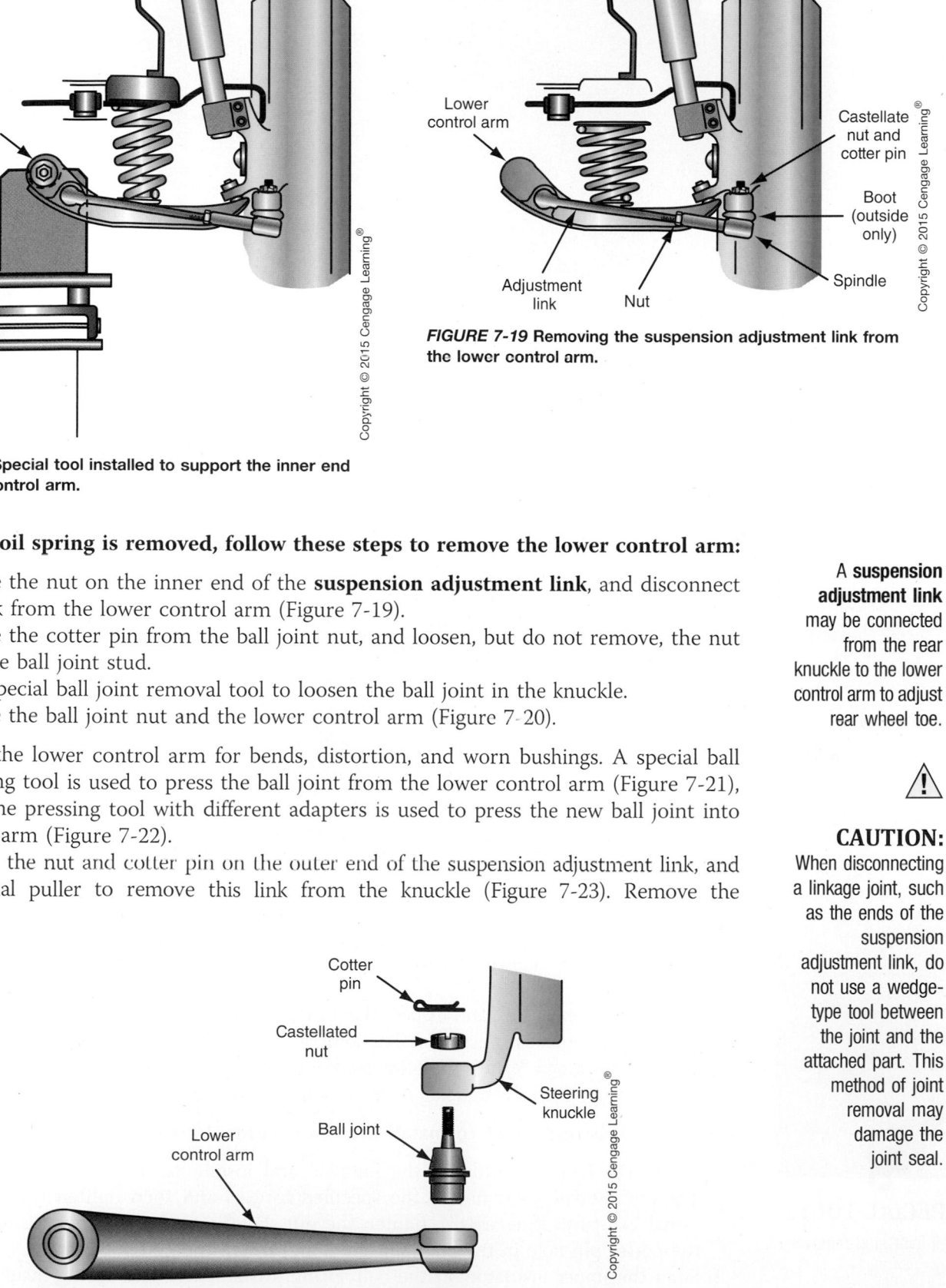

CAUTION:
When disconnecting a linkage joint, such as the ends of the suspension adjustment link, do not use a wedge-type tool between the joint and the attached part. This method of joint removal may damage the joint seal.

FIGURE 7-20 **Removing the lower control arm and ball joint.**

Copyright © 2015 Cengage Learning®

FIGURE 7-21 **Removing the ball joint from the lower control arm.**

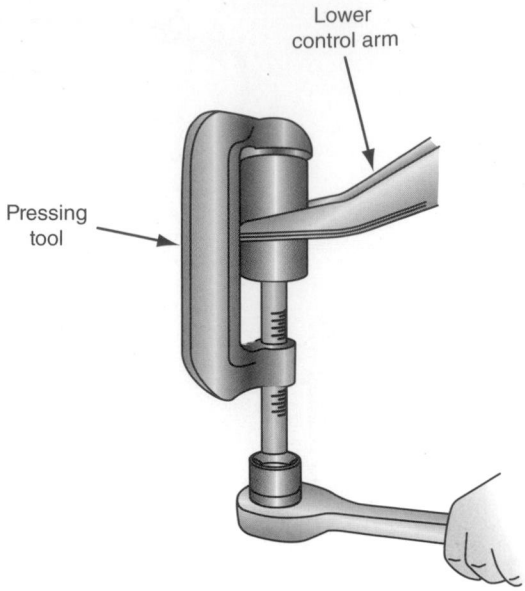

Copyright © 2015 Cengage Learning®

FIGURE 7-22 **Installing the ball joint in the lower control arm.**

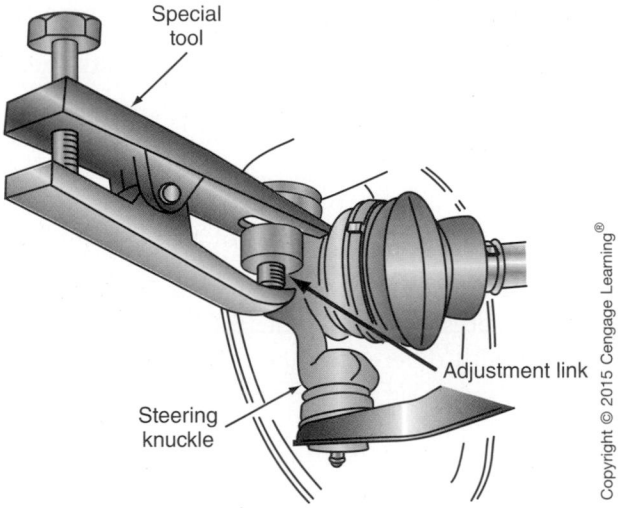

Copyright © 2015 Cengage Learning®

FIGURE 7-23 **Removing the suspension adjustment link from the knuckle.**

suspension adjustment link and inspect the joints. If the joints are loose or the seals are damaged, joint replacement is necessary. The joint studs must fit snugly in the knuckle and lower control arm openings. When the joint studs are worn, joint replacement is necessary, and worn stud openings in the knuckle or lower control arm require component replacement.

SPECIAL TOOLS

Ball joint stud removal tool

Ball joint pressing tool

Follow this procedure to install the lower control arm and spring:

1. Install the ball joint stud in the knuckle and install the nut on the ball joint stud. Tighten the ball joint nut to the specified torque, and then tighten the nut an additional 2/3 turn. If necessary, tighten the nut slightly to align the nut castellations with the cotter pin hole in the ball joint stud, and install the cotter pin.
2. Snap the upper insulator on the coil spring. Install the lower spring insulator and the spring in the lower control arm (Figure 7-24).

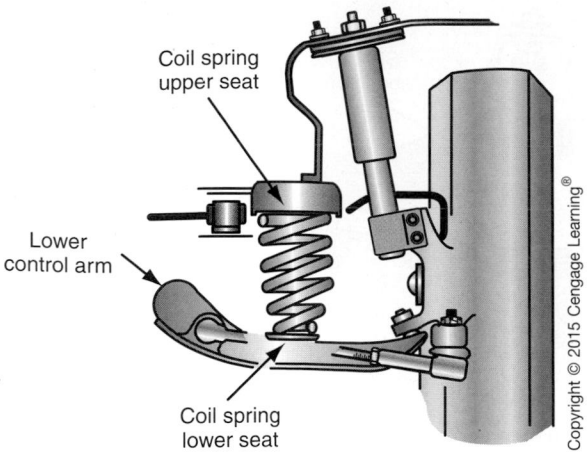

FIGURE 7-24 **Installing the spring and insulators.**

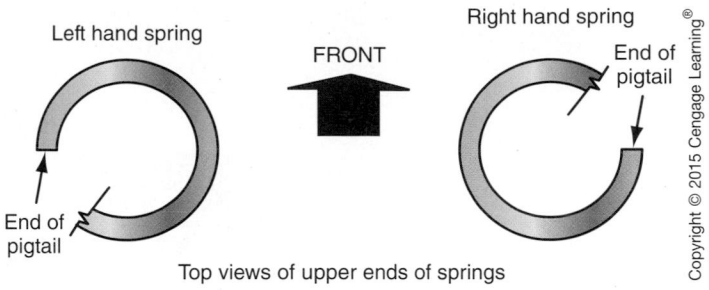

FIGURE 7-25 **Proper location of upper coil spring ends in relation to the front of the vehicle.**

3. Be sure the top of the coil spring is properly positioned in relation to the front of the vehicle (Figure 7-25).
4. Install the special tool on the inner ends of the control arm, and place the transmission jack under the special tool.
5. Slowly raise the transmission jack until the control arm bushing openings are aligned with the openings in the chassis.
6. Install the bolts and nuts in the inner ends of the control arm. Do not torque these bolts and nuts at this time.
7. Install the stabilizer-bar-to-knuckle bracket and tighten the fasteners to the specified torque.
8. Install the parking brake retaining clip.
9. If the vehicle has ELC, install the height sensor link, and tighten the fastener to the specified torque.
10. Install the suspension adjustment link and tighten the retaining nuts to the specified torque. Install cotter pins as required.
11. Remove the transmission jack and install the tire and wheel assembly.
12. Lower the vehicle onto the floor and tighten the wheel hub nuts and lower control arm bolts and nuts to the specified torque.

SPECIAL TOOL

Suspension adjustment link removal tool

CAUTION:

Never loosen a ball joint nut to install a cotter pin, because this action causes improper torquing of the nut.

CAUTION:

The pivot bolts and nuts in the inner ends of the lower control arm must be tightened to the specified torque with the vehicle weight supported on the wheels and the suspension at normal curb height. Failure to follow this procedure may adversely affect ride quality and steering characteristics.

Rear Leaf-Spring Diagnosis and Replacement

This leaf-spring discussion applies to multiple-leaf springs on rear suspension systems that have two springs mounted longitudinally in relation to the chassis. Many leaf springs have plastic **spring silencers** between the spring leafs. If these silencers are worn out, creaking and squawking noises are heard when the vehicle is driven over road irregularities at low speeds.

When the silencers require checking or replacement, lift the vehicle with a floor jack and support the frame on safety stands so the rear suspension moves downward. With the vehicle weight no longer applied to the springs, the spring leafs may be pried apart with a pry bar to remove and replace the silencers.

Worn shackle bushings, brackets, and mounts cause excessive chassis lateral movement and rattling noises. With the normal vehicle weight resting on the springs, insert a pry bar between the rear outer end of the spring and the frame. Apply downward pressure on the bar and observe the rear shackle for movement. Shackle bushings, brackets, or mounts must be replaced if there is movement in the shackle. The same procedure may be followed to check the front bushing in the main leaf. A broken spring center bolt may allow the rear axle assembly to move rearward on one side. This movement changes rear wheel tracking, which results

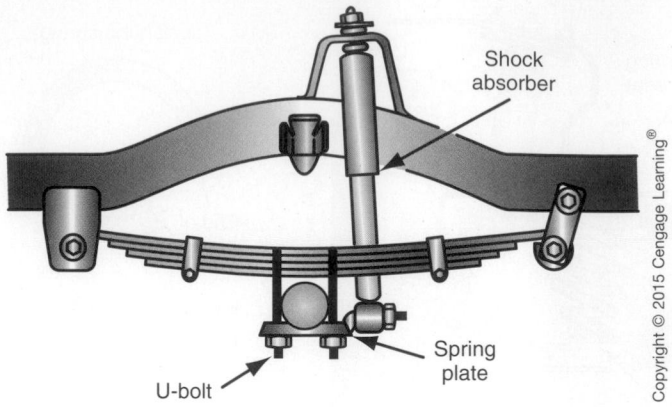

FIGURE 7-26 **Leaf-spring rear suspension.**

in handling problems, tire wear, and reduced directional stability. Sagged rear springs reduce the curb riding height. Spring replacement is necessary if the springs are sagged.

When rear leaf-spring replacement is necessary, proceed as follows:

1. Lift the vehicle with a floor jack and place safety stands under the frame. Lower the vehicle weight onto the safety stands, and leave the floor jack under the differential housing to support the rear suspension weight.
2. Remove the nuts from the spring U-bolts, and remove the U-bolts and lower spring plate (Figure 7-26). The spring plate may be left on the rear shock absorber and moved out of the way.
3. Be sure the floor jack is lowered sufficiently to relieve the vehicle weight from the rear springs.
4. Remove the rear shackle nuts, plate, shackle, and bushings.
5. Remove the front spring mounting bolt and remove the spring from the chassis. Check the spring center bolt to be sure it is not broken.
6. Check the front hanger, bushing, and bolt, and replace as necessary.
7. Check the rear shackle, bushings, plate, and mount; replace the worn components.
8. Reverse steps 1 through 5 to install the spring. Tighten all bolts and nuts to the specified torque.

Classroom Manual
Chapter 7, page 170

Spring silencers are plastic spacers mounted between the spring leafs to reduce spring noise.

Track Bar Diagnosis and Replacement

Some rear suspension systems have a track bar to control **lateral chassis movement**. Rubber mounting bushings insulate the track bar from the chassis components. Worn track bar mounts and bushings may cause rattling and excessive lateral chassis movement.

When the track bar is inspected, lift the vehicle on a hoist or floor jack with the rear suspension in the normal riding height position. If the vehicle is lifted with a floor jack, place safety stands under the rear axle to support the vehicle weight. Grasp the track bar firmly and apply vertical and horizontal force. If there is movement in the track bar mountings, track bar, or bushing, replacement is essential (Figure 7-27).

Directional stability refers to the tendency of the vehicle steering to remain in the straight-ahead position when driven straight ahead on a reasonably smooth, level road surface.

Another track bar checking method is to leave the vehicle on the shop floor and observe the track bar mounts as an assistant applies side force to the chassis or rear bumper. If there is lateral movement in the track bar bushings or brackets, replace the bushings and check the bracket bolts. Bent track bars must be replaced.

When the track bar is replaced, remove the mounting bolts, bushings, grommets, and track bar. Inspect the mounting bolt holes in the chassis for wear. After the track bar is installed with the proper grommets and bushings, tighten the mounting bolts to the specified torque.

Lateral chassis movement refers to sideways movement.

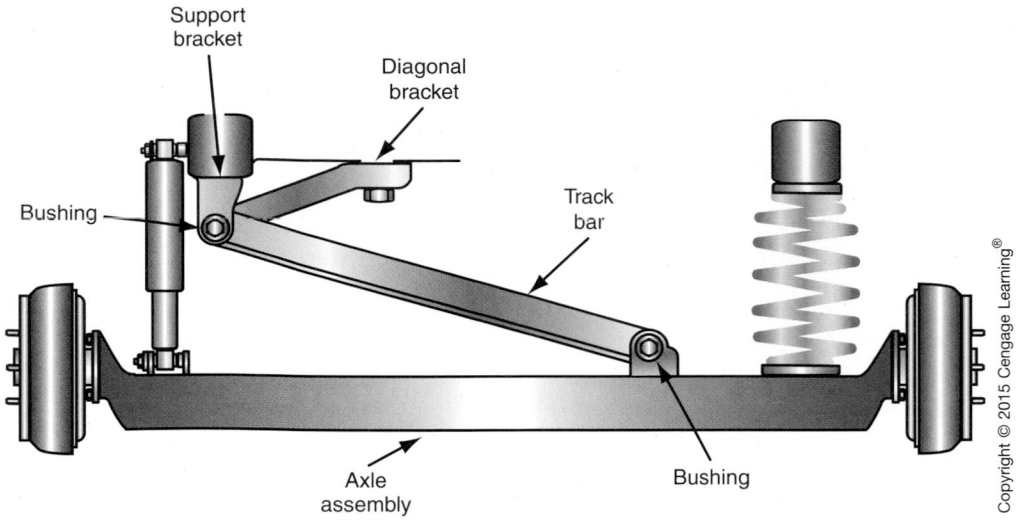

FIGURE 7-27 **Checking track bar bushings.**

Stabilizer Bar Diagnosis and Service

Worn stabilizer bar mounting bushings, grommets, or mounting bolts cause a rattling noise as the vehicle is driven on irregular road surfaces. A weak stabilizer bar or worn bushings and grommets cause harsh riding and excessive body sway while driving on irregular road surfaces. Worn or very dry stabilizer bar bushings may cause a squeaking noise on irregular road surfaces. All stabilizer bar components should be visually inspected for wear. Stabilizer bar removal and replacement procedures vary depending on the vehicle. Always follow the vehicle manufacturer's recommended procedure in the service manual.

A stabilizer bar may be referred to as a sway bar.

Following is a typical rear stabilizer bar removal and replacement procedure:

1. Lift the vehicle on a hoist and allow both sides of the rear suspension to drop downward as the vehicle chassis is supported on the hoist.
2. Remove the mounting bolts at the outer ends of the stabilizer bar and remove the bushings, grommets, brackets, or spacers (Figure 7-28).

SERVICE TIP:
On rear suspension systems with an inverted U-channel, the stabilizer bar inside the U-channel sometimes breaks away where it is welded to the end plate in the U-channel. This results in a rattling, scraping noise when the car is driven over road irregularities.

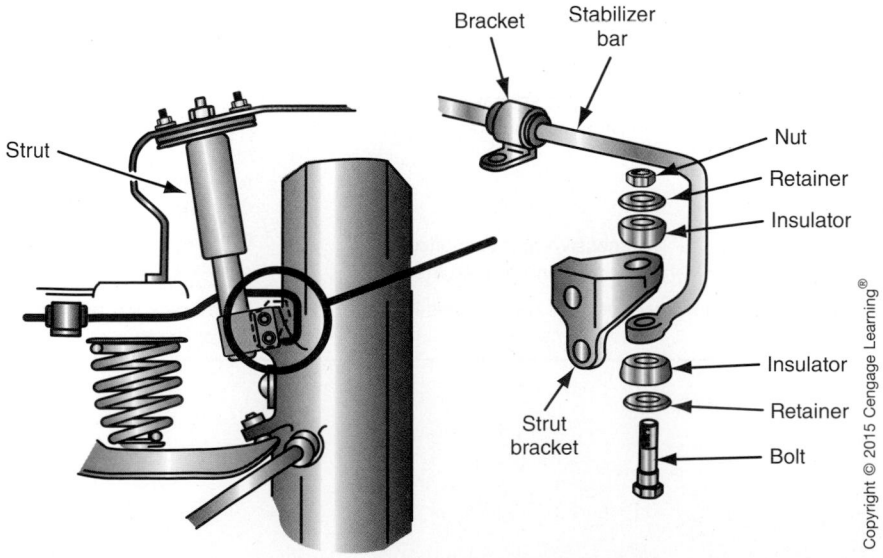

FIGURE 7-28 **Stabilizer bar, bushings, grommets, and brackets.**

3. Remove the mounting bolts in the center area of the stabilizer bar.
4. Remove the stabilizer bar from the chassis.
5. Visually inspect all stabilizer bar components, such as bushings, bolts, and spacer sleeves. Replace the stabilizer bar, grommets, bushings, brackets, or spacers as required. Split bushings may be removed over the stabilizer bar. Bushings that are not split must be pulled from the bar.
6. Reverse steps 2 through 4 to install the stabilizer bar. Make sure all stabilizer bar components are installed in the original position, and tighten all fasteners to the specified torque.

Rear Suspension Tie Rod Inspection and Replacement

Classroom Manual
Chapter 7, page 179

Rear tie rods should be inspected for worn grommets, loose mountings, and bent conditions. Loose tie rod bushings or a bent tie rod will change the rear wheel tracking and result in reduced directional stability. Worn tie rod bushings also cause a rattling noise on road irregularities. When the rear tie rod is replaced, remove the front and rear rod mounting nuts. The lower control arm or rear axle may have to be pried rearward to remove the tie rod. Inspect the tie rod grommets and mountings for wear, and replace parts as required. When the tie rod is reinstalled, tighten the mounting bolts to specifications, and measure the rear wheel toe.

TABLE 7-1 **REAR SUSPENSION DIAGNOSIS**

Problem	Symptoms	Possible Causes
Low riding height	Harsh ride quality, worn strikeout bumpers	Weak springs, worn control arm bushings, bent trailing arms or control arms
Steering pull	Steering pulls to the right or left when driving straight ahead	Improper rear wheel toe
Rear tire wear	Excessive rear tire tread wear	Improper rear wheel toe or camber
Rear chassis vibration	Rear chassis vibration when driving at a certain speed	Improper rear wheel, rotor, or drum balance; improper drive shaft balance or angles on rear-wheel-drive vehicles
Rear suspension noise	Rattling or squeaking noise when driving on road irregularities	Worn or dry stabilizer bar links and bushings, worn shock absorber or strut rod bushings, broken tubular rod in the rear axle inverted U-channel
Excessive rear suspension vertical oscillations	Excessive rear suspension bouncing when driving or road irregularities	Worn out shock absorbers or struts
Excessive rear chassis waddle	Excessive rear chassis lateral oscillations	Worn track bar bushings, loose track bar or brace

Copyright © 2015 Cengage Learning®

CASE STUDY

A customer complained about steering pull to the left on a Chevrolet Impala. The customer also said the problem had just occurred in the last few days. The technician road tested the car and found the customer's description of the complaint to be accurate. A careful inspection of the tires indicated there was no abnormal wear on the front or rear tires. The vehicle was inspected for recent collision damage, but there was no evidence of this type of damage. After lifting the vehicle on a hoist, the technician checked all the front suspension components, including ball joints, control arms, control arm bushings, and wheel bearings. However, none of these front suspension components indicated any sign of wear or looseness.

Realizing that improper rear wheel tracking causes steering pull, the technician inspected the rear suspension. A pry bar was used to apply downward and rearward force to the trailing arms on the rear suspension. When this action was taken on the right rear trailing arm, the technician discovered the trailing arm bushing was very loose. This defect had allowed the right side of the rear axle to move rearward a considerable amount, which explained why the steering pulled to the left.

After installing a new trailing arm bushing, the alignment was checked on all four wheels, and the wheel alignment was within specifications. A road test indicated no evidence of steering pull or other steering problems.

TERMS TO KNOW

Body sway

Lateral chassis movement

Spring silencers

Suspension adjustment link

Wear indicators

ASE-STYLE REVIEW QUESTIONS

1. While discussing lateral movement of the rear chassis:
   Technician A says this problem may be caused by loose stabilizer bar bushings.
   Technician B says this problem may be caused by loose track bar bushings.
   Who is correct?
   A. A only
   B. B only
   C. Both A and B
   D. Neither A nor B

2. While discussing excessive rear body sway or roll when driving on road irregularities:
   Technician A says this problem may be caused by worn spring insulators.
   Technician B says this problem may be the result of sagged rear springs.
   Who is correct?
   A. A only
   B. B only
   C. Both A and B
   D. Neither A nor B

3. While discussing rear suspension curb riding height:
   Technician A says reduced rear curb riding height causes harsh riding.
   Technician B says reduced rear curb riding height increases steering effort.
   Who is correct?
   A. A only
   B. B only
   C. Both A and B
   D. Neither A nor B

4. While discussing curb riding height:
   Technician A says the curb riding height is measured at the same location on each vehicle.
   Technician B says the curb riding height has no effect on other alignment angles.
   Who is correct?
   A. A only
   B. B only
   C. Both A and B
   D. Neither A nor B

5. While discussing spring removal on a MacPherson strut rear suspension:
   Technician A says the entire spring tension must be removed from the upper mount before the upper strut nut is loosened.
   Technician B says a vinyl-coated spring should be taped in the compressing tool contact area.
   Who is correct?
   A. A only
   B. B only
   C. Both A and B
   D. Neither A nor B

6. The most likely cause of excessive rear chassis lateral movement is (are):
   A. A weak stabilizer bar.
   B. Worn track bar bushings.
   C. Worn-out struts.
   D. Sagged rear coil springs.

7. All these statements about rear suspension systems with lower control arms and ball joints are true EXCEPT:
   A. Worn lower control arm bushings do not affect tire tread wear.
   B. Some rear suspension ball joints have wear indicators.
   C. On some rear suspensions, the ball joint is pressed into the lower control arm.
   D. On some vehicles, the vehicle weight must be resting on the tires when the inner, lower control arm bolts are tightened.
8. Rear suspension adjustment links:
   A. May be lengthened or shortened to adjust rear wheel camber.
   B. Are connected from the lower control arm to the knuckle.
   C. May be loosened on each end with a wedge-type tool.
   D. Prevent fore-and-aft control arm and wheel movement.
9. On a rear suspension system with longitudinally mounted leaf springs:
   A. Improper vehicle tracking may be caused by a broken spring center bolt.
   B. Worn spring silencers may cause reduced curb riding height.
   C. A shackle is mounted between the front of the spring and the chassis.
   D. Longitudinal rear leaf springs are usually installed on independent rear suspension systems.
10. While diagnosing improper rear wheel tracking:
   A. Improper rear wheel tracking may be caused when both rear springs are sagged the same amount.
   B. A bent rear suspension tie rod does not affect rear wheel tracking.
   C. Improper rear wheel tracking may result in steering pull when driving straight ahead.
   D. Improper rear wheel tracking may cause front wheel shimmy.

ASE CHALLENGE QUESTIONS

1. A squeak in the rear suspension could be caused by all of the following EXCEPT:
   A. The suspension bushing.
   B. Weak spring leaves.
   C. Worn spring antifriction pads.
   D. A defective shock absorber.
2. A vehicle with a live axle coil spring rear suspension has become hard to steer with harsh ride quality. Which of the following could be the cause of this problem?
   A. Worn lateral link bushings.
   B. Weak rear coil springs.
   C. Bent rear shock rod.
   D. Worn stabilizer bar bushings.
3. A truck with a solid axle and parallel leaf spring rear suspension darts and acts erratic during turns. Which of the following is the most likely cause?
   A. Worn rear sway bar bushings.
   B. Incorrect driveline angle
   C. Incorrect ride height
   D. Loose rear axle U-bolts
4. A car with independent rear suspension has excessive rear tire wear. An inspection of the rear tires shows they are worn on the inside edge and the tread is feathered.

   Technician A says the problem could be the tires are toeing out during acceleration.

   Technician B says worn control arm bushings could be the cause of the problem.

   Who is correct?

   A. A only
   B. B only
   C. Both A and B
   D. Neither A nor B
5. The steering on a front-wheel-drive car pulls to the right.

   Technician A says the strut on the right rear suspension assembly could be the problem.

   Technician B says worn bushings of the left rear lower arm assembly could be the problem.

   Who is correct?

   A. A only
   B. B only
   C. Both A and B
   D. Neither A nor B

JOB SHEET 24

Name _____ **Date** _____

Remove and Service Rear Suspension Strut and Coil Spring Assembly

Upon completion of this job sheet, you should be able to remove and service rear suspension strut and coil spring assemblies.

NATEF AST/MAST Task Correlation:

Suspension and Steering. Suspension Systems Diagnosis and Repair,

Task #C-10. Inspect, remove and install strut cartridge or assembly, strut coil spring, insulators (silencers), and upper strut bearing mount. (P-3)

Tools and Materials

Floor jack
Hoist
Safety stands
Coil spring compressing tool

Describe the Vehicle Being Worked On:

Year _____ Make _____ Model _____

VIN _____ Engine type and size _____

Procedure — Task Completed

1. Remove the rear seat and the package trim tray. ☐
2. Remove the wheel cover, and loosen the wheel nuts. ☐
3. Lift the vehicle with a floor jack, and lower the chassis onto safety stands so the rear suspension is allowed to drop downward.
   Is the rear suspension properly supported on the safety stands? ☐ Yes ☐ No
   Instructor check _____
4. Place the floor jack lift pad under the rear spindle on the side where the strut and spring removal is taking place. Raise the floor jack to support some of the suspension system weight.
   Is the floor jack placed under the rear spindle and supporting some of the vehicle weight? ☐ Yes ☐ No
   Instructor check _____
5. Remove the rear wheel, disconnect the nut from the small spring in the lower arm, and remove the brake hose and antilock brake system (ABS) wire from the strut. ☐
6. Remove the stabilizer bar link from the strut, and loosen the strut-to-spindle mounting bolts. ☐
7. Remove the three upper support nuts under the package tray trim, and lower the floor jack to remove the strut from the knuckle. Remove the strut from the chassis. ☐

Task Completed

8. Following the vehicle and equipment manufacturers' recommended procedures, install a spring compressor on the coil spring, and tighten the spring compressor until the entire spring tension is removed from the upper support.
   Is the spring compressor properly installed on the coil spring? ☐ Yes ☐ No
   Is the spring compressor tightened so all tension is removed from the upper support?
   Instructor check _____

☐ 9. Install a bolt in the upper strut-to-spindle bolt hole and tighten the two nuts on the end of this bolt. One nut must be on each side of the strut bracket.
   Clamp this bolt in a vise to hold the strut, coil spring, and spring compressor.

10. Using the special tool to hold the spring and strut from turning, loosen the nut from the strut rod nut.
   Is the entire coil spring tension removed from the upper strut mount by the spring compressor? ☐ Yes ☐ No
   Is the strut rod nut loosened? ☐ Yes ☐ No
   Instructor check _____

☐ 11. Remove the strut rod nut, upper support, and upper insulator.

☐ 12. Remove the strut from the lower end of the spring.

☐ 13. If the spring is to be replaced, rotate the compressor bolt until the entire spring tension is removed from the compressing tool, then remove the spring from the tool.

14. Inspect the lower insulator and spring seat on the strut. If the spring seat is warped or damaged, replace the strut. A new cartridge may be installed in some rear struts.
   Lower insulator condition: ☐ Satisfactory ☐ Unsatisfactory
   Spring seat condition: ☐ Satisfactory ☐ Unsatisfactory
   Strut condition: ☐ Satisfactory ☐ Unsatisfactory
   List all the components that require replacement and explain the reasons for your diagnosis.
   _____
   _____

SERVICE TIP:
If the plastic coating on a coil spring is chipped, the spring may break prematurely. The spring may be taped in the compressing tool contact areas to prevent chipping.

15. Visually inspect the coil spring, upper mount, insulator, and spring bumper. If any of these components are damaged, worn, or distorted, replacement is necessary.
   Coil spring condition: ☐ Satisfactory ☐ Unsatisfactory
   Upper mount condition: ☐ Satisfactory ☐ Unsatisfactory
   Insulator condition: ☐ Satisfactory ☐ Unsatisfactory
   Spring bumper condition: ☐ Satisfactory ☐ Unsatisfactory
   List all the components that require replacement and explain the reasons for your diagnosis.
   _____

Instructor's Response _____
_____
_____

JOB SHEET 25

Name _____ **Date** _____

Remove Rear Suspension Lower Control Arm and Ball Joint Assembly

Upon completion of this job sheet, you should be able to remove rear suspension lower control arm and ball joint assemblies.

NATEF AST/MAST Task Correlation:

Suspension and Steering. Suspension Systems Diagnosis and Repair,

Task #C-3. Inspect, remove and install upper and lower control arms, bushings, shafts, and rebound bumpers. (P-3)

Task #C-5. Inspect, remove and install upper and/or lower ball joints (with or without wear indicators). (P-2)

Tools and Materials

Floor jack
Control arm removing tool
Safety stands
Transmission jack
Hoist
Ball joint removal and replacement tools

Describe the Vehicle Being Worked On:

Year _____ Make _____ Model _____

VIN _____ Engine type and size _____

Procedure — Task Completed

1. Lift the vehicle on a hoist with the chassis supported in the hoist and control arms dropped downward. The vehicle may be lifted with a floor jack and the chassis supported on safety stands. ☐
2. Remove the tire and wheel assembly. ☐
3. Remove the stabilizer bar from the knuckle bracket. ☐
4. Remove the parking brake cable retaining clip from the lower control arm. ☐
5. If the car has electronic level control (ELC), disconnect the height sensor link from the control arm. ☐
6. Install a special tool to support the lower control arm in the bushing areas. ☐
7. Place a transmission jack under the special tool and raise the jack enough to remove the tension from the control arm bushing retaining bolts. If the car was lifted with a floor jack and supported on safety stands, place a floor jack under the special tool.
   Is the special control arm support tool properly installed and supported?
   ☐ Yes ☐ No
   Instructor check _____

Task Completed

8. Place a safety chain through the coil spring and around the lower control arm.
   Is the safety chain properly installed? ☐ Yes ☐ No
   Instructor check _____

☐ 9. Remove the bolt from the rear control arm bushing.

☐ 10. Be sure the jack is raised enough to relieve the tension on the front bolt in the lower control arm and remove this bolt.

☐ 11. Lower the jack slowly and allow the control arm to pivot downward. When all the tension is released from the coil spring, remove the safety chain, coil spring, and insulators.

12. Inspect the coil spring for distortion and proper free length. If the spring has a vinyl coating, check this coating for scratches or nicks. Check the spring insulators for cracks and wear.
    Coil spring condition: ☐ Satisfactory ☐ Unsatisfactory
    List all the components that require replacement and explain the reasons for your diagnosis.

    _____

    _____

    _____

☐ 13. Remove the nut on the inner end of the suspension adjustment link and disconnect this link from the lower control arm.

14. Remove the cotter pin from the ball joint nut, and loosen, but do not remove, the nut from the ball joint stud.
    Is the ball joint nut loosened? ☐ Yes ☐ No
    Instructor check _____

☐ 15. Use a special ball joint removal tool to loosen the ball joint in the knuckle.

☐ 16. Remove the ball joint nut and the control arm.

17. Inspect the lower control arm for bends, distortion, and worn bushings.
    Lower control arm condition: ☐ Satisfactory ☐ Unsatisfactory
    List the control arm and related parts that require replacement and explain the reasons for your diagnosis.

    _____

    _____

    Instructor check _____

☐ 18. Use a special ball joint pressing tool to press the ball joint from the lower control arm.

☐ 19. Use the same pressing tool with different adapters to press the new ball joint into the control arm.

Instructor's Response _____

_____

_____

JOB SHEET 26

Name _____ **Date** _____

Install Rear Suspension Lower Control Arm and Ball Joint Assembly

Upon completion of this job sheet, you should be able to install rear suspension lower control arm and ball joint assemblies.

NATEF AST/MAST Task Correlation:

Suspension and Steering. Suspension Systems Diagnosis and Repair,

Task #C-3. Inspect, remove and install upper and lower control arms, bushings, shafts, and rebound bumpers. (P-3)

Task #C-5. Inspect, remove and install upper and/or lower ball joints (with or without wear indicators). (P-2)

Tools and Materials

Floor jack
Control arm removing tool
Safety stands
Transmission jack
Hoist

Describe the Vehicle Being Worked On:

Year _____ Make _____ Model _____

VIN _____ Engine type and size _____

Procedure

_____
Task Completed

1. Install the ball joint stud in the knuckle and install the nut on the ball joint stud. Tighten the ball joint nut to the specified torque, and then tighten the nut an additional 2/3 turn. If necessary, tighten the nut slightly to align the nut castellations with the cotter pin hole in the ball joint stud, and install the cotter pin.

   Specified ball joint stud nut torque _____

   Actual ball joint stud nut torque _____

2. Snap the upper insulator on the coil spring. Install the lower spring insulator and the spring in the lower control arm.

   Is the coil spring and insulator properly installed in lower control arm?

   ☐ Yes ☐ No

   Instructor check _____

3. Be sure the top of the coil spring is properly positioned in relation to the front of the vehicle.

   Is the top of the spring properly positioned? ☐ Yes ☐ No

   Instructor check _____

Task Completed

☐ 4. Install the special tool on the inner ends of the control arm, and place the transmission jack or floor jack under the special tool.

Is the special tool properly supported on the control arm? ☐ Yes ☐ No

Instructor check _____

Is the special tool properly supported on the transmission or floor jack?

Instructor check _____

☐ 5. Slowly raise the transmission jack until the control arm bushing openings are aligned with the openings in the chassis.

6. Install the bolts and nuts in the inner ends of the control arm. Do not torque these bolts and nuts at this time.

7. Install the stabilizer-bar-to-knuckle bracket fasteners to the specified torque.

Specified torque on stabilizer-bar-to-knuckle bracket fasteners

Actual torque on stabilizer-bar-to-knuckle bracket fasteners

☐ 8. Install the parking brake retaining clip.

☐ 9. If the vehicle has ELC, install the height sensor link, and tighten the fastener to the specified torque.

10. Install the suspension adjustment link and tighten the fastener to the specified torque. Install cotter pins as required.

Specified adjustment link retaining nut torque _____

Actual adjustment link retaining nut torque _____

Are the cotter pins properly installed in adjustment link retaining nuts?

☐ Yes ☐ No

Instructor check _____

☐ 11. Remove the transmission jack or floor jack, and install the tire and wheel assembly.

12. Lower the vehicle onto the floor. Tighten the wheel hub nuts and lower control arm bolts and nuts to the specified torque.

Specified wheel nut torque _____

Actual wheel nut torque _____

Specified control arm retaining nut torque _____

Actual control arm retaining nut torque _____

CAUTION:
The pivot bolts and nuts in the inner ends of the lower control arm must be tightened to the specified torque with the vehicle supported on the wheels and the suspension at normal curb height. Failure to follow this procedure may adversely affect ride quality and steering characteristics.

Instructor's Response _____

_____

_____

Chapter 8

Steering Column and Linkage Diagnosis and Service

BASIC TOOLS

Basic technician's tool set
Service manual
Steering wheel puller
Foot-pound torque wrench
Machinist's rule
1/4" Electric drill
Set of drill bits
Screw extractors

Upon completion and review of this chapter, you should be able to:

- Diagnose steering columns.
- Remove and replace steering wheels on air-bag-equipped vehicles and non-air-bag-equipped vehicles.
- Remove and replace air bag deployment modules and clock spring electrical connectors.
- Remove and replace steering columns.
- Inspect collapsible steering columns for damage.
- Disassemble steering columns.
- Inspect steering column components and replace necessary parts.
- Assemble steering columns.
- Diagnose and service flexible couplings and universal joints.
- Diagnose steering linkage mechanisms.
- Diagnose, remove, and replace tie rod ends.
- Diagnose, remove, and replace pitman arms.
- Diagnose, remove, and replace center links.
- Diagnose, remove, and replace idler arms.
- Diagnose, remove, and replace steering dampers.
- Diagnose steering arms.

When servicing or replacing steering wheels, columns, and linkages, technicians actually have the customer's life in their hands. If steering components are not serviced properly and not tightened to the specified torque, the steering may become disconnected, resulting in a complete loss of steering control. This condition may cause a collision, resulting in vehicle damage, personal injury, and an expensive lawsuit for technicians and the shops where they are employed. Therefore, when performing any automotive service, always be sure the vehicle manufacturer's recommended service procedures and torque specifications are followed.

Air Bag Deployment Module, Steering Wheel, and Clock Spring Electrical Connector Removal and Replacement

Prior to working on an air bag system, always disconnect the negative battery cable and wait 1 minute before proceeding with the diagnostic or service work. Many air bag systems have a **back-up power supply** circuit designed into the air bag computer or located in a separate module. This back-up power supply provides power to deploy the air bag for a specific length of time after the battery power is disconnected in a collision. One minute

CAUTION:
Always disconnect the negative battery terminal and wait one minute before diagnosing or servicing any air bag system component. Failure to observe this precaution may result in accidental air bag deployment. Air bag service precautions vary depending on the vehicle. Always follow the vehicle manufacturer's air bag service precautions in the service manual.

after the negative battery terminal is disconnected, this back-up power supply is powered down, and no power is available to deploy the air bag while the battery remains disconnected.

CUSTOMER CARE: Before disconnecting the negative battery cable, note how the customer has the radio stations programmed, and reset the radio and clock after the negative battery cable is reconnected.

An air bag warning light in the instrument panel indicates the status of the air bag system. On some vehicles, this warning light should be illuminated for a few seconds when the ignition switch is turned on. The warning light should remain off while cranking the engine, and it should be on for a few seconds after the engine starts. After the engine has been running for a few seconds, the air bag warning light should remain off. On other vehicles, the air bag warning light flashes seven to nine times when the ignition switch is turned on and after the engine is started. The air bag warning light operation varies depending on the make and model year of the vehicle. Always check the vehicle manufacturer's service manual for the exact air bag warning light operation. When the air bag warning light does not operate as specified by the vehicle manufacturer, the air bag system is defective, and the air bag or bags will probably not deploy if the vehicle is involved in a collision.

Air bag module and steering wheel removal and replacement procedures vary depending on the vehicle. Always follow the vehicle manufacturer's recommended procedure in the service manual.

Classroom Manual
Chapter 8, page 193

The following is a typical air bag module and steering wheel removal and replacement procedure:

1. Turn the ignition switch to the Lock position and place the front wheels facing straight ahead.
2. Remove the negative battery terminal and wait for the specified time period.
3. Loosen the three air-bag-retaining Torx screws under the steering wheel (Figure 8-1).
4. Loosen the other two air-bag-retaining Torx screws under the steering wheel (Figure 8-2). Loosen all five Torx screws until the groove along the screw circumference catches on the screw case. Some air bag deployment modules are retained on top of the steering wheel with a spring-loaded clip. To release this clip, insert a flat-tipped screw driver into the slot at the lower edge of the steering wheel and turn the screw driver 1/4 turn to release the clip.

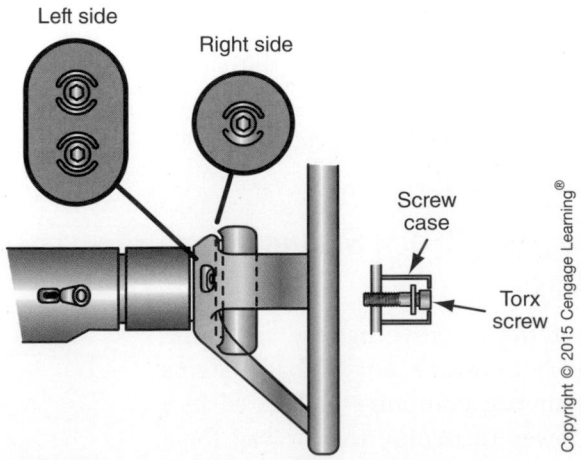

FIGURE 8-1 **Three air-bag–retaining Torx screws.**

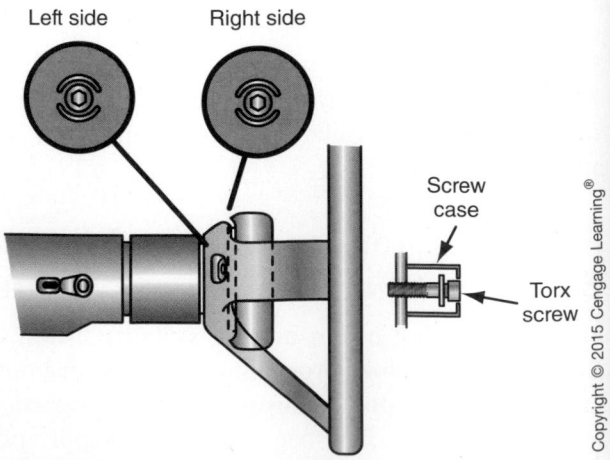

FIGURE 8-2 **Two air-bag–retaining Torx screws.**

WARNING: When an air bag deployment module is temporarily stored on the workbench, always place this module face upward. If the air bag deployment module accidentally deployed when facing downward, the module would become a projectile, and personal injury might result.

An air bag deployment module may be referred to as a steering wheel pad.

5. Pull the **air bag deployment module** from the steering wheel and disconnect the air bag module electrical connector (Figure 8-3). Do not pull on the air bag wires in the steering column. Place the air bag deployment module face upward on the workbench.
6. Disconnect the air bag wiring retainer in the steering wheel (Figure 8-4).
7. Use the proper size socket and a ratchet to remove the steering wheel retaining nut.
8. Observe the matching alignment marks on the steering wheel and the steering shaft. If these alignment marks are not present, place alignment marks on the steering wheel and steering shaft with a center punch and a hammer.

The **air bag deployment module** contains the air bag, inflation chemicals, and an igniting device.

WARNING: Photo Sequence 13 shows a typical procedure for removing a steering wheel. Do not pull on the steering wheel in an attempt to remove it from the steering shaft. The steering wheel may suddenly come off, resulting in personal injury, or the steering wheel may be damaged by the pulling force.

CAUTION: On an air-bag-equipped vehicle, the wait time prior to servicing electrical components after the negative battery terminal is disconnected varies depending on the vehicle make and model year. Always follow the wait time and all other service precautions recommended in the vehicle manufacturer's service manual.

9. Install a steering wheel puller with the puller bolts threaded into the bolt holes in the steering wheel. On some vehicles, the steering wheel has slots for the pulley adapters to fit into rather than threaded bolt holes. This design prevents the possibility of the technician turning the bolts too far into the steering wheel and damaging components below the steering wheel. Tighten the puller nut to remove the steering wheel (Figure 8-5). Visually check the steering wheel condition. If the steering wheel is bent or cracked, replace the wheel.
10. Disconnect the four retaining screws and remove the **clock spring electrical connector** (Figure 8-6).
11. Be sure the front wheels are facing straight ahead. Turn the clock spring electrical connector counterclockwise by hand until it becomes harder to turn as it becomes fully wound in that direction.

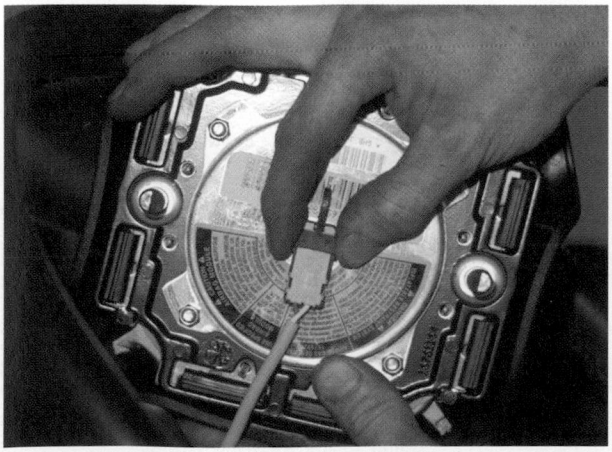

Copyright © 2015 Cengage Learning®

FIGURE 8-3 **Disconnecting the air bag module electrical connector.**

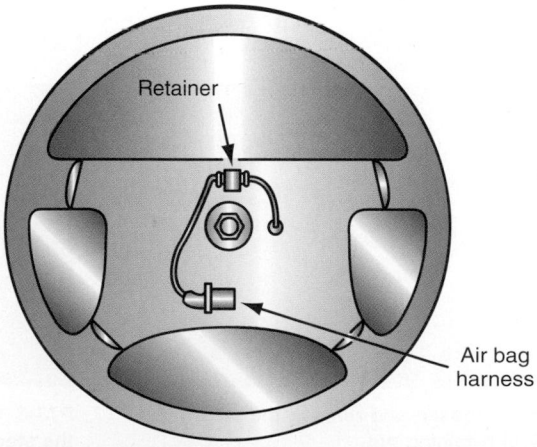

Copyright © 2015 Cengage Learning®

FIGURE 8-4 **Disconnecting the air bag wiring retainer.**

PHOTO SEQUENCE 13

Typical Procedure For Removing a Steering Wheel

P13-1 **Check and record the radio stations programmed in the stereo system.**

Copyright © 2015 Cengage Learning®

P13-2 **Look up the car manufacturer's steering wheel removal procedure in the service manual.**

Copyright © 2015 Cengage Learning®

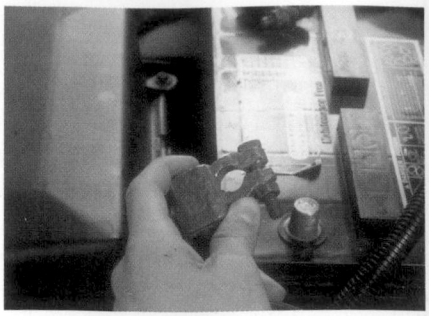

P13-3 **Disconnect the negative battery cable, and wait for the car manufacturer's specified length of time.**

Copyright © 2015 Cengage Learning®

P13-4 **Remove the air bag deployment module retaining screws under the steering wheel.**

Copyright © 2015 Cengage Learning®

P13-5 **Lift the air bag deployment module upward from the steering wheel, and disconnect the module wiring connector.**

Copyright © 2015 Cengage Learning®

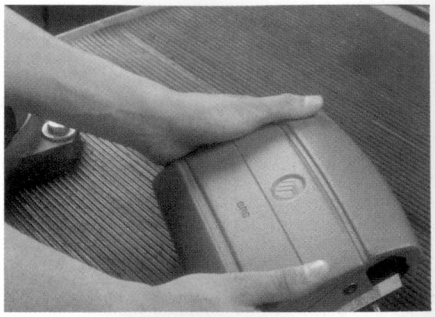

P13-6 **Set the air bag deployment module face upward on the workbench.**

Copyright © 2015 Cengage Learning®

P13-7 **Loosen and remove the steering wheel retaining nut.**

Copyright © 2015 Cengage Learning®

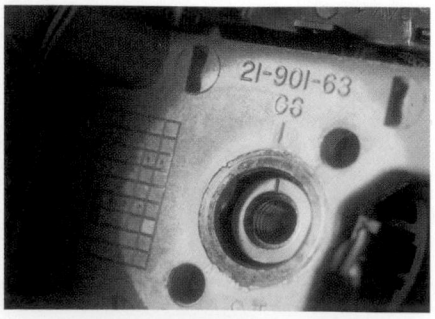

P13-8 **Observe the alignment marks on the steering wheel and shaft.**

Copyright © 2015 Cengage Learning®

P13-9 **Connect the proper steering wheel puller to the steering wheel.**

Copyright © 2015 Cengage Learning®

PHOTO SEQUENCE 13 (CONTINUED)

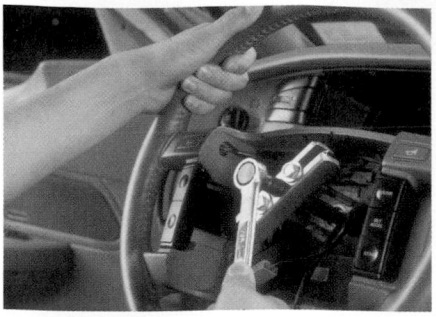

P13-10 **Turn the puller screw to loosen the steering wheel on the shaft.**

Copyright © 2015 Cengage Learning®

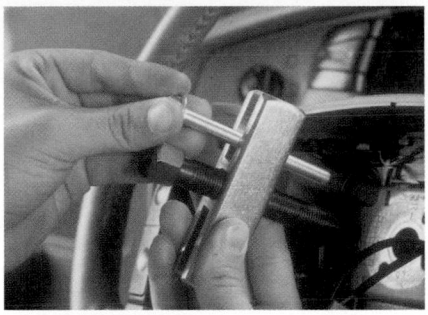

P13-11 **Remove the puller from the steering wheel.**

Copyright © 2015 Cengage Learning®

P13-12 **Lift the steering wheel off the shaft.**

Copyright © 2015 Cengage Learning®

Copyright © 2015 Cengage Learning®

FIGURE 8-5 **Removing steering wheel with the proper steering wheel puller.**

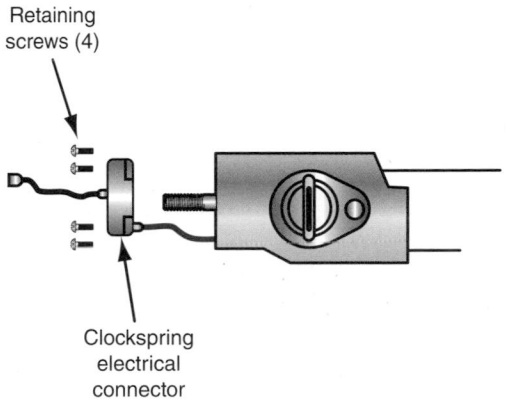

Copyright © 2015 Cengage Learning®

FIGURE 8-6 **Removing clock spring electrical connecter.**

CAUTION: Always use the vehicle manufacturer's recommended tools when diagnosing or servicing an air bag system or the electrical system on an air-bag-equipped vehicle. Do not use a 12-V test light or a self-powered test light to diagnose air bag systems or electrical systems on air-bag-equipped vehicles. **Accidental air bag deployment** may result from improper use of tools.

12. Turn the clock spring electrical connector clockwise three turns and align the red mark on the center part of the spring face with the notch in the cable circumference (Figure 8-7). This action centers the clock spring electrical connector.
13. Install the clock spring electrical connector and tighten the four retaining screws to the specified torque.
14. Align the marks on the steering wheel and the steering shaft, and install the steering wheel on the shaft.
15. Install the steering wheel retaining nut and tighten this nut to the specified torque.
16. Install the air bag wiring retainer in the steering wheel.
17. Hold the air bag deployment module near the top of the steering wheel and connect the air bag module connector.
18. Install the air bag deployment module in the top of the steering wheel and tighten the five retaining Torx screws.
19. Reconnect the negative battery cable.
20. Reset the clock and radio.

If the vehicle is not equipped with an air bag, the steering wheel removal and replacement procedure is basically the same, but all steps pertaining to the air bag module and clock spring are not required. On a non-air-bag-equipped vehicle, the center steering wheel cover must be removed to access the steering wheel retaining nut.

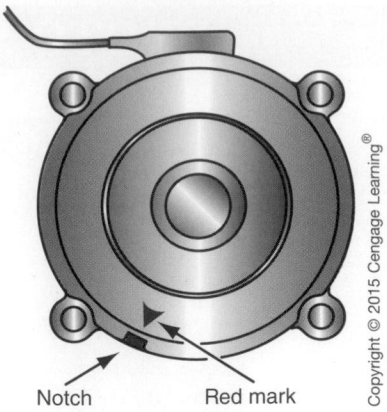

FIGURE 8-7 **Centering clock spring electrical connecter.**

A **clock spring electrical connector** may be called a coil, spiral cable, or cable reel. The clock spring electrical connector connects the air bag electrical system to the air bag deployment module and allows steering wheel rotation.

CUSTOMER CARE: While servicing a vehicle, always inspect the operation of the indicator lights or gauges in the instrument panel. These lights or gauges may indicate a problem that the customer has been ignoring. For example, if the air bag warning light is not operating properly, the air bag or bags may not deploy in a collision, resulting in serious injury to the driver and/or passenger. If the air bag warning light is not working properly, always advise customers that they will not be protected by the air bag in a collision, and the vehicle should not be driven under this condition.

SERVICE TIP: When servicing air bag components on some recent model vehicles, the vehicle manufacturer recommends disconnecting the air bag components only in the zone or area on the vehicle where service work is required rather than disconnecting the negative battery cable.

STEERING COLUMN SERVICE

Some steering column service can be performed with the column installed in the vehicle. In some steering columns removal and replacement of the various switches in the column is possible with the column installed in the vehicle. Always follow the recommended service procedure in the vehicle manufacturer's service manual.

STEERING COLUMN REMOVAL AND REPLACEMENT

Steering column removal and replacement procedures vary depending on the vehicle make, type of steering column, and gearshift lever position. Always follow the vehicle manufacturer's recommended procedure in the service manual.

The following is a typical steering column removal and replacement procedure:

1. Disconnect the negative battery cable. If the vehicle is equipped with an air bag, wait 1 minute.
2. Install a seat cover on the front seat.
3. Place the front wheels in the straight-ahead position and remove the ignition key from the switch to lock the steering column.
4. Remove the cover under the steering column and remove the lower finish panel if necessary.
5. Disconnect all wiring connectors from the steering column.
6. If the vehicle has a column-mounted gearshift lever, disconnect the gearshift linkage at the lower end of the steering column. If the vehicle has a floor-mount gearshift, disconnect the shift interlock.

Classroom Manual
Chapter 8, page 196

7. Remove the retaining bolt or bolts in the lower universal joint or flexible coupling.
8. Remove the steering-column-to-instrument-panel mounting bolts.
9. Carefully remove the steering column from the vehicle. Be careful not to damage the upholstery or paint.
10. Install the steering column under the instrument panel and insert the steering shaft into the lower universal joint.
11. Install the steering-column-to-instrument-panel mounting bolts. Be sure the steering column is properly positioned, and tighten these bolts to the specified torque.
12. Install the retaining bolt or bolts in the lower universal joint or flexible coupling, and tighten the bolts to the specified torque.
13. Connect the gearshift linkage if the vehicle has a column-mounted gearshift.
14. Connect all the wiring harness connectors to the steering column connectors.
15. Install the steering column cover and the lower finish panel.
16. Reconnect the negative battery cable.
17. Road test the vehicle and check for proper steering column operation.

CAUTION: Do not hammer on top of the steering shaft to remove the steering wheel. This action may damage the shaft.

CAUTION: If the steering wheel puller bolts are too long, they will extend through the steering wheel and damage the clock spring.

CAUTION: Failure to center a clock spring prior to installation may cause a broken conductive tape in the clock spring.

Collapsible Steering Column Inspection

Since steering column design varies depending on the vehicle, the **collapsible steering column** inspection procedure should be followed in the vehicle manufacturer's service manual.

The following is a typical collapsible steering column inspection procedure:

1. Measure the clearance between the capsules and the slots in the steering column bracket (Figure 8-8). If this measurement is not within specifications, replace the bracket.
2. Inspect the contact between the bolt head and the bracket (Figure 8-9). If the bolt head contacts the bracket, the shear load is too high, and the bracket must be replaced.
3. Inspect the steering column jacket for **sheared injected plastic** in the openings on the side of the jacket (Figure 8-10). If sheared plastic is present, the column is collapsed. Measure the distance from the end of the bearing assembly to the lower edge of the upper steering column jacket (Figure 8-11). If this distance is not within the vehicle manufacturer's specification, a new jacket must be installed.
4. Visually inspect the gearshift tube for sheared injected plastic. If the gearshift tube indicates sheared plastic, replace the tube.
5. Remove the intermediate steering shaft between the column and the steering gear. Position a dial indicator stem against the lower end of the steering shaft, and rotate the steering wheel. If the runout on the dial indicator exceeds the vehicle manufacturer's specification, the steering shaft is bent and must be replaced.
6. If the steering shaft is not bent but shows sheared injected plastic (Figure 8-12), this shaft may be repaired with a service repair package.

A **collapsible steering column** allows the column to move away from the driver if the driver impacts the column during a collision.

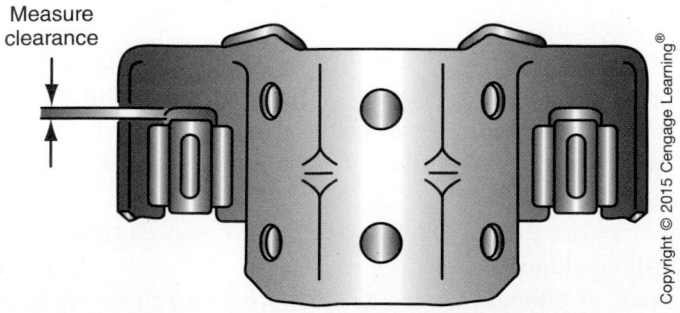

FIGURE 8-8 **Capsules in steering column bracket.**

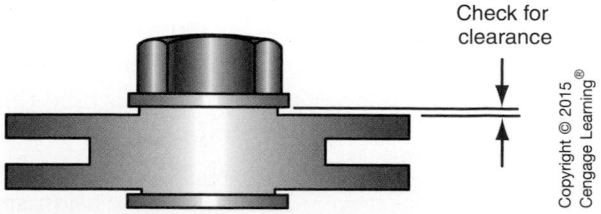

FIGURE 8-9 **Bolt head to bracket clearance.**

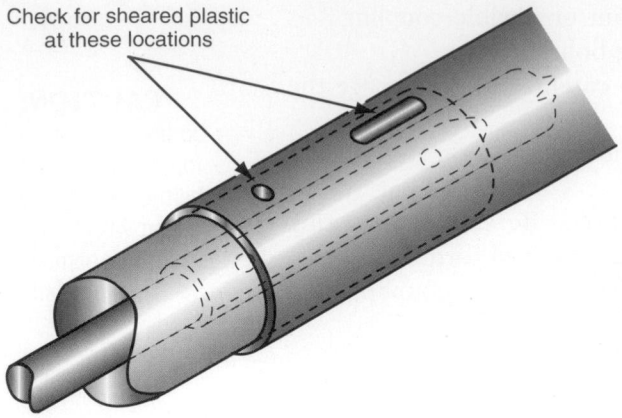

Copyright © 2015 Cengage Learning®

FIGURE 8-10 **Inspecting for sheared plastic in jacket openings.**

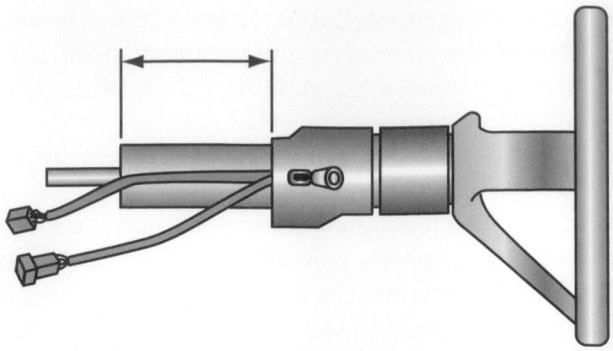

Copyright © 2015 Cengage Learning®

FIGURE 8-11 **Measuring distance from the end of the bearing assembly to the upper steering column jacket.**

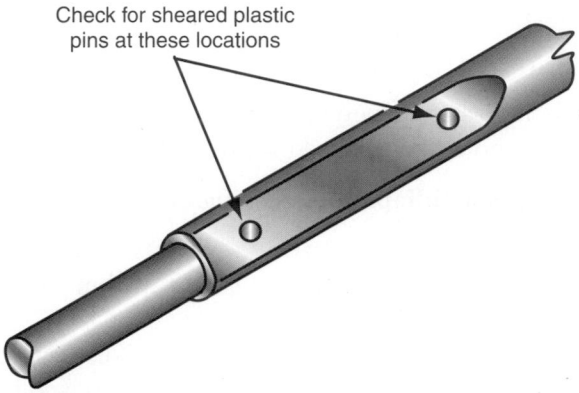

Copyright © 2015 Cengage Learning®

FIGURE 8-12 **Inspecting for sheared injected plastic in steering shaft.**

Classroom Manual
Chapter 8, page 197

Tilt Steering Column Disassembly

Since there are many variations in steering column design, always follow the vehicle manufacturer's recommended steering column disassembly procedure in the service manual.

The following is a typical tilt steering column disassembly procedure:

1. Remove the air bag deployment module, steering wheel, and spiral cable, as mentioned previously in this chapter.
2. Remove the ignition key cylinder illumination mounted on top of the ignition switch cylinder (Figure 8-13).
3. Remove the universal joint from the lower end of the steering shaft.
4. Remove the steering column protector and the wiring harness clamp mounted under the steering column.
5. Remove the steering damper mounted near the top of the column.
6. Remove the retaining screw and the combination switch with the wiring harness (Figure 8-14).
7. Mark the center of the **tapered-head bolts** on the steering column with a center punch. Use a 0.120 to 0.160 in. (3 to 4 mm) drill bit to drill into the tapered-head bolts (Figure 8-15).
8. Remove the tapered-head bolts with a screw extractor, and separate the upper bracket and column tube (Figure 8-16).
9. Grasp the compression spring seat with a pair of pliers and turn the seat to release the seat, compression spring, and bushing (Figure 8-17).
10. Grasp the tension spring with a pair of pliers and extend this spring to remove it from the column (Figure 8-18).

Tapered-head bolts have a tapered head with no provision for connecting any type of wrench to the bolt.

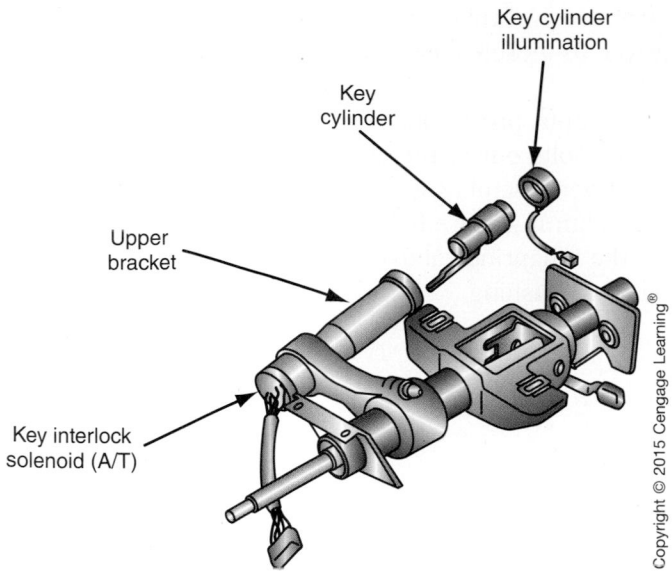

FIGURE 8-13 **Ignition key cylinder illumination mounted above ignition switch cylinder.**

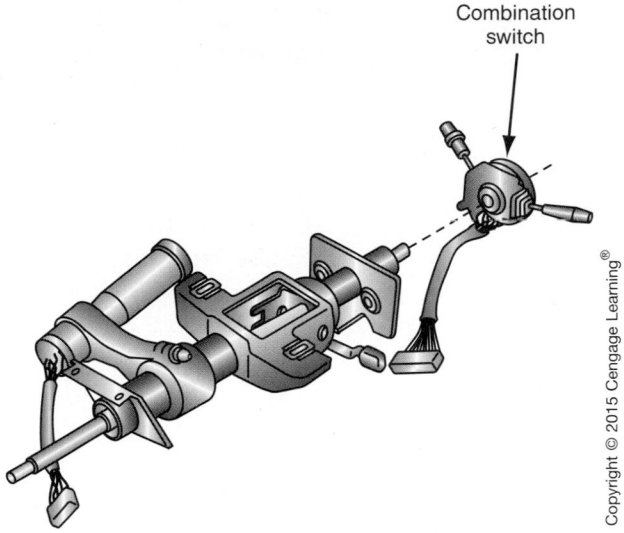

FIGURE 8-14 **Removing combination switch and wiring harness.**

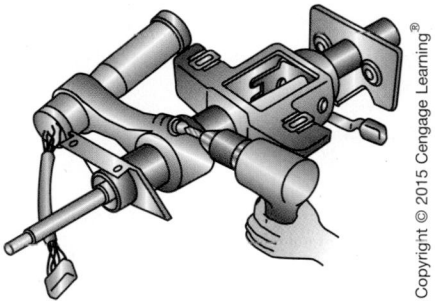

FIGURE 8-15 **Drilling tapered-head bolts.**

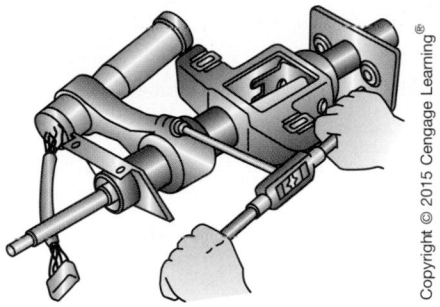

FIGURE 8-16 **Removing tapered-head bolts with a screw extractor.**

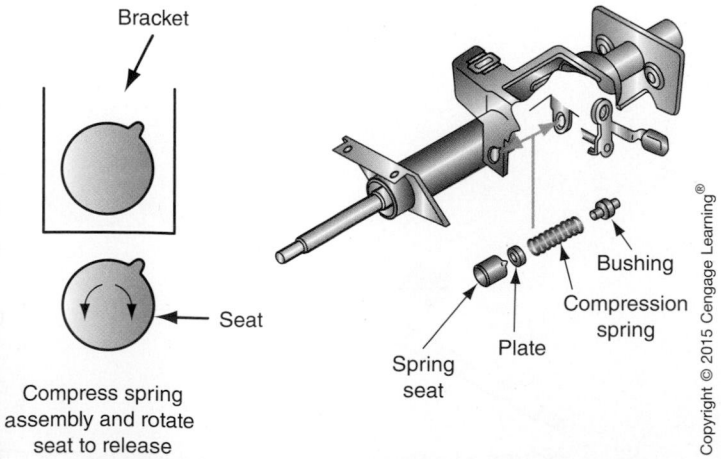

FIGURE 8-17 **Removing compression spring, seat, and bushing.**

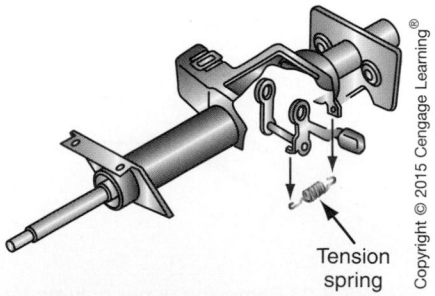

FIGURE 8-18 **Removing tension spring.**

11. Remove the snapring from the upper column tube with snapring pliers (Figure 8-19).
12. Use a soft hammer and a screwdriver to loosen the staked parts of the upper column tube (Figure 8-20).
13. Remove the two nuts on the upper column pivot bolts (Figure 8-21).
14. Use a plastic hammer to tap the pivot bolts out of the column (Figure 8-22).
15. Remove the upper column tube, tilt lever assembly, and plate washers (Figure 8-23).
16. Remove the steering shaft from the column (Figure 8-24).
17. Use snapring pliers to remove the snapring above the steering shaft bushing (Figure 8-25). Remove the steering shaft bushing.

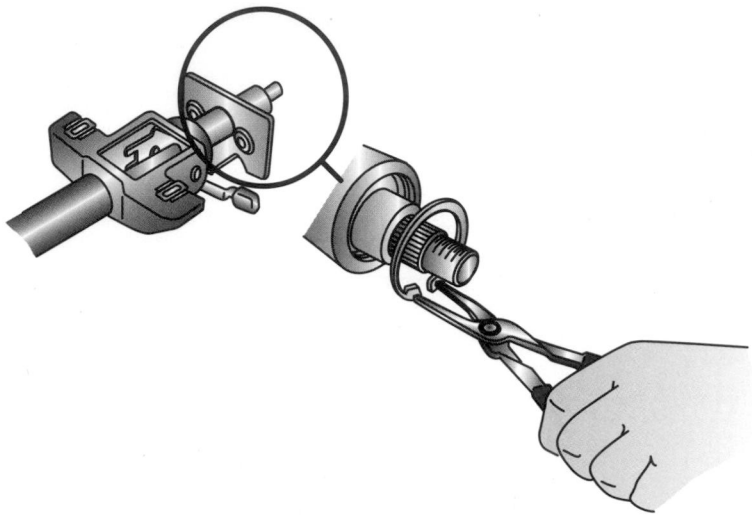

Copyright © 2015 Cengage Learning®

FIGURE 8-19 **Removing snapring from upper column tube.**

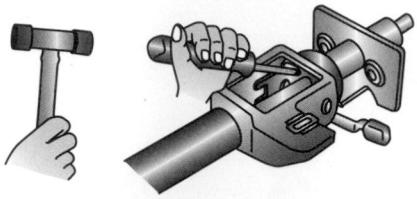

Copyright © 2015 Cengage Learning®

FIGURE 8-20 **Loosening staked parts in upper column tube.**

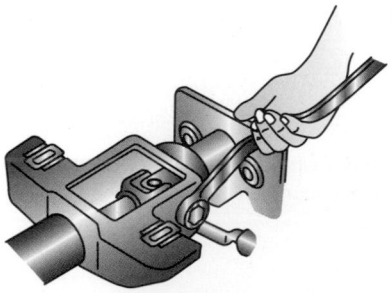

Copyright © 2015 Cengage Learning®

FIGURE 8-21 **Removing nuts on upper column pivot bolts.**

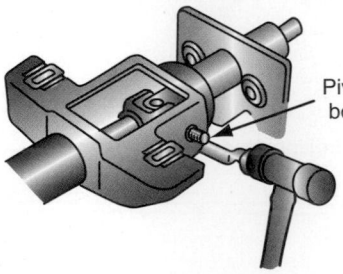

Copyright © 2015 Cengage Learning®

FIGURE 8-22 **Removing pivot bolts from steering column.**

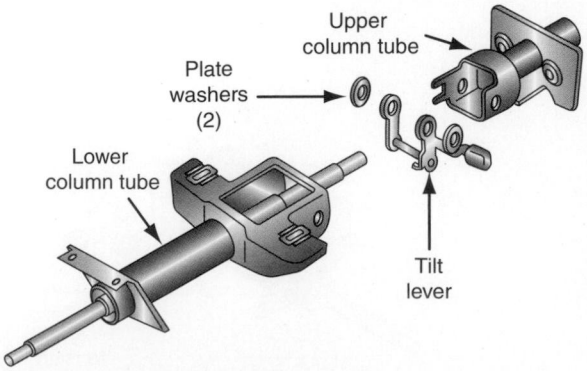

Copyright © 2015 Cengage Learning®

FIGURE 8-23 **Removing upper column tube, tilt lever assembly, and plate washers.**

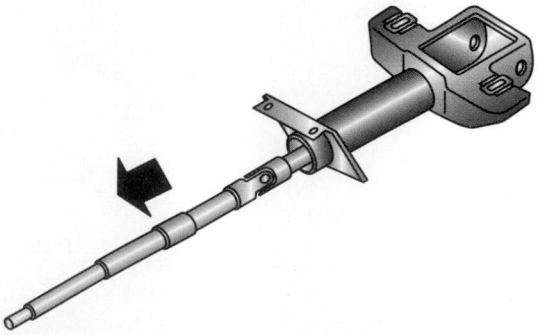

Copyright © 2015 Cengage Learning®

FIGURE 8-24 **Removing steering shaft from the steering column.**

Copyright © 2015 Cengage Learning®

FIGURE 8-25 **Removing snapring above steering shaft bushing.**

Tilt Steering Column Inspection and Parts Replacement

1. Place the ignition key in the switch and move the key through all the switch positions. Be sure the operating pin moves properly in the switch mechanism (Figure 8-26). If the ignition key cylinder must be replaced, turn the key to the accessory (ACC) position. Use a small steel rod to push down on the stop pin near the bottom of the key cylinder (Figure 8-27) and pull the cylinder out of the housing. Install the new key cylinder with the key in the ACC position.
2. Rotate the upper steering shaft bearing and inspect for noise, looseness, and wear (Figure 8-28). If any of these conditions are present, replace the upper tube.
3. Inspect the lower steering shaft bearing for noise, looseness, and wear (Figure 8-29). Replace this bearing if necessary.

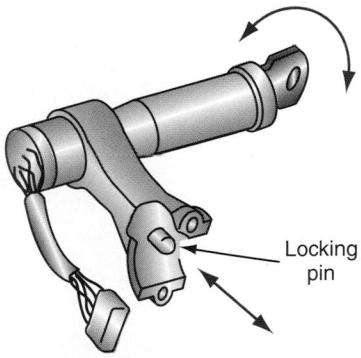

Copyright © 2015 Cengage Learning®

FIGURE 8-26 **Inspecting ignition switch operation.**

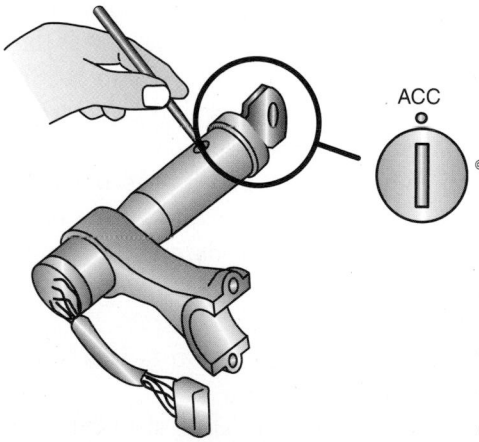

Copyright © 2015 Cengage Learning®

FIGURE 8-27 **Depressing stop pin in release ignition switch cylinder.**

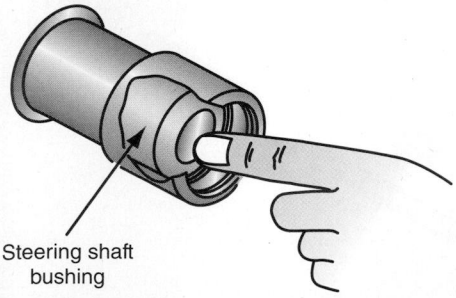

Copyright © 2015 Cengage Learning®

FIGURE 8-28 **Inspecting upper steering shaft bearing.**

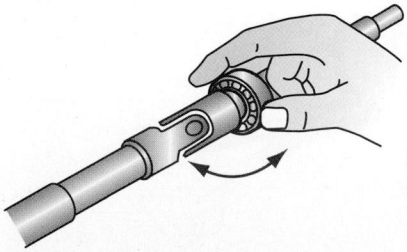

Copyright © 2015 Cengage Learning®

FIGURE 8-29 **Inspecting lower steering shaft bearing.**

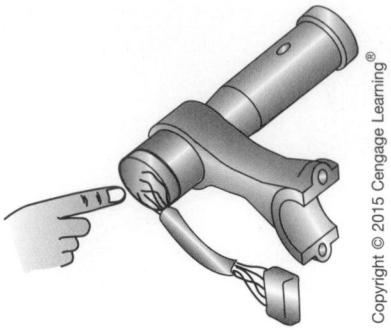
Copyright © 2015 Cengage Learning®

FIGURE 8-30 **Inspecting ignition key interlock solenoid.**

4. Inspect the ignition key interlock solenoid for damaged wires and loose mounting screws (Figure 8-30). Repair or replace this solenoid as required.

Tilt Steering Column Assembly

Molybdenum disulphide lithium-based grease is a special grease that is applied to contacting parts in a tilt steering column.

1. Coat all rubbing parts with **molybdenum disulphide lithium-based grease**. Install the steering shaft bushing and the bushing snapring (Figure 8-31).
2. Install the steering shaft in the lower tube (Figure 8-32).
3. Install the upper column tube, tilt lever mechanism, and two plate washers (Figure 8-33). Install the two pivot bolts.
4. Install the pivot bolt nuts and tighten these nuts to the specified torque (Figure 8-34).
5. Use a brass bar and a hammer to tap the steering shaft into the upper column tube (Figure 8-35).
6. Stake the upper column tube with a pin punch and a hammer (Figure 8-36).
7. Install the snapring in the top of the upper column tube (Figure 8-37).
8. Install the tension spring (Figure 8-38). Assemble the compression spring, bushing, plate, and seat. Use a vise to install the compression spring and related components (Figure 8-39).
9. Install the upper bracket with two new tapered-head bolts. Tighten the tapered-head bolts until the tapered head breaks off (Figure 8-40).
10. Install the protector and wiring harness clamp.
11. Install the steering damper and ignition key illumination.
12. Install the universal joint on the bottom of the steering shaft and tighten the retaining bolt to the specified torque.
13. Install the combination switch and wiring harness and tighten the retaining screw to the specified torque.
14. Install the spiral cable, steering wheel, and air bag deployment module as mentioned earlier in this chapter.

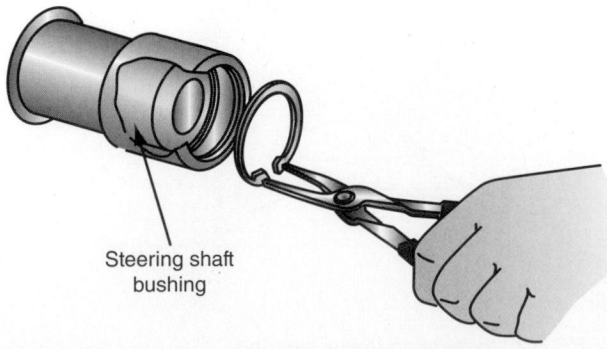

Copyright © 2015 Cengage Learning®

FIGURE 8-31 **Installing steering shaft bushing and snapring.**

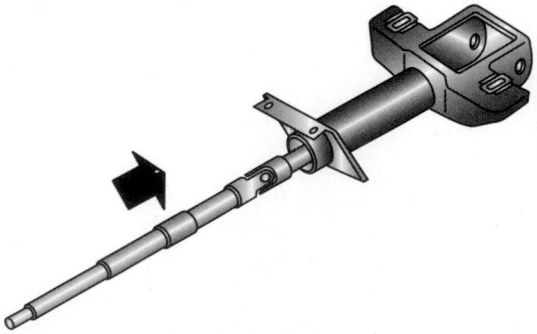
Copyright © 2015 Cengage Learning®

FIGURE 8-32 **Installing the steering shaft in the lower tube.**

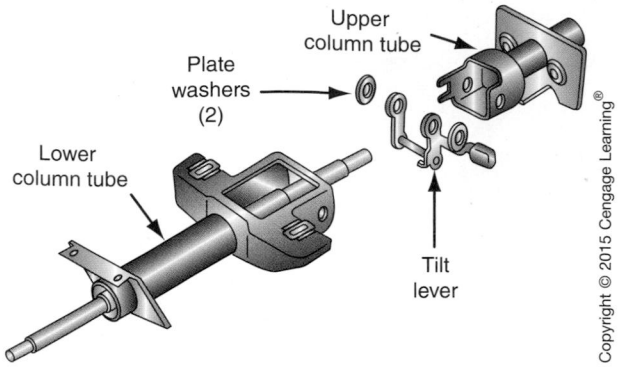

Copyright © 2015 Cengage Learning®

FIGURE 8-33 **Installing upper column tube, tilt lever mechanism, and two plate washers.**

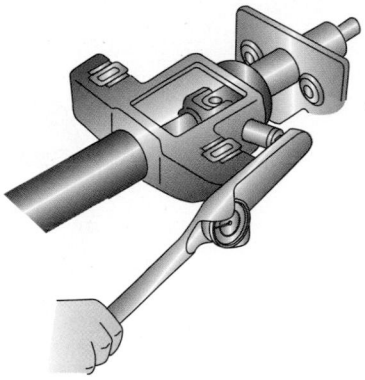

Copyright © 2015 Cengage Learning®

FIGURE 8-34 **Tightening pivot bolt nuts.**

Copyright © 2015
Cengage Learning®

FIGURE 8-35 **Tapping steering shaft into upper column tube.**

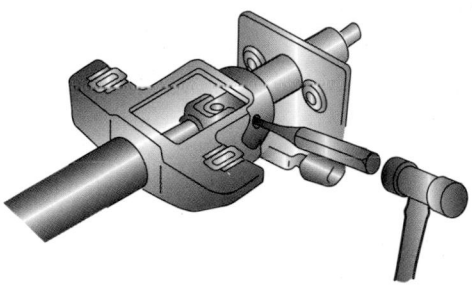

Copyright © 2015 Cengage Learning®

FIGURE 8-36 **Staking upper column tube.**

Copyright © 2015 Cengage Learning®

FIGURE 8-37 **Installing snapring in upper column tube.**

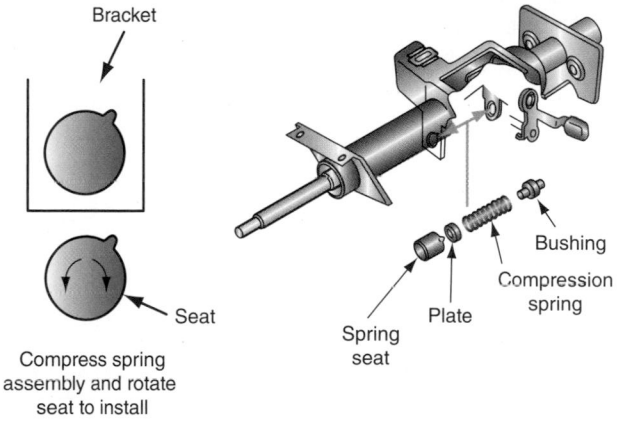

Copyright © 2015 Cengage Learning®

FIGURE 8-38 **Installing tension spring.**

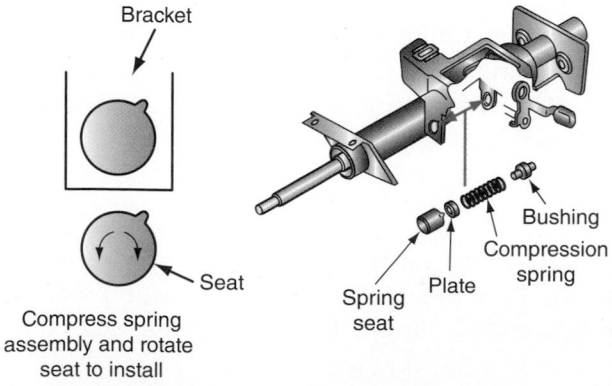

Copyright © 2015 Cengage Learning®

FIGURE 8-39 **Installing compression spring.**

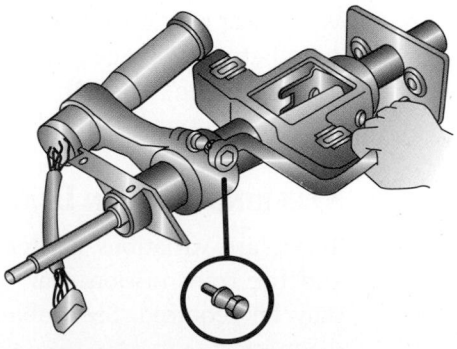

Copyright © 2015 Cengage Learning®

FIGURE 8-40 **Tightening tapered-head bolts in upper bracket.**

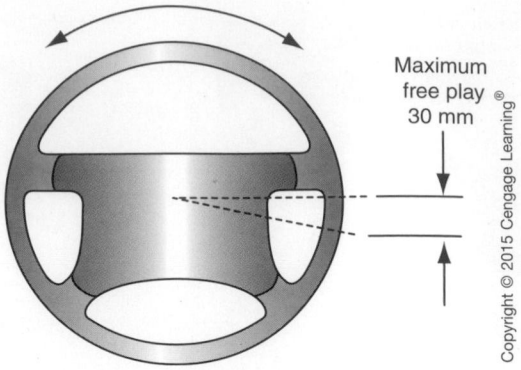

FIGURE 8-41 **Measuring steering wheel free play.**

Steering Column Flexible Coupling and Universal Joint Diagnosis and Service

Checking Steering Wheel Free Play

With the engine stopped and the front wheels in the straight-ahead position, move the steering wheel in each direction with light finger pressure. Measure the amount of steering wheel movement before the front wheels begin to turn (Figure 8-41). This movement is referred to as **steering wheel free play**. On some vehicles, this measurement should not exceed 1.18 in. (30 mm). Always refer to the vehicle manufacturer's specifications. Excessive steering wheel free play is caused by worn steering shaft universal joints or flexible coupling. Other causes of excessive steering wheel free play include worn steering linkage mechanisms and a worn or out of adjustment steering gear.

Steering wheel free play is the amount of steering wheel movement before the front wheels start to turn.

A worn universal joint or flexible coupling in the steering column may also cause rattling noises. The rattling noises may occur while driving the vehicle straight ahead on irregular road surfaces. With the normal vehicle weight resting on the front suspension, observe the flexible coupling or universal joint as an assistant turns the steering wheel 1/2 turn in each direction. If the vehicle has power steering, the engine should be running with the gear selector in Park. The flexible coupling or universal joint must be replaced if there is free play in this component.

Flexible Coupling Replacement

If the flexible coupling must be replaced, loosen the coupling-to-steering-gear-stub-shaft bolt. Disconnect the steering column from the instrument panel, and move the column rearward until the flexible coupling can be removed from the steering column shaft. Remove the coupling-to-steering-shaft bolts, and disconnect the coupling from the shaft. When the new coupling and the steering column are installed on some vehicles, the clearance between the coupling clamp and the steering gear adjusting plug should be 1/16 in. (1.5 mm) (Figure 8-42). This specification may vary depending on the vehicle. Always use the vehicle manufacturer's specifications in the service manual.

Steering Column Diagnosis

There are variations in steering columns depending on the vehicle, type of transmission, and the transmission gear selector position. Thus different column diagnostic procedures may be required. See Table 8-1, Table 8-2, and Table 8-3 for a typical steering column diagnosis.

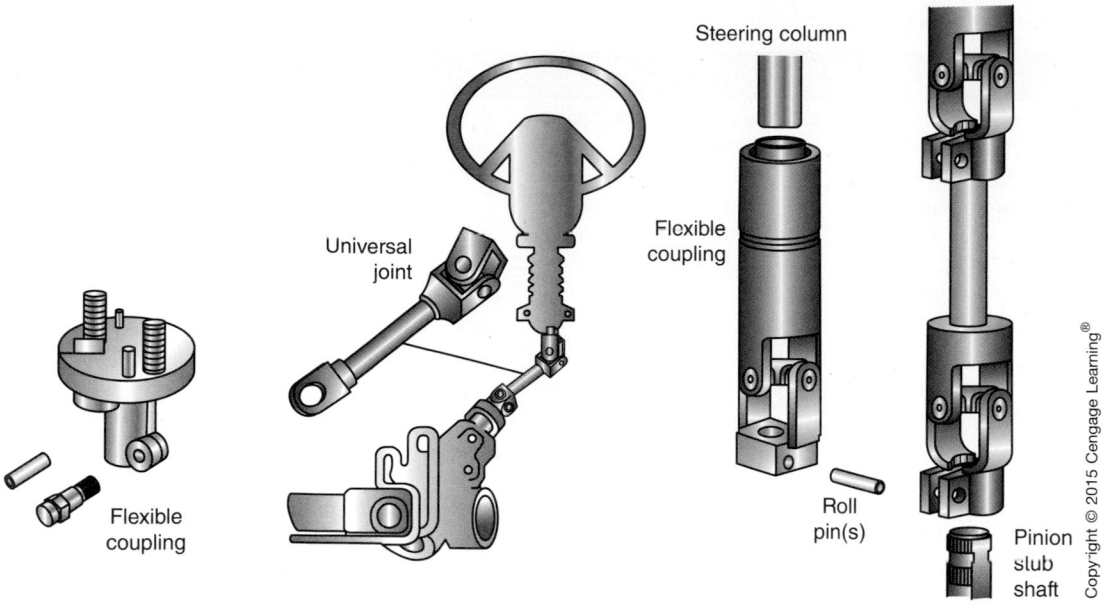

FIGURE 8-42 **Flexible couplers.**

TABLE 8-1 **AUTOMATIC TRANSMISSION—STEERING COLUMN DIAGNOSIS**

Condition	Possible Cause	Correction
Lock system—will not unlock	1. Lock bolt damaged	1. Replace lock bolt.
	2. Defective lock cylinder	2. Replace or repair lock cylinder.
	3. Damaged housing	3. Replace housing.
	4. Damaged or collapsed sector	4. Replace sector.
	5. Damaged rack	5. Replace rack.
	6. Shear flange on sector shaft collapsed	6. Replace.
Lock system—will not lock	1. Lock bolt spring broken or defective	1. Replace spring.
	2. Damaged sector tooth, or sector installed incorrectly	2. Replace, or install correctly.
	3. Defective lock cylinder	3. Replace lock cylinder.
	4. Burr on lock bolt or housing	4. Remove burr.
	5. Damaged housing	5. Replace housing.
	6. Transmission linkage adjustment incorrect	6. Readjust.
	7. Damaged rack	7. Replace rack.
	8. Interference between bowl and coupling (tilt)	8. Adjust or replace as necessary.
	9. Ignition switch stuck	9. Readjust or replace.
	10. Actuator rod restricted or bent	10. Readjust or replace.
Lock system—high effort	1. Lock cylinder defective	1. Replace lock cylinder.
	2. Ignition switch defective	2. Replace switch.
	3. Rack preload spring broken or deformed	3. Replace spring.
	4. Burr on sector, rack, housing, support, tang of shift gate, or actuator rod coupling	4. Remove burr.
	5. Bent sector shaft	5. Replace shaft.
	6. Distorted rack	6. Replace rack.
	7. Misalignment of housing to cover (tilt)	7. Replace either or both.
	8. Distorted coupling slot in rack (tilt)	8. Replace rack.
	9. Bent or restricted actuator rod	9. Straighten, remove restriction, or replace.
	10. Ignition switch mounting bracket bent	10. Straighten or replace.

(*Continued*)

TABLE 8-1 (Continued)

Condition	Possible Cause	Correction
Lock cylinder—high effort between Off and Off-lock positions	1. Burr on tang of shift gate 2. Distorted rack	1. Remove burr. 2. Replace rack.
Sticks in Start position	1. Actuator rod deformed 2. Any high effort condition	1. Straighten or replace. 2. Check items under high effort section.
Key cannot be removed in Off-lock position	1. Ignition switch not set correctly 2. Defective lock cylinder	1. Readjust ignition switch. 2. Replace lock cylinder.
Lock cylinder can be removed without depressing retainer	1. Lock cylinder with defective retainer 2. Lock cylinder without retainer 3. Burr over retainer slot in housing cover	1. Replace lock cylinder. 2. Replace lock cylinder. 3. Remove burr.
Lock bolt hits shaft lock in Off and Park positions	Ignition switch not set correctly	Readjust ignition switch.
Ignition system—electrical system does not function	1. Defective fuse in "accessory" circuit 2. Connector body loose or defective 3. Defective wiring 4. Defective ignition switch 5. Ignition switch not adjusted properly	1. Replace fuse. 2. Tighten or replace. 3. Repair or replace. 4. Replace ignition switch. 5. Readjust ignition switch.
Switch does not actuate mechanically	Defective ignition switch	Replace ignition switch.
Switch cannot be set correctly	1. Switch actuator rod deformed 2. Sector to rack engaged in wrong tooth (tilt)	1. Repair or replace switch actuator rod. 2. Engage sector to rack correctly.
Noise in column	1. Coupling bolts loose 2. Column not correctly aligned 3. Coupling pulled apart 4. Sheared intermediate shaft plastic joint 5. Horn contact ring not lubricated 6. Lack of grease on bearings or bearing surfaces 7. Lower shaft bearing tight or frozen 8. Upper shaft tight or frozen 9. Shaft lock plate cover loose 10. Lock plate snapring not seated 11. Defective buzzer dog cam on lock cylinder 12. One click when in Off-lock position and the steering wheel is moved	1. Tighten pinch bolts to specified torque. 2. Realign column. 3. Replace coupling and realign column. 4. Replace or repair steering shaft and realign column. 5. Lubricate with Lubriplate. 6. Lubricate bearings. 7. Replace bearing. Check shaft and replace if scored. 8. Replace housing assembly. 9. Tighten three screws or, if missing, replace. CAUTION: Use specified screws (15 in.-lbs). 10. Replace snapring. Check for proper seating in groove. 11. Replace lock cylinder. 12. Normal condition: lock bolt is seating.
Steering shaft—high effort	1. Column assembly misaligned in vehicle 2. Improperly installed or deformed dust seal 3. Tight or frozen, upper or lower bearing 4. Flash on ID of shift tube from plastic joint	1. Realign. 2. Remove and replace. 3. Replace affected bearing or bearings. 4. Replace shift tube.

Condition	Possible Cause	Correction
High shift effort	1. Column not aligned correctly in car	1. Realign.
	2. Improperly installed dust seal	2. Remove and replace.
	3. Lack of grease on seal or bearing areas	3. Lubricate bearings and seals.
	4. Burr on upper or lower end of shift tube	4. Remove burr.
	5. Lower bowl bearing not assembled properly (tilt)	5. Reassemble properly.
	6. Wave washer with burrs (tilt)	6. Replace wave washer.
Improper transmission shifting	1. Sheared shift tube joint	1. Replace shift tube assembly.
	2. Improper transmission linkage adjustment	2. Readjust linkage.
	3. Loose lower shift lever	3. Replace shift tube assembly.
	4. Improper gate plate	4. Replace with correct part.
	5. Sheared lower shift lever weld	5. Replace tube assembly.
Lash in mounted column assembly	1. Instrument panel mounting bolts loose	1. Tighten to specifications (20 ft.-lbs).
	2. Broken weld nuts on jacket	2. Replace jacket assembly.
	3. Instrument panel bracket capsule sheared	3. Replace bracket assembly.
	4. Instrument panel to jacket mounting bolts loose	4. Tighten to specifications (15 ft.-lbs).
	5. Loose shoes in housing (tilt)	5. Replace.
	6. Loose tilt head pivot pins (tilt)	6. Replace.
	7. Loose shoe lock pin in support (tilt)	7. Replace.
Miscellaneous	1. Housing loose on jacket noticed with ignition in Off-lock position and a torque applied to the steering wheel	1. Tighten four mounting screws (60 in.-lbs).
	2. Shroud loose on shift bowl	2. Bend tabs on shroud over lugs on bowl.

Copyright © 2015 Cengage Learning®

TABLE 8-2 **MANUAL TRANSMISSION—STEERING COLUMN DIAGNOSIS**

Condition	Possible Cause	Correction
Shift lever sticking	1. Defective upper shift lever	1. Replace shift lever.
	2. Defective shift lever gate	2. Replace shift lever gate.
	3. Loose relay lever on shift tube	3. Replace shift tube assembly.
	4. Wrong shift lever	4. Replace with current lever.
High shift effort	1. Column not aligned correctly	1. Realign column.
	2. Lower bowl bearing not assembled correctly	2. Reassemble correctly.
	3. Improperly installed seal	3. Remove and replace.
	4. Wave washer in lower bowl bearing defective	4. Replace wave washer.
	5. Improper adjustment of lower shift levers	5. Readjust.
	6. Lack of grease on seal, bearing areas, or levers	6. Lubricate seal, levers, and bearings.
	7. Damaged shift tube in bearing areas	7. Replace shift tube assembly.
Improper transmission shifting	Loose relay lever on shift tube	Replace shift tube assembly.

Copyright © 2015 Cengage Learning®

TABLE 8-3 **MANUAL TRANSMISSION—TILT COLUMN DIAGNOSIS**

Condition	Possible Cause	Correction
Housing scraping on bowl	Bowl bent or not concentric with hub	Replace bowl.
Steering wheel loose	1. Excessive clearance between holes in support or housing and pivot pin diameters 2. Defective or missing antilash spring in spheres 3. Upper bearing seat not seating in bearing 4. Upper bearing inner race seat missing 5. Loose support screws 6. Bearing preload spring missing or broken	1. Replace either or both. 2. Add spring or replace both. 3. Replace both. 4. Install seat. 5. Tighten to 60 in.-lbs. 6. Replace preload spring.
Steering wheel loose every other tilt position	Loose fit between shoe and shoe pivot pin	Replace both.
Noise when tilting column	1. Upper tilt bumper worn 2. Tilt spring rubbing in housing or dirt	1. Replace tilt bumper. 2. Lubricate.
Steering column not locking in any tilt position	1. Shoe seized on its pivot pin 2. Shoe grooves might have burrs 3. Shoe lock spring weak or broken	1. Replace shoe and pivot pin. 2. Replace shoe. 3. Replace lock spring.
Steering wheel fails to return to top tilt position	1. Pivot pins bound up 2. Wheel tilt spring defective 3. Turn signal switch wires too tight	1. Replace pivot pins. 2. Replace tilt spring. 3. Reposition wires.

Copyright © 2015 Cengage Learning®

Rack and Pinion Steering Gear Tie Rod Service

Follow these steps for rack and pinion steering inner tie rod replacement:

1. Remove the outer tie rod end from the steering arm.
2. Insert a short piece of wire through each outer tie rod end cotter pin hole and connect a pull scale to this wire. Pull upward on the scale to check the tie rod articulation effort (Figure 8-43). If this effort is not within specifications, replace the inner tie rod end.
3. Mark the outer tie rod and jam nut position with masking tape wrapped around the tie rod threads next to the jam nut, or place a dab of paint on the jam nut and tie rod threads (Figure 8-44). Loosen the jam nuts and remove the outer tie rod ends.

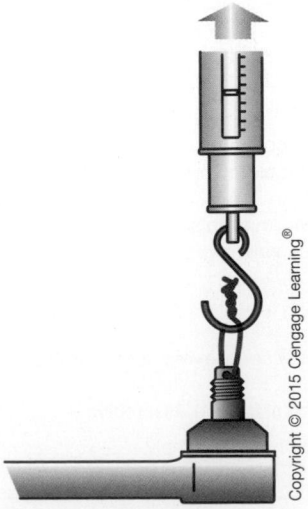

FIGURE 8-43 **Measuring inner tie rod articulation effort.**

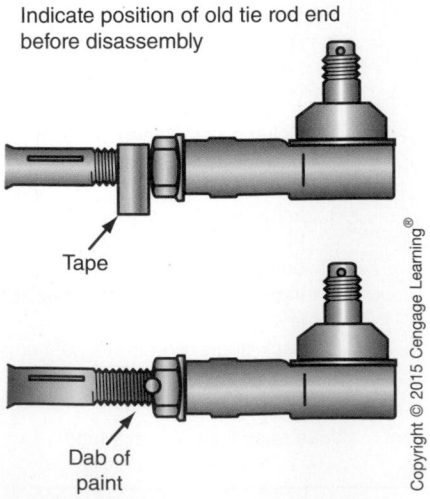

FIGURE 8-44 **Before removing outer tie rod mark the jam nut location or count thread turns to remove.**

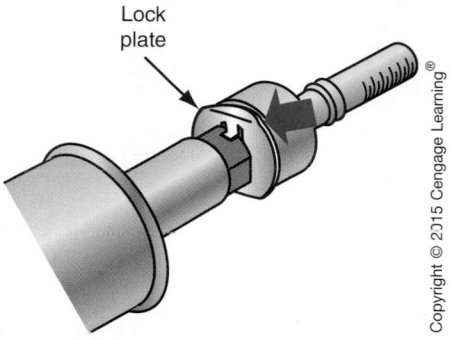

Copyright © 2015 Cengage Learning®

FIGURE 8-45 **Straighten the lock plate to allow removal of inner tie rod end.**

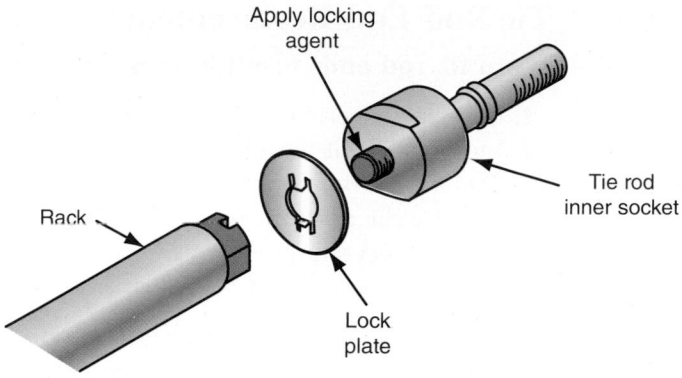

Copyright © 2015 Cengage Learning®

FIGURE 8-46 **Removing inner tie rod end from rack.**

4. Remove the inner and outer bellows boot clamps and pull the bellows boots from the tie rods.
5. Straighten the lock plate where it is bent over the inner tie rod end (Figure 8-45). Hold the rack with a wrench, and use the proper size wrench to remove the inner tie rod end from the rack (Figure 8-46).A jam nut is used in place of the lock ring on some inner tie rod ends, and a roll pin retains some inner tie rod ends to the rack. Various inner tie rod socket removal procedures are required depending on the socket design.
6. Rotate the inner tie rod ends onto the rack until they bottom. Install the inner tie rod pins, or stake these ends as required. Always use a wooden block to support the opposite side of the rack and inner tie rod end while staking. If jam nuts are located on the inner tie rod ends, be sure they are tightened to the specified torque.
7. Place a large bellows boot clamp over each end of the gear housing. Install the bellows boots, and be sure the boots are seated in the housing and the tie rod undercuts. Install and tighten the large inner boot clamps.
8. Install the outer bellows boot clamps on the tie rods, but do not install these clamps on the boots until the steering gear is installed and the toe is adjusted.
9. Install the jam nuts and the outer tie rod ends. Align the marks placed on these components during the disassembly procedure. Leave the jam nuts loose until the steering gear is installed and the toe is adjusted.
10. Install the steering gear in the vehicle and check the front wheel toe. Tighten the outer bellows boot clamps and the outer tie rod end jam nuts.

Classroom Manual
Chapter 8, page 202

Steering Linkage Diagnosis and Service

Diagnosis of Center Link, Pitman Arm, and Tie Rod Ends

The vehicle should be raised and safety stands positioned under the lower control arms to support the vehicle weight. Use vertical hand force to check for looseness in all the pivots on the tie rod ends and the center link. Inspect the seals on each tie rod end and pivot on the center link or pitman arm for damage and cracks. Cracked seals allow dirt to enter the pivoted joints, which results in rapid wear. If looseness or damaged seals are found on any pivoted joint on the tie rods and center link, these components must be replaced.

The second part of this diagnosis is done with the front wheels resting on the shop floor. If the vehicle is equipped with power steering, start the engine and allow the engine to idle with the transmission in Park and the parking brake applied. While someone turns the steering wheel one-quarter turn in each direction from the straight-ahead position, observe all the pivoted joints on the tie rod ends and center link. This test allows the technician to check the steering linkage pivots under load. If any of the pivoted joints show a slight amount of play, they must be replaced.

Front wheel shimmy may be defined as a consistent, fast, side-to-side movement of the front wheels and steering wheel. This movement is usually experienced at speeds above 40 mph (25 km/h), and it may occur more frequently while driving on irregular road surfaces.

Tie Rod End Replacement

Worn tie rod ends result in these problems:

1. Excessive steering wheel free play
2. Incorrect front wheel toe setting
3. Tire squeal on turns
4. Tread wear on front tires
5. Front wheel shimmy
6. Rattling noise on road irregularities

SPECIAL TOOL
Tie rod end puller

The cotter pin and nut must be removed prior to tie rod end replacement. A puller is used to remove the tie rod end from the steering arm (Figure 8-47). Tie rod ends with rubber-encapsulated ball studs require special inspection and diagnostic procedures (Figure 8-48). On this type of tie rod end, inspect for looseness of the ball stud in the rubber capsule and looseness of the stud and rubber capsule in the outer housing. If any looseness is present, replace the tie rod end. Tie rod end replacement is also necessary if there is any indication of the rubber capsule starting to come out of the outer housing.

After the tie rod end is removed from the steering arm, install two nuts on the stud threads and tighten these nuts against each other. Use the proper size of socket and a torque wrench to rotate the ball stud through a 40 degree arc (Figure 8-49). If the ball stud turning torque is less than 20 ft.-lb. (27 N•m), replace the tie rod end.

The tie rod clamp must be loosened before the tie rod end is removed from the sleeve. Count the number of turns required to remove the tie rod end from the sleeve, and install the new tie rod with the same number of turns. Even when this procedure is followed, the toe must be checked after the steering linkage components are replaced. Before the new tie rod end is installed, center the stud in the tie rod end. When the tie rod end stud is installed in the steering arm opening, only the threads should be visible above the steering arm surface. If the machined surface of the tie rod end stud is visible above the steering arm surface, or if the stud fits loosely in the steering arm opening, this opening is worn or the tie rod end is not correct for that application. The tie rod end nut must be torqued to the manufacturer's specifications, and the cotter pin must be installed through the tie rod end and nut openings (Figure 8-50). Photo Sequence 14 illustrates the procedure for removing and replacing an outer tie rod end.

The tie rod end nut must never be loosened from the specified torque to install the cotter pin. Another method of positioning replacement tie rod ends is to measure the

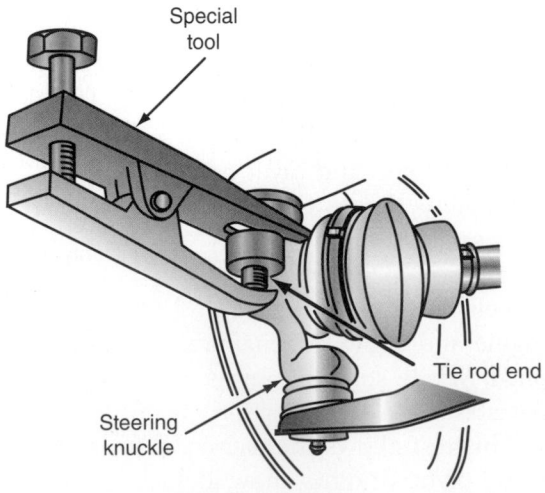

FIGURE 8-47 **Removing tie rod end.**

Copyright © 2015 Cengage Learning®

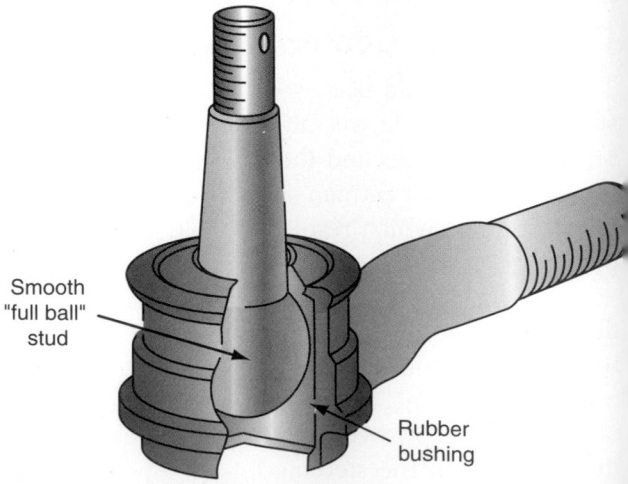

FIGURE 8-48 **Tie rod end with rubber encapsulated ball stuc**

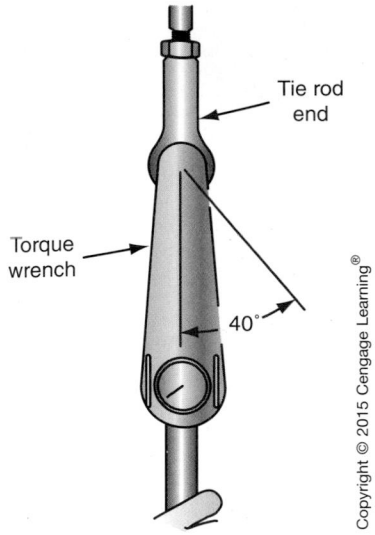

FIGURE 8-49 **Measuring turning torque on a rubber-encapsulated tie rod end.**

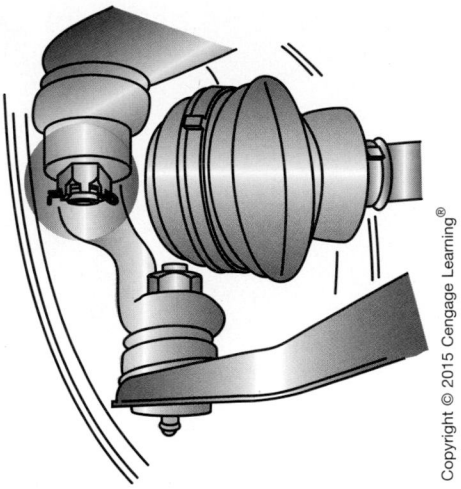

FIGURE 8-50 **Tie rod nut and cotter pin installation.**

CAUTION:
If rubber-encapsulated tie rod ends are tightened with the front wheels in any position but straight ahead, steering pull and wander may occur.

distance from the center of the tie rod stud to the end of the sleeve prior to removal. When the new tie rod end is installed, be sure this measurement is the same. The slots in the tie rod sleeve must be positioned away from the opening in the sleeve clamps (Figure 8-51). Leave the sleeve clamps loose until the front wheel toe is checked, and then tighten the sleeve clamp bolts to the specified torque. A special tool is available to rotate the tie rod sleeves and set the front wheel toe (Figure 8-52).

When rubber-encapsulated tie rod ends are installed and tightened to the specified torque, the front wheels must be straight ahead. Tightening this type of tie rod end with the front wheels in any other position but straight ahead may cause steering pull or wander.

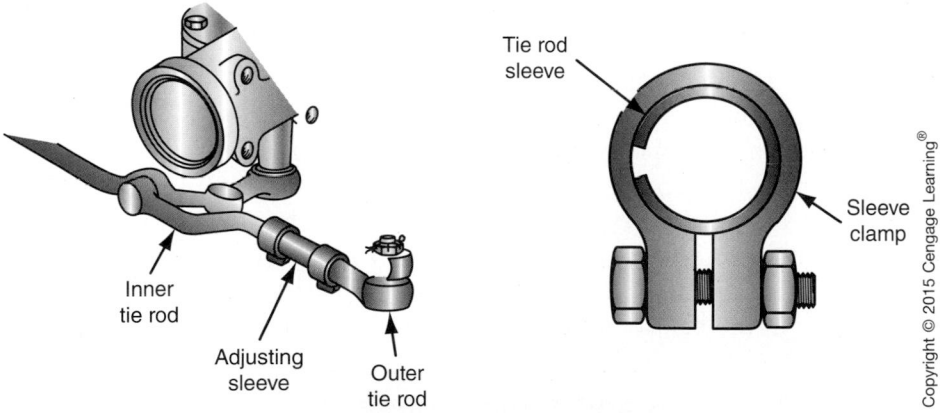

FIGURE 8-51 **Proper slot and clamp position on tie rod sleeves.**

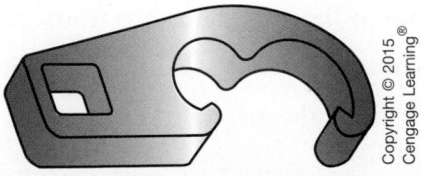

FIGURE 8-52 **Tie rod sleeve-adjusting tool.**

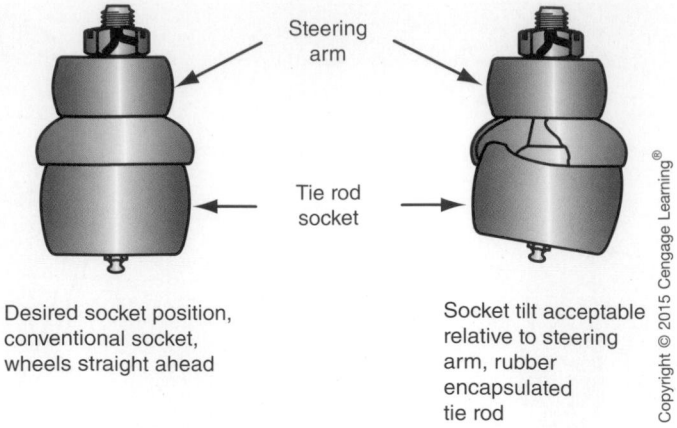

FIGURE 8-53 **Rubber-encapsulated tie rod end housing may be tilted to one side after it is installed and tightened.**

After rubber-encapsulated tie rod ends are tightened to the specified torque, it is acceptable for the tie rod end housing to be tilted to one side (Figure 8-53).

Pitman Arm Diagnosis and Replacement

Some pitman arms contain a ball socket joint on the outer end. The threaded extension on this ball socket fits into the center link. On other steering linkages, the ball socket joint is in the center link, and the threaded extension fits into the pitman arm opening. If the pitman arm is bent, it must be replaced because the tie rod is not parallel to the lower control arm. Under this condition, excessive front wheel toe change occurs on road irregularities, and front tire wear may be excessive.

The following is a typical pitman arm replacement procedure:

SPECIAL TOOL
Pitman arm puller

1. Position the front wheels straight ahead, and remove the cotter pin and nut from the ball socket joint on the outer end of the pitman arm.
2. Remove the ball socket extension from the pitman arm or center link with a tie rod end puller.
3. Loosen the pitman arm-to-pitman shaft nut.
4. Use a puller to pull the pitman arm loose on the shaft.
5. Remove the nut, lock washer, and pitman arm.
6. Check the pitman shaft splines. If the splines are damaged or twisted, the shaft must be replaced.
7. Reverse steps 1 through 5 to install the pitman arm. The pitman arm-to-shaft nut and the ball socket extension nut must be tightened to the manufacturer's specified torque. Be sure the pitman arm is installed in the correct position on the shaft splines. Install the cotter pin in the ball socket extension.

Center Link Diagnosis and Replacement

A bent center link must be replaced. Do not attempt to straighten this rod. If the ball socket joints are loose on either end of the rod, center link assembly replacement is necessary. If the ball stud openings in the center link are worn, replace the center link.

Follow these steps for a typical center link replacement:

1. Remove the cotter pins from the tie rod-to-center-link nuts, and the idler arm and pitman-arm-to-center-link nuts.
2. Remove the nuts on the tie rod inner ends, idler-arm-to-center-link ball socket extension, and the pitman-arm-to-center-link ball socket extension.

PHOTO SEQUENCE 14

Diagnosing, Removing, and Replacing an Outer Tie Rod End on a Vehicle with a Parallelogram Steering Linkage

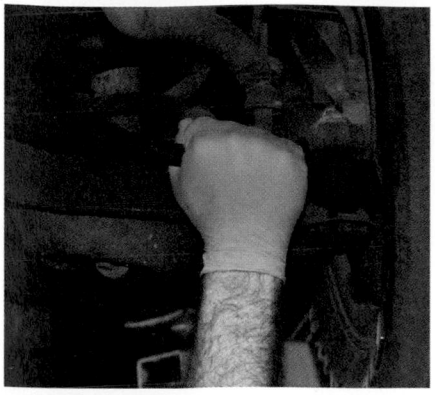

P14-1 **Raise the vehicle on a lift and check for excessive vertical movement in the left outer tie rod end.**

Copyright © 2015 Cengage Learning®

P14-2 **Check for lateral movement in the left outer tie rod end.**

Copyright © 2015 Cengage Learning®

P14-3 **Visually inspect the left outer tie rod end seal for cracks and damage.**

Copyright © 2015 Cengage Learning®

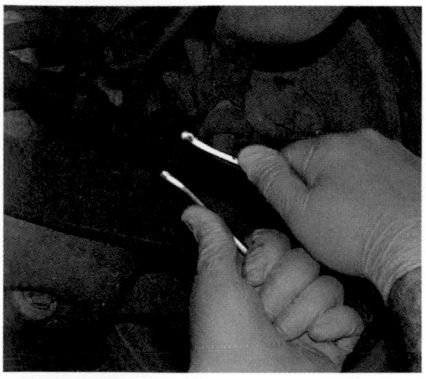

P14-4 **Loosen the nut on the outer bolt in the left tie rod sleeve.**

Copyright © 2015 Cengage Learning®

P14-5 **Use a pair of side cutters to remove the cotter pin in the left outer tie rod end retaining nut.**

Copyright © 2015 Cengage Learning®

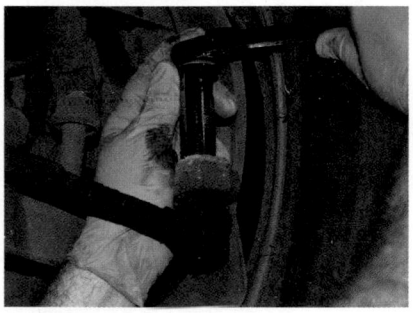

P14-6 **Use the proper size socket and ratchet to remove the retaining nut on the outer tie rod end.**

Copyright © 2015 Cengage Learning®

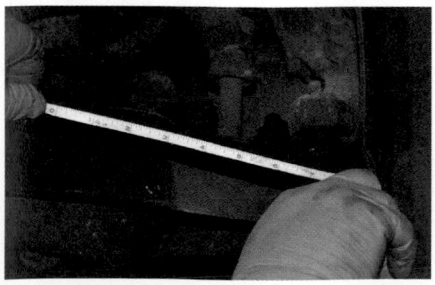

P14-7 **Use a tape measure to carefully measure the distance from the outer edge of the left tie rod sleeve to the outer edge of the tie rod end, and record this distance.**

Copyright © 2015 Cengage Learning®

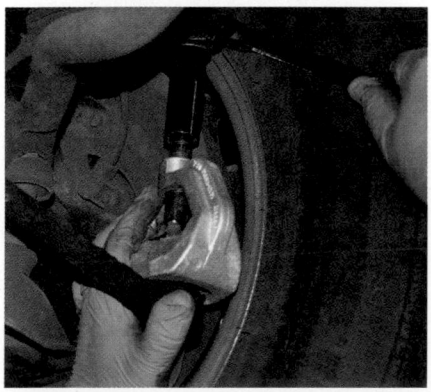

P14-8 **Install a tie rod end puller on the left outer tie rod end and tighten the puller bolt to remove the tie rod end from the steering arm.**

Copyright © 2015 Cengage Learning®

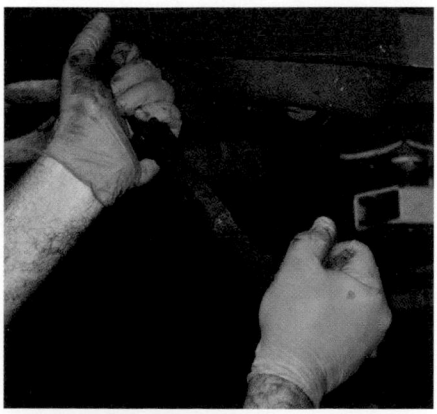

P14-9 **After the tie rod end is removed from the steering arm, rotate the outer tie rod end to remove it from the tie rod sleeve.**

Copyright © 2015 Cengage Learning®

P14-10 **Inspect the outer tie rod sleeve for thread damage.**

Copyright © 2015 Cengage Learning®

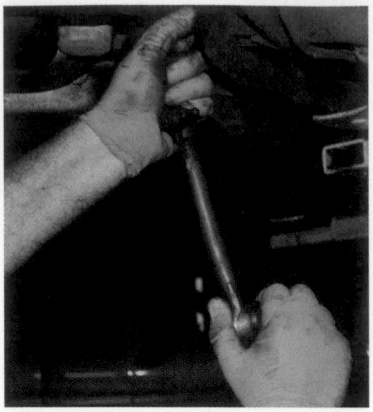

P14-11 **Thread the new tie rod end into the outer tie rod sleeve until the distance from the outer edge of the sleeve to the outer edge of the tie rod end is exactly the same as the original distance measured and recorded previously.**

Copyright © 2015 Cengage Learning®

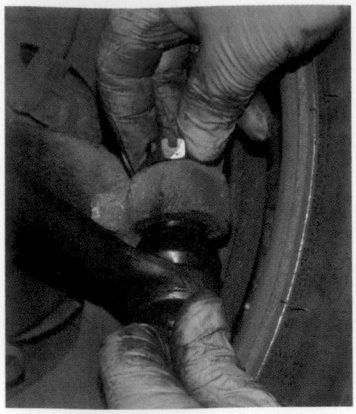

P14-12 **Be sure the seal is properly installed on the new tie rod end, and install the outer tie rod end in the steering arm opening and install the nut to retain the tie rod end.**

Copyright © 2015 Cengage Learning®

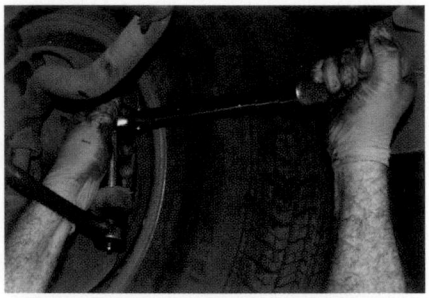

P14-13 **Use a torque wrench and the proper size socket to tighten the left outer tie rod nut to the specified torque. Be sure the serrated openings in the nut are aligned with the hole in the tie rod stud.**

Copyright © 2015 Cengage Learning®

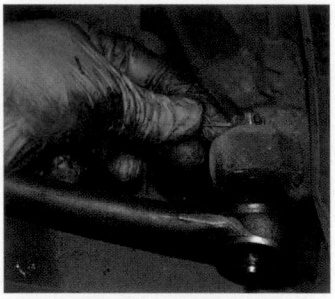

P14-14 **Install the cotter pin in the left outer tie rod nut and bent the ends of this cotter pin around the nut.**

Copyright © 2015 Cengage Learning®

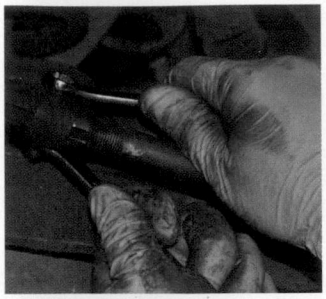

P14-15 **Position the outer tie rod sleeve clamp so the clamp opening is positioned away from the slots in the tie rod sleeve, and tighten the clamp bolt nut to the specified torque.**

Copyright © 2015 Cengage Learning®

P14-16 **Inflate all the vehicle tires to the specified pressure, and position the vehicle properly on a wheel alignment ramp. Use the wheel aligner to check the front wheel toe. Adjust the front wheel toe if necessary.**

Copyright © 2015 Cengage Learning®

3. Use a tie rod end puller to pull the inner tie rods from the center link. Follow the same procedure to remove the center-link-to-pitman-arm ball socket extension.
4. Remove the idler arm from the center link; then remove the center link.
5. Reverse steps 1 through 4 to install the center link. Tighten all the ball socket nuts to the manufacturer's specified torque, and install cotter pins in all the nuts. If the ball sockets have grease fittings, lubricate the ball sockets with a grease gun and chassis lubricant.

SERVICE TIP: Never attempt to straighten steering linkage components. This action may weaken the metal and cause sudden component failure, vehicle damage, and personal injury.

Idler Arm Diagnosis

To measure idler arm vertical movement, attach the magnetic base of a dial indicator to the frame near the idler arm. Position the dial indicator stem against the upper side on the outer end of the idler arm. Preload the dial indicator stem and zero the dial. Use a pull scale to apply 25 lbs. (11.34 kg) of force downward and upward on the idler arm (Figure 8-54). Observe the total vertical idler arm movement on the dial indicator. If this vertical movement exceeds the vehicle manufacturer's specifications, replace the idler arm. Typical maximum vertical idler arm movement from the downward to the upward position is 0.250 in. (63.5 mm). If idler arm vertical movement is excessive, the tie rod is not parallel to the lower control arm.

Excessive idler arm vertical movement causes these steering problems:

1. Excessive toe change and front tire tread wear
2. Excessive steering wheel free play and reduced steering control
3. Front end shimmy

Binding idler arm bushings result in these complaints:

1. Hard steering
2. Squawking noise when the front wheels are turned
3. Poor steering wheel returnability

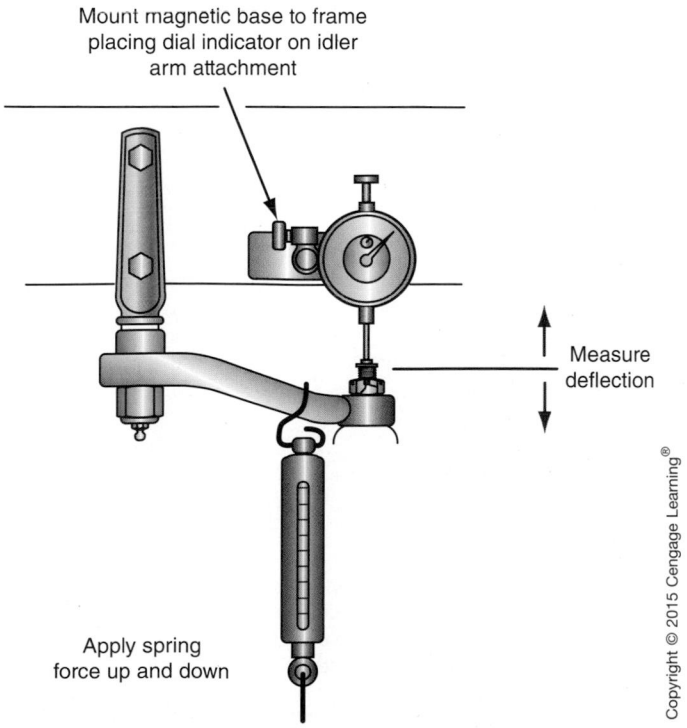

FIGURE 8-54 **Measuring idler arm vertical movement.**

Some idler arms contain steel bushings, whereas others are equipped with rubber bushings.

SERVICE TIP:
A binding idler arm may suddenly break off and cause complete loss of steering control, vehicle damage, and personal injury.

The following is a typical idler arm removal and replacement procedure:

1. Remove the idler-arm-to-center-link cotter pin and nut.
2. Remove the center link from the idler arm.
3. Remove the idler arm bracket mounting bolts, and remove the idler arm.
4. If the idler arm has a steel bushing, thread the bracket into the idler arm bushing until the specified clearance is obtained between the center of the lower bracket bolt hole and the upper idler arm surface (Figure 8-55).
5. Install the idler arm bracket to frame bolts, and tighten the bolts to the specified torque. Be sure that lock washers are installed on the bolts.
6. Install the center link into the idler arm, and tighten the mounting nut to the specified torque. Install the cotter pin in the nut.
7. If the idler arm contains a grease fitting, lubricate as required.

The idler arm adjustment is very important. If this adjustment is incorrect, front wheel toe is affected. After idler arm replacement, the front wheel toe should be checked.

Steering Damper Diagnosis and Replacement

Some steering systems have a damper connected between the center link and the chassis. A damper is similar to a small shock absorber. The purpose of the damper is to prevent the transfer of steering shock and vibrations to the steering wheel. A worn-out steering damper may cause excessive steering shock and vibration on the steering wheel, especially on irregular road surfaces. A rattling noise occurs if the damper mounting bolts or brackets are loose.

The following is a typical steering damper checking and replacement procedure:

1. Lift the vehicle on a hoist, and grasp the damper firmly. Apply vertical and horizontal pressure to the damper, and check for movement in the damper mounts. If movement exists, tighten or replace the damper mounting bushings or brackets.
2. Visually inspect the damper for oil leaks. A slight film of oil on the damper body near the shaft seal is acceptable. If there is any indication of oil dripping from the damper, the unit must be replaced.

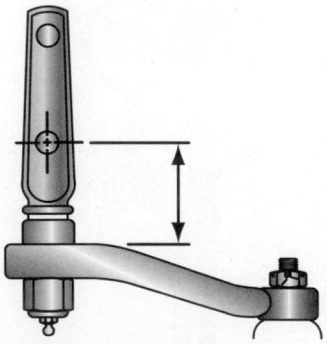

Copyright © 2015 Cengage Learning®

FIGURE 8-55 **Specified clearance between center of lower bracket bolt hole and upper idler arm surface.**

3. Disconnect one end of the damper, and pull the damper back and forth horizontally. The damper should offer a slight equal resistance to movement in either direction. When this resistance is not felt in one or both directions, replace the damper.
4. To replace the damper, remove the mounting bolts from the chassis and the center link.
5. When the new damper is installed, tighten the mounting bolts to the specified torque. Turn the steering wheel fully in each direction, and be sure the damper does not restrict linkage movement.

Steering Arm Diagnosis

Classroom Manual
Chapter 8, page 204

If the front rims have been damaged, the steering arms should be checked for a bent condition. Measure the distance from the center of the tie rod end stud to the edge of the rim on each side. Unequal readings may indicate a bent steering arm. Bent steering arms must be replaced. Steering linkage diagnosis is summarized in Table 8-4.

TABLE 8-4 **STEERING LINKAGE DIAGNOSIS**

Condition	Possible Cause	Correction
Excessive play or looseness in steering system	1. Front wheel bearings loosely adjusted 2. Worn steering shaft couplings 3. Worn upper ball joints 4. Steering wheel loose on shaft or loose pitman arm, tie rods, steering arms, or steering linkage ball studs 5. Steering gear thrust bearings loosely adjusted 6. Excessive over-center lash in steering gear 7. Worn intermediate rod or tie rod sockets	1. Adjust bearings to obtain proper end play. 2. Replace part. 3. Check and replace if necessary. 4. Tighten to specified torque, or replace if necessary. 5. Adjust preload to specifications. 6. Adjust preload to specifications per shop manual. 7. Replace worn part.
Excessive looseness in tie rod or intermediate rod pivots, or excessive vertical lash in idler support	Seal damage and leakage resulting in loss of lubricant, corrosion, and excessive wear	Replace damaged parts as necessary. Properly position upon reassembly.
Hard steering—excessive effort required at steering wheel	1. Low or uneven tire pressure 2. Steering linkage or bolt joints need lubrication 3. Tight or frozen intermediate rod, tie rod, or idler socket 4. Steering gear-to-column misalignment 5. Steering gear adjusted too tightly 6. Front wheel alignment incorrect (manual gear)	1. Inflate to specified pressures. 2. Lube with specified lubricant. 3. Lube, replace, or reposition as necessary. 4. Align column. 5. Adjust over-center and thrust bearing preload to specification. 6. Check alignment and correct as necessary.
Poor returnability	1. Steering linkage or ball joints need lubrication 2. Steering gear adjusted too tightly 3. Steering gear-to-column misalignment 4. Front wheel alignment incorrect (caster)	1. Lube with specified lubricant. 2. Adjust over-center and thrust bearing preload to specifications. 3. Align columns. 4. Check alignment and correct as necessary.

Copyright © 2015 Cengage Learning®

TERMS TO KNOW

Accidental air bag deployment

Air bag deployment module

Back-up power supply

Clock spring electrical connector

Collapsible steering column

Molybdenum disulphide lithium-based grease

Sheared injected plastic

Steering wheel free play

Tapered-head bolts

CASE STUDY

A customer complained about excessive steering effort and poor steering wheel returnability after a turn on a 2009 General Motors Sierra truck with power steering. The technician road tested the car and found the customer's description of the problems to be accurate except for one point. The steering continually required excessive steering effort, and the steering wheel did not return properly after a turn. However, the customer did not mention, or possibly did not notice, that a squawking and creaking noise was sometimes heard during a turn.

The technician checked the power steering fluid level and condition, and found this fluid to be in good condition and at the proper level. Next, the technician checked the power steering belt condition and the power steering pump pressure. The belt tension and condition were satisfactory, and the power steering pump pressure was normal. A check of the power steering pump mounting bolts and brackets indicated they were in good condition. The technician disconnected the center link from the pitman arm and rotated the steering wheel with the engine running. Under this condition, the steering wheel turned very easily. Therefore, the technician concluded the excessive steering effort and poor returnability problems were not in the steering column or steering gear. Next, the technician disconnected the idler arm from the center link. When the technician attempted to move the idler arm back and forth, the idler arm had a severe binding problem. After an idler arm replacement, a road test revealed the excessive steering effort and poor returnability problems had disappeared.

ASE-STYLE REVIEW QUESTIONS

1. While servicing a steering column on an air-bag-equipped vehicle the system must be disarmed by:
   A. Locking the crash sensor in the disarmed position.
   B. Rewinding the clock spring.
   C. Using a scan tool to disable the system.
   D. Disconnecting the battery for manufacturer specified amount of time.

2. A steering wheel on an air-bag-equipped vehicle moves from side to side in the column. Which of the following is the most likely cause?
   A. A loose rack mounting bushing.
   B. A worn upper steering shaft bearing.
   C. A worn pitman (sector) shaft.
   D. Loose steering gear box mounting bolts.

3. On a rack and pinion type steering system, the best way to inspect the inner tie rod for wear with the wheels:
   A. Removed from the vehicle and the outer tie rod disconnected.
   B. Resting on the alignment rack turn plates.
   C. With the vehicle lifted off the ground by the frame or unibody and the suspension in the full rebound position.
   D. Steering the wheel to the left.

4. While discussing steering wheel removal:
   Technician A says a steering wheel should be marked in relation to the steering shaft.
   Technician B says the steering wheel should be grasped firmly with both hands and pulled from the shaft.
   Who is correct?
   A. A only
   B. B only
   C. Both A and B
   D. Neither A nor B

5. While discussing collapsible steering column damage:
   Technician A says if the vehicle has been in a frontal collision, the injected plastic in the column jacket, gearshift tube, and steering shaft may be sheared.
   Technician B says after a collision, some steering columns may be shifted on the bracket.
   Who is correct?
   A. A only
   B. B only
   C. Both A and B
   D. Neither A nor B

6. On a rack and pinion steering system when the inner tie rod is replaced the technician should:
   A. Set the steering rack preload.
   B. Center the steering rack piston prior to removing the inner tie rod.
   C. After the repair the power steering system should be bled.
   D. The rack piston should be prevented from rotating.

7. A rattling noise in the steering column and linkage may be caused by:
   A. A bent center link.
   B. A binding idler arm.
   C. A worn steering shaft U-joint.
   D. Loose tie rod sleeve clamps.

8. When diagnosing and servicing tie rod ends:
   A. Rubber-encapsulated tie rod ends should be tightened with the wheels turned fully to the right or left.
   B. The machined part of the tie rod stud should be visible above the surface of the steering arm.
   C. The nut on the tie rod stud may be loosened to install the cotter pin.
   D. If the turning torque on a rubber-encapsulated tie rod end is less than specified, the tie rod end must be replaced.

9. On the steering arm the tapered attachment hole for the tie rod is out of round. The correct repair for this condition is to:
   A. Replace the steering arm and knuckle assembly.
   B. Place a shim under the retaining nut and torque to specifications.
   C. Measure stud movement with a dial indicator and compare to specifications.
   D. Tighten the tie rod retaining nut with an impact gun until all play is removed.

10. An idler arm has 5/8 in. (15.8 mm) vertical movement. To correct this problem it is necessary to:
    A. Install thicker shims between the idler arm and the bracket.
    B. Install an oversize bushing in the idler arm.
    C. Replace the idler arm assembly.
    D. Replace the Belleville washer and spacer under the bushing.

ASE CHALLENGE QUESTIONS

1. A preliminary steering inspection reveals that the steering wheel has nearly 1/2 inch of movement in either direction before the pitman arm begins to move. All of the following could cause this problem EXCEPT:
   A. Coupling joints.
   B. U-joints.
   C. Tight or seized shaft bearing.
   D. Worn or out-of-adjustment steering gear.

2. A customer complains that sometimes the ignition switch will not unlock and the key will not turn. A preliminary inspection reveals that turning the steering wheel does not help but repositioning the key, after several attempts, unlocks the ignition.

   Technician A says the problem could be the column assembly is misaligned.

   Technician B says the problem could be the ignition key is worn.

   Who is correct?

   A. A only
   B. B only
   C. Both A and B
   D. Neither A nor B

3. If the clock spring electrical connector is not properly centered before it is installed:
   A. The steering may pull to one side.
   B. The clock spring electrical connector may be broken.
   C. The air bag may deploy without the vehicle being in an accident.
   D. The steering may have a binding condition.

4. A customer says the steering on his or her vehicle is very hard to turn in both directions. A preliminary inspection reveals no problems with the manual rack and pinion steering gear.

   Technician A says the problem could be a misaligned steering column.

   Technician B says the problem could be tight or frozen tie rod ends.

   Who is correct?

   A. A only
   B. B only
   C. Both A and B
   D. Neither A nor B

5. A customer says his or her steering has a shimmy problem above 50 mph (80 km/h). Preliminary inspection indicates excessive wear on the right front tire.

   Technician A says the problem could be a worn tie rod end.

   Technician B says the problem could be a worn idler arm ball joint.

   Who is correct?

   A. A only
   B. B only
   C. Both A and B
   D. Neither A nor B

JOB SHEET 27

Name _____ **Date** _____

Remove and Replace Air Bag Inflator Module and Steering Wheel

Upon completion of this job sheet, you should be able to remove and replace an air bag inflator module and steering wheel.

NATEF AST/MAST Task Correlation

Suspension and Steering. Steering Systems Diagnosis and Repair,

Task #B-1. Disable and enable supplemental restraint system (SRS). (P-1)

Task #B-2. Remove and replace steering wheel; center/time supplemental restraint system (SRS) coil (clock spring). (P-1)

Tools and Materials

Steering wheel puller
Torque wrench

Describe the Vehicle Being Worked On:

Year _____ Make _____ Model _____

VIN _____ Engine type and size _____

Procedure — Task Completed

1. Turn the ignition switch to the Lock position and place the front wheels facing straight ahead. ☐

2. Remove the negative battery terminal and wait the time period specified by the vehicle manufacturer before servicing air bag components.

   Specified waiting period _____

   After the negative battery cable is disconnected, explain why a waiting period is necessary before working on the air bag system.

   _____

   _____

   _____

   _____

3. Loosen the air bag retaining screws under the steering wheel. ☐

4. Loosen the other two air bag retaining screws under the steering wheel. ☐

   Loosen all the screws until the groove along the screw circumference catches on the screw case.

WARNING: When an air bag deployment module is temporarily stored on the workbench, always place this module face upward. If the air bag deployment module accidentally deployed when facing downward, the module would become a projectile and personal injury may result.

Task Completed

☐ 5. Pull the air bag deployment module from the steering wheel, and disconnect the air bag module electrical connector. Do not pull on the air bag wires in the steering column. Place the air bag deployment module face upward on the workbench.

☐ 6. Disconnect the air bag wiring retainer in the steering wheel.

☐ 7. Use the proper size socket and a ratchet to remove the steering wheel retaining nut.

☐ 8. Observe the matching alignment marks on the steering wheel and the steering shaft. If these alignment marks are not present, place alignment marks on the steering wheel and steering shaft with a center punch and a hammer.

Alignment marks on steering wheel and steering shaft:

☐ Satisfactory ☐ Unsatisfactory

WARNING: Do not pull on the steering wheel in an attempt to remove it from the steering shaft. The steering wheel may suddenly come off, resulting in personal injury, or the steering wheel may be damaged by the pulling force.

9. Install a steering wheel puller with the puller bolts threaded into the bolt holes in the steering wheel. Tighten the puller nut to remove the steering wheel. Visually check the steering wheel condition. If the steering wheel is bent or cracked, replace the wheel.

Steering wheel condition: ☐ Satisfactory ☐ Unsatisfactory

☐ 10. Disconnect the four retaining screws, and remove the clock spring electrical connector.

☐ 11. Be sure the front wheels are facing straight ahead. Turn the clock spring electrical connector counterclockwise by hand until it becomes harder to turn as it becomes fully wound in that direction.

CAUTION: Do not hammer on the top of the steering shaft to remove the steering wheel. This action may damage the shaft.

12. Turn the clock spring electrical connector clockwise three turns, and align the red mark on the center part of the spring face with the notch in the cable circumference. This action centers the clock spring electrical connector.

Is the clock spring centered? ☐ Yes ☐ No

Instructor check _____

Explain why the clock spring electrical connector must be centered before it is installed.

_____

_____

_____

☐ 13. Install the clock spring electrical connector, and tighten the four retaining screws to the specified torque.

14. Align the marks on the steering wheel and the steering shaft, and install the steering wheel on the shaft.

Are the marks on the steering wheel and steering shaft aligned? ☐ Yes ☐ No

Instructor check _____

Task Completed

15. Install the steering wheel retaining nut, and tighten this nut to the specified torque.
    Specified steering wheel retaining nut torque _____
    Actual steering wheel retaining nut torque _____

16. Install the air bag wiring retainer in the steering wheel. ☐

17. Hold the air bag deployment module near the top of the steering wheel, and connect the air bag module connector. ☐

18. Install the air bag deployment module in the top of the steering wheel, and tighten the retaining screws. ☐

19. Reconnect the negative battery cable. ☐

20. Reset the clock and radio. ☐

21. Turn on the ignition switch and start the vehicle. Check the operation of the air bag system warning light. ☐
    Explain the air bag system warning light operation that indicates normal air bag system operation.

    _____

    _____

    _____

CAUTION:
Failure to center a clock spring prior to installation may cause a broken conductive tape in the clock spring.

Instructor's Response _____

_____

_____

JOB SHEET 28

Name _____ **Date** _____

Remove and Replace Steering Column

Upon completion of this job sheet, you should be able to remove and replace a steering column.

NATEF AST/MAST Task Correlation

Suspension and Steering. Steering Systems Diagnosis and Repair,

Task #B-3. Diagnose steering column noises, looseness, and binding concerns (including tilt mechanisms); determine necessary action. (P-2)

Tools and Materials

Seat cover
Torque wrench

Describe the Vehicle Being Worked On:

Year _____ Make _____ Model _____

VIN _____ Engine type and size _____

Procedure — Task Completed

1. Disconnect the negative battery cable. If the vehicle is equipped with an air bag, wait for the time period specified by the vehicle manufacturer.
   Specified waiting period prior to servicing air bag components _____
   Explain why it is necessary to disconnect the battery before removing the steering column.
   _____
   _____
   _____
2. Install a seat cover on the front seat. ☐
3. Place the front wheels in the straight-ahead position and remove the ignition key from the switch to lock the steering column.
   Are the front wheels straight? ☐ Yes ☐ No
   Is the ignition key removed?
   Instructor check _____ ☐
4. Remove the cover under the steering column and remove the lower finish panel if necessary. ☐
5. Disconnect all wiring connectors from the steering column. ☐
6. If the vehicle has a column-mounted gearshift lever, disconnect the gearshift linkage at the lower end of the steering column. ☐
7. Remove the retaining bolt or bolts in the lower universal joint or flexible coupling. ☐

Task Completed

☐ 8. Remove the steering-column-to-instrument-panel mounting bolts.

☐ 9. Carefully remove the steering column from the vehicle. Be careful not to damage the upholstery or paint.

☐ 10. Install the steering column under the instrument panel and insert the steering shaft into the lower universal joint.

11. Install the steering column to instrument panel mounting bolts. Be sure the steering column is properly positioned, and tighten these bolts to the specified torque.
Specified torque, steering column mounting bolts _____
Actual torque, steering column mounting bolts _____

12. Install the retaining bolt or bolts in the lower universal joint or flexible coupling, and tighten the bolt(s) to the specified torque.
Specified torque, lower U-joint retaining bolt(s) _____
Actual torque, lower U-joint retaining bolt(s) _____

☐ 13. Connect the gearshift linkage if the vehicle has a column-mounted gearshift.

☐ 14. Connect all the wiring harness connectors to the steering column connectors.

☐ 15. Install the steering column cover and the lower finish panel.

☐ 16. Reconnect the negative battery cable.

☐ 17. Road test the vehicle, and check for proper steering column operation.
Steering column operation: ☐ Satisfactory ☐ Unsatisfactory

Instructor's Response _____

_____

_____

JOB SHEET 29

Name _____ **Date** _____

Inspect Manual or Power Rack and Pinion Steering Gear and Tie Rods

Upon completion of this job sheet, you should be able to inspect manual or power rack and pinion steering gears and tie rods.

NATEF AST/MAST Task Correlation

Suspension and Steering. Steering Systems Diagnosis and Repair,

Task #B-8. Inspect rack and pinion steering gear inner tie rod ends (sockets) and bellows boots; replace as needed. (P-2)

Task #B-17. Inspect, replace, and adjust tie rod ends (sockets), tie rod sleeves, and clamps. (P-1)

Tools and Materials

Tape measure

Describe the Vehicle Being Worked On:

Year _____ Make _____ Model _____

VIN _____ Engine type and size _____

Make of steering gear _____

Procedure

1. With the front wheels straight ahead and the engine stopped, rock the steering wheel gently back and forth with light finger pressure, and measure the maximum steering wheel free play with a tape measure.

   Specified steering wheel free play _____

   Actual steering wheel free play _____

   State the necessary repairs to correct steering wheel free play.

   _____

   _____

   _____

   _____

2. With the vehicle sitting on the shop floor and the front wheels straight ahead, have an assistant turn the steering wheel about 1/4 turn in both directions. Watch for looseness in the steering shaft and the flexible coupling or universal joints.

   Looseness in U-joints or flexible coupling: ☐ Satisfactory ☐ Unsatisfactory

   Looseness in steering shaft: ☐ Satisfactory ☐ Unsatisfactory

   State the necessary repairs and explain the reasons for your diagnosis.

   _____

   _____

   _____

   _____

3. While an assistant turns the steering wheel about 1/2 turn in both directions, watch for movement of the steering gear housing in the mounting bushings.

   Looseness in steering gear mounting bushings: ☐ Satisfactory ☐ Unsatisfactory

   State the necessary repairs and explain the reasons for your diagnosis.

   _____

   _____

   _____

   _____

4. Grasp the pinion shaft extending from the steering gear and attempt to move it vertically. If there is steering shaft vertical movement, a pinion bearing preload adjustment may be required. When the steering gear does not have a pinion bearing preload adjustment, replace the necessary steering gear components.

   Excessive pinion shaft vertical endplay: ☐ Satisfactory ☐ Unsatisfactory

   State the necessary repairs and explain the reasons for your diagnosis.

   _____

   _____

   _____

   _____

5. Road test the vehicle and check for excessive steering effort. A bent steering rack, tight rack bearing adjustment, or damaged front drive axle joints in front-wheel-drive cars may cause excessive steering effort.

   Steering effort: ☐ Satisfactory ☐ Excessive

   State the necessary repairs and explain the reasons for your diagnosis.

   _____

   _____

   _____

   _____

6. Visually inspect the bellows boots for cracks, splits, leaks, and proper clamp installation. Replace any boot that indicates any of these conditions.

   Condition of tie rod bellows boots and clamps:

   Left side boot: ☐ Satisfactory ☐ Unsatisfactory

   Right side boot: ☐ Satisfactory ☐ Unsatisfactory

   State the necessary repairs and explain the reasons for your diagnosis.

   _____

   _____

   _____

   _____

7. Loosen the inner bellows boot clamps and move each boot toward the outer tie rod end until the inner tie rod end is visible. Push outward and inward on each front tire, and watch for movement in the inner tie rod end. An alternate method of checking the inner tie rod ends is to squeeze the bellows boots and grasp the inner tie rod end socket. Movement in the inner tie rod end is then felt as the front wheel is moved inward and outward. Hard plastic bellows boots may be found on some applications. With this type of bellows boot, remove the ignition key from the switch to lock the steering column, and push inward and outward on the front tire while observing any lateral movement in the tie rod.

   Condition of inner tie rod ends:

   Left side inner tie rod end: ☐ Satisfactory ☐ Unsatisfactory

   Right side inner tie rod end: ☐ Satisfactory ☐ Unsatisfactory

   State the necessary repairs and explain the reasons for your diagnosis.

   _____

   _____

   _____

   _____

WARNING: Bent steering components must be replaced. Never straighten steering components because this action may weaken the metal and result in sudden component failure, serious personal injury, and vehicle damage.

8. Grasp each outer tie rod end and check for vertical movement. While an assistant turns the steering wheel 1/4 turn in each direction, watch for looseness in the outer tie rod ends. Check the outer tie rod end seals for cracks and proper installation of the nuts and cotter pins. Cracked seals must be replaced. Inspect the tie rods for a bent condition.

   Tie rod and outer tie rod end condition:

   Left side tie rod and outer tie rod end condition: ☐ Satisfactory ☐ Unsatisfactory

   Right side tie rod and outer tie rod end condition: ☐ Satisfactory ☐ Unsatisfactory

   List the necessary repairs and explain the reasons for your diagnosis.

   _____

   _____

   _____

   _____

Instructor's Response _____

_____

_____

JOB SHEET 30

Name _____ **Date** _____

Diagnose, Remove, and Replace Idler Arm

Upon completion of this job sheet, you should be able to diagnose, remove, and replace an idler arm.

NATEF AST/MAST Task Correlation

Suspension and Steering. Steering Systems Diagnosis and Repair,

Task #B-16. Inspect and replace pitman arm, relay (centerlink/intermediate) rod, idler arm and mountings, and steering linkage damper. (P-2)

Tools and Materials

Pull scale
Dial indicator
Torque wrench
Grease gun

Describe the Vehicle Being Worked On:

Year _____ Make _____ Model _____

VIN _____ Engine type and size _____

Procedure — Task Completed

1. Attach the magnetic base of a dial indicator to the frame near the idler arm. ☐
2. Position the dial indicator stem against the upper side on the outer end of the idler arm. Preload the dial indicator stem, and zero the dial. ☐
3. Use a pull scale to apply 25 lbs. (11.34 kg) of force downward and then upward on the idler arm. Observe the total vertical idler arm movement on the dial indicator. If this vertical movement exceeds the vehicle manufacturer's specifications, replace the idler arm.

   Specified vertical idler arm movement _____

   Actual vertical idler arm movement _____

   Recommended idler arm service _____

   Explain the effect of idler arm wear on steering quality.

   _____

   _____

   _____
4. Remove the idler arm to center link cotter pin and nut. ☐
5. Remove the center link from the idler arm. ☐
6. Remove the idler arm bracket mounting bolts and remove the idler arm. ☐

Task Completed

7. If the idler arm has a steel bushing, thread the bracket into the idler arm bushing until the specified clearance is obtained between the center of the lower bracket bolt hole and the upper idler arm surface.

   Specified distance from center of lower bracket hole to upper idler arm surface _____

   Actual distance from center of lower bracket hole to upper idler arm surface _____

8. Install the idler arm bracket to frame bolts and tighten the bolts to the specified torque. Be sure that lock washers are installed on the bolts.

   Specified idler arm retaining bolt torque _____

   Actual idler arm retaining bolt torque _____

9. Install the center link into the idler arm and tighten the mounting nut to the specified torque. Install the cotter pin in the nut.

   Specified center link to idler arm retaining nut torque _____

   Actual center link to idler arm retaining nut torque _____

☐ 10. If the idler arm has a grease fitting, lubricate as required.

Instructor's Response _____

_____

_____

JOB SHEET 31

Name _____ **Date** _____

Remove and Replace Outer Tie Rod End, Parallelogram Steering Linkage

Upon completion of this job sheet, you should be able to remove and replace outer tie rod ends in a parallelogram steering linkage.

NATEF AST/MAST Task Correlation

Suspension and Steering. Steering Systems Diagnosis and Repair,
Task #B-17. Inspect, replace, and adjust tie rod ends (sockets), tie rod sleeves, and clamps. (P-1)

Tools and Materials

Tie rod puller

Describe the Vehicle Being Worked On:

Year _____ Make _____ Model _____
VIN _____ Engine type and size _____

Procedure

1. Raise the vehicle on a lift with the tires supported on the lift. If the vehicle has rubber-encapsulated tie rod ends, be sure the front wheels are straight ahead.
2. Loosen the nut on the tie rod sleeve bolt that retains the sleeve to the tie rod ends.
3. Measure from the top center of the tie rod end ball stud to the outer end of the tie rod sleeve, and record the measurement.
   Distance from the center of the tie rod end ball stud to the outer end of the tie rod sleeve _____
4. Remove the cotter pin from the tie rod end retaining nut.
5. Loosen the tie rod end retaining nut.
6. Use a tie rod end puller to loosen the tie rod end taper in the steering arm.
7. Remove the tie rod end retaining nut and remove the tie rod end from the steering arm.
8. Count the number of turns required to thread the tie rod end out of the tie rod sleeve.
   Number of turns required to remove the tie rod end from the tie rod sleeve
   _____
   Describe the effect of worn, loose outer tie rod ends on steering quality.
   _____
   _____
9. Install the new tie rod end in the tie rod sleeve using the same number of turns recorded in step 8.

10. Push the tie rod end ball stud fully into the steering arm opening. Only the threads on the tie rod end ball stud should be visible above the steering arm surface.

    Is the taper on the tie rod end ball stud visible above the steering arm surface?

    ☐ Yes ☐ No

    If the answer to this question is yes, state the necessary repairs required to correct this problem.

    _____

    _____

    Instructor check _____

11. With the tie rod end ball stud pushed fully upward into the steering arm, measure the distance from the center of the ball stud to the outer end of the tie rod sleeve.

    Distance from the center of the tie rod end ball stud to the outer end of the tie rod sleeve

    _____

12. If this distance is not the same as recorded in step 3, remove the tie rod end ball stud from the steering arm, and rotate the tie rod end until this distance is the same as recorded in step 3.

13. Install the tie rod end ball stud retaining nut, and tighten this nut to the specified torque.

    Specified ball stud nut torque _____

    Actual ball stud nut torque _____

WARNING: Never loosen a tie rod end ball stud nut to align the openings in the ball stud and nut to allow cotter pin installation. This action may cause the nut to loosen during operation of the vehicle, and this will result in excessive steering free play. Complete loss of steering control will occur if the nut comes off the tie rod end ball stud, and this may cause a collision.

14. Install a new cotter pin through the tie rod end ball stud and nut openings, and bend the cotter pin legs separately around this nut.

15. Tighten the nut on the outer tie rod sleeve clamp bolt to the specified torque. Be sure the slot in the tie rod sleeve is positioned away from the opening in the clamp.

16. Measure the front wheel toe with a toe bar.

    Specified front wheel toe _____

    Actual front wheel toe _____

    Explain the type of front tire wear that is caused by improper front wheel toe.

    _____

Instructor's Response _____

_____

_____

Chapter 9

Four Wheel Alignment Procedure, Part 1

BASIC TOOLS
Basic technician's tool set
Service manual

Upon completion and review of this chapter, you should be able to:

- Perform a prealignment inspection.
- Diagnose wheel alignment and tire wear problems.
- Position a vehicle on the alignment rack and connect the wheel units on a computer wheel aligner to the wheel rims.
- Select the specifications in the computer wheel aligner for the vehicle being aligned.
- Perform a preliminary wheel alignment inspection with the preliminary inspection screen.
- Perform a ride height inspection and measurement with the ride height screen.
- Perform an automatic or manual wheel runout compensation procedure.
- Measure front wheel camber and caster.
- Recognize the symptoms of improper rear wheel alignment.
- Measure rear wheel camber and toe.
- Diagnose the causes of improper rear wheel alignment.
- Perform rear wheel camber adjustments.
- Perform rear wheel toe adjustments.
- Adjust front wheel camber on various front suspension systems.
- Adjust front wheel caster on various front suspension systems.
- Adjust front wheel toe.
- Center steering wheel.
- Measure toe change during front wheel jounce and rebound to check steering linkage height.

Proper front and rear wheel alignment is extremely important because it affects directional stability, tire tread wear, and vehicle safety! Technicians must know how to check front and rear wheel alignment angles and diagnose the causes of steering and alignment problems. It is also essential for technicians to know how to adjust front and rear suspension angles while maintaining vehicle safety. On certain vehicles, some wheel alignment angles are considered nonadjustable by the vehicle manufacturer, but aftermarket suppliers often provide parts kits to provide adjustments on these suspension systems. This chapter provides various adjustment procedures of front and rear suspension alignment angles as summarized in Photo Sequence 16.

Wheel Alignment Preliminary Diagnosis and Inspection

Customer Complaints Related to Suspension or Brakes

Drivers may experience a variety of symptoms related to incorrect wheel alignment or defective brakes (Figure 9-1). Not all the symptoms are related to wheel alignment, but the customer may request an alignment to correct these problems. The technician must diagnose

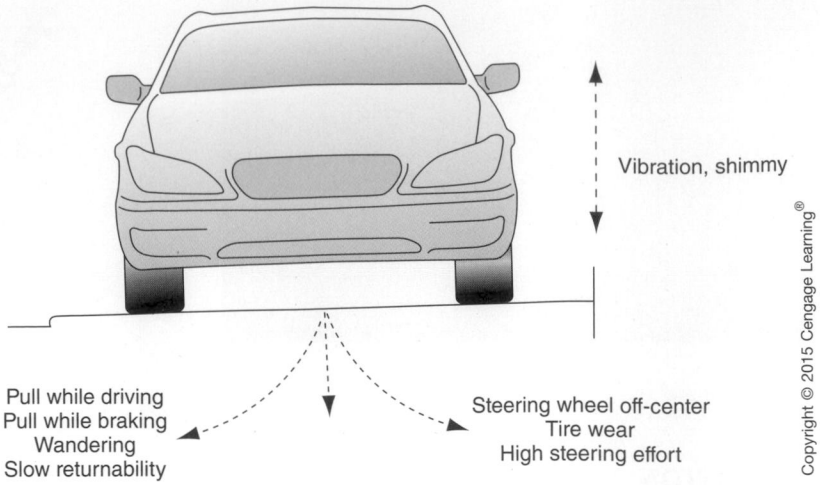

FIGURE 9-1 **Customer complaints related to wheel alignment or brakes.**

and correct the cause of the specific complaint. Wheel alignment angles that affect tire wear are toe and camber. The most likely symptoms of incorrect toe or camber settings are pull to one side while driving, wander, feathered tire wear, or tire wear on one side.

Road Test

Classroom Manual
Chapter 9, page 231

In many cases, a **road test** is necessary to verify customer complaints. During a road test, the vehicle should be driven under the conditions when the customer complaint occurred. While road-testing the vehicle, the technician should listen for any unusual noises related to suspension and steering.

The technician should check for these steering problems during a road test:

1. Excessive vertical chassis oscillations
2. Chassis waddle
3. Steering wander, pull, or drift
4. High steering effort and binding
5. Tire squeal while cornering
6. Bump steer
7. Torque steer
8. Excessive steering wheel free play
9. Suspension bottoming on rough road surfaces
10. Wheel hop during hard acceleration
11. Nose dive while braking or rear suspension squat when accelerating
12. Excessive body sway when cornering
13. Memory steer
14. Steering wheel return

Memory steer occurs when the steering does not want to steer straight ahead after a turn because the steering does not return to the straight-ahead position.

Chassis waddle may be caused by excessive tire or wheel runout. Excessive vertical chassis oscillations are the result of defective shock absorbers or struts. (Refer to Chapter 3 for tire and wheel runout diagnosis and Chapter 5 for shock absorber and strut diagnosis.)

Most steering systems are designed so the steering wheel returns to within 30 to 60 degrees of center after a 180 degree turn (Figure 9-2). If the steering wheel returnability is not satisfactory and **memory steer** is evident, the steering shaft dash seal or the steering shaft universal joints may be binding. When memory steer occurs on a front-wheel-drive car, the upper strut mounts may be binding. High steering effort, or binding, can also be caused by a loose power steering belt, low power steering fluid level, worn front tires, or a defective steering gear.

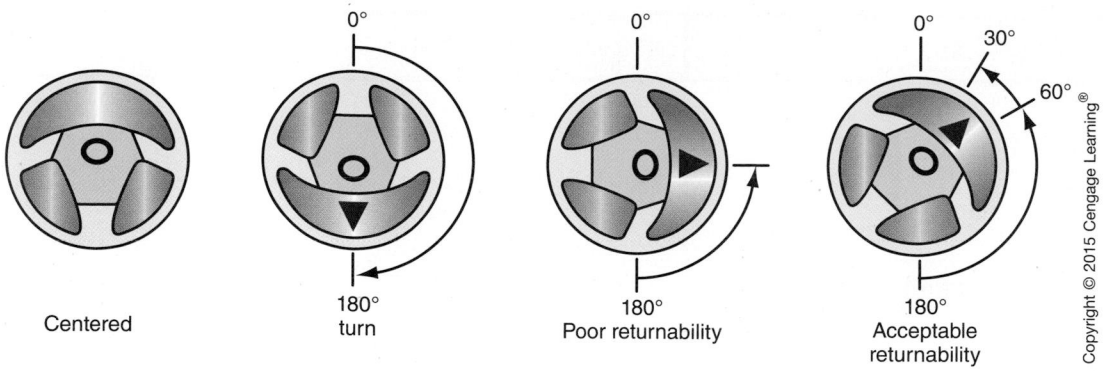

FIGURE 9-2 **Steering wheel returnability.**

Torque steer is the tendency of the steering to pull to one side during hard acceleration.

A front-wheel-drive vehicle with unequal drive axles produces some **torque steer** on hard acceleration. Some front-wheel-drive vehicles, especially those with higher horsepower, have equal front drive axles to reduce torque steer. On a front-wheel-drive vehicle, torque steer is aggravated by different tire tread designs on the front tires or uneven wear on the front tires. Torque steer may also be aggravated by worn and sagged transaxle and engine mounts that cause improper drive axle angles.

Bump Steer is the tendency of the steering to veer suddenly in one direction when one or both front wheels strike a bump.

Bump steer occurs if the tie rods are not the same height, meaning one of the tie rods is not parallel to the lower control arm. This condition may be caused by a bent or worn idler arm or pitman arm on a parallelogram steering linkage. On a rack and pinion steering system, worn steering gear mounting bushings may cause this unparallel condition between one of the tie rods and the lower control arm.

Steering pull may be caused by improper caster or camber adjustments.

SERVICE TIP: If steering pull or drift is evident during the road test, drive the vehicle on the same road in the opposite direction to determine if a crosswind is causing the steering pull.

During a road test, any of the following symptoms may indicate improper rear wheel alignment:

1. The front wheel **toe-in** is set correctly with the steering wheel centered on the alignment rack, but the steering wheel is not centered when the vehicle is driven straight ahead.
2. When there are no worn suspension parts or defective tires and all front suspension angles are within specifications, but the vehicle wanders or drifts to one side.
3. The vehicle is not overloaded, all front suspension angles are within specifications, and there are no worn suspension parts, but tire wear recurs.

Prealignment Inspection

Toe-in occurs when the distance between the front edges of the front or rear tires is less than the distance between the rear edges of the tires.

The technician should identify the exact suspension or steering complaint before a wheel alignment or other suspension work is performed. The technician must diagnose the cause or causes of the complaint and correct this problem during the wheel alignment (Figure 9-3). Because collision damage may affect wheel alignment angles, an inspection for collision damage on the vehicle should be completed prior to a wheel alignment (Figure 9-4).

Worn suspension and steering components must be replaced before a wheel alignment is performed. A wheel alignment should be performed after suspension components such as struts have been replaced.

Worn suspension and steering components must be replaced before a wheel alignment is performed.

The following components and measurements should be checked in a prealignment inspection:

1. Curb weight
2. Tires

Condition	Probable Cause
Tire wear:	
Outer shoulder	Too much positive camber
Inner shoulder	Too much negative camber
Sawtooth pattern:	
Sharp edge towards center	Too little toe-in
Sharp edge away from center	Too much toe-in
Both shoulders	Underinflation, high-speed cornering, overloading
Tread roll under	High speed in curves
Cupping, dishing	Wheel imbalance, radial runout
Diagonal wipe	Incorrect rear toe FWD Vehicle
Shimmy:	
With or without vehicle instability	Too much positive caster, unequal caster between wheels
Without instability	Tire imbalance, tire runout, driveline vibration
Vibration: (Caster/camber not a probable cause. The condition is listed because it is sometimes misdiagnosed as shimmy.)	Driveline misalignment, driveline imbalance, vehicle shake (accompanied by a characteristic moan), tire runout, unequal weight distribution between wheels
Steering wander/pull	Incorrect camber Unequal caster Overload, which elevates front end
Brake pull	Too much negative caster Unequal tire pressure, brake line damage that impedes hydraulic action on one side
Hard steering	Incorrect caster Damaged steering linkage, worn steering linkage, damaged spindles, rear-end overload, bent steering arm causing incorrect turning angle

FIGURE 9-3 **Diagnosis of wheel alignment problems.**

Copyright © 2015 Cengage Learning®

3. Suspension height
4. Steering wheel free play
5. Shock absorbers or struts
6. Wheel bearing adjustment
7. Ball joint condition
8. Control arms and bushings

Look at the complete vehicle

If you suspect collision damage to the inner body panels, measure the reference points or body and frame alignment as described in the service manual.

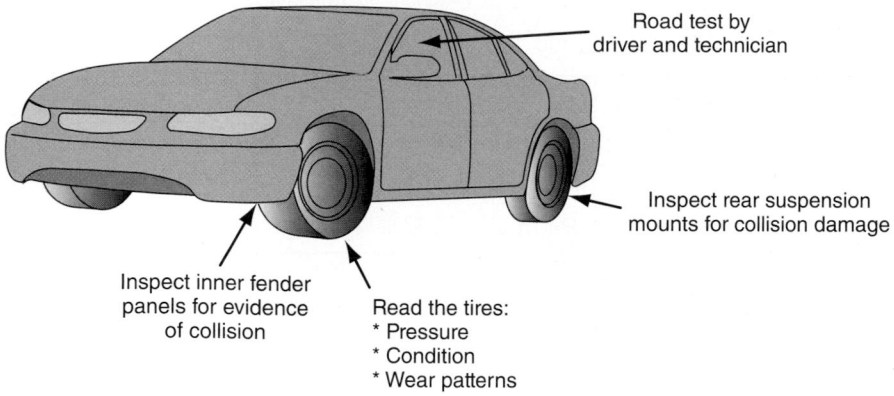

Copyright © 2015 Cengage Learning®

FIGURE 9-4 **Collision damage inspection.**

9. Steering linkages and tie rod ends
10. Stabilizers and bushings
11. Full fuel tank

After suspension components such as struts have been replaced, a wheel alignment should be performed.

Components to be checked in a **prealignment inspection** and wheel alignment measurements are summarized in Figure 9-5.

A **prealignment inspection** of all steering and suspension components must be completed before a wheel alignment procedure.

These checks should be completed with the car on the floor:

1. Inspect for excessive mud adhered to the chassis. Remove heavy items from the trunk and passenger compartment that are not considered in the vehicle curb weight. If heavy items such as tool display cases or merchandise are normally carried in the vehicle, these items should be left in the car during a wheel alignment.
2. Inflate tires to the recommended pressure and note any abnormal tread wear or damage on each tire. Be sure all tires are the same size.
3. Check the front tires and wheels for radial runout.
4. Check the suspension **ride height**. If this measurement is not within specifications, check for broken or sagged springs. On a torsion bar suspension system, check the torsion bar adjustment.
5. Inspect front and rear suspension bumpers. Worn bumpers may indicate weak springs or worn-out shock absorbers or struts.
6. When the wheels are in the straight-ahead position, rotate the steering wheel back and forth to check for play in the steering column, steering gear, or linkages.
7. Inspect the shock absorbers or struts for loose mounting bushings and bolts. Examine each shock absorber or strut for leakage.

Ride height is the distance from a specified location on the front or rear suspension to the road surface.

SERVICE TIP:
Decreasing the rear ride height 1 in. (25.4 mm) moves the front wheel caster 1 degree more positive.

The following checks should be performed with the vehicle raised and the suspension supported:

1. Inspect the front wheel bearings for lateral movement. (This inspection is described in Chapter 4.) On front-wheel-drive vehicles, perform this inspection on all four wheel bearings. Wheel bearings must be properly adjusted prior to a wheel alignment. Clean, repack, or adjust the wheel bearings as necessary.
2. Measure the ball joint radial and axial movement. If excessive movement exists in either direction, ball joint replacement is required. (Ball joint diagnosis and

PREALIGNMENT INSPECTION CHECKLIST

Owner _____ Phone _____ Date _____

Address _____ VIN _____

Make _____ Model _____ Year _____ Lic. number _____ Mileage _____

1. Road test results	Yes	No	Right	Left
Above 30 MPH				
Below 30 MPH				
Bump steer				
When braking				
Steering wheel movement Stopping from 2 to 3 MPH(Front)				
Vehicle steers hard				
Strg wheel returnability normal				
Strg wheel position				
Vibration	Yes	No	Frnt	Rear

2. Tire pressure	Specs Frnt ____ Rear ____
Record pressure found	
RF ____ LF ____ RR ____ LR ____	

3. Chassis height	Specs Frnt ____ Rear ____
Record height found	
RF ____ LF ____ RR ____ LR ____	

	Yes	No
Springs sagged		
Torsion bars adjusted		

4. Rubber bushings	OK
Upper control arm	
Lower control arm	
Sway bar / stabilizer link	
Strut rod	
Rear bushing	

5. Shock absorbers/struts	Frnt	Rear

6. Steering linkage	Frnt OK	Rear OK
Tie rod ends		
Idler arm		
Center link		
Sector shaft		
Pitman arm		
Gearbox/rack adjustment		
Gearbox/rack mounting		

7. Ball joints			OK
Load bearings			
Specs		Readings	
Right ____ Left ____		Right ____ Left ____	
Follower			
Upper strut bearing mount			
Rear			

8. Power steering	OK
Belt tension	
Fluid level	
Leaks/hose fittings	
Spool valve centered	

9. Tires/wheels	OK
Wheel runout	
Condition	
Equal tread depth	
Wheel bearing	

10. Brakes operating properly	

11. Alignment	Spec		Initial reading		Adjusted reading	
	R	L	R	L	R	L
Camber						
Caster						
Toe						

Bump steer	Toe change right wheel		Toe change left wheel	
	Amount	Direction	Amount	Direction
Chassis down 3"				
Chassis up 3"				

	Spec		Initial reading		Adjusted reading	
	R	L	R	L	R	L
Toe-out on turns						
SAI						
Rear camber Rear total toe						
Rear indiv. toe						
Wheel balance						
Radial tire pull						

Copyright © 2015 Cengage Learning®

FIGURE 9-5 **Prealignment inspection and wheel alignment checklist.**

replacement is explained in Chapter 6.) Be sure the suspension is supported correctly during ball joint diagnosis.

3. Inspect the control arms for damage and inspect the control arm bushings for wear.
4. Inspect all the steering linkages and tie rod ends for looseness.
5. Inspect for worn stabilizer mounting links and bushings.
6. Inspect for loose steering gear mounting bolts and worn mounting brackets and bushings.

Improper camber settings cause wear on one side of the front tire treads.

Tire Wear Diagnosis

Various types of tire tread wear patterns indicate specific alignment or balance defects (Figure 9-6).

An **alignment ramp** is a special steel ramp on which the vehicle is positioned for a wheel alignment.

Four-Wheel Alignment with Computer Alignment Systems

Preliminary Procedure

WARNING: When using a computer wheel aligner, always follow the equipment manufacturer's recommended procedures to provide accurate readings and to avoid equipment and vehicle damage or personal injury.

SPECIAL TOOL
Computer wheel aligner

The vehicle should be on an **alignment ramp** with a *turntable* under each front tire and conventional **slip plates** under the rear tires (Figure 9-7). Slip plates under the rear tires allow unrestricted movement of the rear wheels during rear wheel camber and toe adjustments (Figure 9-8). If suspension adjustments are made with the tires contacting the

Conditions	Rapid wear at shoulders	Rapid wear at center	Cracked treads	Wear on one edge	Feathered edge	Diagonal wipe rear tire FWD vehicles	Scalloped wear
Effect							
Causes	Underinflation or lack of rotation	Overinflation or lack of rotation	Underinflation or excessive speed	Excessive camber	Incorrect toe	Incorrect wheel toe	Lack of rotation of tires or worn or out-of-alignment suspension
Corrections	Adjust pressure to specifications when tires are cool. Rotate tires.			Adjust camber to specs	Adjust toe to specs	Perform rear wheel alignment	Rotate tires and inspect suspension

Copyright © 2015 Cengage Learning®

FIGURE 9-6 **Tire tread wear patterns and causes.**

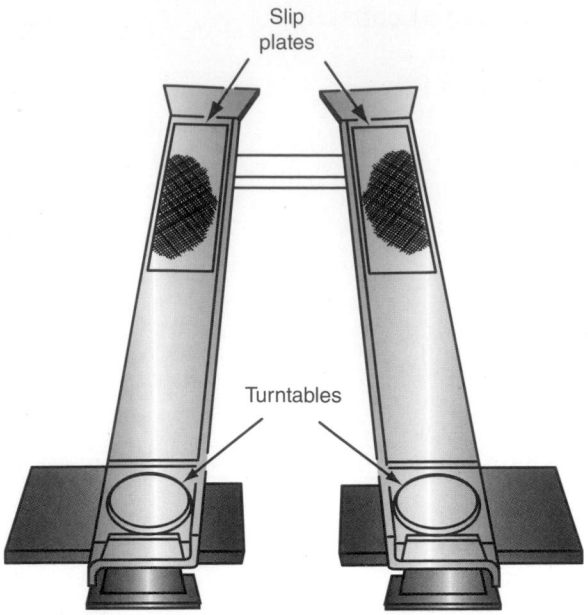

Copyright © 2015 Cengage Learning®

FIGURE 9-7 **Alignment ramp with turntables.**

Copyright © 2015 Cengage Learning®

FIGURE 9-8 **Rear wheel slip plates.**

A **turntable** is placed under each front wheel during a wheel alignment. The turntables allow the wheels to turn, and they also allow the wheels to move during alignment adjustments.

Rim clamps mount each wheel sensor to the rim.

Slip plates under the rear wheels allow the wheels to move during alignment adjustments.

A **digital signal processor (DSP)** in each wheel sensor provides digital signals indicating wheel alignment angles.

alignment ramp or the shop floor, the tires cannot move when the adjustment is completed. This action causes inaccurate suspension adjustments. The center of the front wheel spindles should be in line with the zero mark on the turning plates. The locking pins must be left in these plates and the parking brake applied.

Mount the **rim clamps** on each wheel. It may be necessary to remove the chrome discs or hub caps before these clamps can be installed. The adjustment knob on the rim clamp should be pointing toward the top of the wheel, and the bubble on the wheel unit should be centered (Figure 9-9). A set knob on the wheel unit allows the service technician to lock the wheel unit in this position. Some computer wheel aligners have a display on the monitor that indicates if any of the wheel sensors are not level (Figure 9-10). These sensors may be leveled by pressing a control knob on each wheel sensor. Some **digital signal processor (DSP)** wheel sensors contain a microprocessor and a **high-frequency transmitter** that

Copyright © 2015 Cengage Learning®

FIGURE 9-9 **Rim clamp and wheel sensor.**

Copyright © 2015 Cengage Learning®

FIGURE 9-10 **Wheel sensor level indicator on the monitor screen.**

acquire measurements and process data and then send these data to a *receiver* mounted on top of the wheel alignment monitor. This type of wheel sensor does not require any cables connected between the sensors and the computer wheel aligner. The data from this type of wheel sensor signal are virtually uninterruptible, even by solid objects. When these wheel sensors are stored on the computer wheel aligner, a "docking station" feature charges the batteries in the wheel sensors.

Some wheel sensors contain infrared emitter/detectors or light-emitting diodes (LEDs) between the front and rear wheel sensors to perform alignment measurements. Older wheel sensors have strings between the front and rear wheel sensors. Each wheel sensor contains a microprocessor with a preheat circuit that stabilizes the readings in relation to temperature. Some wheel sensors have touch keypads on the wheel units to allow the entry of commands from each unit. The front wheel units have arms that project toward the front of the vehicle to transmit signals between these units. When a blocked beam prompt appears on any screen, the beam between two wheel units is blocked. This could be caused by a person standing in the beam, an open car door, or a suspension jack. The blocked beam prompt must be eliminated before any tests are completed.

Classroom Manual
Chapter 9, page 239

The manufacturers of various types of computer alignment systems publish detailed operator's manuals for their specific equipment. Our objective here is to discuss some of the general screens that a technician uses while measuring wheel alignment angles with a typical computer aligner.

A **high-frequency transmitter** in each wheel sensor processes data from the DSP and transmits this data to the computer wheel aligner.

Main Menu

One of the first items displayed on the computer aligner screen is the *main menu* screen. From this screen, the technician makes a selection by touching the desired selection on the monitor screen with a mouse (Figure 9-11) or by pressing the number on the keypad that matches the number beside the desired procedure. A cursor may also be used to select the type of alignment.

A **receiver** mounted on or in the computer wheel aligner receives data from the high-frequency transmitters in the wheel sensors.

Specifications Menu

Some computer wheel aligners have specifications contained on CDs. The appropriate CD must be placed in the aligner. Several methods may be used to enter the vehicle specifications. The technician may scroll through the *specifications menu* and then press Enter when the appropriate vehicle is highlighted on the screen (Figure 9-12). The technician may also select the proper vehicle specifications by pointing with the light pen or by typing in the vehicle identification number (VIN) or the first letter in the vehicle make.

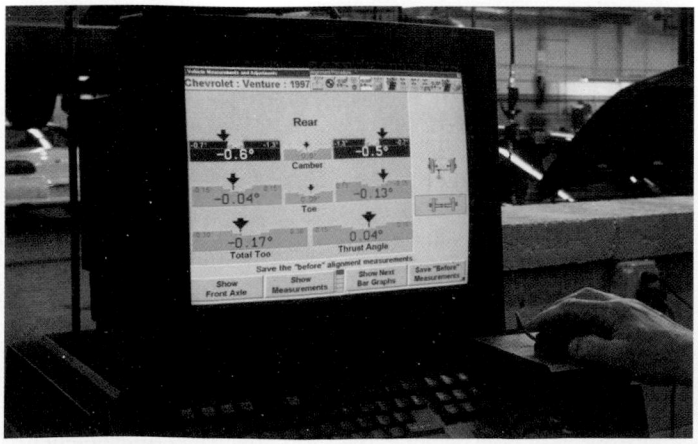

Copyright © 2015 Cengage Learning®

FIGURE 9-11 **Making screen selections with a mouse.**

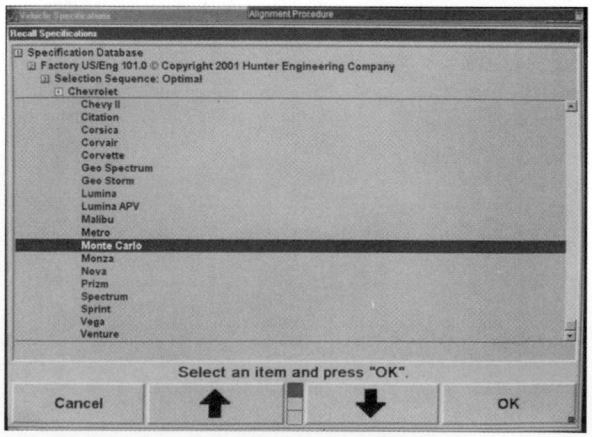

Copyright © 2015 Cengage Learning®

FIGURE 9-12 **Entering vehicle specifications.**

On other computer wheel aligners, the specifications are contained on a floppy disk and the technician selects the appropriate make, model, and year of vehicle with the cursor on the screen.

On early computer alignment systems, the vehicle specifications were contained on bar graphs. The technician moved a wand across the appropriate bar graph to enter the specifications for the vehicle being aligned.

CAUTION:
The ride height must be within specifications before proceeding with the wheel alignment. Improper curb riding height affects many of the other suspension angles.

Preliminary Inspection Screen

The **preliminary inspection screen** allows the technician to enter the condition of all front and rear suspension components that must be checked during a prealignment inspection. On some computer alignment systems, the technician may enter checked, marginal, and repair or replace for each component (Figure 9-13). These entered conditions may be printed out at any time or with the complete wheel alignment results.

Ride Height Screen

Some computer wheel aligners have optical encoders in the wheel sensors that measure ride height when this selection is displayed on the screen and the ride height attachment on each wheel sensor is lifted until it touches the lower edge of the fender (Figure 9-14). The ride height measurement is displayed on the screen. If the screen display is green, the ride height is within specifications. A red ride height display indicates this measurement is not within specifications. Improper ride height may be caused by sagged or broken springs, bent components such as control arms, or worn components such as control arm bushings.

Some computer wheel aligners provide a ***ride height screen*** with a graphic display indicating the exact location where the ride height should be measured (Figure 9-15). The computer aligner compares the ride height measurements entered by the technician to the specifications.

Tire Condition Screen

When the **tire inspection screen** is displayed, the technician may enter various tire wear conditions for each tire (Figure 9-16). The cursor on the screen is moved to the tire being inspected. The tire conditions are printed out with the preliminary inspection results and the alignment report. On other computer wheel aligners, tire condition is included in the preliminary inspection screens.

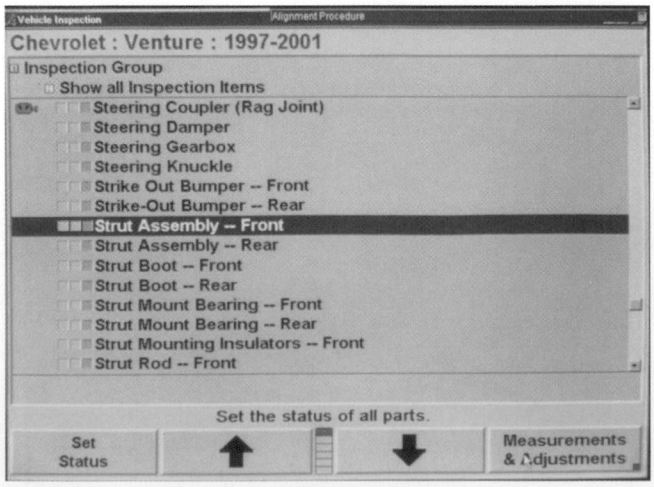

Copyright © 2015 Cengage Learning®

FIGURE 9-13 **Preliminary inspection screen.**

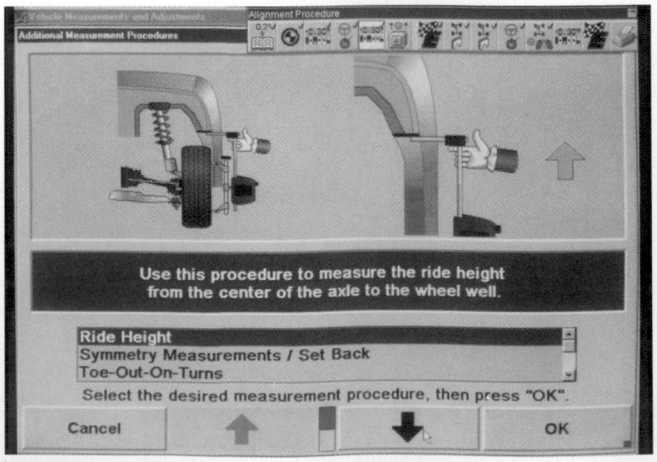

Copyright © 2015 Cengage Learning®

FIGURE 9-14 **Ride height screen and ride height attachment in wheel sensor.**

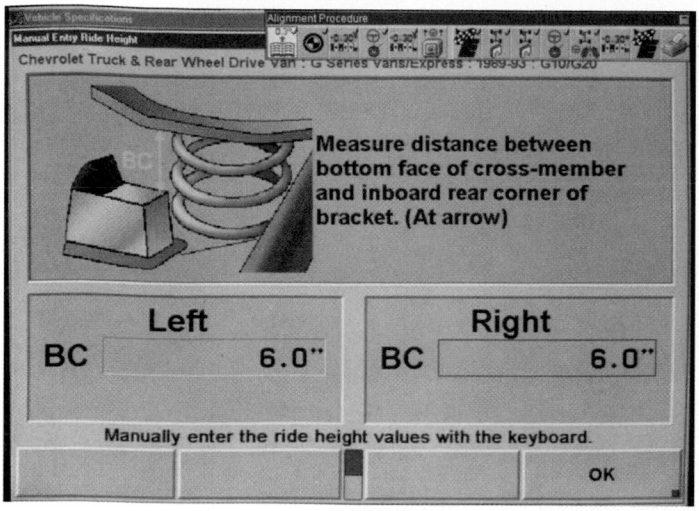

Copyright © 2015 Cengage Learning®

FIGURE 9-15 **Ride height screen.**

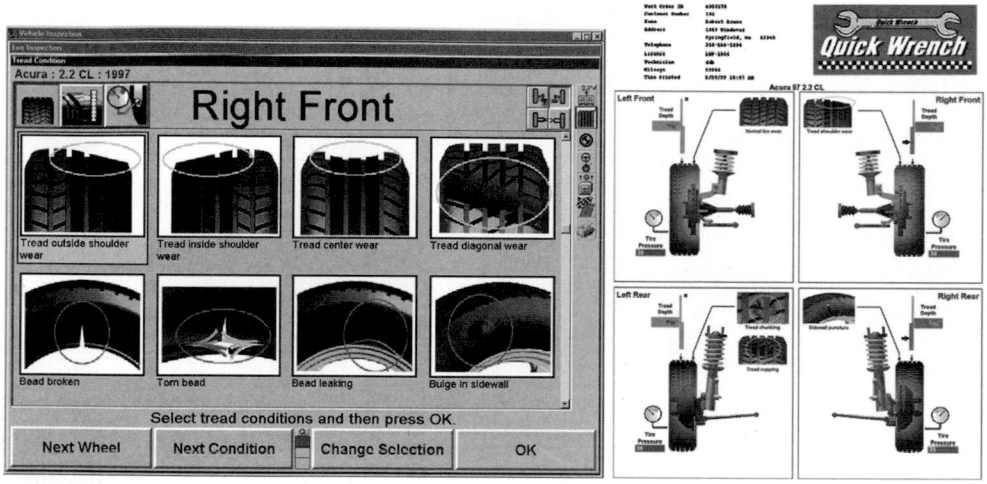

Copyright © 2015 Cengage Learning®

FIGURE 9-16 **Tire inspection screen.**

Classroom Manual
Chapter 9, page 242

SERVICE TIP:
Many computer wheel aligners can print out any single screen. It may be helpful to print out the wheel alignment angle screen before and after the alignment angles are adjusted to the vehicle manufacturer's specifications. These printouts may be presented to the customer; many customers appreciate this service.

Wheel Runout Compensation

As mentioned previously, screen indicators on some computer wheel aligners inform the technician if any of the wheel sensors require leveling or compensating for wheel runout. Runout compensation is accomplished by pressing the appropriate button on each wheel sensor. Some wheel sensors provide continuous compensation. This feature provides accurate alignment angles even when a wheel is rotated after the compensation button on the wheel sensor has been pressed. When the wheel runout screen is displayed, the technician is directed to level and lock each wheel unit and then press the compensation button on each wheel sensor to provide automatic wheel runout compensation (Figure 9-17).

If the computer aligner does not have automatic wheel runout compensation, a manual wheel runout procedure must be followed. This type of computer aligner displays a wheel runout measurement screen. During this procedure, the wheel being checked for runout is lifted with the hydraulic jack on the alignment rack and the wheel is rotated until the rim clamp knob faces downward. Level and lock the wheel unit in this position; then push Yes on the wheel unit as instructed on the screen. Rotate the wheel until the rim clamp knob faces upward; then level and lock the wheel unit. After this procedure, press Yes on the wheel unit. This same basic procedure is followed at each wheel.

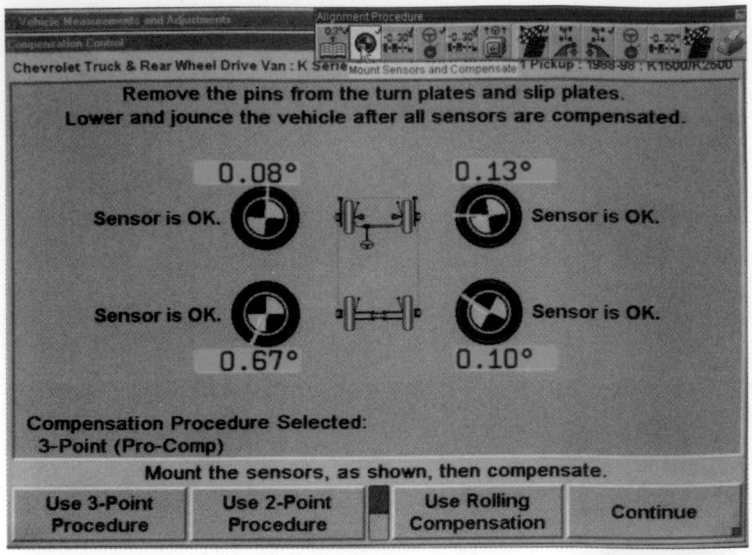

Copyright © 2015 Cengage Learning®

FIGURE 9-17 **Wheel runout compensation screen.**

Camber is the tilt of a line through the tire and wheel centerline in relation to the true vertical centerline of the tire and wheel.

Toe-out is present when the distance between the front edges of the front or rear wheels is more than the distance between the rear edges of the front or rear wheels.

Total toe is the sum of the toe settings on the front wheels.

Setback is the distance that one front or rear wheel is moved rearward in relation to the opposite front or rear wheel.

The **thrust line** is positioned at 90 degrees in relation to the rear axle, and this line projects forward.

Wheel Alignment Screens

Front and Rear Wheel Alignment Angle Screen

Prior to a display of the **front and rear wheel alignment angle screen** on some computer wheel aligners, the screen display directs the technician to position the front wheels straight ahead, lock the steering wheel, apply the **brake pedal depressor**, and level and lock the wheel units. See Photo Sequence 15. The brake pedal depressor is an adjustable rod installed between the front edge of the front seat and the brake pedal (Figure 9-18). If the vehicle has power brakes, the engine should be running when the depressor is used to apply the brakes. Some steering wheel holders are installed between the steering wheel and the top of the front seat (Figure 9-19). A ratchet and handle on the steering wheel holder allow extension of this holder.

After the wheel runout compensation procedure is completed at each wheel sensor, some computer wheel aligners automatically display *camber*, toe-in or **toe-out**, **total toe**, **setback**, and **thrust line** measurements on the front and rear suspension. After this procedure, the screen directs the technician to swing the front wheels outward a specific amount (Figure 9-20). The steering wheel holder must be released and the lock pins removed from the front turntables before the wheel swing procedure. This front wheel swing may be referred to as a caster/SAI swing. The amount of front wheel swing varies depending on the make of the computer aligner. Older computer wheel aligners required the technician to read the degrees of wheel swing on the turntable degree scales. On some newer

Copyright © 2015 Cengage Learning®

FIGURE 9-18 **Brake pedal depressor.**

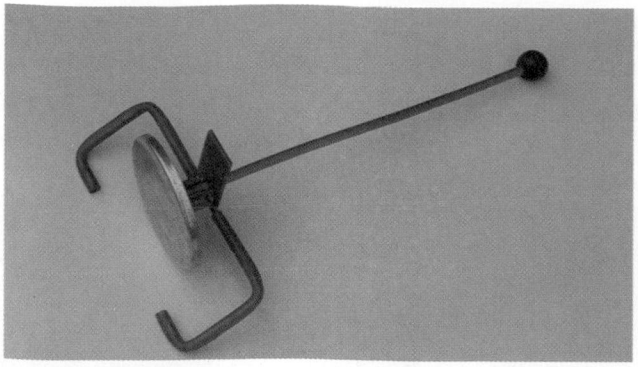

Copyright © 2015 Cengage Learning®

FIGURE 9-19 **Steering wheel holder.**

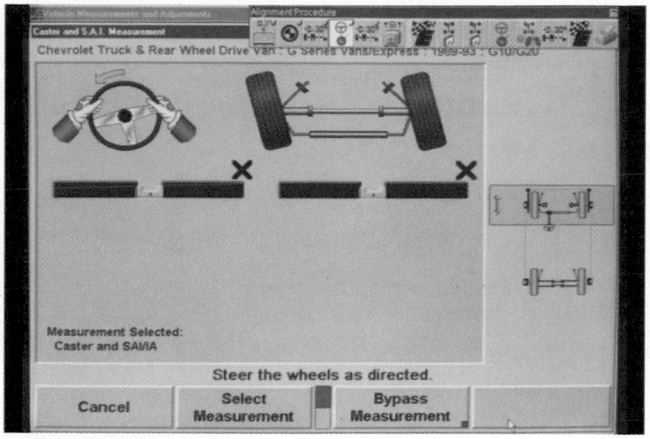

Copyright © 2015 Cengage Learning®

FIGURE 9-20 **Wheel swing instructions on the screen.**

computer wheel aligners, the required amount of wheel swing is illustrated on the screen (Figure 9-21).

After the front wheel swing procedure, all the front and rear wheel alignment angles are displayed, including *caster, steering axis inclination (SAI)*, and **included angle** (Figure 9-22), which are described in more detail in Chapter 10 of the classroom manual. Some computer wheel aligners highlight any wheel alignment angles that are not within specifications. If cross camber and cross caster are displayed on the screen, these readings indicate the maximum difference allowed between the right and left side readings. Alignment angles within specifications are highlighted in green; alignment angles that are not within specifications are highlighted in red.

Caster is the tilt of a line through the centers of the lower ball joint and upper strut mount in relation to a vertical line through the center of the wheel and spindle as viewed from the side.

WinAlign Software with Advanced Vehicle Handling (AVH)

Advanced vehicle handling (AVH) is a standard feature of WinAlign computer alignment software. AVH allows the technician to locate the cause of hidden suspension, chassis, and body problems prior to alignment adjustments. All chassis and suspension components should be inspected during the preliminary alignment inspection, and any worn or defective components must be replaced prior to using the AVH alignment software. AVH measurements include the following:

1. **Caster offset**—Caster offset is the distance from the caster line to the center of the front spindle (Figure 9-23). Caster offset may be called **caster trail**. Assuming the right and left caster angles are equal, if the right and left caster offset measurements

The **steering axis inclination (SAI)** line is an imaginary line through the centers of the upper and lower ball joints or through the center of the lower ball joint and the upper strut mount.

The SAI angle is the angle between the SAI line and the true vertical centerline of the tire and wheel.

The **included angle** is the sum of the SAI angle and the camber angle if the camber is positive. If the camber is negative, the camber setting must be subtracted from the SAI angle to obtain the included angle.

Copyright © 2015 Cengage Learning®

FIGURE 9-21 **Wheel swing procedure illustrated on the screen.**

PHOTO SEQUENCE 15

Typical Procedure for Performing Four-Wheel Alignment with a Computer Wheel Aligner

P15-1 **Position the vehicle on the alignment ramp.**

Copyright © 2015 Cengage Learning®

P15-2 **Be sure the front tires are positioned properly on the turntables.**

Copyright © 2015 Cengage Learning®

P15-3 **Position the rear wheels on slip plates.**

Copyright © 2015 Cengage Learning®

P15-4 **Attach the wheel units.**

Copyright © 2015 Cengage Learning®

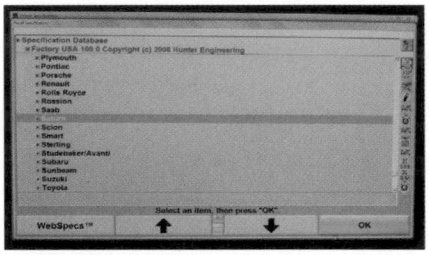

P15-5 **Select the vehicle make and model year.**

Copyright © 2015 Cengage Learning®

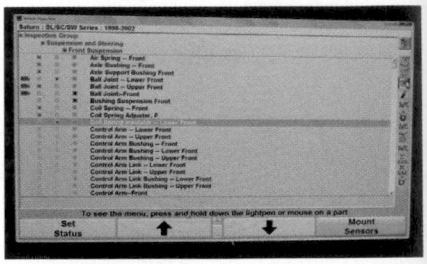

P15-6 **Check items on the screen during preliminary inspection.**

Copyright © 2015 Cengage Learning®

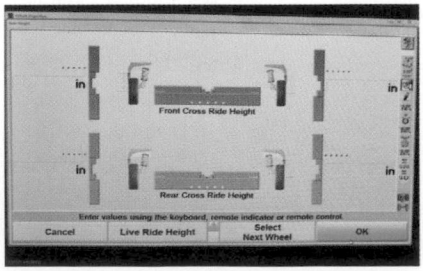

P15-7 **Display the ride height screen.**

Copyright © 2015 Cengage Learning®

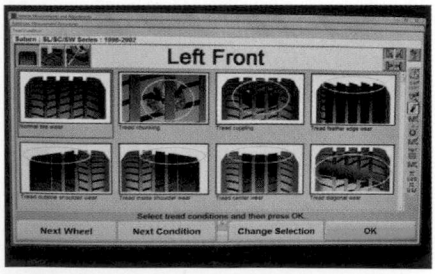

P15-8 **Check the tire condition for each tire on the tire condition screen.**

Copyright © 2015 Cengage Learning®

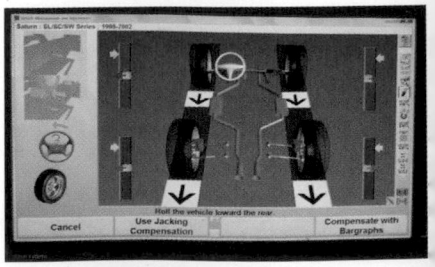

P15-9 **Display the wheel runout compensation screen.**

Copyright © 2015 Cengage Learning®

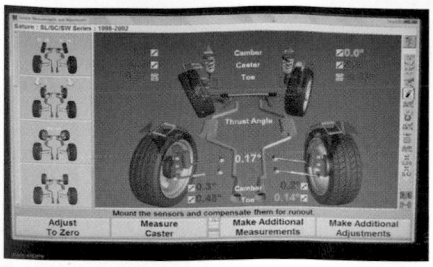

P15-10 **Display the front and rear wheel alignment angle screen.**
Copyright © 2015 Cengage Learning®

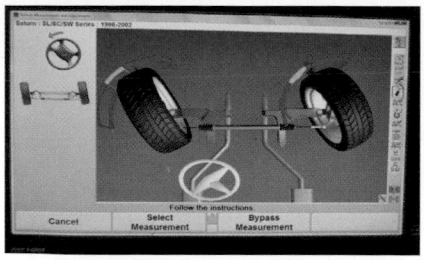

P15-11 **Display the turning angle screen and perform the turning angle check.**
Copyright © 2015 Cengage Learning®

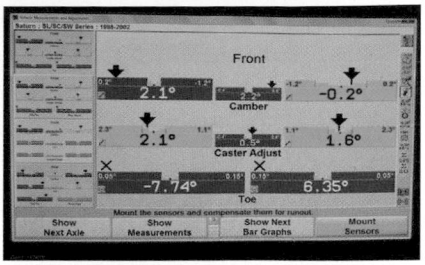

P15-12 **Display the adjustment screens.**
Copyright © 2015 Cengage Learning®

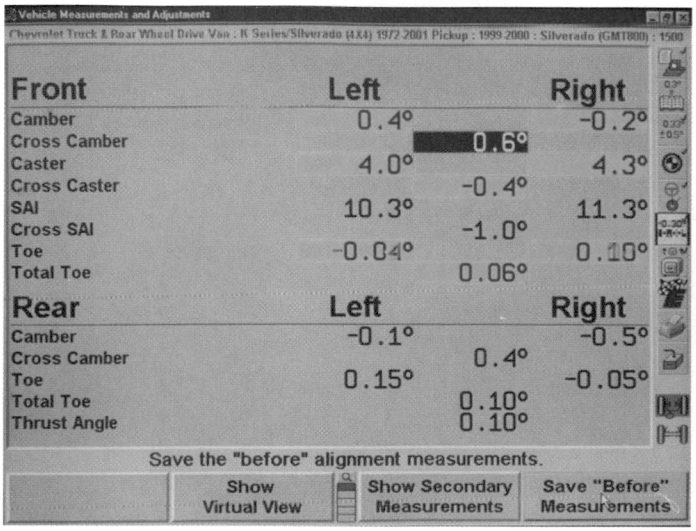

Copyright © 2015 Cengage Learning®

FIGURE 9-22 **Front and rear wheel alignment screen.**

are not equal, suspension components such as the strut or spindle may be worn or bent. Unequal caster offset may cause steering pull or drift.

2. Rolling force lever geometry—The rolling force lever distance is measured from the caster line viewed from the front or rear to a line parallel to the caster line and intersecting the outer end of the spindle (Figure 9-24). If the left and right rolling force lever measurements are unequal, steering pull or drift may occur. Assuming that all suspension and steering components are satisfactory, if the left and right rolling force lever measurements are not equal, one of the spindles is likely bent.
3. Body centerline angle and body centerline offset—Body centerline angle and offset isolates body and chassis centerline. Body centerline and/or body centerline offset problems may indicate collision damage that has not been properly corrected.
4. Body overhang—Body overhang indicates unequal left and right body overhang, which usually indicates worn or damaged body mounts or collision damage.
5. Excessive body roll angle when vehicle is steered—This usually indicates defective or worn suspension components such as the stabilizer bar and related parts.
6. Track circle and curb-to-curb distance measurements—If these measurements are not correct, suspension components such as control arms may be defective or bent.
7. Bump steer—Measures suspension angles in relation to suspension movement.

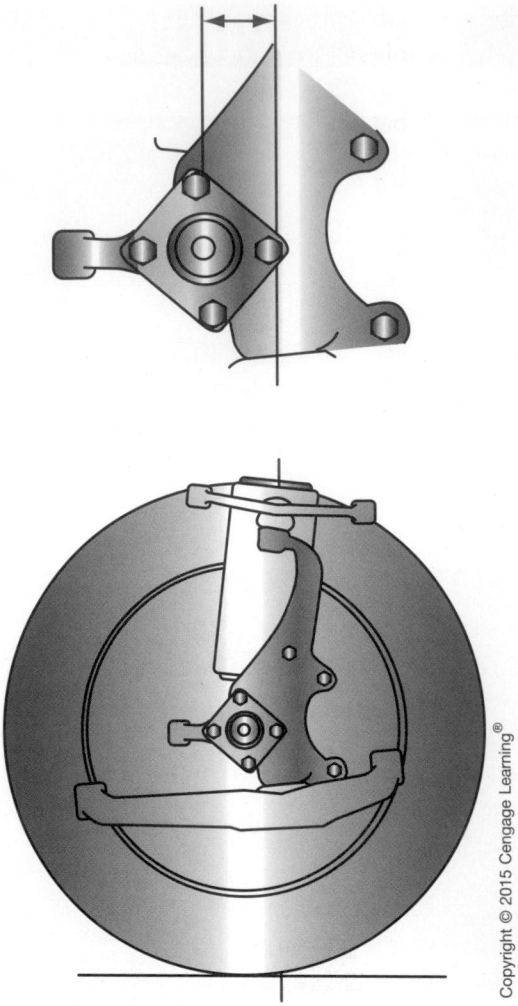
Copyright © 2015 Cengage Learning®

FIGURE 9-23 **Caster offset measurement.**

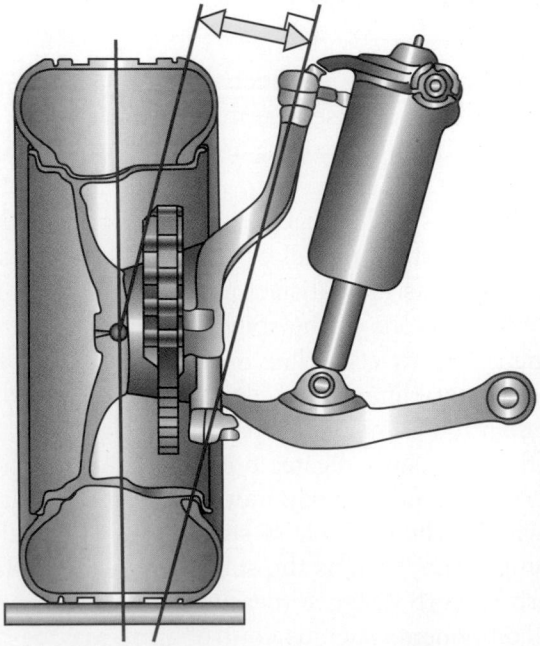
Copyright © 2015 Cengage Learning®

FIGURE 9-24 **Rolling force lever measurement.**

AVH alignment software enables the technician to find the cause(s) of steering problems such as unusual pull complaints including steering pull while braking, excessive torque steer, and bump steer. Locating and correcting the root cause of these problems eliminates steering related comebacks.

Adjustment Screens

The technician may select adjustment screens that provide bar graph readings of camber, caster, and toe. An arrow on the bar graph shows the amount and direction the actual measurement is from the preferred specification. As the alignment angle is adjusted, the arrow moves. If an alignment angle moves from out of the specification range to within the specification range, the bar graph color changes from red to green (Figure 9-25). When the arrow on the bar graph is in the zero position, the alignment angle is at the preferred specification. A zoom feature on some computer wheel aligners provides enlarged bar graphs so they may be seen at a distance while performing the actual suspension adjustments (Figure 9-26). Some computer aligners have a jack and hold feature that allows the suspension to be lifted on the alignment rack to perform an adjustment while maintaining accurate displays on the adjustment screen. Other computer wheel aligners have a remote display that may be connected to the aligner and taken under the car for close-up viewing while performing suspension adjustments (Figure 9-27). The remote display duplicates the bar graphs shown on the monitor screen.

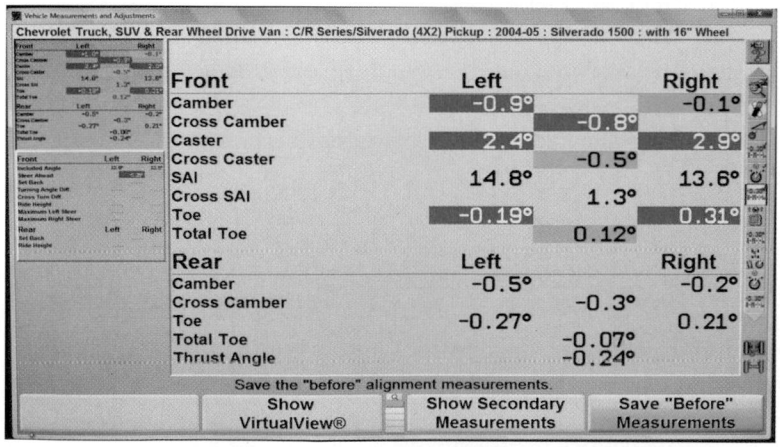

Copyright © 2015 Cengage Learning®

FIGURE 9-25 **Alignment angle adjustment screens.**

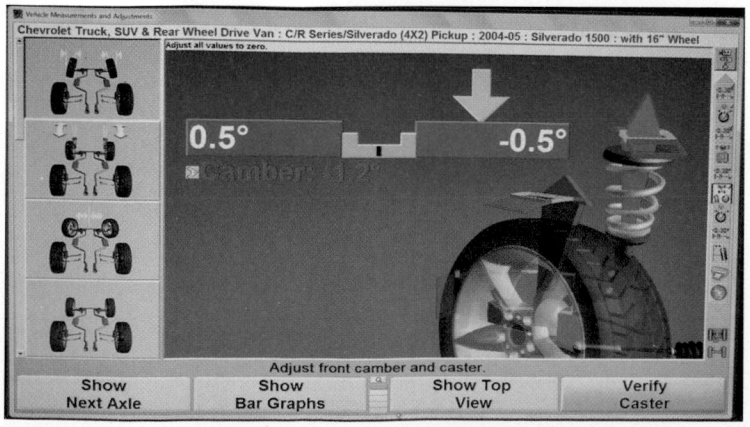

Copyright © 2015 Cengage Learning®

FIGURE 9-26 **Zoom feature on adjustment screens.**

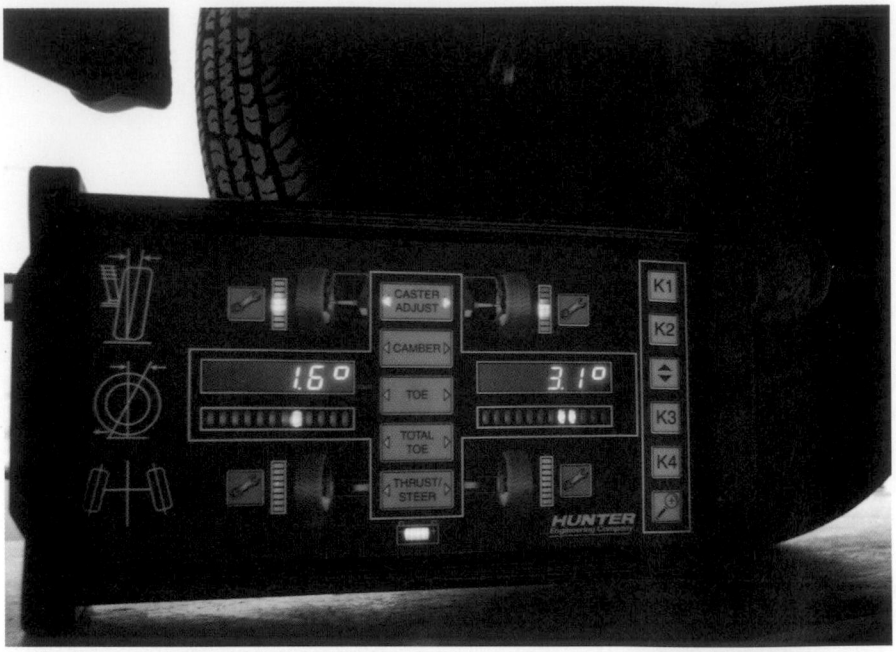

Copyright © 2015 Cengage Learning®

FIGURE 9-27 **Remote display for computer wheel aligner.**

Diagnostic Drawing and Text Screens

The technician may select tools and kits for the vehicle being serviced from the tools and kits database. When this feature is selected, the monitor screen displays the necessary wheel alignment adjustment tools for the vehicle selected at the beginning of the alignment (Figure 9-28). The kits displayed on the monitor screen are special components such as adjustment shims that are available for alignment adjustments on the car being serviced.

The technician may select **digital adjustment photos** that indicate how to perform wheel alignment adjustments (Figure 9-29). These digital photos include photos for cradle inspection and correction of cradle-to-body alignment (Figure 9-30). Live-action videos can also be selected. These CD videos provide suspension component inspection procedures (Figure 9-31).

A **part-finder database** is available in some computer wheel aligners. This database allows the technician to access part numbers and prices from many under-car parts manufacturers.

On many front-wheel-drive cars, the rear wheel camber and toe are adjusted with shims. Some computer aligners have a **shim display screen** that indicates the thickness

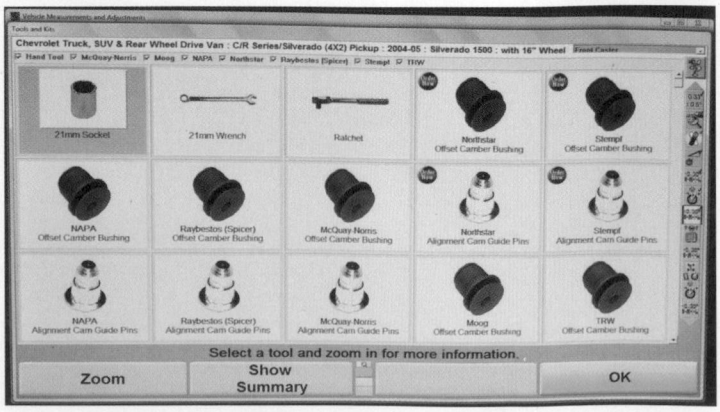

Copyright © 2015 Cengage Learning®

FIGURE 9-28 **Wheel alignment tools and kits display.**

Copyright © 2015 Cengage Learning®

FIGURE 9-29 **Wheel alignment adjustment photos.**

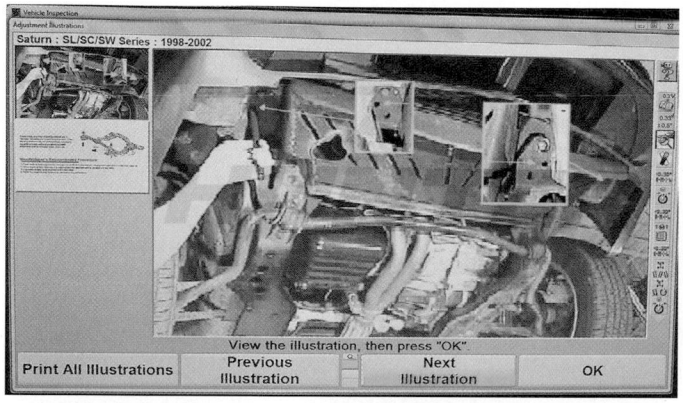

Copyright © 2015 Cengage Learning®

FIGURE 9-30 **Cradle inspection and correction photos.**

Copyright © 2015 Cengage Learning®

FIGURE 9-31 **Live-action videos of suspension inspection and service.**

of shim required and the proper position of the shim. On some computer wheel aligners, the technician may use the light pen to change the orientation angle of the shim on the monitor screen while observing the resulting change in camber and toe (Figure 9-32). Some computer wheel aligners provide an automatic bushing calculator screen (Figure 9-33). This screen shows the required bushing and the proper bushing position to obtain the specified camber and caster on twin I-beam front suspension systems.

Classroom Manual
Chapter 9, page 218

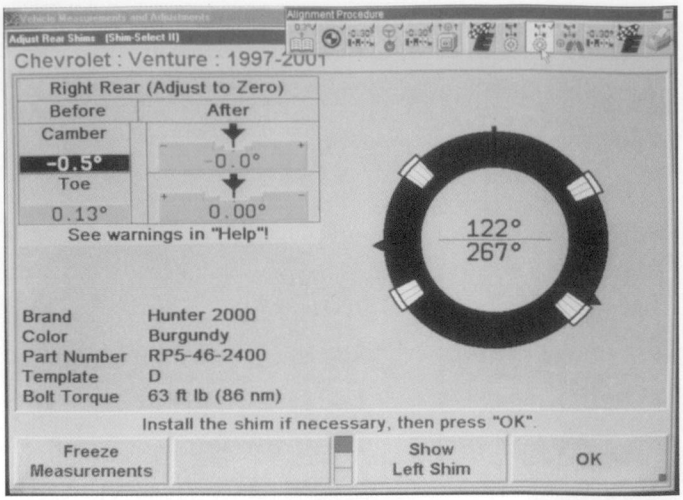

Copyright © 2015 Cengage Learning®

FIGURE 9-32 **Shim selection screen.**

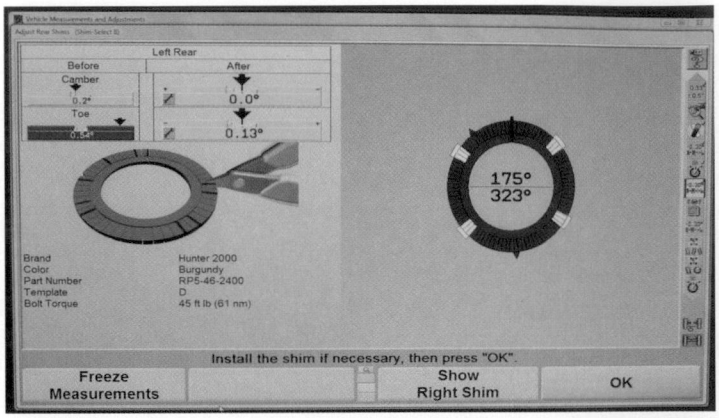

Copyright © 2015 Cengage Learning®

FIGURE 9-33 **Automatic bushing calculation screen.**

A **control arm movement monitor** is available on some computer wheel aligners. On short-and-long arm front suspensions, this feature indicates the required shim thickness to provide the specified camber and caster (Figure 9-34).

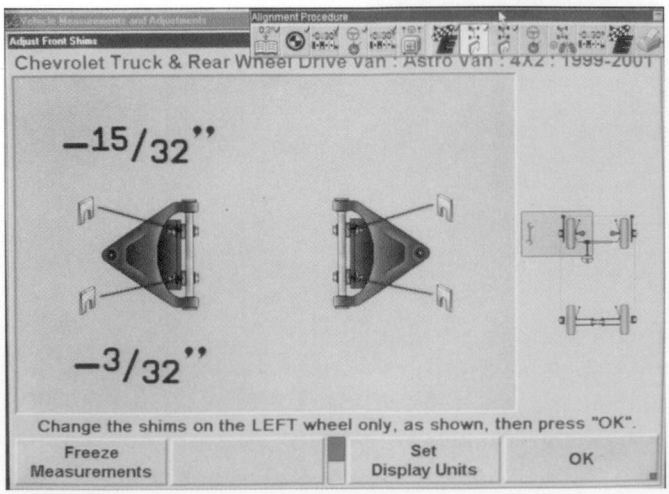

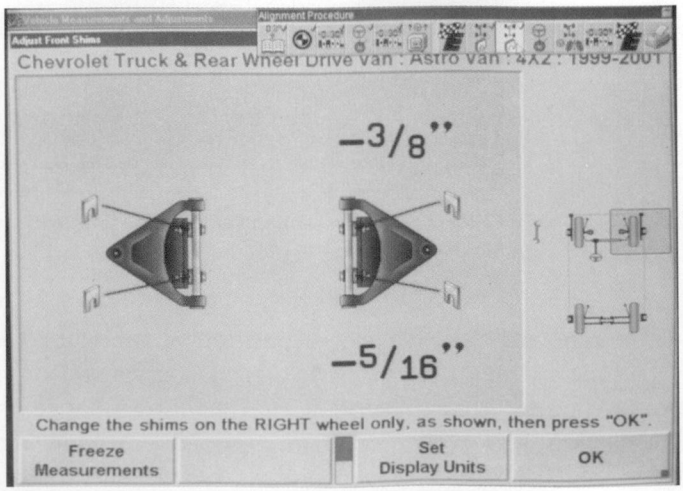

Copyright © 2015 Cengage Learning®

FIGURE 9-34 **Control arm movement monitor indicates the required shim thickness to provide specified camber and caster.**

On some computer wheel aligners, the technician may select Print at any time, and print out the displayed screen, including diagnostic drawings. Another optional procedure is to print out the wheel alignment report before and after the adjustment of wheel alignment angles.

SERVICE TIP:
Because rear wheel alignment plays a significant role in aiming or steering the vehicle, the rear wheel alignment should be corrected before adjusting front suspension angles. The rear wheel camber, toe, and thrust line must be within specifications before adjusting the front suspension angles.

Wheel Alignment Procedure

The proper procedure for front and rear wheel alignment is important because adjusting one wheel alignment angle may change another angle. For example, adjusting front wheel caster changes front wheel toe. The wheel alignment adjustment procedure is especially critical on four-wheel independent suspension systems. A front wheel adjustment procedure is provided in Figure 9-35, and a typical rear wheel adjustment procedure is given in Figure 9-36. Always follow the wheel alignment procedure in the vehicle manufacturer's service manual.

On vehicles that have a provision for camber and toe adjustments in the rear of the vehicle, a four-wheel alignment may be performed. The alignment would begin at the rear of the vehicle to bring rear camber and toe into specifications. If you recall from the classroom manual rear toe errors will impact the thrust angle of the vehicle that is used for setting the front wheel toe angle. If you were to start in the front of the vehicle and then move to the rear, the changes you make to the rear toe would also change your front toe measurements proportionally. It is critical that you always begin by making all necessary adjustments to the rear alignment first and then move to the front.

Causes of Improper Rear Wheel Alignment

The following suspension or chassis defects will cause incorrect rear wheel alignment:

1. Collision damage that results in a bent frame or distorted unitized body
2. A leaf-spring eye that is unwrapped or spread open
3. Leaf-spring shackles that are broken, bent, or worn
4. Broken leaf springs or leaf-spring center bolts
5. Worn rear upper or lower control arm bushings
6. Worn trailing arm bushings or dislocated trailing arm brackets
7. Bent components such as radius rods, control arms, struts, and rear axles

CAUTION:
Always follow the vehicle manufacturer's recommended rear suspension adjustment procedure in the service manual to avoid improper alignment angles and reduced steering control.

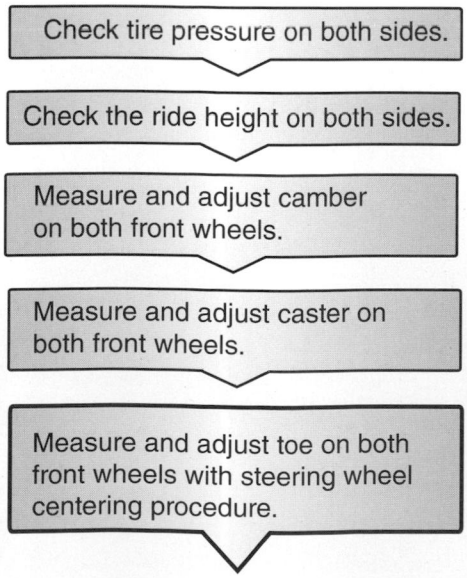

Copyright © 2015 Cengage Learning®

FIGURE 9-35 **Front wheel alignment adjustment procedure.**

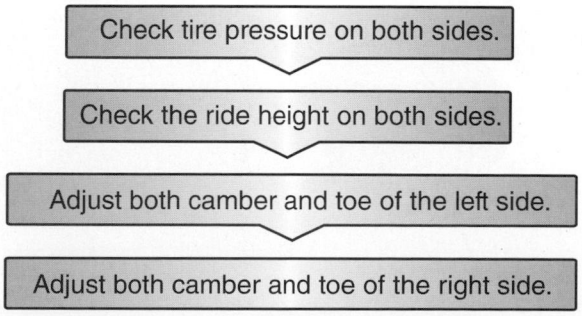

Copyright © 2015 Cengage Learning®

FIGURE 9-36 **Rear wheel alignment adjustment procedure.**

Rear Suspension Adjustments

Rear Wheel Camber Adjustment

Rear wheel camber is the tilt of a line through the center of a rear tire and wheel in relation to the true vertical centerline of the tire and wheel.

Improper **rear wheel camber** causes excessive wear on the edges of the rear tire treads.

Cornering stability may be affected by improper rear wheel camber, especially while cornering at high speeds. Improper rear wheel camber on a front-wheel-drive vehicle may change the understeer or oversteer characteristics of the vehicle. Because a tilted wheel rolls in the direction of the tilt, rear wheel camber may cause steering pull. For example, excessive positive camber on the right rear wheel causes the rear suspension to drift to the right and steering to pull to the left.

Rear suspension adjustments vary depending on the type of suspension system. Some manufacturers of wheel alignment equipment and tools provide detailed diagrams for front and rear wheel alignment adjustments on all makes of domestic and imported vehicles. Our objective is to show a few of the common methods of adjusting rear wheel alignment angles on various types of rear suspension systems.

SPECIAL TOOL

Rear suspension camber tool

On some semi-independent rear suspension systems, camber and toe are adjusted by inserting different sizes of shims between the rear spindle and the spindle mounting surface. These shims are retained by the spindle mounting bolts. The shim thickness is changed between the top and bottom of the spindle to adjust camber (Figure 9-37).

Many rear camber shims are now circular and are available in a wide variety of configurations to fit various rear wheels. Some computer wheel aligners indicate the thickness of shim required and the proper shim position. These same shims also adjust rear wheel toe (Figure 9-38).

On some transverse leaf rear suspension systems such as those on a later model Oldsmobile Cutlass Supreme, the spindle must be removed from the strut, and the lower strut bolt hole elongated to allow a camber adjustment (Figure 9-39). Once this bolt hole is elongated, assemble the knuckle and strut and leave the retaining bolts loose. The top of the tire and the spindle may be moved outward or inward to adjust camber. After the camber is adjusted to specifications, the strut-to-knuckle nuts must be tightened to the specified torque.

On some independent rear suspension systems with a lower control arm and ball joint, the strut-to-knuckle bolts are loosened and a special camber tool must be installed on the lower strut-to-knuckle bolt and the rear side of the strut (Figure 9-40). Rotate the adjusting

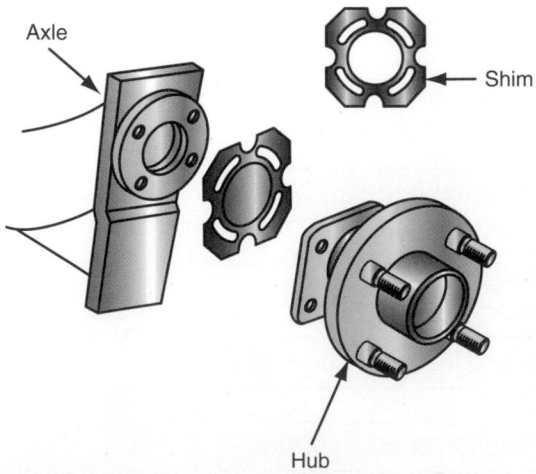

Copyright © 2015 Cengage Learning®

FIGURE 9-37 **Shim adjustment for rear wheel camber, semi-independent rear suspension system.**

Copyright © 2015 Cengage Learning®

FIGURE 9-38 **Circular shim for rear wheel camber and toe adjustment.**

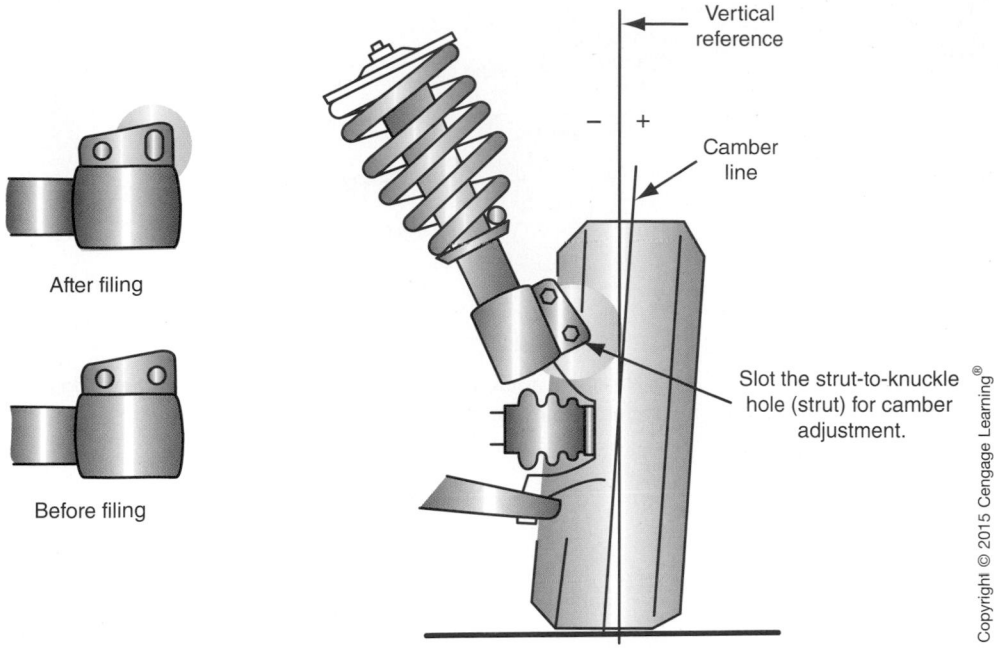

FIGURE 9-39 **Elongating the lower strut bolt hole provides camber adjustment.**

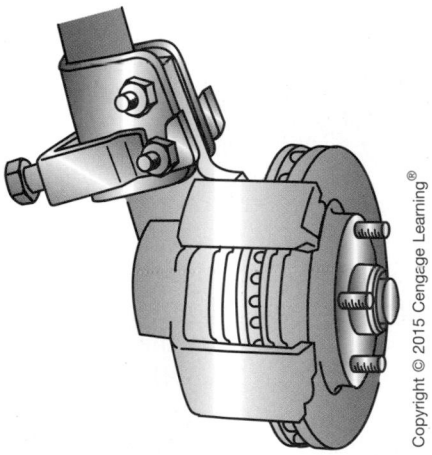

FIGURE 9-40 **Camber adjustment with special tool inserted on rear lower strut-to-knuckle bolt.**

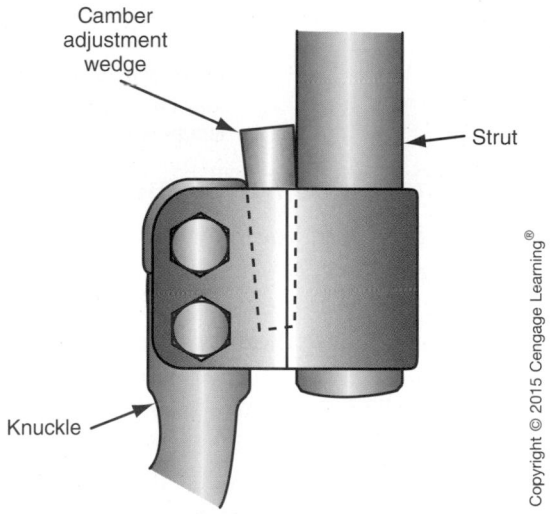

FIGURE 9-41 **Camber adjusting wedge installed between the top of the rear knuckle and strut.**

bolt in the back of the tool to adjust the camber; then tighten the strut-to-knuckle nuts to the specified torque.

On other rear suspension systems, a camber adjustment wedge may be installed between the top of the knuckle and the strut to adjust camber (Figure 9-41). The nuts on the strut-to-knuckle bolts must be loosened prior to wedge installation, and these nuts must be tightened to the specified torque after the camber is adjusted to specifications. Aftermarket adjustment cam kits are available to adjust the rear wheel camber on many cars (Figure 9-42). On some double-wishbone rear suspension systems, graduated cams on the inner ends of the lower control arms provide camber adjustment (Figure 9-43). Improper rear wheel camber may be caused by sagged rear springs on some cars or station wagons (Figure 9-44). If the improper rear wheel camber is caused by sagged rear springs, the springs should be replaced and the camber adjusted to specifications.

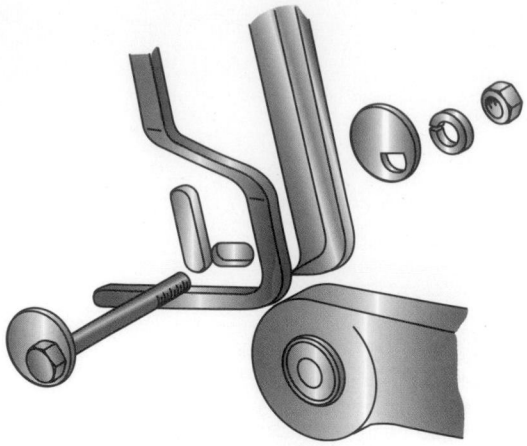

FIGURE 9-42 **Rear wheel camber adjusting kit.**

Copyright © 2015 Cengage Learning®

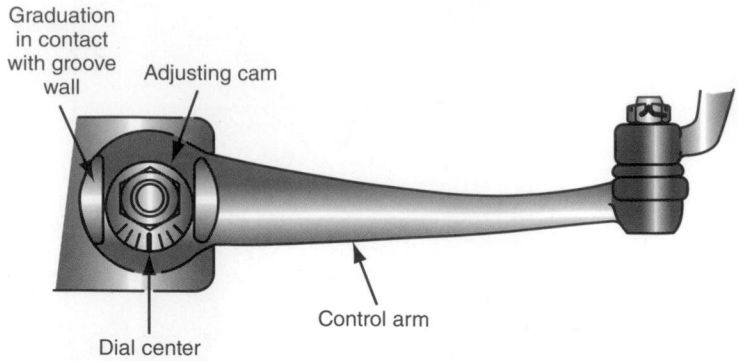

FIGURE 9-43 **Graduated cam for rear camber adjustment, double-wishbone suspension.**

Copyright © 2015 Cengage Learning®

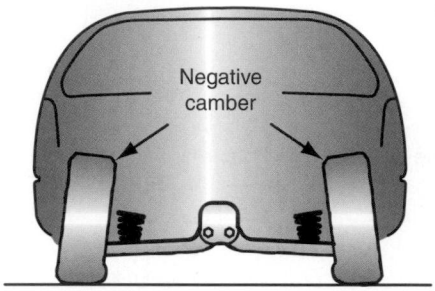

FIGURE 9-44 **Improper rear wheel camber may be caused by sagged rear springs on independent rear suspension systems.**

Copyright © 2015 Cengage Learning®

Rear Wheel Toe Adjustment

Rear wheel toe is the distance between the front edges of the rear tires in relation to the distance between the rear edges of the rear tires.

Improper **rear wheel toe** moves the thrust line away from the geometric centerline, which causes steering pull and feathered tire tread wear. If the left rear wheel has excessive toe-out, the thrust line is moved to the left of the geometric centerline. Excessive toe-in on the left rear wheel moves the thrust line to the right of the geometric centerline. When toe-in or toe-out is excessive on the right rear wheel, it has the opposite effect on the thrust line compared to the left rear wheel. When the thrust line is moved to the left of the geometric centerline, the steering pulls to the right, whereas a thrust line positioned to the right of the *geometric centerline* results in steering pull to the left. Photo sequence 16 shows rear wheel alignment adjustment procedures.

The **thrust line** is a line at 90 degrees to the rear axle and projecting forward.

Improper rear wheel toe adjustment on a nondriving rear axle causes diagonal wipe tire tread wear (Figure 9-45). The toe may be adjusted on many rear suspension systems used

Geometric centerline is the true centerline between the front and rear axles.

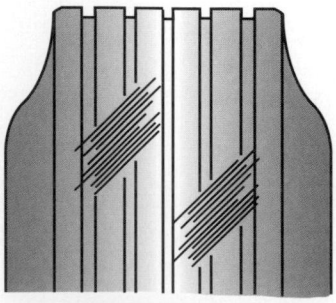

FIGURE 9-45 **Diagonal wipe tire tread wear caused by improper rear wheel toe on a front-wheel-drive vehicle.**

Copyright © 2015 Cengage Learning®

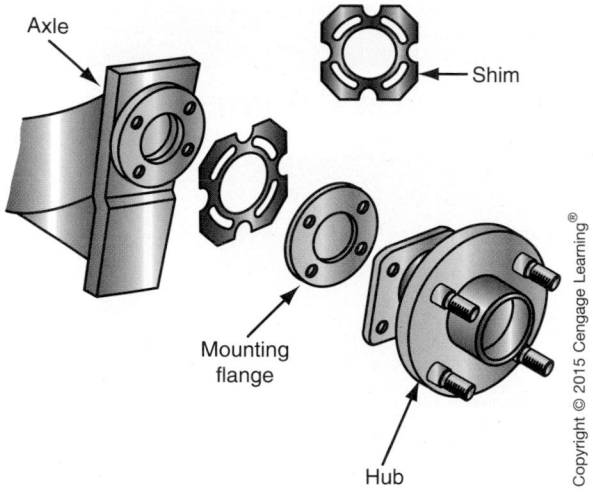

FIGURE 9-46 **Adjusting rear wheel toe by changing shim thickness between rear spindle and mounting flange.**

on front-wheel-drive cars by changing the tapered shim thickness between the front and rear edges of the rear spindle mounting flange (Figure 9-46). As mentioned previously, these shims are also used to adjust rear wheel camber.

On some transverse leaf rear suspension systems, a special toe adjusting tool is inserted in openings in the jack pad and the rear lower control rod (Figure 9-47). This tool has a turnbuckle in the center of the tool. When the nut on the inner end of the rear lower control rod is loosened, the turnbuckle is rotated to lengthen or shorten the tool and move the rear lower control rod to adjust the rear wheel toe (Figure 19-48). After the toe is adjusted to specification, the nut on the rear lower control rod bolt must be tightened to the specified torque. Photo Sequence 17 illustrates front and rear suspension adjustments.

On some independent rear suspension systems, an eccentric star wheel on the rear lower control rod may be rotated to adjust the rear wheel toe (Figure 9-49).

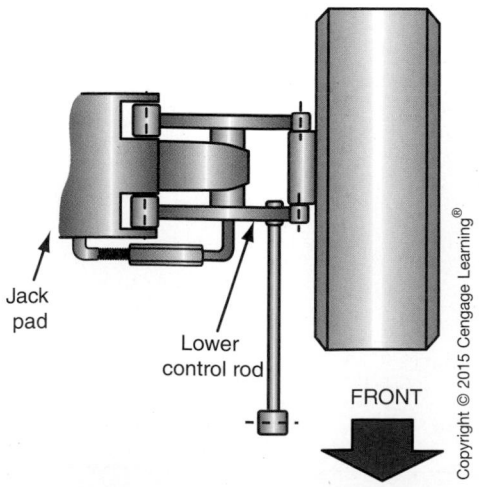

FIGURE 9-47 **Special toe adjusting tool for transverse leaf-spring rear suspension.**

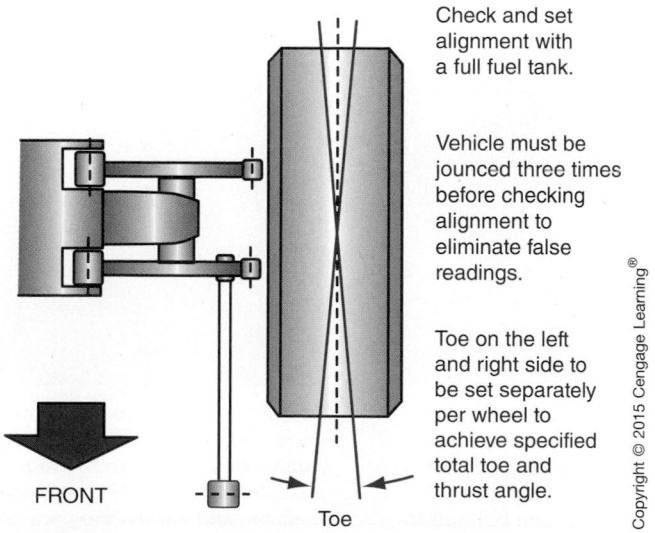

FIGURE 9-48 **Moving lower rear control rod position to adjust rear wheel toe.**

PHOTO SEQUENCE 16

Typical Procedure for Adjusting Rear Wheel Camber and Toe

P16-1 **Be sure the vehicle is properly positioned on an alignment ramp.**

Copyright © 2015 Cengage Learning®

P16-2 **Perform a prealignment inspection.**

Copyright © 2015 Cengage Learning®

P16-3 **Be sure the wheel aligner wheel sensors are properly connected to the vehicle.**

Copyright © 2015 Cengage Learning®

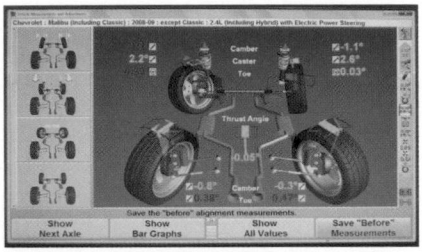

P16-4 **Obtain the front and rear alignment angles using the procedures recommended by the equipment manufacturer. Display the front and rear wheel alignment angles on the aligner monitor.**

Copyright © 2015 Cengage Learning®

P16-5 **Adjust the camber on the right rear wheel by loosening the cam bolt lock nut on the inner end of the right rear lower control arm and turning the eccentric cam until the camber setting is within specifications. Maintain the cam setting and tighten the lock nut to the specified torque.**

Copyright © 2015 Cengage Learning®

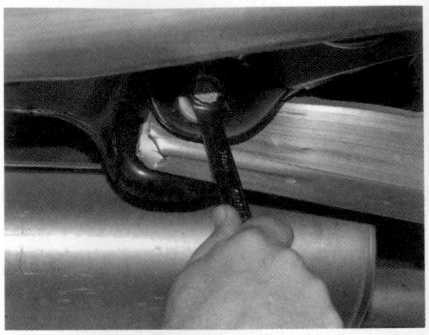

P16-6 **Adjust the camber on the left rear wheel by loosening the cam bolt lock nut on the inner end of the right rear lower control arm and turning the eccentric cam until the camber setting is within specifications. Maintain the cam setting and tighten the lock nut to the specified torque.**

Copyright © 2015 Cengage Learning®

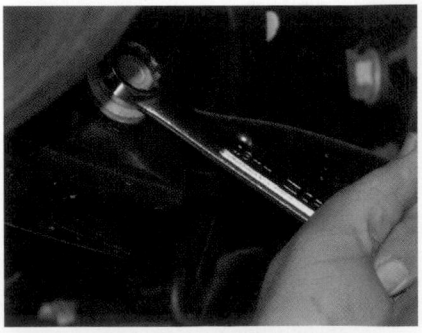

P16-7 **Loosen the cam bolt nut on the inner end of the left rear toe link, and rotate the cam bolt to the required toe specification. Maintain the cam bolt position, and tighten the cam bolt nut to the specified torque.**

Copyright © 2015 Cengage Learning®

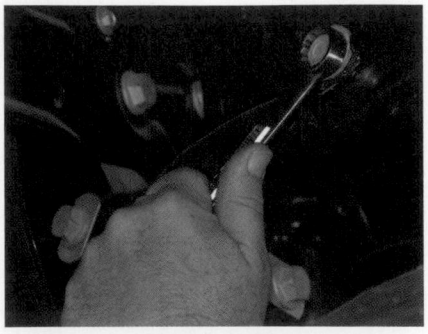

P16-8 **Loosen the cam bolt nut on the inner end of the right rear toe link, and rotate the cam bolt to the required toe specification. Maintain the cam bolt position, and tighten the cam bolt nut to the specified torque.**

Copyright © 2015 Cengage Learning®

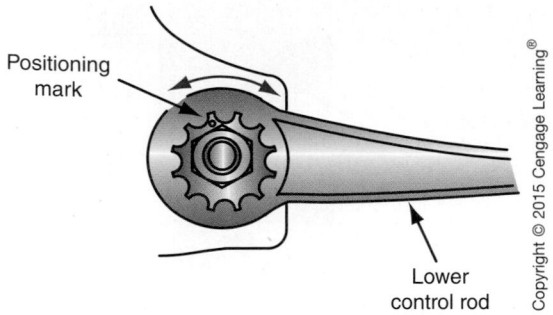

FIGURE 9-49 **Eccentric star wheel on inner end of lower control rod for toe adjustment.**

On some rear suspension systems with a lower control arm and ball joint, the nut on the tie rod is loosened, and the tie rod is rotated to adjust the rear wheel toe (Figure 9-50). After the toe is adjusted to specifications, the tie rod nut must be tightened to the specified torque.

On some front-wheel-drive cars, the rear wheel toe is adjusted by loosening the toe link bolt and moving the wheel in or out to adjust the toe (Figure 9-51). After the rear wheel toe is properly adjusted, the toe link bolt must be tightened to the specified torque. On other front-wheel-drive cars, the rear wheel toe is adjusted by loosening the nut on the inner adjustment link cam nut (Figure 9-52). The cam bolt is rotated to adjust the rear wheel toe. After the rear wheel toe is properly adjusted, the adjustment link cam nut must be tightened to the specified torque. Photo Sequence 17 illustrates front and rear suspension adjustments.

A variety of aftermarket bushing kits are available to adjust rear wheel toe. Some of these kits also adjust rear wheel comber (Figure 9-53).

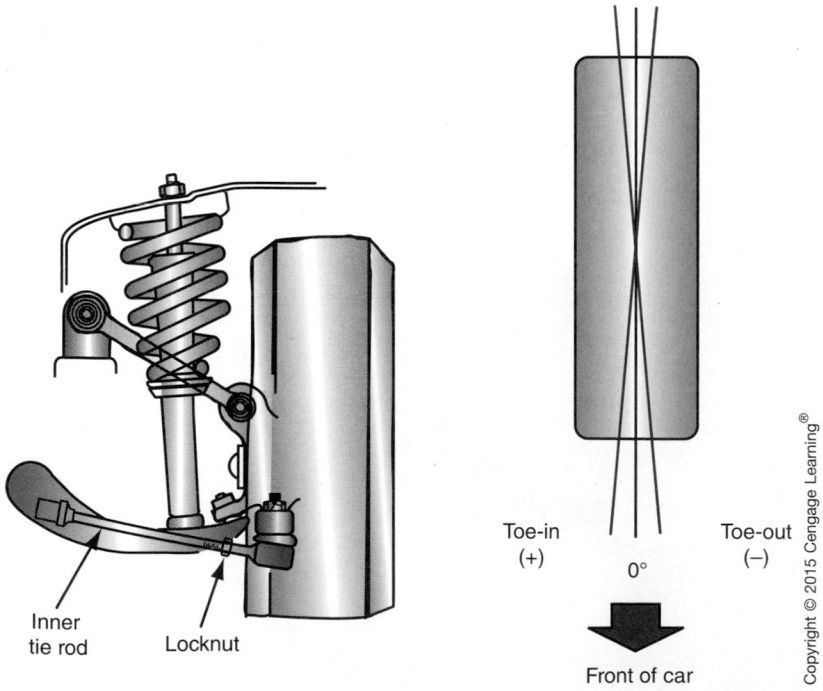

FIGURE 9-50 **Rotating rear tie rod to change tie rod length and adjust rear wheel toe.**

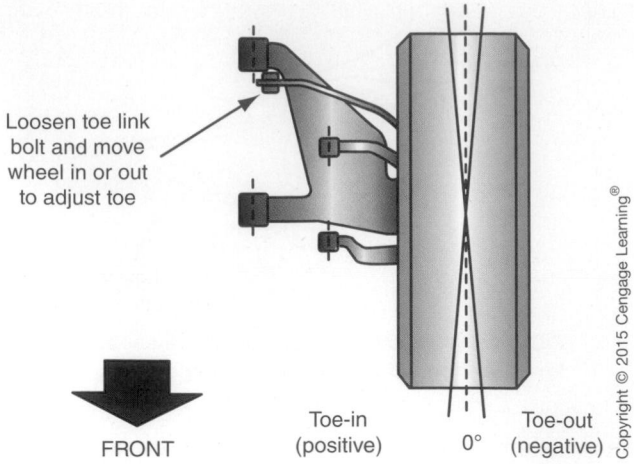

FIGURE 9-51 **Loosening toe link bolt to adjust rear wheel toe.**

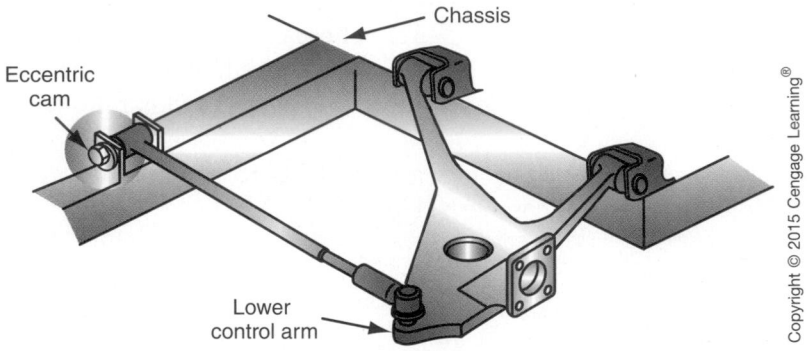

FIGURE 9-52 **Cam bolt for rear wheel toe adjustment.**

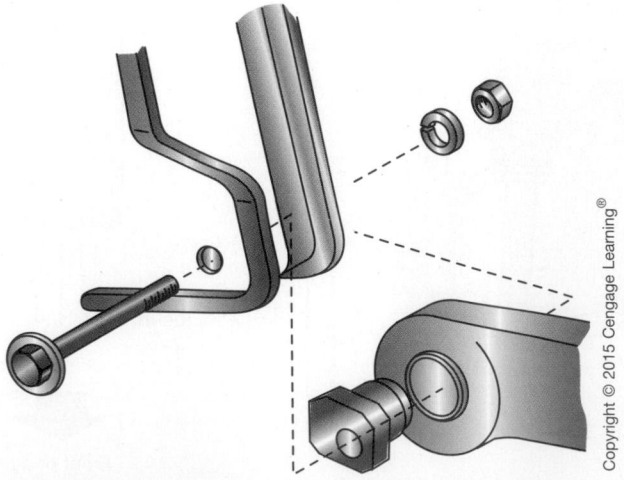

FIGURE 9-53 **Aftermarket bushing kit to adjust rear wheel toe and camber.**

PHOTO SEQUENCE 17

Typical Procedure for Performing Front and Rear Suspension Alignment Adjustments

P17-1 **Position vehicle properly on alignment rack with front wheels centered on turntables and rear wheels on slip plates.**

Copyright © 2015 Cengage Learning®

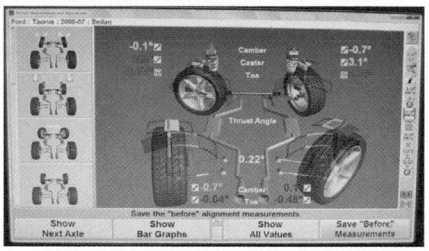

P17-2 **After prealignment inspection and front and rear ride height measurement, measure front and rear suspension angles with computer wheel aligner.**

Copyright © 2015 Cengage Learning®

P17-3 **Drill spot welds to remove alignment plate on top of the strut tower. After strut upper mount retaining nuts and alignment plate are removed, shift the strut inward or outward to adjust front wheel camber. Shift the upper strut mount forward or rearward to adjust front wheel caster.**

Copyright © 2015 Cengage Learning®

P17-4 **Loosen the locknuts and adjust the tie rod length to adjust front wheel toe.**

Copyright © 2015 Cengage Learning®

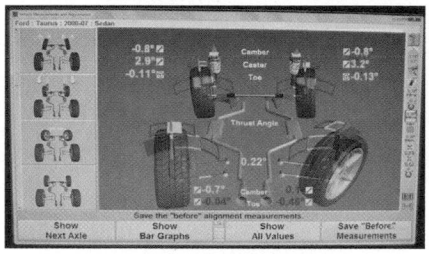

P17-5 **Recheck front suspension alignment readings on the monitor screen.**

Copyright © 2015 Cengage Learning®

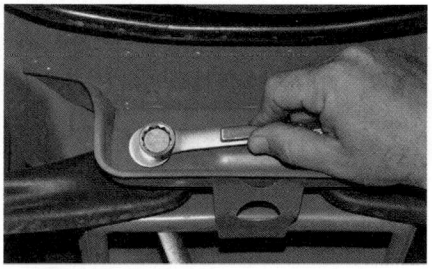

P17-6 **Adjust the left rear wheel camber and toe by rotating the cam bolt on the inner end of the lower control arm. If the rear wheel camber and toe have not been adjusted previously, a rear camber adjusting kit may have to be installed in the inner end of the lower control arm.**

Copyright © 2015 Cengage Learning®

P17-7 **Adjust the right rear wheel camber and toe by rotating the cam bolt on the inner end of the lower control arm. If the rear wheel camber and toe have not been adjusted previously, a rear camber adjusting kit may have to be installed in the inner end of the lower control arm.**

Copyright © 2015 Cengage Learning®

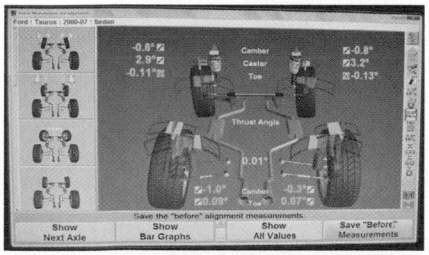

P17-8 **Recheck all front and rear alignment readings on the monitor screen. Be sure all readings are within specifications, including the thrust angle.**

Copyright © 2015 Cengage Learning®

Classroom Manual
Chapter 9, page 214

Front Camber Adjustment

Shims

A **camber adjustment** is usually a shim-type or eccentric cam-type mechanism that is used to correct the camber angle on front or rear wheels.

Various methods are provided by car manufacturers for camber adjustment. On older platforms some car manufacturers provide a shim-type **camber adjustment** between the upper control arm mounting and the inside of the frame (Figure 9-54). In this type of camber adjustment, increasing the shim thickness moves the camber setting to a more negative position, whereas decreasing shim thickness changes the camber toward a more positive position. Shims of equal thickness should be added or removed on both upper control arm mounting bolts to change the camber setting without affecting the caster setting.

Other vehicle manufacturers provide a shim-type camber adjustment between the upper control arm mounting and the outside of the frame (Figure 9-55). On this shim-type camber adjustment, increasing the shim thickness increases positive camber.

Older vehicles have shim-type camber adjustments, but newer vehicles have eccentric cams for this adjustment.

Eccentric Cams

Eccentric cams are out-of-round pieces of metal mounted on a retaining bolt with the shoulder of the cam positioned against a component. When the cam is rotated, the component contacting the cam shoulder is moved.

Some vehicles have **eccentric cams** on the inner ends of the upper control arm to provide camber adjustment (Figure 9-56), whereas other suspension systems have eccentric cams on the inner ends of the lower control arms (Figure 9-57). If the original cam adjustment on the inner ends of the upper control arms does not provide enough camber adjustment, aftermarket upper control arm shaft kits are available that provide an extra 1 1/2 degree of camber adjustment (Figure 9-58). The suspension adjustments shown in Figure 9-54 through Figure 9-58 are used on rear-wheel-drive cars with short-and-long arm front suspension systems.

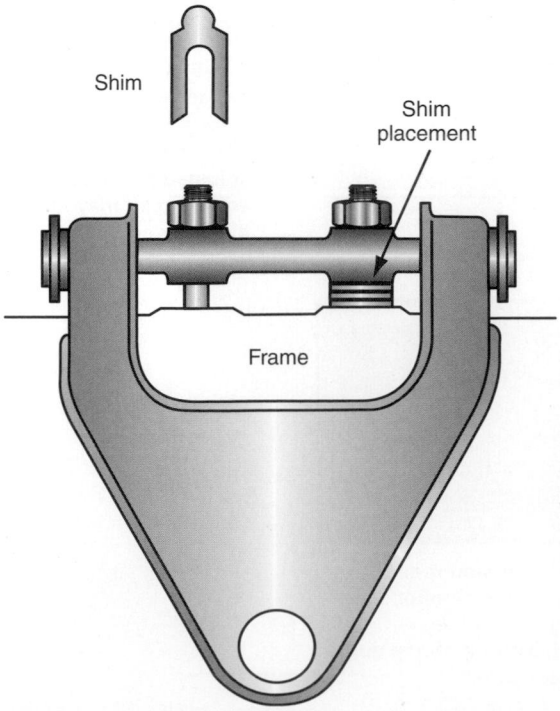

Copyright © 2015 Cengage Learning®

FIGURE 9-54 **Shim-type camber adjustment between upper control arm mounting and the inside of the frame.**

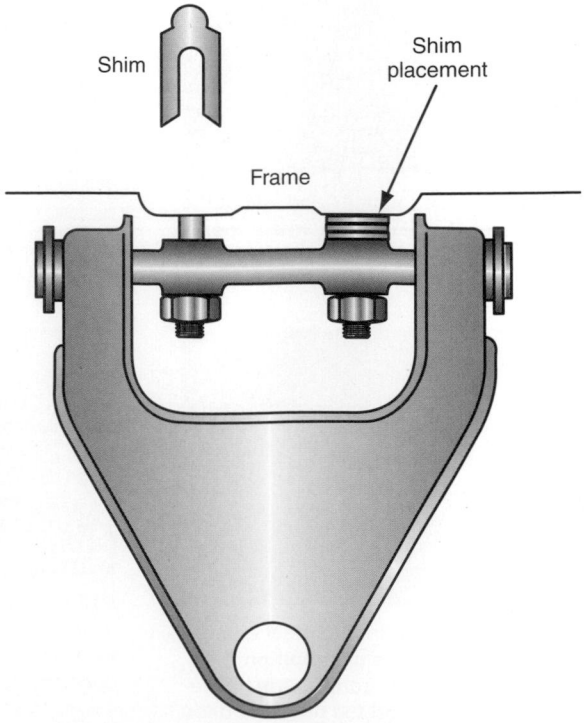

Copyright © 2015 Cengage Learning®

FIGURE 9-55 **Shim-type camber adjustment between upper control arm mounting and the outside of the frame.**

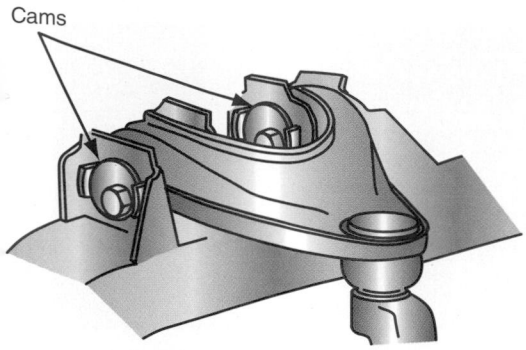

Copyright © 2015 Cengage Learning®

FIGURE 9-56 **Cam adjustment on the inside ends of the upper control arm.**

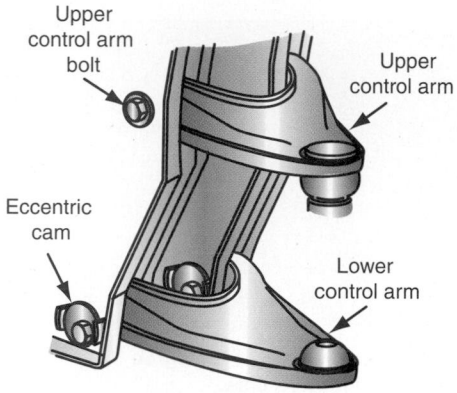

Copyright © 2015 Cengage Learning®

FIGURE 9-57 **Cam adjustment on the inner ends of the lower control arms.**

Copyright © 2015 Cengage Learning®

FIGURE 9-58 **Upper control arm shaft kit provides an extra 1 1/2 degree camber adjustment.**

Some MacPherson strut front suspension systems on front-wheel-drive cars have a cam on one of the steering-knuckle-to-strut bolts to adjust camber (Figure 9-59). Aftermarket camber adjustment kits that provide 2 1/2 degree of camber adjustment are available for many MacPherson strut suspension systems (Figure 9-60 and Figure 9-61). Similar camber adjustment kits are available for many cars with MacPherson strut front suspension systems. On some double-wishbone front suspension systems, a graduated cam on the inner end of the lower control arm provides a camber adjustment (Figure 9-62).

On late model Ford F-150 light-duty truck 4 × 2 front suspension systems, an alignment plate is installed on the upper control arm attaching nuts when the vehicle is manufactured (Figure 9-63). If a camber or caster adjustment is required, remove the control arm attaching nuts one at a time and install eccentric cams behind the adjusting nuts (Figure 9-64). Turn the eccentric cams equally on the front and rear upper control arm attaching nuts to adjust the camber. To increase the positive caster, move the front of the

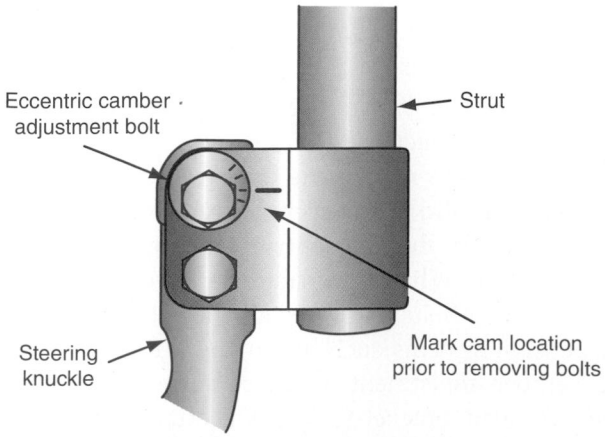

Copyright © 2015 Cengage Learning®

FIGURE 9-59 **Eccentric steering knuckle-to-strut camber adjusting bolt, MacPherson strut front suspension.**

Copyright © 2015 Cengage Learning®

FIGURE 9-60 **Camber adjustment kits provide 2 1/2 degree camber adjustment on some General Motors cars.**

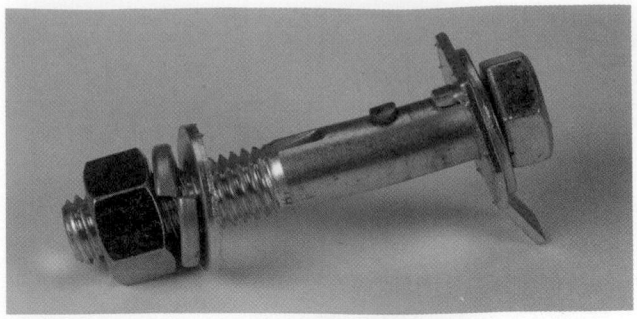

Copyright © 2015 Cengage Learning®

FIGURE 9-61 **Camber adjustment kits provide 2 1/2 degree camber adjustment on some vehicles.**

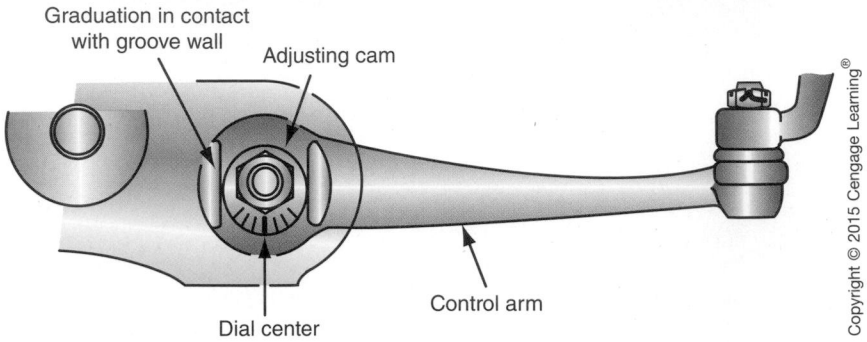

Copyright © 2015 Cengage Learning®

FIGURE 9-62 **Camber adjustment double-wishbone front suspension.**

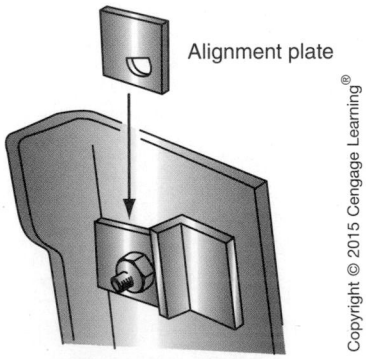

Copyright © 2015 Cengage Learning®

FIGURE 9-63 **Alignment plate on upper control arm attaching nut.**

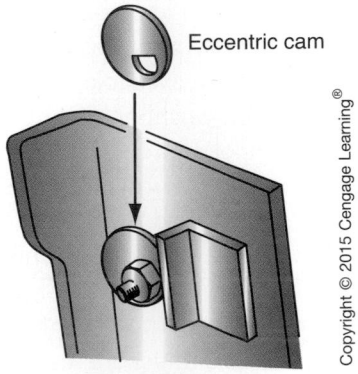

Copyright © 2015 Cengage Learning®

FIGURE 9-64 **Eccentric cam installed behind upper control arm attaching nut.**

upper control arm outboard and move the rear of the upper control arm inboard (Figure 9-65). If it is necessary to decrease the positive caster, move the front of the upper control arm inboard and move the rear of the upper control arm outboard (Figure 9-66). These front suspension systems also have eccentric cams on the lower control arm attaching bolts (Figure 9-67). However, these eccentric cams are only for factory use, and they must not be adjusted in service. When removing the lower control arms, always mark the eccentric cams on the attaching bolts and reinstall these cams in their original position.

In some front suspension systems such as on the PT Cruiser, eccentric cams are not installed to provide a camber adjustment. In these suspension systems, the manufacturer supplies slightly undersize strut bracket-to-knuckle attaching bolts that allow a 2 degree camber adjustment on each side of the front suspension. If a camber adjustment is required, remove the strut bracket-to-knuckle attaching bolts one at a time and install the

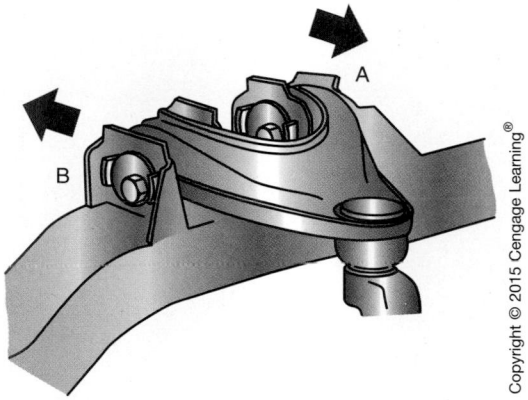

Copyright © 2015 Cengage Learning®

FIGURE 9-65 **To increase the positive caster, move the front of the upper control arm (A) outboard and move the rear of the upper control arm (B) inboard.**

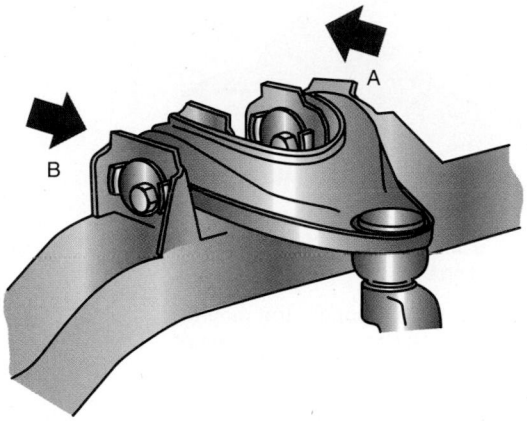

Copyright © 2015 Cengage Learning®

FIGURE 9-66 **To decrease the positive caster move the front of the upper control arm (A) inboard and move the rear of the upper control arm (B) outboard.**

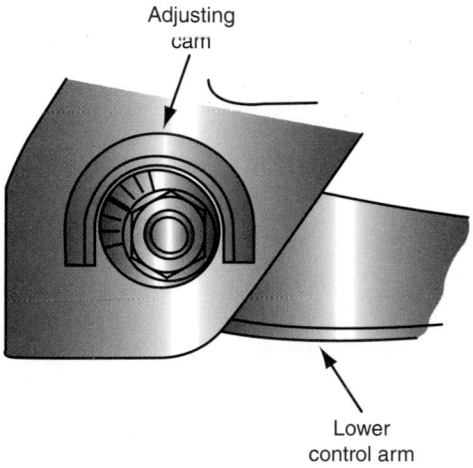

Copyright © 2015 Cengage Learning®

FIGURE 9-67 **Factory-set eccentric cam on lower control arm attaching bolts.**

undersize bolts. Install the nuts snugly on these bolts, but do not torque the nuts. Install each bolt from the rear of the knuckle. With the car on a wheel alignment rack, pull or push the top of the tire to obtain the specified camber. After the camber is set to specifications, tighten the strut bracket-to-knuckle attaching bolts to the specified torque.

Slotted Strut Mounts and Frames

On some MacPherson strut front suspension systems, the upper strut mounts may be loosened and moved inward or outward to adjust camber (Figure 9-68). Other MacPherson strut front suspension systems do not provide a camber adjustment. For these cars, aftermarket manufacturers sell slotted upper strut bearing plates that provide camber adjustment capabilities (Figure 9-69). If the front wheel camber adjustment is not correct, check for worn components such as ball joints or upper strut mounts. Also check the engine cradle/subframe mounts. Because the inner ends of the lower control arms are attached to the engine cradle on many front-wheel-drive cars, a cradle that is out of position may cause improper camber readings.

For example, if the vehicle has been impacted on the left front corner in a collision, the cradle may be shifted to the right. This action moves the bottom of the left front wheel inward and the bottom of the right front wheel outward. Therefore, the left front wheel has excessive positive camber and the right front wheel has too much negative camber.

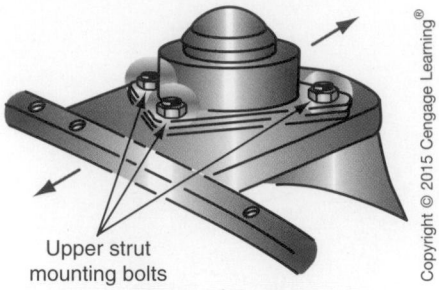

Copyright © 2015 Cengage Learning®

FIGURE 9-68 **The upper strut mounting bolts can be loosened and the mount moved inward or outward to adjust camber on some MacPherson strut front suspension systems.**

Copyright © 2015 Cengage Learning®

FIGURE 9-69 **Slotted upper strut bearing plates provide camber adjustment on some MacPherson strut front suspension systems.**

If the vehicle has been impacted on the left front corner, the cradle may be squashed to some extent with the left side being moved toward the right side without moving the right side of the cradle out of position. In this case the left front wheel has excessive positive camber, and the side-to-side cradle measurements are less than specified.

A camber plate is locked in place with a pop rivet on some upper strut mounts. The pop rivet may be removed to release the camber plate (Figure 9-70). After the upper strut mount bolts are loosened, the mount may be moved inward or outward to adjust the camber. When the camber is adjusted to the vehicle manufacturer's specification, the strut mount bolts should be tightened to the correct torque and a new pop rivet installed.

On some short-and-long arm front suspension systems, the bolt holes in the frame are elongated for the inner shaft on the upper control arm. When these bolts are loosened, the upper control arm may be moved inward or outward to adjust camber (Figure 9-71).

When a camber adjustment is required on some MacPherson strut front suspensions, the vehicle manufacturer recommends removing the strut-to-steering knuckle retaining bolts and removing the strut from the knuckle. A round file is then used to elongate the lower bolt opening in the strut (Figure 9-72). Reinstall the strut and retaining bolts, but leave the nuts slightly loose on the bolts. With the vehicle in the proper position on the wheel aligner rack, grasp the tire and move the top of the tire inward or outward to obtain the desired camber adjustment, and then tighten the nuts on the strut-to-knuckle bolts to the specified torque.

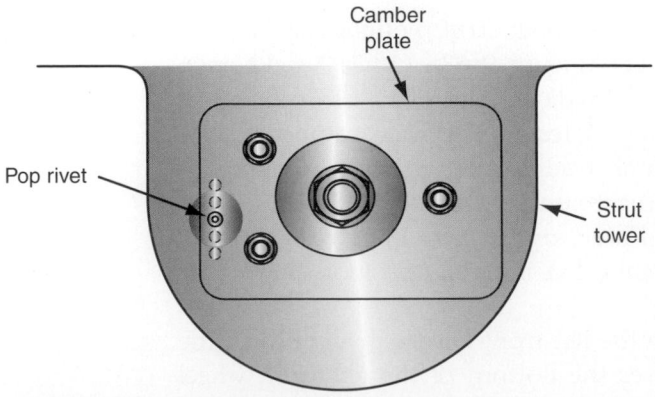

Copyright © 2015 Cengage Learning®

FIGURE 9-70 **Upper strut mount with camber plate and rivet, modified MacPherson strut front suspension.**

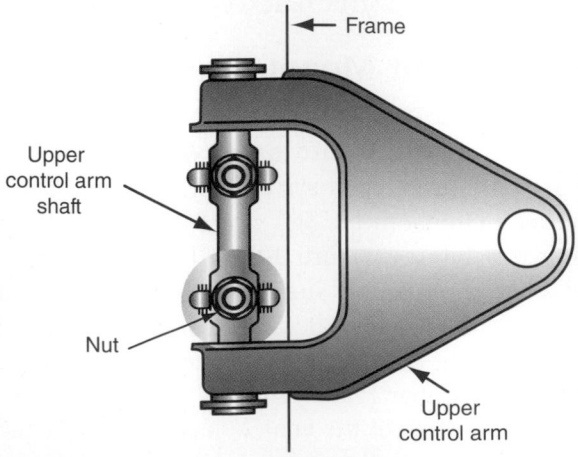

Copyright © 2015 Cengage Learning®

FIGURE 9-71 **Elongated bolt holes for upper control arm shaft bolts provide camber adjustment on some short-and-long arm front suspension systems.**

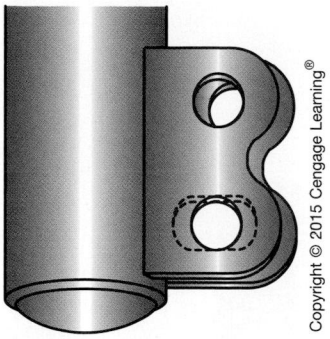

FIGURE 9-72 **Elongated lower strut-to-knuckle bolt hole for camber adjustment.**

Special Adjustment Bolts

On other vehicles such as the Dodge Magnum, the vehicle manufacturer recommends loosening the cradle/subframe mounts and shifting the cradle to provide minor camber or caster adjustments (Figure 9-73). If the desired camber adjustment cannot be obtained by shifting the cradle, the vehicle manufacturer provides an adjustment bolt package for camber and caster adjustment. These bolts are designed to replace the inboard bolts in the lower control arm and tension strut. To adjust camber use both bolts, and to adjust caster only use a replacement bolt in the tension strut.

The wheel alignment adjustment bolts have offset grooves cut into the length of the bolts (Figure 9-74). The grooves in the adjustment bolts force the bolt to be installed in one or two ways depending on whether more positive or negative camber is required. The bolt grooves work with "bat wing" holes in the lower control arm and tension strut bushings. The original nongrooved bolt is installed between the "bat wings" in the center of the tension strut or lower control arm bushing opening (Figure 9-75, view C). After the original bolt is removed from the tension strut or control arm, move the control arm or tension

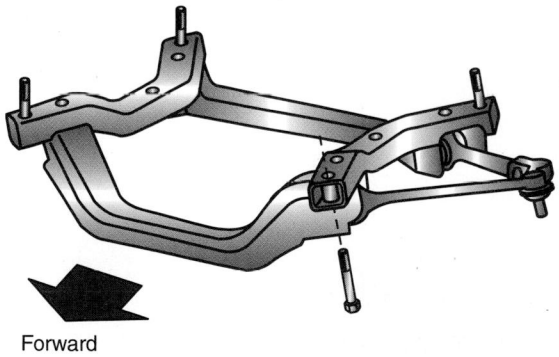

FIGURE 9-73 **Loosening cradle mounting bolts and shifting the cradle to perform minor camber adjustments.**

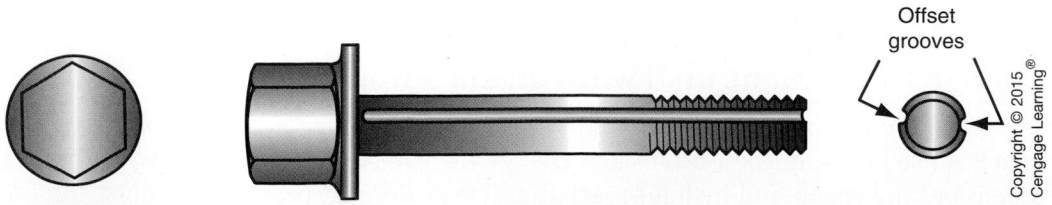

FIGURE 9-74 **Camber/caster adjustment bolt with offset grooves.**

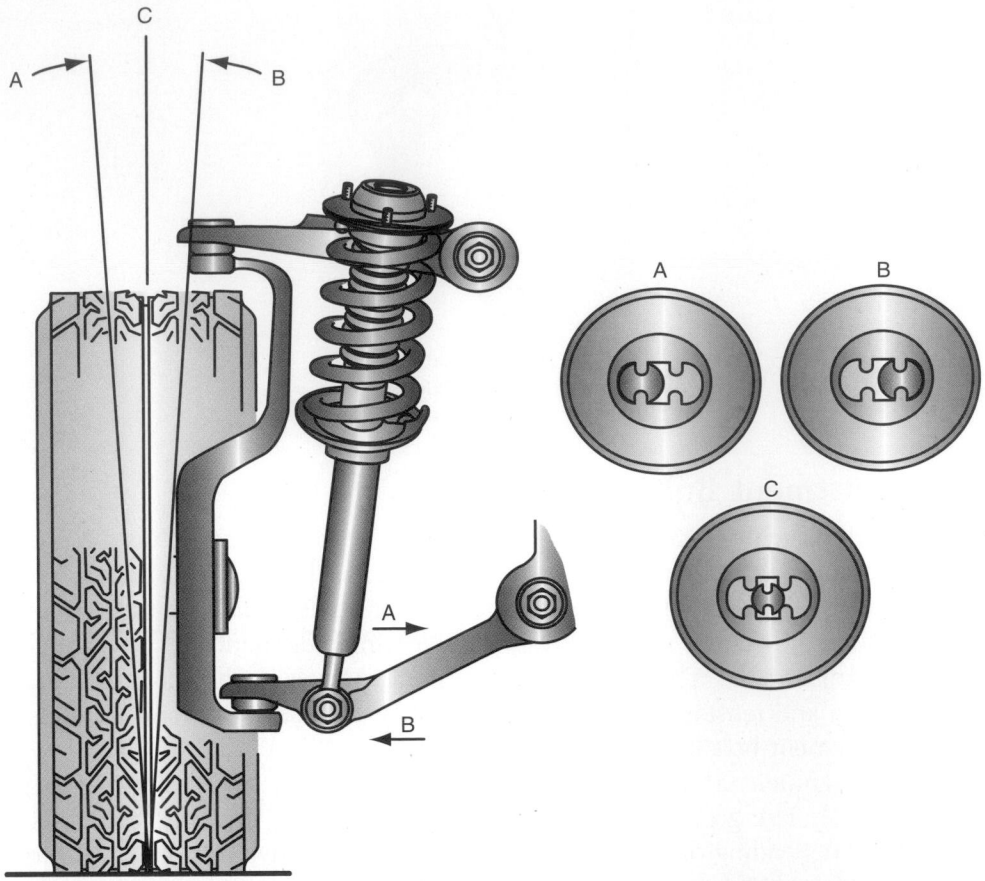

Copyright © 2015 Cengage Learning®

FIGURE 9-75 **Adjustment bolt positions in tension strut or lower control arm bushing to provide camber adjustment.**

SERVICE TIP:
The bolt in the lower control arm bushing must be installed from the rear and the tension strut bushing bolt must be installed from the front to avoid contact with other suspension components.

CAUTION:
After an adjustment bolt is installed, do not attempt to rotate the bolt. This action will damage the "bat wing" in the control arm or tension strut bushing.

CAUTION:
Some vehicle manufacturers do not recommend bending I-beams and other suspension components.

strut inward or outward as desired. To provide more positive camber, install the adjustment bolt grooves in the "bat wings" (Figure 9-75, view A). The adjustment bolts provide an average of 3 degree camber adjustment in either direction. Always install a washer under the head of the bolt. Installing the adjustment bolts in the "bat wings" (Figure 9-75, view b) provides more negative camber. After an adjustment bolt is installed, hold the head of the bolt with the proper size tool, and tighten the nut to the specified torque.

To adjust the front wheel caster, follow the same procedure, but install the adjustment bolt only in the tension strut.

Special Tools for Camber Adjustment

Some vehicle manufacturers provide a special tool for adjusting front wheel camber and caster. This tool has threaded projections on each end extending at a right angle from a turnbuckle in the center of the tool. When the turnbuckle is rotated, the tool is lengthened or shortened. The tool is installed into openings in the lower control arm and the chassis (Figure 9-76). After the bolt is loosened in the inner end of the control arm, the tool can be lengthened or shortened to adjust the camber or caster.

Camber Adjustment on Twin I-Beam Front Suspension System

On twin I-beam front suspension systems, a special clamp-type bending tool with a hydraulic jack is used to bend the I-beams and change the camber setting. On a twin I-beam front suspension, the clamp and hydraulic jack must be positioned properly to adjust the camber to a more positive setting (Figure 9-77) or a more negative setting (Figure 9-78).

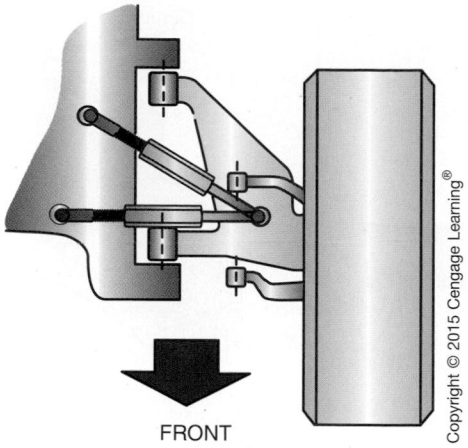

FIGURE 9-76 **Tool installed in lower control arm and chassis openings to adjust camber/caster.**

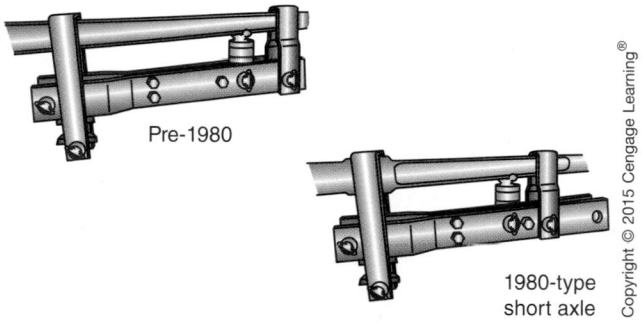

FIGURE 9-77 **Adjusting camber to a more positive setting, twin I-beam front suspension.**

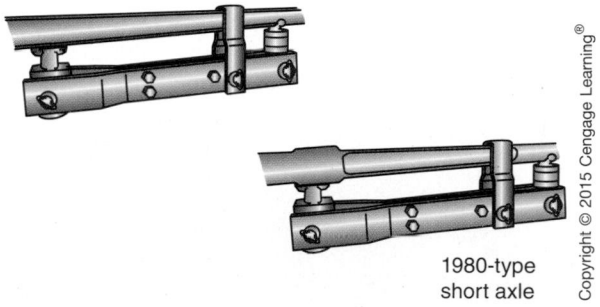

FIGURE 9-78 **Adjusting camber to a more negative setting, twin I-beam front suspension.**

Front Caster Adjustment Procedure

Strut Rod Length Adjustment

On some suspension systems, the nuts on the forward end of the strut rod may be adjusted to lengthen or shorten the strut rod, which changes the caster setting (Figure 9-79). Shorten the strut rod to increase the positive caster.

Eccentric Cams

The same eccentric cams on the inner ends of the upper or lower control arms may be used to adjust camber or caster (Figure 9-80). If an eccentric bushing in the outer end of the upper control arm is rotated to adjust camber, this same eccentric also adjusts caster. On some double-wishbone front suspension systems, the pivot adjuster mounting bolt nuts must be loosened under the compliance pivot, and the graduated cam may then be rotated to adjust caster (Figure 9-81).

Slotted Strut Mounts and Frames

If a **caster adjustment** is required and the suspension does not have an adjustment for this purpose, check for worn components such as upper strut mounts and ball joints. Also check the engine cradle mounts. Because the inner ends of the lower control arms are attached to the engine cradle on many front-wheel-drive cars, a cradle that is out of position may cause improper caster readings.

SERVICE TIP: The vehicle manufacturer does not recommend bending the I-beams to adjust front wheel camber. However, this procedure is suggested by some equipment manufacturers.

Classroom Manual
Chapter 9, page 217

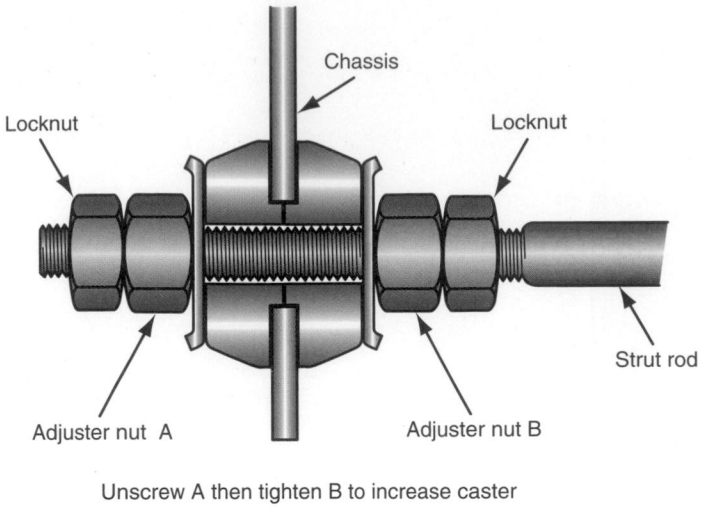

Copyright © 2015 Cengage Learning®

FIGURE 9-79 **Strut rod nuts can be adjusted to lengthen or shorten the strut rod to adjust caster.**

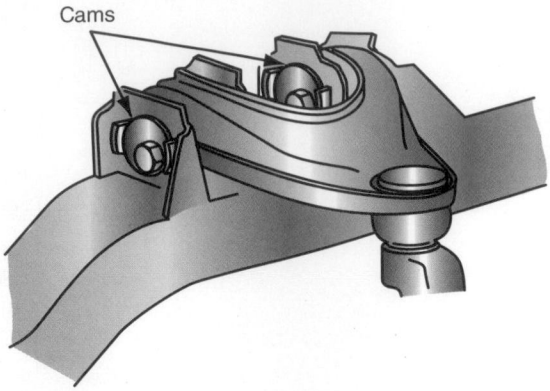

Copyright © 2015 Cengage Learning®

FIGURE 9-80 **Eccentric cams on the inner ends of the upper or lower control arms can be used to adjust camber and caster.**

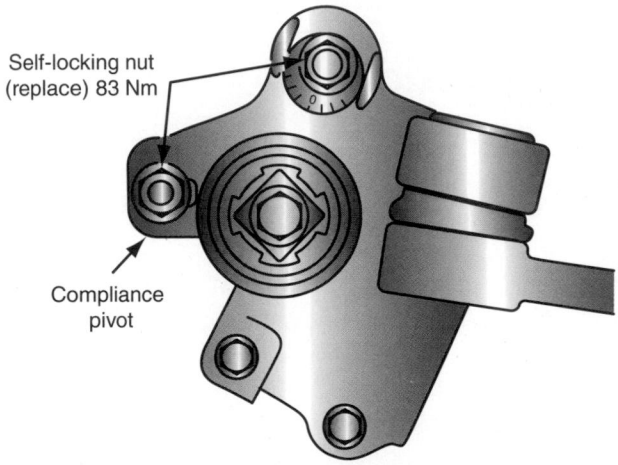

Copyright © 2015 Cengage Learning®

FIGURE 9-81 **Graduated cam for caster adjustment, double-wishbone front suspension.**

Copyright © 2015 Cengage Learning®

FIGURE 9-82 **Elongating upper strut mount bolt openings to provide caster adjustment.**

The slots in the frame at the upper control arm shaft mounting bolts provide camber and caster adjustment. If the upper strut mount is adjustable on a MacPherson strut suspension, the mount retaining bolts may be loosened and the mount moved forward or rearward to adjust caster.

A strut rod may be called a radius rod.

When a caster adjustment is necessary, some vehicle manufacturers recommend removing the upper strut mount bolts and elongating the bolt hole openings in the strut tower with a round file to provide a caster adjustment (Figure 9-82).

WARNING: After any alignment angle adjustment, always be sure the adjustment bolts are tightened to the vehicle manufacturer's specified torque. Loose suspension adjustment bolts will result in improper alignment angles, steering pull, and tire wear. Loose suspension adjustment bolts may also cause reduced directional stability, which could result in a vehicle accident and possible personal injury.

Caster Adjustment on Twin I-Beam Front Suspension Systems

WARNING: Improper use of bending equipment may result in personal injury. Always follow the equipment and vehicle manufacturer's recommended procedures.

SPECIAL TOOL

Twin I-beam radius rod bending tool

The same bending tool and hydraulic jack may be used to adjust camber and caster on twin I-beam front suspension systems. This equipment is installed on the radius rods connected from the I-beams to the chassis to adjust caster. If the hydraulic jack is placed near the rear end of the radius rod and the clamp arm is positioned over the center area of this rod, the rod is bent downward in the center when the jack is operated (Figure 9-83). This bending action tilts the I-beam and caster line rearward to increase positive caster.

Aftermarket parts manufacturers supply a special radius arm bushing to provide a caster adjustment on some twin I-beam front suspension systems (Figure 9-84).

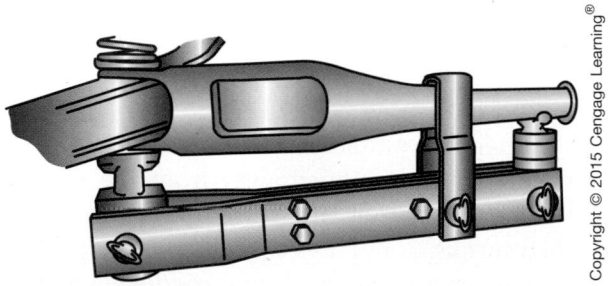

Copyright © 2015 Cengage Learning®

FIGURE 9-83 **Hydraulic jack and bending tool positioned on the radius rod to increase positive caster on a twin I-beam front suspension.**

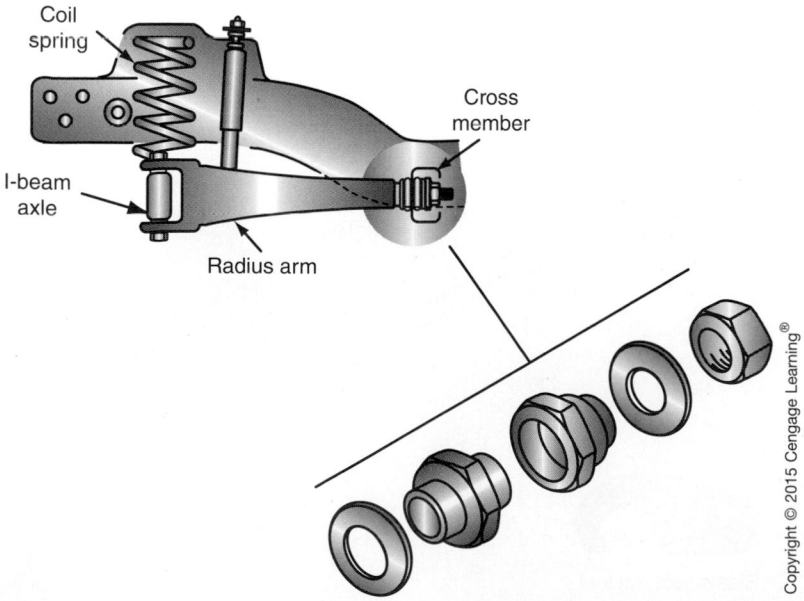

Copyright © 2015 Cengage Learning®

FIGURE 9-84 **Radius arm bushing for caster adjustment on some twin I-beam suspension systems.**

Toe Adjustment

Classroom Manual
Chapter 9, page 229

WARNING: Do not heat tie rod sleeves to loosen them. This action may weaken the sleeves, resulting in sudden sleeve failure, loss of steering control, personal injury, and vehicle damage.

SPECIAL TOOL
Tie rod sleeve rotating tool

WARNING: Do not use a pipe wrench to loosen tie rod sleeves. This tool may crush and weaken the sleeve, causing sleeve failure, loss of steering control, personal injury, and vehicle damage. A tie rod rotating tool must be used for this job.

Total toe is the sum of the toe settings on the front wheels.

When a front or rear wheel toe adjustment is required, apply penetrating oil to the tie rod adjusting sleeves and sleeve clamp bolts. Loosen the tie rod adjusting sleeve clamp bolts enough to allow the clamps to partially rotate. One end of each tie rod sleeve contains a right-hand thread; the opposite end contains a left-hand thread. These threads match the threads on the tie rod and outer tie rod end. When the tie rod sleeve is rotated, the complete tie rod, sleeve, and tie rod end assembly is lengthened or shortened. Use a tie rod sleeve rotating tool to turn the sleeves until the toe-in on each front wheel is equal to one-half the total toe specification (Figure 9-85).

After the toe-in adjustment is completed, the adjusting sleeve clamps must be turned so the openings in the clamps are positioned away from the slots in the adjusting sleeves (Figure 9-86). Replace clamp bolts that are rusted, corroded, or damaged, and tighten these bolts to the specified torque.

On some rack and pinion steering gears, the tie rod locknut is loosened and the tie rod is rotated to adjust toe on each front wheel (Figure 9-87). Prior to rotating the tie rod, the small outer bellows boot clamp should be removed to prevent it from twisting during tie rod rotation. After the toe-in adjustment, the tie rod locknut must be torqued to specifications, and the small bellows boot clamp must be reinstalled.

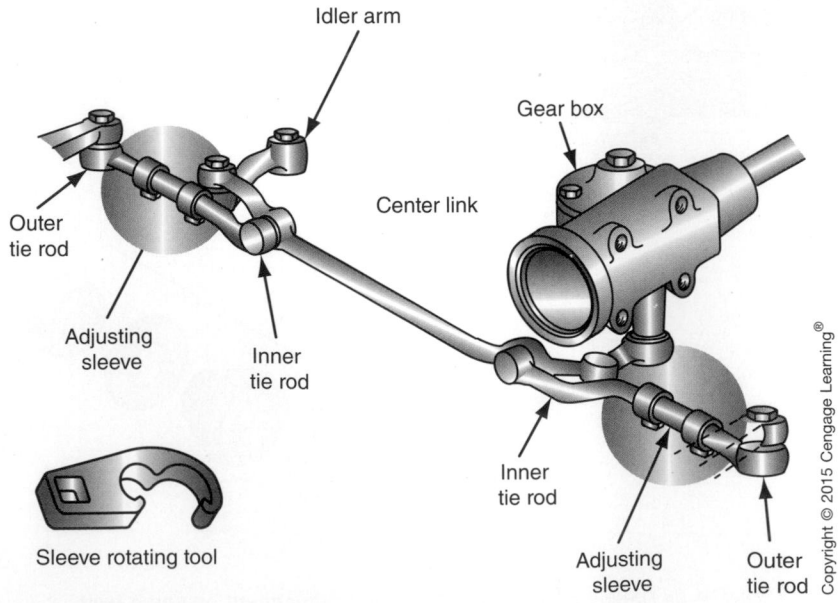

FIGURE 9-85 **Rotating tie rod sleeves to adjust front wheel toe.**

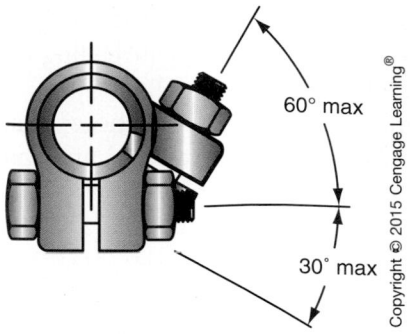

FIGURE 9-86 **Proper tie rod adjusting sleeve and clamp position.**

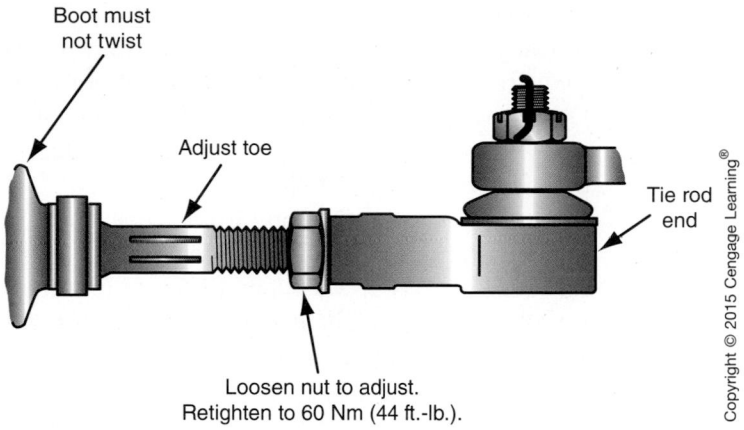

FIGURE 9-87 **Rotating tie rod to adjust the toe on a rack and pinion steering gear.**

On other rack and pinion steering gears, the tie rod and outer tie rod end have internal threads, and a threaded adjuster is installed in these threads. One end of the adjuster has a right-hand thread, and the opposite end has a left-hand thread. Matching threads are located in the outer tie rod end and tie rod. Clamps on the outer tie rod end and tie rod secure these components to the adjuster. A hex-shaped nut is designed into the center of the adjuster (Figure 9-88). After the clamps are loosened, an open-end wrench is placed on this hex nut to rotate the adjuster and change the toe setting.

When the toe-in adjustment is completed, the toe must be set within the manufacturer's specifications with the steering wheel in the centered position. If the steering wheel is not centered, a centering adjustment is required.

Checking Toe Change and Steering Linkage Height

If the steering linkage on each side is not the same height, a condition called bump steer may occur. When bump steer occurs, the steering suddenly swerves to one side when one of the front wheels strikes a road irregularity. When this steering linkage condition is present, abnormal toe changes occur during wheel jounce and rebound.

Follow this procedure to check for abnormal toe changes during wheel jounce and rebound:

1. Be sure the toe is adjusted to specifications with the front wheels centered and the steering wheel locked.

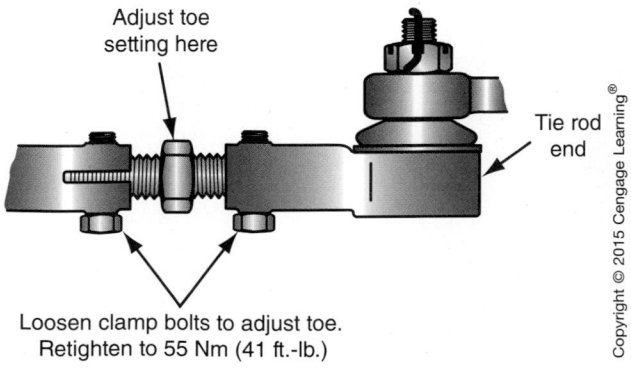

FIGURE 9-88 **Rack and pinion steering gear with externally threaded toe adjuster and internal threads on the tie rod and outer tie rod end.**

2. Pull the chassis downward approximately 3 in. (7.6 cm) and observe the toe change. It is acceptable if the toe setting on each front wheel remains at the original reading or if each front wheel toes in or toes out an equal small amount (Figure 9-89).
3. During wheel jounce, if one front wheel toes inward or outward while the opposite wheel remains at the original setting, the steering linkage height is not equal and must be corrected (Figure 9-90).
4. During wheel jounce, if one front wheel toes in and the opposite front wheel toes out, the steering linkage height is unequal and must be corrected (Figure 9-91).
5. Push the chassis upward 3 in. (7.6 cm) and inspect the toe change (Figure 9-92). If the steering linkage height is equal, the toe setting on each front wheel remains the same or moves the same amount to a toe-in or toe-out position.
6. If the toe on one front wheel remains at the original setting while the opposite front wheel toe changes to a toe-in or toe-out setting, the steering linkage height is unequal. When one front wheel moves to a toe-in position, and the opposite front wheel moves to a toe-out setting, unequal steering linkages are indicated.

On some imported vehicles, the steering linkage height can be adjusted with shims between the rack and pinion steering gear and the chassis (Figure 9-93). The idler arm may be moved upward or downward on some domestic vehicles to adjust the steering

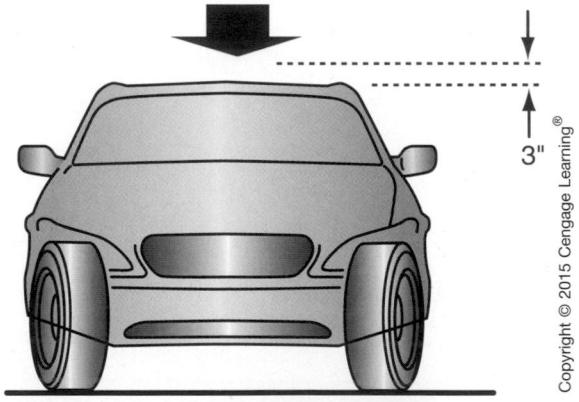

FIGURE 9-89 **Normal toe change during wheel jounce.**

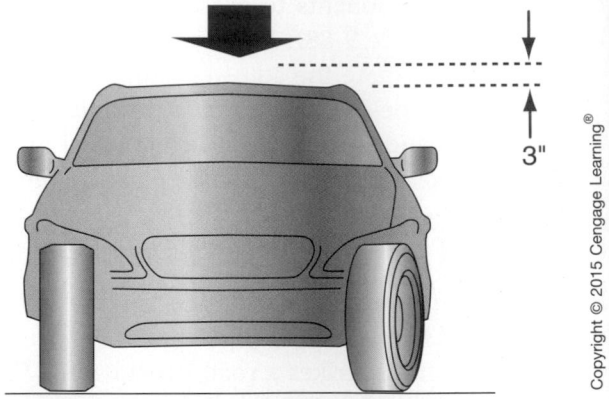

FIGURE 9-90 **If the toe change on one front wheel remains at the original setting while the toe on the opposite wheel toes in or out during wheel jounce, unequal steering linkage height is indicated.**

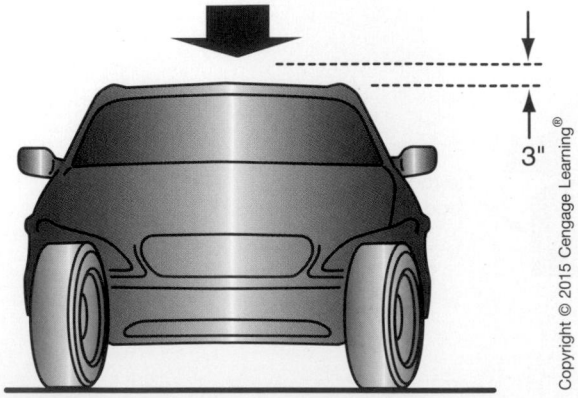

FIGURE 9-91 **If the toe on one front wheel toes in and the toe on the opposite front wheel toes out during wheel jounce, unequal steering linkage height is indicated.**

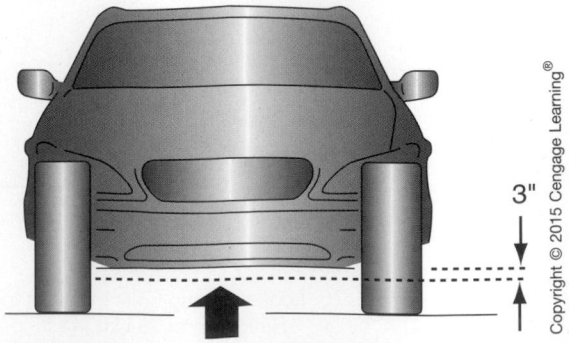

FIGURE 9-92 **Pull the chassis upward 3 in. (7.6 cm) to check for improper toe change indicating unequal steering linkage height.**

linkage height. On other vehicles, if the steering linkage height is unequal, steering components are worn or bent. These components include tie rods, tie rod ends, idler arms, pitman arms, and rack and pinion steering gear mounting bushings.

Another method of inspecting for equal tie rod height on rack and pinion steering gears is to measure the distance from the center of the inner retaining bolt on the lower control arm to the alignment ramp on each side of the front suspension (distances A and B in Figure 9-94). If this distance is not equal on each side of the front suspension, one of the front springs may be sagged or such components as control arm bushings may be worn. When distances A and B are not equal, this problem must be corrected before performing the steering gear measurements. If distances A and B are equal, measure from the outer ends of the rack and pinion steering gear to the alignment ramp (distances C and D). If these distances are not equal, the rack and pinion steering gear mountings may be worn or distorted.

SERVICE TIP:
The most accurate check of a properly centered steering wheel is while driving the vehicle straight ahead during a road test.

Manual Steering Wheel Centering Procedure

Road test the vehicle and determine if the steering wheel spoke is centered when the vehicle is driven straight ahead.

If steering wheel centering is necessary, follow this procedure:

1. Lift the front end of the vehicle with a hydraulic jack and position safety stands under the lower control arms. Lower the vehicle onto the safety stands and place the front wheels in the straight-ahead position.

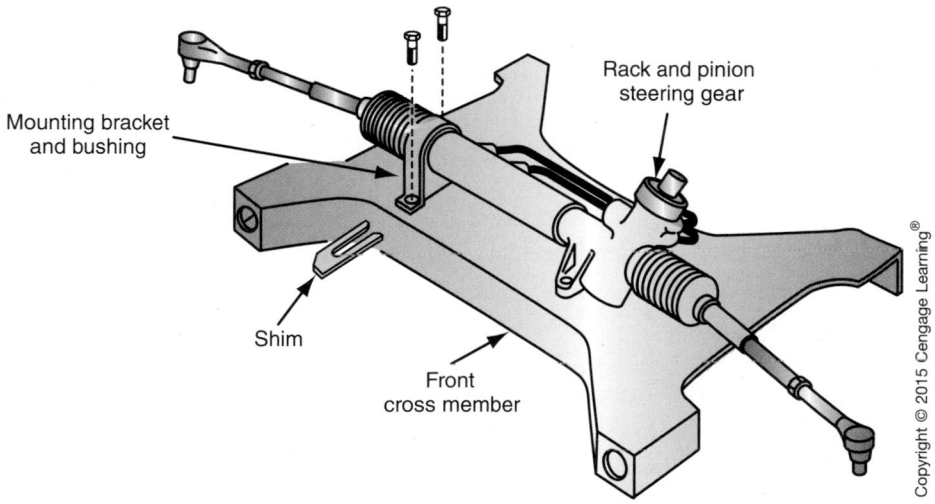

FIGURE 9-93 **On some imported vehicles, shims between the rack and pinion steering gear and the chassis adjust the steering linkage height.**

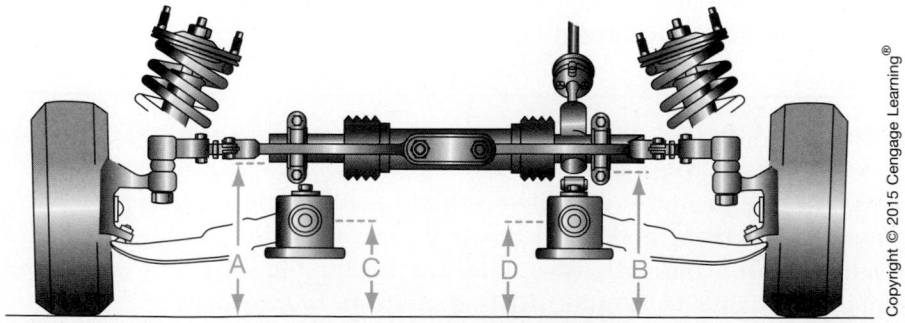

FIGURE 9-94 **Measuring rack and pinion steering gear height at various locations.**

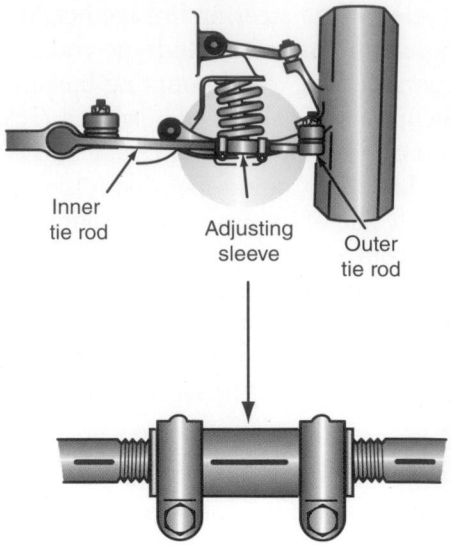

Copyright © 2015 Cengage Learning®

FIGURE 9-95 **Marking tie rod sleeves in relation to the tie rods and outer tie rod ends.**

2. Use a piece of chalk to mark each tie rod sleeve in relation to the tie rod, and loosen the sleeve clamps (Figure 9-95).
3. Position the steering wheel spoke in the position it was in while driving straight ahead during the road test. Turn the steering wheel to the centered position and note the direction of the front wheels.
4. If the steering wheel spoke is low on the left side while driving the vehicle straight ahead, use a tie rod sleeve rotating tool to shorten the left tie rod and lengthen the right tie rod (Figure 9-96). One-quarter turn on a tie rod sleeve moves the steering wheel position approximately one inch. Turn the tie rod sleeves the proper amount to bring the steering wheel to the centered position. For example, if the steering wheel spoke is two inches off-center, turn each tie rod sleeve one-half turn.
5. If the steering wheel spoke is low on the right side while driving the vehicle straight ahead, lengthen the left tie rod and shorten the right tie rod.
6. Mark each tie rod sleeve in its new position in relation to the tie rod. Be sure the sleeve clamp openings are positioned properly, as indicated earlier in this chapter. Tighten the clamp bolts to the specified torque.
7. Lift the front chassis with a floor jack and remove the safety stands. Lower the vehicle onto the shop floor, and check the steering wheel position during a road test.

SERVICE TIP: Some computer wheel aligners have a steering wheel centering display that provides an easier method of steering wheel centering compared to the manual method.

Classroom Manual
Chapter 9, page 233

When all the front suspension alignment angles are adjusted within the manufacturer's specifications, road test the vehicle. When the vehicle is driven on a relatively smooth, straight road, there should be directional stability with no drift to the right or left, and the steering wheel must be centered (Table 9-1).

CUSTOMER CARE: Always concentrate on quality workmanship and customer satisfaction. Most customers do not mind paying for vehicle repairs if the work is done properly and their vehicle problem is corrected. To determine customer satisfaction, make follow-up phone calls a few days after repairing their vehicle. This indicates that you are interested in their vehicle and that you consider quality work and satisfied customers a priority.

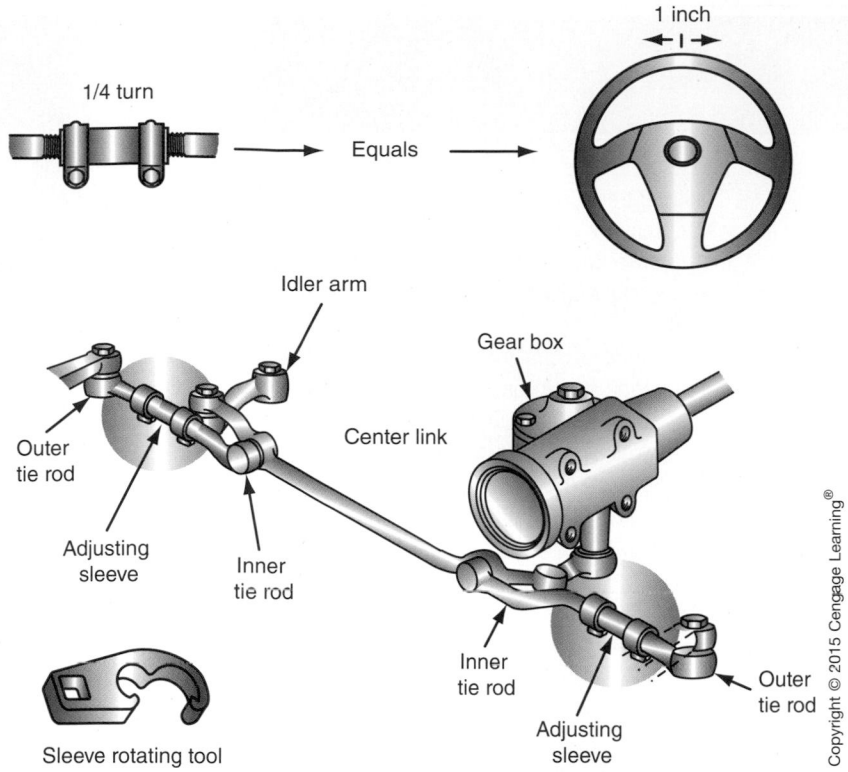

FIGURE 9-96 **Rotating tie rod sleeves to center steering wheel.**

TABLE 9-1 **STEERING AND WHEEL ALIGNMENT DIAGNOSIS**

Problem	Symptoms	Possible Causes
Tire tread wear, inside edge	Steering pull, premature tire replacement	Excessive negative camber
Tire tread wear, outside edge	Steering pull, premature tire replacement	Excessive positive camber
Tire tread wear, feathered	Premature tire replacement	Improper toe setting
Tire tread wear, cupped	Wheel vibration	Improper wheel balance
Steering wander	Reduced directional stability	Reduced positive caster or negative caster
Steering pull	Steering pull to the right when driving straight ahead	Reduced positive caster, right front wheel Excessive positive camber, right front wheel Excessive toe-out, left rear wheel
Steering pull	Steering pull to the left when driving straight ahead	Reduced positive caster, right front wheel Excessive positive camber, right front wheel Excessive toe-out, right rear wheel
Steering wheel return	Excessive steering wheel returning force after a turn	Excessive positive caster on front wheels
Steering wheel return	Steering wheel does not return properly after a turn	Binding column or linkage Reduced positive caster on front wheels
Harsh riding	Reduced ride quality when driving on road irregularities	Worn shock absorbers or struts Excessive positive caster, front wheels Reduced ride height

Copyright © 2015 Cengage Learning®

TERMS TO KNOW

Advanced vehicle handling (AVH)
Alignment ramp
Brake pedal depressor
Bump steer
Camber adjustment
Caster adjustment
Caster offset
Caster trail
Control arm movement monitor
Digital adjustment photos

TERMS TO KNOW
(continued)

Digital signal processor (DSP)
Eccentric cams
Front and rear wheel alignment angle screen
High-frequency transmitter
Included angle
Memory steer
Part-finder database
Prealignment inspection
Preliminary inspection screen
Rear wheel camber
Rear wheel toe
Ride height
Rim clamps
Road test
Setback
Shim display screen
Slip plates
Thrust line
Toe-in
Toe-out
Torque steer
Total toe

CASE STUDY

A customer complained about erratic steering on a front-wheel-drive Dodge Intrepid. A road test revealed the car steered reasonably well on a smooth road surface, but while driving on irregular road surfaces, the steering would suddenly swerve to the right or left.

The technician performed a preliminary wheel alignment inspection and found the right tie rod end was loose; all the other suspension and steering components were in satisfactory condition. The technician replaced the loose tie rod end, but a second road test indicated that the bump steer problem was still present. After advising the customer that a complete wheel alignment was necessary, the technician drove the vehicle on the wheel aligner and carefully checked all front and rear alignment angles. Each front and rear wheel alignment angle was within specifications. The technician realized that somehow he had not diagnosed this problem correctly.

While thinking about this problem, the technician remembered a general diagnostic procedure he learned while studying automotive technology. This procedure stated: Listen to the customer complaints, be sure the complaint is identified, think of the possible causes, test to locate the exact problem, and be sure the complaint is eliminated. The technician realized he had not thought much about the causes of the problem, and so he began to recall the wheel alignment theory he learned in college. He remembered that the tie rods must be parallel to the lower control arms, and if the tie rod height is unequal, this parallel condition no longer exists. The technician also recalled that unequal tie rod height causes improper toe changes during wheel jounce and rebound, which result in bump steer.

An inspection of the toe during front wheel jounce and rebound indicated the toe on the right front wheel remained the same during wheel jounce and rebound, but the toe on the left front wheel moved to a toe-out position. Because the tie rods had been inspected during the preliminary alignment inspection, the technician turned his attention to the rack and pinion steering gear mounting. He found the bushing on the right end of the steering gear was worn and loose. This bushing was replaced and all the steering gear mounting bolts were tightened to the specified torque. An inspection of the toe change during wheel jounce and rebound revealed a normal toe change.

From this experience the technician learned the following two things:

1. His understanding of wheel alignment theory was very important in diagnosing steering problems.
2. Always be thorough! During a prealignment inspection, check all suspension and steering components, including rack and pinion steering gear mountings.

ASE-STYLE REVIEW QUESTIONS

1. While discussing a front suspension height that is 1 in. (2.54 cm) less than specified:
   Technician A says the suspension height must be correct before a wheel alignment is performed.
   Technician B says the lower front suspension height may be caused by worn lower control arm bushings.
   Who is correct?
   A. A only
   B. B only
   C. Both A and B
   D. Neither A nor B

2. While performing a prealignment inspection:
   Technician A says improper front wheel bearing adjustment may affect wheel alignment angles.
   Technician B says worn ball joints have no effect on wheel alignment angles.
   Who is correct?
   A. A only
   B. B only
   C. Both A and B
   D. Neither A nor B

3. While discussing a front suspension system in which the right front wheel has 2 degree positive camber and the left front wheel has 1/2 degree positive camber:

   Technician A says when the vehicle is driven straight ahead, the steering will pull to the left.

   Technician B says there will be excessive wear on the inside edge of the left front tire tread.

   Who is correct?

   A. A only
   B. B only
   C. Both A and B
   D. Neither A nor B

4. When discussing unsatisfactory steering wheel returnability:

   Technician A says the rack and pinion steering gear mounts may be worn.

   Technician B says this problem may be caused by interference between the dash seal and the steering shaft.

   Who is correct?

   A. A only
   B. B only
   C. Both A and B
   D. Neither A nor B

5. While discussing turning radius measurement:

   Technician A says a bent steering arm will cause the turning radius to be out of specification.

   Technician B says if the turning radius is not within specification, tire tread wear is excessive while cornering.

   Who is correct?

   A. A only
   B. B only
   C. Both A and B
   D. Neither A nor B

6. While measuring and adjusting front wheel toe:

   A. If the positive caster is increased on the right front wheel, this wheel moves toward a toe-in position.
   B. Improper front wheel toe setting causes steering wander and drift.
   C. The front wheel toe should be checked with the front wheels straight ahead.
   D. Improper front wheel toe setting causes wear on the inside edge of the tire tread.

7. All of these statements about front wheel toe change during wheel jounce and rebound are true EXCEPT:

   A. If one front wheel toes in and the opposite front wheel toes out during front wheel jounce and rebound, the tie rod height is unequal.
   B. If both front wheels toe in or toe out a small, equal amount during front wheel jounce and rebound, the tie rods are parallel to the lower control arms.
   C. The improper toe change during front wheel jounce and rebound may cause bump steer.
   D. Improper toe change during front wheel jounce and rebound may be caused by a worn upper strut mount.

8. While using a computer wheel aligner:

   A. The technician may select defective suspension components from a list on the screen.
   B. It is not necessary to check or compensate for wheel runout.
   C. If the camber bar graph is red, the camber setting is within specifications.
   D. A wheel sensor containing a high-frequency transmitter requires a cable connected to the aligner computer.

9. While using computer wheel aligners:

   A. On many computer wheel aligners, the technician may only print out the four-wheel alignment results.
   B. Some wheel sensors have the capability to measure ride height and display this reading on the screen.
   C. A front wheel swing is necessary before reading the front wheel camber.
   D. If the computer aligner contains a control arm movement monitor, the technician has to estimate the necessary shim thickness.

10. While using computer wheel aligners:

   A. Symmetry angle measurements display thrust angle, rear wheel toe, and rear wheel camber.
   B. Setback is the angle formed by a line at 90 degrees to the vehicle centerline at the axle attachment point and a line through the left and right wheel centers.
   C. Wheelbase difference is an angle created by a line through the rear wheel centers and the thrust line.
   D. Right or left lateral offset is an angle between a line through the left front and left rear tires and a line between the right front and right rear tires.

ASE CHALLENGE QUESTIONS

1. While discussing turning radius:

   Technician A says incorrect turning radius may often be noted by tire squeal while cornering.

   Technician B says to adjust turning radius toe-out on turns, turn the inner wheel to stop.

   Who is correct?

   A. A only
   B. B only
   C. Both A and B
   D. Neither A nor B

2. The customer says that sometimes her car suddenly swerves to one side on a bump. All of the following could cause this problem EXCEPT:

   A. Loose steering gear.
   B. Worn tie rods.
   C. Sagging front springs.
   D. Steering gear lash adjustment.

3. While discussing steering diagnosis:

   Technician A says uneven half-shaft axle lengths may cause a vehicle to pull to one side when accelerating.

   Technician B says abnormal toe changes can cause a vehicle to pull to one side on road irregularities.

   Who is correct?

   A. A only
   B. B only
   C. Both A and B
   D. Neither A nor B

4. After you have completed a front end alignment, a customer returns to your shop and complains of continued vehicle drift. To correct this problem, you should:

   A. Inspect the manual steering gear for possible miscentering of the sector gear.
   B. Inspect rear wheel alignment.
   C. Ask the customer to fill the fuel tank.
   D. Inspect the steering column flex coupling.

5. While performing a prealignment inspection:

   Technician A says a prealignment inspection should include checking the vehicle interior for heavy items.

   Technician B says tools and other items normally carried in the vehicle should be included during an alignment.

   Who is correct?

   A. A only
   B. B only
   C. Both A and B
   D. Neither A nor B

JOB SHEET 32

Name _____ **Date** _____

Road Test Vehicle and Diagnose Steering Operation

Upon completion of this job sheet, you should be able to road test a vehicle and diagnose steering operation.

NATEF AST/MAST Task Correlation

Suspension and Steering. Wheel Alignment Diagnosis, Adjustment, and Repair,
Task #E-1. Diagnose vehicle wander, drift, pull, hard steering, bump steer, memory steer, torque steer, and steering return concerns; determine necessary action. (P-1)

Describe the Vehicle Being Worked On:

Year _____ Make _____ Model _____

VIN _____ Engine type and size _____

Procedure

1. Road test vehicle under such driving conditions as slow-speed driving, cornering, and normal cruising speed while driving on a straight, level road surface. Check for the following abnormal steering conditions:
2. Vertical chassis oscillations: ☐ Satisfactory ☐ Unsatisfactory
3. Chassis lateral waddle: ☐ Satisfactory ☐ Unsatisfactory
4. Steering pull to right: ☐ Satisfactory ☐ Unsatisfactory
5. Steering pull to left: ☐ Satisfactory ☐ Unsatisfactory
6. Steering effort: ☐ Satisfactory ☐ Unsatisfactory
7. Tire squeal while cornering: ☐ Satisfactory ☐ Unsatisfactory
8. Bump steer: ☐ Satisfactory ☐ Unsatisfactory
9. Torque steer: ☐ Satisfactory ☐ Unsatisfactory
10. Memory steer: ☐ Satisfactory ☐ Unsatisfactory
11. Steering wheel return: ☐ Satisfactory ☐ Unsatisfactory
12. Steering wheel free play: ☐ Satisfactory ☐ Unsatisfactory

13. Return the vehicle to the shop and inspect suspension and steering to determine the cause of abnormal conditions. List the necessary repairs or adjustments to correct all abnormal conditions that occurred during the road test.

_____

_____

Instructor's Response _____

_____

_____

JOB SHEET 33

Name _____ **Date** _____

Measure Front and Rear Wheel Alignment Angles with a Computer Wheel Aligner

Upon completion of this job sheet, you should be able to measure front and rear wheel alignment with a computer wheel aligner.

NATEF AST/MAST Task Correlation

Suspension and Steering. Wheel Alignment Diagnosis, Adjustment, and Repair,
Task #E-2. Perform prealignment inspection and measure vehicle ride height; perform necessary action. (P-1)
Task #E-3. Prepare vehicle for wheel alignment on alignment machine; perform four-wheel alignment by checking and adjusting front and rear wheel caster, camber; and toe as required; center steering wheel. (P-1)
Task #E-4. Check toe-out-on-turns (turning radius); determine necessary action. (P-2)
Task #E-5. Check SAI (steering axis inclination) and included angle; determine necessary action. (P-2)
Task #E-6. Check rear wheel thrust angle; determine necessary action. (P-1)
Task #E-7. Check for front wheel setback; determine necessary action. (P-2)
Task #E-8. Check front and/or rear cradle (subframe) alignment; determine necessary action. (P-3)

Tools and Materials

Computer wheel aligner

Describe the Vehicle Being Worked On:

Year _____ Make _____ Model _____
VIN _____ Engine type and size _____

Procedure

WARNING: When you drive a vehicle onto an alignment ramp, be sure no one is standing in front of the vehicle to avoid causing personal injury.

1. Lock the front turntables and drive the vehicle onto the alignment ramp. Apply the parking brake.
   Are the front tires properly positioned on the front turntables? ☐ Yes ☐ No
   Are the rear tires properly positioned on the slip plates? ☐ Yes ☐ No
   Instructor check _____

2. Install the rim clamps and wheel sensors. Perform wheel sensor leveling and wheel runout compensation procedures.
   Are the wheel sensors level? ☐ Yes ☐ No
   Is the wheel runout compensation procedure completed? ☐ Yes ☐ No
   Instructor check _____

3. Select the specifications for the vehicle being aligned.
   Are the specifications selected? ☐ Yes ☐ No
   Instructor check _____

4. Perform a prealignment inspection using the checklist in the computer wheel aligner. List any components that must be repaired or replaced, and explain the reasons for your diagnosis.

   _____
   _____
   _____
   _____
   _____

5. Measure the ride height.

Left front ride height	_____	Specified ride height	_____
Right front ride height	_____	Specified ride height	_____
Left rear ride height	_____	Specified ride height	_____
Right rear ride height	_____	Specified ride height	_____

   State the necessary repairs to correct ride height and explain the reasons for your diagnosis.

   _____
   _____
   _____
   _____

6. Measure the front and rear suspension alignment angles following the prompts on the computer wheel aligner screen.

Left front camber _____	Right front camber _____
Cross camber _____	Specified front wheel camber _____
Specified cross camber _____	
Left front caster _____	Right front caster _____
Cross caster _____	Specified front wheel caster _____
Specified cross caster _____	
Left front SAI _____	Right front SAI _____
Specified SAI _____	Included angle _____
Thrust angle _____	
Specified thrust angle _____	
Left front toe _____	Right front toe _____
Total toe _____	Specified front wheel toe _____
Left rear camber _____	Right rear camber _____
Specified camber _____	
Left rear toe _____	Right rear toe _____
Total toe _____	Specified rear wheel toe _____

State the necessary adjustments and repairs to correct front and rear suspension alignment angles and explain the reasons for your diagnosis.

_____

_____

_____

_____

_____

7. Measure the turning radius.

Left turn: Turning radius right front wheel _____

Turning radius left front wheel _____

Specified turning radius _____

Right turn: Turning radius left front wheel _____

Turning radius right front wheel _____

Specified turning radius _____

State the necessary repairs to correct the turning radius, and explain the reasons for your diagnosis.

_____

_____

_____

_____

_____

_____

Instructor's Response _____

_____

_____

JOB SHEET 34

Name _____ **Date** _____

Check Proper Steering Linkage Height by Measuring Toe Change

Upon completion of this job sheet, you should be able to check steering linkage height by measuring toe change.

NATEF MAST Task Correlation

Suspension and Steering. General: Suspension and Steering Systems,

Task #A-2. Identify and interpret suspension and steering system concerns; determine necessary action. (P-1)

Tools and Materials

Computer wheel aligner

Describe the Vehicle Being Worked On:

Year _____ Make _____ Model _____

VIN _____ Engine type and size _____

Procedure

1. Be sure the toe is adjusted to specifications with the front wheels centered and the steering wheel locked.
   Toe, left front _____
   Toe, right front _____
   Total toe, front wheels _____
2. Pull the chassis downward approximately 3 in. (7.6 cm) and record the toe change.
   Toe change left front from _____ to _____
   Toe change right front from _____ to _____
3. During the wheel jounce in step 2, did the toe remain at the original setting on the left front? Amount of toe change from the original setting on the left front wheel _____
   During the wheel jounce in step 2, did the toe remain at the original setting on the right front? Amount of toe change from the original setting on the right front wheel _____
4. During the wheel jounce in step 2, did the right and the left front wheels move an equal amount toward a toe-in position? Amount of toe change toward a toe-in position on both front wheels _____
5. During the wheel jounce in step 2, did one front wheel move toward a toe-in or toe-out position while the opposite front wheel remained at the original position? Amount that one front wheel moved toward a toe-in or toe-out position while the opposite front wheel remained at the original position _____

6. During the wheel jounce in step 2, did one front wheel move toward a toe-in and the opposite front wheel move toward a toe-out position?

   Amount that one front wheel moved toward a toe-in position _____

   Amount the opposite front wheel moved toward a toe-out position _____

7. Push the chassis upward 3 in. (7.6 cm) and inspect the toe change.

   Left front toe change: from _____ to _____

   Right front toe change: from _____ to _____

8. During the wheel rebound in step 7, did one front wheel remain at the original setting and the opposite front wheel move toward a toe-in or toe-out position?

   Amount that one front wheel moved toward a toe-in or toe-out position while the opposite front wheel remained at the original setting _____

9. During the wheel rebound in step 7, did one front wheel move toward a toe-in position while the opposite front wheel moved toward a toe-out position?

   Amount that one front wheel moved toward a toe-in position, and the amount the opposite front wheel moved toward a toe-out position _____

10. List all the abnormal conditions that indicate unequal steering arm height.

    _____

    _____

    _____

    _____

11. Inspect the front steering linkages and steering gear mountings, and list the necessary repairs to correct the improper toe change during wheel jounce and rebound. Explain the reasons for your diagnosis.

    _____

    _____

Instructor's Response _____

_____

_____

JOB SHEET 35

Name _____ **Date** _____

CENTER STEERING WHEEL

Upon completion of this job sheet, you should be able to center a steering wheel.

NATEF AST/MAST Task Correlation

Suspension and Steering. Wheel Alignment Diagnosis, Adjustment, and Repair,
Task #E-3. Prepare vehicle for wheel alignment on alignment machine; perform four-wheel alignment by checking and adjusting front and rear wheel caster, camber; and toe as required; center steering wheel. (P-1)

Tools and Materials

Tie rod sleeve adjusting tool

Describe the Vehicle Being Worked On:

Year _____ Make _____ Model _____

VIN _____ Engine type and size _____

Procedure

1. List the suspension and steering service procedures that may require steering wheel centering after these procedures are completed.

   _____

   _____

   _____

   _____

2. The steering wheel is centered properly with the front wheels straight ahead in the shop, but the steering wheel is not centered when driving the vehicle straight ahead. List the causes of this problem.

   _____

   _____

   _____

   _____

3. Lift the front end of the vehicle with a hydraulic jack, and position safety stands under the lower control arms. Lower the vehicle onto the safety stands, and place the front wheels in the straight-ahead position.

   Are the lower control arms securely supported on safety stands? ☐ Yes ☐ No

   Are the front wheels straight ahead? ☐ Yes ☐ No

   Instructor check _____

4. Use a piece of chalk to mark each tie rod sleeve in relation to the tie rod, and loosen the sleeve clamps.

   Are the tie rod sleeves marked in relation to the tie rods? ☐ Yes ☐ No

   Instructor check _____

5. Position the steering wheel spoke in the position it was in while driving straight ahead during the road test. Turn the steering wheel to the centered position and note the direction of the front wheels.

   Direction the front wheels are turned with the steering wheel centered:
   _____ right _____ left

6. If the steering wheel spoke is low on the left side while driving the vehicle straight ahead, use a tie rod sleeve rotating tool to shorten the left tie rod and lengthen the right tie rod. A one-quarter turn on a tie rod sleeve moves the steering wheel position approximately one inch. Turn the tie rod sleeves the proper amount to bring the steering wheel to the centered position. For example, if the steering wheel spoke is two inches off-center, turn each tie rod sleeve one-half turn.

   Left tie rod sleeve: lengthened _____ shortened _____

   Right tie rod sleeve: lengthened _____ shortened _____

   Amount of left tie rod sleeve rotation _____

   Amount of right tie rod sleeve rotation _____

   Is the steering wheel centered with front wheel straight ahead? ☐ Yes ☐ No

   Instructor check _____

7. If the steering wheel spoke is low on the right side while driving the vehicle straight ahead, lengthen the left tie rod and shorten the right tie rod.

8. Mark each tie rod sleeve in its new position in relation to the tie rod. Be sure the sleeve clamp openings are positioned properly, as indicated in this chapter. Tighten the clamp bolts to the specified torque.

   Are the tie rod sleeves marked in relation to tie rods? ☐ Yes ☐ No

   Are the tie rod sleeve clamps properly installed? ☐ Yes ☐ No

   Tie rod sleeve bolt specified torque _____

   Tie rod sleeve bolt actual torque _____

   Instructor check _____

9. Lift the front chassis with a floor jack, and remove the safety stands. Lower the vehicle onto the shop floor, and check the steering wheel position during a road test.

   Steering wheel position while driving straight ahead during a road test:
   ☐ Satisfactory ☐ Unsatisfactory

Instructor's Response _____

_____

_____

JOB SHEET 36

Name _____ **Date** _____

Adjust Front Wheel Alignment Angles

Upon completion of this job sheet, you should be able to adjust front wheel alignment angles.

NATEF AST/MAST Task Correlation

Suspension and Steering. Wheel Alignment Diagnosis, Adjustment, and Repair,
Task #E-2. Perform prealignment inspection and measure vehicle ride height; perform necessary action. (P-1)
Task #E-3. Prepare vehicle for wheel alignment on alignment machine; perform four-wheel alignment by checking and adjusting front and rear wheel caster, camber; and toe as required; center steering wheel. (P-1)
Task #E-4. Check toe-out-on-turns (turning radius); determine necessary action. (P-2)
Task #E-5. Check SAI (steering axis inclination) and included angle; determine necessary action. (P-2)
Task #E-6. Check rear wheel thrust angle; determine necessary action. (P-1)
Task #E-7. Check for front wheel setback; determine necessary action. (P-2)
Task #E-8. Check front and/or rear cradle (subframe) alignment; determine necessary action. (P-3)

Tools and Materials

A vehicle with improperly adjusted front wheel camber, caster, and toe.

Describe the Vehicle Being Worked On:

Year _____ Make _____ Model _____
VIN _____ Engine type and size _____

Procedure

1. Position the vehicle properly on the alignment ramp with the front wheels properly positioned on the turntables and the rear wheels on slip plates.
   Are the front wheels properly positioned on turntables? ☐ Yes ☐ No
   Are the rear wheels properly positioned on slip plates? ☐ Yes ☐ No
   Is the parking brake applied? ☐ Yes ☐ No
   Instructor check _____

2. Perform a prealignment inspection and correct any defective conditions or components found during the inspection. List the defective conditions or components, and explain the reasons for your diagnosis.
   _____
   _____
   _____

3. Measure front and rear ride height and correct it if necessary.
   Specified front suspension ride height _____
   Actual front suspension ride height _____
   Specified rear suspension ride height _____
   Actual rear suspension ride height _____
   List the necessary repairs to correct the front and rear suspension ride height and explain the reasons for your diagnosis.
   _____
   _____
   _____
   _____

4. Measure and adjust front wheel camber.
   Specified front wheel camber _____
   Actual front wheel camber: left _____ right _____
   Front wheel camber after adjustment: left _____ right _____
   Camber adjustment method _____

5. Measure SAI and included angle.
   Specified SAI _____
   Actual SAI: left _____ right _____
   Specified included angle _____
   Actual included angle: left _____ right _____
   If the SAI and included angle are not within specifications, state the necessary repairs and explain the reasons for your diagnosis.
   _____
   _____
   _____
   _____

6. Measure and adjust front wheel caster.
   Specified front wheel caster _____
   Actual front wheel caster: left _____ right _____
   Front wheel caster after adjustment: left _____ right _____
   Caster adjustment method _____

7. Inspect camber measurements and adjust if necessary.
   Camber adjustment: ☐ Satisfactory ☐ Unsatisfactory
   Left front camber readjusted to _____
   Right front camber readjusted to _____

8. Measure and adjust front wheel toe.
   Specified front wheel toe _____
   Actual front wheel toe: left _____ right _____
   Total toe _____
   Front wheel toe after adjustment: left _____ right _____
   Total front wheel toe _____

9. Inspect steering wheel centering with front wheels straight ahead and center the wheel if necessary.

Is steering wheel centered? ☐ Yes ☐ No

Instructor check _____

10. Measure turning angle.

Specified turning angle _____

Actual turning angle left front wheel _____

Actual turning angle right front wheel _____

Turning angle correction required: ☐ Yes ☐ No

Left front wheel turned inward 20 degrees, turning angle on right front wheel _____

Right front wheel turned inward 20 degrees, turning angle on left front wheel _____

If the turning angle is not within specifications, state the necessary repairs and explain the reasons for your diagnosis.

_____

_____

_____

_____

Instructor's Response _____

_____

_____

JOB SHEET 37

Name _____ **Date** _____

Adjust Rear Wheel Alignment Angles

Upon completion of this job sheet, you should be able to adjust rear wheel alignment angles.

NATEF AST/MAST Task Correlation

Suspension and Steering. Wheel Alignment Diagnosis, Adjustment, and Repair,
Task #E-2. Perform prealignment inspection and measure vehicle ride height; perform necessary action. (P-1)
Task #E-3. Prepare vehicle for wheel alignment on alignment machine; perform four-wheel alignment by checking and adjusting front and rear wheel caster, camber; and toe as required; center steering wheel. (P-1)
Task #E-6. Check rear wheel thrust angle; determine necessary action. (P-1)

Tools and Materials

A vehicle with improperly adjusted rear wheel camber and toe.

Describe the Vehicle Being Worked On:

Year _____ Make _____ Model _____
VIN _____ Engine type and size _____

Procedure

1. Position the vehicle properly on the alignment ramp with the front wheels properly positioned on the turntables and the rear wheels on slip plates.
   Are the front wheels properly positioned on turntables? ☐ Yes ☐ No
   Are the rear wheels properly positioned on slip plates? ☐ Yes ☐ No
   Is the parking brake applied? ☐ Yes ☐ No
   Instructor check _____

2. Perform a prealignment inspection, and correct any defective conditions or components found during the inspection. List the defective conditions or components, and explain the reasons for your diagnosis.
   _____
   _____
   _____

3. Measure front and rear ride height and correct if necessary.
   Specified front suspension ride height _____
   Actual front suspension ride height _____
   Specified rear suspension ride height _____
   Actual rear suspension ride height _____

If the rear ride height is not within specifications, state the necessary repairs and explain the reasons for your diagnosis.

_____

_____

_____

_____

4. Measure and adjust camber on left rear wheel.
   Specified left rear wheel camber _____
   Actual left rear wheel camber _____
   Left rear wheel camber after adjustment _____
   Camber adjustment method _____

5. Measure left rear wheel toe.
   Specified left rear wheel toe _____
   Actual left rear wheel toe _____
   Left rear wheel toe after adjustment _____
   Toe adjustment method _____

6. Measure and adjust camber on right rear wheel.
   Specified right rear wheel camber _____
   Actual right rear wheel camber _____
   Right rear wheel camber after adjustment _____
   Camber adjustment method _____

7. Measure right rear wheel toe.
   Specified right rear wheel toe _____
   Actual right rear wheel toe _____
   Right rear wheel toe after adjustment _____
   Toe adjustment method _____

8. Measure thrust angle.
   Specified thrust angle _____
   Actual thrust angle _____
   If the thrust angle is not within specifications, state the necessary repairs and explain the reasons for your diagnosis.

   _____

   _____

   _____

   _____

9. Road test vehicle for satisfactory steering and suspension operation.

Steering directional control: ☐ Satisfactory ☐ Unsatisfactory

Complete steering operation: ☐ Satisfactory ☐ Unsatisfactory

Suspension operation: ☐ Satisfactory ☐ Unsatisfactory

If the steering or suspension operation is unsatisfactory, state the necessary repairs and explain the reasons for your diagnosis.

_____

_____

_____

_____

Instructor's Response _____

_____

_____

Chapter 10

Diagnostic Alignment Angles and Frame Diagnosis

BASIC TOOLS

Basic technician's tool set
Service manual
Tape measure
Chalk
Plumb bob
Floor jack
Safety stands

Upon completion and review of this chapter, you should be able to:

- Measure front and rear wheel setback.
- Measure steering axis inclination (SAI).
- Measure front wheel toe and turning radius.
- Diagnose bent front struts.
- Correct setback conditions.
- Check and correct front engine cradle position.
- Correct SAI angles that are not within specifications.
- Use a track gauge to measure rear wheel tracking.
- Diagnose rear wheel tracking problems from the track gauge measurements.
- Take the necessary precautions to avoid frame damage.
- Diagnose the causes of frame damage.
- Follow safety precautions when measuring and welding frames.
- Visually inspect frames.
- Measure frames with a plumb bob and diagnose frame damage.
- Measure frames with a tram gauge and diagnose frame damage.
- Perform unitized body measurements with a tram gauge.
- Perform unitized body measurements with a dedicated bench system.

Today's computerized wheel alignment system performs many vehicle geometric angle measurements beyond the standard alignment angles of camber, caster, and toe. What used to take many hours using mechanical measuring devises are now done fast and efficiently with today's computerized alignment systems. While not all of these measurements are displayed as part of a routine wheel alignment, they are available as additional measurements in your computerized wheel alignment system, such as that produced by Hunter Engineering, to aid you in diagnosing vehicle geometry issues.

Turning Angle Screen

Some computer wheel aligners have a turning angle screen. When this screen is displayed, the technician removes the locking pins from the turntables. Each front wheel must be turned outward a specified amount, and the turning angle on the opposite front wheel must be entered with the keypad as directed on the screen (Figure 10-1). A toe-out on turns option is available on some DSP wheel sensors. This option has optical encoders in the wheel sensors that measure the turning angle electronically rather than reading the degree scales on the front turntables. Photo Sequence 18 shows a manual procedure for measuring turning angle.

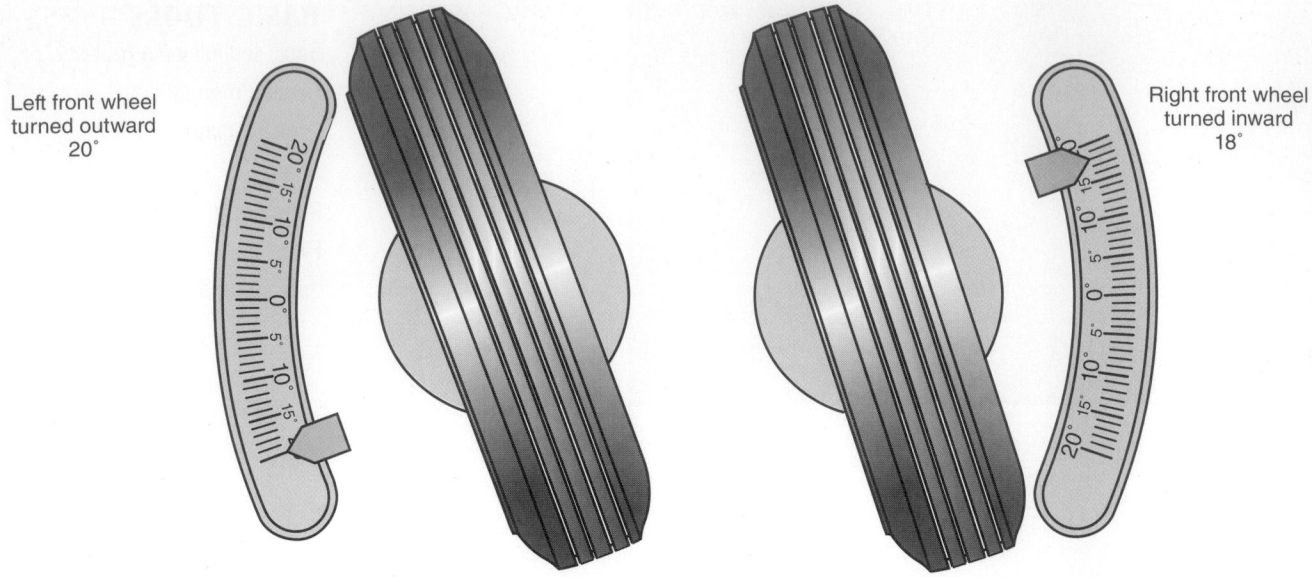

Copyright © 2015 Cengage Learning®

FIGURE 10-1 **Turning angle screen.**

PHOTO SEQUENCE 18

Typical Procedure for Front Wheel Turning Radius Measurement

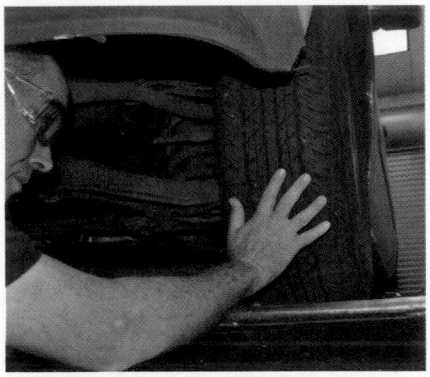

P18-1 **Perform a prealignment inspection.**

Copyright © 2015 Cengage Learning®

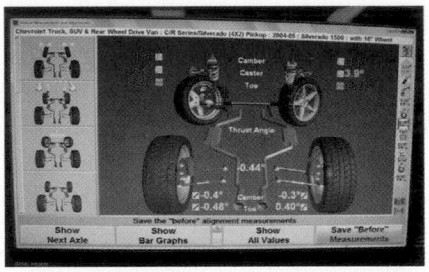

P18-2 **Measure and adjust the other front and rear suspension angles.**

Copyright © 2015 Cengage Learning®

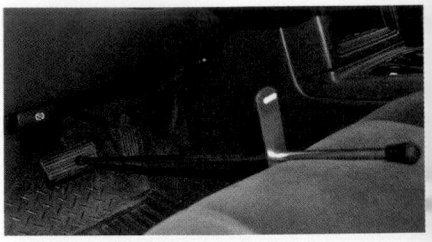

P18-3 **Be sure the front wheels are centered on the turntables and the brake pedal jack is installed to apply the brakes.**

Copyright © 2015 Cengage Learning®

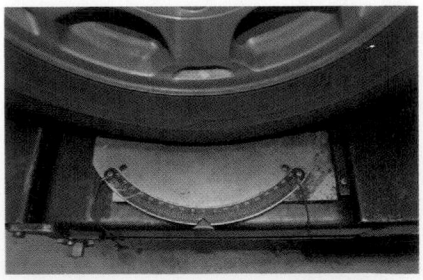

P18-4 **Remove the turntable locking pins and be sure the turning radius gauges are in the zero position with the front wheels straight ahead.**

Copyright © 2015 Cengage Learning®

P18-5 **Turn the right front wheel inward toward the center of the vehicle until the turning radius gauge indicates 20 degrees.**

Copyright © 2015 Cengage Learning®

P18-6 **With the front wheels positioned as described in step 5, read and record the reading on the left turning radius gauge. The reading on the left turning radius gauge should be 22 degrees, or 2 degrees more than the reading on the right turning radius gauge.**

Copyright © 2015 Cengage Learning®

P18-7 **Turn the left front wheel inward toward the center of the vehicle until the turning radius gauge indicates 20 degrees.**

Copyright © 2015 Cengage Learning®

P18-8 **With the front wheels positioned as described in step 7, read and record the reading on the right turning radius gauge. The reading on the right turning radius gauge should be 22 degrees, or 2 degrees more than the reading on the left turning radius gauge.**

Copyright © 2015 Cengage Learning®

Additional Diagnostic Screens

Some computer wheel aligners provide **symmetry angle measurements** that help the technician determine if out-of-specification readings may have been caused by collision or frame damage. These symmetry angle measurements display **axle offset**, right or left **lateral axle sideset** and track-width difference, front and rear wheel setback, and wheelbase difference (Figure 10-2). Setback is an angle formed by a line drawn at a 90 degree angle to the centerline and a line connecting the centers of the front or rear wheels (Figure 10-3). Wheelbase difference is an angle created by a line through the rear wheel centers and a line through the front wheel centers (Figure 10-4). Right or left lateral offset is an angle between the thrust line and a line connecting the centers of the left front and left rear wheels or right front and right rear wheels (Figure 10-5). Track-width difference is an angle created by the line connecting the centers of the right rear and right front wheels and the line connecting the centers of the left rear and left front wheels (Figure 10-6). Axle offset is an angle formed by the line that is projected forward at 90 degrees in relation to the center of the rear axle and the thrust line (Figure 10-7).

Axle offset occurs when the rear axle is rotated so it is no longer at a 90 degree angle to the vehicle centerline.

Lateral axle sideset occurs when the rear axle moves inward or outward in relation to the vehicle centerline, but the axle and vehicle centerline remains at a 90 degree angle.

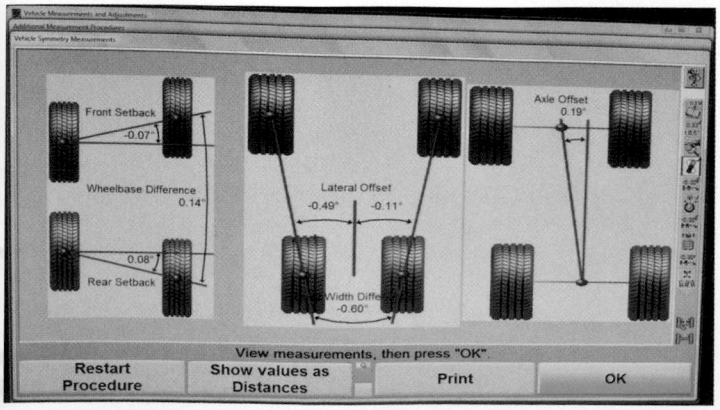

Copyright © 2015 Cengage Learning®

FIGURE 10-2 **Symmetry angle measurements.**

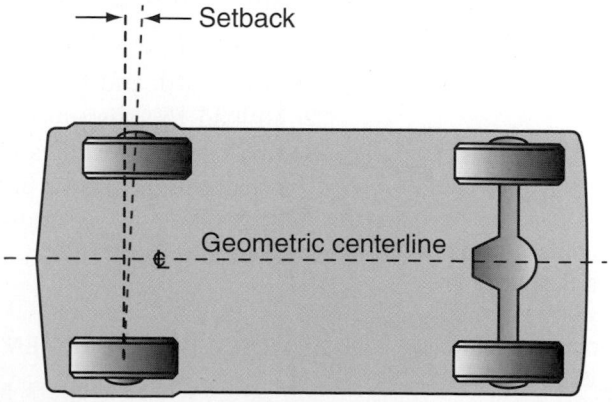

Copyright © 2015 Cengage Learning®

FIGURE 10-3 **Front or rear wheel setback.**

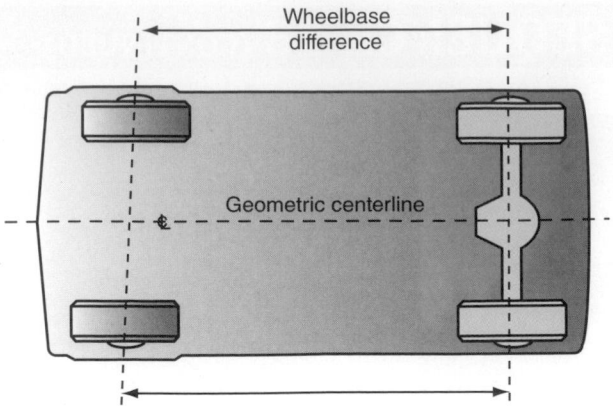

Copyright © 2015 Cengage Learning®

FIGURE 10-4 **Wheelbase difference.**

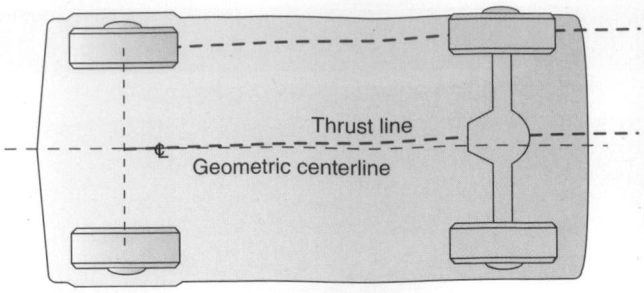

Copyright © 2015 Cengage Learning®

FIGURE 10-5 **Right or left lateral sideset.**

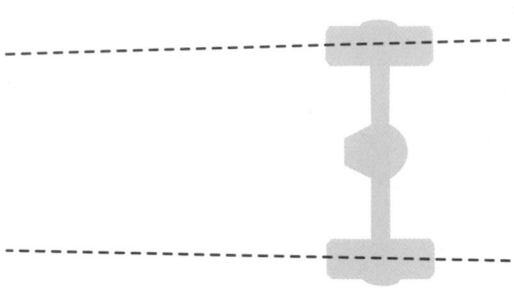

Copyright © 2015 Cengage Learning®

FIGURE 10-6 **Track-width difference.**

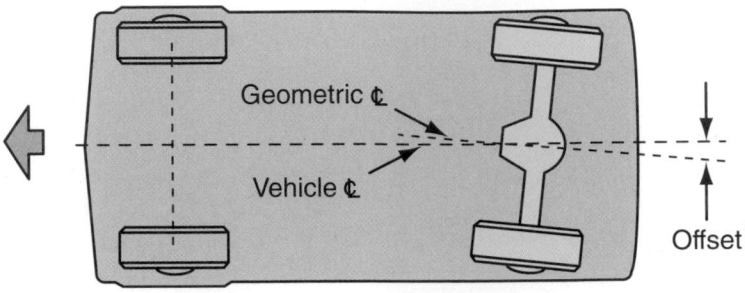

Copyright © 2015 Cengage Learning®

FIGURE 10-7 **Axle offset.**

Bent Front Strut Diagnosis

Diagnostic Procedure

Front struts that are bent forward or rearward may be diagnosed as follows:

1. With the vehicle on the aligning rack and the magnetic gauges in position for the camber measurement, swing one front wheel out 10 degrees and read the camber.
2. Turn the same wheel inward 10 degrees and read the camber. The difference between the two readings is usually less than 4 1/2 degrees. If the strut is bent forward or rearward, the difference in the camber readings will be excessive. A 10 degree difference is not uncommon.
3. Repeat steps 1 and 2 on the other front wheel.

Front struts could also be bent inward or outward. To diagnose this condition, use the following procedure:

1. With the wheel aligner in operation, sit on the front fender to load the suspension downward, and then read the camber.
2. Unload the suspension and lift up on the vehicle while the camber is recorded. The two camber readings should be within 1/2 degree. If the strut is bent inward or outward, the difference in the two camber readings will be excessive. A 4 to 6 degree camber change is not uncommon.
3. Repeat steps 1 and 2 on the other front wheel.

Setback is the distance that one wheel is moved rearward in relation to the opposite wheel.

Setback Measurement and Correction Procedure

Computer wheel aligners have **setback** measuring capabilities on the front wheels. These aligners automatically display setback angles with the other front suspension alignment angles. If a front-wheel-drive vehicle has experienced collision damage on the left front, the left side of the cradle or subframe may be driven rearward (Figure 10-8). This type of collision damage also moves the bottom of the left front strut rearward to a setback condition and shifts the caster to a negative position.

WARNING: Prior to any service operation where the body is going to be moved in relation to the cradle or subframe, the intermediate shaft on the rack and pinion gear must be disconnected. If this shaft is not disconnected, it may be damaged, resulting in loss of steering control, serious vehicle damage, and personal injury.

CAUTION: If the engine cradle has to be removed or loosened, the engine support fixture must hold the engine in the normal position to prevent damage to the fan shroud, power steering hoses, and other components.

With this type of collision damage, it may be possible to adjust the caster back to the specified setting, but that does not correct the setback problem. If the caster is adjusted and the setback problem is ignored, the steering may pull to one side while driving straight ahead, especially if the setback condition is severe. The cradle must be straightened and the unitized body must be pulled back to the original measurements to correct this problem. A severely bent, distorted cradle should be replaced.

Collision damage may push a front cradle sideways. This condition may increase the positive camber on one side of the car and move the camber toward a negative position on the opposite side. When this condition is encountered, always inspect and correct the cradle position. If the camber is adjusted in this situation, the steering will not provide proper directional control.

If the engine cradle/subframe has been removed and reinstalled as part of a repair such as transaxle replacement it is possible that it was not correctly lined up during reinstallation. Sometimes a cradle has been erroneously shifted to the left or the right or forward and backward on the left and right side during reinstallation. In this situation, the camber or caster on both the left and right side may be out of specifications, equally or opposite. As an example, a car with a MacPherson strut front suspension and a cradle assembly may

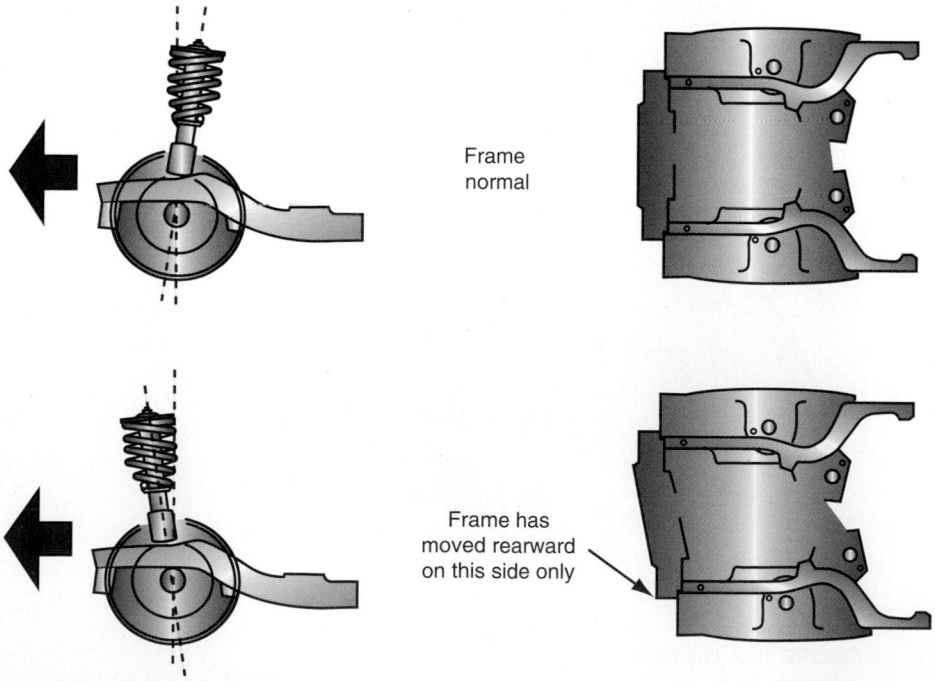

Copyright © 2015 Cengage Learning®

FIGURE 10-8 **Collision damage on the left front moves the cradle and strut rearward to a setback condition and shifts the caster to a negative position.**

Classroom Manual
Chapter 10
page 256

have 2 degree positive camber (caster) on the left front wheel and 2 degree negative camber (caster) on the right front wheel if the cradle was shifted to one side. The vehicle may also have a setback or sideset condition of the front axle assembly. If this is suspected loosen the cradle/subframe assembly and shift it over with a pry bar (Figure 10-9). Often there are alignment holes in the unibody and the subframe that are there for correct positioning of the cradle.

Prior to cradle removal, an engine support fixture must be installed to support the engine and transaxle (Figure 10-10). The engine support fixture must hold the engine in the original position to prevent damage to the power steering hoses or other components.

Before the body is moved in relation to the cradle or subframe, the intermediate shaft on the rack and pinion steering gear must be disconnected to prevent damage to this shaft. Remove all the engine and transaxle mount bolts in the cradle, plus the cradle retaining bolts and insulators to remove the cradle. Perform the front underbody measurements with a tram gauge (Figure 10-11), and pull the unitized body back to these specifications. This pulling operation is usually done in an automotive body shop rather than an automotive repair shop.

The cradle must be restored to the specified measurements (Figure 10-12). Inspect all cradle insulators and replace any that are damaged, oil soaked, or worn. When the cradle is replaced, all the insulators must be installed properly, and all the cradle bolts should be aligned with the bolt holes in the body. Alignment hole A in the cradle must be perfectly

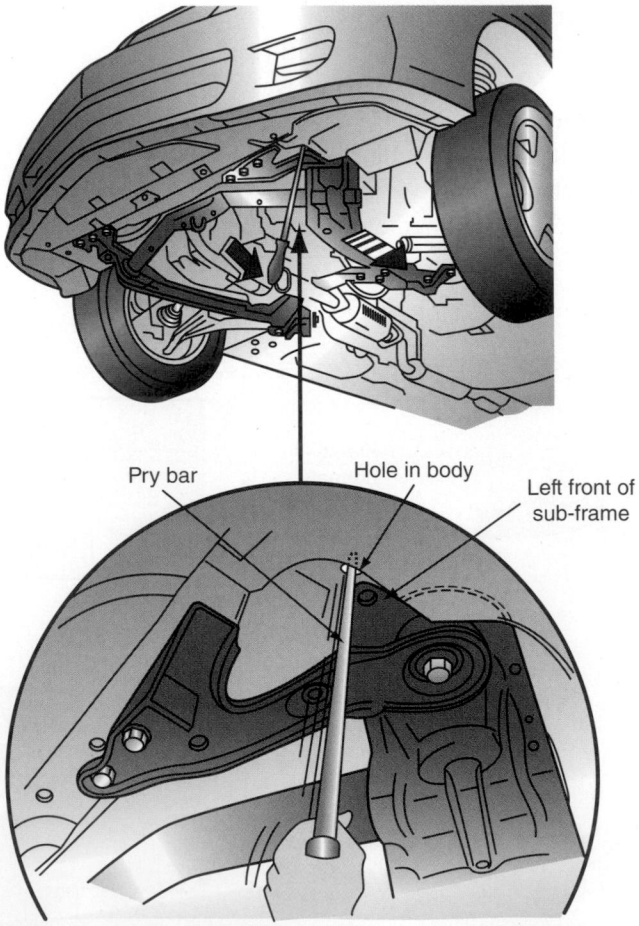

Copyright © 2015 Cengage Learning®

FIGURE 10-9 **The front engine cradle/subframe assembly mounting bolts can be loosened and the cradle can be shifted. Often there are locator holes in the unibody assembly to assist in positioning cradle/subframe assembly.**

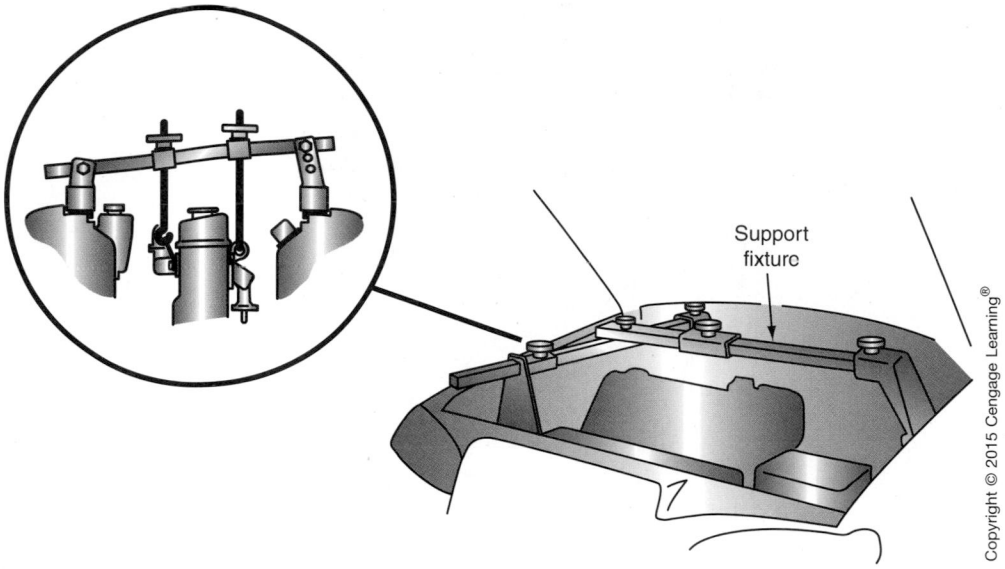

FIGURE 10-10 **Engine support fixture installed to support engine and transaxle prior to cradle removal.**

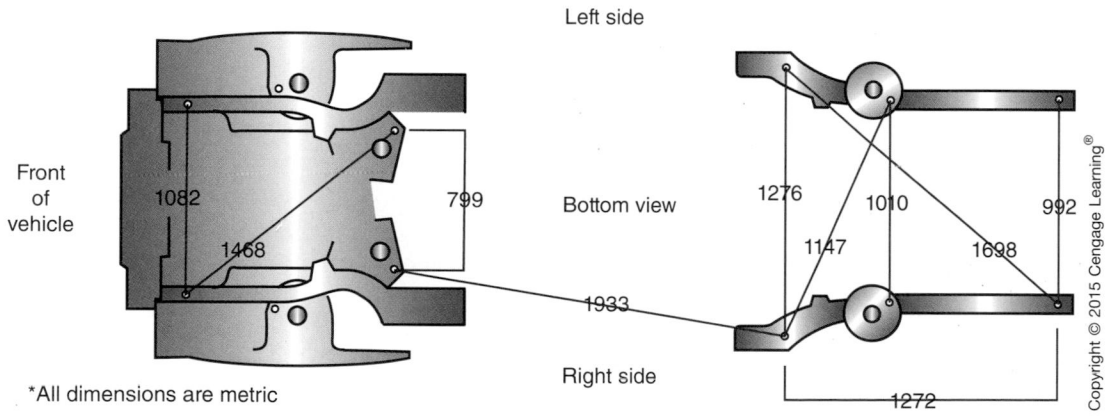

FIGURE 10-11 **Front and rear underbody measurements.**

aligned with the matching alignment hole in the body (Figure 10-13). Observing this alignment hole is a quick way to check cradle position.

Steering Axis Inclination (SAI) Correction Procedure

Steering axis inclination (SAI) is not considered adjustable. If the SAI is not within specifications, the upper strut tower may be out of position, the lower control arm may be bent, or the center cross member, or cradle, could be shifted. In most cases, these defects are caused by collision damage. When the SAI angle is correct, but the camber and included angle are less than specified, the strut or spindle may be bent. The difference in the included angles on the front wheels must not exceed 1 degree.

When the SAI is not within specifications, steering pull may occur while braking or accelerating and steering may be erratic on road irregularities.

The **Steering axis inclination (SAI)** line is an imaginary line through the centers of the upper and lower ball joints, or through the center of the lower ball joint and the upper strut mount. The SAI angle is the angle between the SAI line and the true vertical centerline of the tire and wheel.

Rear Wheel Tracking Measurement with a Track Gauge

The following section will describe rear wheel tracking measurement to give you a better understanding of how today's computer alignment systems perform these checks by

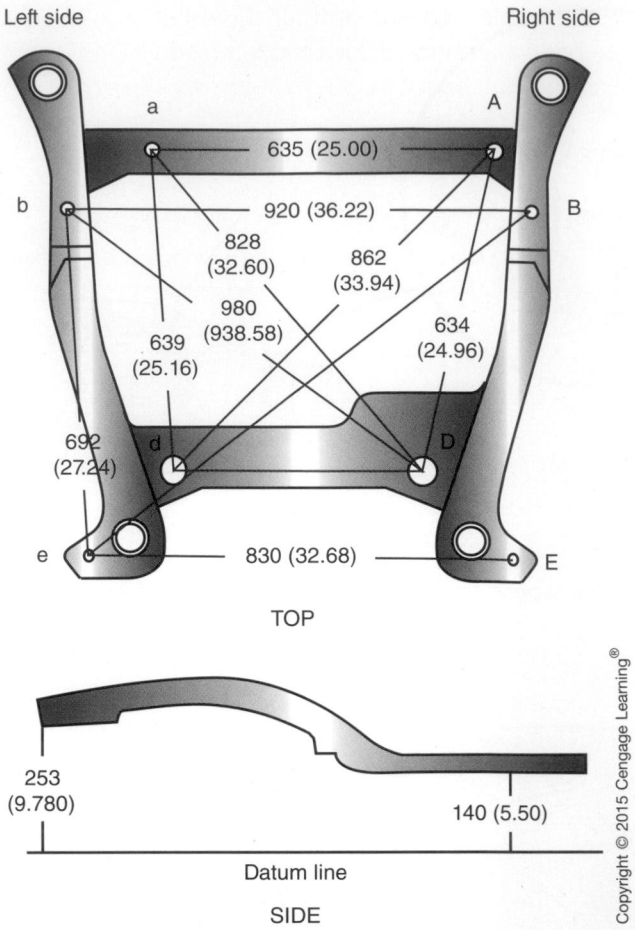

FIGURE 10-12 **Front cradle measurements.**

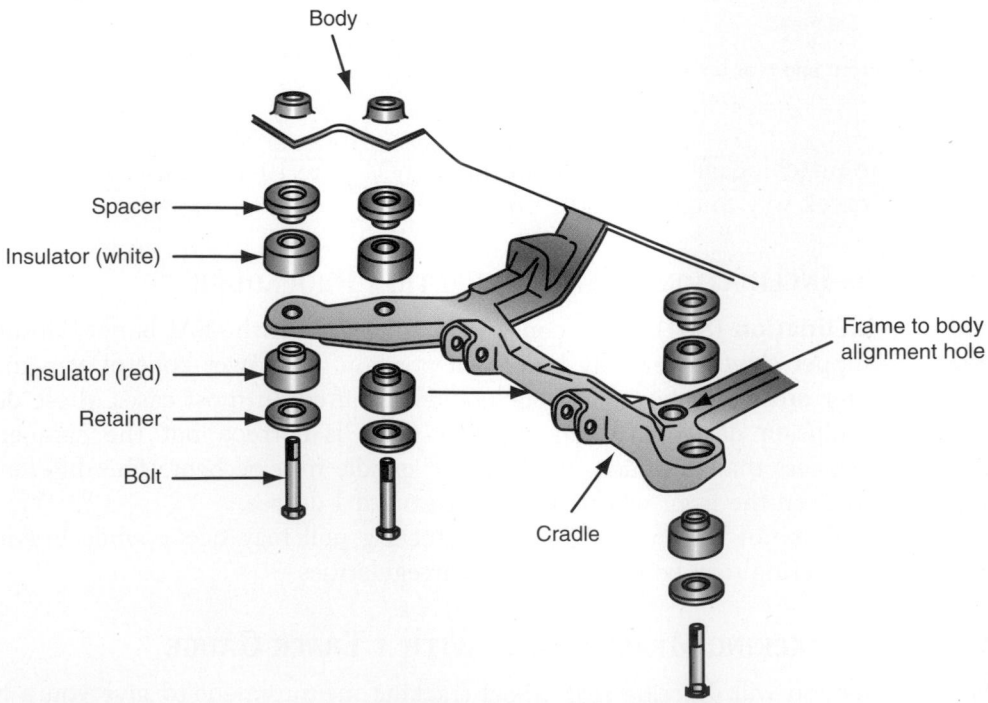

FIGURE 10-13 **Cradle insulators and alignment holes in cradle and body.**

measuring the thrust line angle, setback, and all the other front and rear suspension alignment angles with speed and accuracy. Prior to the introduction of computer alignment systems, wheel alignment equipment performed a two-wheel alignment on the front wheels, and a track gauge was used to measure **rear wheel tracking**.

A **track gauge** is connected between the front and rear wheels to determine the rear wheel tracking or thrust line. When all four wheels are parallel to the geometric centerline of the vehicle, the thrust line is positioned at the geometric centerline. Under this condition, the rear wheels track directly behind the front wheels (Figure 10-14).

If the left front of the vehicle is involved in a severe collision, the left front wheel may be driven rearward to a setback condition, and the rear wheels may be moved straight sideways. Under this condition, the rear wheels no longer track directly behind the front wheels (Figure 10-15). The term sideset is applied to the condition where the rear axle is moved straight sideways.

A track gauge is a straight metal bar graduated in inches or millimeters. The scale on the outside edge of the track bar is similar to a tape measure. Three straight metal pointers are mounted in brackets on the track gauge. These pointers are also graduated much like a tape measure. When the wing nuts in the brackets are loosened, the brackets may be moved lengthwise on the track bar. The metal pointers may be adjusted inward or outward in the brackets.

To inspect rear wheel tracking, one of the track gauge pointers is positioned against the rear outside edge of the front wheel rim at spindle height. The other two pointers are positioned on the outside edges of the rear wheel rim at spindle height. The track gauge must be kept parallel to the side of the vehicle (Figure 10-16).

SPECIAL TOOL

Track gauge

Rear wheel tracking refers to the position of the rear wheels in relation to the front wheels. When all four wheels are parallel to the geometric centerline and the thrust line is positioned on the geometric centerline, the rear wheels track directly behind the front wheels.

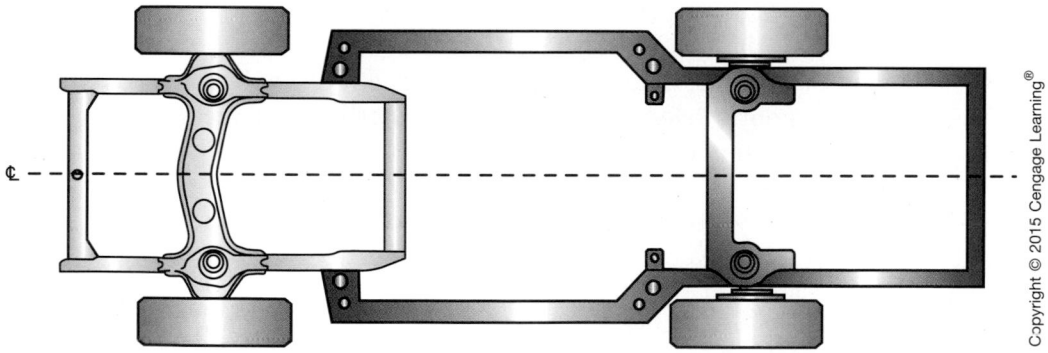

Copyright © 2015 Cengage Learning®

FIGURE 10-14 **When all four wheels are parallel to the geometric centerline, the thrust line is positioned at the geometric centerline and the rear wheels track directly behind the front wheels.**

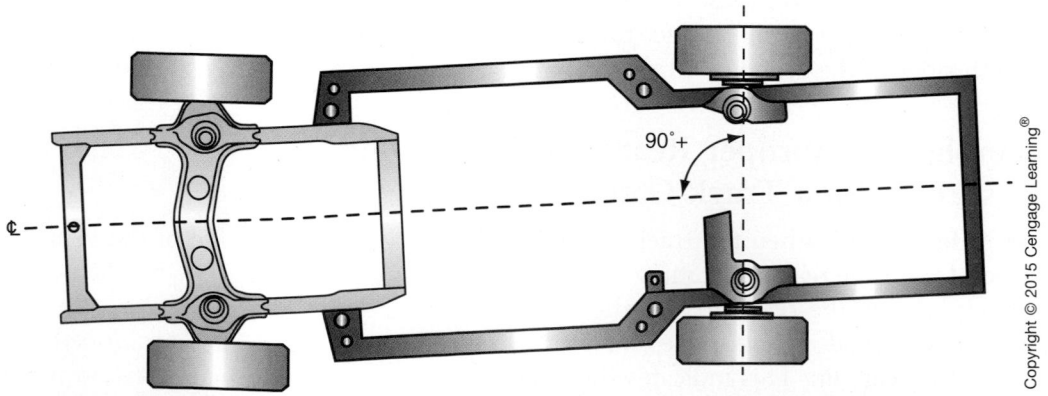

Copyright © 2015 Cengage Learning®

FIGURE 10-15 **Collision damage causes front wheel setback and improper rear wheel tracking with rear wheel offset.**

Copyright © 2015 Cengage Learning®

FIGURE 10-16 **Track bar positioned to measure rear wheel tracking.**

Rear Wheel Tracking Measurement Procedure

Follow these steps to measure rear wheel tracking:

1. Lift one side of the front suspension with a floor jack and lower the suspension onto a safety stand. Be sure the tire is lifted off the shop floor.
2. Place a lateral runout gauge against the outside edge of the rim lip and rotate the wheel to check wheel runout. (Refer to Chapter 3 for wheel runout measurement.) Place a chalk mark at the location of the maximum wheel runout. If the wheel runout is excessive, replace the wheel.
3. Rotate the chalk mark to the exact top or bottom of the wheel, and remove the safety stand and floor jack to lower the wheel onto the shop floor.
4. Repeat steps 1, 2, and 3 at the other three wheels.
5. Hold the single pointer on the track gauge so it contacts the rear edge of the front rim.
6. Adjust the other two pointers so they contact the outer edges of the rear wheel rim.
7. Adjust the pointers on the rear wheel so the track gauge is parallel to the side of the vehicle. Be sure all three pointers are clamped securely.
8. Move the track gauge to the opposite side of the vehicle and position the pointers in the same location on these wheels. If the vehicle has proper tracking, all three pointers on the track gauge will contact the front and rear rims at exactly the same location on both sides of the vehicle, and the track gauge will be parallel to the sides of the vehicle.

Examples of Improper Rear Wheel Tracking Measured with a Track Gauge

Rear axle offset occurs when the rear axle is turned so one rear wheel is moved rearward and the opposite rear wheel is moved forward.

Rear Axle Offset. When the track gauge is positioned on the left side of the vehicle, the rear pointer on the rear wheel must be moved farther inward compared with the pointer at the front edge of the rear rim. On the right side of the vehicle, the front pointer is positioned forward and ahead of the rim edge when the two pointers are positioned on the edges of the rear rim. This indicates the wheelbase is less between the wheels on the right side of the vehicle compared with the left side. The front pointer on the rear rim is some distance from the rim face (Figure 10-17). These pointer positions indicate the right rear

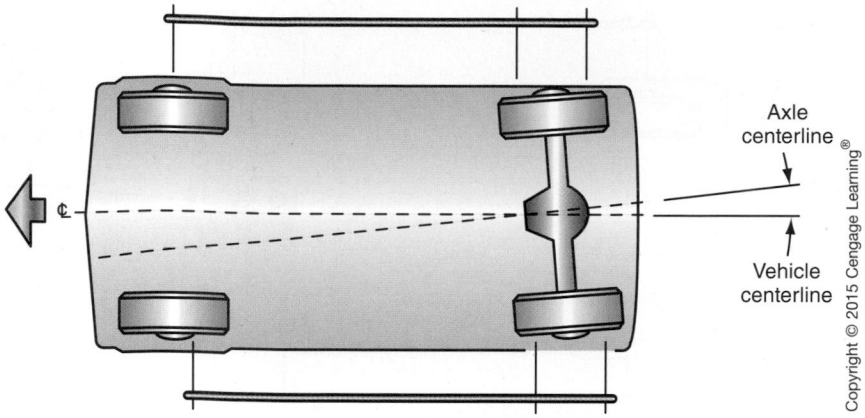

FIGURE 10-17 **Track bar measurement of an offset rear axle.**

wheel has moved forward and the left rear wheel has moved rearward. This **rear axle offset** condition moves the thrust line to the left of the geometric centerline.

Wheel Setback. If a vehicle has been subjected to a severe right front sideways impact, the left front wheel may be driven outward and rearward. The track gauge and pointers are positioned on the right side of the vehicle and then moved to the left side. The front pointer is positioned ahead of the rim edge on the left side, and the track gauge is not parallel to the side of the vehicle (Figure 10-18).

The track gauge must be moved inward on the front pointer to position the bar parallel to the side of the vehicle. When this adjustment is made, the distance on the front pointer is less on the left side of the vehicle compared with the distance on the right side of the vehicle. These track gauge measurements indicate the left front wheel is moved rearward to a setback position as well as being outward.

Diamond-Shaped Frame Condition. When a vehicle experiences severe collision damage on the left front, the entire left side of the vehicle may be driven rearward in relation to the right side of the chassis. Under this condition, the left front wheel is moved to a setback condition, and the rear axle is offset, which moves the thrust line to the left of the geometric centerline. This particular type of damage is most likely to occur on a rear-wheel-drive vehicle with a frame and a one-piece rear axle.

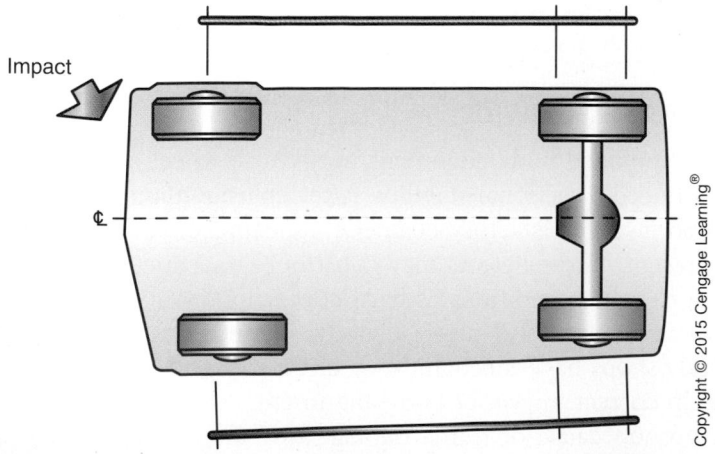

FIGURE 10-18 **Track bar measurement of a left front wheel that is moved rearward to a setback position.**

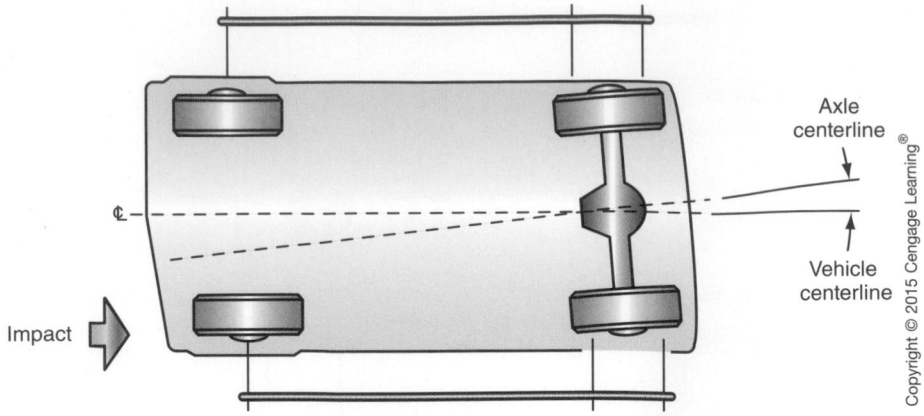

FIGURE 10-19 **A severe left front collision impact moves the entire left side of the vehicle rearward, resulting in front wheel setback and rear axle offset, which provides a diamond-shaped chassis.**

Because the entire left side of the vehicle is driven rearward, the wheelbase is the same on both sides of the vehicle. Therefore, the track gauge pointers indicate the wheelbase is the same between the front and rear wheels on both sides of the vehicle. However, the two rear track gauge pointers must be adjusted to different lengths to contact the rear rim edges. On the left side of the vehicle, the rear track gauge pointer is some distance from the rear rim edge if these two pointers are set at the same measurement (Figure 10-19).

When the track gauge is positioned on the right side of the vehicle, there is some distance between the front pointer and the edge of the rear rim if both rear pointers are set at the same distance.

Measuring vehicle tracking with a track gauge is a two-person operation and requires a considerable amount of time to perform the wheel runout and track bar measurements. The track gauge also lacks precision accuracy.

Computer alignment systems perform the same checks by measuring the thrust line angle, setback, and all the other front and rear suspension alignment angles with speed and accuracy.

CUSTOMER CARE: Those of us involved in the automotive service industry are like everyone else: we do make mistakes. If you make a mistake that results in a customer complaint, always be willing to admit your mistake and correct it. Do not try to cover up the mistake or blame someone else. Customers are usually willing to live with an occasional mistake that is corrected quickly and efficiently.

FRAME DIAGNOSIS AND SERVICE

Improper frame or unitized body alignment may cause wheel alignment defects, rapid tire tread wear, and reduced directional stability. After collision damage, a unitized body vehicle may be repaired so it has satisfactory cosmetic appearance, but body and wheel alignment defects are still present. These defects may contribute to reduced directional stability, especially during such extreme conditions as hard cornering or severe braking. Therefore, diagnosing and correcting frame and unitized body alignment is very important to restoring vehicle safety! Body shops have mechanical or electronic body and chassis alignment equipment that can help correct improper body alignment.

The most common cause of frame damage on cars and light-duty trucks is collision damage. In some cases, the collision damage may be repaired to make the cosmetic appearance of the vehicle satisfactory, but the frame damage may not always be correctable.

When driving the vehicle, some indications of frame damage are:

1. Excessive tire wear when the front suspension alignment angles are correct.
2. Steering pull when the front suspension alignment angles are correct.
3. Steering wheel not centered when driving straight ahead, but the steering wheel was centered in the shop.

Classroom Manual
Chapter 10, page 266

Frame Diagnosis

Preventing Frame Damage

Because frame problems affect wheel alignment, technicians must be able to diagnose frame defects so they are not confused with wheel alignment problems. Refer to Table 10-1 for additional information.

Follow these precautions to minimize frame damage:

1. Do not overload the vehicle.
2. Place the load evenly in a vehicle.
3. Do not operate the vehicle on extremely rough terrain.
4. Do not mount equipment such as a snowplow on a vehicle unless the frame is strong enough to carry the additional load and force.

Diagnosis of Frame Problems

Side Sway. The causes of frame side sway are:

1. Collision damage.
2. Fire damage.
3. Use of equipment on the vehicle for which the frame was not designed.

Sag. The causes of frame sag are:

1. Vehicle loads that exceed the load-carrying capacity of the frame.
2. Uneven load distribution.
3. Sudden changes in **section modulus**.
4. Holes drilled in the **frame flange**.
5. Too many holes drilled in the **frame web**.
6. Holes drilled too close together in the frame web.

Yield strength is a measure of the strength of the material from which the frame is manufactured. Yield strength is the maximum load that may be placed on a material and still have the material retain its original shape. Yield strength is measured in pounds per square inch (psi) or kilopascals (kPa).

Section modulus is a measure of the frame's strength based on its height, width, thickness, and the shape of the side rails. Section modulus does not account for the type of material in the frame.

Frame flange is the horizontal part of the frame on the top and bottom of the web.

Frame web refers to the vertical side of the frame.

TABLE 10-1 **DIAGNOSING FRAME AND BODY DAMAGE**

Problem	Symptoms	Possible Causes
Excessive front tire tread wear	Front tire tread wear with correct front suspension alignment angles	Frame, front cradle damage
Steering pull when driving straight ahead	Steering pull with correct front and rear alignment angles	Frame, front cradle damage
Steering wheel not centered when driving straight ahead	Steering wheel centered in the shop, but it is not centered when driving straight ahead	Frame, front cradle damage
Improper strut tower position	Incorrect front suspension alignment angles, steering pull, tire tread wear	Collision damage
Frame buckle	Wrinkles in frame flanges	Collision damage, vehicle abuse
Improper cradle measurements or position	Incorrect front suspension alignment angles, steering pull, tire tread wear	Collision damage, worn or improperly positioned cradle mounts

Copyright © 2015 Cengage Learning®

7. Welds on the frame flange.
8. Cutting holes in the frame with a cutting torch.
9. Cutting notches in the frame rails.
10. A fire involving the vehicle.
11. Collision damage.
12. The use of equipment for which the frame was not designed.

Buckle. The causes of frame buckle are:

1. Collision damage.
2. Using equipment such as a snowplow when the frame was not designed for this type of service.
3. A fire involving the vehicle.

Diamond-Shaped. Diamond-shaped frame damage may be caused by:

1. Collision damage.
2. Towing another vehicle with a chain attached to one corner of the frame.
3. Being towed by another vehicle with a chain attached to one corner of the vehicle frame.

Twist. Frame twist may be caused by:

1. An accident or collision, especially one involving a rollover.
2. Operating the vehicle on extremely rough terrain.

Checking Frame Alignment

Safety Concerns

While servicing vehicle frames, always wear the proper safety clothing and safety items for the job being performed. This includes proper work clothing, safety goggles, ear protection, respirator, proper gloves, welding shield, and safety shoes (Figure 10-20).

If arc welding is necessary on a vehicle frame, follow these precautions:

1. Remove the negative battery cable before welding (Figure 10-21).
2. Remove the fuel tank before welding (Figure 10-22).
3. Protect the interior and exterior of the vehicle as necessary (Figure 10-23).

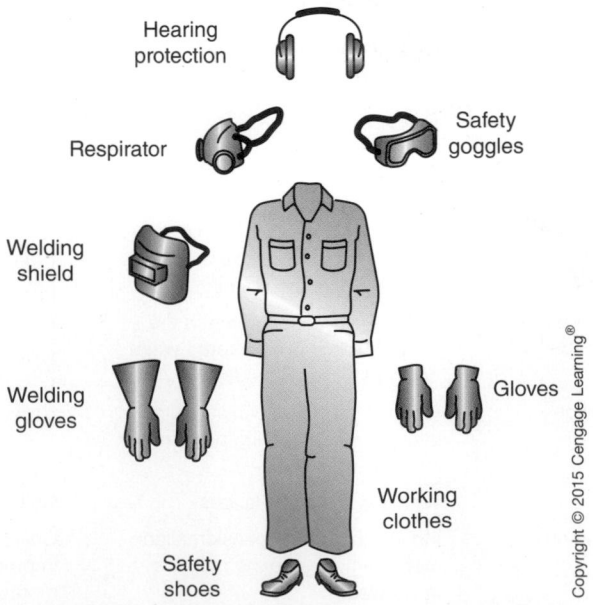

FIGURE 10-20 **Safety items for frame service.**

Copyright © 2015 Cengage Learning®

FIGURE 10-21 **Remove the negative battery cable before arc welding on a vehicle.**

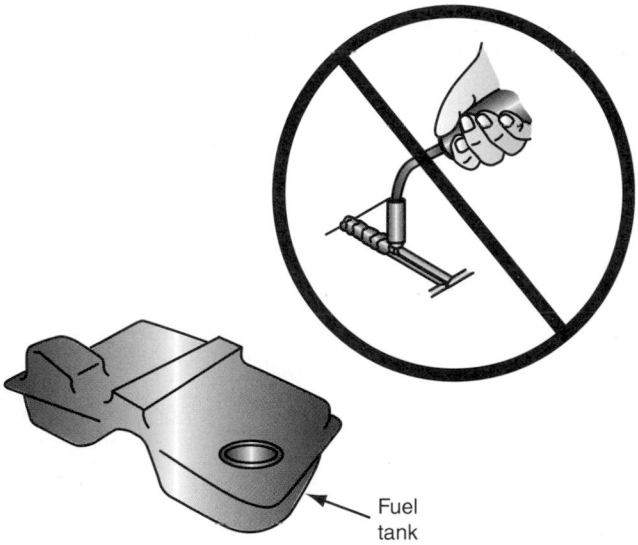

Copyright © 2015 Cengage Learning®

FIGURE 10-22 **Remove the fuel tank before arc welding on a vehicle.**

Copyright © 2015 Cengage Learning®

FIGURE 10-23 **Protect the interior and exterior of the vehicle while arc welding.**

Sunburst cracks radiate outward from a hole in a frame web or cross member.

CAUTION: Always follow the vehicle manufacturer's recommendations in the service manual regarding welding or reinforcing the frame, or mounting additional equipment on the frame. If these recommendations are not followed, the frame may be weakened and damaged.

SERVICE TIP: If the frame is cracked, frame alignment is often necessary.

A **plumb bob** is a weight with a sharp, tapered point that is suspended and centered on a string. Plumbers use this tool for locating pipe openings directly above each other in the tops and bottoms of partitions.

SPECIAL TOOL
Plumb bob

Visual Inspection

Prior to frame measurement, the frame and suspension should be visually inspected. Check for wrinkles on the upper flange of the frame, which indicate a sag problem. Visually inspect the lower flange for wrinkles, which are definite evidence of buckle. Because suspension or axle problems may appear as frame problems, the suspension components should be inspected for wear and damage. For example, an offset rear axle may appear as a diamond-shaped frame. Check all suspension mounting bushings and inspect leaf-spring shackles and center bolts.

The frame should be inspected for cracks, bends, and severe corrosion. Minor frame bends are not visible, but severe bends may be visible. Straight cracks may occur at the edge of the frame flange, and **sunburst cracks** may radiate from a hole in the frame web or cross member (Figure 10-24).

Frame Welding

Follow these steps for a typical frame welding procedure:

1. Remove any components that would interfere with the weld or be damaged by heat.
2. Find the extreme end of the crack and drill a 0.25 in. (6 mm) hole at this point.
3. V-grind the entire length of the crack from the starting point to the drilled hole.
4. The bottom of the crack should be opened 0.062 in. (2 mm) to allow proper weld penetration. A hacksaw blade may be used to open the crack.
5. Arc weld with the proper electrode and welding procedure.

Frame Measurement, Plumb Bob Method

Photo Sequence 19 shows a typical procedure for performing frame measurement using the plumb bob method.

Locating the Frame Centerline. Some vehicle manufacturers recommend measuring the frame with the **plumb bob** method.

Follow these steps to complete a plumb bob measurement for frame damage:

1. Place the vehicle on a level area of the shop floor and use a floor jack to raise the front and rear suspension off the floor. Support the chassis on safety stands at the manufacturer's recommended locations.

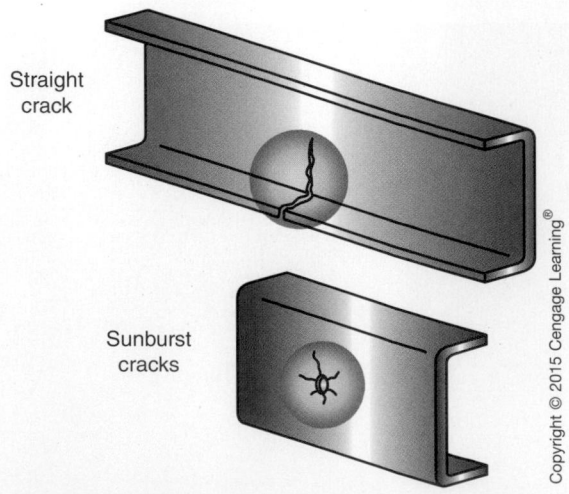

FIGURE 10-24 **Straight cracks may occur in the frame flange; sunburst cracks may radiate from a hole in the frame web or cross member.**

PHOTO SEQUENCE 19

Typical Procedure for Performing Frame Measurement, Plumb Bob Method

P19-1 **Park the vehicle on a level area of the shop floor.**

Copyright © 2015 Cengage Learning®

P19-2 **Raise the front suspension with a floor jack and lower the chassis onto safety stands positioned at the manufacturer's recommended lifting points.**

Copyright © 2015 Cengage Learning®

P19-3 **Raise the rear suspension with a floor jack and lower the chassis onto safety stands positioned at the manufacturer's recommended lifting points.**

Copyright © 2015 Cengage Learning®

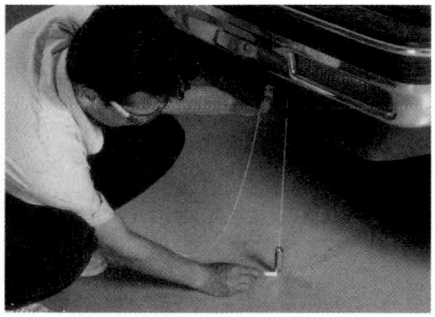

P19-4 **Suspend a plumb bob at the manufacturer's recommended frame measurement locations and place a chalk mark on the floor directly under the plumb bob.**

Copyright © 2015 Cengage Learning®

P19-5 **Use a floor jack to lift the vehicle, remove the safety stands, and lower the vehicle.**

Copyright © 2015 Cengage Learning®

P19-6 **Drive the vehicle away from the chalk-marked area.**

Copyright © 2015 Cengage Learning®

P19-7 **Use a tape measure to measure the vehicle's frame measurements between the chalk marks on the floor.**

Copyright © 2015 Cengage Learning®

P19-8 **Compare the frame measurements obtained with the vehicle manufacturer's specifications in the service manual.**

Copyright © 2015 Cengage Learning®

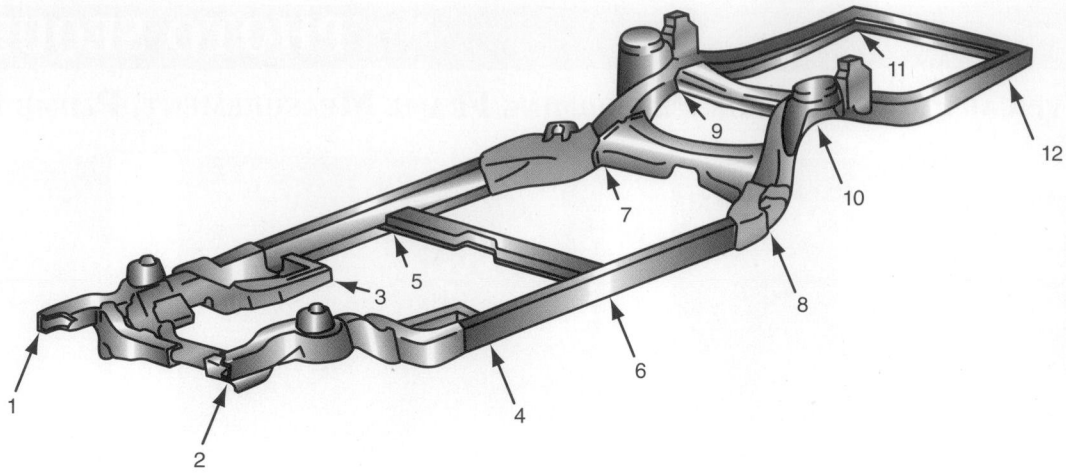

Copyright © 2015 Cengage Learning®

FIGURE 10-25 **Suspend a plumb bob at the locations shown on the frame and mark these locations on the floor with chalk.**

2. Suspend a plumb bob at locations 1, 2, 11, and 12 on the inside of the frame web, and allow the plumb bob to almost touch the floor surface (Figure 10-25). Points 1 and 2 are at the centerline of the slotted hole in each front bumper bracket, and points 11 and 12 are on the inside of the rear frame web. Mark these plumb bob locations on the floor with chalk.
3. Use a plumb bob to transfer points 3 through 10 from the frame to chalk marks on the floor.
4. Raise the vehicle with the floor jack, remove the safety stands, and lower the vehicle onto the floor. Move the vehicle away from the chalk-marked area.
5. Measure the distance between points 1 and 2 with a tape measure, and chalk mark the exact halfway point in this distance. This mark is the frame centerline at the front.
6. Measure the distance between points 11 and 12, and chalk mark the exact center of this measurement. This chalk mark is the frame centerline at the rear.
7. Draw a straight chalk line from the centerline at the front of the frame to the centerline at the rear of the frame. This chalk line is the complete centerline of the frame.

SERVICE TIP:
The points on the left and right frame webs must be at the same location on each web for accurate measurements.

Horizontal Frame Measurements

The procedure used for horizontal frame measurements follows:

1. Measure the distance from the frame centerline to points 3 through 10. Each pair of these points should be equal within 0.125 in. (3 mm). For example, the distance from point 5 to the centerline should be equal to the distance from point 6 to the centerline.
2. Measure diagonally from point 1 to point 6 and point 2 to point 5. These distances should be equal or within the manufacturer's specified tolerance. Place a straight chalk mark from point 1 to point 6 and point 2 to point 5. These diagonal lines should cross each other at the frame centerline.
3. Repeat step 2 at all the other diagonal frame measurements, such as point 3 to 10, point 4 to 9, point 5 to 12, and point 6 to 11. All these diagonal measurements should be equal or within the vehicle manufacturer's specified tolerance. Each pair of diagonal lines must cross each other at the vehicle centerline (Figure 10-26).

SERVICE TIP:
When using a tape measure, avoid twists and bends in the tape to provide accurate readings.

When such frame problems as side sway, buckle, or a diamond-shaped condition are present, some of the horizontal frame measurements are not within specifications, and some of the diagonal chalk lines do not cross at the frame centerline, depending on the location of the damaged area.

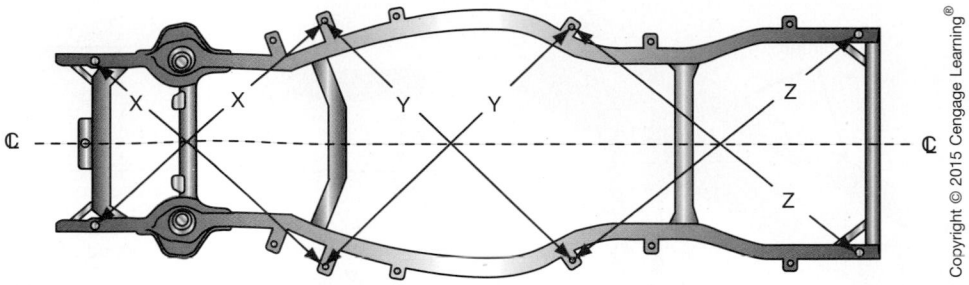

FIGURE 10-26 **Each pair of diagonal measurement lines on the frame must cross each other at the frame centerline.**

Frame Measurement, Tram Gauge Method

A **tram gauge** is a long, straight bar with two adjustable pointers. The distance between the pointers is adjustable, and the height of the pointers is also adjustable (Figure 10-27).

The horizontal frame measurements may be completed with a tram gauge, rather than using the plumb bob method. When measuring a frame, horizontal measurements are completed straight across the frame and diagonally forward and rearward on the frame, as discussed previously.

Vertical measurements from specific locations on the frame to an imaginary standard line are also necessary to detect such conditions as frame sag and twist (Figure 10-28). This imaginary horizontal line used for vertical frame measurements is called the **datum line**.

When vertical frame measurements are completed with the tram gauge, the gauge pointers must be set at the manufacturer's specified distance for each frame location

A **tram gauge** is a long, straight bar with two adjustable pointers that is used for frame measurements. A tram gauge may be called a tracking or track gauge.

An imaginary horizontal line parallel to the frame is used for vertical frame measurements. This line is called the **datum line**.

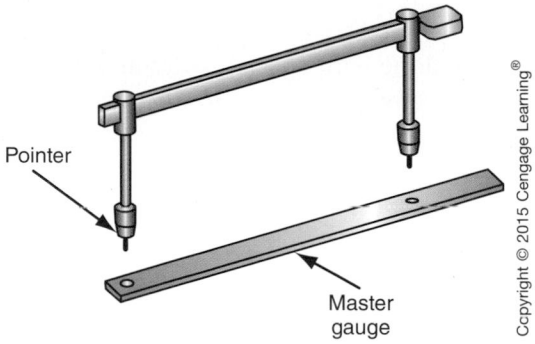

FIGURE 10-27 **Tram gauge for frame measurements.**

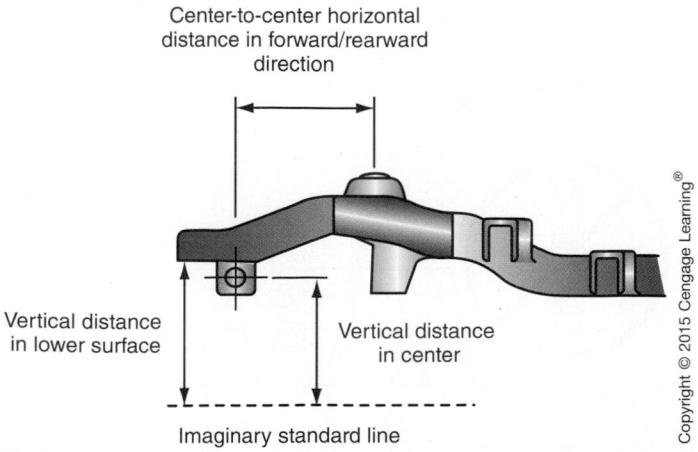

FIGURE 10-28 **Horizontal and vertical frame measurements.**

SERVICE TIP:
When the tram gauge is installed on the frame, be sure the pointers are seated properly in the frame openings (Figure 10-29).

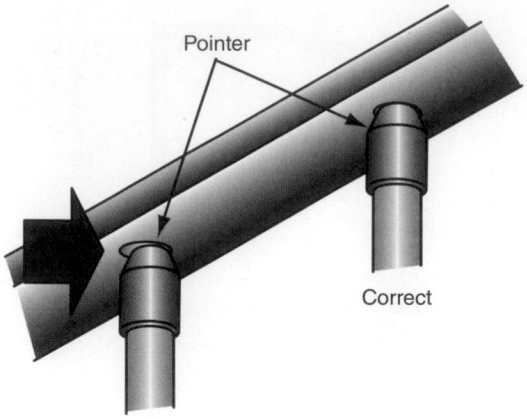

Copyright © 2015 Cengage Learning®

FIGURE 10-29 **Correct and incorrect tram gauge pointer seating in the frame openings.**

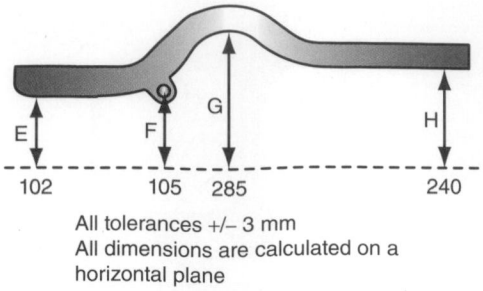

Copyright © 2015 Cengage Learning®

FIGURE 10-30 **Tram gauge setting and locations for vertical frame measurements, rear subframe.**

(Figure 10-30). With the tram gauge pointers set at the specified height and the gauge properly installed across the frame in the recommended frame openings, the tram gauge should be level if the vertical frame measurements are within specifications. If the frame is twisted, the tram gauge is not level when adjusted to specifications and installed in the twisted area.

To measure frame sag, three tram gauges may be installed at various locations for vertical frame measurement. When viewed from the front or the rear, the tram gauges must be level with each other. If the tram gauge near the center of the frame is lower than the tram gauges at the front and rear of the frame, the frame is sagged.

Frame Straightening

SPECIAL TOOL
Tram gauge

Frame straightening is usually done with special hydraulically operated bending equipment. This equipment must be operated according to the equipment manufacturer's recommended procedures. Because frame straightening is usually done by experienced body technicians, we will not discuss this service in detail. All safety precautions must be observed while operating frame-straightening equipment. Never stand in front of pulling equipment when it is in operation (Figure 10-31).

Measuring Unitized Body Alignment

Tram Gauge

Classroom Manual
Chapter 10, page 263

On unitized bodies, a tram gauge may be used to perform such measurements as the upper strut towers (Figure 10-32) and the front cross member (Figure 10-33).

If the underhood body measurements such as the strut tower measurements are not within specifications, the strut towers may not be in the original position. Collision damage

Copyright © 2015 Cengage Learning®

FIGURE 10-31 **Do not stand in front of pulling equipment while it is in operation.**

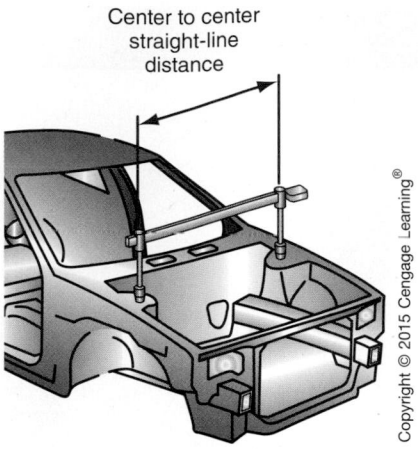

Copyright © 2015 Cengage Learning®

FIGURE 10-32 **Performing an upper strut tower measurement with a tram gauge on a unitized body vehicle.**

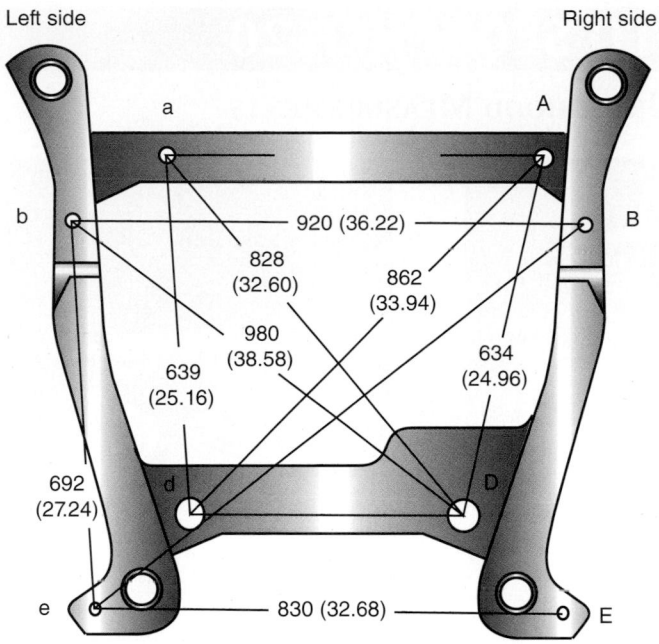

Copyright © 2015 Cengage Learning®

FIGURE 10-33 **Tram gauge measurements on the front cross member of a unitized body vehicle.**

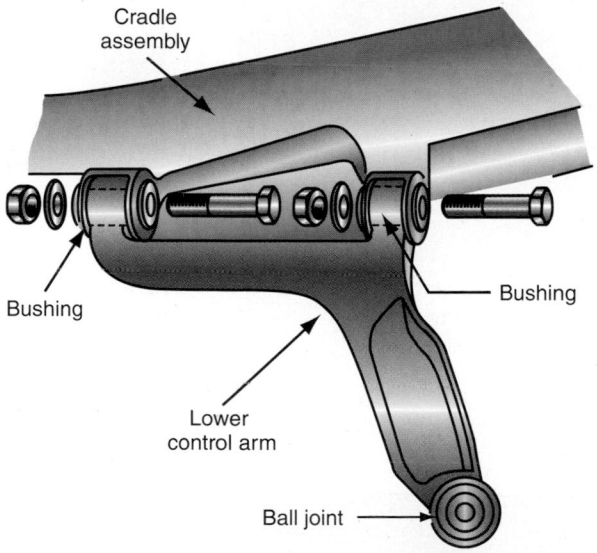

Copyright © 2015 Cengage Learning®

FIGURE 10-34 **Lower control arm-to-cradle attachment.**

may cause the strut towers to move out of the original position. If the strut towers are out of position, the front alignment angles are incorrect. Photo Sequence 20 illustrates the underhood body measurements.

Because the lower control arms are attached to the cross member or cradle and also to the lower end of the steering knuckle, if the cross member is moved from the original intended position, front wheel alignment is adversely affected (Figure 10-34). Improper cross member position may be caused by loose, worn cross member mounts or a bent cross member. Improper front wheel alignment may result in excessive front tire tread wear, steering pull, steering wander, and reduced directional stability. When removing a front cross member, always mark the cradle in relation to the chassis so it can be reinstalled in the original position.

PHOTO SEQUENCE 20

PERFORMING UNDERHOOD MEASUREMENTS

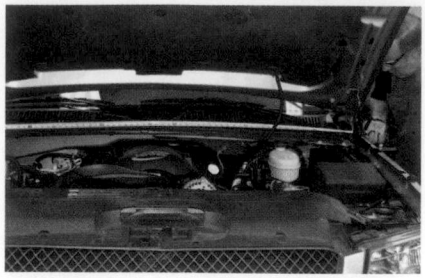

P20-1 **Perform horizontal underhood measurements.**

Copyright © 2015 Cengage Learning®

P20-2 **Record the horizontal underhood measurements.**

Copyright © 2015 Cengage Learning®

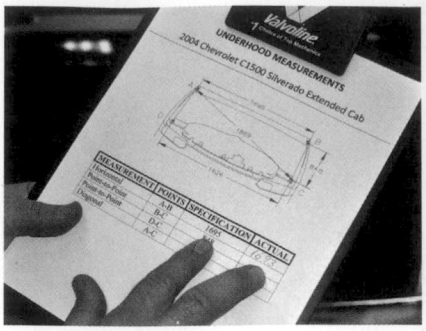

P20-3 **Compare horizontal underhood measurements to specifications and identify measurements that are not equal to the specifications.**

Copyright © 2015 Cengage Learning®

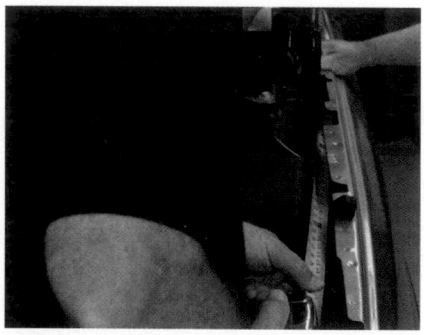

P20-4 **Perform point-to-point underhood measurements.**

Copyright © 2015 Cengage Learning®

P20-5 **Record the point-to-point underhood measurements.**

Copyright © 2015 Cengage Learning®

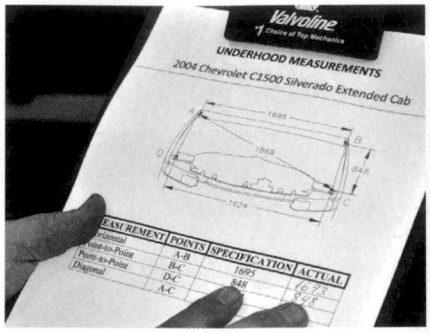

P20-6 **Compare point-to-point underhood measurements to specifications and identify measurements that are not equal to the specifications.**

Copyright © 2015 Cengage Learning®

P20-7 **Perform diagonal underhood measurements.**

Copyright © 2015 Cengage Learning®

P20-8 **Record the diagonal underhood measurements.**

Copyright © 2015 Cengage Learning®

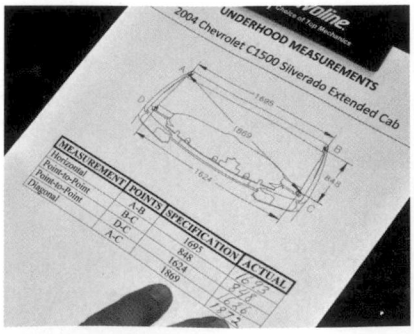

P20-9 **Compare diagonal underhood measurements to specifications and identify measurements that are not equal to the specifications.**

Copyright © 2015 Cengage Learning®

The engine and transaxle mounts are also connected to the front cross member. Therefore, a bent cross member or improperly positioned cross member mounts may cause improper engine and transaxle position, which results in vibration problems. Therefore, when diagnosing and correcting engine, drive axle, or transaxle vibration problems, always be sure the front cross member, engine mounts, transaxle mounts, and cross member mounts are in satisfactory condition. The tram gauge may be used to perform vertical measurements on the front and rear subframes on a unitized body vehicle.

Bench

SPECIAL TOOL
Dedicated bench system

A **dedicated bench system** is necessary to check many unitized body measurements after medium or heavy collision damage. The dedicated bench system has three main parts: the bench, transverse beams, and the dedicated fixtures. When used together, this equipment will perform many undercar unitized body measurements, including length, width, and height at the same time. The bench contains strong steel beams mounted on heavy casters (Figure 10-35). The top of the bench acts as a datum line.

Transverse Beams

The transverse beams are mounted perpendicular to the bench. There are a variety of holes in the bench for proper transverse beam attachment (Figure 10-36). Various holes in the transverse beams provide the correct dedicated fixture positions.

Dedicated Fixtures

The dedicated fixtures are specific to the body style being measured (Figure 10-37). These fixtures are bolted to the transverse beams, and the holes in the upper end of the fixtures must be aligned with specific body openings (Figure 10-38).

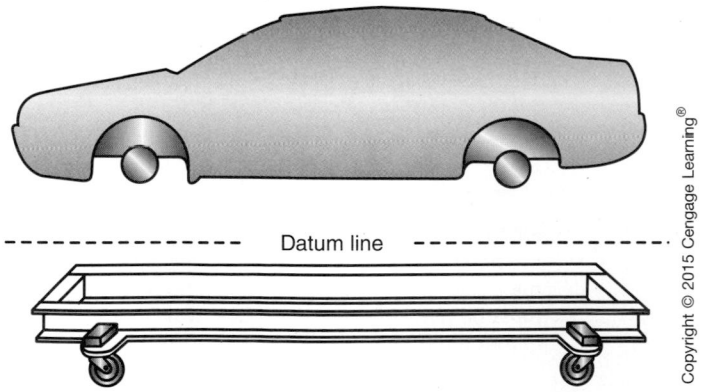

FIGURE 10-35 **Bench from a dedicated bench system.**

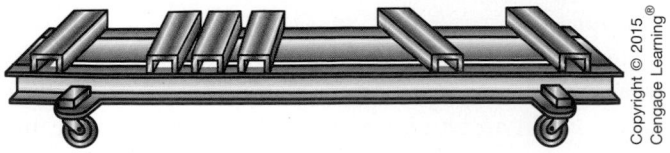

FIGURE 10-36 **Transverse beams placed on the bench.**

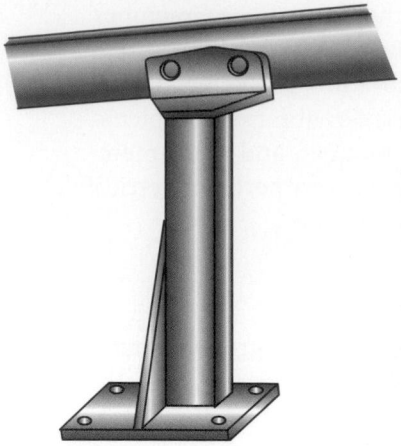

Copyright © 2015 Cengage Learning®

FIGURE 10-37 **Dedicated fixtures are specific to the body style being measured.**

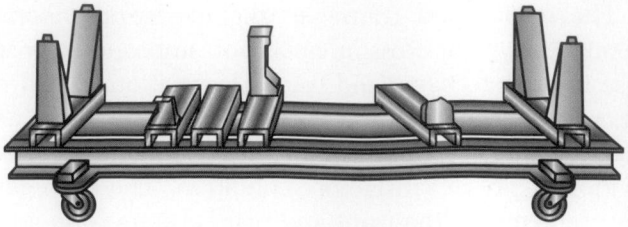

Copyright © 2015 Cengage Learning®

FIGURE 10-38 **Dedicated fixtures are bolted to the transverse beams in specific locations depending on the body style.**

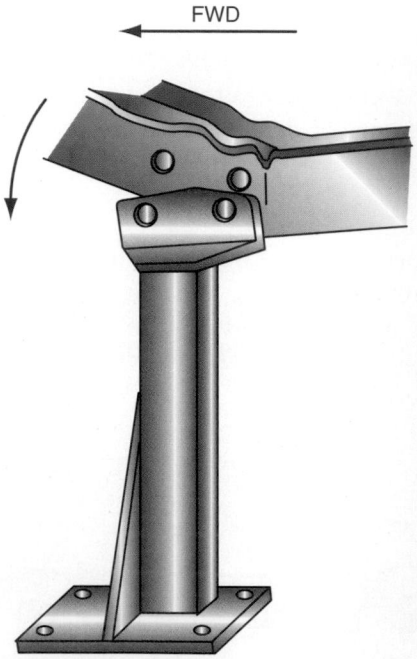

Copyright © 2015 Cengage Learning®

FIGURE 10-39 **When holes in the body are behind and above the openings in the dedicated fixture, the unitized body must be pulled forward and downward in that area.**

The unitized body must be straightened until the holes in the dedicated fixtures and the body openings are aligned. If the holes in the unitized body are behind and above the dedicated fixture openings, the unitized body must be pulled forward and downward in that area (Figure 10-39).

Universal benches are now available with computer measuring systems. Measuring a unitized body with a bench system and straightening these bodies with special body pulling equipment is usually done in an autobody shop rather than an automotive repair shop.

Electronic body and chassis alignment equipment using laser beam technology is used in many body shops during collision repair.

Unitized Body Straightening

WARNING: Always follow all the recommended precautions and procedures in the vehicle manufacturer's service manual and in the equipment manufacturer's operator's manual when straightening a unitized body. Failure to follow these precautions and procedures may result in severe personal injury and property damage.

Because unitized body straightening is usually performed by experienced body technicians, our discussion is brief on this subject. During the straightening process on a unitized body, the body is securely bolted to the dedicated bench at all the possible locations except those where misalignment is present. The dedicated bench is securely chained to special holding fixtures in the shop floor. Hydraulically operated pulling equipment is then attached securely to the unitized body to pull the body in the intended direction. **Single-pull** equipment pulls only one location at a time (Figure 10-40), and **multipull** equipment pulls in several locations at the same time (Figure 10-41). The unitized body is pulled until the holes in the body are aligned with the holes in the dedicated fixtures.

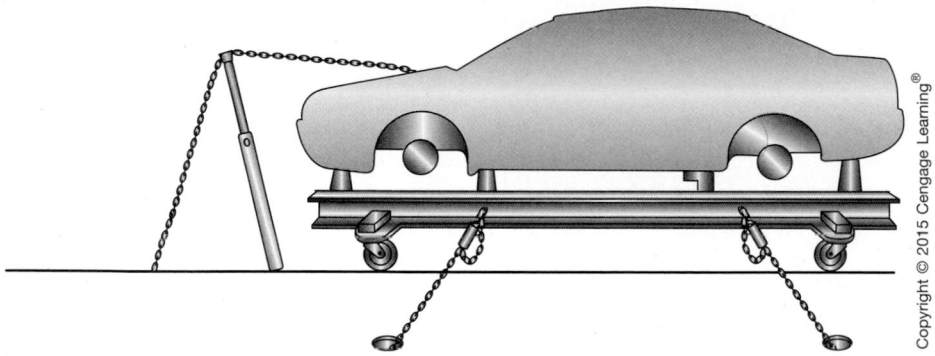

FIGURE 10-40 **Single-pull unitized body straightening system.**

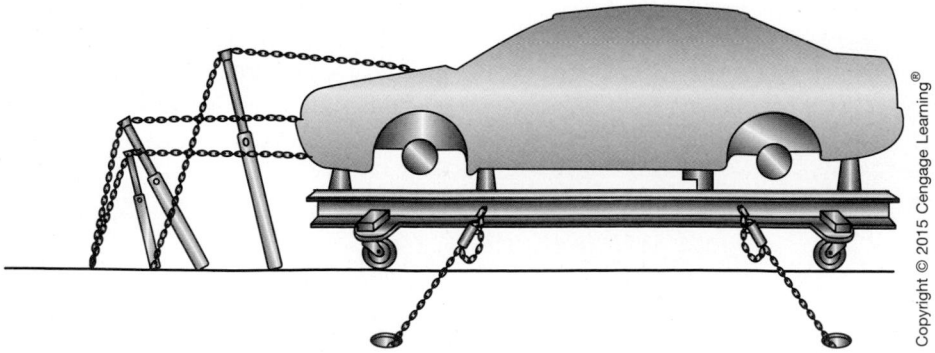

FIGURE 10-41 **Multipull unitized body straightening system.**

CUSTOMER CARE: Always treat customers and their vehicles with respect! Remember that all customers deserve to be treated with respect regardless of their appearance, the mood they are in, or the type of vehicle they drive. When customers notice that you always respect them and their vehicle, they will have a better impression of your business, and they are likely to return to your shop for all their service requirements.

CASE STUDY

A customer complained that the left rear tire on his Cutlass Calais had suffered severe tread wear in the last 2,000 miles. The technician examined this tire and found the customer was not exaggerating. All the tread wear indicators were showing and the other tires showed very little wear. The customer indicated that the left rear tire tread looked like the tread on the other tires 2,000 miles previously. While checking the tread wear on the left rear tire, the technician noticed the car had been newly painted. Further discussion with the customer revealed the car had been recently purchased, and there was no warranty on the vehicle. During a road test, the technician found that the steering pulled to the left while driving straight ahead. The technician suggested a thorough prealignment inspection and complete four-wheel alignment with a computer alignment system.

During the prealignment inspection, all the steering and suspension components except the left rear tire appeared to be in normal condition. However, the technician noticed the left rear quarter panel and other body components had been replaced. Apparently the car had suffered severe collision damage in this area.

The four-wheel alignment results were satisfactory except the left rear wheel, which had a 5 degree toe-in and a thrust angle of 8 degrees. This severe rear toe-in condition had moved the thrust line to the right of the geometric centerline. When the technician referred to the manufacturer's service manual, this information stated the rear wheel toe is not adjustable, and if the rear wheel camber and toe are not within specifications, the rear underbody and suspension should be inspected for damage. The technician located the underbody dimension specifications in the service manual (Figure 10-42 and Figure 10-43).

The technician began checking the rear suspension underbody measurements with a tram gauge. When dimension M was measured between the center of the bolt heads in the trailing arm brackets, this measurement was 1 in. (25.4 mm) less than specified. This measurement indicated the left rear subframe had been forced inward toward the center of the vehicle, which resulted in the improper toe and thrust angle. Further inspection of the rear axle channel with a long straightedge indicated this channel was bent rearward in the center. The car was sent to the body shop to pull the rear subframe and unitized body back to the original position and provide the specified distance between the center of bolt heads on the trailing arms. Because the rear axle channel is a critical component in providing rear suspension ride quality and alignment, the customer was advised this component should be replaced. A new rear tire was also installed.

When the rear suspension was reassembled, all the alignment angles were rechecked and the rear wheel toe and thrust line were within specifications. During a road test, there was no evidence of steering pull or any other steering problems.

From this experience, the technician learned the importance of four-wheel alignment. The technician also discovered that accurate underbody measurement specifications are absolutely essential.

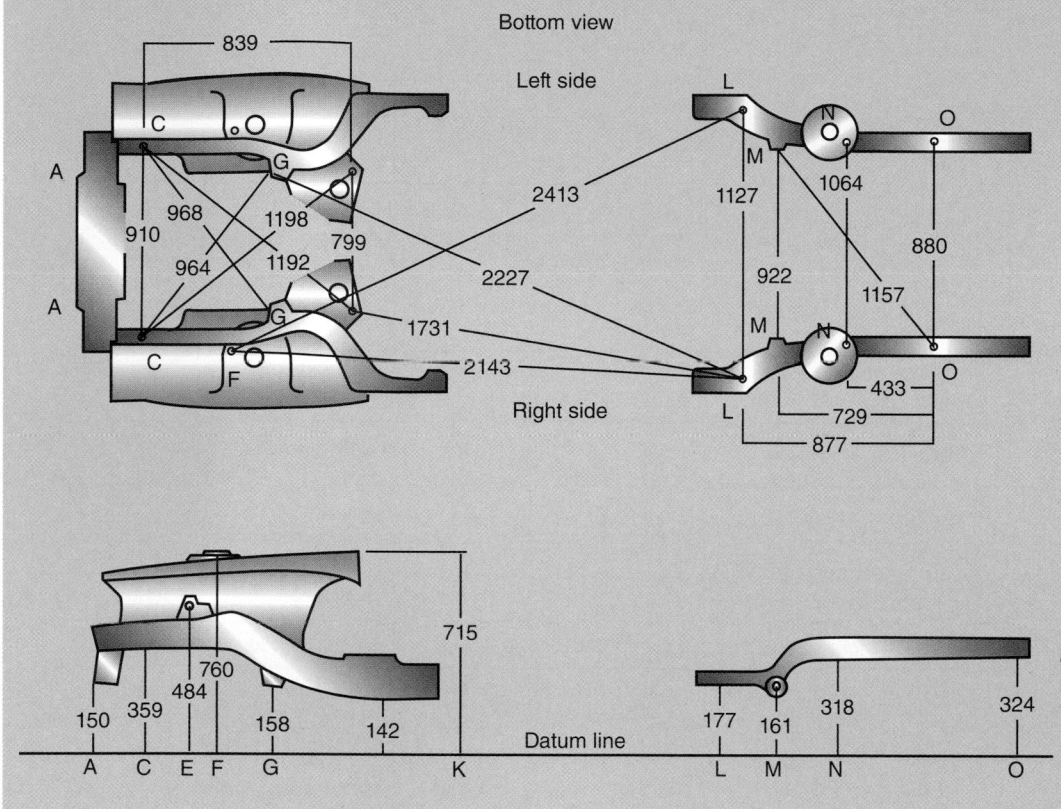

* All dimensions are metric.
* All control points are symmetrical side to side unless otherwise noted.
* All tolerances are +/– 3 mm.

Copyright © 2015 Cengage Learning®

FIGURE 10-42 **Underbody measurement specifications and locations.**

TERMS TO KNOW

Axle offset
Datum line
Dedicated bench system
Frame flange
Frame web
Geometric centerline
Lateral axle sideset
Multipull
Plumb bob
Rear axle offset
Rear wheel tracking
Section modulus
Setback
Single-pull
Steering axis inclination (SAI)
Sunburst cracks
Symmetry angle measurements
Track gauge
Tram gauge
Turntable

REF	HORIZONTAL	VERTICAL	LOCATION
A	Leading edge of 24 mm gauge hole	Lower surface at gauge hole to datum line	Front end lower tie bar
B	Center of 16 mm gauge hole	None	Front end upper tie bar
C	Leading edge of 20 mm gauge hole	Lower surface at gauge hole to datum line	Motor compartment side rail, forward of transaxle anchor support plate on left side rail, and at rear of engine mounting support on right side rail. For access on right side with air conditioning, remove air compressor.
D	Center of 16 mm gauge hole	None	Front upper surface of motor compartment side rail
E	Center of lower attaching hole in transaxle support anchor plate	Datum line to horizontal centerline of transaxle support anchor plate lower attaching hole	Transaxle support anchor plate on left side rail
F	Center of shock tower strut front attaching hole	Upper surface at shock tower strut front attaching hole	Shock tower
G	Leading edge of oblong master gauge hole	Datum line to horizontal leading edge of oblong master gauge hole	Front suspension lower control arm mount support
H	Center of 16 mm gauge hole	None	Gage hole in upper surface of motor compartment side rail
I	Center of 19 mm gauge hole	Datum line to lower surface at gauge hole	Outboard 19 mm gauge hole in front suspension control arm mounting reinforcement support
J	Center of hood hinge pivot pin head	None	Hood hinge
K	Center of front upper hinge pin hole	Upper surface at hinge pin hole	Front upper door hinge, body side
L	Leading edge of 22 mm gauge hole	Lower surface at leading edge of gauge hole to datum line	Compartment pan longitudinal rail, forward of rear suspension spring seat support
M	Center of attaching bolt head	Datum line to horizontal centerline of attaching bolt head of rear suspension control arm support.	Rear suspension control arm support forward of rear suspension spring seat support
N	Center of 16 mm gauge hole	Datum line to lower surface at gauge hole	Outboard gage hole at rear suspension spring seat support
O	Leading edge of 21 mm gauge hole	Lower surface at gauge hole to datum line	Compartment pan longitudinal rail

FIGURE 10-43 **Explanation of underbody horizontal and vertical measurement locations.**

Copyright © 2015 Cengage Learning®

ASE-STYLE REVIEW QUESTIONS

1. While discussing a diamond-shaped frame condition:
   Technician A says this condition may be caused when the vehicle is involved in a fire resulting from an accident.
   Technician B says this condition may be caused by towing another vehicle with the chain attached to one corner of the frame.
   Who is correct?
   A. A only
   B. B only
   C. Both A and B
   D. Neither A nor B

2. When welding a vehicle frame crack:
   A. A 0.25 in. hole should be drilled in the center of the crack.
   B. The crack should be V-ground at both ends of the crack.
   C. Disconnect the negative battery cable before welding the crack.
   D. Use a cutting torch to open the crack for proper weld penetration.

3. A vehicle has a drift to one side, camber is negative on one side and positive on the other but the cross camber is within specification and caster is within specifications. There are no adjustments available for either camber or caster. The technician suggests that the camber may be brought into specifications by:
   A. Rotating the front coil springs.
   B. Bending the lower control arms.
   C. Repositioning the cradle/subframe assembly.
   D. Replacing the front wheel hub assemblies.

4. While inspecting and measuring a front cradle on a front-wheel-drive car:
   A. Front wheel alignment angles may be changed by a bent front cradle.
   B. Engine mounting position is not affected by a bent front cradle.
   C. A bent front cradle has no effect on front drive axle vibration.
   D. A bent front cradle may cause premature front wheel bearing failure.

5. All of these statements about frame problems are true EXCEPT:
   A. Frame sag may be caused by overloading the vehicle.
   B. Frame sag may be caused by the use of equipment for which the frame was not designed.
   C. Frame buckle may be caused by using a snowplow on the front of the vehicle.
   D. A diamond-shaped frame may be caused by operating the vehicle on extremely rough terrain.

6. Unequal SAI angles on the left and right sides of the front suspension may cause:
   A. Tread wear on the front tires.
   B. Brake pull during sudden stops.
   C. Bump steer on a rough road.
   D. Steering wander while driving straight ahead.

7. On many front suspension systems, the maximum variation between the left and right SAI is:
   A. 0.75 degree
   B. 1 degree
   C. 1.5 degrees
   D. 2.5 degrees

8. A bent front cross member affects all of the following alignments EXCEPT:
   A. Centerline.
   B. Suspension.
   C. Axle.
   D. Engine.

9. Referring to OLD Figure 14-14:
   Technician A says the measurement between "e" and "b" should be no more than ±1.0 in. different from the measurement between "E" and "B."
   Technician B says frame damage problems, such as buckling, can be determined if the measurements between points "e" and "E" are more than ±0.25 in. different than between "b" and "B."
   Who is correct?
   A. A only
   B. B only
   C. Both A and B
   D. Neither A nor B

10. When making horizontal frame measurements to determine if the frame or unibody is straight, all of the measurements should be made with reference to which one of the following?
    A. Vertical
    B. Centerline
    C. Horizontal
    D. Body bushings

ASE CHALLENGE QUESTIONS

1. A car with a MacPherson strut front suspension has 1.5 degree positive camber on the left front wheel and 1.5 degree negative camber on the right front wheel. The most likely cause of this problem is:
   A. Worn upper strut mounts.
   B. Improperly positioned steering gear.
   C. Engine cradle/subframe shifted to the right.
   D. Bent strut rods.

2. A pickup truck with a MacPherson strut front suspension pulls to the left. Preliminary inspection shows that the SAI is correct, but the camber of the right front wheel is –1 1/4 degrees.

   Technician A says a bent spindle could be the cause.

   Technician B says the problem may be a worn-out strut.

   Who is correct?

   A. A only
   B. B only
   C. Both A and B
   D. Neither A nor B

3. While discussing frame damage:

   Technician A says a possible indication of frame damage is an uncentered steering wheel when driving straight ahead.

   Technician B says suspension alignment may be impossible if the frame is bent.

   Who is correct?

   A. A only
   B. B only
   C. Both A and B
   D. Neither A nor B

4. Which of the following makes this statement true? Measurement points of a frame or unibody using a tram gauge are:
   A. Seam notches.
   B. Calculated by computer.
   C. Based on a vertical datum plane.
   D. Manufacturer's specified frame openings.

5. The owner of an older pickup says the truck has begun to "wander and shimmy" in the past few months, especially when the bed of the truck is empty.

   Technician A says the twin I-beam front suspension caster is probably out of spec.

   Technician B says the eyes of the rear leaf springs are probably worn.

   Who is correct?

   A. A only
   B. B only
   C. Both A and B
   D. Neither A nor B

JOB SHEET 38

Name _____ **Date** _____

Frame Measurement, Plumb Bob Method

Upon completion of this job sheet, you should be able to perform frame measurements with a plumb bob.

Tools and Materials

Floor jack
Plumb bob
Safety stands
Chalk

Describe the Vehicle Being Worked On:

Year _____ Make _____ Model _____
VIN _____ Engine type and size _____

Procedure

Task Completed

1. Place the vehicle on a level area of the shop floor and use a floor jack to raise the front and rear suspension off the floor. Support the chassis on safety stands at the manufacturer's recommended locations.
   Is the vehicle chassis supported securely on safety stands? ☐ Yes ☐ No
   Instructor check _____
2. Locate the frame measurement locations in the vehicle manufacturer's service manual. ☐
3. Suspend a plumb bob at each frame measurement location and place a chalk mark on the floor directly below the tip of the plumb bob.
   Are all frame measurement locations chalk marked on the shop floor?
   ☐ Yes ☐ No
   Instructor check _____
4. Raise the vehicle with the floor jack, remove the safety stands, and lower the vehicle onto the floor. Move the vehicle away from the chalk-marked area. ☐
5. Measure the distance between the two chalk marks directly across from each other at the front of frame, and chalk mark the exact halfway point in this distance. This mark is the frame centerline at the front.
   Distance between the two chalk marks directly across from each other at the front of the frame _____
   Midpoint in the distance between the two chalk marks directly across from each other at the front of the frame _____

SERVICE TIP: The measurement points on the left and right frame webs must be at the same location on each web to obtain accurate measurements.

6. Measure the distance between the two chalk marks directly across from each other at the rear of the frame, and chalk mark the exact halfway point of this distance. This mark is the frame centerline at the rear.

   Distance between the two chalk marks directly across from each other at the rear of the frame _____

   Midpoint of the distance between the two chalk marks directly across from each other at the rear of the frame _____

SERVICE TIP:
When using a tape measure, avoid twists and bends in the tape to provide accurate readings.

7. Draw a straight chalk line from the centerline at the front of the frame to the centerline at the rear of the frame. This chalk line is the complete centerline of the frame.

   Is the vehicle centerline chalk marked on shop floor? ☐ Yes ☐ No

   Instructor check _____

8. Use a tape measure to measure the distance from the frame centerline to all frame measurement points directly opposite each other. Each pair of these points should be equal within 0.125 in. (3 mm).

   Frame measurements between specified measurement locations that are directly opposite each other and the vehicle centerline, starting at the front of the frame:

   A. left side _____ right side _____ difference _____

   B. left side _____ right side _____ difference _____

   C. left side _____ right side _____ difference _____

   D. left side _____ right side _____ difference _____

   Necessary frame repairs:

   _____

   _____

   _____

9. Use a tape measure to complete all diagonal frame measurements at the vehicle manufacturer's recommended locations. These distances should be equal or within the manufacturer's specified tolerances.

   Distances between diagonal frame measurement points:

   Diagonal frame measurements at the front of the frame:

   diagonal A _____ diagonal B _____

   difference between diagonal A and B _____

   Diagonal frame measurements at the center of the frame:

   diagonal C _____ diagonal D _____

   difference between diagonal C and D _____

   Diagonal frame measurements at the rear of the frame:

   diagonal E _____ diagonal F _____

   difference between diagonal E and F _____

   Necessary frame service:

   _____

   _____

   _____

10. Use a long, straight steel bar to place a straight chalk mark between all frame diagonal measurement locations. These diagonal lines should cross each other at the frame centerline.

If all diagonal lines do not cross the frame centerline at the same location, state the required frame repairs.

_____

_____

Instructor's Response _____

_____

_____

JOB SHEET 39

Name _____ **Date** _____

Inspect and Measure Front Cradle

Upon completion of this job sheet, you should be able to inspect and measure front cradles.

NATEF AST/MAST Task Correlation:

Suspension and Steering. Wheel Alignment Diagnosis, Adjustment, and Repair, Task #E-8. Check front and/or rear cradle (subframe) alignment; determine necessary action. (P-3)

Tools and Materials

Tram gauge

Describe the Vehicle Being Worked On:

Year _____ Make _____ Model _____

VIN _____ Engine type and size _____

Procedure

Task Completed

1. Raise the vehicle on a lift using the vehicle manufacturer's specified lifting points. ☐
2. Inspect front cradle mounts for looseness, damage, oil soaking, wear, and deterioration.
   Cradle mount condition: right front _____ right rear
   left front _____ left rear _____
3. Inspect front cradle for visible bends and damage.
   Front cradle condition indicated in visible inspection:
   _____
   _____
   _____
4. Inspect front cradle alignment.
   Is the front cradle aligning hole(s) properly aligned with the matching hole in the chassis? ☐ Yes ☐ No
   Recommended cradle service:
   _____
   _____
   _____
5. Use a tram gauge to complete all measurements across the width of the cradle starting at the front of the cradle.
   Width measurement A _____
   Specified width measurement A _____
   Width measurement B _____
   Specified width measurement B _____
   Width measurement C _____

Specified width measurement C _____
Width measurement D _____
Specified width measurement D _____
Width measurement E _____
Specified width measurement E _____
Necessary cradle service:

_____

_____

6. Use a tram gauge to complete all front-to-rear measurements on the cradle, starting on the left side of the cradle.
   Front-to-rear measurement A _____
   Specified front-to-rear measurement A _____
   Front-to-rear measurement B _____
   Specified front-to-rear measurement B _____
   Front-to-rear measurement C _____
   Specified front-to-rear measurement C _____
   Front-to-rear measurement D _____
   Specified front-to-rear measurement D _____
   Necessary cradle service:

   _____

   _____

7. Use a tram gauge to complete all diagonal cradle measurements, starting at the left side of the cradle.
   Diagonal measurement A _____
   Specified diagonal measurement A _____
   Diagonal measurement B _____
   Specified diagonal measurement B _____
   Diagonal measurement C _____
   Specified diagonal measurement C _____
   Diagonal measurement D _____
   Specified diagonal measurement D _____
   Necessary cradle service:

   _____

   _____

Instructor's Response _____

_____

_____

JOB SHEET 40

Name _____ **Date** _____

INSPECT AND WELD VEHICLE FRAME

Upon completion of this job sheet, you should be able to inspect and weld a vehicle frame.

CAUTION: Always wear eye protection while working in the shop, and use the specified eye protection while arc welding to prevent eye injury.

Tools and Materials

Hand pump
Welding hammer
Arc welder
Welding rods

Describe the Vehicle Being Worked On:

Year _____ Make _____ Model _____
VIN _____ Engine type and size _____

Procedure

Task Completed

1. Raise the vehicle on a lift using the vehicle manufacturer's specified lifting points. ☐
2. Read the instructions for disconnecting the negative battery cable in the vehicle manufacturer's service manual. These instructions may include connecting a 12-V power source to the cigarette lighter socket to maintain power to computer memories, radio-stereo, and other electronic equipment.
   List the vehicle manufacturer's instructions for disconnecting the negative battery cable.
   _____
   _____
   Have these instructions been completed on the vehicle being serviced?
   ☐ Yes ☐ No
   Instructor check _____
3. If the vehicle is air-bag-equipped, determine the specified waiting period after the negative battery cable is disconnected before performing service work.
   Waiting period _____
4. Disconnect the negative battery cable.
   Is the negative battery cable disconnected? ☐ Yes ☐ No
   Has the specified waiting period been completed before servicing the vehicle?
   ☐ Yes ☐ No
   Instructor check _____
5. Use a hand pump to pump all the fuel from the fuel tank. Place the fuel in the proper fuel safety containers and store this fuel away from the work area and other ignition sources.
   Has all the fuel been pumped from the tank? ☐ Yes ☐ No
   Is this fuel placed in proper fuel safety containers and stored away from the work area? ☐ Yes ☐ No
   Instructor check _____

6. Remove the fuel tank from the vehicle and store the fuel tank away from the work area and other ignition sources.
   Is the fuel tank removed? ☐ Yes ☐ No
   Fuel tank stored away from the work area? ☐ Yes ☐ No
   Instructor check _____

7. Protect the interior and exterior of the vehicle from heat or sparks generated by the welding process.
   Are the vehicle's interior and exterior protected? ☐ Yes ☐ No
   Instructor check _____

8. Inspect the entire frame for cracks, bends, severe corrosion, and wrinkles in the flanges.
   Defective frame conditions:
   _____
   _____
   _____

9. Remove any components that would interfere with the weld or be damaged by heat.
   Have all components been removed that would interfere with the weld or be damaged by the heat? ☐ Yes ☐ No
   Instructor check _____

10. Find the extreme end of the crack and drill a 0.25 in. (6 mm) hole at this point.
    Is a 0.25 in. hole drilled at the end of the crack? ☐ Yes ☐ No
    Instructor check _____

11. V-grind the entire length of the crack from the starting point to the drilled hole.
    Is the entire length of the crack V-ground? ☐ Yes ☐ No
    Instructor check _____

12. The bottom of the crack should be opened 0.062 in. (2 mm) to allow proper weld penetration. A hacksaw blade may be used to open the crack.
    Is the bottom of the crack open 0.062 in. (2 mm)? ☐ Yes ☐ No
    Instructor check _____

13. Arc weld the crack with the electrodes specified by the frame and/or vehicle manufacturer. Complete several passes with the arc weld to fill the crack.
    Is the crack filled with arc weld material? ☐ Yes ☐ No
    Instructor check _____

14. Use a pointed welding hammer to remove slag from the exterior of the weld.
    Is the slag removed from the weld? ☐ Yes ☐ No
    Instructor check _____

Instructor's Response _____
_____
_____

Chapter 11

Computer-Controlled Suspension System Service

BASIC TOOLS

Basic technician's tool set
Service manual
Floor jack
Safety stands
Utility knife
Machinist's ruler or tape measure

Upon completion and review of this chapter, you should be able to:

- Diagnose programmed ride control (PRC) systems.
- Diagnose electronic air suspension systems.
- Remove, replace, and inflate air springs.
- Adjust front and rear trim height on electronic air suspension systems.
- Service and repair nylon air lines.
- Diagnose vehicle dynamic suspension (VDS) systems.
- Remove and replace air springs on VDS systems.
- Perform an on-demand self-test on VDS systems.
- Diagnose electronic suspension control (ESC) systems.
- Display and interpret scan tool data on ESC systems.
- Use the output controls function to diagnose ESC systems.
- Diagnose vehicle networks.

Each year more vehicles are equipped with computer-controlled suspension systems, and these systems are becoming increasingly complex. Therefore, technicians must understand the correct procedures for diagnosing and servicing these systems. When a technician understands computer-controlled suspension systems and the proper diagnostic procedures for these systems, diagnosis becomes faster and more accurate.

Preliminary Inspection of Computer-Controlled Suspension Systems

Prior to diagnosing a computer-controlled suspension system, a preliminary inspection should be performed. The preliminary inspection may locate a minor defect that is the cause of the problem. If the preliminary inspection is not performed, a lot of time may be wasted performing advanced diagnosis when the problem is a minor defect. When the preliminary diagnosis does not locate any minor problems, further diagnosis is required.

Diagnostic Procedures and Steps to Follow:
Any repair procedure requires a systematic approach to diagnose and repair a vehicle. The four principle steps to every repair procedure should always be followed.

1. **Verify the Customer Concern (complainant) and the Involved System:** During this step it is important to gather as much pertinent information as possible.

a. Understand the concern listed on the repair order. Question the service advisor and/or the customer if more detail is required.
b. Road test the vehicle to confirm the customers concern. *Caution* should be advised when trying to repair a vehicle when the concern could not be duplicated or verified.
c. Know how the system or component is supposed to operate. Compare to a known good vehicle when possible.
d. Retrieve any stored DTCs.
   i. Review any Freeze Frame Data stored.
e. Are more than one system malfunction detected. Pay close attention to which systems are functioning properly and which are not.

2. Isolate the Problem:
   f. First, take a *systems based approach* to what could be potentially related to the concern.
   g. Refer to *Technical Service Bulletins* (TSBs) that may be related to the problem area.
   h. Follow manufacturer published diagnostic procedures if available.
   i. Eliminate the easiest to check systems or components.
      i. If the vehicle has a computer-controlled air suspension system, listen for air leaks in the system.
      ii. Measure the vehicle curb (trim) height.
      iii. Check the operation of the suspension warning light or observe the message center in the instrument panel for displayed suspension messages.
   j. Identify related component that are affected by fault and those that are not affected by fault.
   k. Identify any underlying reason that caused the failure to occur (i.e., check all electrical connections and wiring in the computer-controlled suspension system).
3. **Repair the Failure Area (component or circuit)**
   a. *Erase any stored DTC* after all information has been recorded. Remember, freeze frame data will be erased and major monitors will be reset.

   Recheck (verify) Repair Performed
   b. Road test the vehicle *under the same conditions* when the original complaint was detected to verify that repair has been successfully completed.
   c. *Perform any required drive cycle procedures* that may be necessary for OBD II System Monitor to run to completion. Failure to perform proper drive cycle could cause MIL to be illuminated after the customer picks up the vehicle and system monitor detects the same or another system failure.

A diagnostic sequence is very helpful when trying to find the root cause of a customer complaint. The diagnostic paths differ depending on whether or not a diagnostic trouble code (DTC) is present. For faults that do not set DTCs or for intermittent faults always refer to Freeze Frame Data while a road test is being performed. Some enhanced scan tools also have the ability to function as a trip recorder. This function allows the user to trigger the recorder to capture data when the fault is occurring.

Programmed Ride Control System Diagnosis

General Diagnosis

The **programmed ride control (PRC)** module monitors the complete system while the vehicle is driven. If a defect occurs in the PRC system, the mode indicator light starts flashing. When the mode selector switch is moved to the opposite position and returned to the same position, erroneous codes are cleared from the control module memory. If the mode indicator light continues flashing, the self-test should be performed.

Self-Test

The self-test connector is located under the ashtray (Figure 11-1). This connector has two terminals that must be connected during the self-test. A jumper tool should be fabricated to connect these terminals, and this tool should be soldered to the end of a 7-inch slotted screwdriver blade to access the self-test terminals (Figure 11-2).

Follow this procedure to perform the self-test:

1. Turn the ignition switch off and be sure the headlights and parking lights are off. These lights must remain off during the self-test.
2. Position the mode select switch in the Auto position.
3. Remove the ashtray and insert the fabricated tool in the self-test connector terminals.
4. Start the engine and leave the mode select switch in the Auto position.
5. When the engine has been running for 20 seconds or more, remove the tool from the self-test connector.
6. Count the mode indicator light flashes to obtain the trouble codes. For example, if the light flashes six times, code 6 has been provided. The light flashes each code four times at 9 second intervals. The mode indicator light is in the tachometer (Figure 11-3).

SERVICE TIP:
Diagnostic trouble codes (DTCs) together with the vehicle manufacturer's service manual diagnostic procedures will usually locate the problem in a PRC system.

Diagnostic Trouble Codes (DTCs)

Several DTCs are available on the PRC system (Figure 11-4). The DTCs vary depending on the vehicle make and year. Always use the DTC list in the manufacturer's service manual. A trouble code indicates a defect in a specific area. For example, if DTC 2 is received, the defect may be in the right rear actuator or in the connecting wires.

After a PRC system defect has been corrected, the DTCs should be cleared. The system should be checked for DTCs again to make sure there are no other faults in the system.

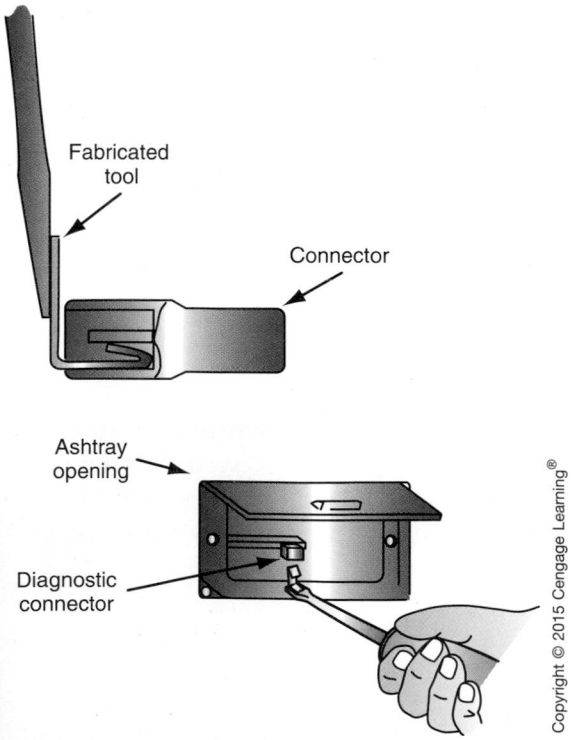

FIGURE 11-1 **Explanation of underbody horizontal and vertical measurement locations.**

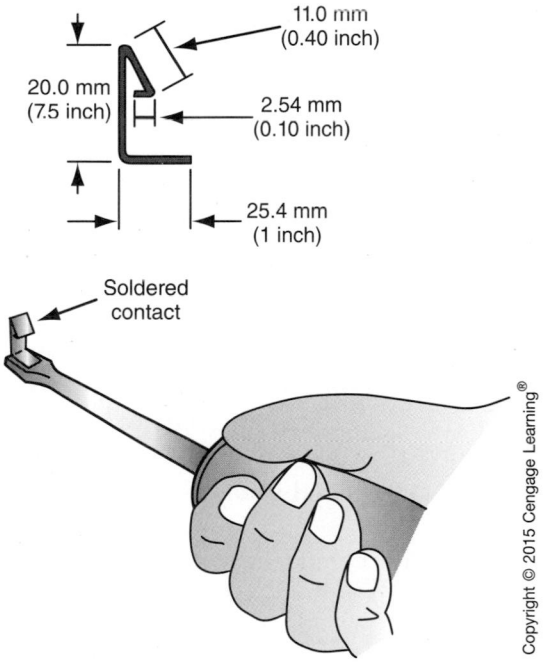

FIGURE 11-2 **Fabricated tool to connect the self-test terminals.**

Copyright © 2015 Cengage Learning®

FIGURE 11-3 **Mode indicator in tachometer flashes trouble codes.**

PRC System Diagnostic Trouble Codes (DTCs)

Code	Defect
6	No problem
1	Left rear actuator circuit
2	Right rear actuator circuit
3	Right front actuator circuit
4	Left front actuator circuit
5	Soft relay control circuit shorted
7	PRC control module
13	Firm relay control circuit shorted
14	Relay control circuit

Copyright © 2015 Cengage Learning®

FIGURE 11-4 **PRC system diagnostic trouble codes (DTCs).**

Diagnostic trouble codes (DTCs) indicate defects in specific areas of computer-controlled systems.

Electronic Air Suspension Diagnosis and Service

Air Spring Removal and Installation

WARNING: The system control switch must be in the Off position when system components are serviced to prevent personal injury and damage to system components.

WARNING: The system control switch must be turned off prior to hoisting, jacking, or towing the vehicle. If the front of the chassis is lifted with a bumper jack, the rear suspension moves downward. The electronic air suspension system will attempt to restore the rear trim height to normal, and this action may cause the front of the chassis to fall off the bumper jack, resulting in personal injury or vehicle damage.

WARNING: When air spring valves are being removed, always rotate the valve to the first stage until all the air escapes from the air spring. Never turn the valve to the second (release) stage until all the air is released from the spring. If this action is taken, the valve may be blown out of the air spring with considerable force, and this may cause personal injury.

WARNING: When removing and replacing many suspension components such as lower control arms in an air spring suspension system, air must be exhausted from the air spring to relieve tension on the components before the removal procedure. Failure to observe this precaution may result in personal injury.

Classroom Manual
Chapter 11, page 288

Many components in an electronic air suspension system, such as control arms, shock absorbers, and stabilizer bars, are diagnosed and serviced in the same way as the components in a conventional suspension system. However, the air spring service procedures are different compared to coil spring service procedures on a conventional suspension system.

The air spring removal and replacement procedure varies depending on the vehicle make and model year. For example, a different air spring removal and replacement procedure is required on vehicles with integral air springs and struts, and vehicles with the air springs mounted separately from the shock absorbers. Front and rear air spring removal and replacement procedures may vary. Always follow the exact air spring removal and replacement procedure in the vehicle manufacturer's service manual. The following is a typical air spring removal and replacement procedure for some vehicles with the air springs mounted separately from the shock absorbers:

Follow these steps for air spring removal:

1. Turn off the electronic air suspension switch in the trunk.
2. Hoist the vehicle and allow the suspension to drop downward, or lift the vehicle with a floor jack and place safety stands under the chassis. Lower the vehicle onto the safety stands and allow the suspension to drop downward.
3. Disconnect the nylon air line from the spring solenoid valve, and rotate the valve to the first stage to allow the air to escape from the spring. Never turn the valve to the second stage until all the air is exhausted from the spring.
4. Disconnect the lower spring retainer and remove the spring from the chassis.
5. Before an air spring is installed, it must be properly folded over the piston at the bottom of the membrane (Figure 11-5).
6. Install the spring in the chassis and connect the lower spring retainer. Be sure the top of the spring is properly seated in the spring seat. When an air spring is installed in the front or rear suspension, the spring must be properly positioned to eliminate folds and creases in the membrane (Figure 11-6).

Removal and Replacement of Air Springs with Integral Struts

Front or rear air springs with integral struts are sold only as an assembly. Do not attempt to service these components. The following is typical front air spring removal and replacement procedure on an automatic air suspension system with air springs containing integral struts:

1. Turn off the air suspension switch.
2. Raise the vehicle on a frame-contact lift and allow the suspension to drop downward.
3. Remove the wheel and tire assembly on the corner of the vehicle that requires air spring and strut removal.
4. Release the metal clip at the top height sensor mount, then remove the height sensor off the ball stud.
5. Remove the air spring solenoid retainer, loosen the air spring solenoid valve to the first position, and allow all the air pressure to be relieved from the air spring.
6. Push the air solenoid line retainer toward the center of the solenoid, and pull the rubber hose and nylon tube out of the solenoid (Figure 11-7). Remove the electrical connector from the solenoid.
7. Remove the air spring solenoid valve.
8. Remove the top strut cover, and remove three strut-to-strut tower retaining nuts (Figure 11-8).
9. Remove the strut-to-suspension arm retaining nut and bolt.
10. Remove the strut and spring assembly from the vehicle.

CAUTION:
Loosening the large center nut on the top of the strut rod may cause a permanent leak through the top of the air spring.

Follow these steps to install a replacement front air spring and integral strut:

1. Install the air spring and strut assembly into the front suspension until the upper strut mount retaining bolts extend through the bolt holes in the strut tower.
2. Install the upper strut mount retaining nuts and tighten these nuts to the specified torque.

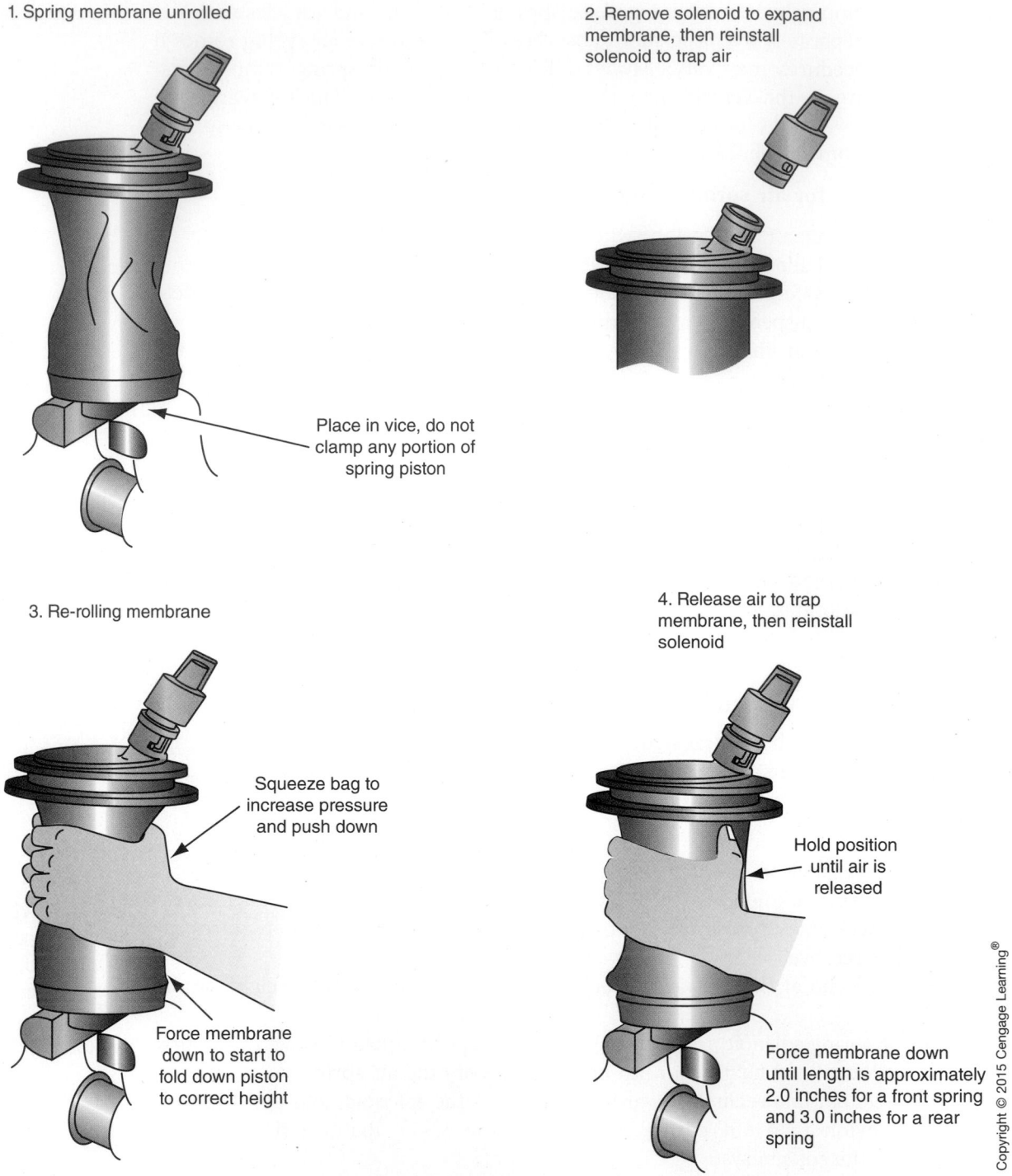

FIGURE 11-5 **Air spring folding.**

3. Install the lower strut-to-suspension retaining bolt and nut, and tighten the nut until it is snug. The nut on this bolt must be tightened to the specified torque after the vehicle weight is applied to the suspension.
4. Install the air spring solenoid valve fully into the air spring, and install the valve retainer.
5. Install the air line and electrical connector into the solenoid valve.
6. Install the wheel and tire assembly and tighten the wheel nuts to the specified torque.
7. Perform an air spring fill procedure described in the next section.

Copyright © 2015 Cengage Learning®

FIGURE 11-6 **When an air spring is installed in the front or rear suspension, the spring must be properly positioned to eliminate folds and creases in the membrane.**

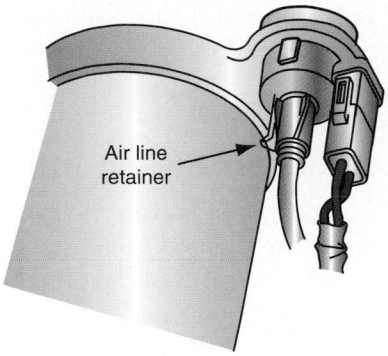

Copyright © 2015 Cengage Learning®

FIGURE 11-7 **Air line removal from the air spring solenoid valve.**

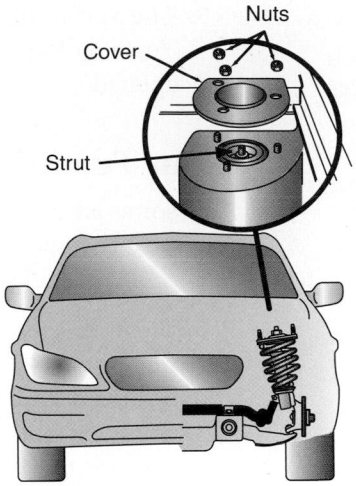

Copyright © 2015 Cengage Learning®

FIGURE 11-8 **Removing the upper strut mount retaining nuts.**

Air Spring Inflation

If an air spring has been replaced, or the air has been completely exhausted from an air spring, a spring fill procedure is required to inflate the air spring. The air spring fill procedure varies depending on the vehicle make and model year. Always follow the air spring fill procedure in the vehicle manufacturer's service manual. The following air spring fill procedure is for a vehicle with the air springs separate from the shock absorbers.

The weight of the vehicle must not be allowed to compress an uninflated air spring.

CAUTION: Do not allow the suspension to compress an air spring until the air spring is inflated. This action may damage the air spring.

When an air spring is being inflated, use this procedure:

1. With the vehicle chassis supported on a hoist, lower the hoist until a slight load is placed on the suspension. Do not lower the hoist until the suspension is heavily loaded.
2. Turn on the air suspension system switch.
3. Turn the ignition switch from Off to Run for 5 seconds with the driver's door open and the other doors shut. Turn the ignition switch off.
4. Ground the diagnostic lead.
5. Apply the brake pedal and turn the ignition to the Run position. The warning lamp will flash every 2 seconds to indicate the fill mode.
6. To fill a rear spring or springs, close and open the driver's door once. After a 6-second delay, the rear spring will fill for 60 seconds.
7. To fill a front spring or springs, close and open the driver's door twice. After a 6-second delay, the front spring will fill for 60 seconds.
8. When front and rear springs require filling, fill the rear springs first. Once the rear springs are filled, close and open the driver's door once to begin filling the front springs.

9. The spring fill mode is terminated if the diagnostic lead is disconnected from ground. Termination also occurs if the ignition switch is turned off or the brake pedal is applied.

A **scan tool** is a digital tester that can be used to diagnose various on-board computer systems.

Scan Tool Procedure for Air Spring Inflation

On some air suspension systems the **scan tool** is used to activate the air compressor and energize the appropriate air spring solenoid valve to inflate an air spring. The following is a typical spring fill procedure using a scan tool on an automatic air suspension system:

The **data link connector (DLC)** is a 16-terminal connector positioned under the left side of the instrument panel. A scan tool is connected to the DLC to diagnose various on-board electronic systems.

1. Turn off the air suspension switch.
2. Be sure the proper module is installed in the scan tool for suspension diagnosis and service on the vehicle being serviced.
3. Connect the scan tool to the **data link connector (DLC)** on the vehicle.
4. Be sure the vehicle is positioned on a frame-contact lift with the suspension dropped downward and no load on the suspension.
5. Connect a battery charger to the battery terminals with the correct polarity, and set the charger at a low charging rate.
6. Select Function Test on the scan tool, and select the desired function test number for the air spring being filled as follows: test number 212—LH front air spring, test number 214—RH front air spring, test number 216—LH rear air spring, and test number 218—RH rear air spring. When the proper function test number is selected, the scan tool commands the compressor and the appropriate air spring solenoid valve on, to allow air spring inflation.
7. When the compressor shuts off, the air spring inflation is complete, and any further air spring inflation or deflation is done during normal operation of the air suspension system.
8. Lower the vehicle onto the shop floor and lower the lift completely. Be sure the lift arms are not contacting the vehicle.
9. Inspect the air spring(s) that were completely deflated, and be sure there are no folds and creases in these springs.
10. Turn on the suspension switch.
11. If the lower front strut-to-control arm bolt and nut were removed and reinstalled during the suspension service procedure, use your knee to push downward and release the front bumper. Repeat this procedure three times.
12. Tighten the lower front strut-to-control arm nut and bolt to the specified torque.
13. Road test the vehicle and verify proper air suspension operation and trim height.

Photo Sequence 21 shows the proper procedure for air spring inflation with a scan tool.

OBD II computer systems became mandatory on cars and light trucks in the 1996 model year. Some vehicles prior to 1996 were equipped with OBD II systems. OBD II systems have some standardized features such as a 16-terminal DLC under the left side of the dash. OBD II systems have increased monitoring capabilities in the PCM. Many engine control systems are monitored, and if any engine system is defective so it allows the exhaust emissions to increase more than 1.5 times the normal limit for that vehicle, the malfunction indicator light (MIL) illuminates in the instrument panel.

Trim Height Mechanical Adjustment

On some vehicles with air suspension, the trim height is adjusted by lengthening or shortening the front or rear height sensors. On other vehicles, the manufacturer recommends using the scan tool to adjust the trim height. Always follow the trim height adjustment procedure in the vehicle manufacturer's service manual.

The **trim height** should be measured on the front and rear suspension at the locations specified by the vehicle manufacturer (Figure 11-10).

If the rear suspension trim height is not within specifications, it may be adjusted by loosening the attaching bolt on the top height sensor bracket (Figure 11-11). When the bracket is moved one index mark up or down, the ride height is lowered or raised to 0.25 in. (6.35 mm).

The front suspension trim height may be adjusted by loosening the lower height sensor attaching bolt. Three adjustment positions are located in the lower front height sensor

PHOTO SEQUENCE 21

INFLATE L/H FRONT AIR SPRING USING A SCAN TOOL

P21-1 **Turn off the air suspension switch.**
Copyright © 2015 Cengage Learning®

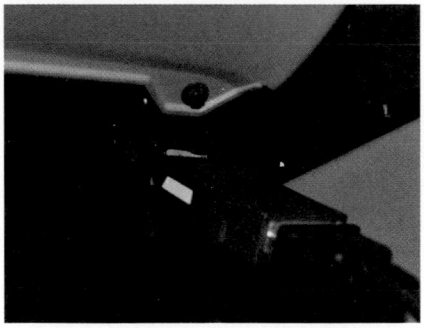

P21-2 **Connect the scan tool to the data link connector (DLC) on the vehicle.**
Copyright © 2015 Cengage Learning®

P21-3 **Be sure the vehicle is positioned on a frame-contact lift with the suspension dropped downward and no load on the suspension.**
Copyright © 2015 Cengage Learning®

P21-4 **Connect a battery charger to the battery terminals with the correct polarity, and set the charger at a low charging rate.**
Copyright © 2015 Cengage Learning®

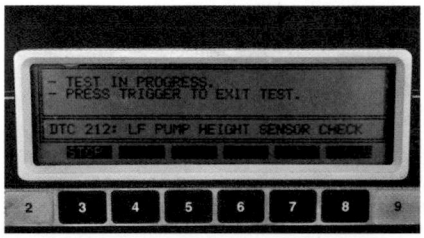

P21-5 **Select Function Test on the scan tool, and select the desired Function Test number 212—LH front air spring. When the proper function test number is selected, the scan tool commands the compressor and the appropriate air spring solenoid valve on, to allow air spring inflation.**
Copyright © 2015 Cengage Learning®

P21-6 **When the compressor shuts off, the air spring inflation is complete. Inspect the L/H front air spring and be sure there are no folds and creases in the spring.**
Copyright © 2015 Cengage Learning®

P21-7 **Shut the battery charger off and disconnect the charger cables. Lower the vehicle onto the shop floor and completely lower the lift. Be sure the lift arms are not contacting the vehicle.**
Copyright © 2015 Cengage Learning®

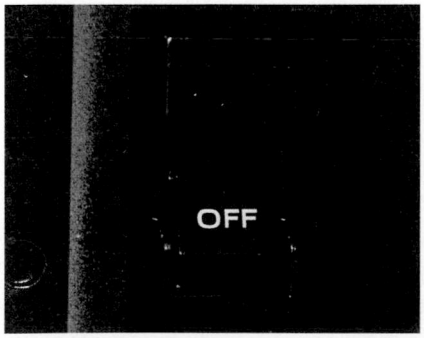

P21-8 **Disconnect the scan tool, and turn on the suspension switch.**
Copyright © 2015 Cengage Learning®

P21-9 **If the lower front strut-to-control arm bolt and nut were removed and reinstalled during the suspension service procedure, use your knee to push downward and release the front bumper. Repeat this procedure three times. Tighten the lower front strut-to-control arm nut and bolt to the specified torque.**
Copyright © 2015 Cengage Learning®

P21-10 **Road test the vehicle and verify proper air suspension operation and trim height.**

Copyright © 2015 Cengage Learning®

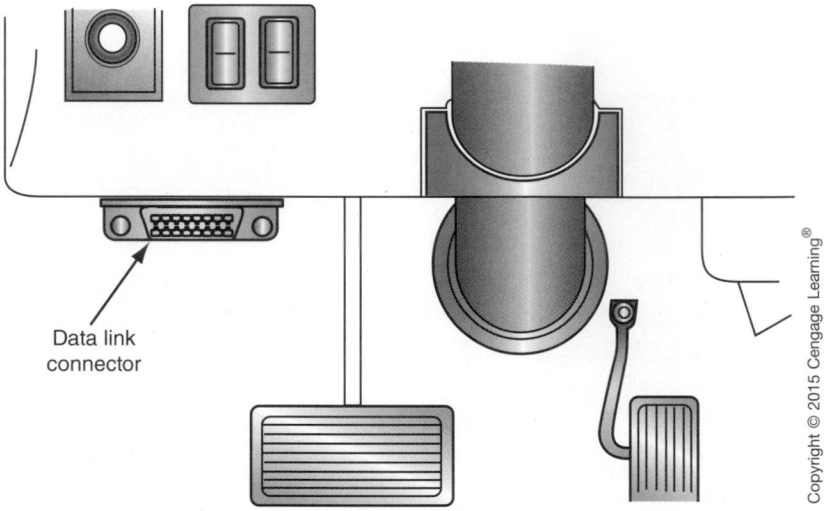

FIGURE 11-9 **Data link connector (DLC), on-board diagnostic II (OBD-II) system.**

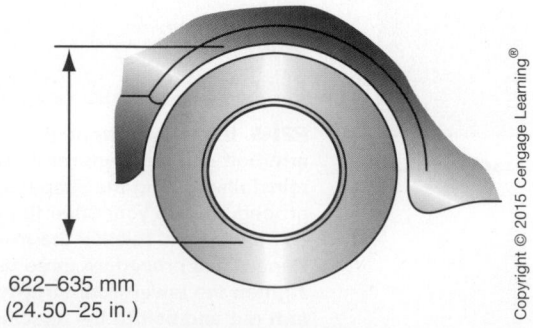

FIGURE 11-10 **Trim height measurement location.**

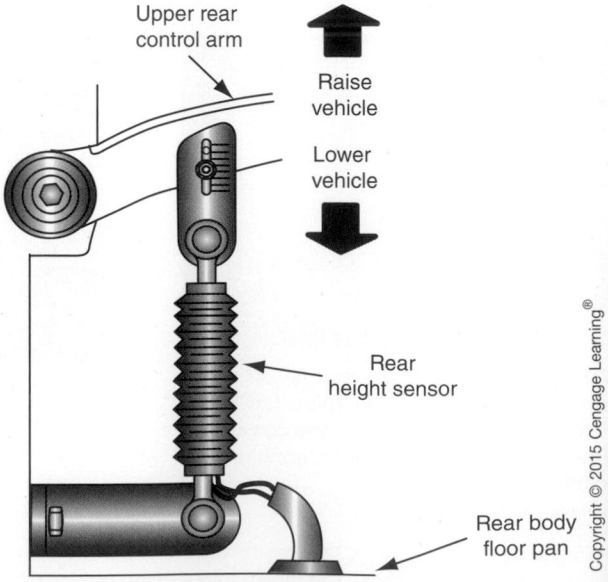

FIGURE 11-11 **Rear trim height adjustment.**

bracket (Figure 11-12). If the height sensor attaching bolt is moved one position up or down, the front suspension height is lowered or raised to 0.5 in. (12.7 mm).

Line Service

Nylon lines on the electronic air suspension system have quick disconnect fittings. These fittings should be released by pushing downward and holding the plastic release ring, and then pulling outward on the nylon line (Figure 11-13). Simply push the nylon line into the fitting until it seats to reconnect an air line.

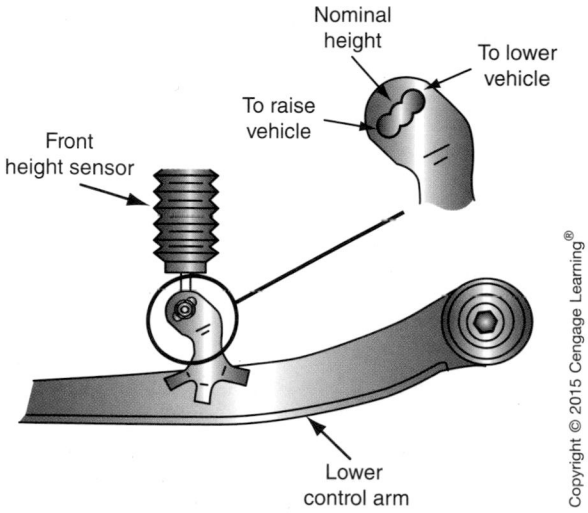

FIGURE 11-12 **Front trim height adjustment.**

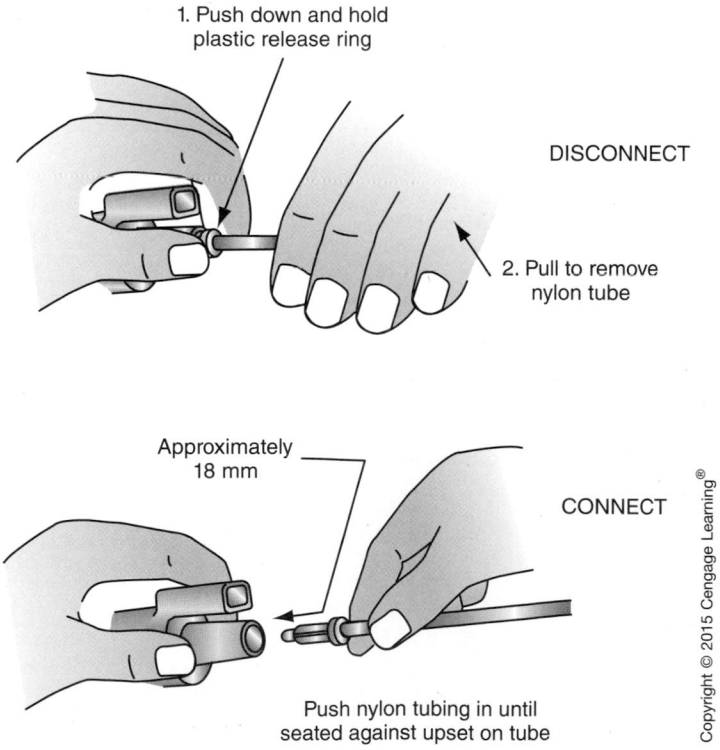

FIGURE 11-13 **Air line removed from the air spring valve or compressor outlets.**

SERVICE TIP: Some on-board diagnostic I (OBD I) vehicles have a separate suspension DLC positioned on the right-front strut tower. Other OBD I vehicles have the suspension DLC in the trunk. **OBD II computer systems** have a universal 16-terminal DLC mounted under the left side of the dash near the steering column (Figure 11-9). Data links are connected from various on-board computers including the suspension computer to the DLC to allow the on-board computers to communicate with the scan tool and vice versa.

Front and rear **trim height** is the distance between the vehicle chassis and the road surface measured at locations specified by the vehicle manufacturer.

If a line fitting is damaged, it may be removed by looping the line around your fingers and pulling on the line without pushing on the release ring (Figure 11-14). When a new collet and release ring are installed, the O-ring under the collet must be replaced. If a leak occurs in a nylon line, a sharp knife may be used to cut the defective area out of the line. A service fitting containing a collet fitting in each end is available for line repairs. After the defective area is cut out of the line, push the two ends of the line into the service fitting.

The **vehicle dynamic suspension** is an air suspension system used on some current model vehicles.

Servicing and Diagnosing Vehicle Dynamic Suspension Systems

On some vehicles the air suspension is referred to as a **vehicle dynamic suspension (VDS)**. Proper diagnosis of the VDS is extremely important to locate the exact root cause of the customer's complaint. Improper diagnosis may lead to wasted time and unnecessary replacement of expensive electronic components. The first step in VDS diagnosis is to visually inspect the system.

SERVICE TIP:
If a door ajar indicator is illuminated even when all the doors, liftgate, and liftgate glass are closed, the VDS will not operate properly. The problem with the door ajar indicator must be repaired before proceeding with the VDS diagnosis.

VDS Inspection and Verification

Follow these steps to inspect the VDS system and verify the symptoms:

1. Be sure the customer complaint is identified; road test the vehicle if necessary.
2. Be sure none of the DOOR AJAR indicators are illuminated.
3. Inspect the suspension system for anything that could result in restricted suspension movement.
4. Inspect the vehicle trunk for excessive loading.
5. Check the front and rear height sensors for loose or worn mountings.
6. Inspect the front and rear air springs for damage, such as punctures and tears.
7. Check the suspension air lines for cuts, splits, or a crimped condition.
8. Check fuses 102, 30, 27, and 111 in the central junction box.
9. Inspect all VDS electrical connections for looseness and corrosion.

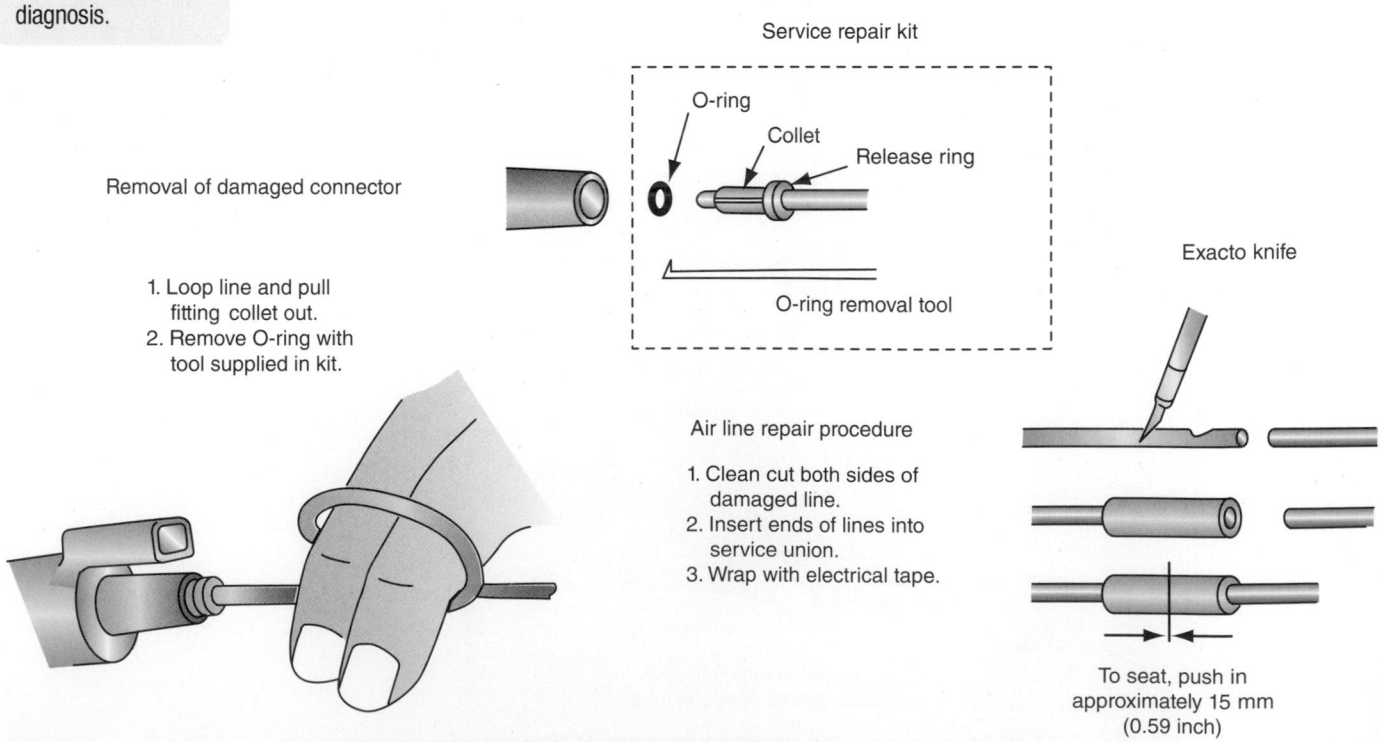

Copyright © 2015 Cengage Learning®

FIGURE 11-14 **Removing the quick disconnect fittings and air line repairs procedure.**

10. Be sure the air suspension switch is on.
11. Inspect the solenoid valves in the air springs for damage.

If the VDS inspection reveals an obvious problem, this defect must be corrected before proceeding with the electronic diagnosis. When the inspection and verification does not indicate any defects, proceed with the pneumatic test and the on-demand self-test.

Ride Height Adjustment or Calibration

Classroom Manual
Chapter 11, page 302

The ride height adjustment procedure with a scan tool may vary depending on the vehicle make and model year. The following is a typical ride height adjustment procedure with a scan tool for a Ford Expedition or Lincoln Navigator equipped with a vehicle dynamic suspension:

1. Be sure the air suspension switch is in the On position.
2. Be sure the battery is fully charged.
3. Turn off the ignition switch and exit the vehicle. Close all doors and wait 45 seconds for the air suspension system to vent down to the Kneel position.
4. Open the driver's door, and turn on the ignition switch.
5. Shift the transmission into drive and back into park.
6. Exit the vehicle, close all the doors, and wait 45 seconds for the suspension system to pump up to the trim height.
7. Use a tape measure to measure the ride height at the locations specified by the vehicle manufacturer (Figure 11-15 and Figure 11-16).
8. If the ride height does not equal the manufacturer's specifications, open the left-front door, and connect the scan tool to the DLC. Be sure the scan tool contains the proper module for the vehicle being tested.
9. Select the correct vehicle year, model, and engine type on the scan tool, and then select *Vehicle Dynamic Module* (VDM).
10. Use the active command mode in the scan tool to vent or lift the vehicle to achieve the specified trim height.
11. When a tape measure indicates the specified front and rear ride height, select *Save Calibration Values* on the scan tool to calibrate the VDM to the specified trim height.
12. Road test the vehicle and verify the proper suspension height.

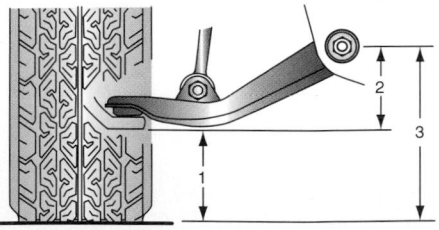

SERVICE TIP: The ride height adjustment procedure must be completed before a wheel alignment is performed.

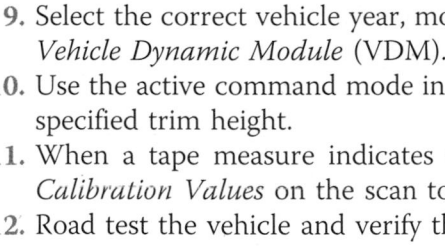

SERVICE TIP: The ride height adjustment must be performed after a VDM is replaced.

Item	Description
1	Distance between the ground and the lower knuckle surface near the ball joint
2	Ride height = 3 - 1
3	Distance between the ground and the center of the lower arm rearward mounting bolt

Copyright © 2015 Cengage Learning®

FIGURE 11-15 **Front ride height measurement.**

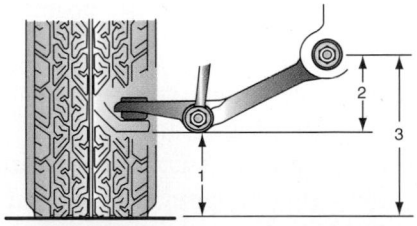

Item	Description
1	Distance between the ground and lower edge of arm below the shock absorber mounting bolt
2	Ride height = 3 - 1
3	Distance between the ground and the center of the lower arm mounting bolt

Copyright © 2015 Cengage Learning®

FIGURE 11-16 **Rear ride height measurement.**

Pneumatic Test

The pneumatic test verifies the air lines are connected and the air compressor, air compressor vent solenoid, and air spring solenoid valves are functioning properly. Follow these steps to complete the pneumatic test:

1. Close all doors, liftgate, and liftgate glass.
2. Place the transmission selector in park.
3. Be sure the air suspension switch is on.
4. Connect the scan tool to the DLC. Be sure the correct module is installed in the scan tool.
5. Turn on the ignition switch and select *Vehicle Dynamic Module* followed by *Pneumatic Test* on the scan tool.
6. This scan tool selection causes the VDM to operate the VDS compressor, spring solenoid valves, and vent valve. While operating the system the VDM checks for proper response from the height sensors. If the VDM does not perform normal air spring fill and vent operations in the required time, the VDM sets diagnostic trouble codes (DTCs) in memory and displays appropriate messages in the message center or in the instrument panel if the vehicle does not have a message center. The results of the pneumatic test are also displayed on the scan tool.

On-Demand Self-Test

When the inspection, verification, and pneumatic tests do not indicate any defects, the on-demand self-test should be performed to locate the problem area in the VDS system.

Complete the following steps to perform the on-demand self-test:

1. Be sure the transmission is in park.
2. Be sure the battery is fully charged. A battery charger may be installed to maintain a fully charged battery.
3. Connect the scan tool to the DLC.
4. Lower the driver's window and close all doors, liftgate, and liftgate glass.
5. Be sure the 4L mode is not selected on four-wheel-drive models.
6. Turn on the ignition switch.
7. Select *Vehicle Dynamic Module* and *On-Demand Self-Test* on the scan tool.
8. Record all DTCs displayed on the scan tool.
9. Turn off the ignition switch.

When DTCs are displayed on the scan tool, interpret the meaning of the DTC and diagnose the root cause of the problem. Possible DTCs on the VDS system are these:

1. B1317—high battery voltage
2. B1318—low battery voltage
3. B1342—VDM is defective
4. B2477—variable assist power steering (VAPS) assist curve not complete
5. C1445—vehicle speed sensor circuit defect
6. C1724—air suspension height sensor power circuit failure
7. C1725—air suspension front pneumatic failure
8. C1726—air suspension rear pneumatic failure
9. C1760—failure in rear height sensor signal circuit
10. C1770—failure in vent solenoid output circuit
11. C1790—L/R air spring solenoid valve output circuit failure
12. C1795—R/R air spring solenoid valve output circuit failure
13. C1830—failure in compressor relay circuit
14. C1840—failure in air suspension On/Off switch

15. C1873—R/F air spring solenoid valve output circuit failure
16. C1877—L/F air spring solenoid valve output circuit failure
17. C1881—failure in R/F height sensor circuit
18. C1889—failure in L/F height sensor circuit
19. C1897—failure in VAPS circuit loop
20. C1964—compressor run time longer than specified
21. C1990—VDM initialization required on replacement VDM
22. C1991—factory trim height setting and/or tests not completed
23. U1900—missing controller area network (CAN) messages
24. U2023—improper CAN messages

If DTCs are displayed on the scan tool, detailed pinpoint tests are provided in the vehicle manufacturer's service manual to locate the exact cause of the problem. These pinpoint tests usually include voltmeter and ohmmeter tests. A DTC indicates the area in which a problem is present, but the pinpoint tests are used to locate the root cause of the defect. When DTC C1990 is displayed, perform the pneumatic test and ride height calibration procedure to initialize the VDM.

Removal and Disassembly of Combined Air Springs and Shock Absorbers on VDS Systems

CAUTION: When removing and replacing components on the VDS, the vehicle manufacturer recommends replacing all the self-locking nuts. Failure to observe this service procedure may cause loose suspension components and component failure.

The removal and disassembly procedures are similar for front and rear air springs. Follow these steps to remove and disassemble a combined air spring and shock absorber:

1. Position the vehicle on a lift and place the lift arms under the specified lift points on the vehicle.
2. Be sure the ignition switch and the VDS switch are turned off.
3. Connect the scan tool to the DLC, and use the scan tool to vent the air spring being removed.

WARNING: Never remove an air spring or a component attached to the air spring without completely deflating the air spring. This action may cause personal injury.

4. Remove the upper strut mount retaining nuts on the air spring to be removed (Figure 11-17). Discard the nuts.
5. Raise the vehicle to a comfortable working height. Disconnect the electrical connector from the air spring solenoid valve (Figure 11-18).

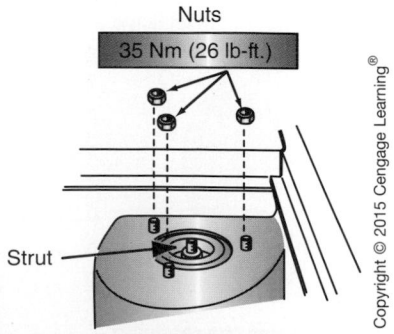

FIGURE 11-17 **Removing the upper mount retaining nuts.**

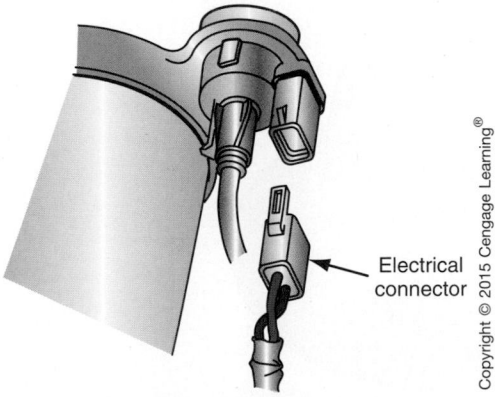

FIGURE 11-18 **Disconnecting the electrical connector from the air spring solenoid valve.**

SERVICE TIP:
If the retainer tabs on the lower end of the air spring are broken, air spring replacement is necessary.

CAUTION:
Never remove or install the wiring connector on a computer or computer system component with the ignition switch on. This action may result in computer damage.

6. Compress the orange quick-disconnect connector on the air line connected to the air spring, and pull downward to disconnect the air line (Figure 11-19).
7. Use the appropriate socket and ratchet to remove the nut from the lower shock absorber mounting bolt, and remove this bolt (Figure 11-20).
8. Remove the air spring and shock absorber from the chassis.
9. Place a wrench on the hex on top of the shock absorber rod, and hold this rod from turning while loosening and removing the nut on the top of the shock absorber rod (Figure 11-21).
10. Place index marks on the lower end of the air spring and shock absorber so these components may be reassembled in the original position. Depress the retainer tabs on the bottom end of the air spring, and remove the air spring from the shock absorber (Figure 11-22).
11. Remove and discard the seal and washer on the upper end of the shock absorber rod (Figure 11-23).
12. Remove and discard the O-ring seals on the lower sealing area between the shock absorber and the air spring (Figure 11-24).
13. Remove and discard the retainer pins that hold the upper mount onto the air spring (Figure 11-25).

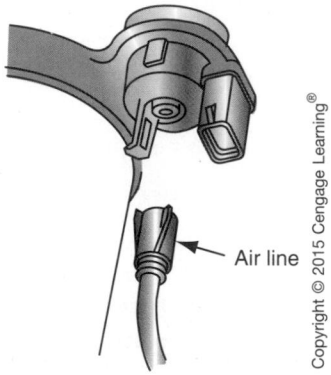

FIGURE 11-19 **Disconnecting the air line from the air spring solenoid valve.**

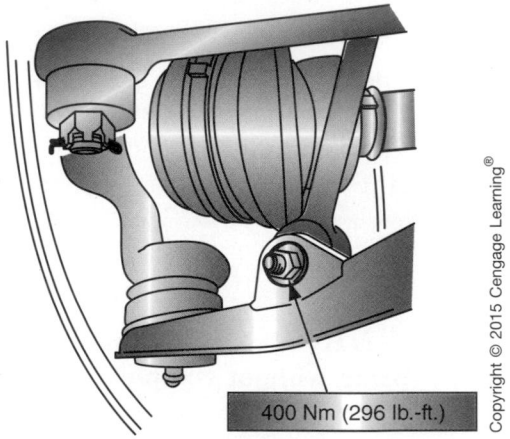

FIGURE 11-20 **Removing the lower shock absorber retaining bolt and nut.**

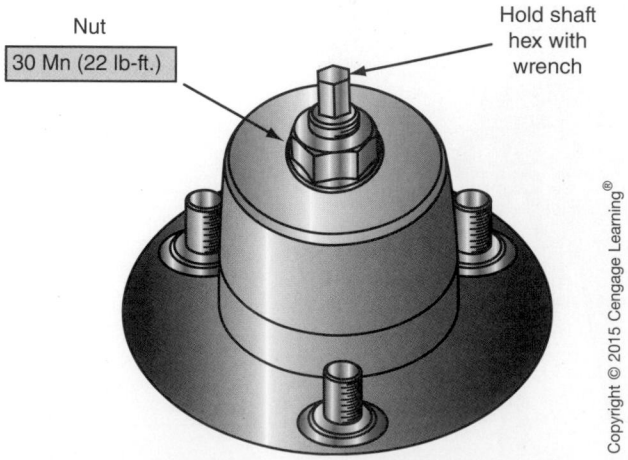

FIGURE 11-21 **Removing the nut from the upper end of the shock absorber rod.**

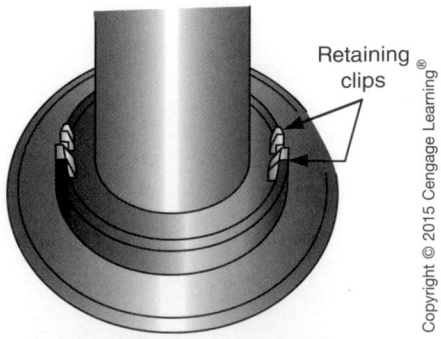

FIGURE 11-22 **Depressing the retaining tabs on the bottom end of the air spring.**

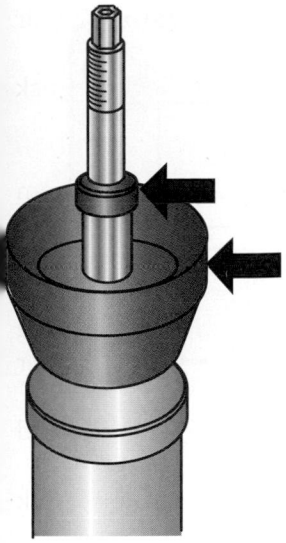

Copyright © 2015 Cengage Learning®

FIGURE 11-23 **Seal and washer on the upper end of the shock absorber rod.**

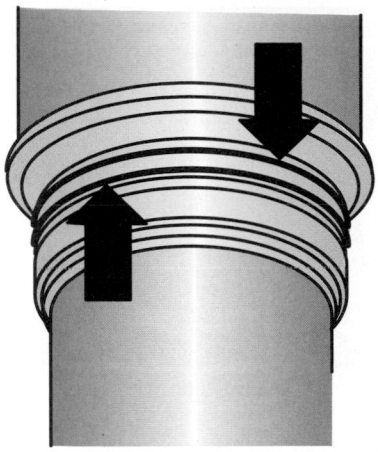

Copyright © 2015 Cengage Learning®

FIGURE 11-24 **O-ring seals on the lower sealing area on the shock absorber.**

CAUTION:
Do not supply voltage to or ground any circuit or component in a computer system unless instructed to do so in the vehicle manufacturer's service manual. This action may damage computer system components.

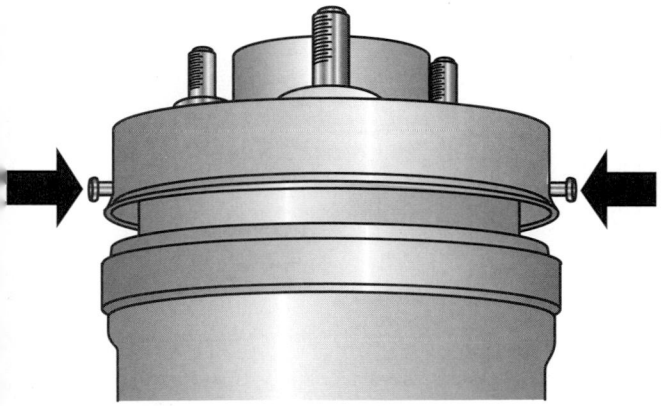

Copyright © 2015 Cengage Learning®

FIGURE 11-25 **Retaining pins on the upper air spring mount.**

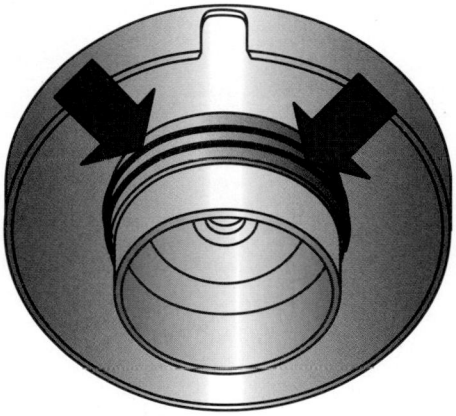

Copyright © 2015 Cengage Learning®

FIGURE 11-26 **O-ring seals on the upper spring mount.**

14. Remove and discard the O-ring seals on the upper mount area that provides a seal between the mount and the air spring (Figure 11-26).

When reassembling the air spring and shock absorber, replace all the O-rings and self-locking nuts, and place a light coating of chassis lubricant on all the O-rings.

CAUTION:
During computer system diagnosis, use only the test equipment recommended in the vehicle manufacturer's service manual to prevent damage to computer system components.

Diagnosis of Electronic Suspension Control Systems

Diagnostic Trouble Code (DTC) Display

Vehicles equipped with the **electronic suspension control (ESC)** system may have a digital or an analog instrument panel cluster (IPC). If the vehicle has a digital IPC, the air-conditioning (A/C) control panel is contained in the IPC (Figure 11-27). A driver message center is located in the center of the IPC underneath the speedometer. This message center is a vacuum fluorescent (VF) display with twenty characters. The same message center is mounted in an analog IPC (Figure 11-28). If the car has an analog IPC, the A/C controls are mounted in a separate display (Figure 11-29).

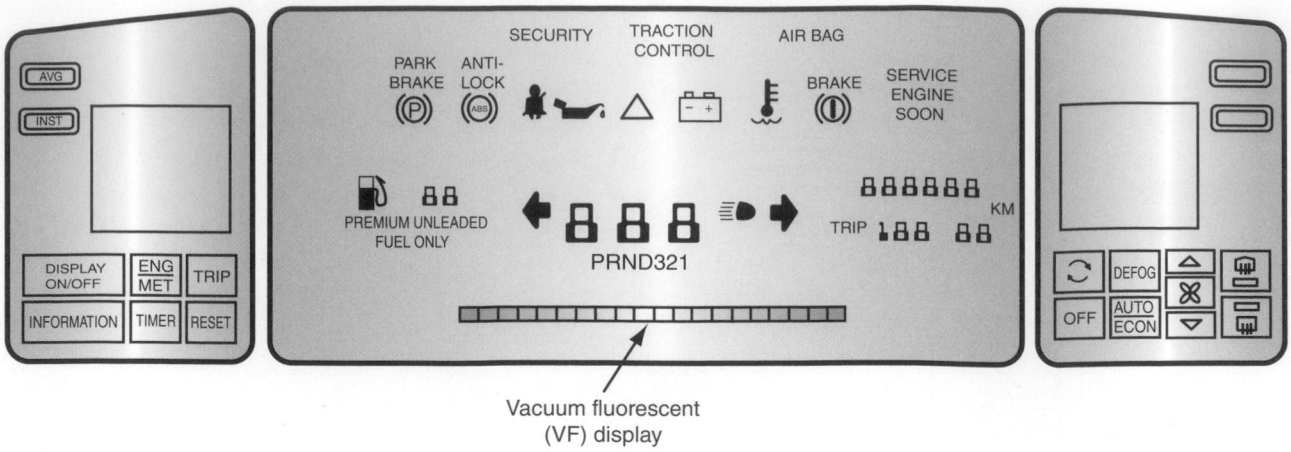

FIGURE 11-27 **Digital instrument panel cluster.**

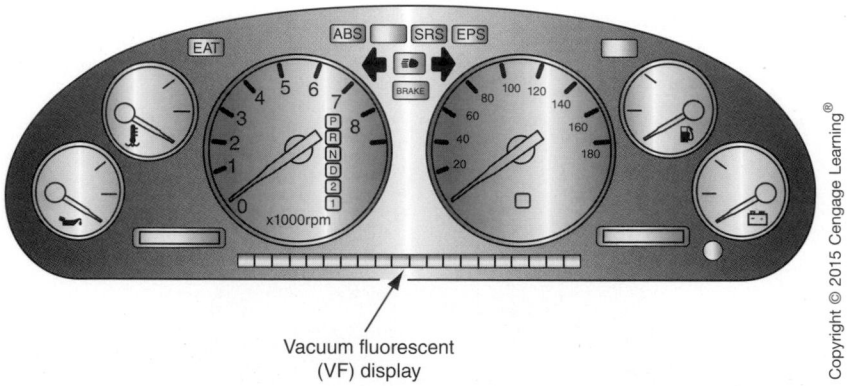

FIGURE 11-28 **Analog instrument panel cluster.**

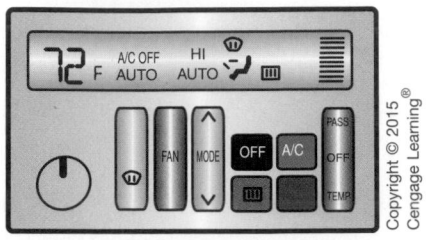

FIGURE 11-29 **A/C controls used with the analog instrument panel cluster.**

SERVICE TIP: When removing, replacing, or servicing an electronic component on a vehicle, always disconnect the negative battery cable before starting the service procedure. If the vehicle is equipped with an air bag or bags, wait 1 minute after the battery negative cable is removed to prevent accidental air bag deployment.

Early versions of the ESC system may be called **continuously variable road sensing suspension (CVRSS)** systems.

When the information (INFO) button is pressed in the IPC, the driver message center cycles through a series of status messages: RANGE, AVG MPG, MPG INST, FUEL USED, OIL LIFE LEFT, ENGINE RPM, BATTERY VOLTS, COOLANT TEMP, ENGLISH/METRIC RESET. The parameters related to fuel usage are not displayed in the message center on a digital IPC, because these readings are displayed in the fuel data center that is contained in the IPC.

If an electrical defect occurs in the ESC system, a SERVICE RIDE CONTROL message is displayed in the driver information center, and a diagnostic trouble code (DTC) is usually stored in the ESC module. The first step in diagnosing the ESC system should be a thorough visual inspection of the system. Inspect the wiring harness for damaged wires and loose or corroded connectors. Be sure the battery is fully charged and all the fuses are in good condition.

If the SERVICE RIDE CONTROL message is displayed, the technician enters the diagnostic mode by pressing the OFF and WARMER buttons simultaneously on the A/C controls with the ignition switch on. When these buttons are pressed, the IPC performs a segment check in which all segments in the IPC and climate control panel (CCP) are illuminated for a brief period of time. Press the low fan speed button to display the abbreviation for the on-board computers, such as PCM and IPC, one after the other. Continue pressing the low fan speed button until ESC appears in the driver message center. When ESC is displayed, press the increase fan speed button to select this parameter. When this action is taken, the ESC DTCs are displayed in the driver message center. Each DTC is designated as CURRENT or HISTORY, and each DTC has a C prefix followed by four digits. If there are no DTCs in the ESC module, NO ESC CODES is displayed. When NO ESC DATA is displayed, there is an electrical defect in the data link, the module ground connection, or in the ESC module battery or ignition inputs. A DTC represents a fault in a certain area of the ESC system. For example, if a C1711 is displayed, there is a short to ground in the left front (LF) damper actuator. The technician has to perform voltmeter or ohmmeter test procedures provided in the appropriate service manual to test the LF damper actuator and the related sensor wiring to locate the exact cause of the defect. When the ohmmeter leads are connected to the damper actuator terminals on the strut, the ohmmeter should indicate 9.5 to 15.5 ohms. The DTCs and ESC system specifications may vary depending on the vehicle make, model, and year. If the ohmmeter reading is infinite, the damper actuator has an open circuit. An ohmmeter reading below this value indicates a shorted damper actuator. If the ohmmeter leads are connected from one of the damper actuator terminals to ground, the ohmmeter should provide an infinite reading indicating this actuator is not grounded. When the ohmmeter indicates a low reading, the damper actuator is shorted to ground. Ohmmeter tests for shorts-to-ground and open circuits must also be performed on the wires from the damper actuator to the ESC module. The DTCs are displayed in the driver message center. The DTCs may be cleared from the ESC module by pressing the increase fan speed button when CLEAR DTCs is displayed in the driver message center. A scan tool may also be used to clear the DTCs. Disconnecting the battery or the ESC module connector will not clear the DTCs.

A scan tool also displays the DTCs and diagnoses the ESC system. The scan tool displays the voltage data from the wheel position sensors, steering sensor, vehicle speed sensor, and battery.

SERVICE TIP:
(Continued)
Many air bag computers have a back-up power supply capable of deploying the air bag for a specific length of time after the battery is disconnected.

Scan Tool Diagnosis of Electronic Suspension Control

When diagnosing an ESC system, the first step is to identify the complaint. Listen to the customer's complaint, and road test the car if necessary to determine the exact problem. The next step in the diagnostic procedure is to visually inspect all ESC components. Check all the ESC wiring connections and wiring harnesses. Be sure all the ESC fuses are satisfactory. Check the shock absorbers, struts, and other suspension components for loose mountings and damage. Inspect the shock absorbers and struts for fluid leaks. Be sure the battery is fully charged.

Classroom Manual
Chapter 11, page 302

Observe the message center for displayed messages related to the ESC system. ESC systems with computer-controlled solenoids in the shock absorbers or struts display SERVICE SUSPENSION SYSTEM or SPEED LIMITED TO XXX if a defect occurs in the ESC. If the vehicle has an ESC system with magneto-rheological fluid in the shock absorbers or struts, the ESC system displays SHOCKS INOPERATIVE, SERVICE RIDE CONTROL, OR MAXIMUM SPEED 129 km/h (80 mph) if a defect occurs in the system. The messages indicating a limited maximum vehicle speed are displayed only when the ESC module has detected a defect and shut down the electronic control of all shock absorbers or struts.

To locate a defect in an ESC system the next step is to perform a diagnostic system check. This check identifies the control modules in the system, and tests the ability of the control modules to communicate through the data link system. The diagnostic system check also displays all DTCs in the various modules related to ESC.

Follow these steps to complete a diagnostic system check:

1. Be sure the ignition switch is off, and connect a scan tool to the DLC.
2. Turn on the ignition switch.
3. Use the scan tool to select the various modules related to ESC control. These modules include the **powertrain control module (PCM)**, ESC module, and the electronic brake control module (EBCM).

The **powertrain control module (PCM)** controls engine functions.

4. Use the scan tool to display all the DTCs stored in each of the modules in step 3. Record all DTCs.
5. If a DTC is displayed that begins with a U, there is a problem with communications through the data links. When one or more of the displayed DTCs begins with a U, a lack of communication between the scan tool and one or more of the modules likely occurred in step 3.
6. The cause of the DTC beginning with a U must be corrected before proceeding with any further diagnosis. After the cause of the data link DTC is corrected, diagnose the cause of all other DTCs.

The following DTCs represent ESC defects:

1. C0550—ESC module malfunction.
2. C0563—ESC module calibration defect.
3. C0577, C0579, C0582, C0584, C0587, C0589, C0592, C0594—the voltage is out of the normal range in one of the strut solenoids or electromagnets. Detailed DTC explanations and diagnostic procedures are provided in the vehicle manufacturer's service manual.
4. C0578, C0583, C0588, C0593—ESC fault in one of the internal ESC module drivers that operate the strut solenoids or electromagnets.
5. C0615, C0620, C0625, C0630—a fault exists in one of the position sensors that causes the sensor voltage to be below or above the normal range of 0.35V to 4.75V for more than 1 second.
6. C0635, C0638, C0640, C0643, C01252, C01253—a defect in the two normal force signal circuits between the ESC module and the electronic brake control module (EBCM). The ESC module controls each shock absorber or strut actuator with a pulse width modulated (PWM) duty cycle or 10 to 90 percent depending on road surface conditions. The ESC module relays this PWM information through a normal force signal circuit to the EBCM, and the EBCM uses this information to provide improved braking on irregular road surfaces.
7. C0665—the ESC module did not detect a low-to-high change in the lift/dive circuit within 31 seconds after it received the ignition switch on signal.
8. C0690, C0691, C0693—a defect in the damper control relay circuit that is integral in the ESC module. If one or more of these DTCs is displayed, ESC module replacement is necessary.
9. C0696—the 5V reference voltage sent from the ESC module to each position sensor is out of the normal range of 3.75V to 5.6V.
10. C0710—the ESC module does not detect a valid steering position signal from the EBCM for 5 seconds. The EBCM controls the ABS and the variable assist steering.
11. C0896—the battery voltage is not within the normal range of 9V to 15.5V.
12. P1652—the output driver module (ODM) in the PCM that controls suspension lift/dive has detected an improper circuit condition.

To set most DTCs in the ESC module, the module must sense a defect on three consecutive ignition cycles, or during the same ignition cycle after clearing the DTC with a scan tool.

Scan Tool Data Display

Select *Special Functions* on the scan tool to access the ESC system data. The ESC data is very helpful to pinpoint the exact cause of DTCs. The following ESC data may be displayed on the scan tool with the ignition switch on, the engine off, and the front wheels straight ahead:

1. Battery voltage signal—typical data 12.6V.

2. DSP software version ID—numerical value.
3. EEPROM calibration ID—numerical value.
4. GM part number ESC module – numerical value.
5. ESC module software version—numerical value.
6. L/F damper actuator command—0 to 100 percent—this percentage indicates the commanded on-time of the damper actuator. If the ESC module commands the damper actuator On, the percentage increases.
7. L/F position sensor—0V to 5V—this voltage is the signal voltage sent from the position sensor to the ESC module. If downward force is supplied to the L/F of the chassis, the voltage reading should change. When the position sensor voltage does not change, the sensor and connecting wires must be tested to locate the exact cause of the problem. After repairs have been completed to correct a problem causing a DTC, the scan tool must be used to clear the DTC from the ESC module.
8. L/F normal force—20 to 80 percent—this percentage indicates a measured level of the road surface condition sent from the ESC module to the EBCM, and the EBCM uses this information to improve antilock braking on irregular road surfaces.
9. The damper actuator command and position sensor data are displayed on the scan tool for the other three corners of the chassis. The normal force data is available only on the front wheels.
10. Lift/Dive status—active, inactive—if there is no change in chassis pitch, the scan tool displays Inactive. When a change in chassis pitch occurs, the scan tool displays Active.
11. Lift/Dive change—changed, unchanged—if there is no change in chassis pitch, Unchanged is displayed on the scan tool. A change in chassis pitch causes the scan tool to display Changed.
12. Steering position PWM—0 to 10 ms—this PWM signal is sent from the EBCM to the ESC module when the steering wheel is turned while cornering. The ESC module commands a firmer shock absorber state when the vehicle is turning a corner.
13. Vehicle speed—0 to 159 mph (0 to 255 km/h)—the vehicle speed sensor signal is relayed on the data links from the PCM to the ESC module.

Scan Tool Output Controls

The output controls are accessed on the scan tool under the Special Functions menu. When output controls are selected on the scan tool, followed by an individual damper solenoid selection, the solenoid is cycled on and off. When an individual damper solenoid actuator is cycled on and off, the technician may perform voltage tests to check the damper actuator and related circuit. The scan tool may also be used to cycle all the damper solenoids simultaneously.

In the Output Controls mode, the scan tool may be used to command the ESC module to send the L/F and R/F normal force percentage readings to be sent to the EBCM, and this percentage is displayed on the scan tool.

ESC Module Recalibration may also be selected on the scan tool in the output controls mode. This recalibration is necessary after a module replacement.

Photo Sequence 22 illustrates the procedure for reading scan tool data on an electronic suspension control system.

Diagnosis of Stability Control System

The stability control system module is combined with the **antilock brake system (ABS)** and traction control system (TCS) modules. Therefore, it is not possible to completely separate the diagnosis of these systems. However, our discussion is mostly concerned with the stability control system, and this diagnostic information pertains to the **Stabilitrak®** system. The Stabilitrak® system may be called a vehicle stability enhancement system (VSES).

The **antilock brake system (ABS)** prevents wheel lockup during a brake application.

The **Stabilitrak®** system is a type of vehicle stability control.

PHOTO SEQUENCE 22

READING SCAN TOOL DATA ON AN ELECTRONIC SUSPENSION CONTROL SYSTEM

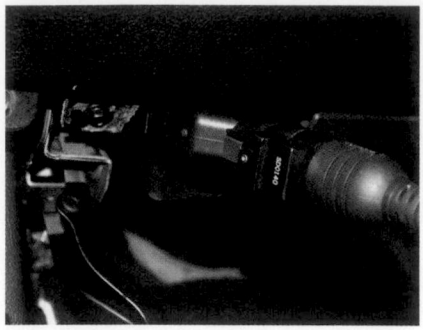

P22-1 **Connect the scan tool to the DLC under the dash.**

Copyright © 2015 Cengage Learning®

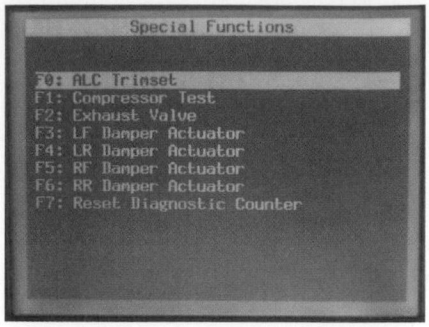

P22-2 **Turn on the ignition switch, and select Special Functions in the ESC menu on the scan tool.**

Copyright © 2015 Cengage Learning®

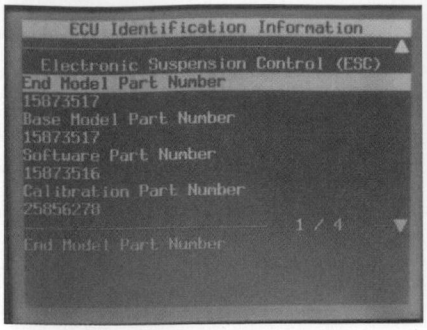

P22-3 **Display and record the DSP software version ID, EEPROM Calibration ID, GM part number ESC module, and ESC module software version on the scan tool.**

Copyright © 2015 Cengage Learning®

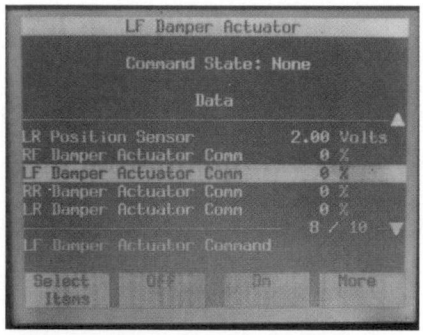

P22-4 **Select the damper actuator command on each damper actuator. This percentage indicates the commanded on-time of the damper actuator.**

Copyright © 2015 Cengage Learning®

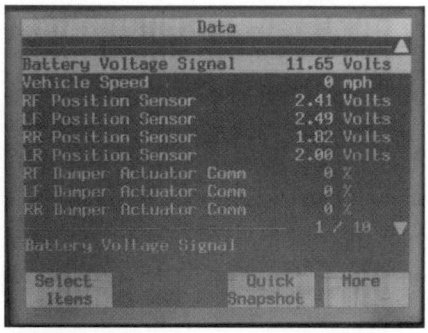

P22-5 **Select the wheel position sensor voltage on each wheel position sensor. This voltage is the signal voltage sent from the position sensor to the ESC module.**

Copyright © 2015 Cengage Learning®

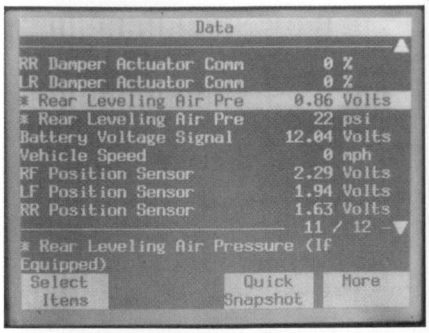

P22-6 **Select the normal force display on the scan tool. This percentage indicates a measured level of the road surface condition sent from the ESC module to the EBCM, and the EBCM uses this information to improve antilock braking on irregular road surfaces.**

Copyright © 2015 Cengage Learning®

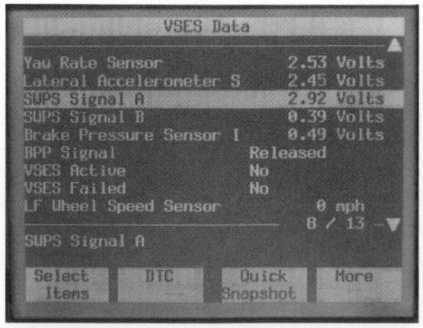

P22-7 **Select the Lift/Dive status on the scan tool.**

Copyright © 2015 Cengage Learning®

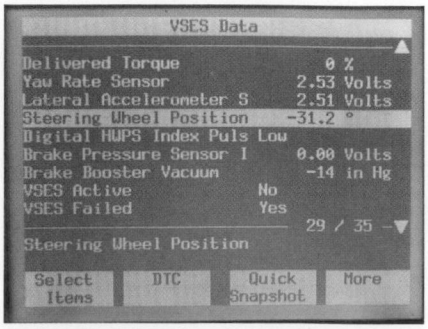

P22-8 **Select Steering Position PWM on the scan tool.**

Copyright © 2015 Cengage Learning®

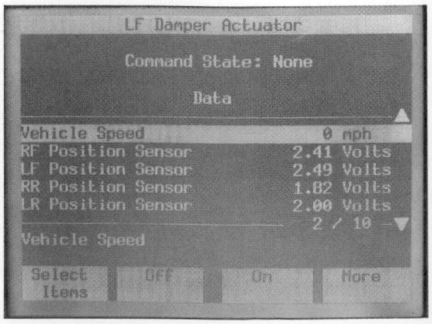

P22-9 **Select Vehicle Speed on the scan tool.**

Copyright © 2015 Cengage Learning®

The **electronic brake and traction control module (EBTCM)** performs an initialization test each time the vehicle is started and the EBTCM does not receive a brake switch input. During the initialization test, the EBTCM cycles each solenoid valve and the pump motor in the **brake pressure modulator valve (BPMV)** for 1.5 seconds. If the EBTCM detects an electrical fault during the initialization test, a DTC is set in the EBTCM, and the amber ABS light and the red brake light may be illuminated in the instrument panel, depending on the severity of the defect. During the initialization test, the customer may hear the solenoids clicking and the pump motor turning in the BPMV.

The **electronic brake and traction control module (EBTCM)** controls ABS, TCS, and Stabilitrak® functions.

The first step in diagnosing the stability control system is a visual inspection of all system components, such as the module connectors, fuses, relays, wiring harness, and sensor mounting and wiring connector.

The **brake pressure modulator valve (BPMV)** controls brake fluid pressure to the wheel calipers or cylinders.

If the visual inspection did not reveal any problems, the next step is to perform a Stabilitrak® diagnostic test drive as follows:

1. Turn the ignition switch off.
2. Connect a scan tool to the data link connector (DLC) under the instrument panel. Be sure the correct module for the system being tested is securely installed in the bottom of the scan tool.
3. Start the engine.
4. Monitor the "steering wheel centered" parameter on the scan tool, and be sure the scan tool displays "Yes" while driving the vehicle straight ahead above 15 mph (24 km/h).
5. Drive the vehicle for at least 10 minutes under a variety of driving conditions: highway driving, rough road driving, and turning maneuvers, including freeway ramps and sharp turns on parking lots.
6. Perform any driving maneuvers under which the customer complaint(s) occurred.
7. With the engine still running, observe and record any DTCs displayed on the scan tool.

If there is a defect in the Stabilitrak® system, "Stability Reduced" may be displayed in the driver information center. When "Display ABS/TCM/ICCS DTCs" is selected on the scan tool, any of the following DTCs may be displayed as indicated in Figure 11-30 and Figure 11-31. The technician usually has to perform voltmeter or ohmmeter tests to locate the exact cause of a problem in the area indicated by the DTC.

Automated Test

An automated test may be selected on the scan tool. During this test, the EBTCM cycles all the solenoid valves and the pump motor in the BPMV, and DTCs are set in the EBTCM if there are defects in these components. This test is the same as the initialization test performed by the EBTCM when the engine is started. The scan tool also performs solenoid tests.

Valve Solenoid/Pressure Hold Test

To perform the valve solenoid/pressure hold test, follow this procedure:

1. Be sure the scan tool is properly connected to the DLC.
2. Turn on the ignition switch.
3. Raise the vehicle on a lift so all four wheels are at least 6 in. (15 cm) off the floor.
4. Select Valve Solenoid Test on the scan tool followed by Hold Pressure on a specific wheel.
5. Hold the brake pedal in the applied position.
6. Have a coworker try to turn the wheel being tested. If the hold pressure solenoid is operating properly, the coworker should be able to turn the wheel.
7. Repeat steps 4 through 6 on the other wheels.

DTC	DESCRIPTION
C1211	ABS Indicator Lamp Circuit Malfunction
C1214	Solenoid Valve Relay Contact or Coil Circuit Open
C1217	BPMV Pump Motor Relay Contact Circuit Open
C1221	LF Wheel Speed Sensor Input Signal - 0
C1222	RF Wheel Speed Sensor Input Signal - 0
C1223	LR Wheel Speed Sensor Input Signal - 0
C1224	RR Wheel Speed Sensor Input Signal - 0
C1225	LF - Excessive Wheel Speed Variation
C1226	RF - Excessive Wheel Speed Variation
C1227	LR - Excessive Wheel Speed Variation
C1228	RR - Excessive Wheel Speed Variation
C1232	LF Wheel Speed Sensor Circuit Open or Shorted
C1233	RF Wheel Speed Sensor Circuit Open or Shorted
C1234	LR Wheel Speed Sensor Circuit Open or Shorted
C1235	RR Wheel Speed Sensor Circuit Open or Shorted
C1236	Low System Voltage
C1237	High System Voltage
C1238	Brake Thermal Model Limit Exceeded
C1241	Magna Steer® Circuit Malfunction
C1242	BPMV Pump Motor Ground Circuit Open
C1243	BPMV Pump Motor Stalled
C1251	RSS Steering Sensor Data Malfunction
C1252	ICCS2 Data Link Left Malfunction
C1253	ICCS2 Data Link Right Malfunction
C1255 xx	EBTCM Internal Malfunction (ABS/TCS/ICCS Disabled)
C1256 xx	EBTCM Internal Malfunction
C1261	LF Hold Valve Solenoid Malfunction

Copyright © 2015 Cengage Learning®

FIGURE 11-30 **DTCs for ABS/TCS/ICCS systems.**

CUSTOMER CARE: While discussing computer-controlled suspension systems with customers, remember that the average customer is not familiar with automotive electronics terminology. Always use basic terms that customers can understand when explaining electronic suspension problems. Most customers appreciate a few minutes spent by service personnel to explain their automotive electronic problems. It is not necessary to provide customers with a lesson in electronics, but it is important that customers understand the basic cause of the problem with their vehicle so they feel satisfied the repair expenditures are necessary. A satisfied customer is usually a repeat customer.

Valve Solenoid/Pressure Release Test

To perform a valve solenoid/pressure release test, follow this procedure:

1. Be sure the scan tool is properly connected to the DLC.
2. Turn on the ignition switch.
3. Raise the vehicle on a lift so all four wheels are at least 6 in. (15 cm) off the floor.
4. Hold the brake pedal in the applied position.
5. Select Valve Solenoid Test on the scan tool followed by Release Pressure on a specific wheel.

DTC	DESCRIPTION
C1262	LF Release Valve Solenoid Malfunction
C1263	RF Hold Valve Solenoid Malfunction
C1264	RF Release Valve Solenoid Malfunction
C1265	LR Hold Valve Solenoid Malfunction
C1266	LR Release Valve Solenoid Malfunction
C1267	RR Hold Valve Solenoid Malfunction
C1268	RR Release Valve Solenoid Malfunction
C1271	LF TCS Master Cylinder Isolation Valve Malfunction
C1272	LF TCS Prime Valve Malfunction
C1273	RF TCS Master Cylinder Isolation Valve Malfunction
C1274	RF TCS Prime Valve Malfunction
C1276	Delivered Torque Signal Circuit Malfunction
C1277	Requested Torque Signal Circuit Malfunction
C1278	TCS Temporarily Inhibited By PCM
C1281	Stabilitrak® Sensors Uncorrelated
C1282	Yaw Rate Sensor Bias Circuit Malfunction
C1283	Excessive Time to Center Steering
C1284	Lateral Accelerometer Sensor Self-Test Malfunction
C1285	Lateral Accelerometer Sensor Circuit Malfunction
C1286	Steering/Lateral Accelerometer Sensor Bias Malfunction
C1287	Steering Sensor Rate Malfunction
C1288	Steering Sensor Circuit Malfunction
C1291	Open Brake Lamp Switch During Deceleration
C1293	DTC C1291 Set In Current or Previous Ignition Cycle
C1294	Brake Lamp Switch Circuit Always Active
C1295	Brake Lamp Switch Circuit Open
C1297	PCM Indicated Brake Extended Travel Switch Failure
C1298	PCM Indicated Class 2 Serial Data Link Malfunction
U1016	Loss of PCM Communications
U1056	Loss of CVRSS Communications
U1255	Generic Loss of Communications
U1300	Class 2 Circuit Shorted to Ground
U1301	Class 2 Circuit Shorted to Battery +

Copyright © 2015 Cengage Learning®

FIGURE 11-31 **DTCs for ABS/TCS/ICCS systems. (continued)**

6. Have a coworker try to turn the wheel being tested. If the release pressure solenoid is operating properly, the coworker should be able to turn the wheel.
7. Repeat steps 4 through 6 on the other wheels.

If any of the hold or release solenoids do not operate properly, the BPMV requires replacement. The solenoids in the BPMV are not serviceable.

Diagnosis of Vehicle Networks

Vehicle networks have various voltages and operating characteristics. Therefore, the vehicle manufacturer's specific diagnostic procedure must be followed for each vehicle network. The most common equipment for diagnosing vehicle networks are a scan tool, a digital volt ohm meter (DVOM), and a lab scope. If a defect occurs in most networks, a DTC(s) is displayed on a scan tool connected to the DLC. The technician must follow the vehicle manufacturer's specified procedure to diagnose the cause of the DTC(s). When using a lab scope to diagnose vehicle networks, the technician must be familiar with the normal lab

Classroom Manual
Chapter 11, page 278

scope pattern for each network. We will discuss the lab scope patterns for two common networks and provide several case studies of network diagnosis.

A Class 2 network is a single-wire system. When the system is at rest with no communication taking place, the system voltage is 0V. When communication takes place on this network, the data are transmitted by a variable pulse width signal of 0V to 7V. The Class 2 network is connected to terminal 2 in the DLC, and it also interconnects with various computers depending on the electronic equipment on the vehicle. To diagnose a Class 2 network with a lab scope, connect the Channel 1 positive lead of the scope to pin 2 in the DLC, and connect the ground lead on the scope to pin 4 in the DLC or to a good chassis ground. Adjust the scope voltage setting to 5V and set the time base setting to 100 milliseconds (ms). With the ignition switch on, a normal network should display a voltage pattern as illustrated in Figure 11-32. A **splice pack** is used in the network on some vehicles. A parts locator manual will help the technician to locate the splice pack. This splice pack may be disconnected to isolate the PCM and truck body computer (TBC) from the rest of the network so the network cannot affect these computers. If the vehicle has a no-start condition, and the engine starts after the splice pack is disconnected, the defect is located in the network. Under this condition, the network may be shorted to ground, the network may be shorted to a 12V source, or a module connected to the network may be shorted internally. The network modules may be disconnected one at a time. If the defect is corrected when one module is disconnected, the network problem is in that module. When none of the modules are causing the network problem, voltage and/or ohmmeter tests must be performed on the network wires to locate the shorted condition.

A **splice pack** is a special connector that may be disconnected to isolate specific components when diagnosing vehicle networks.

The Standard Corporate Protocol (SCP) network is used on many vehicles. The two wires in the SCP system are designated as Bus+ and Bus−. Bus+ is connected to pin 2 in the DLC, and Bus− is connected to pin 10 in the DLC. The network voltages at rest with no communication are 5V on Bus+, and 0V−.2V on Bus−. A DVOM may be used to read these voltages. To read the SCP voltages on a lab scope, connect the Channel 1 lead on the lab scope to pin 2 on the DLC, and connect the Channel 2 lead on the lab scope to pin 10 on the DLC. Connect the ground lead to pin 4 on the DLC or a good chassis ground. When the ignition switch is turned on, the normal SCP pattern shown in Figure 11-33

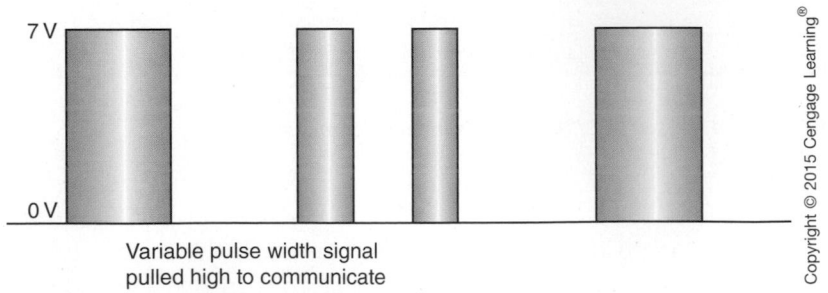

FIGURE 11-32 **Class 2 network voltage signal.**

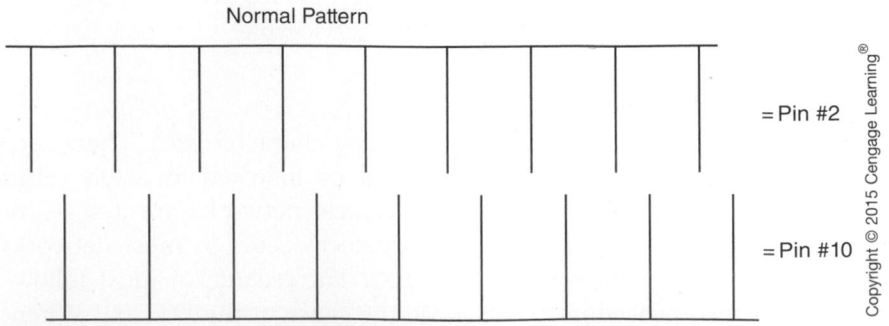

FIGURE 11-33 **SCP network voltage signal.**

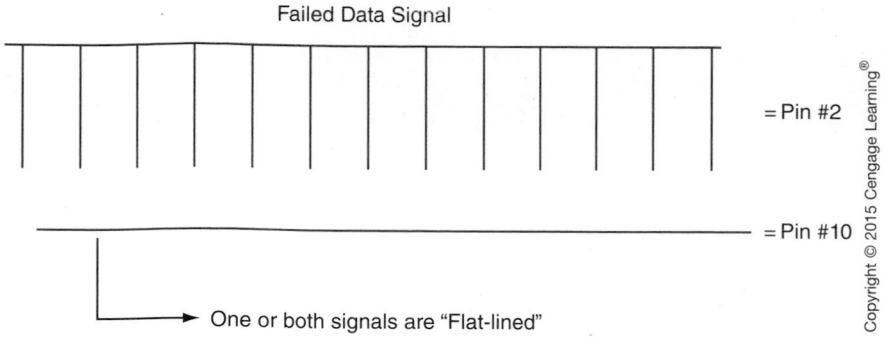

FIGURE 11-34 **SCP network defective voltage signal.**

should be displayed. If one network signal is displayed as a flat line, communication has failed between some of the computers and part of the network (Figure 11-34). When both network signals are displayed as a flat line, no communication is taking place through the network. Some networks remain functional with the ignition switch off, because certain module(s) require information under this condition. Always consult the vehicle manufacturer's specific diagnostic information when diagnosing networks.

CASE STUDY

A 2010 GM 1500 series truck was towed into the shop with a no-start condition. During a preliminary diagnosis, the technician discovered there was 12V on the Class 2 vehicle network with the ignition switched on. The technician used a parts locator manual to locate the splice pack behind the radio. When the splice pack was disconnected, the engine started and ran normally. This proved that the PCM and TBC were not causing the problem, because disconnecting the splice pack isolated these computers from the network. Therefore, the defect must be in the Class 2 network or computers connected in the network. The technician decided to disconnect each computer in the network beginning with the radio/CD player. When the splice pack was reconnected, and the radio disconnected, network voltage was normal, and the engine started and ran normally. Closer internal examination of the radio/CD player indicated a dime lodged between the radio/CD player circuit board and the case. When the dime was removed and the radio reconnected, network voltage and engine operation were normal. The technician used a scan tool to erase all DTCs in the computer memories.

CASE STUDY

A customer brought a 2012 Silverado into the shop with multiple electric problems. The cruise control would cancel when the turn signals were turned on. This only occurred at night when the headlights were on. The customer said several instrument panel readings were randomly intermittent. When a scan tool was connected to the DLC, a U1041 was displayed indicating loss of electronic brake control module (EBCM) data on the network. The technician checked for service bulletins related to this problem and discovered this problem was detailed in a service bulletin. The bulletin indicated this problem was caused by high resistance in ground connection G110 on the vehicle frame below the driver's door. The ground connection was cleaned and tightened and the DTC erased, but the DTC reset again in a short time.

The technician considered the possibility of a defective EBCM. Prior to EBCM replacement, the technician checked the EBCM voltage supply and ground. The EBCM voltage supply was 12V. When a pair of voltmeter leads was connected from the EBCM ground terminal to the battery ground, the voltmeter indicated 3V. The technician inspected all the wiring from the EBCM module to the battery, and discovered that the battery ground cable was connected to the radiator support rather than being connected to the specified location on the left front corner of the vehicle frame. The battery ground cable and the vehicle frame attaching location were thoroughly cleaned, and the ground cable was properly tightened. Now the voltage reading from the EBCM ground terminal to the battery ground was 0.2V. All DTCs were erased with a scan tool, and after driving the vehicle on a road test the DTCs did not reset. All electronic systems operated normally during the road test.

TERMS TO KNOW

Antilock brake system (ABS)

Brake pressure modulator valve (BPMV)

Continuously variable road sensing suspension (CVRSS)

Data link connector (DLC)

Electronic brake and traction control module (EBTCM)

Electronic suspension control (ESC)

Programmed ride control (PRC)

Powertrain control module (PCM)

Scan tool

Splice pack

Stabilitrak®

Trim height

Vehicle dynamic suspension (VDS)

CASE STUDY

A customer complained about the SERVICE RIDE CONTROL light being illuminated on his 2009 Cadillac XLR. When the technician visually inspected the ESC system, no defects were evident. During a diagnostic system check, the scan tool displayed DTC C0577. The detailed DTC explanation in the vehicle manufacturer's service manual indicated this DTC represented a short to ground in the L/F shock absorber damper solenoid circuit.

When the technician disconnected the wiring connector from the L/F damper solenoid and measured the solenoid resistance with an ohmmeter, he discovered the solenoid had the specified resistance of 2 ohms.

The technician disconnected the wiring connectors from the ESC module and identified the L/F damper solenoid wires connected to the module. When the technician connected a pair of ohmmeter leads across the wires from the ESC module terminals to the L/F damper solenoid terminals, each wire had very low resistance. Next, the technician connected the ohmmeter leads from each terminal in the L/F damper solenoid connector to the ground. When connected to one of the damper solenoid terminals, the ohmmeter displayed an infinite reading indicating the wire was not grounded. However, when the ohmmeter leads were connected from the other damper solenoid terminal to the ground, the ohmmeter indicated a very low reading indicating contact between this wire and the chassis.

A closer examination of the L/F damper solenoid wires indicated these wires were jammed against the chassis about 2 ft. from the damper solenoid. The insulation was worn on the wires, and one wire was contacting the chassis. The technician repaired the wiring insulation and repositioned the harness so the wires were not jammed against the chassis. The technician used the scan tool to clear the DTC from the ESC module, and road tested the car to be sure the DTC did not reset and illuminate the SERVICE RIDE CONTROL light.

ASE-STYLE REVIEW QUESTIONS

1. When performing a self-test on a programmed ride control system:
   A. The mode select switch must be in the Auto position.
   B. One of the wires in the self-test connector must be grounded.
   C. The engine must be off and the ignition switch turned on.
   D. The headlights must be on during the self-test.

2. When servicing a vehicle with an air suspension system, the air suspension switch must be turned off:
   A. When changing the engine oil and filter.
   B. When changing the spark plugs.
   C. When jacking the vehicle to change a tire.
   D. During any of the above service procedures.

3. To deflate an air spring prior to removal of the spring:
   A. Disconnect the air line from the air spring.
   B. Turn the air spring solenoid valve to the first stage.
   C. Turn the air spring solenoid valve to the second stage.
   D. Energize the vent solenoid in the air compressor.

4. When using the scan tool to inflate an air spring:
   A. The vehicle must be raised on a tire-contact lift.
   B. The scan tool commands the air compressor on.
   C. The air suspension switch must be on.
   D. The scan tool commands the vent solenoid on.

5. When using a scan tool to perform a ride height adjustment on a vehicle dynamic suspension (VDS),
   Technician A says the VDS switch must be in the Off position.
   Technician B says if the ride height does not equal the manufacturer's specifications, the scan tool is used to command the VDS to lower or lift the vehicle to obtain the specified ride height.
   Who is correct?
   A. A only
   B. B only
   C. Both A and B
   D. Neither A nor B

6. All of these statements about performing an on-demand self-test on a VDS system are true EXCEPT:
   A. The battery must be fully charged.
   B. The ignition switch must be on.
   C. The vehicle must be raised on a lift.
   D. The 4L mode must not be selected on four-wheel-drive vehicles.

7. When diagnosing a VDS system, a U1900 DTC is obtained. This DTC indicates a defect in the:
   A. Controller area network (CAN).
   B. L/R air spring solenoid.
   C. Vent solenoid.
   D. R/F height sensor.

8. When diagnosing an electronic suspension control (ESC) system,
   Technician A says defects represented by a DTC with a U prefix must be repaired before proceeding with further diagnosis or service.
   Technician B says ESC system operation may be affected by low battery voltage.
   Who is correct?
   A. A only
   B. B only
   C. Both A and B
   D. Neither A nor B

9. On an ESC system a normal voltage signal from a wheel position sensor is:
   A. 2.5V.
   B. 5.5V.
   C. 6.1V.
   D. 6.8V.

10. While discussing ESC system diagnosis,
    Technician A says the normal force signal on a scan tool indicates a measured level of road surface condition sent from the ESC module to the EBCM.
    Technician B says the normal force data is sent from the front and rear wheels.
    Who is correct?
    A. A only
    B. B only
    C. Both A and B
    D. Neither A nor B

ASE CHALLENGE QUESTIONS

1. *Technician A* says during air spring inflation the vehicle weight must be applied to the suspension system.

   Technician B says during air spring inflation the vehicle should be positioned on a lift so the wheels are dropped downward.

   Who is correct?

   A. A only
   B. B only
   C. Both A and B
   D. Neither A nor B

2. All of these statements about a programmed ride control system are true EXCEPT:

   A. An air spring is mounted at each corner of the vehicle.
   B. An electric actuator is located in each strut.
   C. A mode indicator light is positioned in the tachometer.
   D. The PRC module provides a firm ride during severe braking.

3. *Technician A* says the suspension switch must be turned off before raising any corner of a car with an electronic air suspension.

   Technician B says the ignition switch must not be turned on while any corner of a car with electronic air suspension is raised.

   Who is correct?

   A. A only
   B. B only
   C. Both A and B
   D. Neither A nor B

4. A vehicle with an electronic air suspension system with mechanical trim height adjustment requires front and rear trim height adjustment.

   Technician A says to adjust the front trim height, rotate the threaded mounting bolt in the upper end of the height sensor.

   Technician B says to adjust the rear trim height, loosen the attaching bolt(s) on the upper height sensor bracket and move the bracket upward or downward.

   Who is correct?

   A. A only
   B. B only
   C. Both A and B
   D. Neither A nor B

5. When removing and replacing an air spring and shock absorber assembly on a VDS:

   A. The self-locking nuts on the upper strut mount can be reused.
   B. Retainer tabs on the lower end of the spring must be depressed to separate the spring and the shock absorber.
   C. The spring must be vented by loosening the spring solenoid valve to the first stage.
   D. The VDS switch and the ignition switch must be in the On position.

JOB SHEET 41

Name _____ **Date** _____

Diagnostic Procedures and Steps to Follow

Upon completion of this job sheet you should be able to follow the four principle steps to every repair procedure. Any repair procedure requires a systematic approach to diagnose and repair a vehicle.

NATEF AST/MAST Task Correlation:

Suspension and Steering. General: Suspension and Steering Systems,

Task #A-1. Research applicable vehicle and service information, vehicle service history, service precautions, and technical service bulletins. (P-1)

MAST Task Correlation:

Task #A-2. Identify and interpret suspension and steering system concerns; determine necessary action. (P-1)

Tools and Materials

Late-model vehicle
Service manual or information system
Safety glasses or goggles
Hand tools, as required

Describe the Vehicle Being Worked On:

Year _____ Make _____ Model _____

VIN _____ Engine type and size _____

Procedure

Always follow the procedures outlined in the service manual. Give a brief description of your procedure following each step. Ensure that the engine is cold and wear eye protection before you begin.

The four principal steps to every repair procedure should always be followed.

1. Verify the Customer Concern (complainant) and the Involved System:
   a. What is the original customer concern with the vehicle? _____

   _____

   b. Road test the vehicle. Was the customer's concern present and verifiable?

   _____

2. Isolate the Problem:
   a. To which system is the concern related to? _____

   _____

   b. Are there any *Technical Service Bulletins* (TSBs) that may be related to the problem area? _____

   _____

   c. Are there manufacturer published diagnostic procedures available? _____

   _____

3. Repair the Failure Area (component or circuit)
   a. *Erase any stored DTC* after all information has been recorded. Remember, freeze frame data will be erased and major monitors will be reset.

4. Recheck (verify) Repair Performed
   a. Road test the vehicle *under the same conditions* in which the original complaint was detected to verify that repair has been successfully completed.
   b. *Perform any required drive cycle procedures,* which may be necessary for OBD II System Monitor to run to completion. Failure to perform proper drive cycle could cause MIL to be illuminated after the customer picks up the vehicle and system monitor detects the same or another system failure.
   c. *Was the customer's concern corrected?* _____

_____

Instructor's Check _____

_____

_____

JOB SHEET 42

Name _____ **Date** _____

Inspection and Preliminary Diagnosis of Computer-Controlled Suspension System

Upon completion of this job sheet, you should be able to inspect computer-controlled suspension systems.

NATEF AST/MAST Task Correlation:

Suspension and Steering. Suspension Systems Diagnosis and Repair,

Task #C-1. Diagnose short and long arm suspension system noises, body sway, and uneven ride height concerns; determine necessary action. (P-1)

Task #C-2. Diagnose strut suspension system noises, body sway, and uneven ride height concerns; determine necessary action. (P-1)

MAST Task Correlation:

Suspension and Steering. Steering Systems Diagnosis and Repair,

Task #B-18. Test and diagnose components of electronically-controlled steering systems using a scan tool; determine necessary action. (P-3)

Tools and Materials

Modern vehicle with electronically controlled suspension system.

Describe the Vehicle Being Worked On:

Year _____ Make _____ Model _____

VIN _____ Engine type and size _____

Procedure _____ Task Completed

1. Discuss the vehicle complaint with the customer.
   Describe the customer complaint(s). _____
   _____
   _____

2. Road test the vehicle and verify the vehicle complaint.
   Describe the vehicle complaint experienced during the road test.
   _____
   _____
   _____

3. Inspect the vehicle for collision damage or other damage that could affect the computer-controlled suspension system. ☐
   Describe any collision or other vehicle damage that could affect the computer-controlled suspension system _____
   _____
   _____

Task Completed

☐ 4. Inspect all electrical connections and wiring harness in the suspension system.
Wiring connection and harness condition, Satisfactory ☐ Unsatisfactory ☐
If the wiring connection and harness condition are unsatisfactory, explain the wiring connection and/or harness defects.

_____

_____

5. If the vehicle has an air suspension system, listen for air leaks near the air-operated components and air lines, and inspect all the air lines for cracks, breaks, damage, and kinks.
Air line condition, Satisfactory ☐ Unsatisfactory ☐
If the air line condition is unsatisfactory, explain the air line defects.

_____

_____

_____

6. Measure the vehicle ride (trim) height.
Ride height, L/F _____, R/F _____, L/R _____, R/R _____
Specified ride height, Front _____ Rear _____

☐ 7. Observe the operation of the suspension warning light or message display in the instrument panel.

8. Suspension warning light or message display operation, Normal ☐ Abnormal ☐

9. Based on your inspection and preliminary diagnosis of the computer-controlled suspension system, list the problems in the system and explain the required further diagnostic and service procedures.

_____

_____

_____

_____

Instructor's Response _____

_____

_____

JOB SHEET 43

Name _____ **Date** _____

ADJUST VEHICLE RIDE (TRIM) HEIGHT WITH A SCAN TOOL

Upon completion of this job sheet, you should be able to adjust the vehicle ride height with a scan tool on computer-controlled suspension systems.

NATEF AST/MAST Task Correlation:

Suspension and Steering. Suspension Systems Diagnosis and Repair,

Task #C-1. Diagnose short and long arm suspension system noises, body sway, and uneven ride height concerns; determine necessary action. (P-1)

Task #C-2. Diagnose strut suspension system noises, body sway, and uneven ride height concerns; determine necessary action. (P-1)

Tools and Materials

Modern vehicle with a vehicle dynamic suspension (VDS) system.
Scan Tool
Tape Measure

Describe the Vehicle Being Worked On:

Year _____ Make _____ Model _____
VIN _____ Engine type and size _____

Procedure

_____
Task Completed

1. Be sure the air suspension switch is in the On position.
   Air suspension switch On, ☐ Yes ☐ No
2. Be sure the battery is fully charged.
   Battery fully charged, ☐ Yes ☐ No
3. Turn off the ignition switch and exit the vehicle. Close all doors and wait for 45 seconds for the air suspension system to vent down to the kneel position.
   VDS suspension in the kneel position, ☐ Yes ☐ No
4. Open the driver's door, and turn on the ignition switch.
   Ignition switch on, ☐ Yes ☐ No
5. Shift the transmission into Drive and back into Park.
   Transmission shifted into Drive and back into Park, ☐ Yes ☐ No
6. Exit the vehicle, close all the doors, and wait 45 seconds for the suspension system to pump up to the trim height.
   VDS system in the trim height position, ☐ Yes ☐ No
7. Use a tape measure to calculate the ride height at the locations specified by the vehicle manufacturer.
   Specified ride height, Front _____ Rear _____
   Actual ride height, L/F _____, R/F _____, L/R _____ R/R _____

Task Completed

8. If the ride height does not equal the manufacturer's specifications, open the left-front door, and connect the scan tool to the DLC.

   Scan tool properly connected to the DLC, ☐ Yes ☐ No

9. Select the correct vehicle year, model, and engine type on the scan tool, and then select *Vehicle Dynamic Module* (VDM).

   Scan tool selections, Vehicle year _____ Model _____

   Engine Size _____

   Vehicle Dynamic Module (VDM) selected on scan tool, ☐ Yes ☐ No

10. Use the active command mode in the scan tool to vent or lift the vehicle to achieve the specified trim height.

    Active command mode on scan tool used to lift or vent VDS system to specified trim height, ☐ Yes ☐ No

11. When a tape measure indicates the specified front and rear ride height, select *Save Calibration Values* on the scan tool to calibrate the VDM to the specified trim height.

    Actual ride height, L/F _____, R/F _____, L/R _____ R/R _____

Instructor's Response _____

_____

_____

JOB SHEET 44

Name _____ **Date** _____

Perform an On-Demand Self-Test on a Vehicle Dynamic Suspension (VDS) System

Upon completion of this job sheet, you should be able to perform an On-Demand Self-Test on a vehicle dynamic suspension (VDS) system.

NATEF MAST Task Correlation:

Suspension and Steering. Steering Systems Diagnosis and Repair,
Task #B-18. Test and diagnose components of electronically-controlled steering systems using a scan tool; determine necessary action. (P-3)

Tools and Materials

Modern vehicle with a vehicle dynamic suspension (VDS) system.
Scan tool

Describe the Vehicle Being Worked On:

Year _____ Make _____ Model _____
VIN _____ Engine type and size _____

Procedure

_____
Task Completed

1. Be sure the transmission is in Park.
   Transmission in Park, ☐ Yes ☐ No
2. Be sure the battery is fully charged.
   Battery fully charged, ☐ Yes ☐ No
3. Connect the scan tool to the DLC.
   Scan tool properly connected to the DLC, ☐ Yes ☐ No
4. Lower the driver's window and close all doors, liftgate, and liftgate glass.
   Driver's window down, ☐ Yes ☐ No
   All doors, liftgate, and liftglass closed, ☐ Yes ☐ No
5. Be sure the 4L mode is not selected on four-wheel-drive models.
   4L switch off, ☐ Yes ☐ No
6. Turn on the ignition switch.
   Ignition switch on, ☐ Yes ☐ No
7. Select *Vehicle Dynamic Module* and *On-Demand Self-Test* on the scan tool.
   Vehicle Dynamic Module selected on scan tool, ☐ Yes ☐ No
   On-Demand Self-Test selected on scan tool, ☐ Yes ☐ No

Task Completed

8. Record all the DTCs displayed on the scan tool.

_____

_____

_____

9. Use the vehicle manufacturer's service information to interpret the meaning of each DTC, and list the meaning of each DTC in the following spaces.

_____

_____

_____

10. Turn off the ignition switch, and disconnect the scan tool.
Ignition switch OFF, ☐ Yes ☐ No
Scan tool disconnected, ☐ Yes ☐ No

Instructor's Response _____

_____

_____

Chapter 12

Power Steering Pump Diagnosis and Service

BASIC TOOLS

Basic technician's tool set
Service manual
Pry bar
Floor jack
Safety stands
Oil drain pan
Crocus cloth

Upon completion and review of this chapter, you should be able to:

- Check power steering belt condition and adjust belt tension.
- Diagnose power steering belt problems.
- Check power steering fluid level and add fluid as required.
- Drain and flush power steering system.
- Bleed air from power steering system.
- Perform power steering pump pressure test.
- Check power steering pump fluid leaks.
- Remove and replace power steering pumps, and inspect pump mounts.
- Remove and replace power steering pump pulleys.
- Remove and replace power steering pump integral reservoirs.
- Remove, replace, and check flow control valve and pressure relief valve.
- Remove, replace, and check power steering pump rotating components.
- Remove and replace power steering pump seals and O-rings.
- Check, remove, and replace power steering lines.
- Inspect, test, diagnose, and service hybrid electric vehicles (HEVs) and electrohydraulic power steering (EHPS) systems.

Power steering pump diagnosis and service is very important to maintain vehicle safety and driver convenience. If a badly worn power steering pump belt is undetected, the belt may suddenly break while the driver is completing a turn. Under this condition, steering effort is greatly increased, and this may result in a collision. Therefore, the power steering belt should be inspected at regular intervals. A loose power steering belt may cause hard steering intermittently, and this condition requires increased driver steering effort, which reduces driver comfort and convenience. Therefore, one of the first checks when diagnosing a power steering system is to inspect the belt condition and test the belt tension.

Power Steering Pump Belt Service

Checking Belt Condition and Tension

Power steering belt condition and tension are extremely important for satisfactory power steering pump operation. A loose belt causes low pump pressure and hard steering. The steering wheel may jerk and surge during a turn if the power steering pump belt is loose. A loose, dry, or worn belt may cause squealing and chirping noises, especially during engine acceleration and cornering.

SPECIAL TOOL

Belt tension gauge

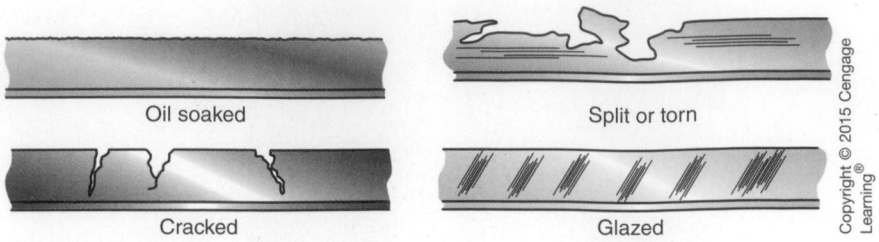

FIGURE 12-1 **Defective belt conditions.**

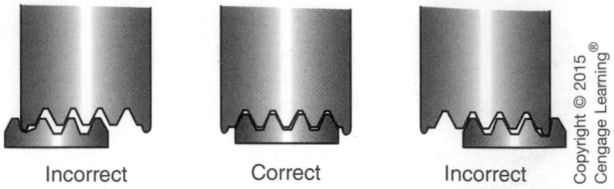

FIGURE 12-2 **Proper and improper installation of multi-ribbed V-belt on a pulley.**

Classroom Manual
Chapter 12, page 327

The power steering pump belt should be checked for tension, cracks, oil-soaking, worn or glazed edges, tears, and splits (Figure 12-1). If any of these conditions are present, belt replacement is necessary.

Most power steering pumps have a **multi-ribbed V-belt** (or **serpentine belt**). Many multi-ribbed V-belts have an automatic tensioning pulley; therefore, a tension adjustment is not required. The multi-ribbed V-belt should be checked to make sure it is installed properly on each pulley in the belt drive system (Figure 12-2). The tension on a multi-ribbed V-belt may be checked with a belt tension gauge. Prior to removal of the serpentine belt locate the routing diagram similar to the one shown in Figure 12-3 (pulley alignment). This diagram is often located under the hood in the engine bay, often on the radiator support cover. The diagram may also be located in the vehicle service information. Some technicians choose to draw a sketch of the belt routing prior to removal.

CAUTION:
Do not pry on the pump reservoir with a pry bar. This action may damage the reservoir.

Many multi-ribbed V-belts have a spring-loaded tensioner pulley that automatically maintains belt tension. As the belt wears or stretches, the spring moves the tensioner pulley to maintain the belt tension. Some of these tensioners have a belt length scale that indicates the new belt range and used belt range (Figure 12-4). If the indicator on the tensioner is out of the used belt length range, belt replacement is required. Many belt tensioners have a 1/2-inch drive opening in which a ratchet or flex handle may be installed to move the tensioner pulley off the belt during belt replacement (Figure 12-5).

SERVICE TIP:
During rainy or wet conditions, a loose power steering belt, which causes low power steering pump pressure and increased steering effort, may be more noticeable because the belt becomes wet and slips more easily.

Belt pulleys must be properly aligned to minimize belt wear. The edges of the pulleys must be in line when a straightedge is placed on the pulleys (Figure 12-6). Repeated belt failure may be caused by a misaligned power steering pump pulley or by extremely high pump pressure caused by a sticking pressure relief valve or a continual or intermittent restriction in the high pressure hose.

On older V-belt systems the friction surfaces are on the sides of the V-type belt, wear occurs in this area. If the belt edges are worn, the belt may be rubbing on the bottom of the pulley. This condition requires belt replacement. Belt tension may be checked by measuring the belt deflection. Press on the belt with the engine stopped to measure the belt deflection, which should be 1/2 inch per foot of free span. The belt tension may be checked with a **belt tension gauge** placed over the belt (Figure 12-7). The tension on the gauge should equal the vehicle manufacturer's specifications.

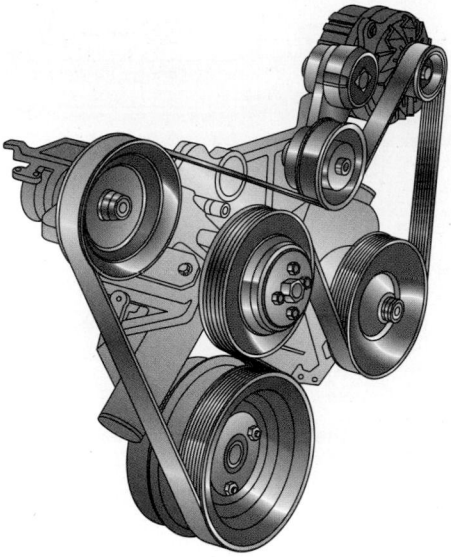

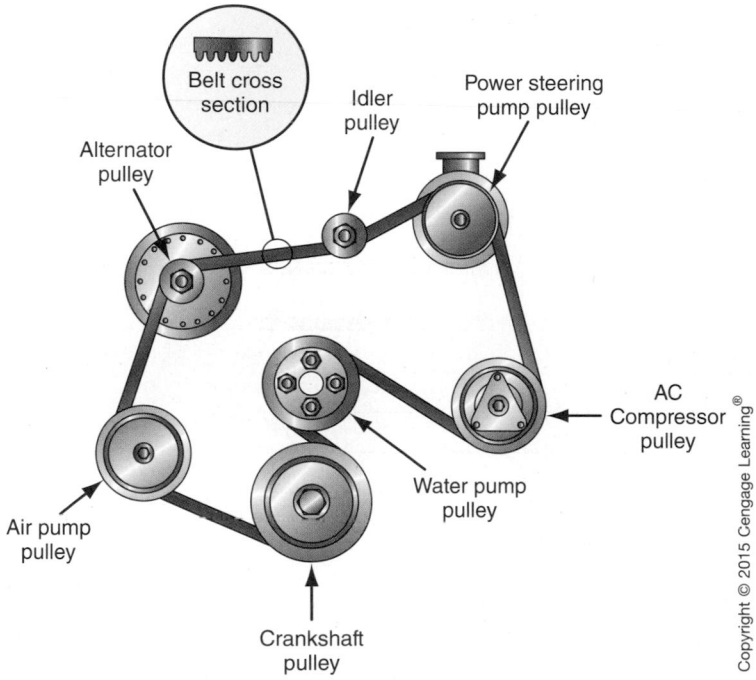

FIGURE 12-3 **A typical serpentine belt routing.**

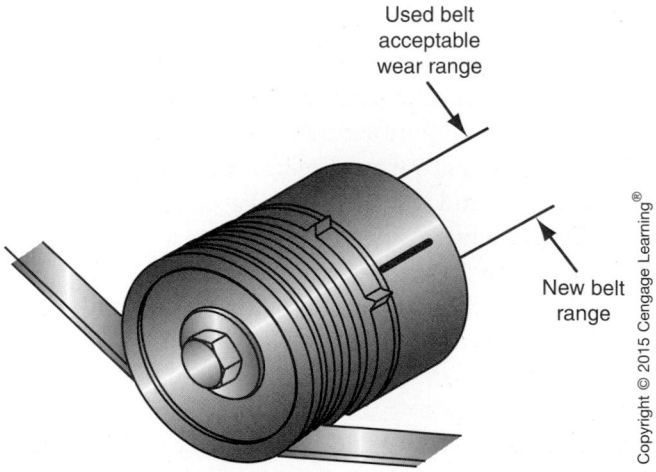

FIGURE 12-4 **Belt tension scale.**

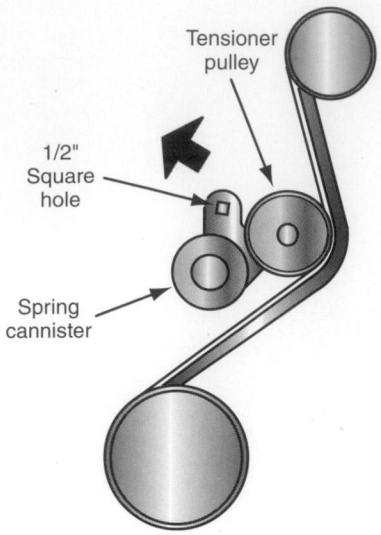

Copyright © 2015 Cengage Learning®

FIGURE 12-5 **One-half inch drive opening in the tensioner pulley.**

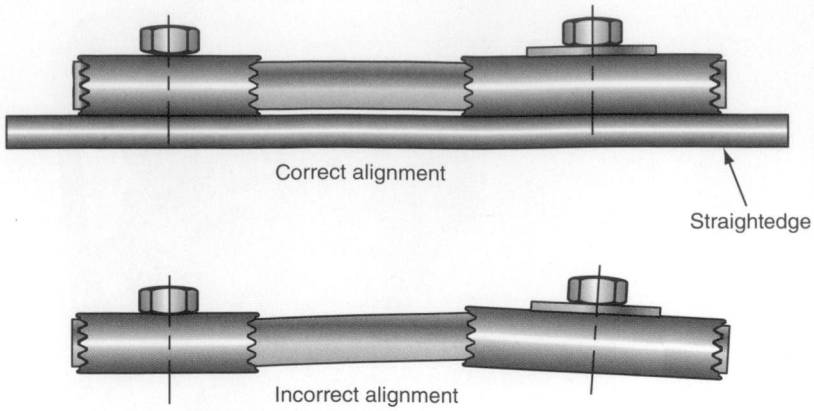

Copyright © 2015 Cengage Learning®

FIGURE 12-6 **Checking pulley alignment.**

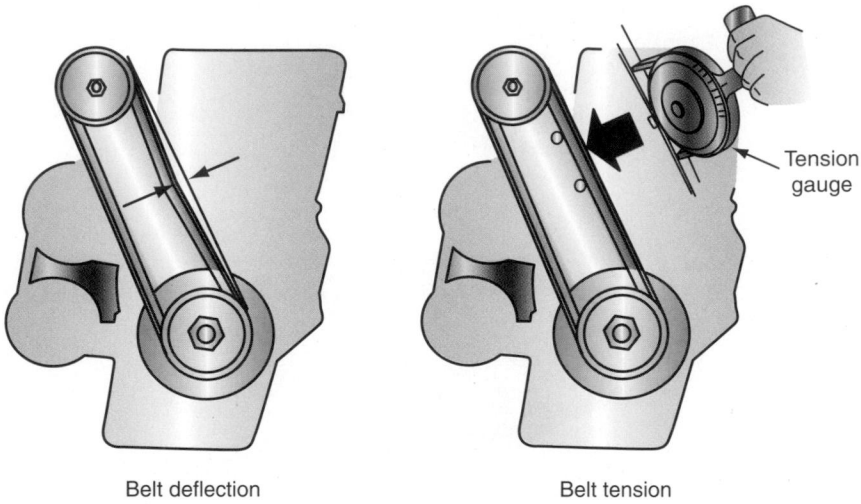

Copyright © 2015 Cengage Learning®

FIGURE 12-7 **Methods of checking belt tension.**

If the belt requires tightening, follow this procedure:

1. Loosen the power steering pump bracket or tension adjusting bolt.
2. Loosen the power steering pump mounting bolts.
3. Check the bracket and pump mounting bolts for wear. If these bolts or bolt openings in the bracket or pump housing are worn, replacement is necessary.
4. Pry against the pump ear and hub with a pry bar to tighten the belt. Some pump brackets have a 1/2-inch square opening in which a breaker bar may be installed to move the pump and tighten the belt.
5. Hold the pump in the position described in step 4, and tighten the bracket or tension adjusting bolt.
6. Recheck the belt tension with the tension gauge. If the belt does not have the specified tension, repeat steps 1 through 5.
7. Tighten the tension adjusting bolt and the mounting bolts to the manufacturer's specified torque.

CUSTOMER CARE: When servicing vehicles, always promote preventive maintenance to the customer. For example, if you are performing some underhood service, such as replacing a battery and starting motor, always make a quick inspection of the belt or belts and cooling system hoses. If you see a multi-ribbed V-belt with chunks out of it, recommend a belt replacement to the customer. The customer will usually appreciate this service and authorize the belt replacement. If you ignore the belt with chunks out of it, and the belt breaks the next day while the customer is driving on the freeway, the customer will be inconvenienced and has to pay a tow bill. Although you did a satisfactory battery and starting motor replacement, the customer will not be happy if this problem occurs the day after you worked on the vehicle. The next time this unhappy customer requires automotive service, he or she may take the vehicle to another repair shop.

Belt Failure Troubleshooting

A variety of critical engine components stop working when a serpentine belt fails. These components may include the water pump, alternator, air-conditioning compressor, and power steering pump to name some of the more common belt driven accessories. It is important to be able to identify the reason for a belt failure to ensure that the replacement belt will last and avoid premature failure. Often by observing the belts appearance we can determine the potential causes for the failure or damage. Serpentine belts may be constructed of either Neoprene or ethylene propylene diene M-class rubber commonly called EPDM. The majority of belts manufactured today are made of EPDM. Discussed next is a list and photos of some common belt failures outlining their cause and solutions recommended by belt manufacturers.

Cracking. Cracks that are small and perpendicular to the ribs of the belt occurring frequently around its circumference (Figure 12-8) is probably one of the most common reasons for belt replacement. These cracks are caused by continuous flexing and bending in high heat conditions found under the hood as the belt bends around pulleys during normal operation. On Neoprene belts, 80 percent of the belt life is gone if three or more cracks are found in a 3 in. segment of the belt, and the belt should be replaced. As a common practice with all belt failures, check belt tensioner and idler pulleys for bearing wear and operation.

Copyright © 2015 Cengage Learning®

FIGURE 12-8 **Multi-ribbed belt cracks are caused by continuous flexing and bending of the material in high heat conditions and is the most common reason for belt replacement.**

Belts made of EPDM should not show cracks for 100,000 miles. But EPDM belts may show rib height material wear.

EPDM Belt Wear. Belts made of EPDM wear with age and often require replacement between 80,000-100,000 miles under normal operating conditions. The EPDM belt is more elastic than a standard Neoprene belt and resists cracking even at higher mileage, so looking for visible cracks is not an effective method for determining belt wear and aging. A better indicator of when to replace EPDM belts is rib wear (Figure 12-9). To aid in determining belt wear a belt wear depth gauge is available (Figure 12-10). One free source

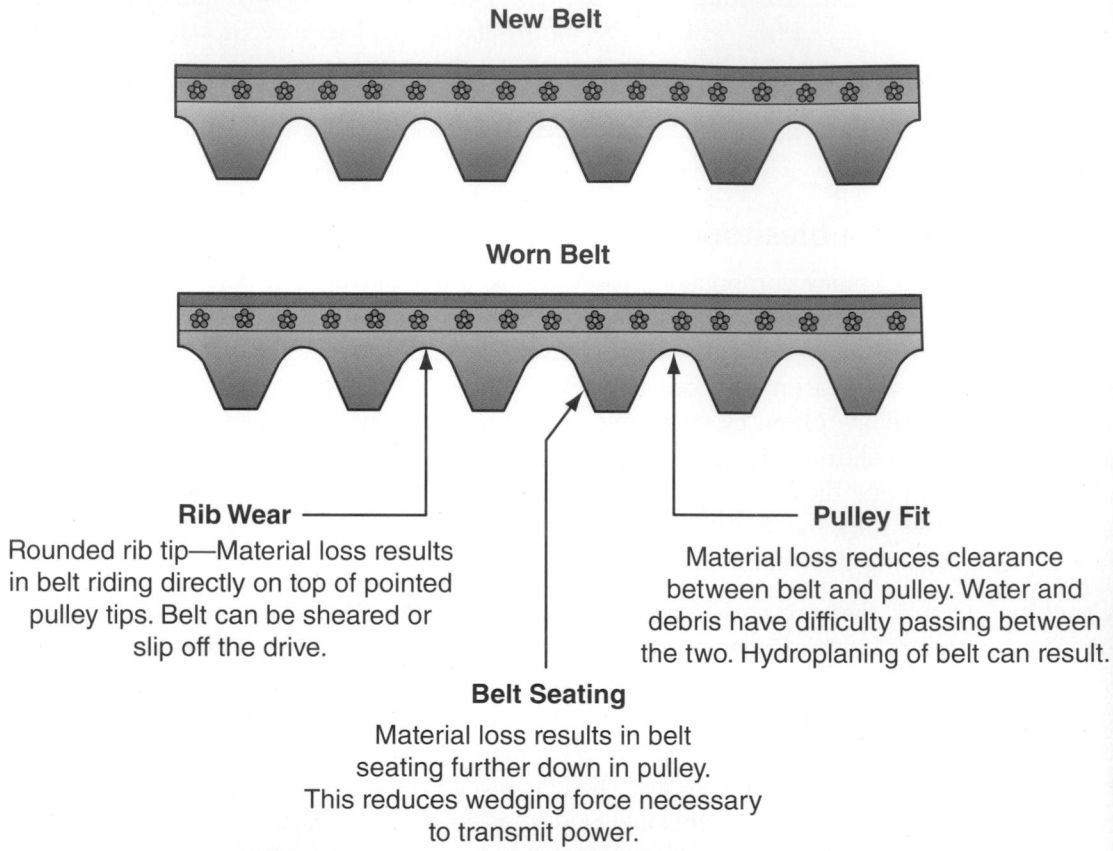

Copyright © 2015 Cengage Learning®

FIGURE 12-9 **Belts made of EPDM are designed to last 100,000 miles, but over time the material on the sides and valleys will wear. A 10 percent loss of material is enough to cause performance issues.**

Copyright © 2015 Cengage Learning®

FIGURE 12-10 **Gates produces a simple belt wear gauge that can be used to detect EPDM belt grove wear.**

Copyright © 2015 Cengage Learning®

FIGURE 12-11 **Belt abrasion is due to a belt making contact with an object such as a flange bolt or foreign object.**

for this belt wear depth gauge is Gates Belts (http://www.gatesprograms.com/beltwear). As little as 5 to 10 percent wear can effect operation if belt wear indicates replacement is required.

Abrasion. Abrasion may cause a shiny or glazed appearance on one or both sides of the belt. In severe cases cord fibers may be visible (Figure 12-11). It is caused by belt making contact with an object, such as a flange bolt or foreign object. It may also be caused by improper belt tension, pulley surface wear or a binding pulley bearing. Inspect belt tensioner for proper operation, replace if necessary.

Improper Installation. A belt rib on either edge begins separating from main body plies (Figure 12-12). If left unnoticed the outer covering may separate causing the belt to unravel. It is caused by improper replacement procedure when one of the belt ribs is placed outside of one of the pulley grooves during installation. Verify that the belt has the same number of ribs as its companion pulley grooves. Once damage has occurred the belt must be replaced. Ensure that all belt ribs fit properly in all pulley groves when replacing.

Pilling. Pilling occurs as belt material is sheared off from the ribs building up in the grooves (Figure 12-13). It is generally caused by insufficient belt tension, pulley misalignment or wear. It is more common on diesel engines, but can happen on any powertrain.

Copyright © 2015 Cengage Learning®

FIGURE 12-12 **A separation of a belt rib on either edge from the main body plies caused by improper replacement procedure.**

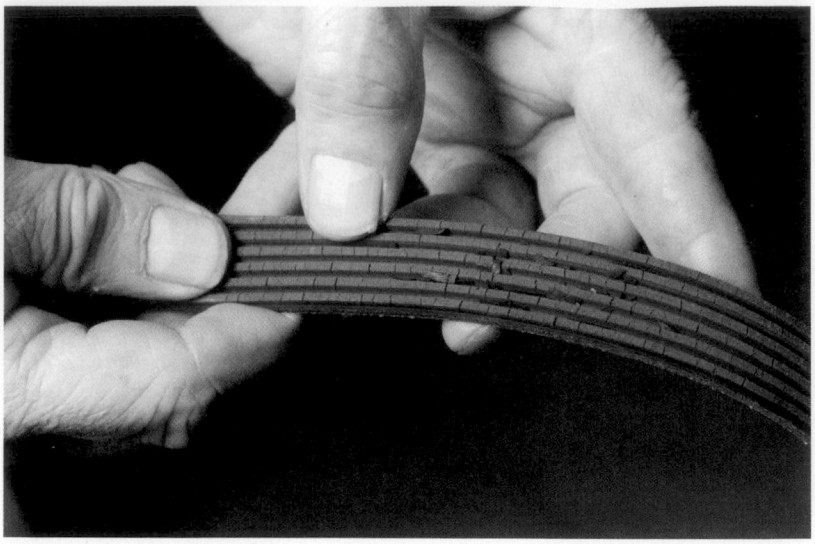

Copyright © 2015 Cengage Learning®

FIGURE 12-13 **A belt pilling occurs as belt material is sheared off from the ribs building up in the grooves.**

Pilling can lead to belt noise and vibration, pulley grooves should be inspected for buildup and alignment should be checked to see if belt needs to be replaced.

It is wise to replace the drive belt along as well as the automatic belt tensioner when any pulley driven component is replaced. By replacing only one system component vibration is increased dramatically (Figure 12-14). This vibration may be both felt and heard leading to customer complaints and the performance and longevity of the new component may be compromised. The green line on the graph indicates vibration minimization with a new belt, tensioner, and component. The red graph line indicates vibration caused by an ageing drive belt, tensioner, and belt driven components at 150,000 miles. The blue line indicates the vibration results of just replacing one belt driven component and reusing the old belt

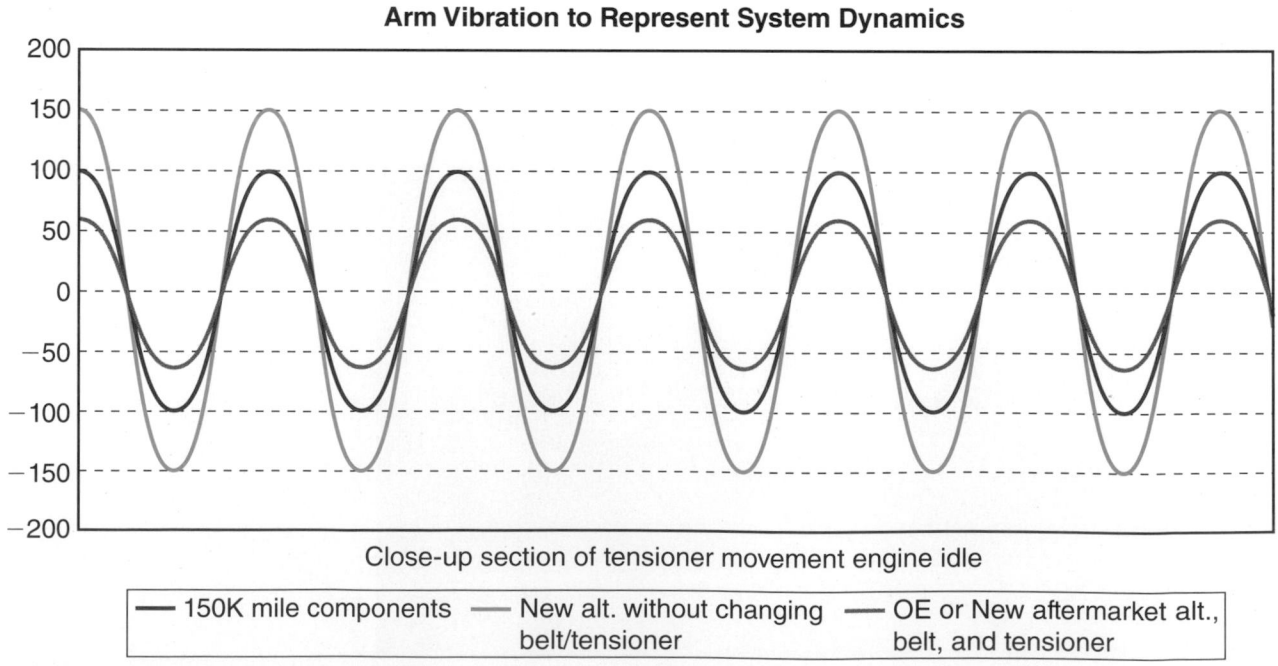

Copyright © 2015 Cengage Learning®

FIGURE 12-14 **When a drive belt component is replaced it is advisable to also replace the belt and automatic tensioner to limit system vibration.**

and tensioner. The old belt and tensioner do not dampen new component vibration effectively. Excessive vibration leads to excessive noise and component bearing wear. Today all belt drive components are closely integrated and each depends on the belt and tensioner to keep the entire system at optimum operating performance. It is critical that you as a technician check every component in the system; this includes idler pulleys if used.

Power Steering Pump Fluid Service

Fluid Level Inspection

Most car manufacturers recommend power steering fluid or automatic transmission fluid in power steering systems. Always use the type of power steering fluid recommended in the vehicle manufacturer's service manual. If the power steering fluid level is low, **steering effort** is increased and may be erratic. A low fluid level may cause a growling noise in the power steering pump. Some car manufacturers now recommend checking the power steering pump fluid level with the fluid at an ambient temperature of 176°F (80°C).

Steering effort is the amount of effort required by the driver to turn the steering wheel.

Follow these steps to check the power steering fluid level:

1. With the engine idling at 1,000 rpm or less, turn the steering wheel slowly and completely in each direction several times to boost the fluid temperature (Figure 12-15).
2. If the vehicle has a remote power steering fluid reservoir, check for foaming in the reservoir, which indicates low fluid level or air in the system.
3. Observe the fluid level in the remote reservoir. This level should be at the hot full mark. Shut off the engine, and remove dirt from the neck of the reservoir with a shop towel. If the power steering pump has an integral reservoir, the level should be at the hot level on the dipstick. When an external reservoir is used, the dipstick is located in the external reservoir (Figure 12-16).
4. Pour the required amount of the car manufacturer's recommended power steering fluid into the reservoir to bring the fluid level to the hot full mark on the reservoir or dipstick with the engine idling.

CAUTION: Always use the vehicle manufacturer's recommended power steering fluid. Using the wrong power steering fluid may cause improper power steering operation.

Power Steering System Draining and Flushing

If the power steering fluid is contaminated with moisture, dirt, or metal particles, the system must be drained and new fluid installed.

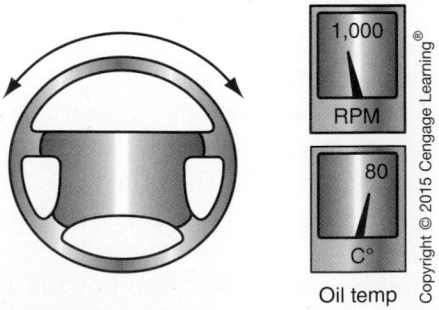

Copyright © 2015 Cengage Learning®

FIGURE 12-15 **Boosting fluid temperature.**

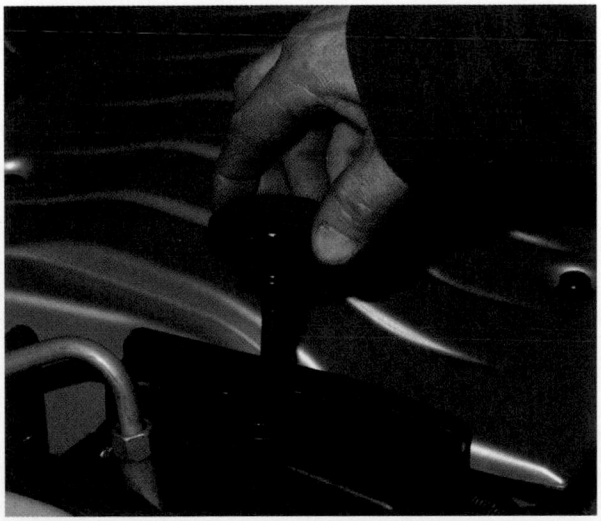

Copyright © 2015 Cengage Learning®

FIGURE 12-16 **Power steering pump dipstick.**

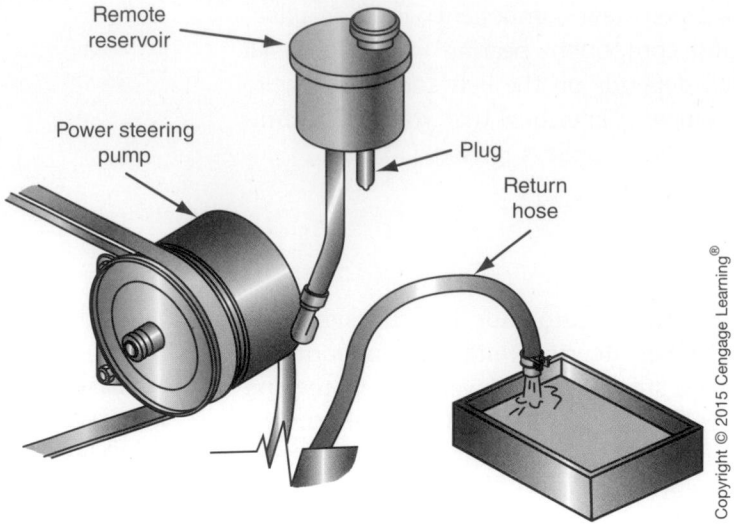

FIGURE 12-17 **Return hose installed in drain pan for power steering draining and flushing.**

FIGURE 12-18 **Draining fluid from the power steering system.**

Follow these steps to drain and flush the power steering system:

A **remote reservoir** is mounted externally from the power steering pump.

1. Lift the front of the vehicle with a floor jack and install safety stands under the suspension. Lower the vehicle onto the safety stands, and remove the floor jack.
2. Remove the return hose from the **remote reservoir** that is connected to the steering gear. Place a plug on the reservoir outlet and position the return hose in an empty drain pan (Figure 12-17).
3. With the engine idling, turn the steering wheel fully in each direction and stop the engine (Figure 12-18).
4. Fill the reservoir to the hot full mark with the manufacturer's recommended fluid.
5. Start the engine, and run the engine at 1,000 rpm while observing the return hose in the drain pan. When fluid begins to discharge from the return hose, shut the engine off.
6. Repeat steps 4 and 5 until there is no air in the fluid discharging from the return hose.
7. Remove the plug from the reservoir and reconnect the return hose. Bleed the power steering system.

Bleeding Air from the Power Steering System

When air is present in the power steering fluid, a growling noise may be heard in the pump and steering effort may be increased or erratic.

When a power steering system has been drained and refilled, follow this procedure to remove air from the system:

1. Fill the power steering pump reservoir as outlined previously.
2. With the engine running at 1,000 rpm, turn the steering wheel fully in each direction three or four times (Figure 12-19). Each time the steering wheel is turned fully to the right or left, hold it there for 2 to 3 seconds before turning it in the other direction.
3. Check for foaming of the fluid in the reservoir. When foaming is present, repeat steps 1 and 2.
4. Check the fluid level with the engine running and be sure it is at the hot full mark. Shut off the engine, and make sure the fluid level does not increase more than 0.020 in. (5 mm) (Figure 12-20).

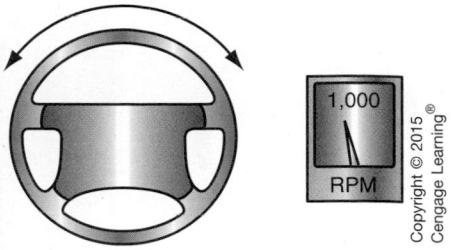

FIGURE 12-19 **Turning the steering wheel slowly to bleed air from the power steering system.**

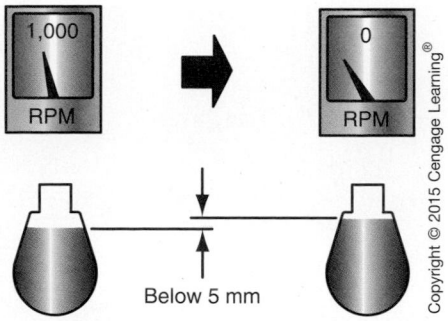

FIGURE 12-20 **Power steering fluid level after bleeding.**

Power Steering Bleeding Procedure with Vacuum Hand Pump

Some vehicle manufacturers recommend using a vacuum hand pump to bleed air from the power steering system. Follow this procedure to bleed the power steering system with a vacuum hand pump:

1. Remove the power steering pump reservoir cap.
2. Install a special fitting to the reservoir cap opening that allows a vacuum hose to be connected to the fitting.
3. Connect the vacuum hand pump hose to the fitting installed on the power steering pump reservoir.
4. Operate the vacuum hand pump until the vacuum gauge on the pump indicates 20 in. Hg (68 kPa).
5. Wait for 5 minutes or vehicle manufacturer's specified time interval.
6. Release the vacuum in the hand pump and reservoir. Remove the hand pump hose and special fitting from the power steering pump.
7. Check the fluid level in the power steering pump reservoir, and correct as necessary.
8. Install the pump reservoir cap and dipstick and check the operation of the power steering system.

Power Steering Pump Diagnosis

Power Steering Pump Pressure Test

Since there are some variations in power steering pump pressure test procedures and pressure specifications, the vehicle manufacturer's test procedures and specifications must be used. If the power steering pump pressure is low, steering effort is increased. Erratic power steering pump pressure causes variations in steering effort, and the steering wheel may jerk as it is turned. Since a power steering pump will never develop the specified pressure if the belt is slipping, the belt tension must be checked and adjusted, if necessary, prior to a pump pressure test.

SPECIAL TOOL
Power steering pressure test gauge

The following is a typical power steering pressure test procedure:

1. With the engine stopped, disconnect the pressure line from the power steering pump and connect the gauge side of the **pressure gauge** to the pump outlet fitting. Connect the valve side of the gauge to the pressure line (Figure 12-21).
2. Start the engine, and turn the steering wheel fully in each direction two or three times to bleed air from the system (Figure 12-22). Be sure the fluid level is correct and the fluid temperature is at least 176°F (80°C). A thermometer may be inserted in the pump reservoir fluid to measure the fluid temperature.

A **pressure gauge** is designed to measure the high pressure in the power steering system.

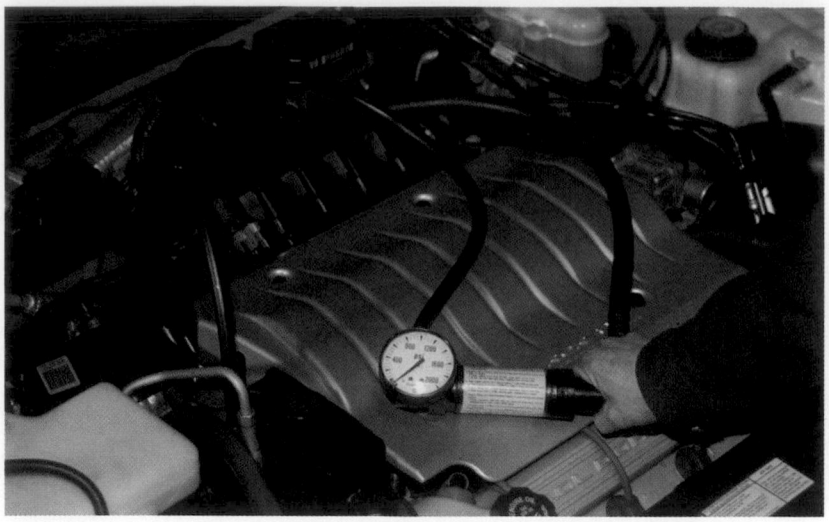

Copyright © 2015 Cengage Learning®

FIGURE 12-21 **Pressure gauge connection to power steering pump.**

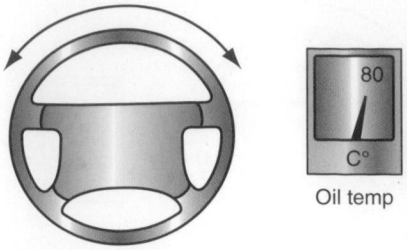

Copyright © 2015 Cengage Learning®

FIGURE 12-22 **Bleeding air from the power steering pump system and checking fluid temperature.**

Classroom Manual
Chapter 12, page 333

WARNING: During the power steering pump pressure test with the engine idling, close the pressure gauge valve for no more than 10 seconds because excessive pump pressure may cause power steering hoses to rupture, resulting in personal injury.

WARNING: Do not allow the fluid to become too hot during the power steering pump pressure test. Excessively high fluid temperature reduces pump pressure. Wear protective gloves, and always shut the engine off before disconnecting gauge fittings, because the hot fluid may cause burns.

3. With the engine idling, close the pressure gauge valve for no more than 10 seconds, and observe the pressure gauge reading (Figure 12-23). Turn the pressure gauge valve to the fully open position. If the pressure gauge reading does not equal the vehicle manufacturer's specifications, repair or replace the power steering pump.
4. Inspect the power steering pump pressure with the engine running at 1,000 rpm and 3,000 rpm, and record the pressure difference between the two readings (Figure 12-24). If the pressure difference between the pressure readings at 1,000 rpm and 3,000 rpm does not equal the vehicle manufacturer's specifications, repair or replace the **flow control valve** in the power steering pump.
5. With the engine running, turn the steering wheel fully in one direction and observe the steering pump pressure while holding the steering wheel in this position (Figure 12-25). If the pump pressure is less than the vehicle manufacturer's specifications, the steering gear housing has an internal leak and should be repaired or replaced.
6. Be sure the front tire pressure is correct and center the steering wheel with the engine idling. Connect a spring scale to the steering wheel and measure the steering effort in both directions (Figure 12-26). If the power steering pump pressure is satisfactory and the steering effort is more than the vehicle manufacturer's specifications, the power steering gear should be repaired. Photo Sequence 23 shows a typical procedure for pressure testing a power steering pump.

The **flow control valve** controls power steering fluid flow from the pump to the steering gear.

SPECIAL TOOL
Spring scale

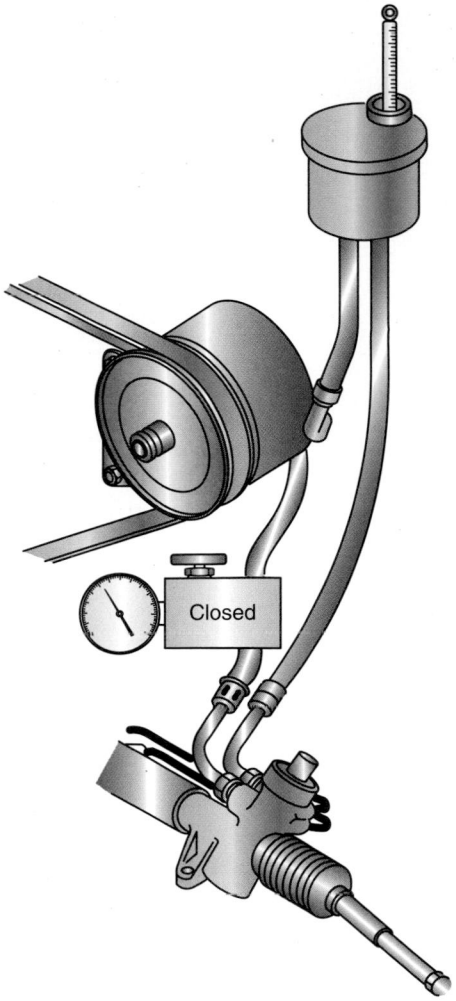

FIGURE 12-23 **Power steering pump pressure test with pressure gauge valve closed.**

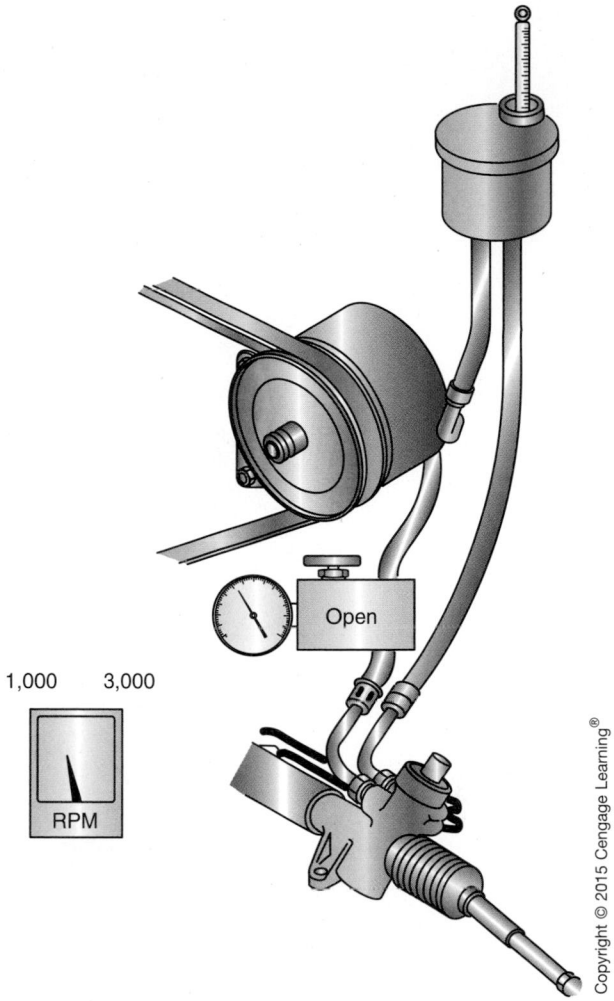

FIGURE 12-24 **Power steering pump pressure test at 1,000 and 3,000 rpm.**

Power Steering Pump Oil Leak Diagnosis

The possible sources of power steering pump oil leaks are the drive shaft seal, reservoir O-ring seal, high-pressure outlet fitting, and the dipstick cap. If leaks occur at any of the seal locations, seal replacement is necessary. When a leak is present at the high-pressure outlet fitting, tighten this fitting to the specified torque (Figure 12-27). If the leak still occurs, replace the O-ring seal on the fitting and retighten the fitting.

Power Steering Pump Service

Power Steering Pump Replacement

WARNING: If the vehicle has been driven recently, the pump, hoses, and fluid could be extremely hot. Use caution when handling components to avoid burns.

If a growling noise is present in the power steering pump after the fluid level is checked and air has been bled from the system, the pump bearings or other components are defective and pump replacement or repair is required. When the power steering pump pressure is lower than specified, pump replacement or repair is necessary.

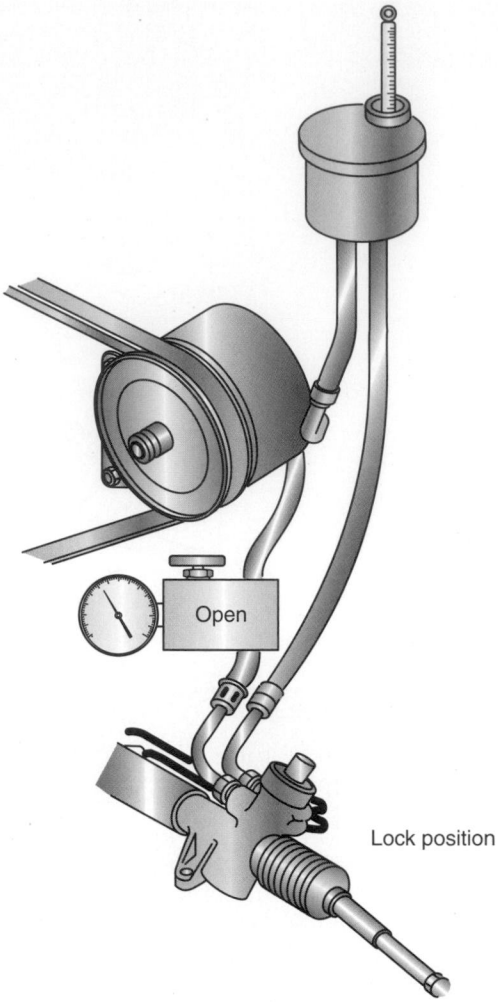

Copyright © 2015 Cengage Learning®

FIGURE 12-25 **Power steering pump pressure test with the front wheels turned fully in one direction.**

Copyright © 2015 Cengage Learning®

FIGURE 12-26 **Steering effort measurement.**

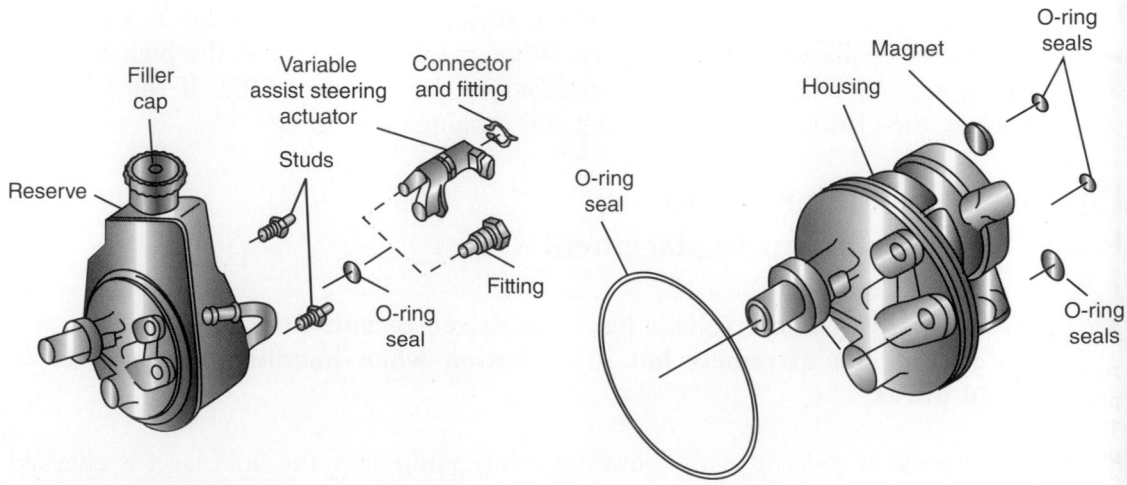

FIGURE 12-27 **Power steering pump oil leak diagnosis.**

PHOTO SEQUENCE 23

Typical Procedure for Pressure Testing a Power Steering Pump

P23-1 **Connect the pressure gauge to the power steering pump.**

Copyright © 2015 Cengage Learning®

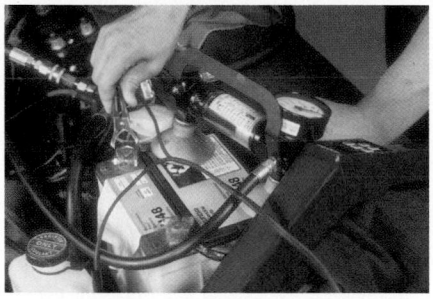

P23-2 **Connect a tachometer to the ignition system.**

Copyright © 2015 Cengage Learning®

P23-3 **Look up the vehicle manufacturer's specified power steering pump pressure in the service manual.**

Copyright © 2015 Cengage Learning®

P23-4 **Start the engine, and turn the steering wheel from lock to lock three times to bleed air from the system and heat the power steering fluid.**

Copyright © 2015 Cengage Learning®

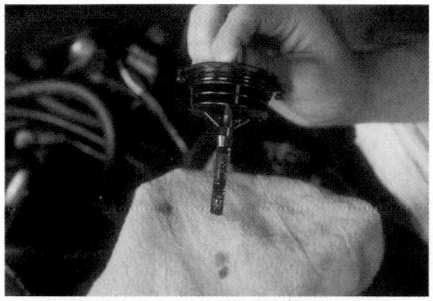

P23-5 **Check the power steering fluid level, and add fluid as required.**

Copyright © 2015 Cengage Learning®

P23-6 **Place a thermometer in the power steering fluid reservoir. Be sure the fluid temperature is at least 176°F (80°C).**

Copyright © 2015 Cengage Learning®

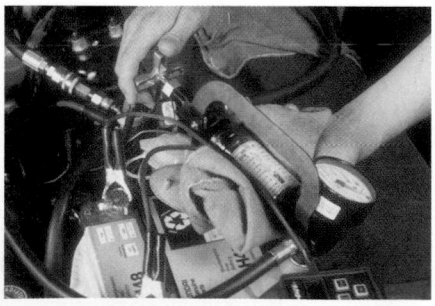

P23-7 **With the engine idling, close the pressure gauge valve for no more than 10 seconds. Observe and record the pressure gauge reading.**

Copyright © 2015 Cengage Learning®

P23-8 **Check and record the power steering pump pressure at 1,000 rpm.**

Copyright © 2015 Cengage Learning®

P23-9 **Check and record the power steering pump pressure at 3,000 rpm.**

Copyright © 2015 Cengage Learning®

P23-10 **Check and record the power steering pump pressure with the steering wheel turned fully in one direction and the engine idling.**

Copyright © 2015 Cengage Learning®

SPECIAL TOOL

Puller for power steering pump pulley, press-on pulley

SPECIAL TOOL

Power steering pump pulley installing tool, press-on pulley

To replace the power steering pump, proceed as follows:

1. Disconnect the power steering return hose from the remote reservoir or pump. Allow the fluid to drain from this hose into a drain pan. Discard the used fluid.
2. Loosen the bracket or belt tension adjusting bolt and the pump mounting bolt.
3. Loosen the belt tension until the belt can be removed. On some cars, it is necessary to lift the vehicle on a hoist and gain access to the power steering pump from underneath the vehicle.
4. Remove the hoses from the pump and cap the pump fittings and hoses.
5. Remove the belt tension adjusting bolt and the mounting bolt, and remove the pump.
6. Check the pump mounting bolts and bolt holes for wear. Worn bolts must be replaced. If the bolt mounting holes in the pump are worn, pump replacement is necessary.
7. Reverse steps 1 through 5 to install the power steering pump. Tighten the belt as described previously, and tighten the pump mounting and bracket bolts to the manufacturer's specifications. If O-rings are used on the pressure hose, replace the O-rings. Be sure the hoses are not contacting the exhaust manifold, catalytic converter, or exhaust pipe during or after pump replacement.
8. Fill the pump reservoir with the manufacturer's recommended power steering fluid and bleed air from the power steering system as described earlier.

A **woodruff key** is a half-moon shaped key that fits snugly in a shaft opening, with the top edge of the key extending out of the shaft. The top edge of the key fits snugly in a groove cut inside the pulley hub. When the pulley is installed on the shaft, the pulley groove slides over the key to prevent pulley rotation on the shaft.

Power Steering Pump Pulley Replacement

If the pulley wobbles while it is rotating, the pulley is probably bent, and pulley replacement is necessary. Worn pulley grooves also require pulley replacement. Always check the pulley for cracks. If this condition is present, pulley replacement is essential. A pulley that is loose on the pump shaft must be replaced. Never hammer on the pump drive shaft during pulley removal or replacement. This action will damage internal pump components. If the pulley is pressed onto the pump shaft, a special puller is required to remove the pulley (Figure 12-28), and a pulley installation tool is used to install the pulley (Figure 12-29).

If the power steering pump pulley is retained with a nut, mount the pump in a vise. Always tighten the vise on one of the pump mounting bolt surfaces. Do not tighten the vise with excessive force. Use a special holding tool to keep the pulley from turning, and loosen the pulley nut with a box end wrench (Figure 12-30). Remove the nut, pulley, and **woodruff key**. Inspect the pulley, shaft, and woodruff key for wear. Be sure the key slots in the shaft and pulley are not worn. Replace all worn components.

Copyright © 2015 Cengage Learning®

RE 12-28 **Press-on power ing pump pulley removal.**

Copyright © 2015 Cengage Learning®

FIGURE 12-29 **Press-on power steering pump pulley installation.**

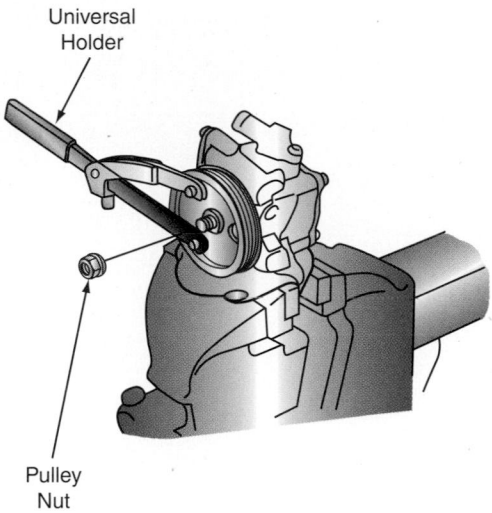

Copyright © 2015 Cengage Learning®

FIGURE 12-30 **Removing power steering pump pulley and nut.**

Remove and Replace the Power Steering Pump Reservoir

Power steering pump service procedures vary depending on the type of pump. Always follow the vehicle manufacturer's recommended service procedures in the service manual. The following service procedures are for a power steering pump with an **integral reservoir**.

An **integral reservoir** is mounted on the power steering pump.

When the power steering pump has an integral reservoir, follow these steps for reservoir removal and replacement:

1. Remove the filler cap and drain the oil from the reservoir.
2. Remove the two clips that retain the reservoir to the housing.
3. Rock the reservoir by hand or tap it gently with a soft hammer to remove it from the housing.
4. Clean all the parts and discard the O-ring. Inspect the O-ring surfaces for damage.
5. Lubricate the new O-ring with the manufacturer's recommended power steering fluid and install it (Figure 12-31).
6. Install the reservoir and retaining clips.

Remove and Replace Flow Control Valve and End Cover

When the flow control valve and end plate are serviced, follow these steps:

1. Remove the retaining ring with a slotted screwdriver and a punch (Figure 12-32).
2. Remove the flow control valve, end cover, spring, and magnet. Inspect the flow control valve for burrs. Remove minor burrs with **crocus cloth** and clean the flow control valve in solvent. Damaged or worn flow control valves must be replaced. Inspect the end cover sealing surface for damage. Inspect the pump driveshaft for corrosion and damage. Remove corrosion with crocus cloth (Figure 12-33). Clean the magnet with a shop towel.
3. Clean all parts and lubricate the end cover with power steering fluid.
4. Install the end cover, retaining ring, and related components.

Crocus cloth is a very fine polishing-type emery paper.

Remove and Replace Pressure Relief Valve

Follow this procedure to service the pressure relief valve:

The **pressure relief valve** limits the maximum power steering pump pressure.

1. Wrap a shop towel around the land end of the flow control valve and clamp this end in a soft-jawed vise. Be very careful not to mark valve lands.
2. Remove the hex-head ball seat (Figure 12-34). Clean the components in solvent. A worn or damaged pressure relief ball, spring, guide, or seat must be replaced.

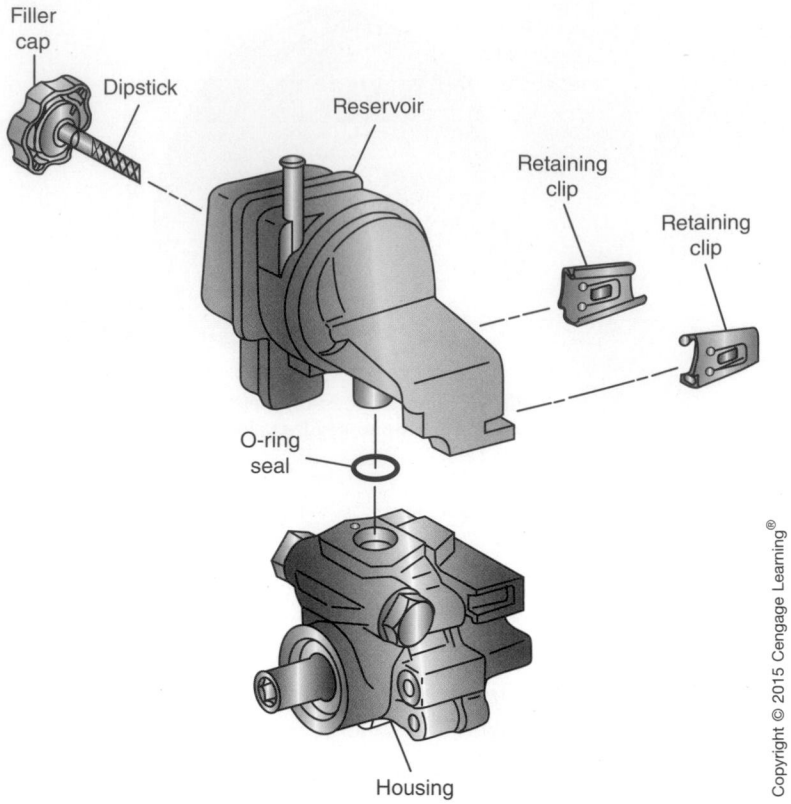

FIGURE 12-31 **Integral power steering pump reservoir seal and O-ring.**

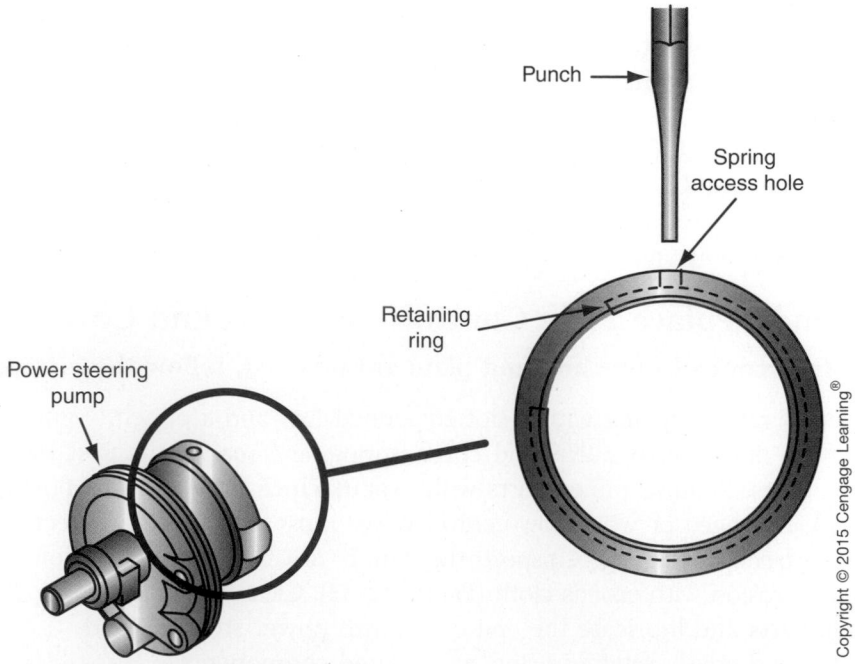

FIGURE 12-32 **Retaining ring removal.**

Inspecting and Servicing Power Steering Lines and Hoses

Power steering lines should be inspected for leaks, dents, sharp bends, cracks, and contact with other components. Lines and hoses must not rub against other components. This action could wear a hole in the line or hose. Many high-pressure power steering lines are made from high-pressure steel-braided hose with molded steel fittings on each end (Figure 12-35).

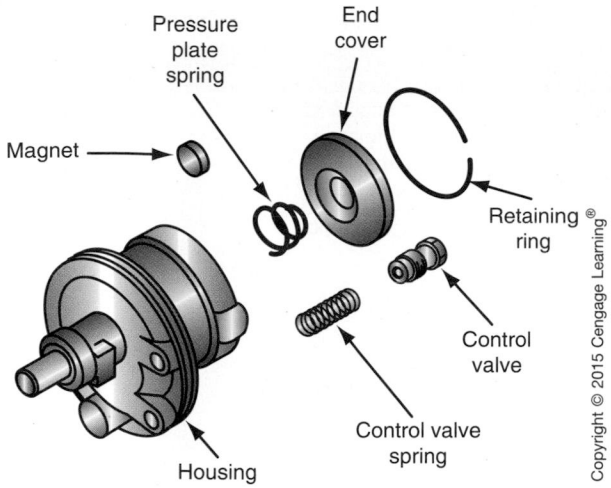

FIGURE 12-33 **End cover and related component service.**

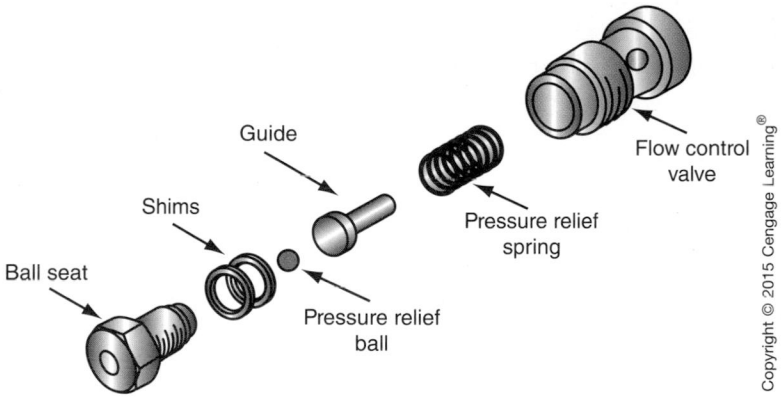

FIGURE 12-34 **Pressure relief valve removal.**

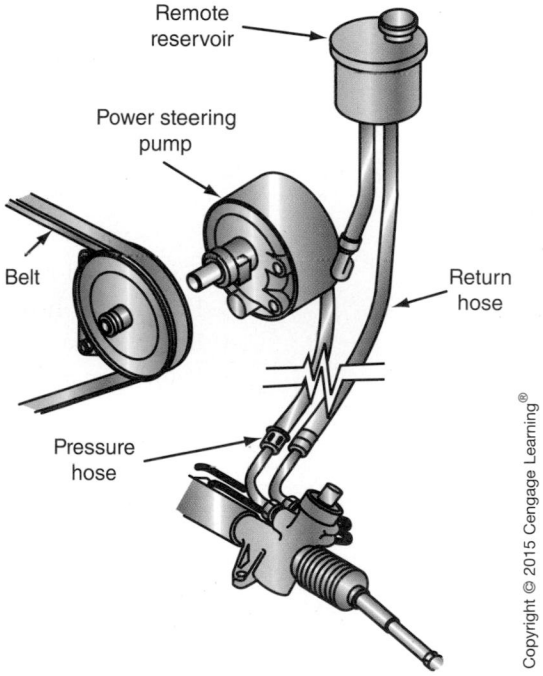

FIGURE 12-35 **Power steering hoses and lines.**

Power Steering Hose Replacement

WARNING: If the engine has been running, the power steering hoses, components, and fluid may be extremely hot. Wear protective gloves and use caution to avoid burns.

When power steering hose replacement is required, follow these steps:

1. With the engine stopped, remove the return hose at the power steering gear, and allow the fluid to drain from this hose into a drain pan. On some cars, the vehicle must be lifted on a hoist to replace the power steering hoses.

TABLE 12-1 **STEERING COLUMN AND STEERING LINKAGE DIAGNOSIS**

Problem	Symptoms	Possible Causes
Underhood noise	Squealing noise during engine acceleration or during steering wheel rotation	Loose power steering pump belt
	Squealing and chirping noise with the engine idling	Dry power steering pump belt friction surfaces
	Growling noise with engine idling	Low power steering fluid level
		Air in the power steering fluid
Improper steering wheel turning effort	Erratic steering wheel turning effort while cornering	Loose power steering pump belt
		Low power steering fluid level
		Air in the power steering fluid
	Excessive steering wheel turning effort	Loose power steering pump drive belt
		Low power steering fluid level
		Low power steering pump pressure
Power steering pump fluid leaks	Fluid leaks at the drive shaft behind the pulley	Worn drive shaft seal
	Fluid leaks between the reservoir and the housing	Leaking reservoir O-ring
	Fluid leaks at the back of the reservoir	Leaking O-ring seals or pump-to-line fittings
Rapid power steering belt wear	Power steering belt wears out prematurely	Misalignment of power steering pump pulley, and other pulleys driven by the same belt
		Excessive power steering belt tension

Copyright © 2015 Cengage Learning®

2. Loosen and remove all hose fittings from the pump and steering gear.
3. Remove all hose-to-chassis clips.
4. Remove the hoses from the chassis and cap the pump and steering gear fittings.
5. If O-rings are used on the hose ends, install new O-rings.
6. Reverse steps 1 through 4 to install the power steering hoses. Tighten all fittings to the manufacturer's specified torque. Be sure all hose-to-chassis clips are in place. Do not position hoses where they rub on other components.
7. Fill the pump reservoir to the full mark with the manufacturer's recommended fluid. Bleed air from the power steering system, as mentioned previously in this chapter. Check the fluid level in the reservoir and add fluid as required.

Refer to Table 12-1 for steering column and steering linkage diagnosis.

Hybrid Electric Vehicle (HEV) and Electrohydraulic Power Steering (EHPS) Diagnosing and Servicing Procedures

HEV Service Precautions

Most HEVs have high-voltage electrical systems that pose a severe electrical shock hazard for anyone servicing these vehicles. When servicing HEV electrical systems, always follow these precautions:

1. Wear high-voltage electrical lineman's gloves rated at 1,000V. Inspect these gloves periodically to be sure they do not contain pin holes or other damage. Immediately replace all damaged gloves.
2. When working on an HEV, always be sure the HEV system is deactivated by removing the electronic or conventional key from the instrument panel. This action prevents the engine from starting.

3. Turn the high-voltage circuit switch off to isolate the battery pack from the electrical system (Figure 12-36). Some HEV manufacturers recommend removing the 20-A yellow HEV fuse in the underhood fuse block (Figure 12-37).
4. Disconnect the negative cable on the 12-V battery (Figure 12-38).
5. After the disabling procedure on some HEVs, power is maintained in the SRS system for 90 seconds and in the high-voltage system for 5 minutes. After disabling the electrical system, always wait for the time period specified by the vehicle manufacturer.
6. Be sure the high- and low-voltage systems have been disabled before using welding equipment on the vehicle. If the HEV system is not disabled, the welding equipment may send high-voltage spikes through the system resulting in component damage.
7. If the battery pack has experienced collision or other damage and battery pack leakage is suspected, wear protective gloves and safety goggles when servicing the battery pack. Nickel–metal hydride batteries contain potassium hydroxide a highly alkaline solution that damages human tissue. Battery pack leakage may be sprayed with vinegar to neutralize the solution.
8. Inspect high-voltage (orange) cables for frayed insulation and damage, which could result in electrical shocks, sparks, and fires. Frayed, damaged orange cables must be replaced.

Classroom Manual
Chapter 12, page 334

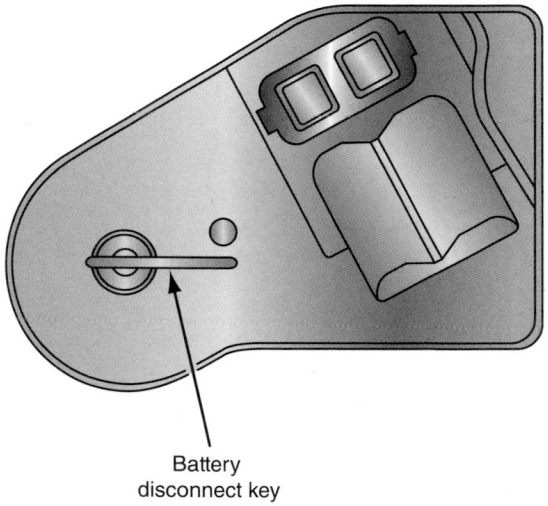

Copyright © 2015 Cengage Learning®

FIGURE 12-36 **Battery disconnect key in the ESB.**

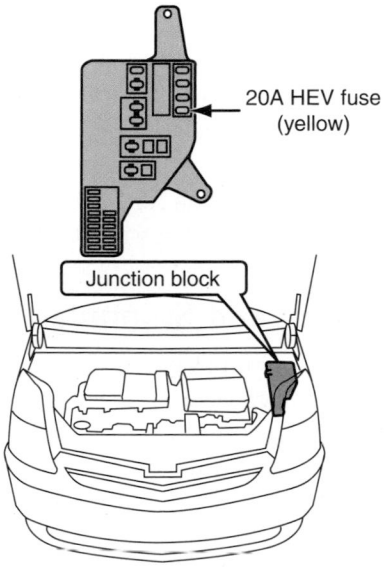

Copyright © 2015 Cengage Learning®

FIGURE 12-37 **Removing HEV fuse from fuse block.**

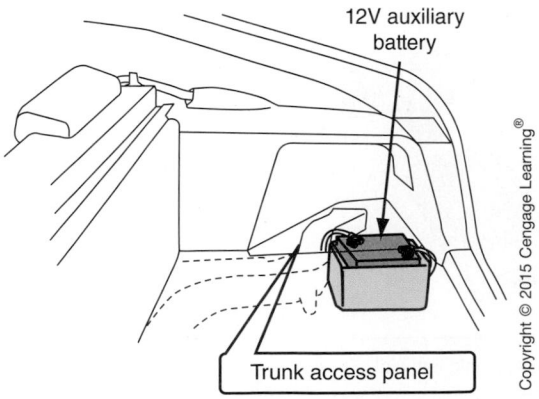

Copyright © 2015 Cengage Learning®

FIGURE 12-38 **Disconnecting the 12-V battery.**

9. If HEVs are towed with the drive wheels on the road surface, the generator may begin producing voltage and current even if the high-voltage electrical system is disabled. This may result in sparks and a fire.
10. The battery packs in many HEVs contain nickel–metal hydride type batteries. These batteries self-discharge quickly compared with a conventional lead acid battery. A nickel–metal hydride battery may lose 30 percent of its charge per month. Do not leave an HEV sitting for two or three months without starting the engine. If an HEV is not being driven, the engine should be started and allowed to run for 30 minutes every two or three weeks to maintain the battery pack state of charge. Some test equipment manufacturers are designing battery chargers for HEV battery packs, but this test equipment is not available at the present time. The engine must be operated to allow the generator to recharge the battery pack. Some HEVs have an auxiliary starter to start the engine if the battery pack is discharged.
11. When working on an HEV with a smart key, always be sure the smart key is removed from the instrument panel, and the disable button under the steering column is pushed to the Off position. This action prevents the HEV system from being activated by pressing the smart key when it is located in close proximity to the vehicle.
12. Use caution when hoisting or lifting HEVs to avoid damage to the high-voltage cable compartments and cables.
13. On HEVs the refrigerant oil specified by the vehicle manufacturer such as SE 10-Y, ND-OIL, 11 must be used in the AC system. Many HEVs have an electric-drive AC compressor that remains operational during the engine stop mode to maintain AC system operation. In many of these compressors, 300V from the inverter is supplied to the compressor drive motor. This higher voltage requires lower amperes of current through the compressor clutch coil. The refrigerant oil in these compressors must have special insulating qualities or the high-voltage insulation inside the compressor may be damaged.

Starting and Stopping Procedures and Warning Indicators

Technicians must be familiar with the proper starting and stopping procedures and warning indicators on HEVs. The engine starting and stopping procedures and warning indicators may vary on hybrid vehicles depending on the vehicle and model year. Always refer to the vehicle manufacturers' information. The following procedures and indicators are typical. Some hybrid vehicles have an electronic key that fits into a key slot in the dash. A power button and an indicator light are located above the electronic key slot in the dash (Figure 12-39). The power button is pressed to cycle through the ignition modes rather than rotating the electronic key in the slot. The electronic key also contains the remote keyless entry buttons and a hidden metal key (Figure 12-40).

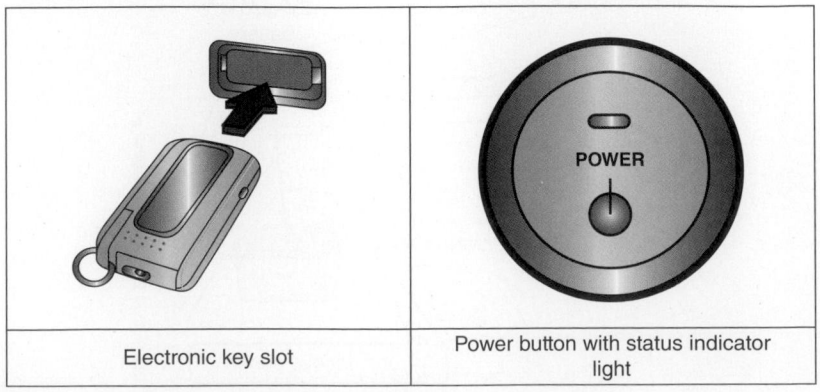

Copyright © 2015 Cengage Learning®

FIGURE 12-39 **Electronic key and power button.**

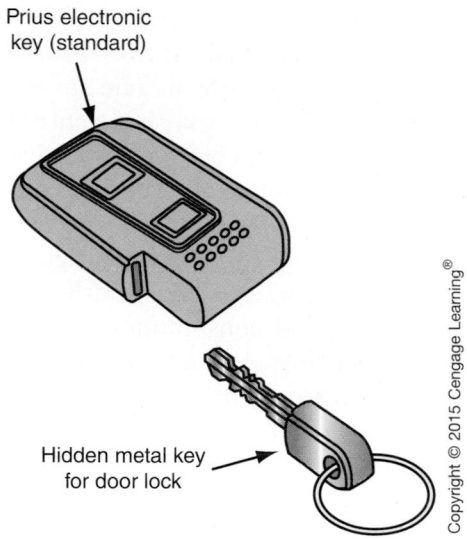

Copyright © 2015 Cengage Learning®

FIGURE 12-40 **Electronic key with remote keyless entry button and hidden metal key.**

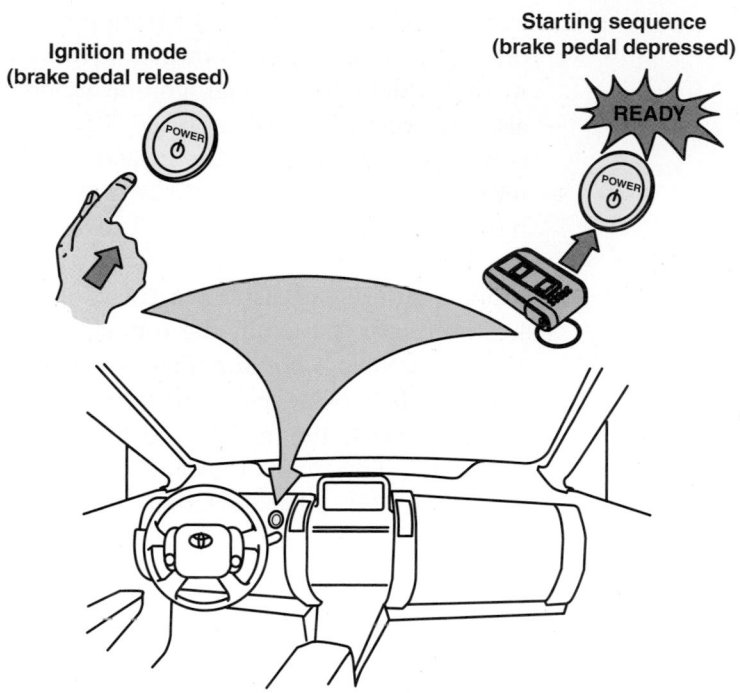

Copyright © 2015 Cengage Learning®

FIGURE 12-41 **Ignition modes and starting sequence.**

The doors are unlocked and locked by pressing the appropriate buttons on the electronic key. Another option for opening the doors is to remove the hidden metal key from the electronic key, and insert the metal key into the key cylinder in the drivers' door. Rotating the metal key once unlocks the drivers' door, and rotating this key a second time unlocks the other doors. There are no lock cylinders in the other doors.

To activate the HEV system and other electronic systems, insert the electronic key into the slot in the dash. With the brake pedal released, pressing the power button once turns on the accessory mode, and pushing the power button again turns on the ignition mode. Pressing the power button a third time turns the ignition off. When the ignition mode is off, the power button indicator light is also off. In the accessory mode the power button indicator light is green, and in the ignition mode this light is amber. When a defect occurs in the electrical system, the amber power button light starts blinking.

With the brake pedal depressed, pressing the power button once activates the HEV and other electronic systems. Starting the vehicle takes priority over all other modes. When the HEV system is activated, the power button indicator light goes off and the Ready light in the instrument panel is illuminated (Figure 12-41). When starting off under light load conditions, the engine does not start because the vehicle is only driven by the propulsion motor. Never work on an HEV with the electronic key installed in the dash and the HEV system activated. Under this condition, the engine may start and cause personal injury and/or vehicle component damage if the battery pack voltage reaches a specific discharged state. To shut off the vehicle, bring the vehicle to a complete stop and press the power button once and remove the electronic key. When the Ready light is illuminated, the electronic key cannot be removed from the key slot.

The vehicle may have an optional smart entry system with a smart electronic key. This key contains a transceiver that communicates bidirectionally and allows the vehicle to recognize the smart key in close proximity to the vehicle. The smart entry and smart key can unlock and lock the doors without pushing the buttons on the smart key. This system can also start the hybrid system without inserting the smart key into the dash slot. If the smart key is in close proximity to the vehicle, the doors may be unlocked by touching the sensor on the back of either exterior front door handle (Figure 12-42). Press the black lock button on either exterior front door handle to lock the doors. If the smart key is in close proximity to the vehicle, the normal ignition modes and start mode are operational. The smart key also contains a hidden metal key. Vehicles equipped with a smart key and smart entry system have a disabling button located in the dash under the steering column. If the disable button is pressed, the smart key must be inserted into the key slot to activate the ignition modes and start the vehicle (Figure 12-43).

The speedometer, gear shift indicator, Ready light, fuel gauge, and warning lights are located in a digital display in the center of the dash near the base of the windshield. An LCD monitor located at the top, center of the dash displays fuel consumption, A/C controls, and energy monitor (Figure 12-44). The electronic gearshift selector is a momentary select, shift-by-wire system that engages the transaxle in Reverse, Neutral, Drive, or engine Brake modes. There are no mechanical connections between the dash indictors and the transaxle. The dash indicator also informs the driver regarding the gear selected (Figure 12-45). After selecting the desired gear position, the transaxle remains in that position as indicated by the dash display, but the shift selector returns to a default mode.

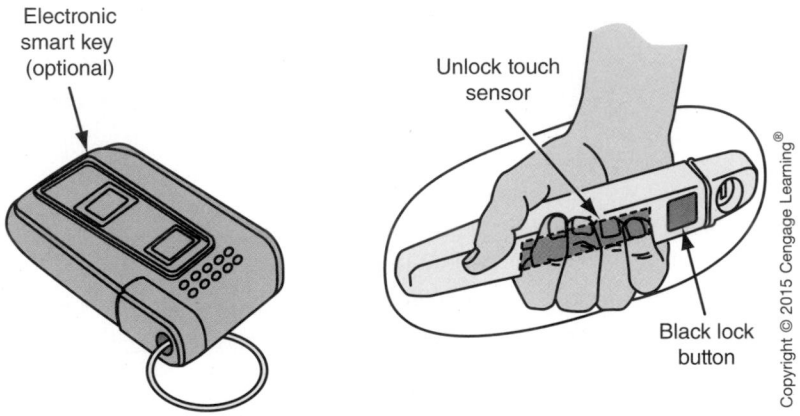

FIGURE 12-42 **Smart key and smart entry system.**

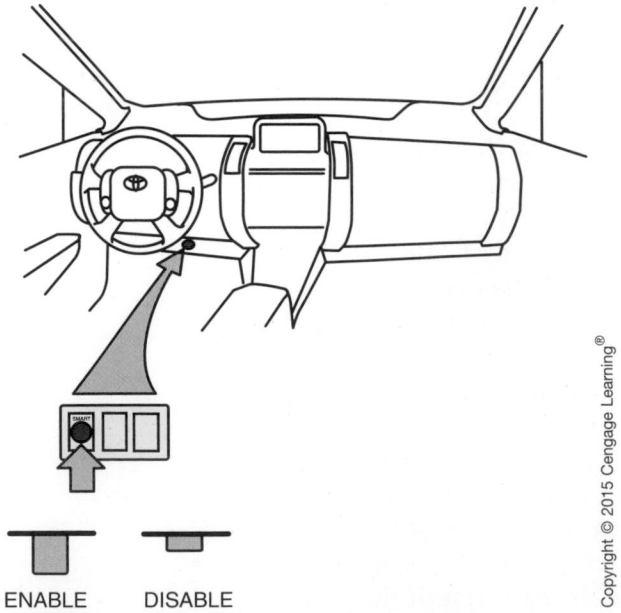

FIGURE 12-43 **Disable button on smart key systems.**

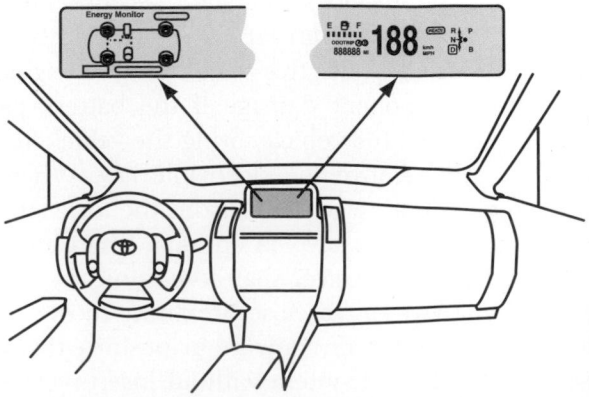

FIGURE 12-44 **Digital dash displays.**

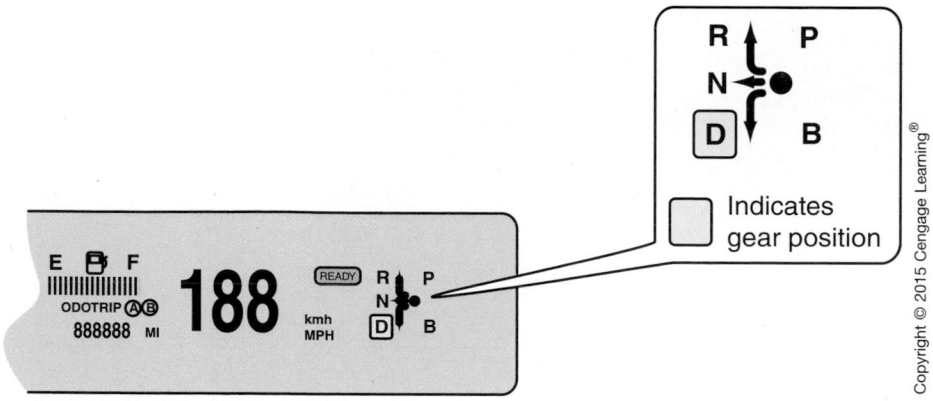

FIGURE 12-45 **Gearshift indicator.**

When the vehicle is stopped, Park may be selected by pressing the P switch in the dash display. If the power button is pressed to shut off the vehicle, Park is automatically engaged. Voltage is supplied to the gearshift selector and the electromechanical parking pawl by the 12-V battery. If the 12-V battery is discharged, or disconnected, the vehicle cannot be started or shifted out of park.

When the Ready light is illuminated, the vehicle is operational even though the engine is not running. The driver simply selects Drive (D) or Reverse (R) on the gearshift selector, releases the brake pedal, and steps on the accelerator pedal to begin driving the vehicle. The vehicle computer decides when to start and stop the engine depending on many factors such as vehicle load and speed, and battery pack state of charge. A hybrid vehicle does not have the usual noise associated with starting motor operation, because the battery simply passes current through the generator windings to rotate and start the engine. The vehicle computer stops the engine during deceleration below a specific speed, and during idle operation, to improve fuel economy.

Scan Tool Diagnosis of HEV and EHPS Systems

To begin the HEV system diagnosis, visually inspect all HEV system components, wiring harnesses, and connectors. Any damaged components, wiring harnesses, or connectors should be replaced. The orange high-voltage cables may be checked for high-voltage leakage by using a digital volt-ohm meter (DVOM). With the DVOM on the highest DC voltage scale, connect the negative DVOM lead to ground on the vehicle chassis, and move the red lead along the high-voltage cable surfaces. High-voltage leakage causes a reading on the voltmeter. A scan tool may be used to diagnose the HEV system. HEV system diagnosis varies depending on the vehicle make and model year. Always use the vehicle manufacturer's specific diagnostic information for the vehicle being diagnosed and serviced. The following is a typical HEV diagnosis. The scan tool and the module installed in the scan tool must be compatible with the HEV system and the vehicle network(s). A special CANdi module must be used with the scan tool on some HEV systems. If the visual inspection does not indicate any HEV system problems, the scan tool may be used to perform a diagnostic system check. Follow these steps to complete the diagnostic system check:

1. Connect the scan tool to the DLC under the dash. If the scan tool does not power up, check the 12-V supply terminal and the ground terminal in the DLC.
2. Turn the ignition switch on with the engine off. Using the scan tool, attempt to establish communication with the energy storage control module (ESCM), hybrid control module (HCM), and the starter and generator control module (SGCM). If the scan tool does not communicate with any of these modules, there is a problem in the LAN

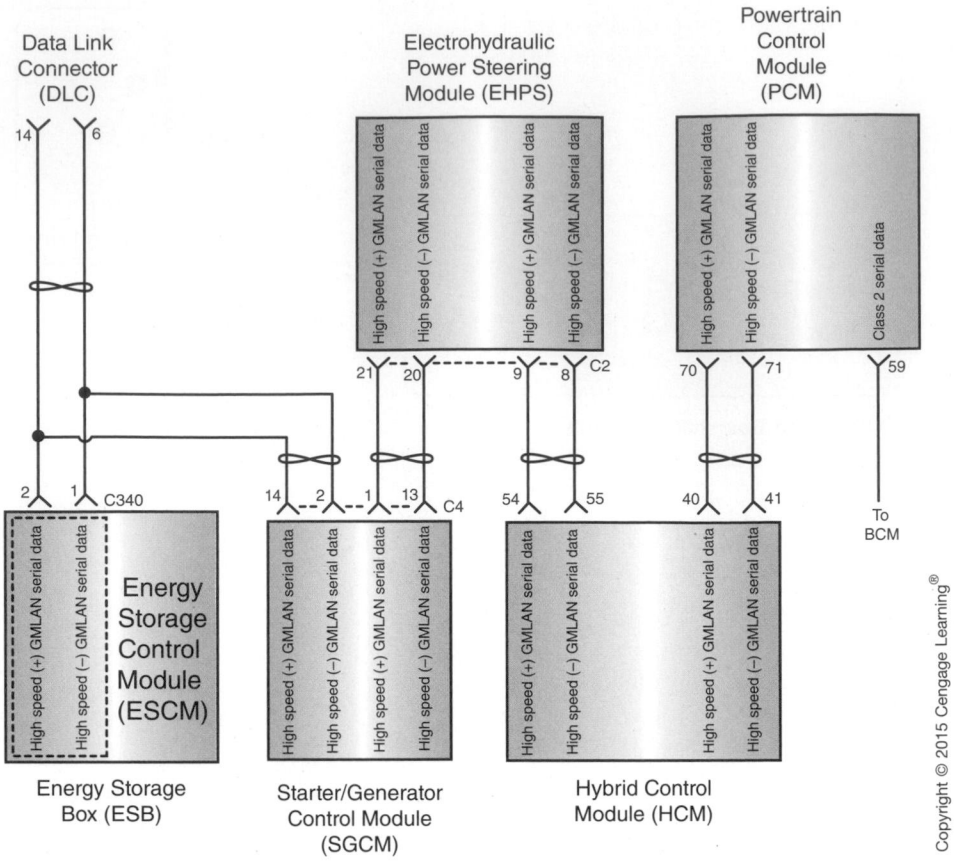

FIGURE 12-46 **The LAN bus interconnects the SGCM, PCM, EHPS modules, ESB, and DLC.**

network and further network diagnosis is required (Figure 12-46). Refer to Chapter 8 in the Shop Manual for network diagnosis. Always follow the vehicle manufacturer's specific diagnostic procedure for the vehicle being diagnosed.

3. Use the scan tool to establish communication with the PCM. If the scan tool does not communicate with the PCM, there is a problem in the Class 2 network and further diagnosis of this network is necessary.
4. Select PCM DTC display on the scan tool. DTCs with a U prefix indicate network problems, and specific DTC diagnosis is required.
5. Check the scan tool display for PCM DTCs. When DTCs are present, the cause of the DTCs must be diagnosed and repaired.

Diagnostic Example for a U0131 Network DTC

Current vehicles have hundreds of DTCs, and thus it is impossible to include an interpretation and diagnostic procedure for all DTCs in this publication. A U0131 DTC is set when the communication is lost between the EHPS module and the PCM. Under this condition, the EHPS operates in a default mode. The possible causes of this DTC are as follows:

1. An open circuit in the EHPS module connector.
2. An open circuit in the LAN network.
3. An open circuit in the voltage supply or ground circuits to the EHPS module.
4. An internal module malfunction. Many modules have specific DTCs for internal module defects.

To locate the root cause of the U0131 DTC, the technician must perform tests with a DVOM or lab scope on the suspected components.

Diagnostic Example for a P0562 DTC

A P0562 DTC is set when the PCM detects an improper voltage below 11V for 5 seconds. When this DTC is set in the PCM memory, the PCM takes these actions:

1. The PCM stores faulty circuit conditions that were present when the DTC set.
2. The PCM disables many outputs.
3. The transmission defaults to a predetermined gear.
4. Torque converter clutch lockup is inhibited.
5. The instrument panel cluster (IPC) displays a warning message.
6. The malfunction indicator light (MIL) will not illuminate.

When this DTC is present in the PCM memory, the technician must use an amp-volt tester to test the charging circuit and use a DVOM to test all the electrical circuits between the charging circuit, battery, and PCM to locate the root cause of the problem. DTCs may be erased with the scan tool.

A scan tool data display may be useful when diagnosing HEV and EHPS systems especially when there is a system problem with no DTCs. Before using the data display, the diagnostic system check should be performed and all DTCs corrected. Available scan tool data from a HEV with EHPS is illustrated in Figure 12-47. Photo Sequence 24 illustrates the safety precautions that must be observed when working on hybrid vehicles.

Scan Tool Parameter	Data List	Units Displayed	Typical Data Value
Park brake set with the ignition switch ON/Ignition Switch in RUN/Automatic Transmission in PARK/Air Conditioning is OFF			
Brake Pedal Travel	Data	mm	−50 to 200 mm
BPP Sensor Rate	Data	ms	0 to +10%/25ms
BPP Sensor Input Volts	Data	Volts	0 to 5 Volts
BPP Sensor Output Volts	Data	Volts	0 to 5 Volts
EHPS 42 V Bus Voltage	Data	Volts	0 to 76 Volts
EHPS Ignition 0	Data	OFF/ON	ON
EHPS Ignition 1	Data	OFF/ON	ON
EHPS Demand	Data	%	−100 to +100%
EHPS Demand Speed	Data	RPM	0 to +7500 RPM
EHPS ECU Temperature	Data	°C	−180 to +180°C (−238 to +302°F)
EHPS Manifold Temperature	Data	°C	−180 to +180°C (−238 to +302°F)
EHPS Motor Current	Data	A	0 to +200A
EHPS Motor RPM	Data	RPM	0 to +7500 RPM
Steering Wheel Rate	Data	°/s	0 to 2048°/s
Steering Wheel Sensor Phase A Volts	Data	Low/High	Low
Steering Wheel Sensor Phase B Volts	Data	Low/High	High
Vehicle Speed	Data	km/h	0

Copyright © 2015 Cengage Learning®

FIGURE 12-47 **EHPS data.**

PHOTO SEQUENCE 24

Observing Safety Precautions When Servicing Hybrid Electric Vehicles (HEVs)

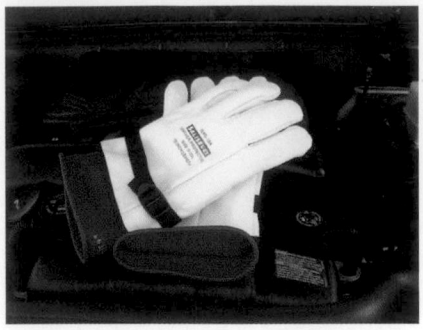

P24-1 **Always wear high-voltage lineman's gloves when servicing HEV electrical systems.**

Copyright © 2015 Cengage Learning®

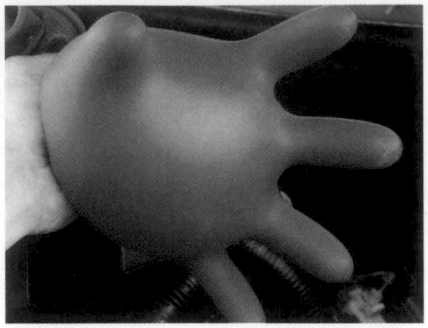

P24-2 **Periodically inspect the high-voltage gloves for pin holes or damage.**

Copyright © 2015 Cengage Learning®

P24-3 **Shut off the high-voltage circuit switch to isolate the battery pack from the electrical system.**

Copyright © 2015 Cengage Learning®

P24-4 **Disconnect the negative cable on the 12-V battery.**

Copyright © 2015 Cengage Learning®

P24-5 **After the high- and low-voltage systems are disabled, wait for the time specified by the vehicle manufacturer to allow these systems to completely power down.**

Copyright © 2015 Cengage Learning®

P24-6 **Inspect the high-voltage (orange) cables for frayed insulation and damage.**

Copyright © 2015 Cengage Learning®

P24-7 **If the HEV is not driven for a few weeks, the engine should be started and run for 30 minutes every two weeks to maintain the state of charge in the high-voltage battery pack.**

Copyright © 2015 Cengage Learning®

P24-8 **Always remove the electronic or conventional key from the instrument panel to prevent the engine from starting when working on an HEV electrical system.**

Copyright © 2015 Cengage Learning®

CASE STUDY

The owner of a Dodge Magnum with a 5.7 L turbocharged engine requested a price on a power steering gear replacement. The service writer questioned the owner regarding the reason for the steering gear replacement, and discovered that the problem was intermittent hard steering. Further discussion with the owner also revealed that the car had been taken to another automotive service center, and the owner was informed that his or her car required a steering gear replacement. The owner was now looking for a second opinion.

The technician road tested the vehicle and found no indication of hard steering. The owner was asked about the exact driving conditions when the hard steering was experienced, and the answer to this question provided the solution to the problem. The owner indicated that the hard steering was usually experienced while cornering and always on a rainy day. Once this information was revealed, the technician suspected a problem with the power steering belt, because a wet belt slips more easily. A check of the power steering belt indicated the belt was oil soaked from engine oil leaks and was also slightly loose. The power steering belt was replaced and adjusted, and the owner was informed that the engine oil leaks should be corrected or the belt would become oil soaked again.

A short while later, after a rain, the owner brought the car to the shop to have the engine oil leaks repaired. The owner reported that the belt corrected the hard steering problem. In this situation, a brief discussion about the customer's complaint led to accurate diagnosis of the problem, and because of this accurate diagnosis, the shop gained a steady customer.

TERMS TO KNOW

Belt tension gauge
Crocus cloth
Flow control valve
Integral reservoir
Multi-ribbed V-belt
Pressure gauge
Pressure relief valve
Remote reservoir
Serpentine belt
Steering effort
Woodruff key

ASE-STYLE REVIEW QUESTIONS

1. *Technician A* says that serpentine belts made of EPDM resist cracking and instead exhibit wear to the belt ribs similar to tire wear and that a depth gauge should be used to assess belt wear.

   Technician B says that as a Neoprene serpentine belt ages and wears, cracks form on the belt ribs, and if there are more than three cracks in a 3 in. span the belt should be replaced.

   Who is correct?

   A. A only
   B. B only
   C. Both A and B
   D. Neither A nor B

2. While discussing power steering belt adjustment:

   Technician A says that all belt drive components are closely integrated and each depends on the belt and tensioner to keep the entire system at optimum operating performance.

   Technician B says a loose power steering pump drive belt may cause the steering wheel to jerk while turning.

   Who is correct?

   A. A only
   B. B only
   C. Both A and B
   D. Neither A nor B

3. In a hybrid electric vehicle (HEV) electrical system with a electrohydraulic power steering system (EHPS), a diagnostic trouble code (DTC) with U prefix represents a fault in the:

   A. EHPS module.
   B. Powertrain control module (PCM).
   C. Inverter.
   D. Vehicle network.

4. While discussing the steering pump pressure test:

   Technician A says the pressure gauge and valve should be connected in the power steering pump return hose to check pump pressure.

   Technician B says the pressure gauge valve should be closed for 30 seconds during the power steering pump pressure test.

   Who is correct?

   A. A only
   B. B only
   C. Both A and B
   D. Neither A nor B

5. While discussing power steering system draining, flushing, refilling, and bleeding:

   Technician A says the return hose from the remote reservoir should be disconnected at the remote reservoir to drain the fluid.

   Technician B says the return hose should be loosened at the remote reservoir with the engine running to bleed air from the power steering system.

   Who is correct?

   A. A only
   B. B only
   C. Both A and B
   D. Neither A nor B

6. A vehicle experiences excessive, erratic steering effort with a squealing noise. The most likely cause of this problem is:

   A. Underinflated and worn front tires.
   B. Power steering pump drive belt slipping on the pulley.
   C. Worn power steering pump mountings.
   D. Continual low power steering pump pressure.

7. A power steering system has the specified pressure with the pressure gauge valve closed or with the pressure gauge valve open and the engine running at 1,000 and 3,000 rpm. The power steering pump pressure is lower than specified with the front wheels turned fully in either direction. The cause of this problem could be:

   A. A defective steering gear.
   B. A stuck flow control valve.
   C. A stuck pressure relief valve.
   D. Worn pump vanes and cam ring.

8. All of these statements about power steering pump pulley replacement are true EXCEPT:

   A. If the pulley is retained with a nut, a woodruff key prevents pulley rotation on the shaft.
   B. If the pulley is pressed onto the shaft, a special puller must be used to remove the pulley.
   C. If the pulley is pressed onto the shaft, a soft hammer may be used to drive the pulley onto the shaft when reinstalling the pulley.
   D. If the pulley is misaligned with the engine running, the power steering pump mountings may be worn.

9. A power steering pump has the specified pressure with the pressure gauge valve closed. However, with this valve open, the pump does not have the specified pressure difference between the pressure readings at 1,000 and 3,000 rpm. The cause of this problem could be:

   A. A sticking pressure relief valve.
   B. A sticking flow control valve.
   C. A slipping pump belt.
   D. A misaligned pump pulley.

10. All of these statements about HEV service precautions are true EXCEPT:

    A. Open the high-voltage switch to isolate the high-voltage battery pack.
    B. High-voltage gloves must be worn when servicing HEV electrical systems.
    C. If the vehicle is not being driven, the high-voltage battery pack must be vented manually.
    D. Remove the electronic or conventional key from the instrument panel.

ASE CHALLENGE QUESTIONS

1. A vehicle has intermittent excessive steering effort.

   Technician A says inspect the fluid because it may have air in it.

   Technician B says inspect the belt because it may be slipping.

   Who is correct?

   A. A only
   B. B only
   C. Both A and B
   D. Neither A nor B

2. While discussing power steering pump leaks:

   Technician A says leaks at the drive shaft seal will leave a wet and oily pump pulley back.

   Technician B says leaks at the drive shaft seal may leave oil on the hood pad above the pump.

   Who is correct?

   A. A only
   B. B only
   C. Both A and B
   D. Neither A nor B

3. A vehicle has continual excessive steering effort, but there is no noise and the fluid level in the reservoir is correct.

   Technician A says the cause of the problem could be a stuck flow control valve.

   Technician B says the cause of the problem could be air in the power steering fluid.

   Who is correct?

   A. A only
   B. B only
   C. Both A and B
   D. Neither A nor B

4. There is air in the power steering fluid. To correct this problem, you should:
   A. With the engine running, crack the pressure line at the pump to release trapped air.
   B. With the engine running, turn the steering wheel full left and right.
   C. With the engine running, drain all fluid from the system and refill.
   D. With the engine stopped, raise wheels and remove the return hose from the gear. Turn the steering wheel full left and right. Refill the reservoir and replace the return hose.

5. All of the following are true EXCEPT:
   A. A loose belt may cause the steering wheel to jerk.
   B. If V-belt edges are worn, the belt may be stretched or the wrong width.
   C. An oil-soaked drive belt may slip only in wet weather.
   D. When air is in the fluid, the fluid will never reach operating temperature.

JOB SHEET 45

Name _____ **Date** _____

Draining and Flushing Power Steering System

Upon completion of this job sheet, you should be able to drain and flush a power steering system.

NATEF AST/MAST Task Correlation:

Suspension and Steering. Steering Systems Diagnosis and Repair,
Task #B-9. Determine proper power steering fluid type; inspect fluid level and condition. (P-1)
Task #B-10. Flush, fill, and bleed power steering system. (P-2)

Tools and Materials

Drain pan
Specified type of power steering fluid
Floor jack
Safety stands

Describe the Vehicle Being Worked On:

Year _____ Make _____ Model _____
VIN _____ Engine type and size _____

Procedure _____ Task Completed

1. Describe the steering problems and noises that occur when there is air in the power steering system.

   _____
   _____
   _____

2. Lift the front of the vehicle with a floor jack and install safety stands under the suspension. Lower the vehicle onto the safety stands, and remove the floor jack. ☐

3. Remove the return hose from the remote reservoir that is connected to the steering gear. Place a plug on the reservoir outlet and position the return hose in an empty drain pan. ☐

4. With the engine idling, turn the steering wheel fully in each direction and stop the engine. ☐

5. Fill the reservoir to the hot full mark with the manufacturer's recommended fluid.
   Specified type of power steering fluid _____
   Power steering reservoir filled to the specified level with the proper power steering fluid? ☐ Yes ☐ No
   Instructor check _____

6. Start the engine, and run the engine at 1,000 rpm while observing the return hose in the drain pan. When fluid begins to discharge from the return hose, shut the engine off. ☐

Task Completed

7. Repeat steps 4, 5, and 6 until there is no air in the fluid discharging from the return hose.
   Have steps 4, 5, and 6 been repeated? ☐ Yes ☐ No
   Instructor check _____

8. Remove the plug from the reservoir and reconnect the return hose.
   Is the return hose connected and tightened? ☐ Yes ☐ No
   Instructor check _____

☐ 9. Fill the power steering pump reservoir to the specified level.

10. With the engine running at 1,000 rpm, turn the steering wheel fully in each direction three or four times. Each time the steering wheel is turned fully to the right or left, hold it there for 2 to 3 seconds before turning it in the other direction.

    Number of times the steering wheel was turned fully in each direction _____

11. Check for foaming of the fluid in the reservoir. When foaming is present, repeat steps 8 and 9.
    Is foaming present in the power steering pump reservoir? ☐ Yes ☐ No
    Have steps 8 and 9 been repeated? ☐ Yes ☐ No
    Instructor check _____

☐ 12. Check the fluid level and be sure it is at the hot full mark.

☐ 13. Raise vehicle with a floor jack, remove safety stands, and lower vehicle.

14. Explain how you know that all the air has been bled from the power steering system.

_____

_____

_____

_____

Instructor's Response _____

_____

_____

JOB SHEET 46

Name _____ **Date** _____

Testing Power Steering Pump Pressure

Upon completion of this job sheet, you should be able to test power steering pump pressure.

NATEF AST/MAST Task Correlation:

Suspension and Steering. Steering Systems Diagnosis and Repair,

Task #B-5. Diagnose power steering gear (rack and pinion) binding, uneven turning effort, looseness, hard steering, and noise concerns; determine necessary action. (P-2)

Tools and Materials

Specified type of power steering fluid
Power steering pressure test gauge
Thermometer

Describe the Vehicle Being Worked On:

Year _____ Make _____ Model _____

VIN _____ Engine type and size _____

Procedure — Task Completed

1. With the engine stopped, disconnect the pressure line from the power steering pump. ☐
2. Connect the gauge side of the pressure gauge to the pump outlet fitting. Connect the valve side of the gauge to the pressure line. ☐
   Is the power steering pressure gauge properly connected? ☐ Yes ☐ No
   Are all power steering gauge fittings tightened to the specified torque?
   ☐ Yes ☐ No
3. Start the engine and turn the steering wheel fully in each direction two or three times to bleed air from the system. ☐
4. Install a thermometer in the pump reservoir fluid to measure the fluid temperature. Be sure the fluid level is correct and the fluid temperature is at least 176°F (80°C).
   Is the power steering fluid at the specified level? ☐ Yes ☐ No
   Is the power steering fluid at the proper temperature? ☐ Yes ☐ No
   Instructor check _____

WARNING: During the power steering pump pressure test with the engine idling, close the pressure gauge valve for no more than 10 seconds because excessive pump pressure may cause power steering hoses to rupture, resulting in personal injury.

WARNING: Do not allow the fluid to become too hot during the power steering pump pressure test. Excessively high fluid temperature reduces

pump pressure. Wear protective gloves, and always shut the engine off before disconnecting gauge fittings, because the hot fluid may cause burns.

5. With the engine idling, close the pressure gauge valve for no more than 10 seconds, and observe the pressure gauge reading. Turn the pressure gauge valve to the fully open position. If the pressure gauge reading does not equal the vehicle manufacturer's specifications, repair or replace the power steering pump.

   Specified power steering pump pressure with pressure gauge valve closed _____

   Actual power steering pump pressure with the pressure gauge valve closed _____

   Recommended power steering pump service.

   _____

6. Check the power steering pump pressure with the engine running at 1,000 rpm and 3,000 rpm. Record the pressure difference between the two readings.

   Specified power steering pump pressure at 1,000 rpm _____

   Specified power steering pump pressure at 3,000 rpm _____

   Actual power steering pump pressure at 1,000 rpm _____

   Actual power steering pump pressure at 3,000 rpm _____

   Recommended power steering pump service. _____

   _____

7. With the engine running, turn the steering wheel fully in one direction. Observe the steering pump pressure while holding the steering wheel in this position.

   Specified power steering pump pressure with the steering wheel turned fully in one direction _____

   Actual power steering pump pressure with the steering wheel turned fully in one direction _____

   Recommended power steering service.

   _____

8. Be sure the front tire pressure is correct, and center the steering wheel with the engine idling. Connect a spring scale to the steering wheel, and measure the steering effort in both directions.

   Specified force required to turn the steering wheel to the right _____

   Specified force required to turn the steering wheel to the left _____

   Actual force required to turn the steering wheel to the right _____

   Actual force required to turn the steering wheel to the left _____

   List all the required power steering service and explain the reasons why this service is necessary.

_____

_____

_____

Instructor's Response _____

_____

_____

JOB SHEET 47

Name _____ **Date** _____

Measure and Adjust Power Steering Belt Tension and Alignment

Upon completion of this job sheet, you should be able to measure and adjust power steering belt tension and alignment.

NATEF AST/MAST Task Correlation:

Suspension and Steering. Steering Systems Diagnosis and Repair,
Task #B-12. Remove, inspect, replace, and adjust power steering pump drive belt. (P-1)

Tools and Materials

Belt tension gauge
Pry bar

Describe the Vehicle Being Worked On:

Year _____ Make _____ Model _____
VIN _____ Engine type and size _____

Procedure — Task Completed

1. Describe the power steering problems and noises that occur when the power steering pump belt is loose.

   _____
   _____
   _____

2. Inspect the power steering belt for fraying, oil soaking, wear on friction surfaces, cracks, glazing, and splits. ☐
   Belt condition: ☐ Satisfactory ☐ Unsatisfactory

3. With the engine stopped, press on the belt at the longest belt span to measure the belt deflection, which should be 1/2 inch per foot of free span. ☐
   Length of belt span where belt deflection is measured _____
   Amount of belt deflection _____
   Belt tension: ☐ Satisfactory ☐ Unsatisfactory

4. Install a belt tension gauge over the belt in the center of the longest span to measure the belt tension.
   Specified belt tension _____
   Actual belt tension _____

5. Loosen the power steering pump bracket or tension adjusting bolt. ☐

6. Loosen the power steering pump mounting bolts. ☐

7. Remove the power steering belt. ☐

Task Completed

☐ **8.** Check the bracket and pump mounting bolts for wear.

Power steering pump mounting bolt holes and bolt condition:

☐ Satisfactory ☐ Unsatisfactory

Power steering pump bracket and bracket bolt condition:

☐ Satisfactory ☐ Unsatisfactory

List the required power steering pump and bracket service and explain the reasons for your diagnosis.

_____

_____

_____

_____

☐ **9.** Install the new power steering pump belt over all the pulleys.

☐ **10.** Pry against the pump ear and hub with a pry bar to tighten the belt. Some pump brackets have a 1/2-inch square opening in which a breaker bar may be installed to move the pump and tighten the belt.

☐ **11.** Hold the pump in the position described in step 9 and tighten the bracket or tension adjusting bolt.

12. Recheck the belt tension with the tension gauge.

Specified power steering belt tension _____

Actual power steering pump belt tension _____

13. Tighten the tension adjusting bolt and the mounting bolts to the manufacturer's specified torque.

Specified power steering pump mounting bolt torque _____

Actual power steering pump mounting bolt torque _____

Specified power steering pump bracket or tension adjusting bolt torque _____

Actual power steering pump bracket or tension adjusting bolt torque _____

14. Check alignment of the power steering pump pulley in relation to the other pulleys surrounded by the power steering belt.

Power steering pump pulley alignment: ☐ Satisfactory ☐ Unsatisfactory

Explain the service required to align the power steering pump pulley with other related pulleys and give the reasons for your diagnosis.

_____

_____

Instructor's Response _____

_____

_____

Chapter 13

Rack and Pinion Steering Gear and Four Wheel Steering Diagnosis and Service

BASIC TOOLS

Basic technician's tool set
Service manual
Floor jack
Safety stands
Machinist's rule

Upon completion and review of this chapter, you should be able to:

- Perform a manual or power rack and pinion steering gear inspection.
- Remove and replace manual or power rack and pinion steering gears.
- Disassemble, inspect, repair, and reassemble manual rack and pinion steering gears.
- Adjust manual rack and pinion steering gears.
- Diagnose manual rack and pinion steering systems.
- Diagnose oil leaks in power rack and pinion steering gears.
- Disassemble, inspect, and repair power rack and pinion steering gears.
- Adjust power rack and pinion steering gears.
- Diagnose Magnasteer.
- Diagnose column-driven electronic power steering systems.
- Diagnose active steering system.
- Perform a preliminary inspection on a four-wheel steering (4WS) system.
- Diagnose Quadrasteer.
- Perform rear wheel steering and rear wheel alignment procedure.
- Diagnose four-wheel active steering systems (4WAS).

Proper rack and pinion steering gear operation is essential to maintain vehicle safety and reduce driver fatigue. Such steering gear conditions as looseness and excessive steering effort may contribute to a loss of steering control, resulting in a vehicle collision. Worn steering gear mountings may cause improper wheel alignment and **bump steer**. Improper wheel alignment increases tire tread wear, and bump steer may increase driver fatigue. Excessive steering gear looseness or high steering effort also contributes to driver fatigue. Therefore, in the interest of vehicle safety and driver alertness, rack and pinion steering gear diagnosis and service are extremely important.

Bump steer occurs when one of the front wheels strikes a road irregularity while driving straight ahead, and the steering suddenly veers to the right or left.

Manual or Power Rack and Pinion Steering Gear On-Car Inspection

The wear points are reduced to four in a rack and pinion steering gear. These wear points are the inner and outer tie rod ends on both sides of the rack and pinion assembly (Figure 13-1).

The first step in manual or power rack and pinion steering gear diagnosis is a very thorough inspection of the complete steering system. During this inspection, all steering

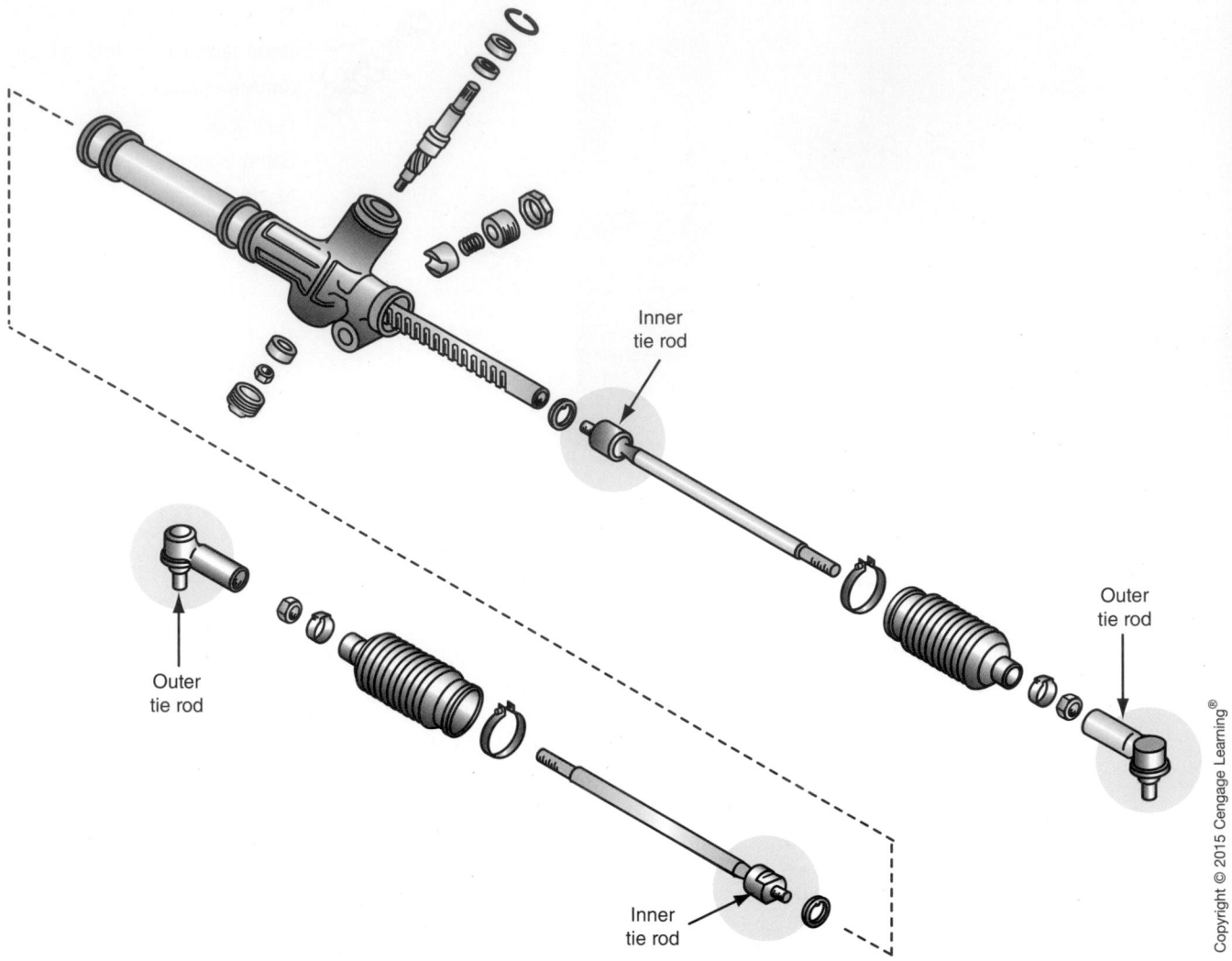

FIGURE 13-1 **Wear points at the inner and outer tie rod ends in a rack and pinion steering gear.**

Bellows boots are clamped over each end of the steering gear housing to keep dirt out of the rack.

system components such as the inner and outer tie rod ends, **bellows boots**, mounting bushings, couplings or universal joints, ball joints, tires, and **steering wheel free play** must be checked.

Follow these steps for manual or power rack and pinion steering gear inspection:

Steering wheel free play is the distance that the steering wheel moves before the front wheels begin to turn when the steering wheel is rocked back and forth with light finger pressure.

1. With the front wheels straight ahead and the engine stopped, rock the steering wheel gently back and forth with light finger pressure (Figure 13-2). Measure the maximum steering wheel free play. The maximum specified steering wheel free play on some vehicles is 1.18 in. (30 mm). Always refer to the vehicle manufacturer's specifications in the service manual. Excessive steering wheel free play indicates worn steering components.
2. With the vehicle sitting on the shop floor and the front wheels straight ahead, have an assistant turn the steering wheel about 1/4 turn in both directions. Watch for looseness in the flexible coupling or universal joints in the steering shaft. If looseness is observed, replace the coupling or universal joint.
3. While an assistant turns the steering wheel about 1/2 turn in both directions, watch for movement of the steering gear housing in the mounting bushings. If there is any movement of the housing in these bushings, replace the bushings. The steering gear mounting bushings may be deteriorated by oil soaking, heat, or age.

Copyright © 2015 Cengage Learning®

FIGURE 13-2 **Checking steering wheel free play.**

4. Grasp the pinion shaft extending from the steering gear and attempt to move it vertically. If there is steering shaft vertical movement, a pinion bearing preload adjustment may be required. When the steering gear does not have a pinion bearing preload adjustment, replace the necessary steering gear components.
5. Road test the vehicle and check for excessive steering effort. A bent steering rack, tight rack bearing adjustment, or damaged front drive axle joints in a front-wheel-drive car may cause excessive steering effort.
6. Visually inspect the bellows boots for cracks, splits, leaks, and proper clamp installation. Replace any boot that indicates any of these conditions. If the boot clamps are loose or improperly installed, tighten or replace the clamps as necessary. Since the bellows boots protect the inner tie rod ends and the rack from contamination, boot condition is extremely important. Boots should be inspected each time under-car service such as oil and filter change or chassis lubrication is performed.
7. Loosen the inner bellows boot clamps and move each boot toward the outer tie rod end until the inner tie rod end is visible. Push outward and inward on each front tire, and watch for movement in the inner tie rod end. If any movement or looseness is present, replace the inner tie rod end. An alternate method of checking the inner tie rod ends is to squeeze the bellows boots and grasp the inner tie rod end socket. Movement in the inner tie rod end is then felt as the front wheel is moved inward and outward. Hard plastic bellows boots may be found on some applications. With this type of bellows boot, remove the ignition key from the switch to lock the steering column and push inward and outward on the front tire while observing any lateral movement in the tie rod. When lateral movement is observed, replace the inner tie rod end.

Classroom Manual
Chapter 13, Page 348

WARNING: Bent steering components must be replaced. Never straighten steering components, because this action may weaken the metal and result in sudden component failure, serious personal injury, and vehicle damage.

8. Grasp each outer tie rod end and check for vertical movement. While an assistant turns the steering wheel 1/4 turn in each direction, watch for looseness in the outer

tie rod ends. If any looseness or vertical movement is present, replace the tie rod end. Check the outer tie rod end seals for cracks and proper installation of the nuts and cotter pins. Cracked seals must be replaced. Inspect the tie rods for a bent condition. Bent tie rods or other steering components must be replaced. Do not attempt to straighten these components.

Manual or Power Rack and Pinion Steering Gear Removal and Replacement

The replacement procedure is similar for manual or power rack and pinion steering gears. This removal and replacement procedure varies depending on the vehicle. On some vehicles, the front crossmember or engine support cradle must be lowered to remove the rack and pinion steering gear. Always follow the vehicle manufacturer's recommended procedure in the service manual.

The following is a typical rack and pinion steering gear removal and replacement procedure:

1. Place the front wheels in the straight-ahead position and remove the ignition key from the ignition switch to lock the steering wheel. Place the driver's seat belt through the steering wheel to prevent wheel rotation if the ignition switch is turned on (Figure 13-3). This action maintains the clock spring electrical connector or spiral cable in the centered position on air-bag-equipped vehicles.
2. Lift the front end with a floor jack, and place safety stands under the vehicle chassis. Lower the vehicle onto the safety stands. Remove the left and right fender apron seals (Figure 13-4).
3. Place punch marks on the lower universal joint and the steering gear pinion shaft so they may be reassembled in the same position (Figure 13-5). Loosen the upper universal joint bolt, remove the lower universal joint bolt, and disconnect this joint.
4. Remove the cotter pins from the outer tie rod ends. Loosen, but do not remove, the tie rod end nuts. Use a tie rod end puller to loosen the outer tie rod ends in the steering arms (Figure 13-6). Remove the tie rod end nuts and remove the tie rod ends from the arms.

SPECIAL TOOL

Tie rod end puller

Copyright © 2015 Cengage Learning®

FIGURE 13-3 **Driver's seat belt wrapped around the steering wheel to prevent wheel rotation.**

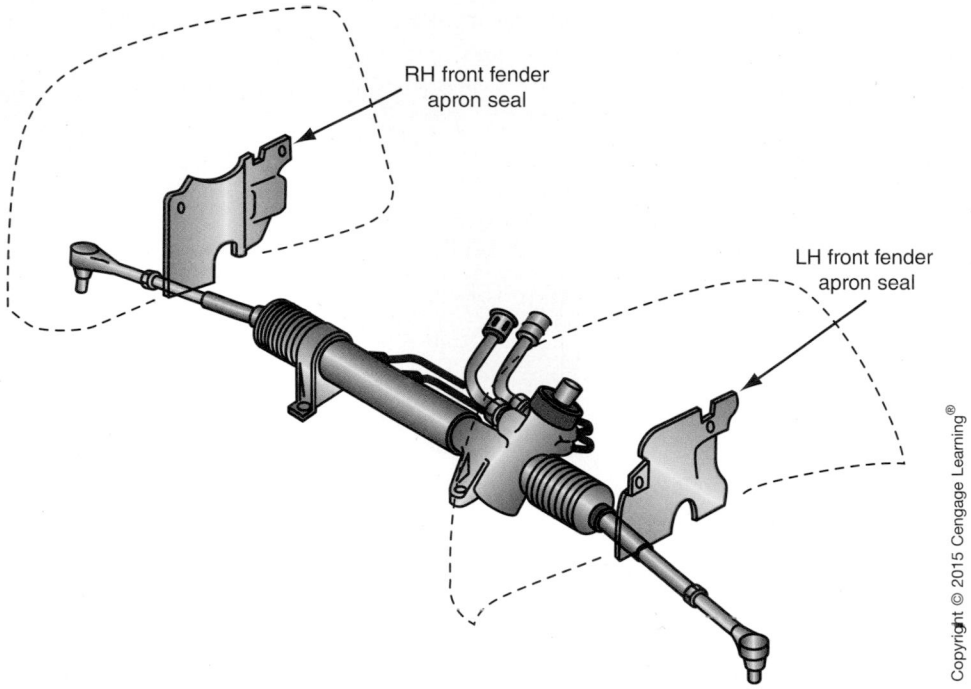

FIGURE 13-4 **Left and right fender apron seals.**

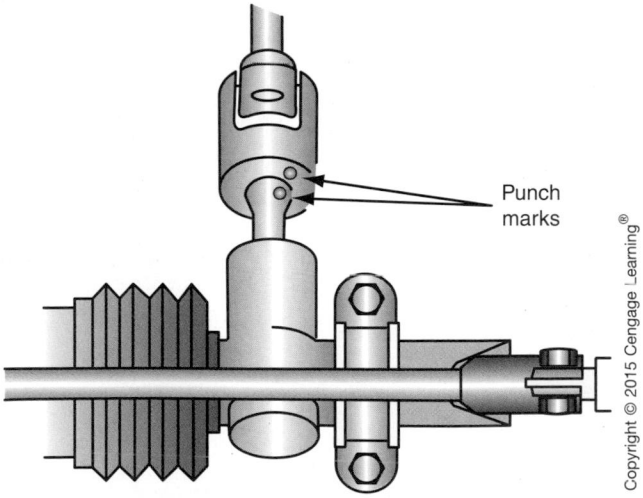

FIGURE 13-5 **Punch marks on universal joint and pinion shaft.**

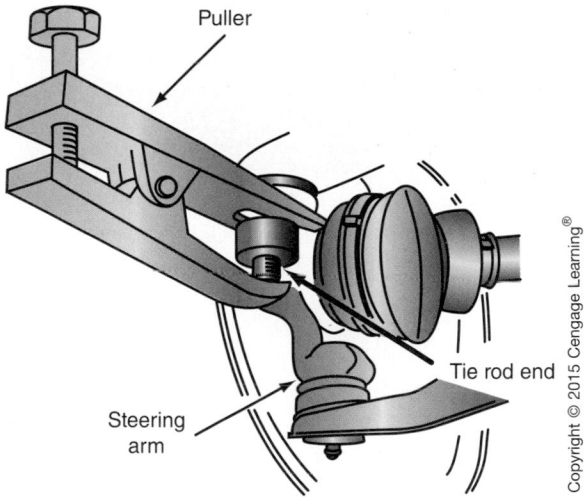

FIGURE 13-6 **Removing outer tie rod ends.**

5. Use the proper wrenches to disconnect the high-pressure hose and the return hose from the steering gear (Figure 13-7). This step is not required on a manual steering gear. The removal procedure for a power rack and pinion steering gear is shown in Photo Sequence 25.
6. Remove the four stabilizer bar mounting bolts (Figure 13-8).
7. Remove the steering gear mounting bolts (Figure 13-9).
8. Remove the steering gear assembly from the right side of the car (Figure 13-10).
9. Position the right and left tie rods the specified distance from the steering gear housing (Figure 13-11). Install the steering gear through the right fender apron.
10. Install the pinion shaft in the universal joint with the punch marks aligned. Tighten the upper and lower universal joint bolts to the specified torque.

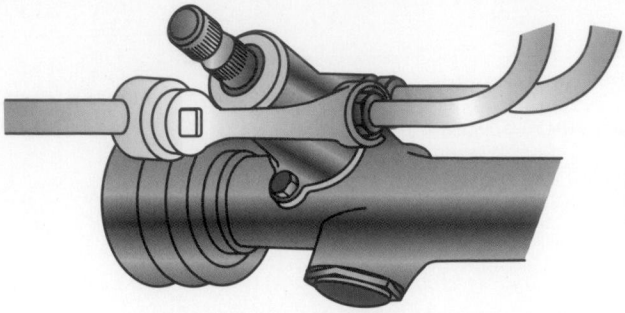

Copyright © 2015 Cengage Learning®

FIGURE 13-7 **Removing high-pressure and return hoses from steering gear.**

FIGURE 13-8 **Removing stabilizer bar mounting bolts.**

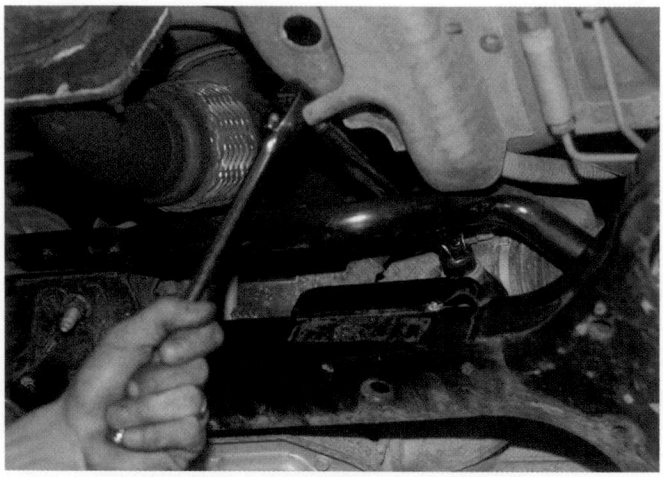

Copyright © 2015 Cengage Learning®

FIGURE 13-9 **Removing steering gear mounting bolts.**

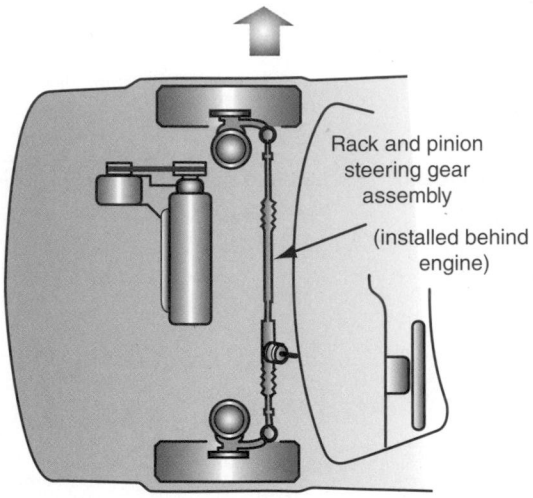

Copyright © 2015 Cengage Learning®

FIGURE 13-10 **Removing steering gear from the car.**

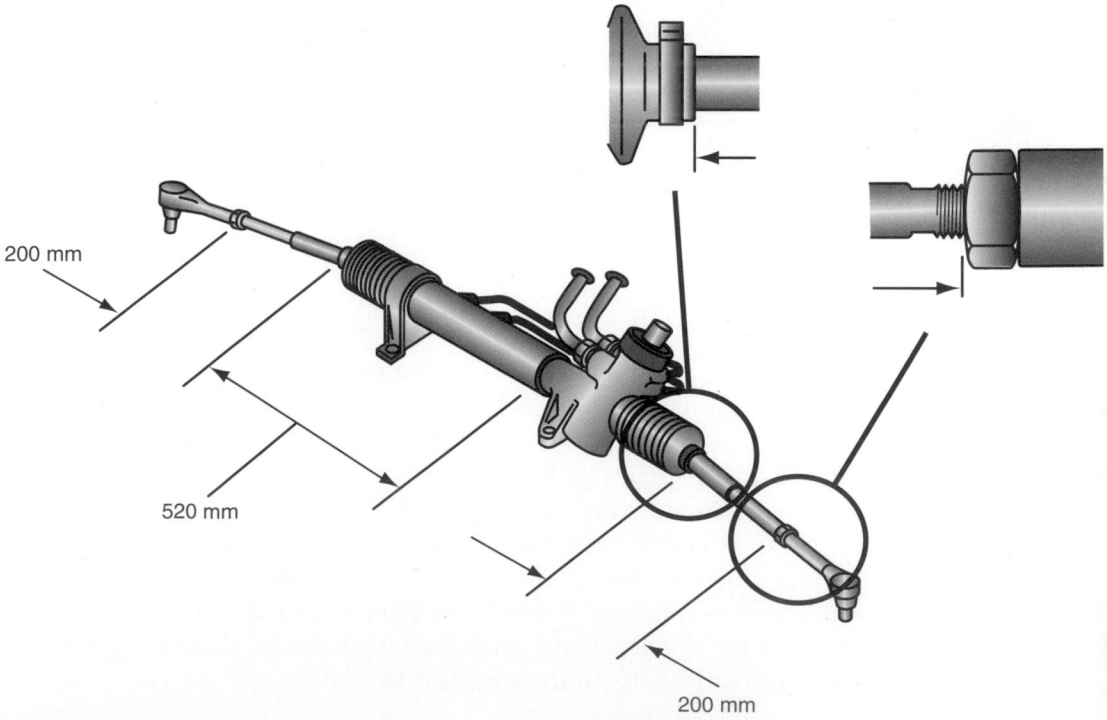

FIGURE 13-11 **Right and left tie rod position in relation to the steering gear housing prior to installation.**

PHOTO SEQUENCE 25

Typical Procedure for Removing and Replacing a Power Rack and Pinion Steering Gear

P25-1 **Position the vehicle on a tire-contact lift, and place the front wheels in the straight-ahead position.**

Copyright © 2015 Cengage Learning®

P25-2 **Remove the key from the ignition switch to lock the steering column, and place the driver's seat belt through the steering wheel opening to prevent wheel rotation if the ignition switch is turned on.**

Copyright © 2015 Cengage Learning®

P25-3 **Raise the vehicle a short distance off the floor, and remove the left and right fender apron seals.**

Copyright © 2015 Cengage Learning®

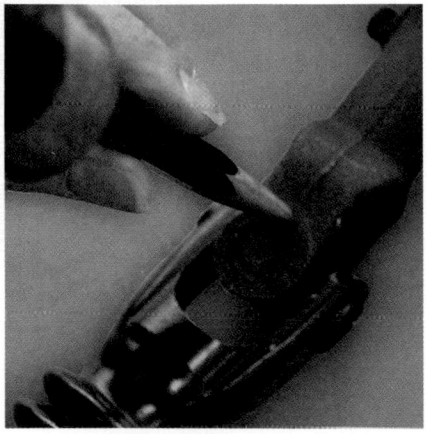

P25-4 **Place punch or chalk marks on the lower universal joint and steering gear pinion shaft so they may be reassembled in the same position.**

Copyright © 2015 Cengage Learning®

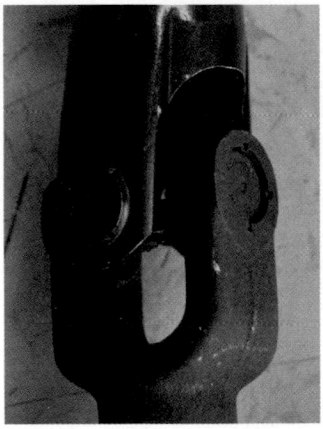

P25-5 **Loosen the upper universal joint bolt, remove the lower universal joint bolt, and disconnect the lower universal joint.**

Copyright © 2015 Cengage Learning®

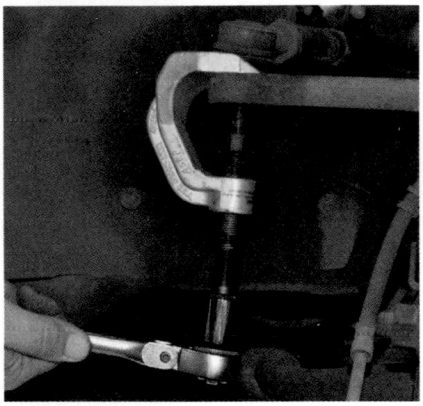

P25-6 **Remove the cotter pins from the outer tie rod ends, and loosen but do not remove the tie rod end nuts. Use a tie rod end puller to loosen the outer tie rod ends in the steering arms. Remove the tie rod end nuts, and remove the tie rod ends from the steering arms.**

Copyright © 2015 Cengage Learning®

P25-7 **Use the proper size wrenches to remove the high-pressure hose and return hose from the steering gear.**

Copyright © 2015 Cengage Learning®

P25-8 **Remove the four stabilizer bar mounting bolts.**

Copyright © 2015 Cengage Learning®

P25-9 **Remove the steering gear mounting bolts.**

Copyright © 2015 Cengage Learning®

P25-10 **Remove the steering gear through one of the fender apron openings.**
Copyright © 2015 Cengage Learning®

P25-11 **Measure the distance from the outer end of the steering gear housing to the outer edge of the outer tie rod end on both sides of the steering gear, and record these measurements.**
Copyright © 2015 Cengage Learning®

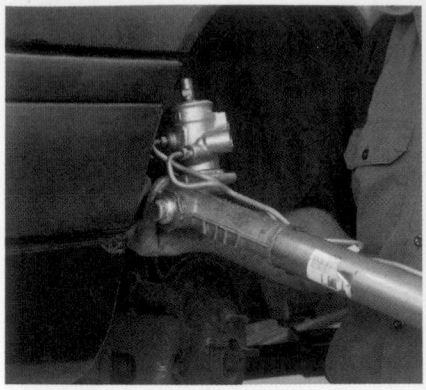

P25-12 **Rotate the pinion shaft to obtain the measurements obtained in step 11, and install the steering gear through one of the fender apron openings.**
Copyright © 2015 Cengage Learning®

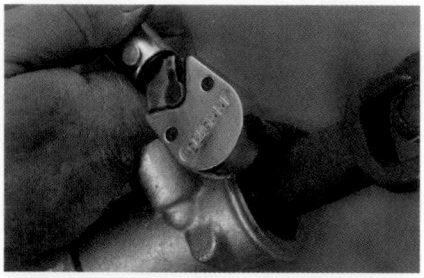

P25-13 **Install the pinion shaft into the universal joint with the punch marks aligned. Tighten the upper and lower universal joint bolts to the specified torque.**
Copyright © 2015 Cengage Learning®

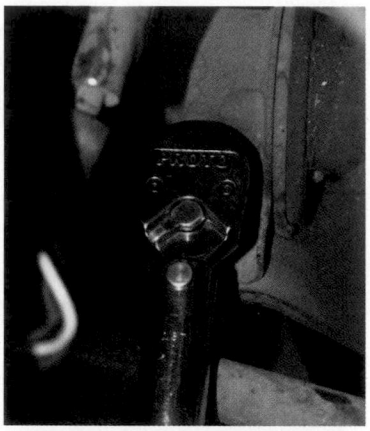

P25-14 **Install the steering gear mounting bolts and tighten these bolts to the specified torque.**
Copyright © 2015 Cengage Learning®

P25-15 **Install the stabilizer mounting bolts and tighten these bolts to the specified torque.**
Copyright © 2015 Cengage Learning®

P25-16 **Install and tighten the high-pressure and return line fittings to the specified torque.**
Copyright © 2015 Cengage Learning®

P25-17 **Install the outer tie rod ends in the steering arms, and tighten the retaining nuts to the specified torque. Install the cotter pins in the nuts.**
Copyright © 2015 Cengage Learning®

P25-18 **Lower the vehicle to a convenient working height, and install the fender apron seals.**
Copyright © 2015 Cengage Learning®

P25-19 **Fill the power steering pump reservoir to the proper level with the specified power steering fluid. Start the engine for a short time, then shut the engine off, and add power steering fluid as required. Run the engine for several minutes, turn the steering wheel fully in both directions, and then add power steering fluid as necessary.**

Copyright © 2015 Cengage Learning®

11. Install the steering gear mounting bolts and tighten these bolts to the specified torque.
12. Install the stabilizer bar mounting bolts and torque these bolts to specifications.
13. Install and tighten the high-pressure and return hoses to the specified torque. (This step is not required on a manual rack and pinion steering gear.)
14. Install the outer tie rod ends in the steering knuckles, and tighten the nuts to the specified torque. Install the cotter pins in the nuts.
15. Check the front wheel toe and adjust as necessary. Tighten the outer tie rod end jam nuts to the specified torque, and tighten the outer bellows boot clamps.
16. Install the left and right fender apron seals, and lower the vehicle with a floor jack.
17. Fill the power steering pump reservoir with the vehicle manufacturer's recommended power steering fluid and bleed the air from the power steering system. (Refer to Chapter 12 for these tasks. This step is not required on a manual rack and pinion steering gear.)
18. Road test the vehicle and check for proper steering gear operation and steering control.

CAUTION: Do not loosen the tie rod end nuts to align cotter pin holes. This action causes improper torquing of these nuts.

Manual Rack and Pinion Steering Gear Diagnosis and Service

Manual Rack and Pinion Steering Gear Tie Rod Service

Follow these steps for manual rack and pinion steering gear disassembly:

1. Clamp the center of the steering gear housing in a **soft-jaw vise**. Do not apply excessive force to the vise.
2. Insert a short piece of wire through each outer tie rod end cotter pin hole and connect a pull scale to this wire. Pull upward on the scale to check the tie rod articulation effort (Figure 13-12). If this effort is not within specifications, replace the inner tie rod end.
3. Mark the outer tie rod and jam nut position with masking tape wrapped around the tie rod threads next to the jam nut, or place a dab of paint on the jam nut and tie rod threads (Figure 13-13). Loosen the jam nuts and remove the outer tie rod ends.
4. Remove the inner and outer bellows boot clamps and pull the bellows boots from the tie rods.
5. Hold the rack in a soft-jaw vise and straighten the lock plate where it is bent over the inner tie rod end (Figure 13-14). Hold the rack with a wrench, and use the proper size wrench to remove the inner tie rod end from the rack (Figure 13-15). A jam nut is used in place of the lock ring on some inner tie rod ends, and a roll pin retains some inner tie rod ends to the rack. Various inner tie rod socket removal procedures are required depending on the socket design.

A **soft-jaw vise** has brass jaws to prevent marking components.

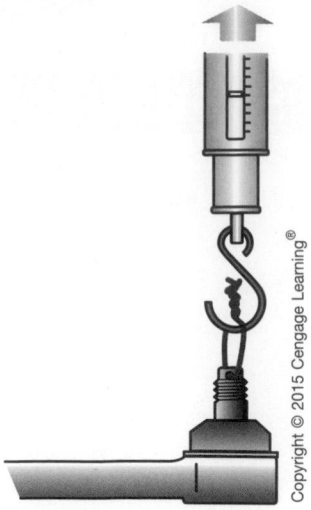
Copyright © 2015 Cengage Learning®

FIGURE 13-12 **Measuring inner tie rod end articulation effort.**

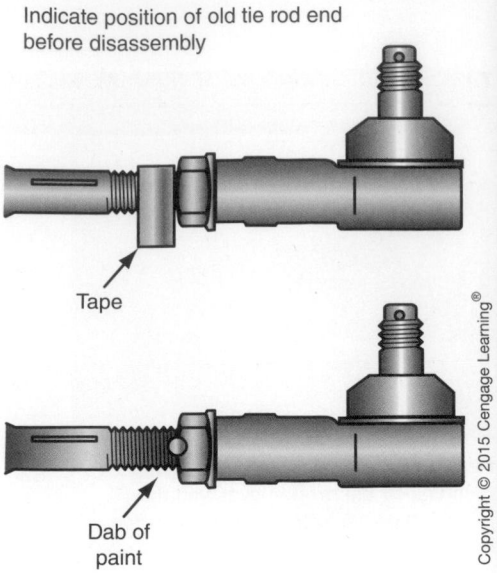

Copyright © 2015 Cengage Learning®

FIGURE 13-13 **Marking outer tie rod end and jam nut position.**

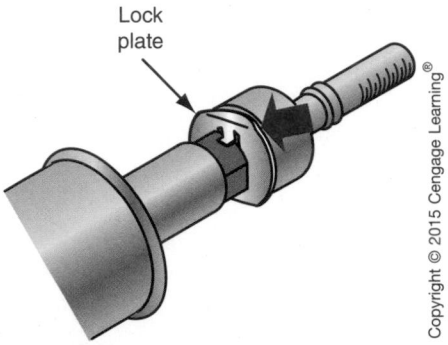

Copyright © 2015 Cengage Learning®

FIGURE 13-14 **Straightening lock plate where it is bent over the inner tie rod end.**

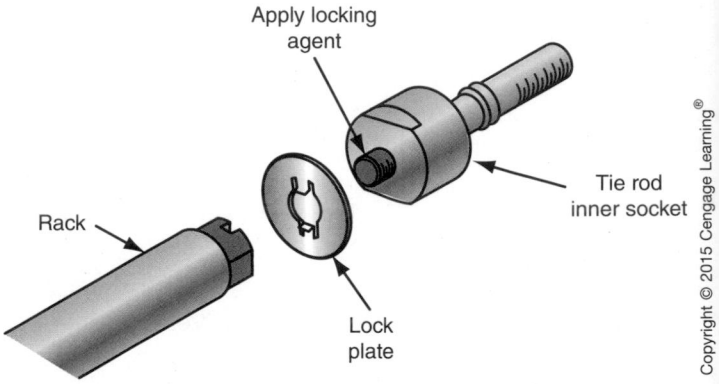

Copyright © 2015 Cengage Learning®

FIGURE 13-15 **Removing inner tie rod end from the rack.**

6. Rotate the inner tie rod ends onto the rack until they bottom. Install the inner tie rod pins, or stake these ends as required. Always use a wooden block to support the opposite side of the rack and inner tie rod end while staking. If jam nuts are located on the inner tie rod ends, be sure they are tightened to the specified torque.
7. Place a large bellows boot clamp over each end of the gear housing. Install the bellows boots, and be sure the boots are seated in the housing and the tie rod undercuts. Install and tighten the large inner boot clamps.
8. Install the outer bellows boot clamps on the tie rods, but do not install these clamps on the boots until the steering gear is installed and the toe is adjusted.
9. Install the jam nuts and the outer tie rod ends. Align the marks placed on these components during the disassembly procedure. Leave the jam nuts loose until the steering gear is installed and the toe is adjusted.
10. Install the steering gear in the vehicle and check the front wheel toe. Tighten the outer bellows boot clamps and the outer tie rod end jam nuts.

Diagnosis of Manual Steering and Suspension Systems

Classroom Manual
Chapter 13, page 348

It is sometimes difficult to separate the diagnosis of manual steering gears, steering columns, and suspension systems because some noises and symptoms may be caused by problems in any of these components or systems. Diagnostic information on three common steering problems is provided in Table 13-1. Refer to the vehicle manufacturer's service manual for complete diagnosis.

TABLE 13-1 **DIAGNOSIS OF MANUAL STEERING GEARS, SUSPENSION SYSTEMS, AND STEERING COLUMNS**

Condition	Possible Cause	Correction
Loose steering	**1.** Steering linkage connections loose	**1.** Tighten as required.
	2. Steering linkage ball stud rubber deteriorated	**2.** Replace as required.
	3. Flex coupling-to-steering-gear attaching bolt loose	**3.** Tighten the bolt to specifications.
	4. Intermediate shaft-to-steering-column-shaft attaching bolt loose	**4.** Tighten the bolt to specifications.
	5. Flex coupling damaged or worn	**5.** Replace as required.
	6. Front and/or rear suspension components loose, damaged, or worn	**6.** Tighten or replace as required.
	7. Steering rack adjustment set too low	**7.** Check the steering rack adjustment and readjust as required.
Noise: Knock, clunk, rapping, squeaking noise when turning	**1.** Steering wheel-to-steering-column shroud interference	**1.** Adjust or replace as required.
	2. Lack of lubrication where speed control brush contacts steering wheel pad	**2.** Lubricate as required.
	3. Steering column mounting bolts loose	**3.** Tighten the bolt to specifications.
	4. Intermediate shaft-to-steering-column-shaft attaching bolt loose	**4.** Tighten the bolt to specifications.
	5. Flex coupling-to-steering-rack attaching bolt loose	**5.** Tighten the bolt to specifications.
	6. Steering rack mounting bolts loose	**6.** Tighten the bolt to specifications.
	7. Tires rubbing or grounding out against body or chassis	**7.** Adjust/replace as required.
	8. Front suspension components loose, worn, or damaged	**8.** Tighten or replace as required.
	9. Steering linkage ball stud rubber deteriorated	**9.** Replace as required.
	10. Normal lash between pinion gear teeth and rack gear teeth when the steering gear is in the off-center position permits gear tooth chuckle noise while turning on rough road surfaces	**10.** This is a normal condition and cannot be eliminated.
Vehicle pulls/drifts to one side *Note:* This condition cannot be caused by the manual steering gear.	**1.** Vehicle overloaded or unevenly loaded	**1.** Correct as required.
	2. Improper tire pressure	**2.** Adjust air pressure as required.
	3. Mismatched tires and wheels	**3.** Install correct tire and wheel combination.
	4. Unevenly worn tires	**4.** Replace as required (check for cause).
	5. Loose, worn, or damaged steering linkage	**5.** Tighten or replace as required.
	6. Steering linkage stud not centered within the socket	**6.** Repair or replace as required.
	7. Bent spindle or spindle arm	**7.** Repair or replace as required.
	8. Broken and/or sagging front and/or rear suspension springs	**8.** Replace as required.
	9. Loose, bent, or damaged suspension components	**9.** Tighten or replace as required.
	10. Bent rear axle housing	**10.** Replace as required.
	11. Excessive camber/caster split (excessive side-to-side variance in the caster/camber settings)	**11.** Adjust camber/caster split as required.
	12. Improper toe setting	**12.** Adjust as required.
	13. Front wheel bearings out of adjustment	**13.** Adjust as required.
	14. Investigate tire variance (conicity for radial tires and unequal circumference for bias-ply and belted bias-ply tires)	**14.** Repair or replace as required.

Copyright © 2015 Cengage Learning®

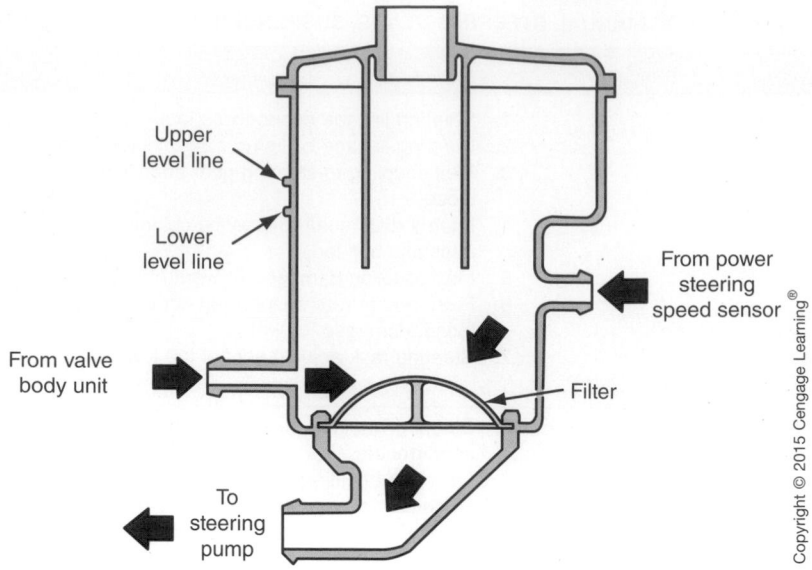

FIGURE 13-16 **Power steering reservoir and filter.**

Power Rack and Pinion Steering Gear Diagnosis and Service

Fluid, Filter, and Fluid Leak Diagnosis

Power steering fluid should be inspected for discoloration and contamination with metal particles, dirt, or water. A suction gun may be used to remove some fluid from the power steering reservoir for inspection. If the fluid is contaminated or discolored, the power steering system must be flushed and then filled to the specified level with the vehicle manufacturer's specified fluid. After refilling the power steering system, an air bleeding procedure is usually necessary. Some power steering systems have a filter that is usually located in the remote power steering reservoir (Figure 13-16). On some systems, the filter and reservoir must be replaced as an assembly. Filter/reservoir replacement is necessary if the power steering system has been opened for repairs or if the fluid is contaminated with water, metal, or dirt particles. The filter/reservoir and fluid should also be changed at the vehicle manufacturer's specified service intervals. A restricted power steering filter may cause erratic steering operation.

WARNING: If the engine has been running for a length of time, power steering gears, pumps, lines, and fluid may be very hot. Wear eye protection and protective gloves when servicing these components.

If power steering fluid leaks from the cylinder end of the power steering gear, the outer rack seal is leaking (Figure 13-17).

The inner rack seal is defective if oil leaks from the pinion end of the housing when the rack reaches the left internal stop (Figure 13-18). An oil leak at one rack seal may result in oil leaks from both boots because the oil may travel through the breather tube between the boots.

If an oil leak occurs at the pinion end of the housing and this leak is not influenced when the steering wheel is turned, the pinion seal is defective (Figure 13-19).

If an oil leak occurs in the pinion coupling area, the input shaft seal is leaking (Figure 13-20). This seal and the pinion seal will require replacement because the pinion seal must be replaced if the pinion is removed.

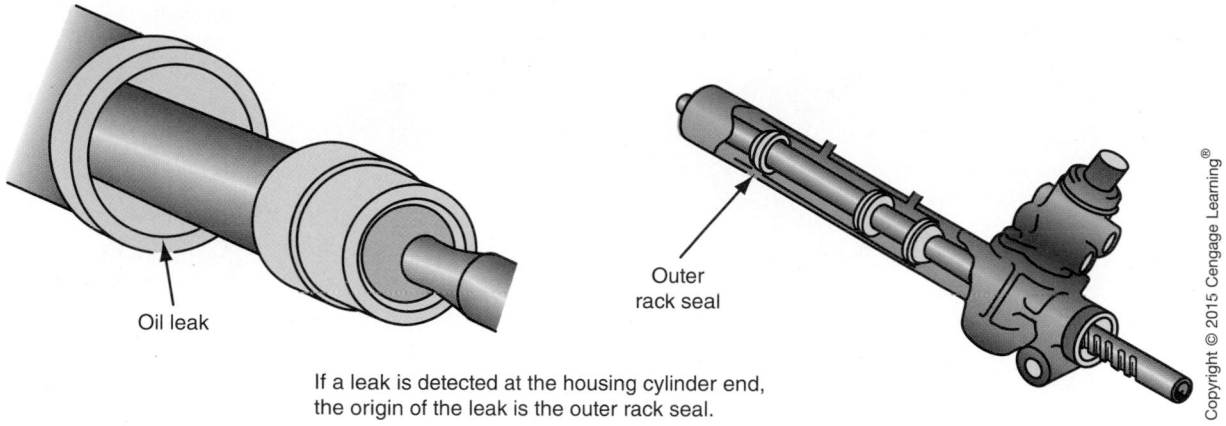

If a leak is detected at the housing cylinder end, the origin of the leak is the outer rack seal.

FIGURE 13-17 **Oil leak diagnosis at outer rack seal.**

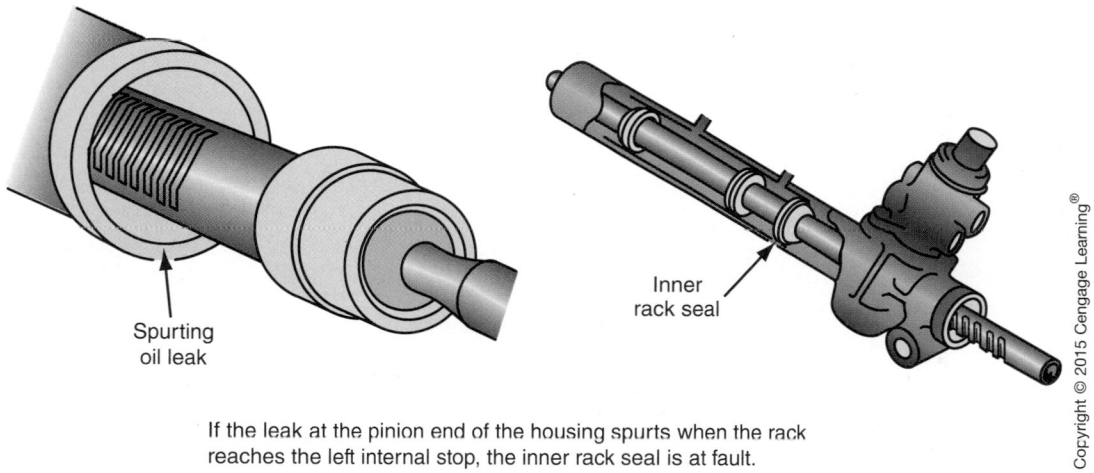

If the leak at the pinion end of the housing spurts when the rack reaches the left internal stop, the inner rack seal is at fault.

FIGURE 13-18 **Inner rack seal leak diagnosis.**

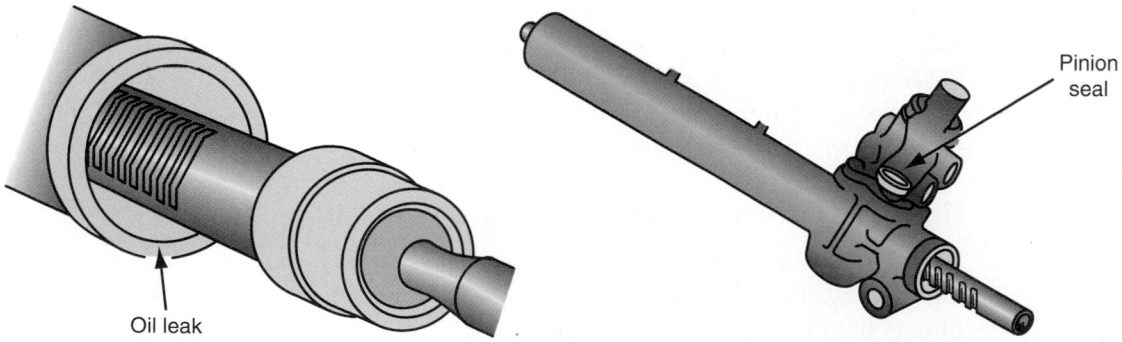

If you detect a leak at the pinion end of the housing and it is not influenced by the direction of the turn, the origin of the leak is the pinion seal.

FIGURE 13-19 **Pinion seal leak diagnosis.**

Steering effort imbalance is the difference in steering effort when making right and left turns.

SPECIAL TOOL

Steering gear holding tool

SERVICE TIP: A common problem with power rack and pinion steering gears is excessive steering effort when the car is first started after sitting overnight. The steering effort becomes normal, however, after the steering wheel is turned several times. This problem usually indicates severe scoring where the pinion sealing rings contact the pinion housing. Steering gear replacement is required to correct this problem.

In a power rack and pinion steering gear, a condition that causes excessive steering effort when the vehicle is first started may be called "morning sickness."

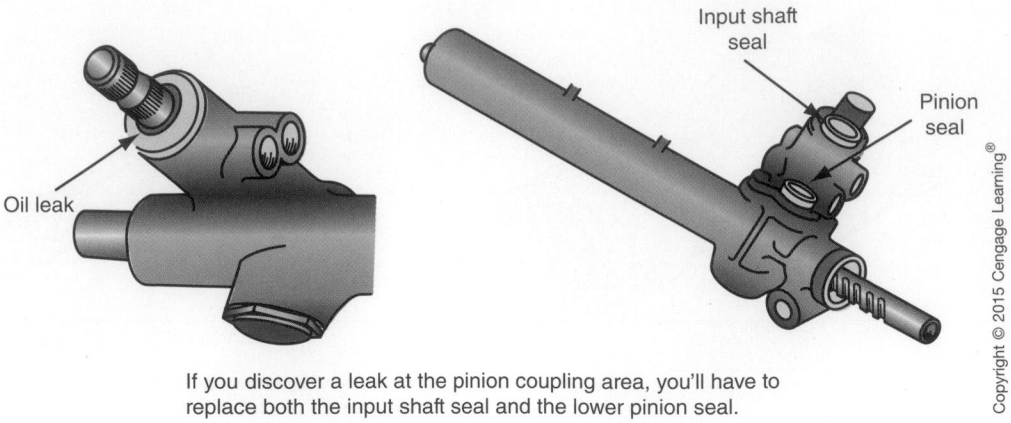

If you discover a leak at the pinion coupling area, you'll have to replace both the input shaft seal and the lower pinion seal.

FIGURE 13-20 **Oil leak diagnosis on pinion coupling area.**

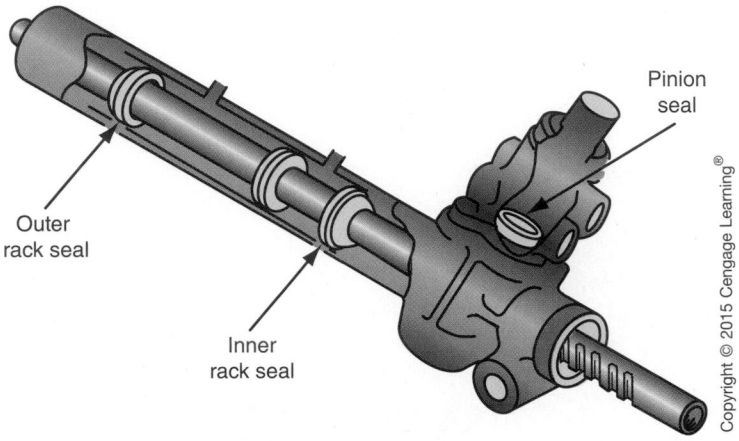

FIGURE 13-21 **Rack seals and pinion seal.**

When the rack is removed, the inner and outer rack seals and the pinion seal must be replaced (Figure 13-21). If oil leaks occur at fittings, these fittings must be torqued to the manufacturer's specifications. If the leak is still present, the line and fitting should be replaced. Leaks in the lines or hoses require line or hose replacement.

Turning Imbalance Diagnosis

The same amount of effort should be required to turn the steering wheel in either direction. A pressure gauge connected to the high-pressure hose should indicate the same pressure when the steering wheel is turned in each direction. **Steering effort imbalance** or lower power assist in each direction may be caused by defective rack seals (Figure 13-22). Steering effort imbalance or low power assist in both directions may also be caused by defective rotary valve rings and seals or restricted hoses and lines (Figure 13-23).

Rotary Valve and Spool Valve Operation

The spool valve and rotary valve are mounted over the upper end of the pinion gear. During a turn, torsion bar twisting moves the spool valve inside the rotary valve and this action directs fluid to the appropriate side of the rack piston to provide steering assist. The pinion gear spool valve directs fluid pressure to the power rack piston right or left chamber for power assist (Figure 13-24) during a turn.

When driving with the front wheels straight ahead, most of the fluid is directed through the rotary valve to the return hose under normal operation. During a left turn,

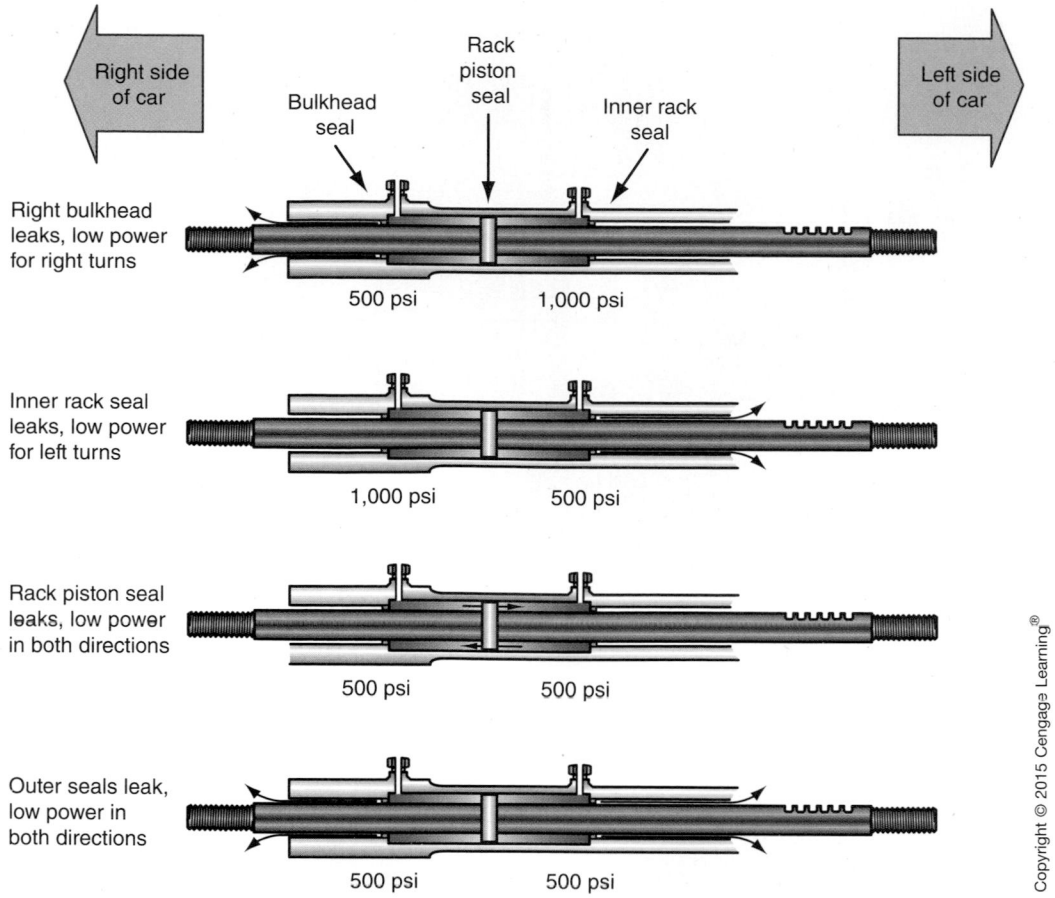

Copyright © 2015 Cengage Learning®

FIGURE 13-22 **Effect of defective rack seals on steering effort imbalance and low power assist.**

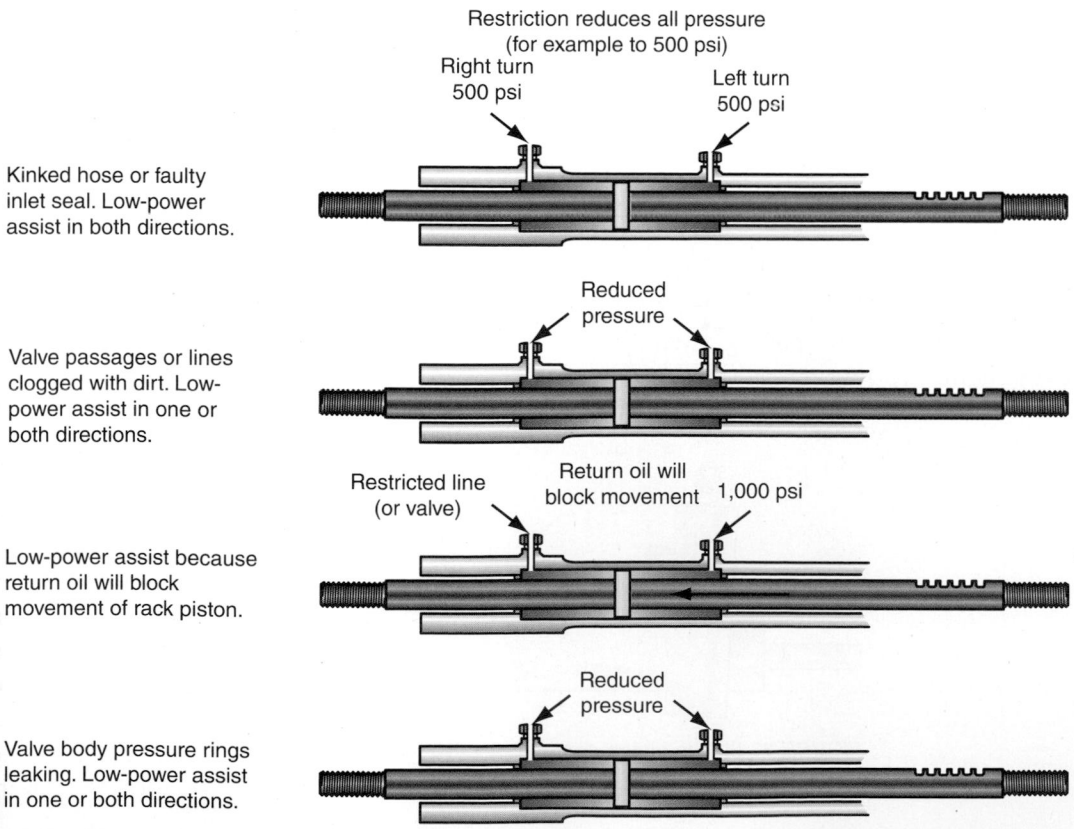

Copyright © 2015 Cengage Learning®

FIGURE 13-23 **Effect of worn rotary valve rings and seals or restricted lines or hoses on steering effort.**

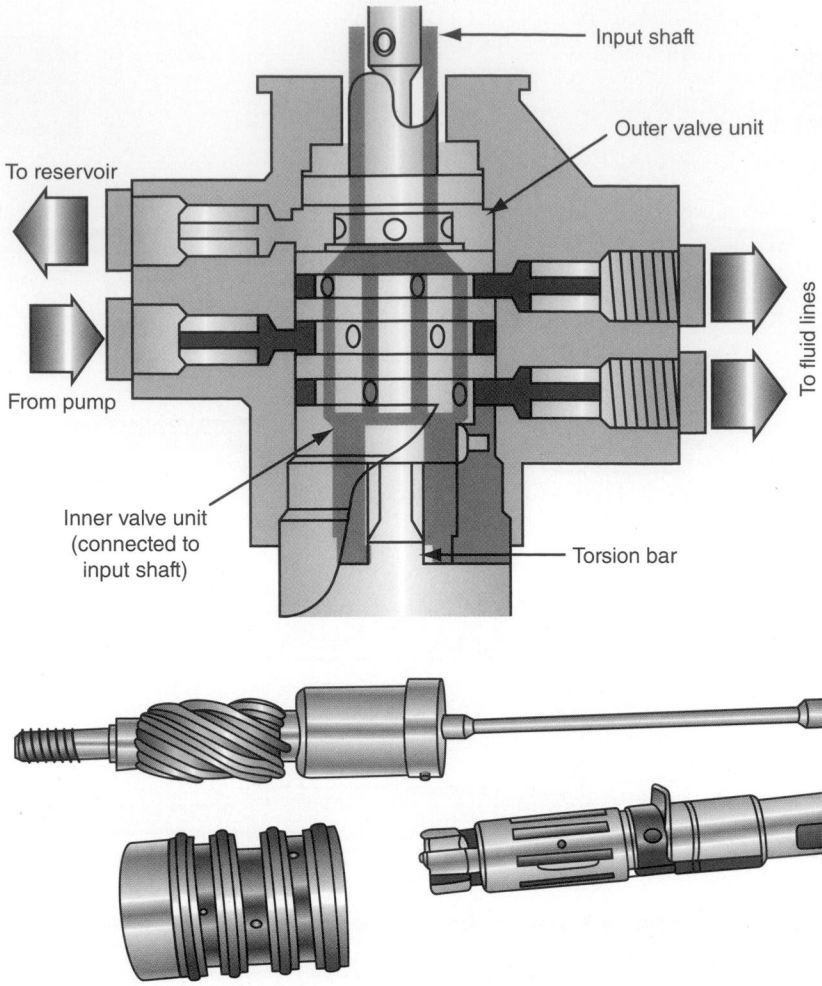

Copyright © 2015 Cengage Learning®

FIGURE 13-24 **The pinion gear spool valve directs fluid pressure to the power rack piston right or left chamber for power assist.**

SERVICE TIP:
To provide equal turning effort in both directions, the rack must be centered with the front wheels straight ahead.

spool valve movement directs fluid to the left side of the rack piston. During a right turn, spool valve movement directs fluid to the right side of the rack piston.

The pinion gear spool valve can bind or stick causing fluid to be directed to the power rack piston chamber causing a drift or pull to one side (Figure 13-25). The symptom for this is a vehicle drift or pull to one side. Alignment angle will be within vehicle specifications but a drift is still present. With both front wheels of the vehicle jacked off the ground start the vehicle while paying attention to steering wheel position. If the steering wheel moves slightly to the left or right suspect a binding spool valve directing pressure to one

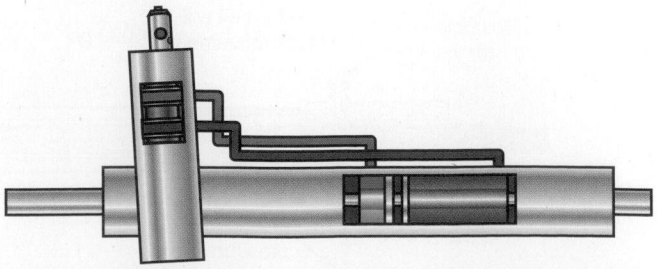

Copyright © 2015 Cengage Learning®

FIGURE 13-25 **The pinion gear spool valve can bind or stick causing fluid to be directed to the power rack piston chamber causing a drift or pull to one side.**

side of the rack piston. Be aware that a front tire with excessive conicity can also cause the vehicle to pull or drift to one side. This is often referred to as a radial tire pull.

Power Rack and Pinion Steering Gear Tie Rod and Rack Bearing Service

1. Install a holding tool on the steering gear housing, and clamp this tool in a vise (Figure 13-26).
2. Mark the outer tie rod ends, jam nuts, and tie rods (Figure 13-27). Loosen the jam nuts and remove the tie rod ends and jam nuts.
3. Loosen the inner and outer clamps on both bellows boots, and remove these boots (Figure 13-28).
4. Clamp the inner tie rod socket lightly in a soft-jaw vise and use a hammer and chisel to unstake the **claw washers** on the inner tie rod sockets (Figure 13-29).
5. Hold the rack with an adjustable wrench and use the proper tool to loosen the inner tie rod ends (Figure 13-30). Remove the inner tie rod ends and the claw washers.
6. Use the proper tool to loosen the adjuster plug locknut (Figure 13-31).
7. Remove the adjuster plug with the proper tool (Figure 13-32).
8. Remove the spring, rack guide, and the seat from the adjuster plug opening (Figure 13-33).

Claw washers are used in some steering gears to lock the tie rods to the rack.

SERVICE TIP:
In some power rack and pinion steering gears, the inner tie rod ends are staked after they are installed on the rack (refer to Photo Sequence 26 later in this chapter). In other power rack and pinion steering gears, a pin retains the inner tie rod ends to the rack, and this pin must be removed before the inner tie rod ends are removed. When the inner tie rod ends are replaced, a new retaining pin must be installed.

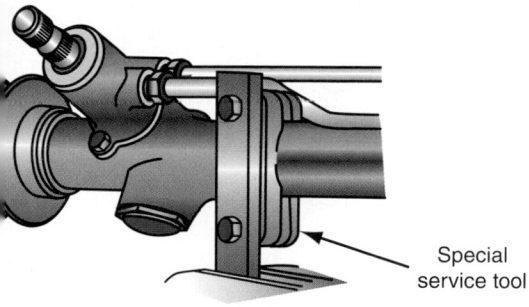

Copyright © 2015 Cengage Learning®

E 13-26 **Steering gear holding tool.**

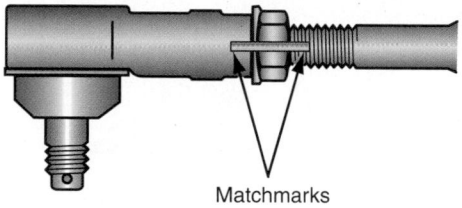

Copyright © 2015 Cengage Learning®

FIGURE 13-27 **Marking outer tie rod ends, jam nuts, and tie rods prior to outer tie rod end removal.**

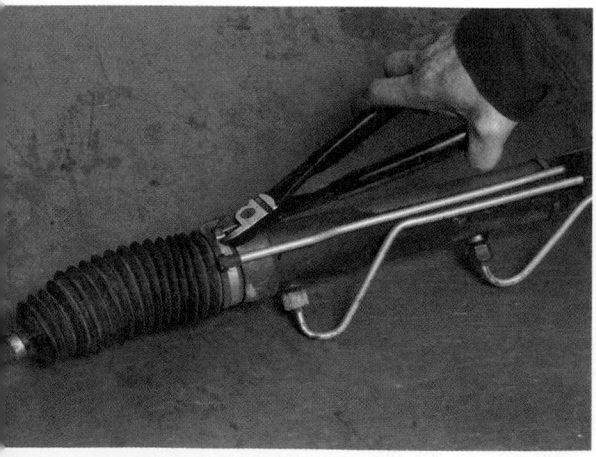

Copyright © 2015 Cengage Learning®

E 13-28 **Removing bellows boot clamps.**

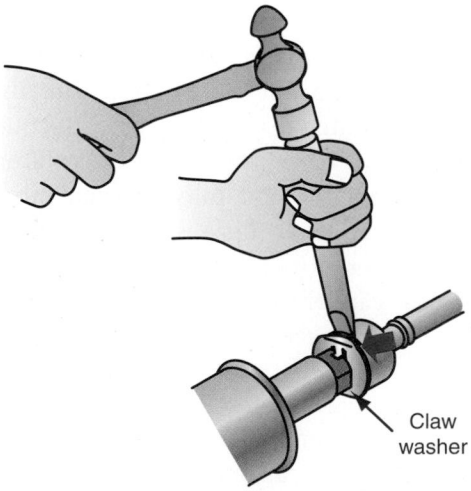

Copyright © 2015 Cengage Learning®

FIGURE 13-29 **Removing claw washers from inner tie rod sockets.**

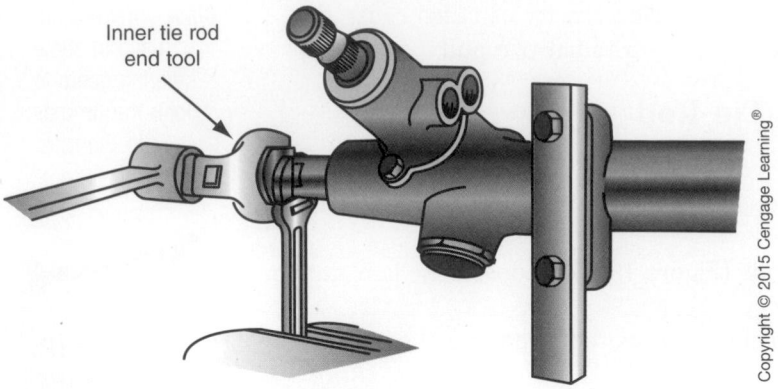

Copyright © 2015 Cengage Learning®

FIGURE 13-30 **Loosening inner tie rod ends.**

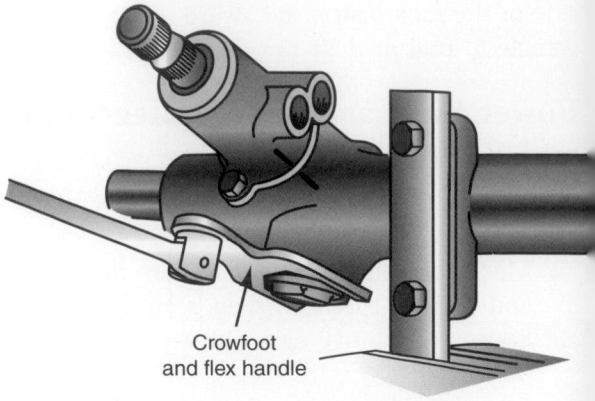

FIGURE 13-31 **Loosening adjuster plug locknut.**

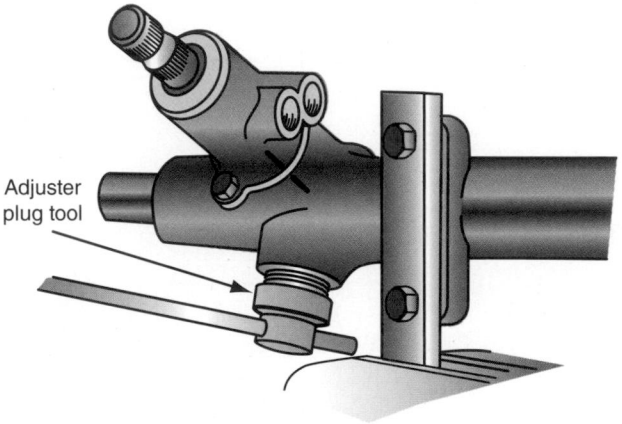

Copyright © 2015 Cengage Learning®

FIGURE 13-32 **Removing adjuster plug.**

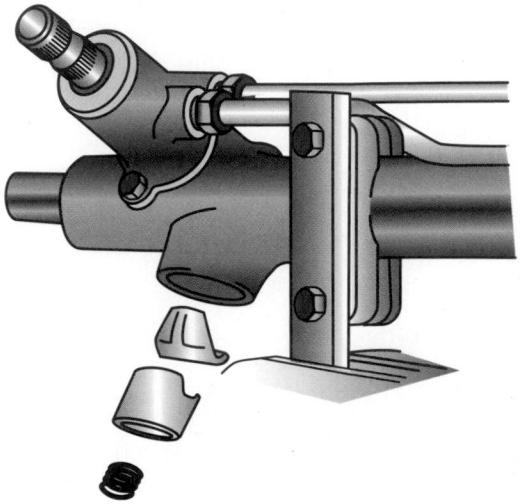

Copyright © 2015 Cengage Learning®

FIGURE 13-33 **Removing spring, rack guide, and seat.**

9. Clean and inspect the spring, rack guide, and seat. Replace only worn components. Lubricate the rack guide and seat with the specified power steering fluid.
10. Coat the rack guide seat and rack guide with power steering fluid. Install the rack guide seat, rack guide, and spring (Figure 13-34). Apply Loctite 242 sealant or its equivalent to the first two or three threads on the rack spring cap. Install the rack spring cap and tighten it to the specified torque (Figure 13-35).
11. Use the proper tool to rotate the rack spring cap 12° counterclockwise. Turn the pinion shaft fully in each direction, and repeat this action. Loosen the rack spring cap until there is no tension on the rack guide spring. Place the proper turning tool and a foot-pound torque wrench on top of the pinion shaft. Tighten the rack spring cap while rotating the pinion shaft back and forth (Figure 13-36). Continue tightening the rack spring cap until the specified turning torque is indicated on the torque wrench.
12. Apply Loctite 242 sealant or its equivalent to the first two or three threads on the locknut. Install the locknut on the rack spring cap. Use the proper tool to hold the rack spring cap, and tighten the locknut to the specified torque (Figure 13-37).

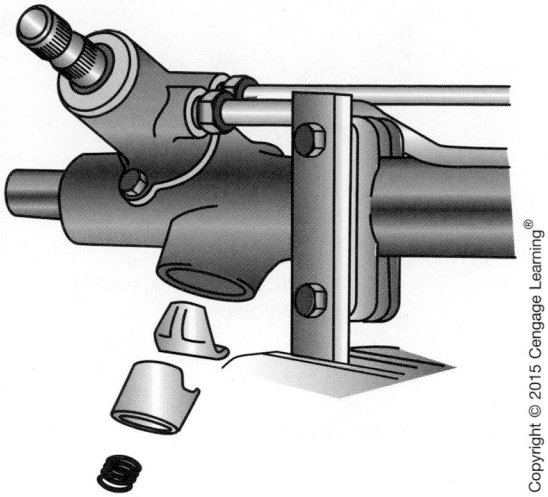
Copyright © 2015 Cengage Learning®

FIGURE 13-34 **Installing rack guide seat, rack guide, and spring.**

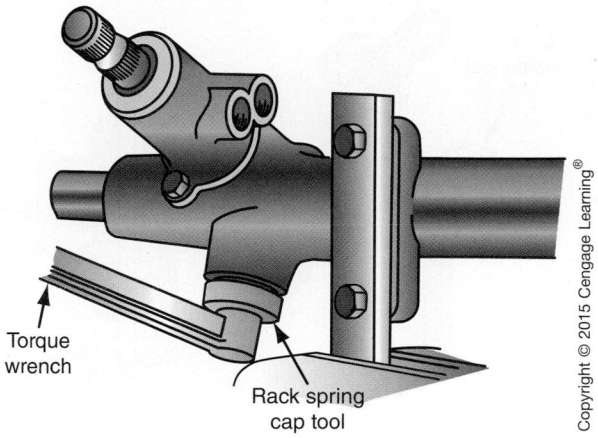

Copyright © 2015 Cengage Learning®

FIGURE 13-35 **Installing rack spring cap.**

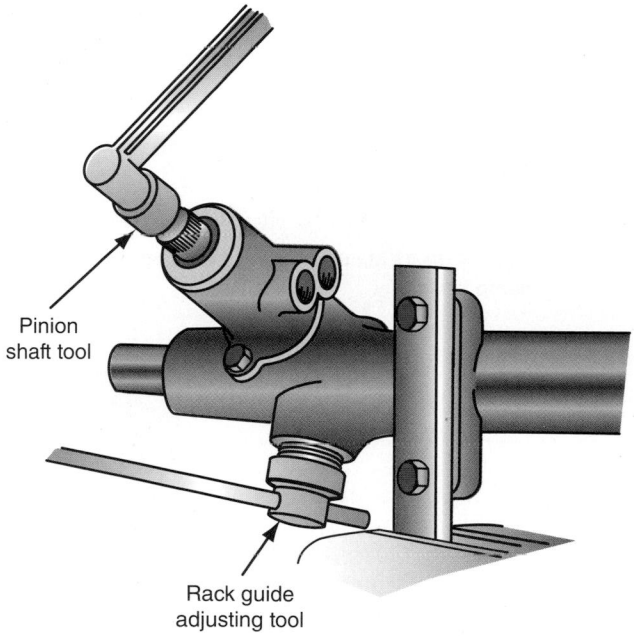

Copyright © 2015 Cengage Learning®

FIGURE 13-36 **Adjusting pinion shaft turning torque.**

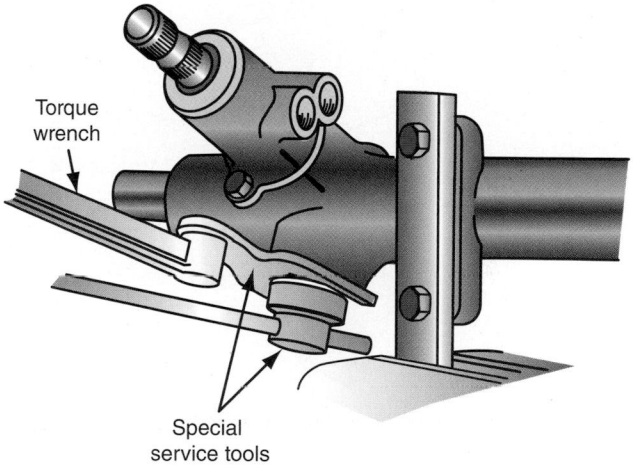

Copyright © 2015 Cengage Learning®

FIGURE 13-37 **Tightening locknut on rack spring cap.**

13. Install a new claw washer on the inner tie rod end, and then install the inner tie rod end on the rack. Hold the rack with an adjustable wrench, and use a torque wrench with the correct tool to tighten the inner tie rod end to the specified torque (Figure 13-38). Clamp the inner tie rod end in a soft-jaw vise and stake the claw washer with a brass bar and a hammer (Figure 13-39). Repeat this procedure on both inner tie rod ends.
14. Check the vent hole in each end of the rack and be sure they are not clogged with grease or other material (Figure 13-40). These vent holes allow air flow from one boot to the other during a turn.
15. Install the bellows boots, clamps, and clips. When the inner boot clamps are tightened, there must be a minimum of 2 mm (0.08 in.) between the ends of the clamps (Figure 13-41).

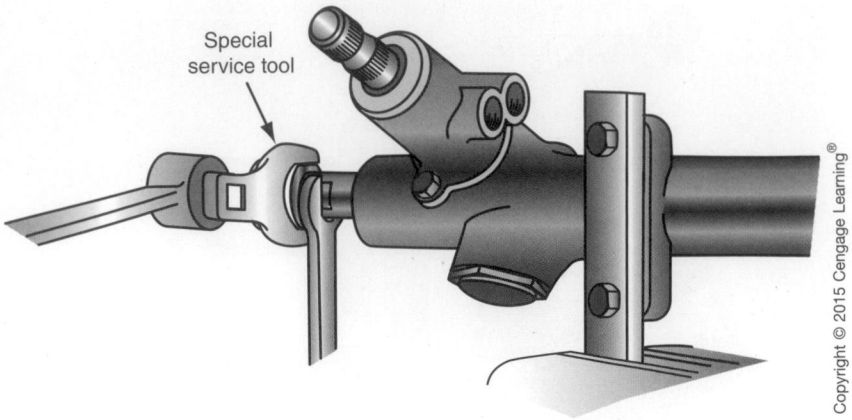

Copyright © 2015 Cengage Learning®

FIGURE 13-38 **Tightening inner tie rod end on the rack.**

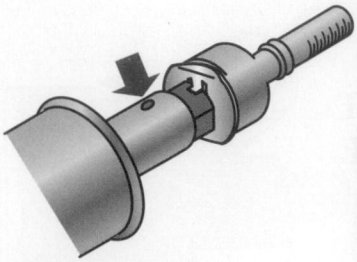

Copyright © 2015

FIGURE 13-40 **Rack vent holes must be open.**

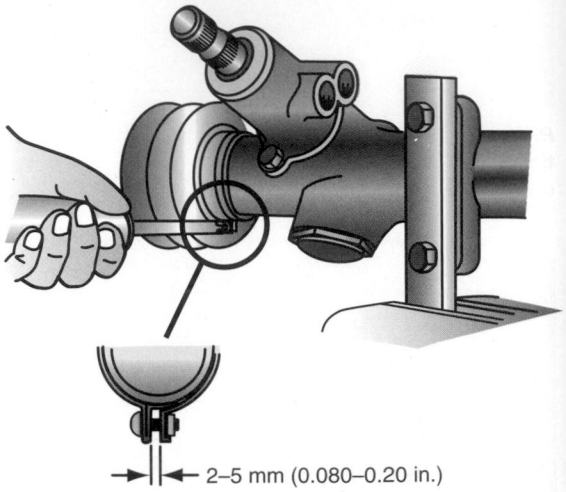

FIGURE 13-41 **Minimum clearance between boot clamp ends.**

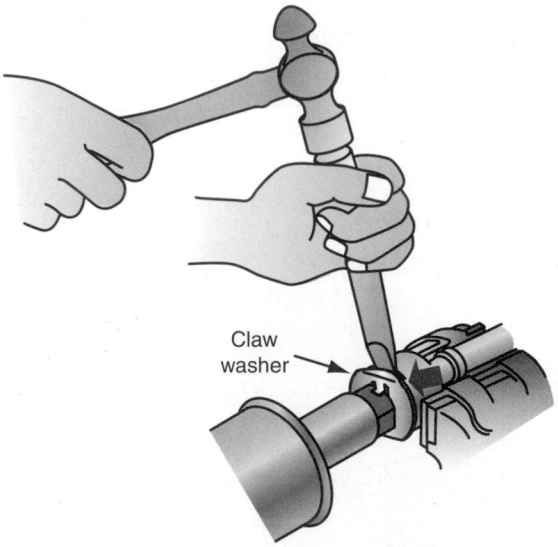

Copyright © 2015 Cengage Learning®

FIGURE 13-39 **Staking claw washer on inner tie rod end.**

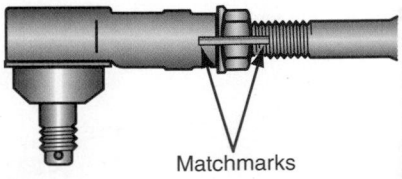

FIGURE 13-42 **Aligning marks on the outer tie rod ends, jam nuts, and tie rods.**

16. Install the outer tie rod ends and jam nuts. Align the marks placed on the tie rod ends, jam nuts, and tie rods during disassembly (Figure 13-42). Leave the jam nuts loose until the steering gear is installed and the front wheel toe is adjusted.
17. Follow the steering gear installation procedure described earlier in this chapter to install the steering gear in the vehicle. Tighten all fasteners to the specified torque. (See Chapter 12 for power steering pump filling and bleeding procedure.) Check the front wheel toe and tighten the outer tie rod end jam nuts to the specified torque. Tighten the outer bellows boot clamps. Road test the vehicle and check for proper steering operation and control.

Photo Sequence 26 shows a typical procedure for removing and replacing an inner tie rod end on a power rack and pinion steering gear.

PHOTO SEQUENCE 26

Typical Procedure for Removing and Replacing Inner Tie Rod End, Power Rack, and Pinion Steering Gear

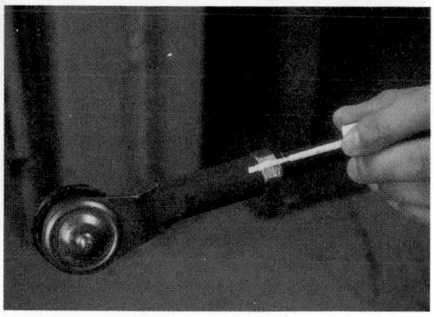

P26-1 **Place an index mark on the outer tie rod end, jam nut, and tie rod.**
Copyright © 2015 Cengage Learning®

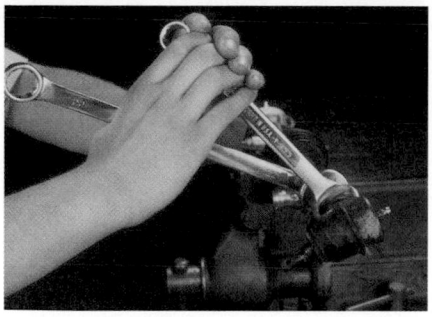

P26-2 **Loosen the jam nut and remove the outer tie rod end and jam nut.**
Copyright © 2015 Cengage Learning®

P26-3 **Remove the inner and outer boot clamps.**
Copyright © 2015 Cengage Learning®

P26-4 **Remove the bellows boot from the tie rod.**
Copyright © 2015 Cengage Learning®

P26-5 **Hold the rack with the proper size wrench and loosen the inner tie rod end with the proper wrench.**
Copyright © 2015 Cengage Learning®

P26-6 **Remove the inner tie rod from the rack.**
Copyright © 2015 Cengage Learning®

P26-7 **Be sure the shock damper ring is in place on the rack.**
Copyright © 2015 Cengage Learning®

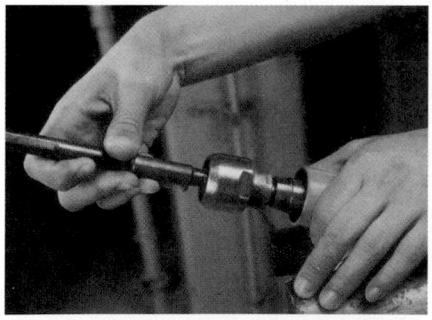

P26-8 **Install the inner tie rod on the rack, and tighten the tie rod to the specified torque while holding the rack with the proper size wrench.**
Copyright © 2015 Cengage Learning®

P26-9 **Support the inner tie rod on a vise, and stake both sides of the inner tie rod joint with a hammer and punch.**
Copyright © 2015 Cengage Learning®

PHOTO SEQUENCE 26 (CONTINUED)

P26-10 **Use a feeler gauge to measure the clearance between the rack and the inner tie rod joint housing stake.**

Copyright © 2015 Cengage Learning®

P26-11 **Install the bellows boot and new clamps.**

Copyright © 2015 Cengage Learning®

P26-12 **Install the jam nut and outer tie rod end with the index marks aligned.**

Copyright © 2015 Cengage Learning®

SERVICE TIP:
A scored cylinder surface in the steering gear housing may cause temporary excessive steering effort after the car sits overnight.

Diagnosis of Power Steering, Steering Column, and Suspension Systems

Some power steering problems may be caused by defects in the steering gear, pump, or column. For example, heavy steering effort may be caused by a defective power steering pump, damaged rack piston and ring, or misalignment of the steering column. Diagnostic information on common power steering problems is provided in Table 13-2. Refer to the vehicle manufacturer's service manual for more complete diagnosis.

TABLE 13-2 **POWER STEERING SYSTEM, STEERING COLUMN, AND SUSPENSION SYSTEM DIAGNOSIS**

Condition	Possible Cause	Correction
Wander: The vehicle wanders side to side on the roadway when it is driven straight ahead, while the steering wheel is held in a firm position. Evaluation should be conducted on a level road (little road crown)	**1.** Loose/worn tie rod ends or ball socket	**1.** Replace tie rod end or tie rod assembly.
	2. Inner ball housing loose or worn	**2.** Replace tie rod assemblies.
	3. Gear assembly loose crossmember	**3.** Tighten the two mounting nuts to specification.
	4. Excessive yoke clearance	**4.** Adjust yoke clearance.
	5. Loose suspension struts or ball joints	**5.** Tighten or replace as required.
	6. Column intermediate shaft connecting bolts loose	**6.** Tighten bolts to specification at gear and column.
	7. Column intermediate shaft joints loose or worn	**7.** Replace intermediate shaft assembly.
	8. Improper wheel alignment	**8.** Set alignment to specification.
	9. Tire size and pressure	**9.** Check tire sizes and adjust tire pressure.
	10. Vehicle unevenly loaded or overloaded	**10.** Adjust load.
	11. Steering gear mounting insulators and/or attachment bolts loose or damaged	**11.** Replace insulators and/or attachment nuts and bolts. Tighten to specification.
	12. Steering gear adjustments	**12.** Refer to steering gear section of *Shop Manual.*
	13. Front end misaligned	**13.** Check and align to specifications.
	14. Worn front end parts or wheel bearings	**14.** Inspect and replace affected parts.
	15. Unbalanced or badly worn steering gear control valve	**15.** Inspect and replace affected parts.

Condition	Possible Cause	Correction
Pulls to one side: The vehicle tends to pull to one side when driven on a level surface.	**1.** Improper tire pressure **2.** Improper tire size or different type **3.** Vehicle unevenly or excessively loaded **4.** Improper wheel alignment **5.** Damaged front suspension components **6.** Damaged rear suspension components **7.** Steering gear valve effort out of balance **8.** Front and rear brakes operation **9.** Bent rear axle housing, or damaged or sagging springs in the front and/or rear suspension **10.** Loose/worn shock absorber or suspension attaching fasteners in rear suspension	**1.** Adjust tire pressure. **2.** Replace as required. **3.** Adjust load. **4.** Adjust as required. **5.** Refer to front suspension section of Shop Manual. **6.** Refer to rear suspension section of Shop Manual. **7.** Place transmission in Neutral while driving and turn engine off (coasting). — If vehicle does not pull, replace the steering gear valve assembly. Refer to steering gear section of Shop Manual. — If vehicle does drift: • Cross-switch front tire/wheel assemblies. • If vehicle pulls to opposite side, cross-switch tire/wheel assemblies that were on the rear to the same on the front. • If vehicle pull direction is not changed, check front suspension components and wheel alignment. **8.** Adjust if necessary. **9.** Replace if necessary. **10.** Tighten all attaching fasteners to specification.

Copyright © 2015 Cengage Learning®

Diagnosis of Magnasteer Systems

The most common customer complaints on Magnasteer systems are continually heavy or light steering effort. When diagnosing a Magnasteer system, the first step is to visually inspect the system. Be sure the power steering fluid level is correct, and check the power steering belt tension. Inspect the wiring harness connected to the actuator solenoid in the steering gear.

If the visual inspection does not locate any defects, connect a scan tool to the DLC near the steering column under the dash (Figure 13-43). Turn the ignition switch on and select the Electronic Brake and Traction Control Module (EBTCM) on the scan tool menu. If there is an electrical defect in the actuator solenoid, connecting wires, or EBTCM, Diagnostic Trouble Code (DTC) C1241 will be displayed on the scan tool. To locate the exact cause of this DTC, perform voltmeter tests on the actuator solenoid and connecting wires. A detailed diagnostic chart is included in the vehicle manufacturer's service manual.

If the customer complains about continually heavy or light steering and there are no defects in the Magnasteer system, reprogramming of the system may be required.

CAUTION: Always be sure the ignition switch is off when connecting or disconnecting the scan tool. Computer system and/or scan tool damage may occur if the ignition switch is on when connecting or disconnecting the scan tool from the data link connector (DLC).

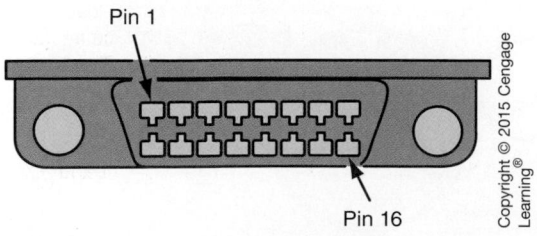

FIGURE 13-43 **Data link connector (DLC) on-board diagnostic II (OBD-II) system.**

SERVICE TIP: Some generic scan tools do not have the capability to reprogram computers.

To reprogram the system, follow these steps:

1. Connect the scan tool to the DLC.
2. Select Magnasteer and Recalibration on the scan tool.
3. Respond to the VIN query on the scan tool. If the displayed VIN is incorrect, enter the correct VIN.
4. When Factory Standard Calibration Will Be Used For This VIN appears on the scan tool, press Enter on the scan tool to install factory calibration.
5. When calibration is complete, disconnect the scan tool.

Diagnosis of Rack-Drive Electronic Power Steering

When the ignition switch is turned on in a vehicle with **rack-drive electronic power steering**, the electronic power steering (EPS) light in the instrument panel should be illuminated (Figure 13-44). This EPS light action proves the bulb and related circuit are functioning normally. When the engine starts, the EPS light is turned off by the EPS control unit if there are no electrical defects in the EPS system. The EPS light should remain off while the engine is running.

If an electrical defect occurs in the EPS system, the EPS control unit turns on the EPS light to alert the driver that a defect exists in the EPS system. Under this condition, the EPS control unit shuts down the EPS system, and electric assist is no longer available. Manual steering with increased steering effort is provided in this mode. When the EPS control unit senses an electrical defect in the EPS system, a diagnostic trouble code (DTC) is stored in the control unit memory. The EPS control unit memory is erased if battery voltage is disconnected from the control unit. The 7.5 ampere clock fuse may be disconnected for 10 seconds to disconnect battery voltage from the EPS control unit and erase the DTCs from the control unit memory.

SERVICE TIP:
If the special jumper wire is left in the EPS service connector and the engine is started, the malfunction indicator light (MIL) for the engine computer system is illuminated.

If the EPS light is on with the engine running, shut the ignition switch off and disconnect the clock fuse for 10 seconds; then road test the car to determine if the EPS light comes back on. If the EPS light comes on again with the engine running, proceed with the diagnosis. Locate the two-wire EPS service check connector under the glove box. With the ignition switch off, connect a special jumper wire to the service check connector (Figure 13-45). This connects the two wires together in the service check connector. When the ignition switch is turned on, the EPS light begins to flash DTCs. Each DTC is indicated by a long light flash or flashes followed by a short flash or flashes. For example, DTC 11 is one long flash followed by a short flash, and DTC 12 is one long flash followed by two short flashes (Figure 13-46).

If more than one DTC is stored in the EPS control unit memory, these DTCs are flashed in numerical order. The DTC sequence continues until the ignition switch is turned off. Be sure to remove the special jumper wire after the ignition switch is turned off.

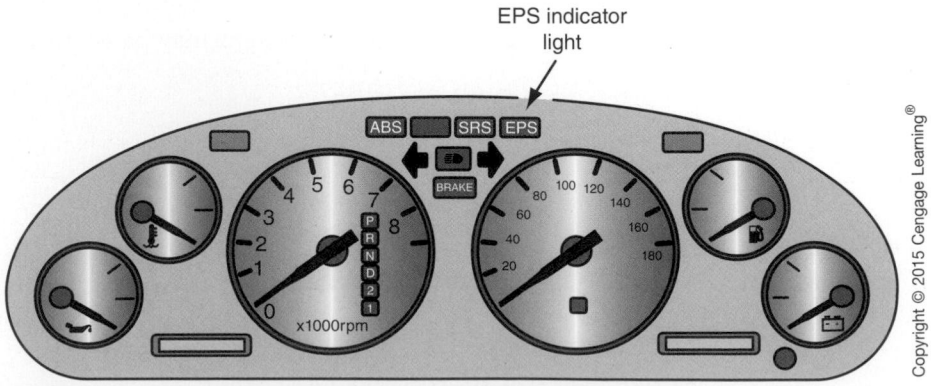

FIGURE 13-44 **Electronic power steering (EPS) warning light.**

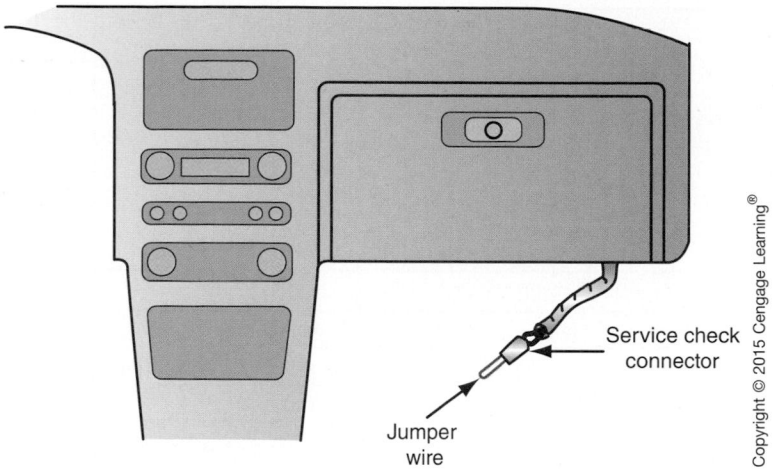

FIGURE 13-45 **EPS system service check connector and special jumper wire.**

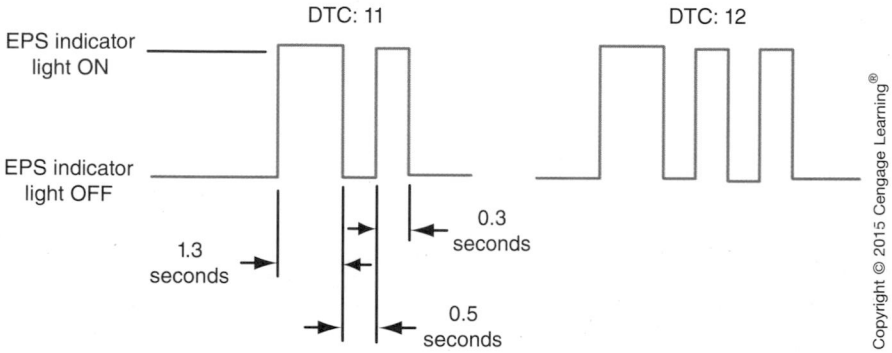

FIGURE 13-46 **DTC 11 is one long flash followed by a short flash, and DTC 12 is one long flash followed by two short flashes.**

All the possible DTCs in the EPS system are provided in Figure 13-47. A DTC indicates a defect in a certain area. For example, DTC 11 indicates high or low voltage from the torque sensor. The defect may be in the sensor itself, or the problem could be in the connecting wires from the sensor to the control unit. Tests usually have to be performed with a voltmeter or ohmmeter to locate the exact cause of the defect. Detailed diagnostic charts are provided in the car manufacturer's service manual.

SERVICE TIP: If DTC 33 is set in the EPS control unit memory, the EPS system remains functional, but the EPS light is illuminated.

CUSTOMER CARE: Diagnosing automotive problems is an extremely important part of a technician's job on today's high-tech vehicles. Always take time to diagnose a customer's vehicle accurately. Fast, inaccurate diagnosis of automotive problems leads to unnecessary, expensive repairs and unhappy customers, who may take their business to another shop. Accurate diagnosis may take more time, but in the long term it will improve customer relations and bring customers back to the shop.

Classroom Manual
Chapter 13, page 362

Diagnostic trouble code (DTC)	EPS indicator light	Description/symptom	Diagnostic period			After detecting for system	Reset	Individual diagnosis
			Individual diagnosis	Individual diagnosis	Regular diagnosis		Individual diagnosis	
—	O	EPS indicator light does not come on when ignition is switched ON						17 – 27
—	O	EPS indicator light does not go off after engine is started				System OFF		17 – 29
3	O	A problem with the current sensor offset	O			^		17 – 36
4	O	A problem with the current sensor offset	O	O		^	O	
5	O	A problem with the current sensor fixed		O		^	O	
6	O	A problem with the current sensor fixed		O		^	O	
11	O	A problem the high voltage with or low voltage of the torque sensor (TRQ1 and TRQ2)		O		^	O	17 – 31
12	O	A problem with the voltage for torque sensor (TRQ3)		O		^	O	
13	O	A problem with the average of voltage on TRQ1 and TRQ2		O		^	O	
14	O	A problem with the 2.5-V reference voltage		O		^	O	
21	O	A problem with the circuit for the input motor voltage in the EPS control unit	O	O	O	^	O	17 – 36
22	O	A problem with the lower current			O	^	O	
23	O	A problem with the circuit for check function in the EPS control unit	O	O		^	O	
24	O	The fail-safe relay or power relay is stuck ON	O			^	O	
25	O	The lower FET is stuck ON	O			^	O	
26	O	The upper FET is stuck ON	O			^	O	
31	O	A problem with the voltage for IG1	O	O		^	O	—
33	O	A problem with the average of VSS1 and VSS2			O	^	O	17 – 38
34	O	A problem with the CPU in the EPS control unit	O	O	O	^	O	Replace EPS control unit

- Initial diagnosis: Performed right after the engine starts until the EPS indicator light goes off.
- Regular diagnosis: Continuously performed (under some conditions) after the EPS indicator goes off and engine is running.
- Individual part/system diagnosis: Diagnoses a specific part/system under its operating conditions.
- CPU: Central Processing Unit.

Copyright © 2015 Cengage Learning®

FIGURE 13-47 **DTC list for EPS system.**

Diagnosis of Column-Drive Electronic Power Steering

When diagnosing a **column-drive EPS**, the first step is to verify the customer complaint. Road test the vehicle if necessary to be sure the complaint is identified. The next step is to perform a visual inspection of the EPS using the following steps:

1. Inspect all EPS electrical connections for corrosion or looseness including ground connections.
2. Inspect EPS wiring for damaged insulation.
3. Be sure the vehicle is equipped with the vehicle manufacturer's specified tire and wheel sizes.
4. Inspect the steering column and EPS components for collision damage.
5. Be sure the battery is fully charged and that the battery connections are clean and tight.
6. Check all the EPS fuses.

7. Inspect the vehicle for aftermarket accessories such as stereo equipment that could affect the EPS operation if they are improperly installed.
8. Search for service bulletins related to the EPS complaint.

Repair any problems discovered during the visual inspection. If the visual inspection did not locate any problems, perform a diagnostic system check using the following procedure:

SERVICE TIP: When connecting certain scan tools to controller area network (CAN) or local area network (LAN) data link systems, a CANdi module must be connected in series between the scan tool and the DLC. Always refer to the vehicle manufacturer's recommended procedure.

1. Be sure the ignition switch is off, and connect a scan tool to the DLC under the left side of the dash.
2. Turn on the ignition switch, and select Vehicle DTC Information on the scan tool. If the scan tool displays No Comm. From Any Computer, repair the data link communication problem by referring to the appropriate diagnostic procedure in the vehicle manufacturer's service manual.
3. Be sure the engine cranks and starts properly.
4. Turn on the ignition switch with the engine stopped, and select List All DTCs on the scan tool.
5. Record all DTCs displayed on the scan tool. Check for DTCs from other computer systems that may affect the EPS system. For example, a DTC representing the VSS is displayed as a PCM DTC, but this sensor signal is transmitted on the data links to the PSCM and affects EPS operation.

Repair the causes of the DTCs displayed on the scan tool. Some typical EPS DTCs are:

1. C0460 steering position sensor
2. C0475 electric steering motor circuit
3. C0545 steering wheel torque input sensor
4. C0870 PSCM voltage range performance

In these **On-Board Diagnostic II (OBD II)** DTCs the first letter C indicates the DTC is related to a chassis problem. If the second digit in the DTC is a 0, the DTC is a universal Society of Automotive Engineers (SAE) code. When the second digit in the DTC is a 1, a vehicle manufacturer's code is indicated. The third digit in the DTC indicates the subsystem to which the DTC belongs, and the last two digits indicate the specific area where the fault exists. A complete list of DTCs is provided in the vehicle manufacturer's service manual. Some scan tools provide an explanation of each DTC indicating the area where the problem is located. A DTC indicates only the area in which a problem exists. Voltmeter and ohmmeter tests may be required to test the circuit and locate the exact cause of the problem. The vehicle manufacturer's service manual also provides a detailed diagnostic procedure for each DTC. Always follow these diagnostic procedures when diagnosing the cause of a DTC.

On-Board Diagnostic II (OBD II) systems are a type of computer system mandated on cars and light trucks since 1996. OBD II systems have a number of mandated standardized features including several monitoring systems in the PCM.

Scan Tool Data Display

The scan tool also displays the following EPS data, which are very helpful when diagnosing EPS defects:

1. Battery voltage—The scan tool displays 6.5 V to 16.5 V as sensed at the power steering control module.
2. Calculated system temperature—The ambient temperature up to 275° F (135° C) as sensed by an internal temperature sensor in the PSCM.
3. Motor command—The amount of EPS assist motor current commanded by the PSCM.
4. Steering shaft torque—The amount of torque being applied to the steering column shaft. A positive value indicates a right turn, and a negative number indicates a left turn.

5. Steering position sensor voltage 1—The amount of steering position sensor voltage from the sensor signal 1 circuit. The voltage range is 0 V to 5 V as the steering wheel is turned.
6. Steering position sensor voltage 2—The amount of steering position sensor voltage from the sensor signal 2 circuit. The voltage range is 0 V to 5 V as the steering wheel is turned.
7. Steering tuning—The scan tool displays 1 or 2 depending on the vehicle. Steering tuning 1 is for 4-cylinder engines, and steering tuning 2 is for 6-cylinder engines. Steering tuning 3 is for the extended sedan.
8. Steering wheel position—The scan tool displays from 562° to −562° indicating the number of degrees the steering wheel has been turned for the center position. A positive number indicates a right turn, and a negative number indicates a left turn.
9. Torque sensor signal 1—The scan tool displays 0.25 V to 4.75 V indicating the voltage from the steering shaft torque sensor 1 circuit.
10. Torque sensor signal 2—The scan tool displays 0.25 V to 4.75 V indicating the voltage from the steering shaft torque sensor 2 circuit.
11. Vehicle speed—The actual vehicle speed from 0 to 158 mph (255 km/h).

SERVICE TIP: After the ignition switch is turned off, wait 25 seconds before performing any procedures that require disconnecting the vehicle battery. If this precaution is not observed, the PSCM memory may be erased. This precaution also applies to torque sensor calibration and steering tuning selection.

Steering Position Sensor Calibration

If the PSCM or steering column is replaced, use the scan tool to calibrate the center position of the steering wheel position sensor as follows:

1. Be sure the ignition switch is off, and connect the scan tool to the DLC.
2. Turn on the ignition switch, but do not start the engine.
3. Turn the steering wheel from lock-to-lock and count the number of turns. From the fully left or fully right position, turn the steering wheel back toward the center position one-half the total number of turns.
4. Select Special Functions on the scan tool.
5. Select Steering Position Sensor Calibration on the scan tool and press the enter key. The scan tool screen indicates Calibration in Progress during the calibration procedure.
6. When the calibration process is completed, and the scan tool displays Calibration Completed, press the exit key.
7. Use the scan tool to clear any DTCs.
8. Turn off the ignition switch and disconnect the scan tool.

SERVICE TIP: The combination steering wheel position sensor and steering shaft torque sensor are not serviced separately; the steering column and these two sensors are replaced as an assembly. The PSCM and motor can be replaced as an assembly separate from the steering column.

Torque Sensor Calibration

If the PSCM or steering column is replaced, a scan tool must be used to perform a center position calibration for the steering shaft torque sensor as described in the following steps:

1. Be sure the ignition switch is off, and connect the scan tool to the DLC.
2. Turn on the ignition switch, but do not start the engine.
3. Turn the steering wheel from lock-to-lock and count the number of turns. From the fully left or fully right position, turn the steering wheel back toward the center position one-half the total number of turns.
4. After the steering wheel is centered, remove your hands or any other objects from the steering wheel. Be sure the suspension is relaxed, and that no uneven force is applied to the steering system.
5. Select Special Functions on the scan tool.
6. Select Torque Sensor Calibration on the scan tool and press the enter key. The scan tool screen indicates Calibration in Progress during the calibration procedure.

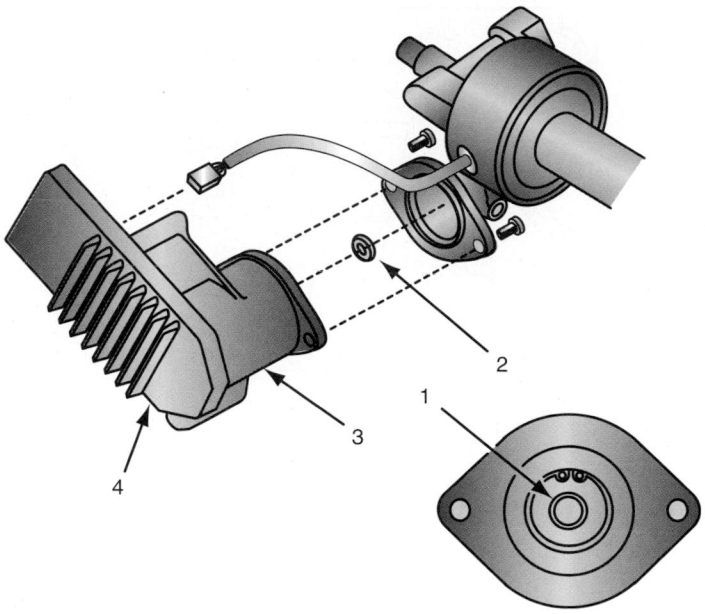

Copyright © 2015 Cengage Learning®

FIGURE 13-48 **Removing and replacing the PSCM and motor assembly.**

7. When the calibration process is completed, and the scan tool displays Calibration Completed, press the exit key.
8. Use the scan tool to clear any DTCs.
9. Turn off the ignition switch and disconnect the scan tool.

If the PSCM and motor assembly is removed from the steering column, inspect the steering column assist mechanism input shaft for the rotor isolator bumper (Figure 13-48). If present, the isolator bumper should be removed from the input shaft, and this bumper must be installed in the rotor isolator in the PSCM and motor assembly. Remove any debris from the steering column assist mechanism housing, but do not remove the remaining grease on the steering column assist mechanism input shaft.

Steering Tuning Selection

If the PSCM is replaced, these steps must be followed to perform a steering tuning procedure:

1. Be sure the ignition switch is off, and connect the scan tool to the DLC.
2. Turn on the ignition switch, but do not start the engine.
3. Select Special Functions on the scan tool.
4. Select Steering Tuning Selection on the scan tool and press the enter key. The scan tool screen indicates Selection in Progress during the selection procedure.
5. When the selection process is completed, and the scan tool displays Selection Completed, press the exit key.
6. Use the scan tool to clear any DTCs.
7. Turn off the ignition switch and disconnect the scan tool.

Active Steering System Preliminary Diagnosis

The active steering system has a warning light and a service light in the instrument panel (Figure 13-49). When the active steering system is operating normally, the active steering warning light and service light go out after the engine starts. If an electronic defect occurs in the active steering system, the warning and service lights are illuminated. The service

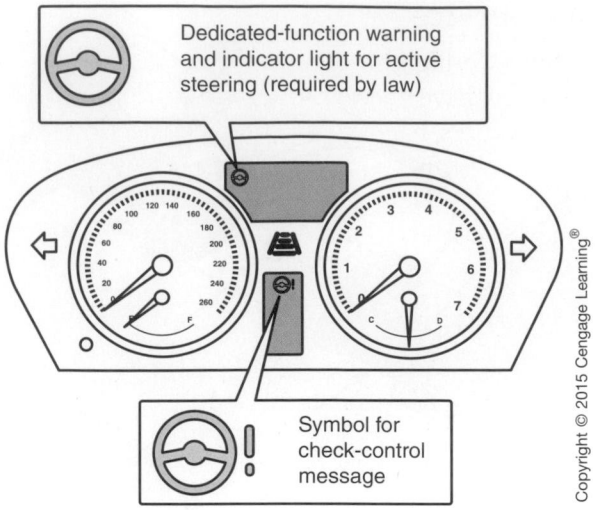

FIGURE 13-49 **Active Steering System light.**

light indicates that a service message is available in the central information display screen in the instrument panel. When "service" is selected in the central information display screen, the following messages may be displayed:

Classroom Manual
Chapter 13, page 376

1. Active steering! Exercise care when steering.
   Active steering fault.
   Steering behavior altered. Steering wheel might be at an angle. Possible to continue journey with caution. Exercise care when steering. Have the problem checked by the nearest BMW service.
2. Active steering inactive.
   Active steering.
   Active steering inactive.
   Steering behavior altered. Steering wheel might be at an angle. Possible to continue journey with caution. Exercise care when steering.
3. Servotronic failure!
   Servotronic failure.
   Possible to continue journey with caution.
   Important: Power steering assistance is no longer automatically adapted to the vehicle's speed.
   Have problem checked by the nearest BMW service.

Preliminary Inspection

If the active steering warning lights are illuminated indicating a defect in the system, the first step in the diagnostic procedure is to perform a visual inspection of all system components, wiring, and fiber-optic cables. During the preliminary inspection, check these items:

1. Check the power steering fluid level and condition.
2. Inspect all wiring harness and electrical connectors in the system for damage, corrosion, and frayed insulation.
3. Inspect all fiber-optic cables for sharp bends, heat damage, punctures, cuts, or looseness.
4. Check all the hydraulic lines and fittings for damage or leaks.
5. Inspect all system mounting bolts for looseness.

Fiber-Optic Cable Service

Many vehicles with active steering have fiber-optic cable networks. Media oriented systems transport (MOST) networks have green fiber-optic cables, and intelligent safety and integration systems (ISIS) networks have yellow fiber-optic cables. Some manufacturers provide orange fiber-optic cables for repair purposes. Fiber-optic cables should always be inspected for sharp bends, heat damage, punctures or cracks, and loose connectors. Repair or replace fiber-optic cables as necessary. To remove a fiber-optic cable from a connector, follow this procedure:

1. To remove the connector, use a small screwdriver in the connector openings (Figure 13-50), and expand the cover catches. Remove the cover.
2. To remove the fiber optic cable from the connector lift the lock (Figure 13-51) and carefully feed the fiber optic cable out of the connector.
3. To install the fiber-optic cable in the connector, push the cable all the way into the connector and be sure the lock secures the cable.
4. If it is necessary to remove the fiber optic cable from its mounting position press to release the lock in the bore (Figure 13-52) and remove the cable connector.

Active Steering Diagnosis System

A diagnosis system is supplied by the vehicle manufacturer (Figure 13-53). The diagnosis system is connected to the vehicle DLC, and this system is used for all coding,

Copyright © 2015 Cengage Learning®

FIGURE 13-50 **Removing cover on a fiber-optic connector.**

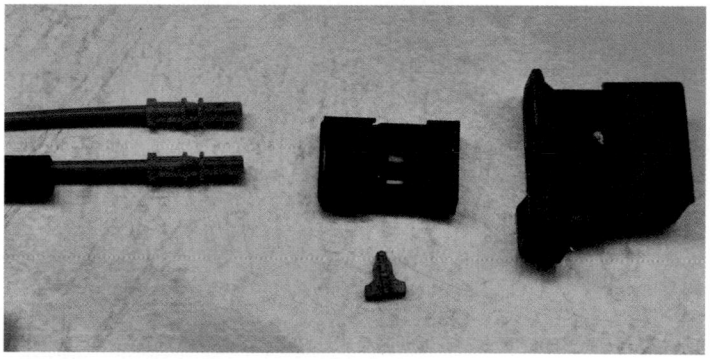

Copyright © 2015 Cengage Learning®

FIGURE 13-51 **Removing a fiber-optic cable from a connector.**

Copyright © 2015 Cengage Learning®

FIGURE 13-52 **Removing a fiber-optic connector from its mounting surface.**

Copyright © 2015 Cengage Learning®

FIGURE 13-53 **Diagnosis system.**

programming, and diagnosis of the active steering system and other systems. ECU and module replacement nearly always require coding/programming of the replacement ECU or module. After the diagnosis system is connected to the DLC, to access the active steering system, select "Service functions," "Suspension," and "Active steering" on the diagnosis system menus. Always follow the vehicle manufacturer's recommended service, diagnosis, or programming procedure when using the diagnosis system.

If an electronic defect occurs in the active steering system, the diagnosis system displays DTCs. As in other electronic systems, the DTCs represent defects in a certain area, and voltmeter or ohmmeter tests may have to be performed to locate the root cause of the problem. The technician must select the appropriate ECU to access the DTCs. For example, the **servotronic valve** and the **electronically controlled orifice (ECO) valve** are controlled by the **safety and gateway module (SGM)** on some models. Therefore, DTCs representing electronic faults in these valves are stored in the SGM, and the technician must access the SGM in the diagnosis system to obtain these DTCs. On other models, the servotronic valve and ECO valve are controlled by the **active steering ECU**, and DTCs representing defects in these components are stored in this ECU.

Certain service procedures and specific adjustments must be performed using the diagnosis system. For example, a steering angle sensor adjustment must be performed after any of the following procedures:

1. Adjustment procedures on the front steering or suspension.
2. Disconnecting the battery.
3. Any replacement of steering components.
4. Replacement of steering column sensors or switches.
5. Replacement of **dynamic stability control (DSC) ECU**, **active roll stabilization (ARS) ECU**, or active steering ECU.

When the active steering system is accessed in the diagnosis system, the steering angle sensor adjustment procedure is entered by selecting "Initial operation/adjustment for active front steering." Follow the vehicle manufacturer's recommended procedure in the diagnosis system to complete the steering angle sensor adjustment. Service and adjustment procedures may vary depending on the model and year of vehicle.

The need for a wheel alignment is indicated by any of the following conditions:

1. Inclination of the steering wheel from the straight ahead position with no faults stored in the active steering ECU memory.
2. Fault(s) stored in the active steering ECU memory and diagnosis system troubleshooting provides "Check wheel alignment" message.
3. Replacement of suspension and/or steering components that requires wheel alignment.

Electronic Four-Wheel Steering Diagnosis and Service

Diagnosis, service, and adjustments on 4WS systems must be performed precisely as explained in the vehicle manufacturer's service manual. On some systems inaccurate sensor adjustments may cause improper rear wheel steering operation, and this may result in reduced steering control. Always follow the diagnostic, service, and adjustment procedures carefully and accurately!

Preliminary Inspection

Prior to any four-wheel steering diagnosis, the following concerns should be considered:

1. Have any suspension modifications been made that would affect steering?
2. Are the tire sizes the same as specified by the vehicle manufacturer?
3. Are the tires inflated to the pressure specified by the vehicle manufacturer?
4. Is the power steering belt adjusted to the vehicle manufacturer's specified tension?
5. Is the power steering pump reservoir filled to the proper level with the type of fluid specified by the vehicle manufacturer?
6. Is the engine idling at the speed specified by the vehicle manufacturer? Is the idle speed steady?
7. Is the steering wheel original equipment?
8. Is the battery fully charged?
9. Are all electrical connections clean and tight?
10. Is there any damaged electrical wiring in the system?
11. Are the rear wheel steering fuses in satisfactory condition?
12. Are there any damaged or worn rear steering gear or rear axle components?
13. Is the Service 4 Wheel Steer indicator illuminated in the instrument panel cluster?

Photo Sequence 27 illustrates a preliminary four-wheel steering inspection.

Quadrasteer Diagnosis

Fast, accurate Quadrasteer diagnosis is very important when correcting steering complaints. If the preliminary inspection does not indicate any problems, the next step in Quadrasteer diagnosis is to perform a diagnostic system check using these steps:

1. With the ignition switch off, connect a scan tool to the DLC under the dash.
2. Turn on the ignition switch, and select Rear Wheel Steering Control Module on the scan tool. If the scan tool does not communicate with this module, check the scan tool electrical connection and the data links. Be sure the scan tool communicates with all the other modules on the vehicle.
3. Access the Class 2 Power Mode on the scan tool, and rotate the ignition switch through all positions while observing the scan tool. The engine may start with the ignition switch in the Start position. The ignition switch position displayed on the scan tool should match the actual ignition switch position.
4. Select the Display DTCs function on the scan tool and then select Rear Wheel Steering Control Module to display the DTCs related to the 4WS system. Select all the other modules on the vehicle, and display any DTCs stored in these modules. Record all DTCs.
5. Are there any DTCs beginning with a U? These DTCs relate to data link problems and must be repaired before proceeding with 4WS diagnosis.
6. Does the scan tool display DTC B1000? This DTC indicates an internal defect in the body control module (BCM), and causes the BCM to refuse all additional inputs. The Quadrasteer system depends on some inputs from the BCM. Therefore, 4WS operation is affected by this BCM problem.

A fail-safe mode may be called a backup mode.

The SAE J1930 terminology is an attempt to standardize electronics terminology in the automotive industry.

In the SAEJ1930 terminology, the term malfunction indicator light (MIL) replaces other terms for computer system indicator lights.

PHOTO SEQUENCE 27

PRELIMINARY INSPECTION, FOUR-WHEEL STEERING DIAGNOSIS

P27-1 **Inspect the front and rear suspension and steering for modifications and damage that could affect steering.**

Copyright © 2015 Cengage Learning®

P27-2 **Check the tire sizes to be sure they are the size specified by the vehicle manufacturer.**

Copyright © 2015 Cengage Learning®

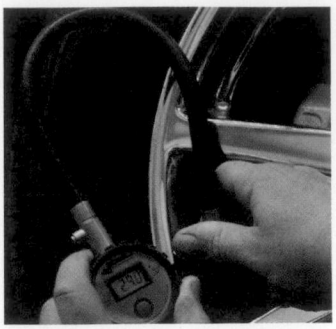

P27-3 **Inflate the tires to the specified pressure.**

Copyright © 2015 Cengage Learning®

P27-4 **Check the power steering belt tension.**

Copyright © 2015 Cengage Learning®

P27-5 **Check the fluid level in the power steering pump reservoir.**

Copyright © 2015 Cengage Learning®

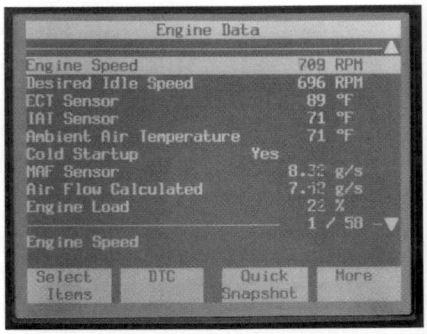

P27-6 **Be sure the engine idle speed is correct.**

Copyright © 2015 Cengage Learning®

P27-7 **Be sure the steering wheel is original equipment.**

Copyright © 2015 Cengage Learning®

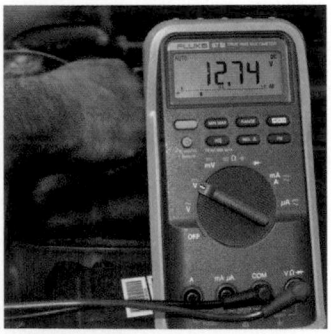

P27-8 **Be sure the battery is fully charged.**

Copyright © 2015 Cengage Learning®

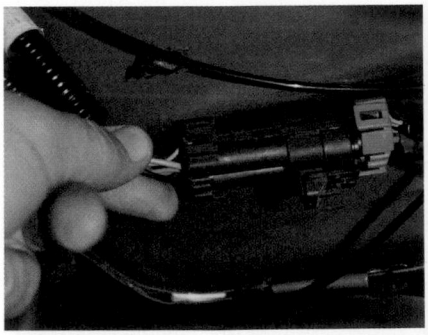

P27-9 **Inspect all electrical connections and harness in the four-wheel steering system to be sure they are in satisfactory condition.**

Copyright © 2015 Cengage Learning®

P27-10 **Check the rear wheel steering system fuses.**
Copyright © 2015 Cengage Learning®

P27-11 **Check the four-wheel steering warning light in the instrument panel for proper operation.**
Copyright © 2015 Cengage Learning®

Rear Wheel Steering Data Display

Select Rear Wheel Steering and Data Display on the scan tool. Observe the data displayed on the scan tool and illustrated in Figures 13-54 and 13-55. Compare the displayed data to the vehicle manufacturer's specified data. Repair the cause of any incorrect data.

Compare the incorrect data to the DTCs recorded previously. If any DTCs and incorrect data are from the same component, repair this component or the related circuit.

Classroom Manual
Chapter 13, page 384

Rear Wheel Steering DTC Interpretation

The possible 4WS system DTCs and causes are the following:

1. B3593—The mode select switch 5 V reference circuit is open, shorted to ground, or shorted to voltage.
2. C0000—The vehicle speed sensor, related circuit, or Class 2 data links are defective.
3. C0253—The steering wheel position sensor phase A and phase B voltage signals may be out of the valid alignment range or the steering wheel position marker pulse points may not be occurring at the proper time in relation to steering wheel rotation (Figure 13-56). The steering wheel position sensor analog voltage may be out of the

SERVICE TIP:
If any of the sensors or the control module is replaced in a Quadrasteer system, a Learn Rear Wheel Alignment Procedure must be performed with a scan tool.

Operating conditions: Ignition ON/Engine ON

Scan tool parameter	Data List	Units displayed	Typical data value
2 wheel steer mode lamp	Output	On/Off	Varies
4 wheel steer mode lamp	Output	On/Off	Varies
4WS-Tow mode lamp	Output	On/Off	Varies
8-digit GM part number	ID information	Numerical	Varies
Actual rear wheel steering angle	Data display	Degrees	0° to 12°
Base model part number	ID information	Numerical	Varies
Battery voltage	Data display	Voltage	9.0-16.0 volts
Calibration part number	ID information	Numerical	Varies
Calibration S/W suffix	ID information	Numerical	Varies
Digital SWPS phase A	Data display	High/Low	Varies
Digital SWPS phase B	Data display	High/Low	Varies
Hall sensor reference	Data display	Voltage	12 volts
Ignition 3	Data display	Voltage	9.0-16.0 volts

Copyright © 2015 Cengage Learning®

FIGURE 13-54 **Rear wheel steering data.**

Operating conditions: Ignition ON/Engine ON

Scan tool parameter	Data List	Units displayed	Typical data value
Ign. cycles since last fault	Data display	Counts	Counts
Marker pulse	Data display	High/Low	1
Motor current	Data display	Amps	20-50A typical 0 to 85A max.
Motor relay commanded state	Data display	On/Off	Varies
Motor relay feedback state	Data display	On/Off	Varies
Rear steer mode actual	Data display	2WS,4WS,4WS tow	Varies
Rear steer mode requested	Data display	2WS,4WS,4WS tow	Varies
Rear steer position 1	Data display	Voltage	.25 to 4.75 volts
Rear steer position 2	Data display	Voltage	.25 to 4.75 volts
Rear steer select switch	Data display	On/Off	Off
Rear wheel centering 1	Data display	Voltage	2.5 volts
Rear wheel centering 2	Data display	Voltage	2.5 volts
Requested rear steer angle	Data display	Degrees	0° to 12°
Sensor supply voltage	Data display	Voltage	4.97 volts
Steering wheel angle	Data display	Degrees	-609° to +609°
Steering wheel angle (TBC)	Data display	Degrees	.25 to 4.75 volts
Steering wheel sensor signal	BCM data	Voltage	Varies
Vehicle speed	Data display	km/h (mph)	

Copyright © 2015 Cengage Learning®

FIGURE 13-55 **Rear wheel steering data**

Copyright © 2015 Cengage Learning®

FIGURE 13-56 **Wiring connection M41 front ECU**

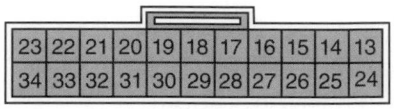

Copyright © 2015 Cengage Learning®

FIGURE 13-57 **Wiring connection M42 front ECU**

valid alignment range (Figure 13-57), the rear wheel position sensor analog voltage may be out of the valid alignment range, or a valid Learn Wheel Alignment Procedure may not have been performed.

4. C0455—The steering wheel position sensor phase A, phase B, or marker pulse signals are defective or the related circuits are open, or shorted to ground or voltage. This DTC may also be caused by a fault that sets a BCM DTC C0472 or C0473.
5. C0472 or C0473—The BCM detects the analog steering wheel position sensor circuit is open or shorted to ground or voltage.
6. C0522—The rear wheel position sensor position 1 or position 2 signals are defective, the 5 V reference circuit to this sensor is open or shorted, or the sensor ground circuit is open.
7. C0527—The Hall Effect switches in the rear steering gear actuator motor are defective, the voltage supply to these switches is open, or the ground circuit is open.
8. C0532—The rear wheel position sensor and the actual rear wheel position varies by more than 1.4°.
9. C0533—The shorting relay in the rear steering actuator motor is stuck open or closed.
10. C0538—The rear steering motor electrical circuit is open, or shorted to ground or voltage.
11. C0543—The difference between the commanded rear wheel steering position and the actual rear wheel steering position exceeds 3°.
12. C0550—When the ignition switch is turned on, the rear wheel steering control module detects an internal malfunction.

The vehicle manufacturer's service manual contains detailed diagnostic procedures for each rear wheel steering DTC. These procedures must be followed when diagnosing the cause of any DTC. Table 13-3 provides a general Quadrasteer diagnostic procedure.

TABLE 13-3 **QUADRASTEER DIAGNOSIS**

Problem	Symptoms	Possible Causes
Inoperative rear wheel steer system	Rear wheels do not steer when front wheels are turned	Open fuse(s) related to 4WS system Defect in data links Defective BCM Defective input sensor(s) Defective rear wheel steering gear
Repeated blowing of fuse(s) related to 4WS system	Inoperative rear wheel steering system	Grounded or shorted circuit in related fuse circuit
Improper 4WS operation	Steering wheel position sensor phase A, phase B, marker pulse, and steering wheel position analog voltage signals are out of alignment with steering wheel rotation; DTC C0253 displayed	Defective steering wheel position sensor Learn Rear Wheel Alignment procedure not performed after component replacement

Copyright © 2015 Cengage Learning®

Rear Wheel Steering Learn Rear Wheel Alignment Procedure

Follow these steps to complete a Learn Rear Wheel Alignment procedure:

1. Turn on the ignition switch and start the engine.
2. Connect the scan tool to the DLC under the dash.
3. Center the steering wheel.
4. Lift the rear of the vehicle so the rear tires are a few inches off the shop floor. Be sure the chassis is securely supported on safety stands, and the rear wheels are centered.
5. Select the Learn Alignment menu on the scan tool.
6. Follow the directions on the scan tool. When directed, the steering wheel must be turned 90° to the left and 90° to the right, and then returned to the center position.
7. Press the Continue button on the scan tool until the alignment procedure is completed.
8. Use the scan tool to erase the rear wheel steering DTCs.
9. Shut off the ignition, disconnect the scan tool, and lower the rear wheels onto the shop floor.

SERVICE TIP:
If step 6 is not completed properly, a false steering wheel position sensor DTC may be set in the rear wheel steering control module memory.

Diagnosis of Four-Wheel Active Steering (4WAS) System

Service Precautions Related to the Supplemental Restraint System (SRS)

Classroom Manual
Chapter 13, page 396

The SRS contains air bags in different locations and seat belt pre-tensioners. Improper service procedures may result in unintended air bag deployment and personal injury. To avoid personal injury and/or rendering the SRS inoperative, all precautions recommended by the vehicle manufacturer must be observed. These precautions include the following:

1. Never use any electrical test equipment on any circuit unless this equipment is recommended in the vehicle manufacturer's service manual.
2. Use only service procedures recommended by the vehicle manufacturer.
3. If heavy vibration occurs near a SRS system sensor while the ignition switch is on or the engine is running, air bag deployment may occur. When using air or electric power tools and hammers, always turn the ignition switch Off, disconnect the negative battery cable followed by the positive battery cable, and wait 3 minutes before proceeding with the service work.
4. To avoid accidental air bag deployment when working on the electrical system, turn the ignition switch Off, disconnect both battery cables, and wait 3 minutes.

Service Precautions Related to Electronic Control Unit (ECU) Replacement

Before removing and replacing any ECU, turn the push-button ignition switch to the Lock position, and then disconnect negative battery cable followed by the positive battery cable. After the replacement control unit is installed, be sure all electrical connectors are properly connected and all related wiring harness is in satisfactory condition. Then reconnect the positive cable followed by the negative cable. Use the vehicle manufacturer's recommended scan tool to perform a self-diagnosis on the replacement control unit.

Safety Precautions Related to the Steering Lock Unit

A number of vehicles are equipped with a push-button ignition switch and a steering lock that engages if the battery is disconnected or discharged. If steering wheel rotation is required with the steering lock engaged, follow these steps:

1. Connect both battery cables.
2. If the battery is discharged, use jumper cables connected from a booster battery to the battery terminals with the correct polarity to provide battery voltage to the electrical system.
3. Turn the push-button ignition switch to the ACC position. This action releases the steering lock.
4. Disconnect the negative battery cable followed by the positive battery cable. The steering lock will remain released with both battery cables disconnected, and steering wheel rotation is possible.
5. When the service work is completed, reconnect the positive battery cable followed by the negative cable. With the brake pedal released, turn the push-button ignition switch from the ACC position to the On position, and then to the Lock position. The steering lock will be engaged when the ignition switch is moved to the Lock position. Use the vehicle manufacturer's recommended scan tool to perform a self-diagnosis of all control units in the system being serviced.

Preliminary Diagnosis and Visual Inspection

When diagnosing the 4WAS system, the first step is to identify the complaint. Questioning the customer regarding the vehicle operation often helps to identify the complaint. A vehicle road test may be necessary to experience and determine the exact complaint. After the complaint is identified, the next step is to visually inspect the system to identify any defects. When diagnosing the 4WAS system, be sure the power steering fluid level and the belt tension are correct. Inspect all the wiring harness and wiring connectors for damage and corrosion. Be sure all system components such as the front and rear actuators are mounted securely. Check the 4WAS warning light. This light should go out after the engine starts. If the warning light remains On with the engine running, the 4WAS system has an electronic defect. If no defects are found during the visual inspection, further diagnosis is required.

Scan Tool Diagnosis

The ignition switch should be in the Off position when connecting and disconnecting the scan tool. A scan tool may be connected to the DLC to perform these tests:

1. Self-diagnostic tests—retrieves DTCs indicating electronic defects in certain electronic areas.
2. Data monitor tests—reads data from the system components to determine component operation and condition.

3. CAN diagnostic monitor—indicates if CAN transmit/receive communication is satisfactory.
4. Active tests—activates specific system components to determine component operation.
5. ECU part number—identifies the front ECU part number.

Diagnostic Example DTC C1631

The electronic systems on current vehicles contain many DTCs and provide a large number of data items. The lists of DTCs and data items for various electronic systems are provided in the vehicle manufacturer's service manual. We will provide some diagnostic examples to illustrate how the DTCs and data items may be used to diagnose the 4WAS system.

If DTC C1631 is displayed on the scan tool, the DTC list for the front control unit indicates this DTC represents an error in the front ECU or in the voltage supply to the front ECU. To locate the root cause of this defect, follow these steps:

1. Turn the ignition switch Off and disconnect terminal M41 and M42 on the front ECU as illustrated in Figures 13-56 and 13-57. A partial wiring diagram is shown in Figure 13-58.

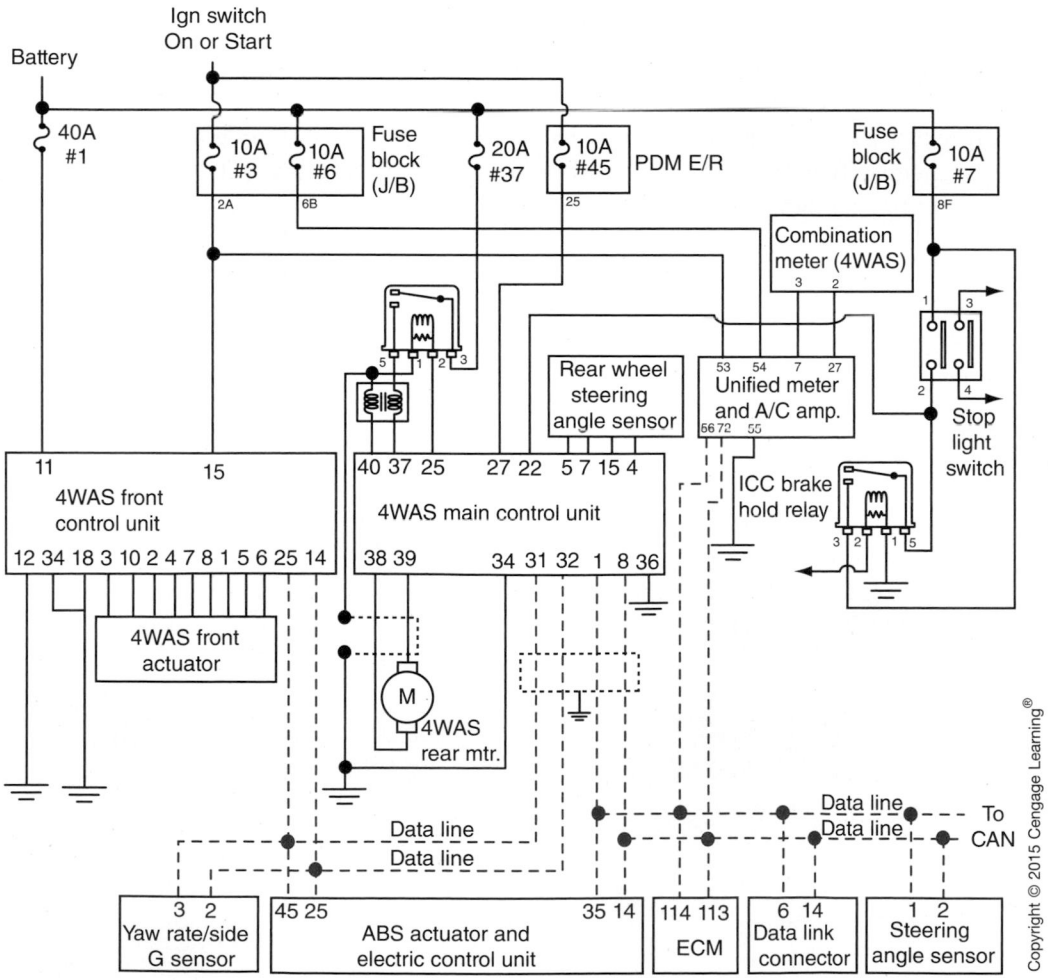

FIGURE 13-58 **4WAS wiring diagram**

2. With the ignition switch Off, connect a voltmeter from terminal 11 to ground in terminal M41. The voltmeter should indicate 12 V. When the voltmeter is connected from terminal 15 in connector M42 to ground, the meter should read 0 V.
3. Turn the ignition switch On and the voltage measured from both terminals 11 and 15 to ground should be 12 V.
4. If the voltage reading is low at terminal 11 to ground, check the 40-A fuse and related circuit, and repair as necessary. When the voltage reading is low at terminal 15, check the 10-A fuse and related circuit, and repair as required.
5. Turn off the ignition switch and connect an ohmmeter from terminal 12 in terminal M41 to ground. The meter should read very low resistance if the ground wire is satisfactory. A high meter reading indicates excessive resistance in this wire or wire terminals and connectors. If a high reading is obtained, repair or replace the ground wire from terminal 12 to ground.
6. Connect the ohmmeter leads from terminals 18 to ground and 34 to ground in connector M42. The ohmmeter should indicate a very low reading at each of these connections. If either connection indicates a high ohmmeter reading, repair or replace the ground wires from terminals 18 and 34 to ground.
7. If the meter readings in steps 1 through 7 are satisfactory, replace the front ECU.

Diagnostic Example DTC C1902

If DTC C1902 is obtained on the scan tool, a rear motor current error is indicated. To pinpoint the root cause of this DTC, follow this procedure:

1. Turn the ignition switch Off and disconnect wiring connector B54 (Figure 13-59) from the rear ECU and remove wiring connector B36 (Figure 13-60) from the rear motor.
2. Connect a pair of ohmmeter leads from terminal 38 in connector B54 to terminal 1 in connector B36. The ohmmeter should indicate a very low reading. If the ohmmeter reading is high, repair or replace the wire from terminal 1 to terminal 38 as required.
3. Connect a pair of ohmmeter leads from terminal 39 in connector B54 to terminal 2 in connector B36. The ohmmeter should indicate a very low reading. If the ohmmeter reading is high, repair or replace the wire from terminal 2 to terminal 39 as required.
4. Connect a pair of ohmmeter leads from terminals 1 and 2 on the rear motor. A low ohmmeter reading is satisfactory. If the ohmmeter reading is high, replace the rear steering actuator.
5. Reconnect all disconnected connectors and connect the scan tool to the DLC. Using the scan tool, select ACTIVE TEST on the rear ECU and rear actuator. This active test forces the rear ECU to operate the motor in the rear actuator. During the active test, select MOTOR VOLTAGE, MOTOR CURRENT, and MTR CRNT OPE on the

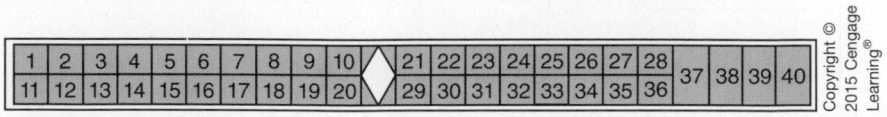

FIGURE 13-59 **Wiring connector B54 rear ECU**

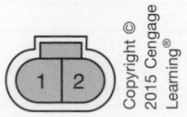

FIGURE 13-60 **Wiring connector B36 rear motor**

scan tool data monitor. The following readings should be obtained on the scan tool during these selections:

MOTOR VOLTAGE 12 V
MOTOR CURRENT 0–20 A

MTR CRNT OPE—with the steering in the straight ahead position, the rear actuator in the neutral condition −2 to 2A, and rear motor running −20 to 20A.

6. If the readings in step 5 are not within specifications, replace the rear actuator.
7. If all the readings in steps 1 through 5 are satisfactory and DTC 1902 continues to appear, replace the rear ECU.

4WAS Adjustments

If the front ECU or front actuator is replaced or a wheel alignment is performed, a front actuator neutral position adjustment is necessary. Proceed as follows to perform the front actuator neutral position adjustment:

1. With the ignition switch Off, connect the scan tool to the DLC.
2. Be sure the front wheels are in the straight ahead position. Start the engine and select DATA MONITOR and ACTR ROTA ANG on the scan tool.
3. Turn the steering wheel until the ACTR ROTA ANG reading on the scan tool indicates −3.5 to 3.5 degrees.
4. Turn the ignition switch Off.

CASE STUDY

A customer complains about excessive steering effort on a 2012 Buick Enclave. The technician road tested the car and found no evidence of hard steering. Further questioning of the customer revealed that the problem only occurred when with the engine idling a 90 degree parking maneuver was attempted in either forward or reverse and it occurred during both right and left turns.

Prior to starting the engine, the technician connected a power steering pressure test gauge in series with the high-pressure hose. When the engine was started, the technician discovered the steering effort seemed normal when the vehicle was driven and only lane change maneuvers where attempted. The line pressure was at the lower end of the normal range compared to specifications when the wheels were not turned. But when a 90 degree turn was attempted at idle speed the line pressure dropped off significantly, and the steering felt much like manual steering while the steering wheel was turned. A deadhead pressure test was performed and line pressure was below specifications.

Since the power steering pump pressure was lower than specified, the technician concluded the power steering pump was faulty. The technician also reasoned that other causes of excessive steering effort, such as binding in the steering column, flexible coupling, or power steering rack spool valve would be constant.

The customer was advised that the power steering pump required replacement. A replacement power steering pump was installed, and the system was flushed and refilled with power steering fluid. When the engine was started, the steering effort and the power steering pump pressure were normal.

TERMS TO KNOW

Active roll stabilization (ARS) ECU
Active steering ECU
Bellows boots
Bump steer
Claw washers
Column-drive EPS
Dynamic stability control (DSC) ECU
Electronically controlled orifice (ECO) valve
On-board diagnostic II (OBD II)
Rack-drive electronic power steering
Safety and gateway module (SGM)
Servotronic valve
Soft-jaw vise
Steering effort imbalance
Steering wheel free play

ASE-STYLE REVIEW QUESTIONS

1. While discussing steering wheel free play:

   Technician A says excessive steering wheel free play may be caused by a worn outer tie rod end.

   Technician B says excessive steering wheel free play may be caused by loose steering gear mounting bushings.

   Who is correct?

   A. A only
   B. B only
   C. Both A and B
   D. Neither A nor B

2. While discussing steering effort:

   Technician A says a tight rack bearing adjustment may cause excessive steering effort.

   Technician B says a loose inner tie rod end may cause excessive steering effort.

   Who is correct?

   A. A only
   B. B only
   C. Both A and B
   D. Neither A nor B

3. While discussing steering gear removal and replacement on air-bag-equipped vehicles:

   Technician A says the steering column should be locked in the centered position to prevent air bag clock spring damage.

   Technician B says that an index mark should be placed on the lower universal joint and steering gear pinion shaft.

   Who is correct?

   A. A only
   B. B only
   C. Both A and B
   D. Neither A nor B

4. While being test driven a vehicle equipped with power rack and pinion steering has a clunking noise, which can also be felt in the steering wheel as it is driven straight ahead on rough pavement. What is the most likely cause?

   A. Loose center link
   B. Loose wheel bearing
   C. Worn flexible steering coupler
   D. Warped brake rotors

5. A steering imbalance or low power assist in both directions may be caused by all of the following EXCEPT:

   A. Defective rack seals
   B. Defective rotary valve rings or seals
   C. Worn lower control arm bushings
   D. Restricted power steering lines or hoses

6. While discussing power rack and pinion steering gears:

   Technician A says power steering fluid should be inspected for discoloration and contamination with metal particles.

   Technician B says after flushing and refilling the power steering system, it is not necessary to perform an air bleeding procedure.

   Who is correct?

   A. A only
   B. B only
   C. Both A and B
   D. Neither A nor B

7. When performing a Learn Rear Wheel Alignment procedure:

   Technician A says the steering wheel must be centered.

   Technician B says during this procedure the front wheels must be lifted off the shop floor.

   Who is correct?

   A. A only
   B. B only
   C. Both A and B
   D. Neither A nor B

8. Technician A says the pinion gear spool valve can bind or stick causing fluid to be directed to the power rack piston chamber causing a drift or pull to one side.

   Technician B says that a front tire with excessive conicity can cause the vehicle to pull or drift to one side and is often referred to as a radial tire pull.

   Who is correct?

   A. A only
   B. B only
   C. Both A and B
   D. Neither A nor B

9. When diagnosing a 4WAS with a scan tool:

   Technician A says the CAN diagnostic monitor indicates if transmit/receive communication is satisfactory.

   Technician B says that during the active tests the scan tool activates specific system components.

   Who is correct?

   A. A only
   B. B only
   C. Both A and B
   D. Neither A nor B

10. When diagnosing an active steering system with a scan tool:

    Technician A says if the DTC's first letter is a C it indicates it is related to a chassis problem.

    Technician B says that when the DTC's second digit is a 1 the DTC is a universal code.

    Who is correct?

    A. A only
    B. B only
    C. Both A and B
    D. Neither A nor B

ASE CHALLENGE QUESTIONS

1. While discussing power rack and pinion steering gears:

   Technician A says inner rack seal leaks can cause power rack and pinion steering to turn easier in one direction than another.

   Technician B says a hissing noise is normal when the steering is turned to the limit on either side.

   Who is correct?

   A. A only
   B. B only
   C. Both A and B
   D. Neither A nor B

2. A pull, wandering, or poor straight-ahead tracking may be caused by all of the following EXCEPT:

   A. Worn or loose rack mounting bushings.
   B. Rack piston seal leaks.
   C. Binding spool valve.
   D. Loose or worn tie rod cnds.

3. A customer says his front-wheel-drive car with power rack and pinion steering is hard to steer for several minutes the first time the car is driven during the day.

   Technician A says the cause of the problem could be a plugged spool valve spring.

   Technician B says the cause of the problem could be a scored steering gear cylinder.

   Who is correct?

   A. A only
   B. B only
   C. Both A and B
   D. Neither A nor B

4. A front-wheel-drive car requires very heavy steering effort in the parking lot but seems not too bad on the highway. Tire pressure is fine. After checking the fluid and finding the level is OK, you should:

   A. Check the engine idle speed.
   B. Look for steering system leaks.
   C. Check the tie rod ends.
   D. Inspect the rack for damage.

5. *Technician A* says an oil leak at one rack seal may result in oil leaking from both boots.

   Technician B says the inner rack seal is defective if oil leaks from the pinion end of the housing when the rack reaches the left internal stop.

   Who is correct?

   A. A only
   B. B only
   C. Both A and B
   D. Neither A nor B

JOB SHEET 48

Name _____ **Date** _____

Remove and Replace Manual or Power Rack and Pinion Steering Gear

Upon completion of this job sheet, you should be able to remove and replace manual or power rack and pinion steering gears.

NATEF AST/MAST Task Correlation:

Suspension and Steering. Steering Systems Diagnosis and Repair,
Task #B-6. Inspect steering shaft universal-joint(s), flexible coupling(s), collapsible column, lock cylinder mechanism, and steering wheel; perform necessary action. (P-2)
Task #B-7. Remove and replace rack and pinion steering gear; inspect mounting bushings and brackets. (P-2)

Tools and Materials

Floor jack
Safety stands
Outer tie rod end puller
Tape measure
Torque wrench
Power steering fluid

Describe the Vehicle Being Worked on:

Year _____ Make _____ Model _____
VIN _____ Engine type and size _____

Procedure

_____
Task Completed

1. Place the front wheels in the straight-ahead position and remove the ignition key from the ignition switch to lock the steering wheel. Place the driver's seat belt through the steering wheel to prevent wheel rotation if the ignition switch is turned on. This action maintains the clock spring electrical connector or spiral cable in the centered position on air-bag-equipped vehicles.
   Are the front wheels straight ahead and steering wheel locked? ☐ Yes ☐ No
   Instructor's Check _____

2. Lift the front end with a floor jack and place safety stands under the vehicle chassis. Lower the vehicle onto the safety stands. Remove the left and right fender apron seals.
   Is the chassis properly supported on safety stands? ☐ Yes ☐ No
   Are the fender apron seals removed? ☐ Yes ☐ No
   Instructor's Check _____

3. Place punch marks on the lower universal joint and the steering gear pinion shaft so they may be reassembled in the same position.
   Are the pinion shaft and U-joint punch marked? ☐ Yes ☐ No
   Instructor's Check _____

Task Completed

4. Loosen the upper universal joint bolt, remove the lower universal joint bolt, and disconnect this joint.
   Inspect the U-joint, is the U-joint in good condition? ☐ Yes ☐ No
   Instructor's Check _____

5. Remove the cotter pins from the outer tie rod ends. Loosen, but do not remove, the tie rod end nuts.
   Are the outer tie rod end nuts loosened? ☐ Yes ☐ No
   Instructor's Check _____

6. Use a tie rod end puller to loosen the outer tie rod ends in the steering arms. Remove the tie rod end nuts and remove the tie rod ends from the arms.
   Are the outer tie rod ends removed from the steering arms? ☐ Yes ☐ No
   Instructor's Check _____

7. Use the proper wrenches to disconnect the high-pressure hose and the return hose from the steering gear. (This step is not required on a manual steering gear.)
   Are the high-pressure and return hoses disconnected from the steering gear?
   ☐ Yes ☐ No
   Instructor's Check _____

8. Remove the four stabilizer bar mounting bolts.
   Are the stabilizer mounting bolts removed? ☐ Yes ☐ No
   Instructor's Check _____

9. Remove the steering gear mounting bolts.
   Are the steering gear mounting bolts removed? ☐ Yes ☐ No
   Instructor's Check _____

☐ 10. Remove the steering gear assembly from the right side of the car.

11. Position the right and left tie rods the specified distance from the steering gear housing. Measure this distance with a tape measure.
    Are the right and left tie rods positioned at equal distances from the steering gear housing? ☐ Yes ☐ No
    Instructor's Check _____

☐ 12. Install the steering gear through the right fender apron.

13. Install the pinion shaft in the universal joint with the punch marks aligned. Tighten the upper and lower universal joint bolts to the specified torque.
    Are the punch marks on the pinion shaft and U-joint properly aligned? ☐ Yes ☐ No
    Instructor's Check _____
    Specified U-joint bolt torque _____
    Actual U-joint bolt torque
    Instructor's Check _____

14. Install the steering gear mounting bolts and tighten these bolts to the specified torque.
    Specified steering gear mounting bolt torque _____
    Actual steering gear mounting bolt torque _____

15. Install the stabilizer bar mounting bolts and torque these bolts to specifications.
Specified stabilizer mounting bolt torque _____
Actual stabilizer mounting bolt torque _____

Task Completed _____

16. Install and tighten the high-pressure and return hoses to the specified torque. (This step is not required on a manual rack and pinion steering gear.)
Specified high-pressure and return hose fitting torque
Actual high-pressure and return hose fitting torque

CAUTION:
Do not loosen the tie rod end nuts to align the cotter pin holes. This action causes improper torquing of these nuts.

17. Install the outer tie rod ends in the steering knuckles, and tighten the nuts to the specified torque. Install the cotter pins in the nuts.
Specified outer tie rod end nut torque _____
Actual outer tie rod end nut torque _____
Are the cotter pins properly installed? ☐ Yes ☐ No

18. Tighten the outer tie rod end jam nuts to the specified torque; then tighten the outer bellows boot clamps.
Specified outer tie rod end jam nut torque _____
Actual outer tie rod end jam nut torque _____
Are the bellows boot clamps properly installed and tightened? ☐ Yes ☐ No
Instructor's Check _____

19. Install the left and right fender apron seals; then lower the vehicle with a floor jack. ☐

20. Fill the power steering pump reservoir with the vehicle manufacturer's recommended power steering fluid, and bleed air from the power steering system. (This step is not required on a manual rack and pinion steering gear.)
Is the power steering pump reservoir filled to the specified level with the proper power steering fluid? ☐ Yes ☐ No
Is the air bled from the power steering system? ☐ Yes ☐ No

21. Perform a front wheel toe measurement, and adjust the toe as required.
Toe satisfactory ☐ Yes ☐ No
Toe adjustment performed ☐ Yes ☐ No
Instructor's Check _____

22. Road test the vehicle and check for proper steering gear operation and steering control.
Steering operation: ☐ Satisfactory ☐ Unsatisfactory
If the steering operation in unsatisfactory, state the necessary repairs and explain the reason for your diagnosis.

_____
_____
_____

Instructor's Response _____
_____
_____

JOB SHEET 49

Name _____ **Date** _____

Diagnose Power Rack and Pinion Steering Gear Oil Leakage Problems

Upon completion of this job sheet, you should be able to diagnose power rack and pinion steering gear oil leakage problems.

NATEF AST/MAST Task Correlation:

Suspension and Steering. Steering Systems Diagnosis and Repair,
Task #B-5. Diagnose power steering gear (rack and pinion) binding, uneven turning effort, looseness, hard steering, and noise concerns; determine necessary action. (P-2)
Task #B-11. Inspect for power steering fluid leakage; determine necessary action. (P-1)

Tools and Materials

Power steering fluid

Describe the Vehicle Being Worked on:

Year _____ Make _____ Model _____
VIN _____ Engine type and size _____

Procedure _____ Task Completed

WARNING: If the engine has been running for a length of time, power steering gears, pumps, lines, and fluid may be very hot. Wear eye protection and protective gloves when servicing these components.

1. Be sure the power steering reservoir is filled to the specified level with the proper power steering fluid. ☐

2. Be sure the power steering fluid is at normal operating temperature. If necessary, rotate the steering wheel several times from lock-to-lock to bring the fluid to normal operating temperature.
   Is the power steering reservoir filled to the specified level? ☐ Yes ☐ No
   Is the power steering fluid at normal temperature? ☐ Yes ☐ No
   Instructor's Check _____

3. Inspect the cylinder end of the steering gear for oil leaks.
   Oil leaks at cylinder end of the power steering gear:
   ☐ Satisfactory ☐ Unsatisfactory
   If there are oil leaks in this area, state the necessary repairs and explain the reasons for your diagnosis.
   _____
   _____
   _____

4. Inspect the steering gear for oil leaks at the pinion end of the housing with the engine running and the steering wheel turned so the rack is against the left internal stop.

   Oil leaks at the pinion end of the steering gear with the rack against the left inner stop:

   ☐ Satisfactory ☐ Unsatisfactory

   If there are oil leaks in this area, state the necessary repairs and explain the reasons for your diagnosis.

   _____

   _____

   _____

SERVICE TIP:
An oil leak at one rack seal may result in oil leaks from both boots because the oil may travel through the breather tube between the boots.

5. Inspect the pinion end of the power steering gear housing for oil leaks with the steering wheel turned in either direction.

   Oil leak at the pinion end of the housing with the steering wheel turned in either direction: ☐ Satisfactory ☐ Unsatisfactory

   If there are oil leaks in the area, state the necessary repairs and explain the reasons for your diagnosis.

   _____

   _____

   _____

6. Inspect the pinion coupling area for oil leaks.

   Oil leaks in the pinion coupling area: ☐ Satisfactory ☐ Unsatisfactory

   If there are oil leaks in the area, state the necessary repairs and explain the reasons for your diagnosis.

   _____

   _____

   _____

7. Inspect all the lines and fittings on the steering gear for oil leaks as an assistant turns the wheels with the engine running.

   Oil leaks at steering gear fittings: ☐ Satisfactory ☐ Unsatisfactory

   Oil leaks in steering gear hoses and lines: ☐ Satisfactory ☐ Unsatisfactory

   If there are oil leaks in the power steering fittings, lines, or hoses, state the necessary repairs and explain the reasons for your diagnosis.

   _____

   _____

   _____

8. With the engine running, turn the steering wheel fully in each direction and check the steering effort.

   Steering effort in both directions: ☐ Normal ☐ Excessive

   If the steering effort is not equal in both directions, state the direction in which steering effort is highest: ☐ Right ☐ Left

   If the steering effort is excessive or unequal, state the necessary repairs and explain the reason for your diagnosis.

   _____

   _____

   _____

Instructor's Response _____

_____

_____

JOB SHEET 50

Name _____ **Date** _____

Perform a Learn Rear Wheel Alignment Procedure on a Quadrasteer System

Upon completion of this job sheet, you should be able to perform a Learn Rear Wheel Alignment procedure on a Quadrasteer system.

NATEF MAST Task Correlation:

Suspension and Steering. Steering Systems Diagnosis and Repair,
Task #B-18. Test and diagnose components of electronically-controlled steering systems using a scan tool; determine necessary action. (P-3)

Tools and Materials

Vehicle with a Quadrasteer system
Compatible scan tool

Describe the Vehicle Being Worked on:

Year _____ Make _____ Model _____
VIN _____ Engine type and size _____

Procedure

_____
Task Completed

1. Turn on the ignition switch and start the engine.
   Engine running: ☐ Yes ☐ No
2. Connect the scan tool to the DLC under the dash.
   Scan tool properly connected to the DLC: ☐ Yes ☐ No
3. Center the steering wheel.
   Steering wheel properly centered: ☐ Yes ☐ No
4. Lift the rear of the vehicle so the rear tires are a few inches off the shop floor. Be sure the chassis is securely supported on safety stands, and the rear wheels are centered.
   Rear wheels raised off the shop floor: ☐ Yes ☐ No
   Chassis properly supported on safety stands: ☐ Yes ☐ No
   Steering wheel centered: ☐ Yes ☐ No
5. Select the Learn Alignment menu on the scan tool.
   Learn Alignment displayed on the scan tool: ☐ Yes ☐ No
6. Follow the directions on the scan tool. When directed, the front wheels must be turned 90° to the left and 90° to the right, and then returned to the center position.
   Steering wheel rotated 90° in each direction: ☐ Yes ☐ No
   Steering wheel centered: ☐ Yes ☐ No

7. Press the Continue button on the scan tool until the Learn Alignment procedure is completed.
   Learn Alignment procedure properly completed: ☐ Yes ☐ No

8. Use the scan tool to erase the rear wheel steering DTCs.
   DTCs erased: ☐ Yes ☐ No

9. Shut off the ignition, disconnect the scan tool, and lower the rear wheels onto the shop floor.
   Ignition shut off and scan tool disconnected: ☐ Yes ☐ No
   Vehicle lowered onto shop floor: ☐ Yes ☐ No

Instructor's Response _____

_____

_____

JOB SHEET 51

Name _____ **Date** _____

Quadrasteer Diagnostic System Check

Upon completion of this job sheet, you should be able to perform a Diagnostic System Check on a Quadrasteer System.

NATEF MAST Task Correlation:

Suspension and Steering. Steering Systems Diagnosis and Repair,
Task #B-18. Test and diagnose components of electronically-controlled steering systems using a scan tool; determine necessary action. (P-3)

Tools and Materials

Vehicle with a Quadrasteer system
Compatible scan tool

Describe the Vehicle Being Worked on:

Year _____ Make _____ Model _____
VIN _____ Engine type and size _____

Procedure

_____
Task Completed

1. With the ignition switch off, connect a scan tool to the DLC under the dash.
   Scan tool properly connected to the DLC: ☐ Yes ☐ No
2. Turn on the ignition switch, and select Rear Wheel Steering Control Module on the scan tool. If the scan tool does not communicate with this module, check the scan tool electrical connection and the data links. Be sure the scan tool communicates with all the other modules on the vehicle.
   Proper scan tool communication with rear wheel steering control module and other modules on the vehicle: ☐ Yes ☐ No
3. Access the Class 2 Power Mode on the scan tool, and rotate the ignition switch through all positions while observing the scan tool. The engine may start with the ignition switch in the Start position. The ignition switch position displayed on the scan tool should match the actual ignition switch position.
   Actual ignition switch positions match displayed ignition switch positions on the scan tool: ☐ Yes ☐ No
4. Select the Display DTCs function on the scan tool and then select Rear Wheel Steering Control Module to display the DTCs related to the 4WS system. Select all the other modules on the vehicle, and display any DTCs stored in these modules. Record all DTCs.
   Rear wheel steering control module DTCs _____
   _____
   DTCs from other modules _____
   _____

5. Are there any DTCs beginning with a U? These DTCs relate to data link problems and must be repaired before proceeding with 4WS diagnosis.
   DTCs beginning with a U _____

6. Does the scan tool display DTC B1000? This DTC indicates an internal defect in the body control module (BCM) and causes the BCM to refuse all additional inputs.
   DTC B1000 displayed: ☐ Yes ☐ No

Instructor's Response _____

_____

_____

Chapter 14

Recirculating Ball Steering Gear Diagnosis and Service

BASIC TOOLS
Basic technician's tool set
Service manual

Upon completion and review of this chapter, you should be able to:

- Diagnose power recirculating ball steering gear problems.
- Remove and replace power recirculating ball steering gears.
- Adjust worm shaft thrust bearing preload in power recirculating ball steering gears.
- Adjust sector lash in power recirculating ball steering gears.
- Diagnose and repair oil leaks in power recirculating ball steering gears.
- Disassemble, repair, and reassemble power recirculating ball steering gears.

Steering gear and linkage inspection and service is extremely important to maintain vehicle safety! For example, if a vehicle has excessive steering wheel free play, this problem should be diagnosed and corrected as soon as possible. If the excessive steering wheel free play is caused by a loose sector shaft lash or worm bearing preload adjustments, follow the proper adjustment procedure to correct the problem. However, if this problem is caused by worn tie rod ends or loose steering gear mounting bolts, these conditions create a safety hazard. If a worn tie rod end becomes disconnected, or the steering gear becomes disconnected from the frame, the result is a complete loss of steering control. This condition may cause a collision resulting in vehicle damage and/or personal injury. Diagnostic Table 14-1 on page 559 contains symptoms and causes of common steering gear issues.

Classroom Manual
Chapter 14, page 404

Power Recirculating Ball Steering Gear Diagnosis

If the steering gear is noisy, check these items:

1. Loose pitman shaft lash adjustment—may cause a rattling noise when the steering wheel is turned.
2. Cut or worn dampener O-ring on the valve spool—when this defect is present, a squawking noise is heard during a turn.
3. Loose steering gear mounting bolts.
4. Loose or worn flexible coupling or steering shaft U-joints.

A hissing noise from the power steering gear is normal if the steering wheel is at the end of its travel, or when the steering wheel is rotated with the vehicle standing still. If the steering wheel jerks or surges when the steering wheel is turned with the engine running, check the power steering pump belt condition and tension. When excessive kickback is felt on the steering wheel, check the poppet valve in the steering gear.

When the steering is loose, check these defects:

1. Air in the power steering system. To remove the air, fill the power steering pump reservoir and rotate the steering wheel fully in each direction several times.
2. Loose pitman lash adjustment.
3. Loose worm shaft thrust bearing preload adjustment.
4. Worn flexible coupling or universal joint.
5. Loose steering gear mounting bolts.
6. Worn steering gears.

A complaint of hard steering while parking could be caused by one of these defects:

1. Loose or worn power steering pump belt.
2. Low oil level in the power steering pump.
3. Excessively tight steering gear adjustments.
4. Defective power steering pump with low-pressure output.
5. Restricted power steering hoses.
6. Defects in the steering gear such as:
   (a) Pressure loss in the cylinder because of scored cylinder, worn piston ring, or damaged backup O-ring.
   (b) Excessively loose spool in the valve body.
   (c) Defective or improperly installed gear check poppet valve.

CAUTION:
Hard steel bolts can be used for steering gear mounting. Ribs on the bolt head indicate the bolt hardness. Harder bolts have five, six, or seven ribs on the bolt heads. Never substitute softer steel bolts in place of the harder original bolts, because these softer bolts may break, allowing the steering gear box to detach from the frame and resulting in a loss of steering control.

Power Recirculating Ball Steering Gear Replacement

When the power steering gear is replaced, proceed as follows:

1. Disconnect the hoses from the steering gear and cap the lines and fittings to prevent dirt from entering the system.
2. Remove the pitman arm nut and washer and mark the pitman arm in relation to the shaft with a center punch. Use a puller to remove the pitman arm.
3. Disconnect the steering shaft from the worn shaft.
4. Remove the steering gear mounting bolts and remove the steering gear from the chassis (Figure 14-1).
5. Reverse steps 1 through 4 to install the steering gear. All bolts must be tightened to the specified torque. Be sure the pitman arm is installed in the original position.

CAUTION:
When the steering linkage is disconnected from the gear, do not turn the steering wheel hard against the stops. This action may damage internal steering gear components.

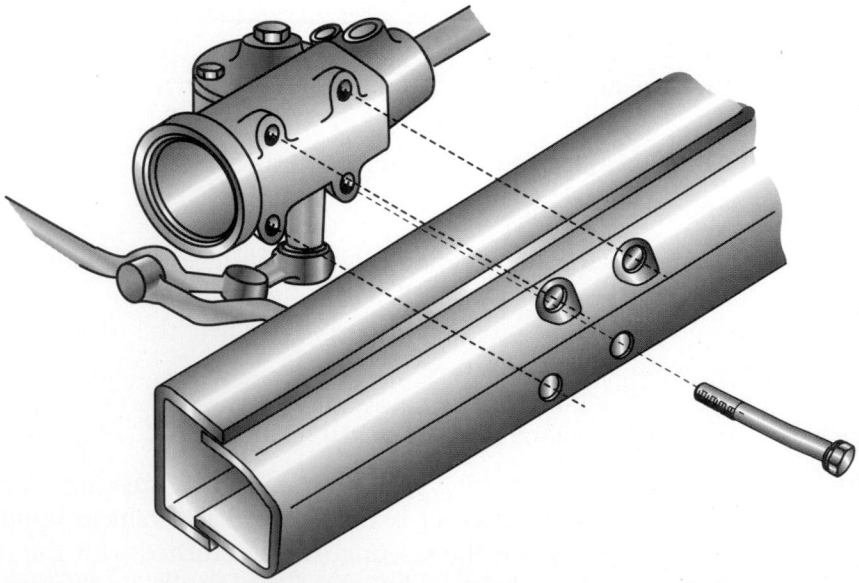

Copyright © 2015 Cengage Learning®

FIGURE 14-1 **Removing steering gear from the chassis.**

WARNING: The steering gear is often mounted near the exhaust system components. Wear protective gloves when working on a steering gear located near the exhaust system.

Power Recirculating Ball Steering Gear Adjustments

Worm Shaft Thrust Bearing Preload Adjustment

Classroom Manual
Chapter 14, page 408

A loose worm shaft thrust **bearing preload** adjustment or sector lash adjustment causes excessive **steering wheel free play** and steering wander. The power recirculating ball steering gear adjustment procedures may vary depending on the vehicle make and model year. Always follow the vehicle manufacturer's recommended procedure in the service manual.

When the worm shaft thrust bearing preload adjustment is performed, use this procedure:

1. Remove the worm shaft thrust bearing adjuster plug locknut with a hammer and brass punch (Figure 14-2).
2. Turn this adjuster plug inward, or clockwise, until it bottoms, and tighten the plug to 20 ft-lb. (27 Nm).
3. Place an index mark on the steering gear housing next to one of the holes in the adjuster plug (Figure 14-3).
4. Measure 0.50 in. (13 mm) counterclockwise from the index mark, and place a second index mark at this position (Figure 14-4).
5. Rotate the adjuster plug counterclockwise until the hole in the adjuster plug is aligned with the second index mark placed on the housing (Figure 14-5).
6. Install and tighten the adjuster plug locknut to the specified torque.

Photo Sequence 28 shows a typical procedure for performing a worm shaft bearing preload adjustment.

Bearing preload is a condition when end play is removed from the bearing and slight tension is applied to the bearing.

Steering wheel free play is the amount of steering wheel movement before the front wheels begin to turn.

Pitman Sector Shaft Lash Adjustment

When the **sector shaft lash** adjustment is too loose, steering wheel free play is excessive, which causes **vehicle wander** when the vehicle is driven straight ahead. A loose sector

SPECIAL TOOL
Worm shaft adjuster plug rotating tool

Sector shaft lash refers to the movement between the sector teeth and ball nut teeth.

Vehicle wander is the tendency of a vehicle to steer right or left as it is driven straight ahead.

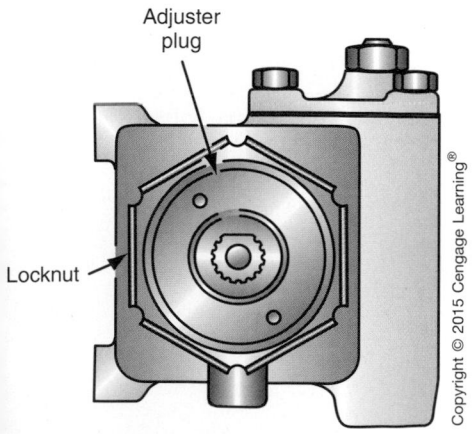

FIGURE 14-2 **Removing worm shaft thrust bearing adjuster plug locknut.**

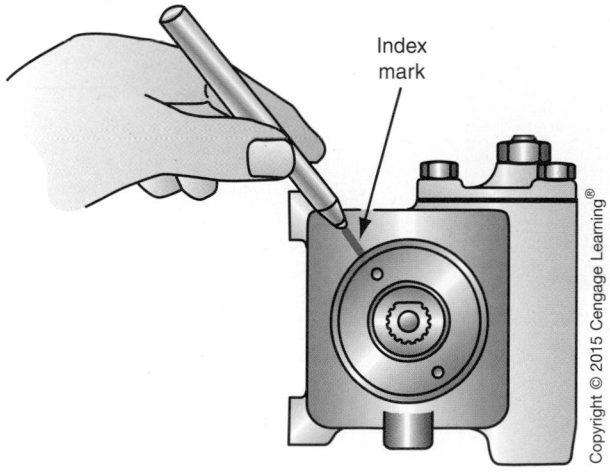

FIGURE 14-3 **Placing index mark on steering gear housing opposite one of the adjuster plug holes.**

PHOTO SEQUENCE 28

Typical Procedure for Performing a Worm Shaft Bearing Preload Adjustment

P28-1 **Retain the steering gear in a vise.**
Copyright © 2015 Cengage Learning®

P28-2 **Remove the worm shaft bearing locknut with a hammer and brass punch.**
Copyright © 2015 Cengage Learning®

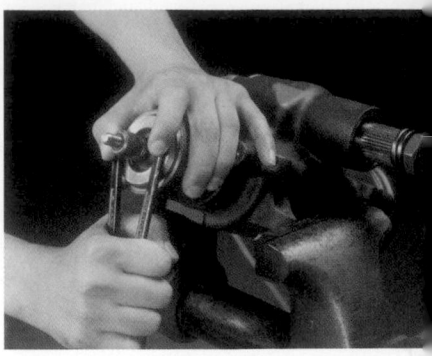

P28-3 **Use the proper tool to turn the worm shaft adjuster plug clockwise until it bottoms, and tighten this plug to 20 ft-lb. (27 Nm).**
Copyright © 2015 Cengage Learning®

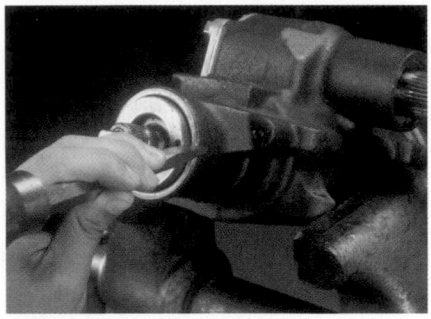

P28-4 **Place an index mark on the steering gear housing next to one of the adjuster plug holes.**
Copyright © 2015 Cengage Learning®

P28-5 **Measure 0.50 in. (13 mm) counterclockwise from the mark placed on the housing in step P11-4, and place a second mark on the housing.**
Copyright © 2015 Cengage Learning®

P28-6 **Rotate the worm shaft adjuster plug counterclockwise until the adjuster plug hole is aligned with the second mark placed on the housing.**
Copyright © 2015 Cengage Learning®

P28-7 **Install and tighten the worm shaft adjuster plug locknut.**
Copyright © 2015 Cengage Learning®

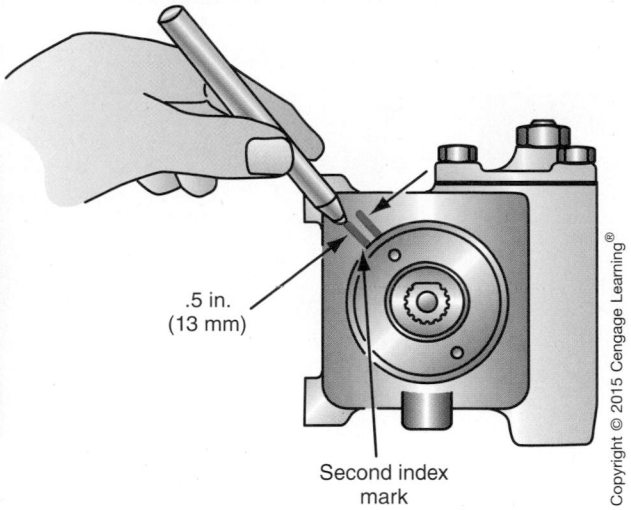

FIGURE 14-4 **Measuring 0.50 in. (13 mm) counterclockwise from the index mark on the steering gear housing.**

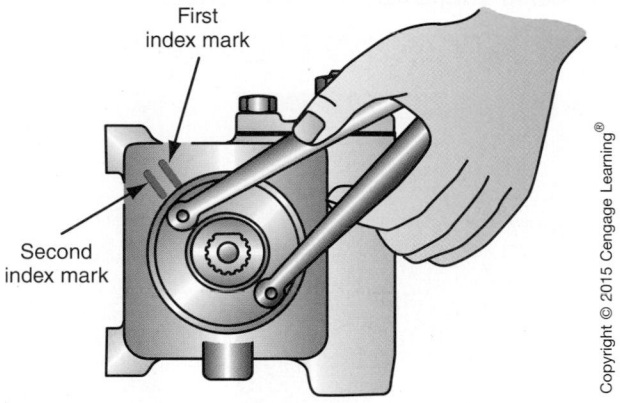

FIGURE 14-5 **Aligning adjuster plug hole with the second index mark placed on steering gear housing.**

shaft lash adjustment decreases driver **road feel**. If the sector lash adjustment is too tight, steering effort is increased, especially with the front wheels in the straight-ahead position.

When the pitman sector shaft lash adjustment is performed, proceed as follows:

1. Rotate the stub shaft from stop to stop and count the number of turns.
2. Starting at either stop, turn the stub shaft back one-half of the total number of turns. In this position, the flat on the stub shaft should be facing upward (Figure 14-6), and the master spline on the pitman shaft should be aligned with the pitman shaft backlash adjuster screw (Figure 14-7).
3. Loosen the locknut and turn the pitman shaft backlash adjuster screw fully counterclockwise, and then turn it clockwise one turn.
4. Use an inch-pound torque wrench to turn the stub shaft through a 45° arc on each side of the position in step 2. Read the over-center torque as the stub shaft turns through the center position (Figure 14-8).
5. Continue to adjust the pitman shaft adjuster screw until the torque is 6 to 10 in-lb. (0.6 to 1.2 Nm) more than the torque in step 4.
6. Hold the pitman shaft adjuster screw in this position, and tighten the locknut to the specified torque as illustrated in Photo Sequence 29.

Road feel is experienced by a driver when the steering wheel is turned and the driver has a positive feeling that the front wheels are turning in the intended direction.

SERVICE TIP:
When tightening the pitman backlash adjuster screw locknut, it is very important to hold the pitman arm lash adjuster screw with a screwdriver to prevent this screw from turning. If the pitman lash adjuster screw turns when the locknut is tightened, the adjustment is changed.

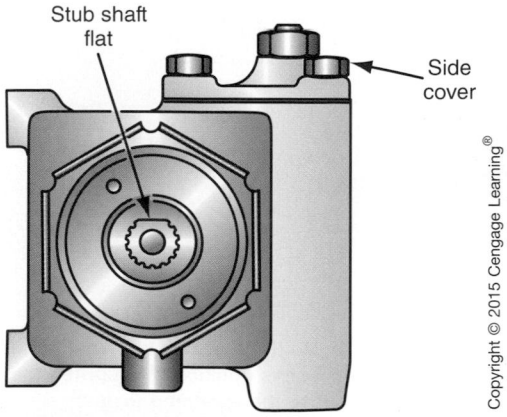

FIGURE 14-6 **Stub shaft flat facing upward and parallel with the side cover.**

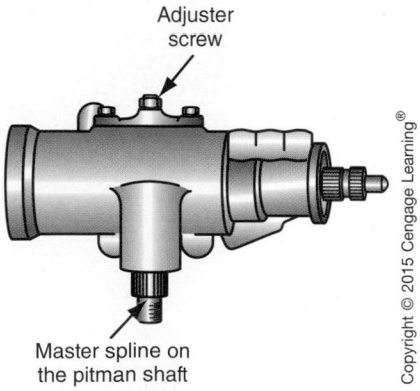

FIGURE 14-7 **Pitman shaft master spline aligned with the pitman backlash adjuster screw.**

PHOTO SEQUENCE 29

PITMAN SECTOR SHAFT LASH ADJUSTMENT

P29-1 **Clamp the steering gear securely in a vise.**

Copyright © 2015 Cengage Learning®

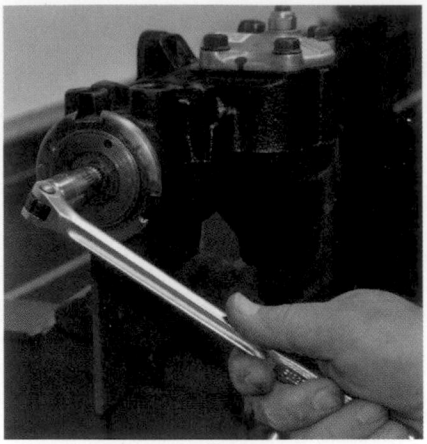

P29-2 **Rotate the stub shaft from stop to stop and count the number of turns.**

Copyright © 2015 Cengage Learning®

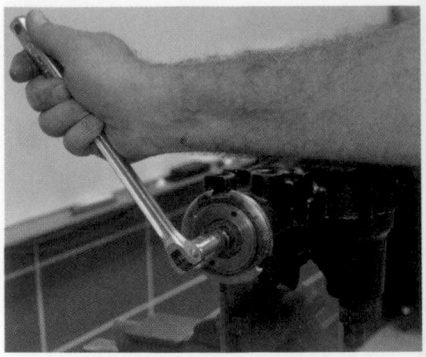

P29-3 **Position the stub shaft at either stop and rotate the shaft back one-half the number of total turns.**

Copyright © 2015 Cengage Learning®

P29-4 **Be sure the master spline on the stub shaft is aligned with the pitman shaft backlash adjuster screw.**

Copyright © 2015 Cengage Learning®

P29-5 **Loosen the locknut and rotate the backlash adjuster screw fully counterclockwise, and then turn this screw clockwise one turn.**

Copyright © 2015 Cengage Learning®

P29-6 **Use an inch-pound torque wrench and the proper size socket to rotate the stub shaft through a 45° on each side of the stub shaft center position.**

Copyright © 2015 Cengage Learning®

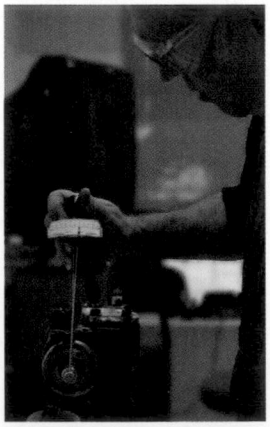

P29-7 **Read the over-center torque reading on the torque wrench as the stub shaft is rotated through the center position.**

Copyright © 2015 Cengage Learning®

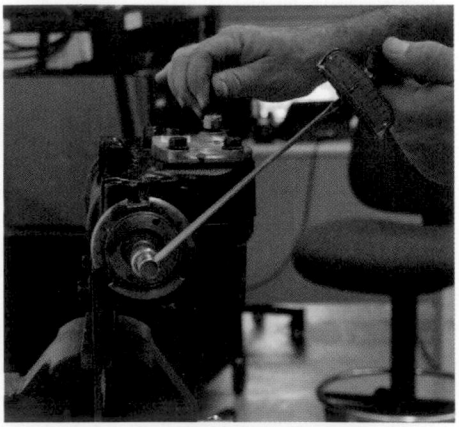

P29-8 **While rotating the stub shaft through the center position, turn the sector backlash adjuster screw until the torque wrench reading is 6–10 in-lb. (0.6–1.2 Nm) more than the torque reading in step 7.**

Copyright © 2015 Cengage Learning®

P29-9 **Hold the sector backlash adjuster screw in this position and tighten the locknut to the specified torque.**

Copyright © 2015 Cengage Learning®

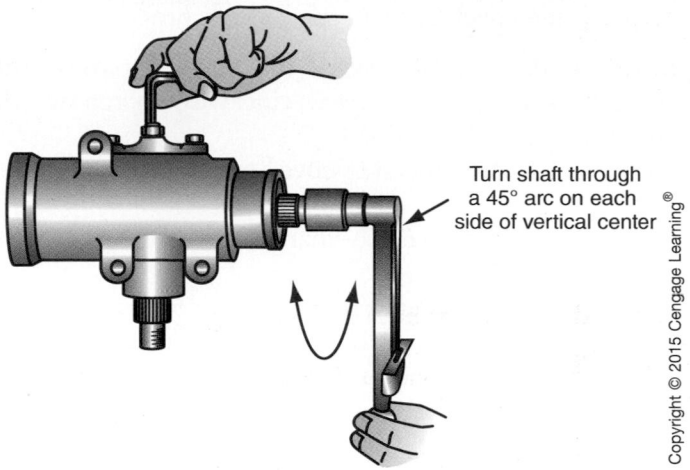

FIGURE 14-8 **Measuring wormshaft turning torque to adjust pitman backlash adjuster screw.**

Power Recirculating Ball Steering Gear Oil Leak Diagnosis

Five locations where oil leaks may occur in a power steering gear are the following:

1. Side cover O-ring seal (Figure 14-9)
2. Adjuster plug seal
3. Pressure line fitting
4. Pitman shaft oil seals
5. End cover seal

If an oil leak is present at any of these areas, complete or partial steering gear disassembly and seal or O-ring replacement is necessary.

Power Recirculating Ball Steering Gear Seal Replacement

Side Cover O-Ring Replacement

Prior to any disassembly procedure, clean the steering gear with solvent or in a parts washer. The steering gear service procedures vary depending on the make of gear. Always follow the vehicle manufacturer's recommended procedure in the service manual.

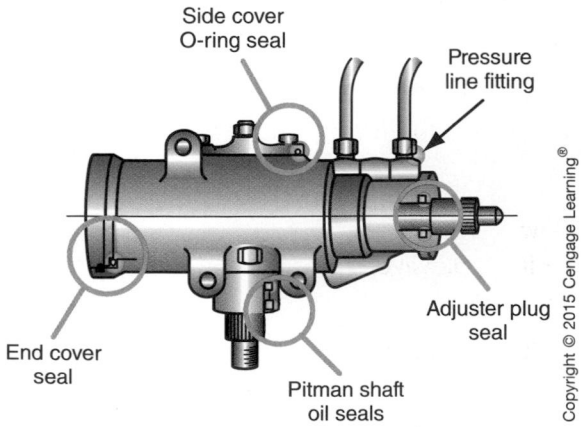

FIGURE 14-9 **Power recirculating ball steering gear oil leak locations.**

Following is a typical side cover O-ring replacement procedure:

1. Loosen the pitman backlash adjuster screw locknut and remove the side cover bolts. Rotate the pitman backlash adjuster screw clockwise to remove the cover from the screw (Figure 14-10).
2. Discard the O-ring and inspect the side cover matching surfaces for metal burrs and scratches.
3. Lubricate a new O-ring with the vehicle manufacturer's recommended power steering fluid and install the O-ring.
4. Rotate the pitman backlash adjuster screw counterclockwise into the side cover until the side cover is properly positioned on the gear housing. Turn this adjuster screw fully counterclockwise and then one turn clockwise. Install and tighten the side cover bolts to the specified torque. Adjust the pitman sector shaft lash as explained earlier.

CAUTION:
The bearing identification number must face the driving tool to prevent bearing damage during installation.

End Plug Seal Replacement

Follow these steps for end plug seal replacement:

1. Insert a punch into the access hole in the steering gear housing to unseat the retaining ring, and remove the ring (Figure 14-11).
2. Remove the end plug and seal.
3. Clean the end plug and seal contact area in the housing with a shop towel.
4. Lubricate a new seal with the vehicle manufacturer's recommended power steering fluid, and install the seal.
5. Install the end plug and retaining ring.

Classroom Manual
Chapter 14, page 408

Worm Shaft Bearing Adjuster Plug Seal and Bearing Replacement

Follow these steps for worm shaft bearing adjuster plug seal and bearing service:

1. Remove the adjuster plug locknut, and use a special tool to remove the adjuster plug.
2. Use a screwdriver to pry at the raised area of the bearing retainer to remove this retainer from the adjuster plug (Figure 14-12).
3. Place the adjuster plug face down on a suitable support, and use the proper driver to remove the needle bearing, dust seal, and lip seal.
4. Place the adjuster plug face up on a suitable support, and use the proper driver to install the needle bearing dust seal and lip seal.
5. Install the bearing retainer in the adjuster plug, and lubricate the bearing and seal with the vehicle manufacturer's recommended power steering fluid.
6. Install the adjuster plug and locknut, and adjust the worm shaft bearing preload as discussed previously.

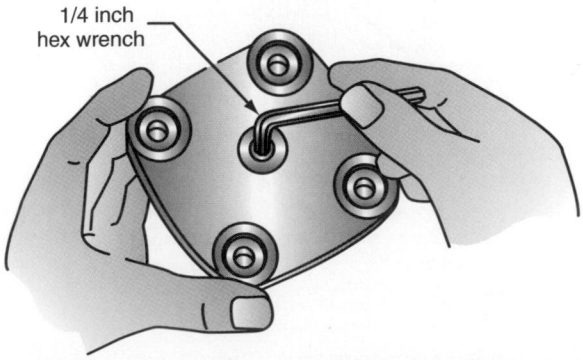

Copyright © 2015 Cengage Learning®

FIGURE 14-10 **Removing steering gear side cover.**

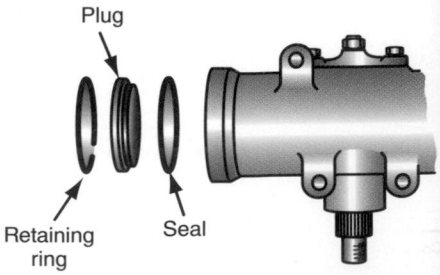

FIGURE 14-11 **Removing steering gear end plug, retaining ring, and seal.**

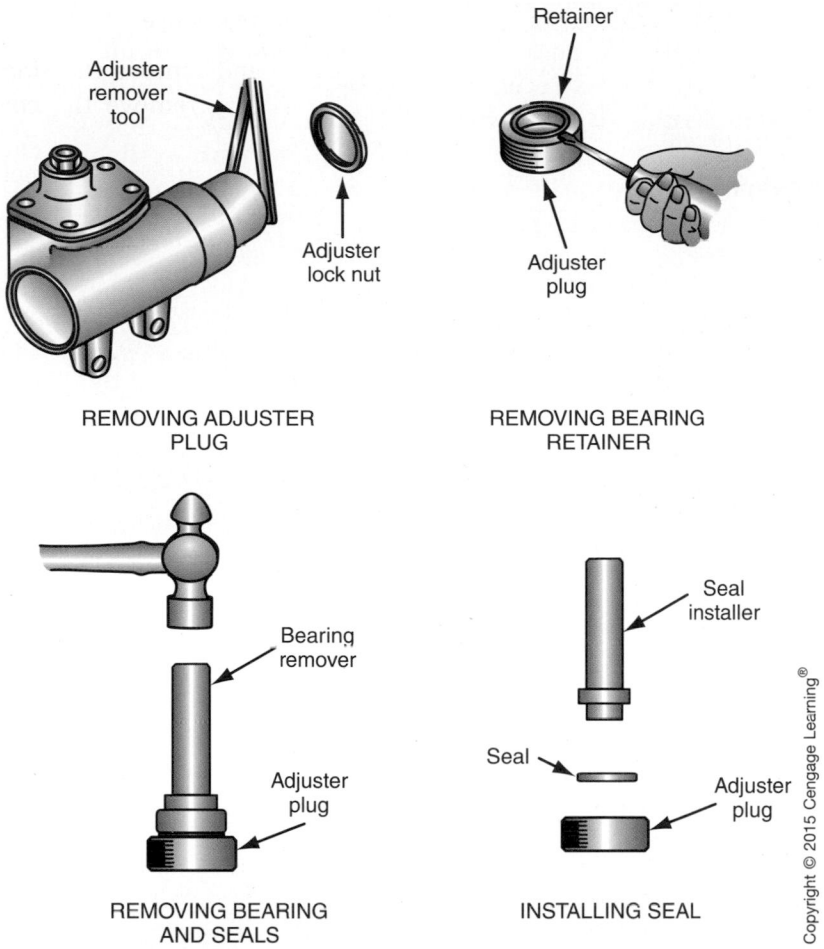

FIGURE 14-12 **Removing and replacing worm shaft adjuster plug, bearing, and seal.**

TABLE 14-1 **STEERING GEAR DIAGNOSIS**

Problem	Symptoms	Possible Causes
Excessive steering wheel free play	Steering wander when driving straight ahead	Loose worm shaft preload adjustment Loose sector lash adjustment Loose steering gear mounting bolts Worn steering linkage components
Excessive steering effort	Excessive steering wheel turning effort when turning a corner or parking	Worn flexible coupling or universal joint in steering shaft.
Under hood noise	Rattling noise when driving over road irregularities Squawking noise while turning a corner	Lack of steering gear lubricant Tight worm shaft bearing preload adjustment Tight sector lash adjustment Worn flexible coupling or universal joint in steering shaft Loose steering gear mounting bolts Cut or worn dampener O-ring on spool valve
Erratic steering effort	Erratic steering effort when turning a corner	Low power steering fluid level Air in power steering system Worn, damaged worm shaft, ball nut, or sector teeth

Copyright © 2015 Cengage Learning®

CUSTOMER CARE: As an automotive technician, you should be familiar with the maintenance schedules recommended by various vehicle manufacturers. Of course, it is impossible to memorize all the maintenance schedules on different makes of vehicles, but maintenance schedule books are available. This maintenance schedule information is available in the owner's manual, but the vehicle owner may not take time to read this manual. If you advise the customer that his or her vehicle requires some service, such as a cooling system flush, according to the vehicle manufacturer's maintenance schedule, the customer will often have the service performed. The customer will usually appreciate your interest in his or her vehicle, and the shop will benefit from the increased service work.

CASE STUDY

The owner of a 2005 Chevrolet Silverado truck complained about increased and somewhat erratic steering effort. The service writer asked the customer about the conditions when this problem occurred, and the customer said that the condition was always present. During a road test, the technician discovered that the customer's description of the problem was accurate.

The technician checked the power steering fluid level and made a careful check of the belt tension and condition without finding any problems. Next, the technician checked the power steering pump pressure and found it to be normal. A check of the power steering hoses did not reveal any hose restrictions.

The technician removed and disassembled the steering gear and found a severely scored cylinder bore in the gear housing. The ball nut piston ring was also worn and scored. A replacement steering gear was installed, and the system filled with the manufacturer's recommended power steering fluid. A road test indicated the new gear worked satisfactorily.

TERMS TO KNOW

Bearing preload
Sector shaft lash
Road feel
Steering wheel free play
Vehicle wander

ASE-STYLE REVIEW QUESTIONS

1. While discussing power recirculating ball steering gear diagnosis:

   Technician A says excessive steering wheel free play may be caused by a worm shaft bearing preload adjustment that is tighter than normal.

   Technician B says excessive steering wheel free play may be caused by loose steering gear mounting bolts.

   Who is correct?

   A. A only
   B. B only
   C. Both A and B
   D. Neither A nor B

2. While discussing power recirculating ball steering gear service:

   Technician A says steering wander may be caused by a loose sector lash adjustment.

   Technician B says a loose sector lash adjustment may cause reduced feel of the road.

   Who is correct?

   A. A only
   B. B only
   C. Both A and B
   D. Neither A nor B

3. While discussing power recirculating steering gear diagnosis:

   Technician A says the sector lash adjustment is performed with the worm shaft halfway between the centered position and the full-right position.

   Technician B says the worm shaft bearing preload adjustment is performed before the sector lash adjustment.

   Who is correct?

   A. A only
   B. B only
   C. Both A and B
   D. Neither A nor B

4. A power recirculating ball steering gear has a rattling noise while driving the vehicle. The most likely cause of this problem is:

   A. Low power steering fluid level.
   B. Worn flexible coupling in the steering shaft.
   C. Loose worm bearing preload adjustment.
   D. Leaking pitman shaft seal.

5. A power recirculating ball steering gear has a fluid leakage problem, and fluid appears on the top of the gear.

   Technician A says the side cover O-ring seal may be leaking.

   Technician B says the high pressure line fitting on the gear may be leaking.

   Who is correct?

   A. A only
   B. B only
   C. Both A and B
   D. Neither A nor B

6. A power recirculating ball steering gear has excessive kickback on the steering wheel. The most likely cause of this problem is:

   A. A loose sector lash adjustment.
   B. A loose worm shaft bearing preload adjustment.
   C. A defective poppet valve in the steering gear.
   D. A worn pitman shaft bearing.

7. All of these statements about power recirculating ball steering gear defects are true EXCEPT:

   A. A worn steering shaft U-joint may cause a growling noise when turning the front wheels.
   B. Excessive steering effort may be caused by a defective power steering pump.
   C. Excessive steering effort may be caused by a scored steering gear cylinder.
   D. Steering wheel jerking when turning may be caused by a slipping power steering belt.

8. When adjusting on a power recirculating ball steering gear:

   A. The worm shaft bearing adjuster plug should be bottomed and then backed off until the specified worm shaft turning torque is obtained.
   B. The sector lash adjustment screw should be tightened one turn before the worm shaft bearing preload is adjusted.
   C. The pitman shaft backlash adjuster screw is turned fully counterclockwise and then one turn clockwise prior to the backlash adjustment.
   D. The pitman shaft is positioned one-half turn from the fully right or left position prior to the pitman shaft backlash adjustment.

9. A power recirculating ball steering gear experiences excessive steering effort while parking. The most likely cause of this problem is:

   A. A restricted high-pressure steering hose.
   B. A misaligned power steering belt.
   C. A leaking pitman shaft seal.
   D. Excessive torque on the worm shaft adjuster plug locknut.

10. A power recirculating ball steering gear experiences hard steering for a short time after the vehicle sits overnight. The most likely cause of this problem is:

    A. Excessively tight worm bearing preload adjustment.
    B. A scored cylinder and worn piston ring in the steering gear.
    C. A restricted power steering return hose.
    D. A binding U-joint in the steering shaft.

ASE CHALLENGE QUESTIONS

1. A car with a power recirculating ball steering gear has excessive steering kickback, and a preliminary inspection shows no abnormal wear in the linkage. Which of the following could be the cause of the problem?

   A. Worn gear piston or bore.
   B. Slipping pump belt.
   C. Worn pump poppet valve.
   D. Sticking valve spool.

2. The complaint is loss of power assist, but there is no mention of any associated noise. All of the following could cause this problem EXCEPT:

   A. Low fluid.
   B. Improperly inflated tires.
   C. Broken pump belt.
   D. Steering column misalignment.

3. While discussing steering problems:

   Technician A says a "jerky" steering wheel and a "clunking" noise could indicate worn steering column U-joints.

   Technician B says lack of assist and a "growling" noise in a fluid-filled steering pump could indicate a hose or pump internal restriction.

   Who is correct?

   A. A only
   B. B only
   C. Both A and B
   D. Neither A nor B

4. A vehicle with a power recirculating ball steering system requires much higher than normal steering effort, especially in the parking lot.

   Technician A says the cause of the problem is a worn pitman shaft seal.

   Technician B says the cause of the problem is a worn worm shaft thrust bearing.

   Who is correct?

   A. A only
   B. B only
   C. Both A and B
   D. Neither A nor B

5. The complaint is excessive steering wheel free play. All of the following could cause this problem EXCEPT:

   A. Loose worm shaft bearing preload.
   B. Worn steering gears.
   C. Steering gear column misalignment.
   D. Worn flex coupling or U joint.

JOB SHEET 52

Name _____ **Date** _____

Power Recirculating Ball Steering Gear Oil Leak Diagnosis

Upon completion of this job sheet, you should be able to diagnose oil leaks in a power recirculating ball steering gear.

NATEF AST/MAST Task Correlation:

Suspension and Steering. Steering Systems Diagnosis and Repair,

Task #B-4. Diagnose power steering gear (non-rack and pinion) binding, uneven turning effort, looseness, hard steering, and noise concerns; determine necessary action. (P-2)

Tools and Materials

Modern vehicle with a power recirculating ball steering gear.

Describe the Vehicle Being Worked On:

Year _____ Make _____ Model _____

VIN _____ Engine type and size _____

Procedure

Task Completed

1. Place fender, seat, and floor mat covers on the vehicle. ☐
2. Clean the outside of the steering gear.
3. Be sure the power steering pump reservoir is filled to the proper level with the specified fluid.
4. Start the engine and observe the power steering gear while an assistant turns the steering wheel fully in each direction several times. Turn off the ignition switch.
5. If the power steering gear is leaking fluid on the lower side, raise the front of the vehicle with a floor jack and support the front suspension on safety stands positioned at the specified vehicle lift points. Repeat steps 2–4.
6. List the fluid leak locations on the power steering gear.

   _____

   _____

7. State the necessary repairs to correct the power steering gear leaks.

   _____

   _____

Instructor's Response _____

_____

_____

JOB SHEET 53

Name _____ **Date** _____

Adjust Power Recirculating Ball Steering Gear Worm Shaft Thrust Bearing Preload, Steering Gear Removed

Upon completion of this job sheet, you should be able to adjust worm shaft thrust bearing preload on power recirculating ball steering gears.

NATEF AST/MAST Task Correlation:

Suspension and Steering. Steering Systems Diagnosis and Repair,

Task #B-4. Diagnose power steering gear (non-rack and pinion) binding, uneven turning effort, looseness, hard steering, and noise concerns; determine necessary action. (P-2)

Tools and Materials

Torque wrench, in-lb.

Describe the Vehicle Being Worked On:

Year _____ Make _____ Model _____

VIN _____ Engine type and size _____

Make of steering gear _____

Procedure

Task Completed

1. Remove the worm shaft thrust bearing adjuster plug locknut with a hammer and brass punch. ☐
2. Turn this adjuster plug inward, or clockwise, until it bottoms and tighten the plug to 20 ft-lb. (27 Nm).

   Actual worm shaft adjuster plug locknut torque _____
3. Place an index mark on the steering gear housing next to one of the holes in the adjuster plug.

   Is an index mark placed on steering gear housing beside one of the adjuster plug holes?

   ☐ Yes ☐ No

   Instructor's Check _____
4. Measure 0.50 in. (13 mm) counterclockwise from the index mark, and place a second index mark at this position.

   Is a second index mark placed 0.50 in. (13 mm) counterclockwise from the first index mark?

   ☐ Yes ☐ No

   Instructor's Check _____

Task Completed

5. Rotate the adjuster plug counterclockwise until the hole in the adjuster plug is aligned with the second index mark placed on the housing.
   Is the hole in adjuster plug properly aligned with second index mark?
   ☐ Yes ☐ No
   Instructor's Check _____

6. Install and tighten the adjuster plug locknut to the specified torque.
   Specified adjuster plug locknut torque _____
   Actual adjuster plug locknut torque _____

Instructor's Response _____

_____

_____

_____

JOB SHEET 54

Name _____ **Date** _____

Adjust Power Recirculating Ball Steering Gear Sector Lash, Steering Gear Removed

Upon completion of this job sheet, you should be able to adjust sector lash on power recirculating ball steering gears.

NATEF AST/MAST Task Correlation:

Suspension and Steering. Steering Systems Diagnosis and Repair,
Task #B-4. Diagnose power steering gear (non-rack and pinion) binding, uneven turning effort, looseness, hard steering, and noise concerns; determine necessary action. (P-2)

Tools and Materials

Torque wrench, in-lb.

Describe the Vehicle Being Worked On:

Year _____ Make _____ Model _____
VIN _____ Engine type and size _____
Make of steering gear _____

Procedure

_____
Task Completed

1. Rotate the worm shaft from stop to stop and count the number of turns.
   Total number of worm shaft turns from stop to stop _____.
2. Starting at either stop, turn the worm shaft back two-thirds of the total number of turns.
   Number of worm shaft turns from fully right or fully left to position the worm shaft properly prior to sector lash adjustment _____.
   Explain the reason for placing the worm shaft in this position prior to the sector lash adjustment.
   _____
   _____
   _____
   _____
3. In this position, the flat on the worm shaft should be facing upward, and the master spline on the pitman shaft should be aligned with the pitman shaft backlash adjuster screw.
   Is the worm shaft flat properly positioned? ☐ Yes ☐ No
   Is the master spline on the pitman shaft properly positioned? ☐ Yes ☐ No
   If the answer is no to either of the above questions, state the necessary corrective action.
   _____
   _____

Task Completed

4. Turn the pitman shaft backlash adjuster screw fully counterclockwise, and then turn it clockwise one turn.

   Is the pitman shaft backlash adjuster screw properly positioned? ☐ Yes ☐ No

   Instructor's Check _____

5. Use an inch-pound torque wrench to turn the worm shaft through a 45° arc on each side of the position in step 2. Read the over-center torque as the worm shaft turns through the center position.

   Stub shaft turning torque _____

6. Continue to adjust the pitman shaft adjuster screw until the torque is 6 to 10 in-lb. (0.6 to 1.2 Nm) more than the torque in step 5.

   Final worm shaft turning torque after adjustment _____

7. Hold the pitman shaft adjuster screw in this position and tighten the locknut to the specified torque.

   Specified pitman shaft adjuster screw locknut torque _____

   Actual pitman shaft adjuster screw locknut torque _____

Instructor's Response _____

_____

_____

1. After new tires and new alloy rims are installed on a sports car, the owner complains about steering wander and steering pull in either direction while braking.

   Technician A says there may be brake fluid on the front brake linings.

   Technician B says the replacement rims may have a different offset than the original rims.

   Who is correct?

   A. Technician A **C.** Both A and B
   B. Technician B **D.** Neither A nor B

2. *Technician A* says when a vehicle pulls to one side; the problem may be caused by a pinion gear binding on the rack gear of a power rack and pinion steering gear.

   Technician B says when a power steering rack and pinion steering gear spool valve binds it may cause a vehicle to pull to one side.

   Who is correct?

   A. Technician A **C.** Both A and B
   B. Technician B **D.** Neither A nor B

3. The outside edge of the left front tire on a front-wheel-drive car is badly worn.

   Technician A says the cause could be worn ball joints.

   Technician B says the cause could be that the camber setting is excessively negative on the left front suspension.

   Who is correct?

   A. Technician A **C.** Both A and B
   B. Technician B **D.** Neither A nor B

4. The owner of a 2-wheel drive pick-up says the front tires squeal loudly during low-speed turns. The most probable cause of this condition is:

   A. Excessive positive camber.
   B. Negative caster adjustment.
   C. Improper steering axis inclination (SAI).
   D. Improper turning angle.

5. A 2-wheel drive pickup has a severe shudder when the vehicle is started from a stop with a load in the bed.

   Technician A says the problem may be worn spring eyes.

   Technician B says the problem may be axle torque wrap-up.

   Who is correct?

   A. Technician A **C.** Both A and B
   B. Technician B **D.** Neither A nor B

6. Steering pull to the right occurs only when braking. The most likely cause of this problem is:

   A. Excessive positive caster on the right-front wheel.
   B. Excessive positive camber on the left-front wheel.
   C. A loose strut rod bushing on the right-front suspension.
   D. Excessive toe-out on the right-front wheel.

7. *Technician A* says hard steering may be caused by low hydraulic pressure due to a stuck flow control valve in the pump.

   Technician B says hard steering may be caused by low hydraulic pressure due to a worn steering gear spool valve piston ring or housing bore.

   Who is correct?

   A. Technician A **C.** Both A and B
   B. Technician B **D.** Neither A nor B

8. *Technician A* says incorrect wheel alignment angles will reduce tire tread life.

   Technician B says wheel alignment is necessary after ball joint replacement.

   Who is correct?

   A. Technician A **C.** Both A and B
   B. Technician B **D.** Neither A nor B

9. While discussing tire tread wear:

Technician A says a scalloped pattern of tire wear indicates an out-of-round wheel or tire.

Technician B says uneven wear on one side of a tire may indicate radial force variation.

Who is correct?

A. Technician A
B. Technician B
C. Both A and B
D. Neither A nor B

10. *Technician A* says excessive front wheel bearing endplay may cause steering wander.

Technician B says a defective front wheel bearing may cause a growling noise while cornering.

Who is correct?

A. Technician A
B. Technician B
C. Both A and B
D. Neither A nor B

11. A vehicle pulls to the right when driving straight ahead.

Technician A says the right-front tire may have excessive radial runout.

Technician B says the right-front tire may have a conicity defect.

Who is correct?

A. Technician A
B. Technician B
C. Both A and B
D. Neither A nor B

12. All of the following could cause shock absorber noise EXCEPT:

A. A bent piston rod.
B. Worn shock absorbers.
C. Shock fluid leaks.
D. Extreme temperatures.

13. When a vehicle pulls to one side, any of the following problems may be the cause EXCEPT:

A. Worn ball joints.
B. Reduced curb height.
C. Bent strut rod.
D. Improper turning angle.

14. A customer says when he applied the brakes hard on his front-wheel-drive car, "the whole car shook."

Technician A says the problem could be worn sway bar bushings.

Technician B says the problem could be worn or loose strut rod bushings.

Who is correct?

A. Technician A
B. Technician B
C. Both A and B
D. Neither A nor B

15. A customer with a sport utility vehicle says after an off-road outing over the weekend, his vehicle pulls to the left on acceleration. Which of the following could cause this problem?

A. Broken leaf spring center bolt.
B. Loose steering unit.
C. Stuck brake pad.
D. Stuck in 4WD.

16. While discussing electronic air suspension:

Technician A says the compressor must be running when a corner of the vehicle is lifted from the ground.

Technician B says on a car with electronic air suspension, the switch should be in the "on" position.

Who is correct?

A. Technician A
B. Technician B
C. Both A and B
D. Neither A nor B

17. Front wheel "shimmy"—a side-to-side movement of the front wheels that is felt in the steering wheel—may be caused by any of the following EXCEPT:

A. Worn tie rod ends.
B. Tire/wheel imbalance.
C. Rack bushings and rack alignment.
D. Tight sector shaft adjustment.

18. A customer says her front-wheel-drive car is hard to steer because the steering wheel no longer returns to center. After the turn, she has to bring it back to center.

Technician A says a corroded or stuck strut bearing plate could be the cause of the problem.

Technician B says a bent tension rod could be the cause of the problem.

Who is correct?

A. Technician A **C.** Both A and B
B. Technician B **D.** Neither A nor B

19. The tire pressure monitoring system is being discussed.

Technician A says that the tire pressure monitoring system must be reinitialized after a wheel pressure sensor has been replaced.

Technician B says that the tire pressure monitoring system must be reinitialized after the tires have been rotated on some vehicles.

Who is correct?

A. Technician A **C.** Both A and B
B. Technician B **D.** Neither A nor B

20. When one of the front wheels strikes a road irregularity, the steering suddenly veers to the right or left on a vehicle with a parallelogram steering linkage. The most likely cause of this problem is:

A. A weak stabilizer bar.
B. Worn-out shock absorbers.
C. Sagged coil springs.
D. Center link that is not level.

21. While discussing suspension height:

Technician A says raising the suspension height in the rear of a vehicle will affect the front suspension geometry.

Technician B says lowering the suspension height of a vehicle will affect the suspension geometry.

Who is correct?

A. Technician A **C.** Both A and B
B. Technician B **D.** Neither A nor B

22. A vehicle with recirculating ball steering has excessive steering wheel free play.

Technician A says a loose worm bearing preload adjustment may cause the problem.

Technician B says loose column U-joints may cause the problem.

Who is correct?

A. Technician A **C.** Both A and B
B. Technician B **D.** Neither A nor B

23. The engine has recently been replaced in a front-wheel-drive car with power rack and pinion steering. The customer now complains about excessive steering effort. A preliminary check of the steering revealed the fluid level was OK and the belts were not slipping.

Technician A says perhaps the rack was knocked out of alignment when the engine was installed.

Technician B says perhaps the pressure line was bent or pinched when the engine was installed.

Who is correct?

A. Technician A **C.** Both A and B
B. Technician B **D.** Neither A nor B

24. While discussing rear suspension systems:

Technician A says semi-independent rear suspension systems momentarily have a slight negative camber and toe-in when the wheel goes over a bump.

Technician B says constantly carrying a lot of weight in the rear of a car with semi-independent rear suspension may cause the rear tires to wear on the inside edge.

Who is correct?

A. Technician A **C.** Both A and B
B. Technician B **D.** Neither A nor B

25. The right front wheel setback is excessive on a Macpherson strut front suspension system. The most likely cause of this problem is:

A. Broken insulators on the right coil spring.
B. A binding right upper strut mount.
C. A broken stabilizer bar.
D. A bent engine cradle.

26. During a wheel alignment, a technician suspects that the front cradle on a front-wheel-drive vehicle may have shifted. If the cradle was shifted from the right to the left, what effect would it have?
    - **A.** Left caster will become more positive and right caster will become more negative evenly.
    - **B.** Right caster will become more positive and left caster will become more negative evenly.
    - **C.** Left camber will become more positive and right camber will become more negative evenly.
    - **D.** Right camber will become more positive and left camber will become more negative evenly.

27. The customer says his vehicle has a rapid "thumping" noise and a vibration in the steering wheel that is most noticeable when the vehicle is at a steady speed in a long curve.

    Technician A says the problem could be caused by a tire defect.

    Technician B says the problem could be improper dynamic wheel balance.

    Who is correct?
    - **A.** Technician A
    - **B.** Technician B
    - **C.** Both A and B
    - **D.** Neither A nor B

28. While discussing power steering fluids:

    Technician A says a foamy, milky power steering fluid is caused by mixing automatic transmission fluid with hydraulic fluid intended for power steering use.

    Technician B says using automatic transmission fluid instead of power steering hydraulic fluids will lower pump pressure and increase steering effort.

    Who is correct?
    - **A.** Technician A
    - **B.** Technician B
    - **C.** Both A and B
    - **D.** Neither A nor B

29. On a car with power rack and pinion steering, the steering suddenly swerves to the right or left when a front wheel strikes a road irregularity. The most likely cause of the problem is:
    - **A.** A loose rack bearing adjustment.
    - **B.** Worn front struts.
    - **C.** Loose steering gear mounting bushings.
    - **D.** Bent steering arm.

30. A customer complains of a steering wheel shake on a rear wheel drive vehicle that occurs at speeds between 50 and 60 mph. What is the most likely cause of this complaint?
    - **A.** Worn control arm bushings.
    - **B.** An out of balance front tire and wheel assembly.
    - **C.** Steering gear is adjusted too tightly.
    - **D.** Worn shocks/struts.

31. When diagnosing an HEV with a scan tool a U0131 is obtained. This DTC indicates a defect in the:
    - **A.** Data network.
    - **B.** Inverter.
    - **C.** Propulsion motor and related circuit.
    - **D.** Generator and related circuit.

32. *Technician A* says front curb riding height is reduced by worn control arm bushings.

    Technician B says front curb riding height is reduced by weak coil springs.

    Who is correct?
    - **A.** Technician A
    - **B.** Technician B
    - **C.** Both A and B
    - **D.** Neither A nor B

33. In a four-wheel active steering system (4WAS), when changing lanes at high speed the:
    - **A.** Rear wheels are steering a few degrees in the same direction as the front wheels.
    - **B.** Rear wheels are steered 5 degrees in the opposite direction to the front wheels.
    - **C.** Camber angle is changed on the rear wheels.
    - **D.** The rear wheels remain in the straight ahead position.

34. While discussing unibody and frame problems:

    Technician A says an indication of possible frame damage is excessive tire wear when the alignment angles are correct.

    Technician B says a worn strut lower ball joint is often an indicator of unibody torsional damage.

    Who is correct?
    - **A.** Technician A
    - **B.** Technician B
    - **C.** Both A and B
    - **D.** Neither A nor B

35. While discussing frame damage:

Technician A says wrinkles in the upper flange of a truck frame indicate sag.

Technician B says diamond-shaped distortion of a 4WD sport utility vehicle frame was possibly caused by towing or being towed from a frame corner.

Who is correct?

A. Technician A
B. Technician B
C. Both A and B
D. Neither A nor B

36. Power steering fluid is leaking from the high-pressure outlet fitting on a nonsubmerged power steering pump. To correct this problem, you should:

A. Replace the cover seal.
B. Replace the O-ring.
C. Replace the fitting.
D. Replace the check valve.

37. During a road test on a front-wheel-drive vehicle, it is determined that the vehicle pulls right after a right turn, and pulls left after a left turn. The most likely cause of this condition is:

A. A misaligned subframe/cradle assembly.
B. Excessive positive caster.
C. Excessive lower ball joint play.
D. A binding upper strut bearing.

38. When the projected steering axis of a front wheel viewed from the side is tilted rearward from the true vertical center of the wheel, this alignment angle is referred to as:

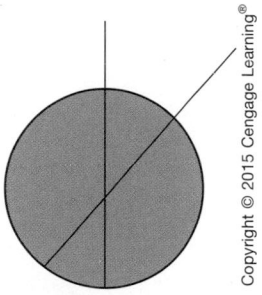

A. Negative camber
B. Positive camber
C. Positive scrub radius
D. Positive caster

39. While discussing unibody and frame damage:

Technician A says a vehicle that has been rear-ended could have later problems with steering and tracking.

Technician B says towing a 2,000 pound trailer on a class II hitch bolted to the frame of a minivan will cause diamond distortion of the frame.

Who is correct?

A. Technician A
B. Technician B
C. Both A and B
D. Neither A nor B

40. *Technician A* says in some short arm long arm (SLA) front suspension systems the springs are mounted between the lower control arm and the chassis.

Technician B says in an SLA type of suspension system the upper ball joints are the load-carrying ball joints.

Who is correct?

A. Technician A
B. Technician B
C. Both A and B
D. Neither A nor B

41. A customer says the steering wheel of her front-wheel-drive car does not return to the center position after a turn. A test drive reveals that the steering wheel is stiff and only returns to within approximately 90° of center after a 180° turn.

Technician A says the problem is memory steer and could be caused by binding in the steering shaft universal joints.

Technician B says memory steer in a front-wheel-drive car may be also caused by binding upper strut mounts.

Who is correct?

A. Technician A
B. Technician B
C. Both A and B
D. Neither A nor B

42. A customer says his front-wheel-drive car shakes and moans when decelerating at low speeds. The most likely cause of the problem is:

A. Worn upper strut bearings.
B. Worn inner drive axle joints.
C. Loose wheel bearings.
D. Excessive tire inflation pressure.

43. A sport utility vehicle with a power recirculating ball steering gear makes a rattling noise when the steering wheel is turned. This noise is noticeable whether the car is in motion or standing still.

Technician A says the cause of the problem could be a loose pitman arm or shaft.

Technician B says the cause of the problem could be the high-pressure power steering hose touching some part of the vehicle.

Who is correct?

A. Technician A
B. Technician B
C. Both A and B
D. Neither A nor B

44. While a vehicle is taken for a test drive, a loud clunking noise is heard from the right front of the vehicle only during hard brake maneuvers. What is the most likely cause?

A. A loose right front shock absorber
B. A worn rack and pinion mounting bushing
C. A loose sway bar link
D. A worn strut rod bushing

45. On a vehicle equipped with rear parallel leaf springs and a solid rear axle, the customer is complaining that the vehicle reacts erratically (it darts) during turns. What is the most likely cause of this complaint?

A. Loose rear axle U-bolts
B. Missing jounce/rebound bumpers
C. Incorrect ride height
D. Incorrect drive line angle

46. Excessive play in the steering wheel is often because of loose steering components. A check of these components should be:

A. Done with vehicle weight on the tires.
B. Done with a computerized alignment system.
C. Completed after a front wheel alignment.
D. Performed with the vehicle raised and the wheels unsupported.

47. While discussing shock absorber diagnosis:

Technician A says the "bounce test" is the sure way to quickly pinpoint bad shocks and struts.

Technician B says oil film on the lower chamber of a shock or strut indicates leakage and requires replacement.

Who is correct?

A. Technician A
B. Technician B
C. Both A and B
D. Neither A nor B

48. While discussing front suspension angles:

Technician A says turning angle is fixed by the steering and suspension system design.

Technician B says toe-out on turns may be adjusted with the steering stopper bolts.

Who is correct?

A. Technician A
B. Technician B
C. Both A and B
D. Neither A nor B

49. While discussing rear suspension alignment:

Technician A says in straight-ahead driving, the rear wheels must track exactly behind the front wheels or the vehicle will not handle correctly.

Technician B says a dynamic tracking rear suspension that lets the rear wheels turn in the same direction as the front wheels makes curve negotiation and lane changes much quicker and safer.

Who is correct?

A. Technician A
B. Technician B
C. Both A and B
D. Neither A nor B

50. The most likely cause of excessive body lean and sway while cornering is:

A. Worn upper control arm bushings.
B. A broken stabilizer link
C. Sagged coil springs
D. Worn strut rod bushings

APPENDIX B

METRIC CONVERSIONS

TO CONVERT THESE	TO THESE	MULTIPLY BY
TEMPERATURE		
Centigrade Degrees	Fahrenheit Degrees	C degrees × 1.8 + 32 = F degrees
Fahrenheit Degrees	Centigrade Degrees	F degrees − 32 × .556 = C degrees
LENGTH		
Millimeters	Inches	0.03937
Inches	Millimeters	25.4
Meters	Feet	3.28084
Feet	Meters	0.3048
Kilometers	Miles	0.62137
Miles	Kilometers	1.60935
AREA		
Square Centimeters	Square Inches	0.155
Square Inches	Square Centimeters	6.45159
VOLUME		
Cubic Centimeters	Cubic Inches	0.06103
Cubic Inches	Cubic Centimeters	16.38703
Cubic Centimeters	Liters	0.001
Liters	Cubic Centimeters	1000
Liters	Cubic Inches	61.025
Cubic Inches	Liters	0.01639
Liters	Quarts	1.05672
Quarts	Liters	0.94633
Liters	Pints	2.11344
Pints	Liters	0.47317
Liters	Ounces	33.81497
Ounces	Liters	0.02957
WEIGHT		
Grams	Ounces	0.03527
Ounces	Grams	28.34953
Kilograms	Pounds	2.20462
Pounds	Kilograms	0.45359
WORK		
Centimeter-Kilograms	Inch-Pounds	0.8676
Inch-Pounds	Centimeter-Kilograms	1.15262
Newton-Meters	Foot-Pounds	0.7376
Foot-Pounds	Newton-Meters	1.3558
PRESSURE		
Kilograms/Sq. Cm	Pounds/Sq. Inch	14.22334
Pounds/Sq. Inch	Kilograms/Sq. Cm	0.07031
Bar	Pounds/Sq. Inch	14.504
Pounds/Sq. Inch	Bar	0.06895

Automotive Group
Kent-Moore Division, SPX Corporation Roseville, MI

Fluke Corporation
Everett, WA

Easco/KD Tools
Lancaster, PA

Hennessy Industries, Inc.
LaVerge, TN

Hunter Engineering Company
Bridgeton, MO

Mac Tools, Inc.
Washington Courthouse, OH

OTC Division, SPX Corporation
Owatonna, MN

Sears Industrial Sales
Cincinnati, OH

Snap-On Tools Corporation
Kenosha, WI

Specialty Products Company
Longmont, CO

Federal Mogul Corporation
Ann Arbor, MI

Audi diagnostic tool and service information
http://www.ebahn.com/audi

Chrysler service information
http://www.techauthority.com

General Motors tool and service information
http://www.gmgoodwrench.com
http://www.acdelco.com

Ford Motor Company vehicle information
http://www.ford.com

Helm Incorporation, distributors of service manuals and training publications for many original equipment manufacturers
http://www.helm.com

Honda service information
http://www.serviceexpress.honda.com

Training materials, tools, and equipment
https://gmspecialservicetools.service-solutions.com/

Special service tools
http://www.otctools.com/product-landing

KIA general information
http://www.kia.com

Mac Tools, tool and equipment information
http://www.mactools.com

Nissan service manuals, 2000 and later models
http://www.nissan-techinfo.com

OTC Division of SPX Corporation, tools and equipment information
http://www.otctools.com

Snap-on Tools Corporation, tools and service equipment
http://www.snapon.com

Toyota Motor Corporation, vehicles and parts information
http://www.toyota.com

Volkswagen service information
http://www.ebahn.com/vw

Volvo service information, 1999 and later models
https://www.volvotechinfo.com/index.cfm

Service information pre-1999
http://www.volvotechinfo.com

Automotive Service Association (ASA)
www.asashop.org
P.O. Box 929
Bedford Texas 76021

International Automotive Technicians Network (IATN)
www.iatn.net

Society of Automotive Engineers (SAE)
www.sae.org
400 Commonwealth Dr.
Warrendale PA 15096

Moog Service Link
www.moogauto.com
Cooper Moog Automotive
P.O. Box 7224
6565 Wells Ave.
St. Louis MO 63177

GLOSSARY
GLOSARIO

Note: **Terms are highlighted in color,** followed by **Spanish translation in bold.**

Accidental air bag deployment An unintended air bag deployment caused by improper service procedures.

Despliegue accidental del Airbag Despliegue imprevisto del Airbag ocasionado por procedimientos de reparación inadecuados.

Active roll stabilization (ARS) ECU A computer that receives input signals and controls output functions in an ARS system.

ECU de estabilización de rodamiento activo (ARS, en inglés) Una computadora que recibe señales de entraday controla las funciones de salida en un sistema ARS.

Active steering ECU A computer that receives input signals and controls output functions in an active steering system.

ECU de dirección activa Una computadora que recibe señales de entrada y controla las funciones de salida en un sistema de dirección activa.

Advanced vehicle handling (AVH) Computer software on CD that allows the technician to locate the cause of hidden suspension, body, and chassis problems.

Manejo avanzado de vehículos (AVH, en inglés) Programa de software en CD que permite que el técnico localice la causa de problemas ocultos de suspensión, carrocería y chasis.

Air bag deployment module The air bag and deployment canister assembly that is mounted in the steering wheel for the driver's side air bag, or in the dash panel for the passenger's side air bag.

Unidad de despliegue del air bag El conjunto del Airbag y elemento de despliegue montado en el volante de dirección para proteger al conductor, o en el tablero de instrumentos para proteger al pasajero.

Air bag system Is designed to protect the driver and/or passengers in a vehicle collision.

Sistema de bolsa de aire Se diseñó para proteger al conductor y/o a los pasajeros en caso de un choque.

Alignment ramp A metal ramp positioned on the shop floor on which vehicles are placed during wheel alignment procedures.

Rampa de alineación Rampa de metal ubicada en el suelo del taller de reparación de automóviles sobre la que se colocan los vehículos durante los procedimientos de alineación de ruedas.

Antilock brake system (ABS) A computer-controlled system that prevents wheel lockup during brake applications.

Sistema de frenado antibloqueo (ABS) Un sistema controlado por computadora que previene el bloqueo de las ruedas durante la aplicación de los frenos.

Antitheft locking wheel covers Locking wheel covers that help to prevent wheel theft.

Cubrerruedas anti-robo Cubrerruedas autobloqueantes que ayudan a evitar el robo de las ruedas.

Antitheft wheel nuts Special wheel nuts designed to prevent wheel and tire theft.

Tuercas antirrobo de fijación de las ruedas Tuercas especiales de fijación de las ruedas diseñadas para evitar el robo de las ruedas y los neumáticos.

Approved gasoline storage can A special gasoline container that meets safety requirements.

Bidón aprobado para almacenamiento de gasolina Un recipiente especial para gasolina que cumple los requisitos de seguridad.

Aqueous parts cleaning tank Uses a water-based, environmentally friendly cleaning solution, such as Greasoff® 2, rather than traditional solvents.

Tanque de limpieza de partes acuosas Usa una solución a base de agua y compatible con el medio ambiente, tal como Greasoff® 2, en lugar delos solventes tradicionales.

ASE blue seal of excellence An ASE logo displayed by automotive service shops that employ ASE-certified technicians.

Sello azul de excelencia de la ASE Logotipo exhibido en talleres de reparación de automóviles donde se emplean mecánicos certificados por la ASE.

Axle offset A condition where the complete rear axle assembly has turned so one rear wheel has moved forward, and the opposite rear wheel has moved rearward.

Eje descentrado Una condición en la cual la asamblea total del eje trasero se ha girado para que una rueda se ha movido hacia afrente, y la rueda opuesta trasera se ha movido hacia atrás.

Axle pullers Are usually slide-hammer type. These pullers attach to the axle studs to remove the axle.

Extractores del eje Son generalmente de tipo martillo de corredera. Estos extractores se vinculan a los espárragos roscados del eje para quitarlo.

Backup power supply A voltage source, usually located in the air bag computer, that is used to deploy an air bag if the battery cables are disconnected in a collision.

Alimentación de reserva Fuente de tensión, por lo general localizada en la computadora del Airbag, que se utiliza para desplegar el Airbag si se desconectan los cables de la batería a consecuencia de una colisión.

Ball joint radial measurement Vertical movement in a ball joint because of internal joint wear.

Medida de radial de la junta esférica Movimiento horizontal en una junta esférica ocasionado por el desgaste interno de la misma.

Ball joint removal and installation tools Special tools required for ball joint removal and replacement.

Herramientas para la remoción y el ajuste de la junta esférica Herramientas especiales requeridas para la remoción y el reemplazo de la junta esférica.

Ball joint vertical movement Vertical movement in a ball joint because of internal joint wear.

Movimiento vertical de la junta esférica Un movimiento vertical en la junta esférica debido al desgaste interior de la junta.

Ball joint wear indicator A visual method of checking ball joint wear.

Indicador de desgaste de la junta esférica Método visual de revisar el desgaste interior de una junta esférica.

Bearing brinelling Straight line indentations on the races and rollers.

Acción de Brinnell en un cojinete Hendiduras en línea recta en los anillos y los rodillos.

Bearing fatigue spalling Flaking of the surface metal on the rollers and races.

Escamación y fatiga del cojinete Condición que ocurre cuando el metal de la superficie de los rodillos y los anillos comienza a escamarse.

Bearing frettage A fine, corrosive wear pattern around the races and rollers, with a circular pattern on the races.

Cinceladura del cojinete Desgaste corrosivo y fino alrededor de los anillos y los rodillos que se hace evidente a través de figuras circulares en los anillos.

Bearing indentations Surface indentations on races and rollers caused by hard, foreign particles.

Depresiones del cojinete Las depresiones en la superficie de los anillos y los rodillos causadas por las partículas duras foraneas.

Bearing preload A tension placed on the bearing rollers and races by an adjustment or assembly procedure.

Carga previa del cojinete Tensión aplicada a los rodillos y los anillos del cojinete a través de un procedimiento de ajuste o de montaje.

Bearing pullers Special tools designed for bearing removal.

Extractores de cojinetes Herramientas especiales diseñadas para la remoción del cojinete.

Bearing smears Metal loss from the races and rollers in a circular, blotched pattern.

Ralladuras en los cojinetes Pérdida del metal de los anillos y los rodillos que se hace evidente a través de una figura circular oxidada.

Bellows boots Accordion-style boots that provide a seal between the tie rods and the housing on a rack and pinion steering gear.

Botas de fuelles Botas en forma de acordeón que proveen una junta de estanqueidad entre las barras de acoplamiento y el alojamiento en un mecanismo de dirección de cremallera y piñón.

Belt tension gauge A special gauge used to measure drive belt tension.

Calibrador de tensión de la correa de transmisión Calibrador especial que se utiliza para medir la tensión de una correa de transmisión.

Body sway Excessive body movement from side to side.

Oscilación de la carrocería Movimiento lateral excesivo de la carrocería.

Bounce test A shock absorber test in which weight is applied to, and released from, the vehicle bumper.

Prueba de rebote Una prueba de los amortiguadores en la cual se aplica y se quita un peso en el parachoques del vehículo.

Brake parts washer A parts washer designed to wash brake parts without releasing asbestos dust into the atmosphere.

Lavador de partes de freno Un lavador de partes diseñada a lavar las partes de freno sín dejar escapar el polvo de amianta al medio ambiente.

Brake pedal depressor A special tool installed between the front seat and the brake pedal to apply the brakes during certain wheel alignment measurements.

Depresor del pedal de frenos Una herramienta especial instalada entre el asiento de afrente y el pedal de frenos para aplicar los frenos durante ciertas medidas de alineación de las ruedas.

Brake pedal jack A special tool installed between the front seat and the brake pedal to apply the brakes during certain wheel alignment measurements.

Gato del pedal de freno Herramienta especial instalada entre el asiento delantero y el pedal de freno que se utiliza para aplicar los frenos durante ciertas medidas de alineación de ruedas.

Brake pressure modulator valve (BPMV) An assembly containing a group of solenoid valves that control brake fluid pressure during antilock brake, traction control, and vehicle stability control functions.

Válvula modulador de presión del freno (BPMV) Una asamblea que contiene un grupo de válvulas de solenoide que controlan la presión del fluido de freno durante las funciones de freno antibloqueo, control de tracción, y control de estabilidad del vehículo.

Brake torque test A test that is performed with the brakes applied.

Prueba de torsión de frenos Una prueba que se lleva acabo mientras que se aplicacan los frenos.

Bump steer The tendency of the steering to veer suddenly in one direction when one or both front wheels strike a bump.

Cambio de dirección ocacionado por promontorios en el terreno Tendencia de la dirección a cambiar repentinamente de sentido cuando una o ambas ruedas delanteras golpea un promontorio.

Camber adjustment A method of adjusting the inward or outward tilt of a front or rear wheel in relation to the true vertical centerline of the tire.

Ajuste de la combadura Método de ajustar la inclinación hacia adentro o hacia afuera de una rueda delantera o trasera con relación a la línea central vertical real del neumático.

Camber angle The inward or outward tilt of a line through the center of a front or rear tire in relation to the true vertical centerline of the tire and wheel.

Ángulos de combaduras Inclinación hacia adentro o hacia afuera de una línea a través del centro de una rueda delantera o trasera con relación a la línea central vertical real del neumáticoy la rueda.

Caster adjustment A method of adjusting the forward or rearward tilt of a line through the center of the upper and lower ball joints, or lower ball joint and upper strut mount, in relation to the true vertical centerline of the tire and wheel viewed from the side.

Ajuste de comba de eje Método de ajustar la inclinación hacia adelante o hacia atrás de una línea a través del centro de las juntas esféricas superior e inferior, o la junta esférica inferior y el montaje del montante superior, con relación a la línea central vertical real del neumático y la rueda vista desde la parte lateral.

Caster angle A line through the center of the upper and lower ball joints, or lower ball joint and upper strut mount, in relation to the true vertical centerline of the tire and wheel viewed from the side.

Ángulo de comba de eje Línea a través del centro de las juntas esféricas superior e inferior, o la junta esférica inferior y el montaje del montante superior, con relación a la línea central vertical real del neumático y la rueda vista desde la parte lateral.

Camber The position of an imaginary line through the centers of the upper and lower ball joints or upper strut mount and lower ball joint in relation to the line through the tire center viewed from the front.

Ángulo de verticalidad La posición de una línea imaginaria que atraviesa los centros de las articulaciones esféricas superiores e inferiores o del montaje del montante superior y la articulación esférica inferior en relación con la línea que atraviesa el centro del neumático visto de frente.

Caster The position of an imaginary line through the centers of the upper and lower ball joints or upper strut mount and lower ball joint in relation to the a line through the wheel and spindle center viewed from the side.

Ángulo de convexión La posición de una línea imaginaria que atraviesa los centros de las articulaciones esféricas superiores e inferiores o del montaje del montante superior y la articulación esférica inferior en relación con la línea que atraviesa la rueda y el centro del palier visto de perfil.

Caster offset The distance between the caster line and the center of the front spindle.

Desfasaje del ángulo de convexión La distancia entre la línea de tiro y el centro del palier delantero.

Caster trail An alternate term for caster offset.

Rastro del ángulo de convexión Un término alternativo para desfasaje del ángulo de convexión.

Circlip A round circular clip used as a locking device on components such as front drive axles.

Anillo de pistón Un anillo circular que se utiliza como dispositivo de traba en componentes como los ejes rotativos delanteros.

Claw washer A special locking washer used to retain the tie rods to the rack in some rack and pinion steering gears.

Arandelas de garras Arandela de bloqueo especial que se utiliza para sujetar las barras de acoplamiento a la cremallera en algunos mecanismos de dirección de cremallera y piñón.

"C" lock A thick metal locking device used to lock components such as the rear drive axles in place.

Retenedores en C Un dispositivo de cerrojo de metal grueso que sirve para enclavar los componentes en su lugar, tal como los ejes propulsores traseros.

Clock spring electrical connector A conductive ribbon in a plastic container mounted on top of the steering column that maintains electrical contact between the air bag inflator module and the air bag electrical system.

Conector eléctrico de cuerda de reloj Cinta conductiva envuelta en una cubierta plástica montada sobre la parte superior de la columna de dirección, que mantiene el contacto eléctrico entre la unidad infladora y el sistema eléctrico del Airbag.

Coil spring compressor tool A special tool required to compress a coil spring prior to removal of the spring from the strut.

Herramienta para la compresión del muelle helicoidal Herramienta especial requerida para comprimir un muelle helicoidal antes de remover el muelle del montante.

Cold parts washer Uses a non-heated solution to clean metal parts.

Lavador de partes en frío usa una solución en frío para limpiar las partes de metal.

Collapsible steering column A steering column that is designed to collapse when impacted by the driver in a collision to help reduce driver injury.

Columna de dirección plegable Columna de dirección diseñada para plegarse al ser impactada por el conductor durante una colisión con el propósito de reducir las lesiones que el conductor pueda recibir.

Column-drive EPS An electronically controlled steering gear in which the electric motor supplies torque to the steering column shaft.

Transmisión de columna de GPE (guiado de propulsión eléctrica) El primer paso para diagnosticarlo es verificar la queja de falla del cliente. Haga una prueba de carretera al vehículo si es necesario para identificar la falla.

Computer four wheel aligner A type of wheel aligner that uses a computer to measure wheel alignment angles at the front and rear wheels.

Alineador computerizado de cuatro ruedas Tipo de alineador de ruedas que utiliza una computadora para medir los ángulos de alineación de las ruedas delanteras y traseras.

Continuously variable road sensing suspension (CVRSS) An updated version of the road sensing suspension (RSS).

Suspensión detector continuo de variedades (CVRSS) Un sistema de suspensión detector más avanzada.

Control arm bushing tools Special tools required for control arm bushing removal and replacement.

Herramientas para el buje del brazo de mando Herramientas especiales requeridas para la remoción y el reemplazo del buje del brazo de mando.

Control arm movement monitor A monitor available in some computer wheel aligners. On short-and-long arm front suspensions this feature indicates the required shim thickness to provide the specified camber and caster.

Monitor del movimiento del brazo de mando Un monitor disponible en algunos alineadores de ruedas computerizadas. En las suspensiones delanteras tipo brazo largo-y-corto este elemento indica el espesor de la cuña requerido para proveer el ángulo de la inclinación especificado.

Corrosive A material that is corrosive causes another material to be gradually worn away by chemical action.

Corrosivo Un material corrosivo provoca el desgaste gradual de otro por una acción química.

Crocus cloth A very fine paper for polishing or removing small abrasions from metal.

Tela fina de esmeril Papel sumamente fino que se utiliza para pulir o remover pequeñas abrasiones del metal.

Curb riding height The distance between the vehicle chassis and the road surface measured at specific locations.

Altura del cotén del viaje Distancia entre el chasis del vehículo y la superficie del camino medida en puntos específicos.

Data link connector (DLC) An electrical connector for computer system diagnosis mounted under the instrument panel or in the engine compartment.

Conector de enlace de datos Conector eléctrico montado debajo del tablero de instrumentos o en el compartimiento del motor, que se utiliza para la diagnosis del sistema informático.

Datum line A straight reference line such as the top of a dedicated bench system.

Línea de datos Línea recta de referencia, como por ejemplo la parte superior de un sistema de banco dedicado.

Dedicated bench system A heavy steel bed with special fixtures for aligning unitized bodies.

Sistema de banco dedicado Asiento pesado de acero con aparatos especiales que se utiliza para la alineación de carrocerías unitarias.

Diagnostic procedure charts Provide step-by-step diagnostic procedures for specific problems.

Cuadros de procedimientos diagnósticos Proveen los procedimientos diagnóticos paso por paso de los problemas específicos.

Diagnostic trouble codes (DTCs) Codes displayed in digital form that represent faults in a computer-controlled system.

Códigos diagnósticos de averías (DTCs) Los códigos manifestados en forma digital que representan las averías en un sistema controlado por computadora.

Dial indicators A precision-measuring device with a stem and a rotary pointer.

Indicador de cuadrante Dispositivo para medidas precisas con vástago y aguja giratoria.

Digital adjustment photos Photos that are available in some computer-controlled wheel aligners to inform the technician regarding suspension adjustment procedures.

Fotos digitales de ajuste Las fotos que son disponibles en algunos alineadores de ruedas controlados por computadoras para informar a los técnicos en cuanto a los procedimientos de ajustes de la suspensión.

Digital signal processor (DSP) An electronic device in some wheel sensors on computer-controlled wheel aligners.

Procesor de señales digitales Un dipositivo electrónico en algunos sensores de ruedas en los alineadores de ruedas controlados por computadora.

Directional stability The tendency of a vehicle steering to remain in the straight-ahead position when driven straight ahead on a smooth, level road surface.

Estabilidad direccional Tendencia de la dirección del vehículo a permanecer en línea recta al ser así conducido en un camino cuya superficie es lisa y nivelada.

Downshift test A test for chassis vibrations performed with the transmission downshifted.

Prueba de cambio descendente Una prueba de las vibraciones del chasis que se efectúan durante un cambio de carrera de transmisión descendente.

Dynamic stability control (DSC) ECU A computer that receives input signals and controls output functions in a DSC system.

ECU de control de estabilidad dinámico (DSC, en inglés) Una computadora que recibe señales de entrada y controla las funciones de salida en un sistema DSC.

Dynamic wheel balance Refers to proper balance of the tire and wheel assembly during tire and wheel rotation.

Equilibrado dinámico de la rueda Se refiere al equilibrado apropiado del ensamblaje de la llanta y la rueda durante su rotación.

Eccentric camber bolt A bolt with an out-of-round metal cam on the bolt head that may be used to adjust camber.

Perno de combadura excéntrica Perno con una leva metálica con defecto de circularidad en su cabeza, que puede utilizarse para ajustar la combadura.

Eccentric cams Out-of-round metal cams mounted on a retaining bolt with the shoulder of the cam positioned against a component. When the cam is rotated, the component position is changed. Delete Eccentric cams or bushings.

Levas eccéntricas Las levas de metal ovaladas montadas en un perno retenedor con lo saliente de la leva posicionado contra un componente. Cuando se gira la leva cambia la posición del componente.

Electronic brake and traction control module (EBTCM) A module that controls antilock brake, traction control, and vehicle stabilty functions.

Módulo electrónico de control del frenado y tracción (EBTCM) Un módulo que controla las funciones del freno antibloqueo, control de tracción, y control de estabilidad del vehículo.

Electronically controlled orifice (ECO) valve A computer-controlled solenoid valve in the power steering pump that controls fluid flow from the pump to the servotronic valve.

Válvula de orificio controlado electrónicamente (ECO, en inglés) Una válvula solenoide controlada por computadora que se encuentra en la bomba de dirección de potencia que controla el flujo de fluidos desde la bomba hasta la válvula servotrónica.

Electronic suspension control (ESC) An electronically controlled suspension system in which the computer controls strut firmness.

Control electrónico de la suspensión (CES) Los vehículos que se equipan con este sistema pueden tener el tablero de control análogo o digital.

Electronic vibration analyzer (EVA) A tester that measures component vibration.

Analizador electrónico de vibraciones Un detector que mide la vibración de un componente.

Electronic wheel balancer A computer-controlled balancer that provides static and dynamic wheel balance.

Equilibrador de ruedas electrónico Un equilibrador controlado por computadora que provee la equilibración estática y dinámico.

Environmental Protection Agency (EPA) A federal government agency that is responsible for air and water quality in the United States.

Agencia de Protección Ambiental (EPA, en inglés) Una agencia del Gobierno Federal responsable de la calidad del aire y del agua en los Estados Unidos.

Fail-safe mode A mode entered by a computer if the computer detects a fault in the system.

Modo de seguridad Un modo iniciado por la computadora si ésta descubre un fallo en el sistema.

Floor jack A hydraulically operated lifting device for vehicle lifting.

Gato de pie Dispositivo activado hidráulicamente que se utiliza para levantar un vehículo.

Flow control valve A special valve that controls fluid movement in relation to system demands.

Válvula de control de flujo Válvula especial que controla el movimiento del fluido de acuerdo a las exigencias del sistema.

Fluorescent trouble lights Contain fluorescent-type bulbs.

Luces de peligro fluorescentes Incluyen bombillas de tipo fluorescente.

Frame flange The upper or lower horizontal edge on a vehicle frame.

Brida del armazón Borde horizontal superior o inferior en el armazón del vehículo.

Frame web The vertical side of a vehicle frame.

Malla del armazón Lado vertical del armazón del vehículo.

Front and rear wheel alignment angle screen A display on a computer wheel aligner that provides readings of the front and rear wheel alignment angles.

Pantalla para la visualización del ángulo de alineación de las ruedas delanteras y traseras Representación visual en una computadora para alineación de ruedas que provee lecturas de los ángulos de alineación de las ruedas delanteras y traseras.

Geometric centerline An imaginary line through the exact center of the front and rear wheels.

Línea central geométrica Línea imaginaria a través del centro exacto de las ruedas delanteras y traseras.

Hazard Communication Standard A forerunner to the Right-To-Know laws published by the Occupational Health and Safety Administration (OSHA).

Estándar de comunicación de riesgos Un precursor de las leyes del derecho a saber publicado por la Administración de Seguridad y Salud Ocupacional (OSHA, en inglés).

Heavy spot Is a location in a tire with excessive weight.

Zona Dura es un lugar en la llanta con peso excesivo.

High-frequency transmitter An electronic device that sends high-frequency voltage signals to a receiver. Some wheel sensors on computer-controlled wheel aligners send high-frequency signals to a receiver in the wheel aligner.

Transmisor de alta frecuencia Un dispositivo electrónico que manda las señales de voltaje de alta frecuencia a un receptor. Algunos sensores de ruedas en los alineadores controlados por computadora mandan señales de alta frecuencia a un receptor en el alineador de ruedas.

History code A fault code in a computer that represents an intermittent defect.

Código histórico Código de fallo en una computadora que representa un defecto intermitente.

Hot cleaning tank Uses a heated solution to clean metal parts.

Tanque de limpieza en caliente Usa una solución caliente para limpiar las partes de metal.

Hybrid vehicle A hybrid vehicle has a power train with two power sources. The most common type of hybrid vehicle has a gasoline engine and an electric drive motor(s).

Vehículo híbrido Un vehículo híbrido posee un tren de potencia con dos fuentes de energía. El tipo más común de vehículo híbrido posee un motor de gasolina y uno o varios motores eléctricos.

Hydraulic press A hydraulically operated device for disassembling and assembling components that have a tight press-fit.

Prensa hidráulica Dispositivo activado hidráulicamente que se utiliza para desmontar y montar componentes con un fuerte ajuste en prensa.

Ignitable A liquid, solid, or a gas that can be set on fire spontaneously or by an ignition source.

Inflamable Un líquido, sólido o gas que se puede encender espontáneamente o mediante una fuente de ignición.

Incandescent bulb-type trouble lights Contain incandescent-type bulbs.

Luces de peligro con bombilla de tipo incandescente Incluyen bombillas de tipo incandescente.

Included angle The sum of the camber and steering axis inclination (SAI) angles.

Ángulo incluido La suma del ángulo de inclinación de la comba y de ángulo de inclinación del eje de dirección (SAI).

Integral reservoir A reservoir that is joined with another component such as a power steering pump.

Tanque integral Tanque unido a otro componente, como por ejemplo una bomba de la dirección hidráulica.

International system (SI) A system of weights and measures.

Sistema internacional Sistema de pesos y medidas.

Lateral axle sideset The amount that the rear axle is moved straight sideways in relation to the front axle.

Desplazamiento lateral del eje La cantidad que el eje trasero está desplazado directamente hacia un lado en relación al eje delantero.

Lateral chassis movement Movement from side to side.

Movimiento lateral del chasis Movimiento de un lado a otro.

Lateral tire runout The variation in side-to-side movement.

Desviación lateral del neumático Variación del movimiento de un lado a otro.

Machinist's rule A steel ruler used for measuring short distances. These rulers are available in USC or metric measurements.

Regla para mecánicos Regla de acero que se utiliza para medir distancias cortas. Dichas reglas están disponibles en medidas del USC o en medidas métricas.

Main menu A display on a computer wheel aligner from which the technician selects various test procedures.

Menú principal Representación visual en una computadora para alineación de ruedas de la que el mecánico puede eligir varios procedimientos de prueba.

Magnetic wheel alignment gauge A gauge that may be attached magnetically to the wheel hub to indicate some wheel alignment angles.

Indicador de alineación de rueda magnética Un indicador que se puede adjuntar magnéticamente al buje de la rueda para indicar algunos ángulos de alineación de la rueda.

Manual test A shock absorber test in which the lower end of the shock is disconnected from the suspension.

Prueba manual Una prueba del amortiguador en la cual el extremo inferior del amortiguador se disconecta de la suspensión.

Memory steer The tendency of the vehicle steering not to return to the straight-ahead position after a turn but to keep steering in the direction of the turn.

Dirección de memoria Tendencia de la dirección del vehículo a no regresar a la posición de línea recta después de un viraje, sino a continuar girando en el sentido del viraje.

Molybdenum disulphide lithium-based grease A special lubricant containing molybdenum disulphide and lithium that may be used on some steering components.

Grasa de bisulfuro de molibdeno con base de litio Lubricante especial compuesto de bisulfuro de molibdeno y litio que puede utilizarse en algunos componentes de la dirección.

Multipull unitized body straightening system A hydraulically operated system that pulls in more than one location when straightening unitized bodies.

Sistema de tiro múltiple para enderezar la carrocería unitaria Sistema activado hidráulicamente que tira hacia más de una dirección al enderezar carrocerías unitarias.

Multipurpose dry chemical fire extinguisher Contains a dry chemical powder used to extinguish various types of fires.

Extintor de incendios multipropósito de polvo químico seco Contiene un polvo químico seco que se utiliza para apagar diversos tipos de incendios.

Multi-ribbed V-belt A belt containing a series of small "V" grooves on the underside of the belt.

Correa trapezoidal de ranuras múltiples Una correa que posee varias ranuras pequeñas en forma de "V" en la cara interna.

National Institute for Automotive Service Excellence (ASE) An organization that provides voluntary automotive technician certification in eight areas of expertise.

Instituto Nacional para la excelencia en la reparación de automóviles Organización que provee una certificación voluntaria de mecánico de automóviles en ocho áreas diferentes de especialización.

Neutral coast-down test A chassis vibration test performed while the vehicle is coasting in neutral.

Prueba de marcha desembragado Una prueba de vibración del chasis que se efectúa mientras que el vehículo marcha desembragado.

Neutral run-up test A vibration test performed by accelerating the engine with the transmission in neutral.

Prueba de aceleración desembragado Una prueba de vibración que se efectúa acelerando el motor con la transmisión desembragada.

OBD II computer systems Have monitoring capabilities and inform the driver if emission levels exceed 1.5 time the normal levels for that vehicle year.

Sistemas de computación OBD II Cuentan con capacidades de monitoreo e informan al conductor si los niveles de emisiones superan 1,5 veces los niveles normales para ese año vehicular.

Occupational Safety and Health Act (OSHA) A federal government agency and related laws in the United States that are responsible for ensuring safe and healthful working conditions.

Administración de Seguridad y Salud Ocupacional (OSHA, en inglés) Una agencia del Gobierno Federal, y sus leyes relacionadas, en los Estados Unidos, responsable de garantizar condiciones de trabajo seguro y saludable.

Onboard Diagnostic II (OBD II) Systems are a type of computer system mandated on cars and light trucks since 1996. OBD II systems have a number of mandated standardized features, including several monitoring systems in the PCM.

Sistemas de Capacidad de diagnóstico a bordo II (CDB II) Son un tipo de sistema computarizado obligatorio en automóviles y camionetas ligeras desde 1996. Los sistemas CDB II tienen un número de características estandarizadas obligatorias que incluyen varios sistemas de monitoreo en el módulo de control del motor (MCM).

On-demand DTCs Diagnostic trouble codes that represent intermittent faults in some computer-controlled systems.

DTCs a solicitud Los códigos de fallos diagnósticos que representan los fallos intermitentes en algunos sistemas controlados por computadora.

Oversteer The tendency of a vehicle to turn sharper than the turn selected by the driver.

Sobreviraje Tendencia de un vehículo a girar más de lo que el conductor desea.

Part-finder database Information available in some computer wheel aligners that allows the technician to access part numbers and prices from many undercar parts manufacturers.

Datos para localización de partes La información disponible en algunos alineadores de ruedas electrónicos que permite que el técnico tenga acceso a los números y precios de partes de muchos de los fabricantes de partes del carro inferior.

Pitman arm puller A special puller required to remove a pitman arm.

Extractor del brazo pitman Extractor especial requerido para la remoción de un brazo pitman.

Plumb bob A weight with a sharp, tapered point that is suspended and centered on a string.

Plomada Balanza con un extremo cónico puntiagudo, suspendida y centrada en una hilera.

Ply separation A parting of the plies in a tire casing.

Separación de estrias División de las estrias en la cubierta de un neumático.

Power steering pressure gauge A gauge and valve with connecting hoses for checking power steering pump pressure.

Calibrador de presión de la dirección hidráulica Calibrador y válvula con mangueras de conexión utilizadas para verificar la presión de la bomba de la dirección hidráulica.

Powertrain control module (PCM) A computer that controls engine and possibly transmission functions.

Unidad de control del tren transmisor de potencia Computadora que controla las funciones del motor y posiblemente las de la transmisión.

Prealignment inspection A check of steering and suspension components prior to a wheel alignment.

Inspección antes de una alineación Verificación de los componentes de la dirección y de la suspensión antes de llevarse a cabo una alineación de ruedas.

Preliminary inspection screen A display on a computer wheel aligner that allows the technician to check and record the condition of many suspension and steering components.

Pantalla para la visualización durante una inspección preliminar Representación visual en una computadora para alineación de ruedas que le permite al mecánico revisar y anotar la condición de un gran número de componentes de la suspensión y de la dirección.

Pressure gauge A gauge that may be connected to a pressure source, such as a power steering pump, to determine pump condition.

Indicador de presión Un manómetro que se puede conectar a una fuente de presión, tal como la bomba de la dirección hidráulica, para determinar la condición de la bomba.

Pressure relief valve A valve designed to limit pressure on a pump, such as a power steering pump.

Válvula de relieve de presión Una válvula diseñada para limitar la presión en una bomba, tal como en una bomba de dirección hidráulica.

Programmed ride control (PRC) A computer-controlled suspension system in which the computer operates an actuator in each strut to control strut firmness.

Control programado del viaje Sistema de suspensión controlado por computadora en el que la computadora opera un accionador en cada uno de los montantes para controlar la firmeza de los mismos.

Rack-drive electronic power steering The electronic power steering (EPS) light in the instrument panel should be illuminated when the ignition switch is turned on, in vehicles with rack-drive electronic power steering.

Cremallera de dirección GPE (guiado de propulsión electrónica) La luz del guiado de propulsión electrónica (GPE) en el tablero de control debe iluminarse cuando el selector de encendido se prende en vehículos con cremallera de dirección GPE.

Radial runout The variations in diameter of a round object such as a tire.

Desviación radial Variaciones en el diámetro de un objeto circular, como por ejemplo un neumático.

Reactive A material that is reactive reacts with some other chemicals and gives off gas(es) during the reaction.

Reactivo Un material reactivo es aquel que reacciona con otros químicos y emana gases durante la reacción.

Resource Conservation and Recovery Act (RCRA) Provides specific laws regarding hazardous waste disposal.

Acta de Conservación y Recuperación de Recursos (RCRA, en inglés) Proporciona leyes específicas respecto de la eliminación de desechos peligrosos.

Rear axle offset A condition in which the complete rear axle assembly has turned so one rear wheel has moved forward and the opposite rear wheel has moved rearward.

Desviación del eje trasero Condición que ocurre cuando todo el conjunto del eje trasero ha girado de manera que una de las ruedas traseras se ha movido hacia adelante y la opuesta se ha movido hacia atrás.

Rear axle sideset A condition in which the rear axle assembly has moved sideways from its original position.

Resbalamiento lateral del eje trasero Condición que ocurre cuando el conjunto del eje trasero se ha movido lateralmente desde su posición original.

Rear main steering angle sensor An input sensor in the rear steering actuator of an electronically controlled four-wheel steering system.

Sensor principal del ángulo de la dirección trasera Sensor de entrada en el accionador de la dirección trasera de un sistema de dirección en las cuatro ruedas controlado electrónicamente.

Rear steering center lock pin A special pin used to lock the rear steering during specific service procedures in an electronically controlled four-wheel steering system.

Pasador de cierre central de la dirección trasera Pasador especial que se utiliza para bloquear la dirección trasera durante procedimientos de reparación específicos en un sistema de dirección en las cuatro ruedas controlado electrónicamente.

Rear sub steering angle sensor An input sensor in the rear steering actuator of a four-wheel steering system.

Sensor auxiliar del ángulo de la dirección trasera Sensor de entrada en el accionador de la dirección trasera de un sistema de dirección en las cuatro ruedas.

Rear wheel camber The tilt of a line through the center of a rear tire and wheel in relation to the true vertical centerline of the tire and wheel.

Comba de la rueda trasera La inclinación de una línea que atraviesa el centro del neumático y la rueda en relación al eje mediano verdadero del neumático y la rueda.

Rear wheel toe The toe setting on the rear wheels.

Ángulo de inclinación de la rueda trasera El ajuste del ángulo de inclinación en las ruedas traseras.

Rear wheel tracking Refers to the position of the rear wheels in relation to the front wheels.

Encarrilamiento de las ruedas traseras Se refiere a la posición de las ruedas traseras con relación a las ruedas delanteras.

Rebound bumpers Rubber stops that prevent metal-to-metal contact between the suspension and the chassis when a wheel strikes a road irregularity and the suspension moves fully upward.

Paragolpes de rebotes Topes de cauchos que previene el contacto de metal a metal entre la suspensión y el chasis cuando una rueda golpea una irregularidad en el camino y la suspensión se mueva totalmente hacia arriba.

Receiver An electronic device on an aligner that receives signals from the wheel sensors.

Receptor Un dispositivo electrónico de una alienador computerizado de cuatro ruedas que recibe las señales de los sensores de las ruedas.

Remote reservoir A container containing fluid that is mounted separately from the fluid pump.

Tanque remoto Recipiente lleno de líquido que se monta separado de la bomba del fluido.

Ribbed V-belt A belt containing a series of small "V" grooves on the underside of the belt.

Correa nervada en V Correa con una serie de ranuras pequeñas en forma de "V" en la superficie inferior de la misma.

Ride height The distance from a specified chassis location to the road surface.

Altura del viaje La distancia de un punto específico del chasis a la superficie del camino.

Ride height screen A display on a computer wheel aligner that illustrates the locations for ride height measurement.

Pantalla para la visualización de la altura del viaje Representación visual en una computadora para alineación de ruedas que muestra los puntos donde se mide la altura del viaje.

Right-to-Know Laws List the worker's rights regarding the handling of hazardous waste materials.

Leyes del Derecho a saber Lista de los derechos de los trabajadores respecto del manejo de materiales de desecho peligrosos.

Rim clamps Special clamps designed to clamp on the wheel rims and allow the attachment of alignment equipment such as the wheel units on a computer wheel aligner.

Abrazaderas de la llanta Abrazaderas especiales diseñadas para sujetar las llantas de las ruedas y así permitir la fijación del equipo de alineación, como por ejemplo las unidades de la rueda en una computadora para alineación de ruedas.

Road feel A feeling experienced by a driver during a turn when the driver has a positive feeling that the front wheels are turning in the intended direction.

Sensación del camino Sensación experimentada por un conductor durante un viraje cuando está completamente seguro de que las ruedas delanteras están girando en la dirección correcta.

Road test A driving procedure used to diagnose specific vehicle problems.

Prueba de camino Un procedimiento de conducción que se usa para diagnosticar los problemas específicos del vehículo.

Safety and gateway module (SGM) Operates the servotronic valve and the electronically controlled orifice valve in some power steering systems.

Módulo de puerta de enlace y seguridad (SGM, en inglés) Opera la válvula servotrónica y la válvula de orificio controlado electrónicamente en algunos sistemas de dirección de potencia.

Safety data sheets (SDS) Safety data sheets (SDS) provide all the necessary data about hazardous materials. They were previously called Material Safety Data Sheets (MSDS).

Hojas de datos de seguridad (SDS) Las hojas de datos de seguridad (SDS) brindan todos los datos necesarios sobre los materiales peligrosos. Se conocían previamente como Hojas de datos de seguridad del material (MSDS).

Safety stands May be called jack stands.

Bases de seguridad Pueden también llamarse bases del gato.

Scan tool A digital computer system tester used to read trouble codes and perform other diagnostic functions.

Exploradore Un probador del sistema de computadora digital que sirve para leer los códigos de errores y para executar otras funciones diagnósticas.

Seal drivers Designed to maintain even contact with a seal case and prevent seal damage during installation.

Empujadores para sellar Diseñados para mantener un contacto uniforme con el revestimiento de la junta de estanqueidad y evitar el daño de la misma durante el montaje.

Section modulus The measurement of a frame's strength based on height, width, thickness, and the shape of the side rails.

Coeficiente de sección Medida de la resistencia de un armazón, basada en la altura, el ancho, el espesor y la forma de las vigas laterales.

Sector shaft lash Refers to the movement between the sector teeth and ball nut teeth.

Coletazo del eje de sector Se refiere al movimiento entre el dentado del sector y el de la tuerca esférica.

Serpentine belt A ribbed V-belt drive system in which all the belt-driven components are on the same vertical plane.

Correa serpentina Sistema de transmisión con correa nervada en V en el que todos los componentes accionados por una correa se encuentran sobre el mismo plano vertical.

Service check connector A diagnostic connector used to diagnose computer systems. A scan tool may be connected to this connector.

Conector para la revisión de reparaciones Conector diagnóstico que se utiliza para diagnosticar sistemas informáticos. A dicho conector se le puede conectar un verificador de exploración.

Servotronic valve A computer-controlled valve in an active steering system that controls power steering pump pressure in relation to vehicle speed.

Válvula servotrónica Una válvula controlada por computadora en un sistema de dirección activo que controla la presión de la bomba de dirección de potencia de acuerdo con la velocidad del vehículo.

Setback Occurs when one front or rear wheel is driven rearward in relation to the opposite front or rear wheel.

Retroceso Condición que ocurre cuando una de las ruedas delanteras se mueve hacia atrás con relación a la rueda delantera opuesta.

Sheared injected plastic Is inserted into steering column shafts and gearshift tubes to allow these components to collapse if the driver is thrown against the steering column in a collision.

Plástico cortado inyectado Insertado en los árboles de la columna de dirección y en los tubos del cambio de engranajes de velocidad para permitir que estos componentes se pleguen si la columna de dirección es impactada por el conductor durante una colisión.

Shim display screen A display on a computer wheel aligner that shows the technician the thickness and location of the proper shims for rear wheel alignment.

Pantalla de laminillas de ajuste Una presentación en un alineador de ruedas que demuestra al técnico el espesor y localización de las laminillas apropriadas para alineación de las ruedas traseras.

Single-pull unitized body straightening system Hydraulically operated equipment that pulls in one location while straightening unitized bodies.

Sistema enderezador de tiro único para la carrocería unitaria Sistema activado hidráulicamente que tira hacia una dirección al enderezar carrocerías unitarias.

Slip plates Plates that are placed under the rear wheels during a wheel alignment to allow rear wheel movement during rear suspension adjustments.

Placas de patinaje Las placas que se colocan debajo de las ruedas traseras durante un alineación de ruedas traseras para permitir el movimiento de las ruedas traseras durante el ajuste de la suspensión trasera.

Slow acceleration test A chassis vibration test performed by slowly accelerating the vehicle.

Prueba de aceleración lenta Una prueba de vibración del chasis que se efectúa por medio de una aceleración lenta del vehículo.

Soft-jaw vise A vise equipped with soft metal, such as copper, in the jaws.

Tornillo con tenacilla maleable Tornillo equipado con un metal blando, como por ejemplo cobre, en las tenacillas.

Specifications menu A display on a computer wheel aligner that allows the technician to select, enter, alter, and display vehicle specifications.

Menú de especificaciones Representación visual en una computadora para alineación de ruedas que le permite al mecánico seleccionar, introducir datos, cambiar e indicar especificaciones referentes al vehículo.

Splice pack A special connector in a vehicle network that may be disconnected to isolate specific components for diagnostic purposes.

Paquete de empalmes Un conector especial en una red de vehículos que se puede desconectar para aislar componentes específicos con fines de diagnóstico.

Spring compressing tools Tools that are used to compress a coil spring during the removal and replacement procedure.

Herramientas de compresión de resortes Las herramientas que se usan para comprimir un resorte helicoidal durante el procedimiento de remoción y refacción.

Spring insulators Rings usually made from plastic and positioned on each end of a coil spring to reduce noise and vibration transfer to the chassis.

Aisladores de muelles Anillos, por lo general fabricados de plástico y colocados a ambos extremos de un muelle helicoidal, que se utilizan para disminuir la transferencia de ruido y de vibración al chasis.

Spring silencers Plastic spacers placed between spring leaves to reduce noise.

Amortiguadores de los resortes Las arandelas de plástico puestas entre los muelles de hojas para disminuir el ruido.

Stabilitrak® The vehicle stability control system used on General Motors vehicles.

Stabilitrak® El sistema de control de la estabilidad del vehículo que se utiliza en los vehículos General Motors.

Stabilizer bar A round steel bar connected between the front or rear lower control arms that reduces body sway.

Barra estabilizadora Barra circular de acero contectada entre los brazos de mando delantero o trasero que disminuye la oscilación lateral de la carrocería.

Standing acceleration test A vibration test performed during moderate vehicle acceleration.

Prueba de acceleración fijo Una prueba de vibración que se efectúa durante una aceleración moderada del vehículo.

Static balance Refers to the balance of a tire and wheel at rest.

Equilibrio estático Refiere al equilibrio de una rueda y de un neumático y una rueda cuando se ha detenido la marcha del vehículo.

Steering axis inclination (SAI) The tilt of a line through the center of the upper and lower ball joints or through the center of the lower ball joint and the upper strut mount in relation to the true vertical centerline of the tire and wheel.

Inclinación del eje de dirección (SAI) La inclinación de una línea que atraviesa el centro de las juntas esféricas superiores e inferiores o por el centro de la junta esférica inferior y el montaje del tirante superior en relación al eje mediano verdadero del neumático y de la rueda.

Steering effort The amount of effort required by the driver to turn the steering wheel.

Esfuerzo de dirección Amplitud de esfuerzo requerido por parte del conductor para girar el volante de dirección.

Steering effort imbalance Occurs when more effort is required to turn the steering wheel in one direction than in the opposite direction.

Desequilibrio del esfuerzo en dirección Ocurre cuando se requiere más esfuerzo para girar la rueda en una dirección que en la dirección opuesta.

Steering input test A vibration test performed while driving the vehicle through turns.

Prueba de entrada de dirección Una prueba de vibración que se efectúa mientras que el vehículo completa los virajes.

Steering pull The tendency of the steering to pull to the right or left when the vehicle is driven straight ahead on a smooth, straight road surface.

Tiro de la dirección Tendencia de la dirección a desviarse hacia la derecha o hacia la izquierda mientras se conduce el vehículo en línea recta en un camino cuya superficie es lisa y nivelada.

Steering wheel free play The amount of steering wheel movement before the front wheels begin to turn.

Juego libre del volante de dirección Amplitud de movimiento del volante de dirección antes de que las ruedas delanteras comiencen a girar.

Steering wheel locking tool A special tool used to lock the steering wheel during certain wheel alignment procedures.

Herramienta para el cierre del volante de dirección Herramienta especial que se utiliza para bloquear el volante de dirección durante ciertos procedimientos de alineación de ruedas.

Stethoscope A special tool that amplifies sound to help diagnose noise location.

Estetoscopio Herramienta especial que amplifica el sonido para ayudar a diagnosticar la procedencia de los ruidos.

Struts Similar to shock absorbers, but struts also support the steering knuckle.

Montantes Parecidos a los amortiguadores, pero los montantes tambien sostiene el muñon de dirección.

Strut cartridge The inner components in a strut that may be replaced rather than replacing the complete strut.

Cartucho del montante Componentes internos de un montante que pueden ser reemplazados en vez de tener que reemplazarse todo el montante.

Strut chatter A chattering noise as the steering wheel is turned, often caused by a binding upper strut mount.

Vibración del montante Rechinamiento producido mientras se gira el volante de dirección. A menudo ocasionado por el trabamiento del montaje del montante superior.

Strut rod A rod connected from the lower control arm to the chassis to prevent forward and rearward control arm movement.

Varilla del montante Varilla conectada del brazo de mando inferior al chasis que se utiliza para evitar el movimiento hacia adelante o hacia atrás del brazo de mando.

Strut tower A circular, raised, reinforced area inboard of the front fenders that supports the upper strut mount and strut assembly.

Torre del montante Área circular, elevada y reinforzada en la parte interior de los guardafangos delanteros que apoya el conjunto del montante y montaje del montante superior.

Sunburst cracks Cracks that radiate outward from an opening in the vehicle frame.

Grietas radiales Las grietas que radian hacia afuera desde una apertura en el armazón del vehículo.

Suspension adjustment link An adjustable link connected horizontally from the suspension to the chassis.

Biela de ajuste de la suspensión Una biela ajustable conectada horizontalmente de la suspensión al chasis.

Symmetry angle measurements On some computer-controlled wheel aligners, they help the technician determine if out-of-specifications readings may have been caused by collision or frame damage.

Medidas de simetría de los ángulos En algunos alineadores de ruedas controlados por computadora, ayudan al técnico en determinar si las medidas fuera de specificación puedan haber sido causadas por una colisión o un daño al bastidor.

Tapered-head bolts A bolt with a taper on the underside of the head.

Pernos de cabeza cónica Perno con una cabeza cuya superficie inferior es cónica.

Thrust line A line positioned at a 90° angle to the rear axle and projected toward the front of the vehicle.

Línea de empuje Línea colocada a un ángulo de 90° con relación al eje trasero y proyectada hacia la parte frontal del vehículo.

Tie-rod end and ball joint puller A special puller designed for removing tie-rod ends and ball joints.

Extractor para la junta esférica y el extremo de la barra de acoplamiento Extractor especial diseñado para la remoción de las juntas esféricas y los extremos de barras de acoplamiento.

Tie-rod sleeve adjusting tool A special tool required to rotate tie-rod sleeves without damaging the sleeve.

Herramienta para el ajuste del manguito de la barra de acoplamiento Herramienta especial requerida para girar los manguitos de las barras de acoplamiento sin estropearlos.

Tire changer Equipment used to dismount and mount tires on wheel rims.

Cambiador de neumáticos Equipo que se utiliza para desmontar y montar los neumáticos en las llantas de las ruedas.

Tire conicity Occurs when the tire belt is wound off-center in the manufacturing process creating a cone-shaped belt, which results in steering pull.

Conicidad del neumático Condición que ocurre cuando la correa del neumático se devana fuera del centro durante el proceso de fabricación creando así una correa en forma cónica y trayendo como resultado el tiro de la dirección.

Tire inspection screen A screen on a computer-controlled wheel aligner that allows the technician to enter the condition of each vehicle tire.

Pantalla de inspección de los neumáticos Una pantalla en un alineador de ruedas controlado por computadora que permite que el técnico anote la condición de cada neumático del vehículo.

Tire rotation Involves moving each tire and wheel to a different location on the vehicle to increase tire life.

Rotación de los neumáticos Involucra moviendo cada neumático y rueda a una posición diferente en el vehículo para prolongar la vida del neumático.

Tire thump A pounding noise as the tire and wheel rotate, usually caused by improper wheel balance.

Ruido sordo del neumático Ruido similar al de un golpe pesado que se produce mientras el neumático y la rueda están girando. Por lo general este ruido lo ocasiona el desequilibrio de la rueda.

Tire tread depth gauge A special tool required to measure tire tread depth.

Calibrador de la profundidad de la huella del neumático Herramienta especial requerida para medir la profundidad de la huella del neumático.

Tire vibration Vertical or sideways tire oscillations.

Vibración del neumático Oscilaciones verticales o laterales del neumático.

Toe gauge A special tool used to measure front or rear wheel toe.

Calibrador del tope Herramienta especial que se utiliza para medir el tope de las ruedas delanteras o traseras.

Toe-in A condition in which the distance between the front edges of the tires is less than the distance between the rear edges of the tires.

Convergencia Condición que ocurre cuando la distancia entre los bordes frontales de los neumáticos es menor que la distancia entre los bordes traseros de los mismos.

Toe-out A condition in which the distance between the front edges of the tires is more than the distance between the rear edges of the tires.

Divergencia Condición que ocurre cuando la distancia entre los bordes delanteros de los neumáticos es mayor que la distancia entre los bordes traseros de los mismos.

Torque steer The tendency of the steering to pull to one side during hard acceleration on front wheel drive vehicles with unequal length front drive axles.

Dirección de torsión Tendencia de la dirección a desviarse hacia un lado durante una aceleración rápida en un vehículo de tracción delantera con ejes de mando desiguales.

Total toe The sum of the toe angles on both wheels.

Tope total Suma de los ángulos del tope en ambas ruedas.

Track gauge A long straight bar with adjustable pointers used to measure rear wheel tracking in relation to the front wheels.

Calibrador del encarrilamiento Barra larga y recta con agujas ajustables que se utiliza para medir el encarrilamiento de las ruedas traseras con relación a las ruedas delanteras.

Tram gauge A long, straight bar with adjustable pointers used to measure unitized bodies.

Calibrador del tram Barra larga y recta con agujas ajustables que se utiliza para medir carrocerías unitarias.

Tread wear indicators Raised portions near the bottom of the tire tread that are exposed at a specific tread wear.

Indicadores del desgaste del neumático Secciones elevadas cerca de la parte inferior de la huella del neumático que quedan expuestas cuando la huella alcanza una cantidad de desgaste específica.

Trim height The normal chassis riding height on a computer-controlled air suspension system.

Altura equilibrada Altura normal de viaje del chasis en un sistema de suspensión controlado por computadora.

Turning radius gauge A gauge with a degree scale mounted on turntables under the front wheels.

Calibrador del radio de giro Calibrador con una escala de grados montado sobre plataformas giratorias debajo de las ruedas delanteras.

Turntable A mechanical device placed under each front wheel during a wheel alignment to allow the front wheels to be turned in each direction to measure various front suspension angles.

Plataforma giratoria Un dispositivo mecánico colocado debajo de cada rueda delantera durante la alineación de ruedas para permitir girar las ruedas delanteras en cada dirección para medir varios ángulos de la suspensión delantera.

Understeer The tendency of a vehicle not to turn as much as desired by the driver.

Dirección pobre Tendencia de un vehículo a no girar tanto como lo desea el conductor.

U.S. Customary (USC) system A system of weights and measures patterned after the British system.

Sistema usual estadounidense (USC) Sistema de pesos y medidas desarrollado según el modelo del Sistema Imperial Británico.

Vacuum hand pump A mechanical pump with a vacuum gauge and hose used for testing vacuum-operated components.

Bomba de vacío manual Bomba mecánica con un calibrador de vacío y una manguera que se utiliza para probar componentes a depresión.

Vehicle Dynamic Suspension (VDS) Is a computer-controlled air suspension system used on some current model vehicles.

Suspensión dinámica del vehículo (SDV) Es un sistema de suspensión de aire que se usa en algunos vehículos de modelo reciente.

Vehicle lift A hydraulically or air-operated mechanism for lifting vehicles.

Levantamiento del vehículo Mecanismo activado hidráulicamente o por aire que se utiliza para levantar vehículos.

Vehicle wander The tendency of the steering to pull to the right or left when the vehicle is driven straight ahead on a straight road.

Desviación de la marcha del vehículo Tendencia de la dirección a desviarse hacia la derecha o hacia la izquierda cuando se conduce el vehículo en línea recta en un camino cuya superficie es lisa.

V-type belt A drive belt with a V shape.

Correa en V Una correa de transmisión en forma de V.

Wear indicators Wear indicators show ball joint wear by the position of the grease fitting in the ball joint.

Indicadores de desgaste Los indicadores de desgaste muestran el desgaste de las rótulas de la suspensión por la posición de la válvula engrasadora en la rótula de suspensión.

Woodruff key A half-moon-shaped metal key used to retain a component, such as a pulley, on a shaft.

Chaveta woodruff Chaveta de metal en forma de media luna que se utiliza para sujetar un componente, como por ejemplo una roldana, en un árbol.

Workplace hazardous materials information systems (WHMIS) Are similar to MSDS sheets and they provide information regarding the handling of hazardous waste materials.

Sistemas de información acerca de materiales peligrosos en el lugar de trabajo (WHMIS, en inglés) Son similares a las hojas de datos MSDS y brindan información acerca del manejo de materiales de desecho peligrosos.

INDEX

A

ABS (antilock brake system), 441
Accessory drive frequencies, 132
Active roll stabilization (ARS), 528
Actuators, 194–196
Adjustment bolts, 351–352
Adjustment screens, of computer alignment system, 333–334
Advanced vehicle handling (AVH), 329
Air bag deployment modules, 273–278
  avoiding accidental deployment, 277
  removal and replacement, 274–275
Air bag safety, 17
Air bag system, 17
Air quality, 5–6
Air shock absorbers, diagnosis and replacement, 181
Air springs
  inflation, 427–428
  with integral struts, removal and replacement, 425–427
  removal and disassembly, on VDS systems, 435–437
  removal and installation, 424–425
Air suspension systems, 424–432
  air spring inflation, 427–428
  air spring removal and installation, 424–425
  air springs with integral struts, removal and replacement, 425–427
  line service, 431–432
  trim height mechanical adjustment, 428–431
Alcohol use, 12–13
Alignment. *See* Wheel alignment
Alignment angles, 324
Alignment ramps, 323
Antilock brake system (ABS), 441
Antitheft locking wheel covers, 99
Antitheft wheel nuts, 99
Aqueous parts cleaning tanks, 25
Asbestos, 3, 10
ASE blue seal of excellence, 79
Automatic transaxles (ATX), 160–161
Automatic transmissions, steering column diagnosis, 287–289
Automotive shops
  cleanliness of, 16–17
  layout, 9–10
  safety. *See* Safety
Automotive technicians
  job responsibilities, 78
  obligations of, 76–78
Automotive training, 76
Axle offset, 385, 392–393
Axle pullers, 49

B

Backup mode, 529
Back-up power supply, 273
Balancing, tire and wheel servicing and, 95–132
  dynamic wheel balance procedure, 119–121
  electronic wheel balancers, 121–124
  on-car balancing, 125–126
  preliminary wheel balancing checks, 117–118
  static wheel balance procedure, 119–120
  tire and wheel runout measurement, 114–116
  tire and wheel service, 98–100
  tire and wheel service precautions, 100–104
  tire inflation pressure, 118–119
  tire inspection and repair, 105
  tire noises and steering problems, 95–98
  tire pressure monitoring systems, 108–110
  tire remounting procedure, 107–108
  tire rotation, 97–98
  tread wear measurement, 116–117
  wheel rim service, 106–107
Ball joint removal and installation tools, 54
Ball joints
  checking, on twin I-beam axles, 220
  diagnosis and replacement, 223–225, 256
  horizontal measurement, 217
  inspection, 213
  unloading, 214–217
  vertical measurement, 217
Ball joint vertical movement, 217
Ball joint wear indicator, 213
Batteries, 2, 5
  disposal, 26
Bearing brinelling, 148
Bearing defects, diagnosis of, 147–148
Bearing etching, 148
Bearing fatigue spalling, 148
Bearing frettage, 148
Bearing galling, 147
Bearing preload, 553, 554
Bearing pullers, 48–49
Bearing races, 151
Bearing smears, 148
Bearings. *See* Wheel bearings
Bellows boots, 498
Belts, power steering, 459–462
Belt tension gauge, 460
Bench grinders, 61
Bench systems, 405
Bodies. *See* Frame diagnosis and service
Body sway, 250
Bolt mountings, 179
Bounce tests, 181
Brake parts washer, 10
Brake pedal depressors, 328
Brake pedal jack, 58
Brake pressure modulator valve (BPMV), 443
Brake problems, 317–319
Brake torque test, 129
Buckle, 395
Bump steer, 318–319, 497
Bushing condition, 179
Bushings, front lower control arm bushing removal and replacement, 230–232

C

Camber, 328
Camber adjustment, 346–352
  eccentric cams, 353
  rear wheel, 338–340
  shims, 346
  slotted strut mounts and frames, 349–350
  special adjustment bolts, 351–352
  special tools for, 352
  on twin I-beam front suspension systems, 352
Camber angles, 217
Camber bolts, 184
Camber settings, 318
Carbon monoxide, 2, 5, 12
Carrying objects, 20–21
Caster, 329
Caster adjustment, 353–355
Caster angles, 217
Caster offset, 329
Caster trail, 329
Caustic liquids, 2
Center link, diagnosis and replacement, 291–297
Certification, ASE, 78–79

Chassis vibration, 264
Chassis waddle, 116, 264
Chattering noise, 196
Circlips, 158, 160
Claw washers, 513
Cleanliness, 16–17
"C" locks, 163
Clock spring electrical connector, 273–275, 277–278
Coil spring compressor tool, 55, 68–70
Coil spring interference problems, 183–184
Coil springs
  installation, on strut, 187–189
  installation of strut-and-spring assembly, 189
  lower control arm and spring replacement, 228–229
  rear, diagnosis and service, 251–256
  removal of strut from, 186–187
Cold parts washers, 25
Collapsible steering columns, 279
Column-drive electronic power steering, 522–525
Compressed-air equipment safety, 23–24
Compressor, coil springs, 55
Computer alignment systems, 60, 73–74, 323–337
  adjustment screens, 333
  diagnostic drawing and test screens, 334–337
  main menu, 325
  preliminary inspection screen, 326
  preliminary procedure, 323–325
  ride height screen, 326
  specifications menu, 325–326
  tire condition screen, 326–327
  wheel alignment screens, 328–329
  wheel runout compensation screen, 327–328
Computer-controlled suspension system service, 421–447
  electronic air suspension diagnosis and service, 424–432
  electronic suspension control systems, 437–439
  preliminary inspection, 421–422
  programmed ride control system diagnosis, 422–423
  scan tool diagnosis of ESC, 439–447
  vehicle dynamic suspension systems, 432–437
Computer four wheel aligner, 60, 74
Continuously variable road sensing suspension (CVRSS) systems, 438
Control arm bushing tools, 54
Control arm diagnosis and service, 225–234
  front leaf spring inspection and replacement, 234
  front lower control arm bushing removal and replacement, 230–231
  lower control arm and ball joint diagnosis and replacement, 256–261
  lower control arm and spring replacement, 228–229
  lower control arm replacement, 226
  rebound bumpers, 230
  stabilizer bar diagnosis and replacement, 232–233
  strut rod diagnosis and replacement, 233
  upper control arm removal and replacement, 229–230
Control arm movement monitor, 336
Cords, frayed, 2, 13
Corrosive materials, 27
Crocus cloth, 475
Curb riding height measurement, 211–212, 250–251
Customer satisfaction, 360
Customer service, 48
  attitude toward, 256
  careful vehicle operation and, 16
  preventive maintenance promotion, 463
  terminology use, 444
  time estimates, 83

D

Data link connector (DLC), 428
Datum line, 401
Dedicated fixtures, 405–407
Diagnostic alignment angles and frame diagnosis, 383–410
  bent front strut diagnosis, 386
  frame alignment, checking, 396–402
  rear wheel tracking measurement, 389–394
  SAI correction procedure, 389
  and service, 394–395
  setback measurement and correction procedure, 386–389
  unitized body aligment measurement, 402–408
Diagnostic drawing and test screens, of computer alignment system, 334–337
Diagnostic procedure charts, 81
Diagnostic procedures, basic, 47–48
Diagnostic trouble codes (DTCs)
  electronic suspension control systems, 437–441
  Magnasteer systems, 519
  programmed ride control system, 422–424
  rack-drive EPS, 520–521
  rear wheel steering, 531–533
Dial indicators, 50, 66–68
Diamond-shaped frames, 393–396
Digital adjustment photos, 334
Digital signal processor (DSP), 324
Directional stability, 211, 262
  lack of, 148
Distance conversions, 47
Downshift test, 128
Drain covers, 2
Drive axles
  diagnosis, 158
  removal, 158–159
Drive display, 110
Drug use, 12–13
Dry chemical fire extinguishers, 6
Dust, 3
Dynamic stability control (DSC), 528
Dynamic wheel balance, 119

E

Eccentric camber bolts, 184
Eccentric cams, 346–349
Electrical safety, 13
Electrohydraulic power steering (EHPS), 478–485
Electronic air suspension systems, 424–432
  air spring inflation,427–428
  air spring removal and installation, 424–425
  air springs with integral struts, removal and replacement, 425–427
  line service, 431–432
  trim height mechanical adjustment, 428–431
Electronically controlled orifice (ECO) valve, 528
Electronically controlled shock absorbers, diagnosis, 194–196
Electronic brake and traction control module (EBTCM), 443, 519

Electronic four-wheel steering, 528–537
diagnosis, 529–531
preliminary inspection, 529
Quadrasteer diagnosis, 529
Electronic power steering (EPS), 520
column-drive, 522–525
rack-drive, 520–522
Electronic suspension control (ESC) systems, 437
diagnosis, 437–439
scan tool diagnosis, 439–447
Electronic training, 76
Electronic vibration analyzer (EVA), 127, 129–131
Electronic wheel balancers, 52, 73, 121–124
with LFM and radial force variation capabilities, 121–124
Employer obligations, 77
End plug seal replacement, 558
Engine and engine firing frequency, 132
Engine coolant, disposal, 26
Environmental Protection Agency (EPA), 27
Equipment
floor jacks, 62–64
hydraulic presses, 61–62
safety stands, 63–64
vehicle lifts, 62–65
Equipment cleaning, 24–27
Eye injuries, 7
Eyewash fountains, 7

F

Face shields, 7–8
Fail-safe mode, 529
Falls, 3
Fire extinguishers, 6–7, 15–16
Fire safety, 14–15
First-aid kits, 9
Flammable liquids, 2
Flammable materials, 2
Flexible coupling installation, 286
Floor jacks, 62–64
Flow control valves, 475
Fluids
disposal, 26
power steering, 467
Fluorescent trouble lights, 12
Four wheel alignment procedure, 317–362
camber adjustment, 346–353
caster adjustment, 353–355
checking toe change and steering linkage height, 357–359
with computer alignment systems, 323–328
preliminary inspection, 317–319
rear suspension adjustments, 338–345
steering wheel centering procedure, 359–362
toe adjustment, 356–359
Four-wheel steering, electronic, 528–537
diagnosis, 528–531
preliminary inspection, 529, 530–531
Quadrasteer diagnosis, 529
Frame damage
indications of, 395
prevention, 395
Frame diagnosis and service, 394–410
checking frame alignment, 396–402
measurement of unitized body alignment, 402–407
Frame flange, 395
Frame measurement
plumb bob method, 398–402
tram gauge method, 400–402
of unitized bodies, 402–407
Frame straightening, 402
Frame web, 395
Frame welding, 398
Free play, steering wheel, 286, 498, 553
Front and rear wheel alignment angle screens, 328–329
Front bearing hub tool, 50
Front drive axles
diagnosis, 158
removal, 158–161
Front spring noise diagnosis, 236
Front struts, bent, 386
Front suspension adjustments, 341
Front suspension system service, 211–237
ball joint checking, 220
ball joint inspection, 213
ball joint measurement, 217–219
ball joint replacement, 223–225
ball joint unloading, 214–217
control arm diagnosis and service, 225–234
curb riding height measurement, 211–212
diagnosis and service, 212–225, 233
front steering knuckle diagnosis and service, 220–222
front steering knuckle replacement, 222–223
torsion bar adjustment, 212–213
torsion bars, removing and replacing, 235–236
Front wheel alignment, 337. *See also* Wheel alignment
Front wheel bearings, hub unit removal and replacement, 161–163
Front wheel shimmy, 291–292

G

Gasoline safety, 13–14
Gasoline storage cans, 13
Geometric centerline, 340

H

Hand tool safety, 21
Harsh ride quality, 196
Hazard Communication Standard, 28
Hazardous waste disposal, 27–30
Right-to-Know laws, 2
safety considerations, 27–28
Hazardous waste material, 2
Heavy objects, lifting and carrying, 20–21
Heavy spots, 120
HEV service precautions, 478–487
High-frequency transmitters, 324–325
High-pressure air, 2
Hoists, 63
Hot cleaning tanks, 24
Housekeeping safety, 16–17
Hub nut torque, 162
Hybrid electric vehicle (HEV) safety, 30–34
high-voltage service plug, 33–34
insulated glove integrity test, 31–33
Hybrid vehicle, 4
Hydraulic jack safety, 22–23
Hydraulic presses, 61–62

I

Idler arms, 297–299
Ignitable materials, 27
Incandescent bulbs, 3
Incandescent bulb-type trouble lights, 12
Included angle, 329
Integral reservoirs, 475
International system (IS), 45
Invalid display, 110

J

Jerry cans, 14
Job responsibilities, 78

L

Lateral axle sideset, 385
Lateral chassis movement, 262

Lateral force measurement (LFM), 121–123
Lateral movement, 250
Lateral tire runout, 116
Layout, shop, 9
Leaf springs
  front, inspection and replacement, 234
  noise, 236
  rear, diagnosis and replacement, 261–262
Leaks
  oil, 26, 180, 196, 471–472, 557–558
  wheel rim leak repair, 107
Learn display, 110
Length, 47
Lifting equipment, 61–62
Lifting objects, 20–21
Lift safety, 21–22
Line fittings, 557
Liquids, disposal of, 26
Lock systems, 287
Loose clothing/hair, 2
Loose steering, 507
Low battery display, 110
Low riding height, 236, 264
Lubrication, bearing, 152

M

Machinist's rule, 52
MacPherson strut suspension systems
  lower control arm replacement, 226–228
  strut and spring removal and replacement, 251–255
  strut removal and replacement, 184
Magnasteer systems, 519–520
Magnetic learning tool, 112
Magnetic wheel alignment gauge, 58
Main menu, of computer alignment system, 325
Maintenance schedules, 560
Malfunction indicator light (MIL), 485
Manual steering and suspension systems, diagnosis, 506
Manual tests, of shock absorbers, 181
Manual transmissions
  steering column diagnosis, 286–290
  tilt column diagnosis, 290
Mass, 46
Material safety data sheet (MSDS), 28–30
Measurement equivalents, 46
Measuring systems, 45–47
Memory steer, 318
Metric Conversion Act, 46
Metric system, 45–46
Molybdenum disulphide lithium-based grease, 284
Morning sickness, 510
Mounting bolts, 179
Mounting bushings, 179
Multipull equipment, 407
Multipurpose dry chemical fire extinguishers, 6–7
Multi-ribbed V-belt, 460

N

National Institute for Automotive Service Excellence (ASE) certification, 78–79
Neutral coast-down test, 128
Neutral run-up test, 128–129
Noise levels, 3
Noises
  chattering, 196
  diagnostics for, 148
  front spring, 183–184
  and harshness diagnosis, 126–127
  leaf spring, 234
  rattling, 196, 250
  rear suspension, 249
  strut chatter, 184
  suspension, 234, 236, 254
  tire, 95
  when turning, 559

O

OBD II computer systems, 428, 431
Occupational Safety and Health Act (OSHA), 2
Occupational Safety and Health Administration (OSHA), 28
Oil, disposal, 26
Oil filters, disposal, 26
Oil leakage, 26, 180, 196, 471–472, 557–558
On-board diagnostic II (OBD II) systems, 72, 431, 523
On-car balancing, 125–126
O-ring replacement, 557–558

P

Part-finder database, 334
Parts washers
  with agitation immersion tanks, 25
  with electromechanical agitation, 24–25
Patch installation, 106
Personal safety, 11–12
Pitman arm puller, 56
Pitman arms, 291, 294
Pitman sector shaft lash adjustment, 553, 555
Plug installation, 105
Plumb bob method, of frame measurement, 398
Plumb bobs, 57, 398
Ply separation, 105
Pneumatic test, 434
Power steering diagnosis, 518
Power steering lines and hoses
  hose replacement, 477
  inspection and service, 476
Power steering pressure gauge, 55, 65–66
Power steering pumps, 459–486
  belt service, 459–462
  diagnosis, 469–471
  flow control valve and end cover removal and replacement, 475
  fluid service, 467
  oil leak diagnosis, 471–472
  pressure tests, 469, 473–474
  pulley replacement, 474–475
  replacement, 471
  reservoir removal and replacement, 475
  servicing, 476
Power tool safety, 23
Powertrain control module (PCM), 439
Prealignment inspection, 319–323
Preliminary inspection screen, of computer alignment system, 326
Pressure gauges, 55, 469
Pressure relief valves, 475
Programmed ride control (PRC) systems, 422–423
Protective clothing/equipment, 3

Q

Quadrasteer diagnosis, 529–533

R

Rack bearing service, 513–516
Rack-drive electronic power steering, 520–521
Rack and pinion steering gear, 497–537
  column-drive EPS, 522
  inner tie rod end, removing and replacing, 517–518
  Magnasteer systems, 519–520
  manual, diagnosis and service, 505–506

Rack and pinion steering gear (*continued*)
on-car inspection, 497–500
power, diagnosis and service, 508–516
oil leak diagnosis, 508–509
tie rod and rack bearing service, 513–516
turning imbalance diagnosis, 510–511
power, removal and replacement, 518
power steering, steering column, and suspension system diagnosis, 518
rack-drive electronic power steering, 520–521
removal and replacement, 500–505
tie rod service, 290–291, 505–506
Radial force variations, 121
Radial runout, 95–96
Radial tire runout, 114
Rattling noise, 196, 234
Reactive materials, 27
Rear-axle bearing and seal service, 163–166
Rear axle offset, 392–393
Rear shock absorber visual inspection and bounce test, 182
Rear suspension adjustments, 338–344
rear wheel camber adjustment, 338–340
rear wheel toe adjustment, 340
Rear suspension service, 249–264
calculating vibration frequencies, 131–132
curb riding height measurement, 250–251
diagnosis, 264
leaf-spring diagnosis and replacement, 261–262
lower control arm and ball joint diagnosis and replacement, 256, 258–261
noise diagnosis, 219
rear strut, coil spring, and upper mount diagnosis and service, 251, 253–254
ride harshness diagnosis, 249
road tests, 127–129
stabilizer bar diagnosis and service, 263–264
sway and lateral movement diagnosis, 250
tie rod inspection and replacement, 264
track bar diagnosis and replacement, 262
vibration diagnosis, 126–127
vibration diagnosis with EVA, 129–131
Rear wheel alignment. *See also* Wheel alignment
causes of improper, 337
procedure, 337
Rear wheel bearings, adjusting, 155
Rear wheel camber, 338
Rear wheel camber adjustment, 338–340
Rear-wheel-drive vehicles
front steering knuckle replacement, 222–223
rear-axle bearing and seal service, 163–166
Rear wheel steering data, 531
Rear wheel toe, 340
Rear wheel toe adjustment, 340–344
Rear wheel tracking, 391
measurement, 389, 391–394
Rebound bumpers, 196, 214, 230
Receivers, 325
Recirculating ball steering gear, 551–559
adjustments, 553–555
pitman sector shaft lash, 553, 555
worm shaft thrust bearing preload, 553
diagnosis, 551–552
oil leak diagnosis, 557
replacement, 552
seal replacement, 557–558
Refrigerants, disposal, 27
Remeasure display, 110
Remote reservoirs, 468
Resource Conservation and Recovery Act (RCRA), 27
Respect, for customers, 408
Ribbed V-belt, 460
Ride harshness, 219, 249
Ride height, 321
Ride height adjustment, 433
Ride height screen, of computer alignment system, 326
Right-to-Know laws, 2, 28
Rim clamps, 58, 324
Road feel, 555
Road tests, 127–129, 318

S

SAE J1930 terminology, 529
Safety, 1–34
air bag, 17–20
air quality, 5–6
in automotive shop, 10–11
carrying, 20–21
in checking frame alignment, 396
compressed-air equipment, 23–24
electrical, 13
equipment cleaning and, 24–27
fire, 14–16
gasoline, 13–14
hand tool, 21
with hazardous wastes, 27–29
housekeeping, 16–17
hydraulic jack, 22–23
importance of, 2
lift, 21–22
lifting, 20–21
OSHA, 2
personal, 11–12
power tool, 23
safety stand, 22–23
shop hazards and, 2–3
shop safety rules, 3–5
vehicle operation, 16
Safety data sheets (SDS), 28
Safety equipment, 6–9
eyewash fountains, 7
face shields, 7–8
fire extinguishers, 6–7
first-aid kits, 9
safety glasses, 7–8
Safety and gateway module (SGM), 528
Safety glasses, 7–8
Safety stands, 63–64
Safety stand safety, 22–23
Sag, 395
SAI angle, 329
Scan tool, 60, 71–73, 428
for air spring inflation, 428
diagnosis of column-drive EPS, 523
diagnosis of ESC, 439–440
diagnosis of HEV and EHPS systems, 483–484
diagnosis of Magnasteer systems, 519
inflate L/H front air spring using, 429–430
Quadrasteer diagnosis, 529
reading data on electronic suspension control system, 442
Seal drivers, 48
Section modulus, 395
Sector shaft lash, 553, 555
Self-test
of PRC systems, 423
of VDS systems, 434–435
Serpentine belts, 460
Service check connector, 520–521
Service manuals, 80–83
Servotronic valve, 528
Setback, 328, 386, 393
Setback measurement and correction procedure, 386–389

Sewer covers, 2
Sheared injected plastic, 279
Shim display screen, 334–335
Shims, 346
Shock absorbers
  air, 181
  bounce test, 181
  diagnosis of electronically controlled, 194–196
  manual tests, 181
  removal and disassembly, on VDS systems, 435–437
  replacement, 183–184
  visual inspections, 179–180
Shop hazards, 2–3
Shop layout, 9–10
Shop safety equipment, 6–9
Shop safety rules, 3–5, 10–12
Shop towels, disposal, 26
Short-and-long arm (SLA) suspension systems
  lower control arm and spring replacement, 228–229
  upper control arm removal and replacement, 229–230
Side cover O-ring replacement, 557–558
Side sway, 395
Single-pull equipment, 407
Slip plates, 323
Slips, 3
Slow acceleration test, 127–128
Smoking, 12
Society of Automotive Engineer (SAE), 523
Sodium hydroxide, 20
Soft-jaw vise, 505
Solenoids, 194
Solvents, disposal, 26
Specifications menu, of computer alignment system, 325–326
Splice pack, 446
Spring compressing tools, 186
Spring insulators, 184
Spring silencers, 261–262
Stabilitrak system, 441, 443
Stability control system, diagnosis, 441, 443
Stabilizer bars, 232
  diagnosis and replacement, 232–233
  diagnosis and service, 263–264
Standing acceleration test, 129
Static balance, 119
Stationary display, 110
Steering arms, 299
Steering axis inclination (SAI), 329
  correction procedure, 389
Steering column and linkage service, 273–299
  air bag module removal and replacement, 273–275
  collapsible steering column inspection, 279
  diagnosing, removing, and replacing outer tie rod end, 295–296
  flexible coupling and universal joint diagnosis and service, 286
  manual, diagnosis, 289
  steering column diagnosis, 286–290
  steering column removal and replacement, 278–279
  steering linkage diagnosis and service, 291–299
  steering wheel removal and replacement, 274
  tilt steering column assembly, 284
  tilt steering column disassembly, 280–282
  tilt steering column inspection and parts replacement, 283–284
Steering control problems, 236
Steering dampers, 298
Steering diagnosis, 527–528
Steering effort, 211, 467, 510, 518
Steering effort imbalance, 510
Steering gear
  diagnosis, 551–552
  rack and pinion, 497–528
  recirculating ball, 551–558
Steering input test, 129
Steering knuckles
  front, diagnosis and service, 220, 222
  replacement, rear-wheel-drive vehicles, 222–223
Steering linkage height, 357–359
Steering position sensor calibration, 524
Steering problems, 95
Steering pull, 96–97, 127, 264
Steering system preliminary diagnosis, active, 525–526
  active steering diagnosis system, 527–528
  fiber-optic cable service, 527
  preliminary inspection, 526
Steering tools, 48–61
Steering tuning selection, 525
Steering wander, 236
Steering wheel centering procedure, 359–360
Steering wheel free play, 286, 297, 498, 553
Steering wheel holders, 328
Steering wheel locking tool, 59
Steering wheel pad. *See* Air bag deployment modules
Stethoscopes, 49, 163
Strut cartridge, 190
Strut cartridge installation
  off-car, 190, 192–193
  on-car, 193–194
Strut chatter, 183
Strut mounts, 349–350
Strut rod, 233
  diagnosis and replacement, 233
  length adjustment, 353
Struts, 180
  bent front, 386
  bounce test, 181
  diagnosis, 196
  disposal procedure, 187
  installation of coil spring on, 187–189
  installation of strut-and-spring assembly, 189
  noise diagnosis, 219
  rear, diagnosis and service, 251, 253–254
  rear strut replacement, 190
  removal and replacement, 184–185
  removal from coil spring, 186–187
  visual inspections, 179–180
Strut tower, 183
Substance abuse, 13
Sulfuric acid, 5
Sunburst cracks, 396–398
Suspension adjustment link, 259
Suspension noise, 236
Suspension problems, 396
Suspension systems, 79, 160, 211, 250
  computer-controlled, 421–447
  front, 211–236
  manual, diagnosis, 506
  power, diagnosis, 518
  rear, 249–264
  SLA, 228–230
  VDS, 432–437
Suspension tools, 48–58
Symmetry angle measurement, 385

T

Tapered-head bolts, 280
Tapered roller bearings, service and adjustment of, 148–152
Thrust line, 328, 340

Tie rod adjusting tool, 59
Tie rod end and ball joint puller, 54
Tie rod ends
diagnosis, 291
replacement, 292–294
Tie rods
rack and pinion steering gear tie rod service, 505–506
rear, inspection and replacement, 264
steering gear, 513–514
Tie rod sleeve adjusting tool, 59
Tilt steering columns
assembly, 284–285
diagnosis, 290
disassembly, 280–283
inspection and parts replacement, 283–284
Tire changers, 51–52, 70–71, 100–101, 107, 111–112
Tire condition screen, of computer alignment system, 326–327
Tire conicity, 96, 123
Tire diagnosis, 79
Tire dismounting, 100–105
Tire inflation pressure, 83
Tire inspection screen, 326
Tire noises, 95
Tire plug installation, 105–106
Tire pressure monitoring system (TPMS), 108
Tire remounting, 107
Tire removal, 115–116
Tire rotation, 97–98
Tire runout measurement, 114–116
Tire service, 104–105
inspection and repair, 105
precautions, 105
Tire thump, 95, 132
Tire tread depth gauge, 51
Tire tread wear, 116, 220, 234
Tire vibration, 95
Tire waddle, 116
Tire wear diagnosis, 323
Tire/wheel frequency, 131
Toe adjustment, 356–357
rear wheel, 340–345
Toe change, checking, 357–359
Toe gauge, 59
Toe-in, 319, 358
Toe-out, 328
Toe, rear wheel, 340
Tools
for camber adjustment, 352
coil spring compressing tool, 68–70
computer four wheel aligners, 73–74
dial indicators, 66–68
electronic wheel balancers, 73
power steering pressure gauge, 65–66
scan. *See* Scan tool
spring compressing, 186
suspension and steering, 48–60
tire changes, 70–71
Torque sensor calibration, 524–525
Torque steer, 319
Torque sticks, 56–57
Torque wrenches, 56–57
Torsion bar adjustment, 212–213
Torsion bars, removing and replacing longitudinally mounted, 235–236
Total toe, 328, 356
Toxic substances, 27
Track bar diagnosis and replacement, 262
Track gauge, 59, 391, 400
Tracking, rear wheel, 389–393
Track-width difference, 386
Tram gauge, 57, 400
Tram gauge method, of frame measurement, 400–402
Transverse beams, 405
Tread wear, 118
Tread wear indicators, 116
Tread wear measurement, 116–117
Trim height, 428
mechanical adjustment, 428–432
Turning angle screen, of computer alignment system, 383
Turning imbalance diagnosis, 510–511
Turning radius gauges, 57
Turntables, 324, 328
Twin I-beam axles, checking ball joints on, 220
Twin I-beam front suspension systems
camber adjustment, 352–353
caster adjustment, 355
Twist, frame, 396

U

U-channels, 263
Underhood measurements, 404
Unitized bodies
measurement, 402–403, 405–407
straightening, 407
Universal joints, 286
Upward wheel movement, 181
U.S. customary (USC) system, 45

V

Vacuum hand pump, 56
Valid display, 110
Valve solenoid/pressure hold test, 443
Valve solenoid/pressure release test, 444–445
Valve stem removal, 111–112
Vapors, 3
Vehicle dynamic suspension (VDS) systems, 432–437
air spring and shock absorbers removal and disassembly, 435–437
inspection and verification, 432–433
on-demand self-test, 434–435
pneumatic test, 434
ride height adjustment or calibration, 433
Vehicle exhaust, 5
Vehicle identification number (VIN), 80
Vehicle lifts (hoists), 62–63
Vehicle operation, safety rules for, 16
Vehicle pulling/drifting, 507
Vehicle wander, 553
Ventilation, 5
Vertical ball joint measurement, 218
Vibration, 158
chassis, 264
diagnosis, 126–129
diagnosis with EVA, 129–131
tire, 95
wheel, 121, 127
Vibration frequencies, calculating, 131–132
Volume, 47
V-type belts, 460

W

Wake display, 110
Wander, 553
Wear indicators, 213, 256
Wheel alignment. *See also* Computer alignment systems; Four wheel alignment procedure
diagnosis, 361
procedure, 337, 384–385
Wheel alignment screens, of computer alignment system, 328–333
Wheel balancer, 52
Wheel balancing
on-car balancing, 125–126
preliminary checks, 117–118
procedure, 122–123
rear wheel balancing, 126

Wheelbase difference, 386
Wheel bearings
adjusting, 155
cleaning, 148–152
defects, 147–148
front, removal and replacement, 161–162
lubrication and assembly, 152–153
tapered roller, service and adjustment, 153–154
Wheel diagnosis, 127
Wheel hub unit, 154, 156, 158
Wheel jounce, 357–359
Wheel nut tightening, 99, 163
Wheel removal, 98–100
Wheel rim leak repair, 107
Wheel rim service, 106
Wheel runout compensation, 327–328
Wheel runout measurement, 114–116
Wheel sensors, 324
Wheel service, 98–105
precautions, 100–105
Wheel setback, 393–394
Wheel shimmy, 291–292
Wheel vibration, 121, 125
Wiring harness colors, 196
Woodruff key, 474
Workplace hazardous materials information system (WHMIS), 28
Worm shaft bearing adjuster plug seal replacement, 558–559

Y

Yield strength, 395